Representation Theory and Automorphic Forms

Reprints from the

BULLETIN

of the

American Mathematical Society

Volume 2

Representation Theory and Automorphic Forms

Paul J. Sally, Jr.
Nolan R. Wallach
Editors

American Mathematical Society
Providence, Rhode Island

This volume consists of 11 papers focusing on the common theme of the representation theory of locally compact groups and its numerous applications. These articles appeared in the Bulletin of the American Mathematical Society between 1955–1984.

1991 *Mathematics Subject Classification.* Primary 22–XX.

Library of Congress Cataloging-in-Publication Data

Representation theory and automorphic forms/edited by Paul J. Sally ,Jr. and Nolan R. Wallach.
p. cm. – (Reprints from the Bulletin of the American Mathematical Society, ISSN 1064-9662; v. 2)
Bibliography: p.
ISBN 0-8218-0720-X (alk. paper)
1. Representations of groups. 2. Automorphic forms. I. Sally, Paul. II. Wallach, Nolan R. III. Series.
QA171.R427 1989 89-35474
512′.2—dc20 CIP

Printed in the United States of America.
The paper used in this book is acid-free and falls within the guidelines established to ensure permanence and durability. ♾
Printed on recycled paper. ♻

10 9 8 7 6 5 4 3 2 1 98 97 96 95 94 93

Table of Contents

Introduction

The collection of articles reprinted in this volume appeared in the *Bulletin of the American Mathematical Society* during the years 1955 to 1984. There are 11 papers focusing on the common theme of the representation theory of locally compact groups and its numerous applications. The choice of the papers presented here was dicated by the wish of the Editors to provide a glimpse at the historical development of a subject which has expanded into many areas of mathematics over the past 40 years. In addition, this volume is intended to give easy access to a useful set of references for specialists in representation theory and in those parts of mathematics in which representation theory plays an essential role.

The first eight papers printed here were based on Invited Lectures presented by the authors at meetings of the American Mathematical Society. Prior to the establishment of the *Bulletin* (*New Series*), this was the standard route for papers published in the *Bulletin*. The remaining three papers were submitted by the authors for publication as Research Expository Articles in the *Bulletin* (*New Series*). As a consequence, some of the earlier papers have a narrower focus and are directed more towards the research results of the authors. Of particular note in this category are the papers of Mackey (1963) and Harish-Chandra (1970) which contain expositions of the authors' Colloquium Lectures delivered at the Summer Meetings of the American Mathematical Society in 1961 and 1969 respectively. As such, they contain a wealth of material related to the work of two of the major figures of the past half century in the field of Representation Theory.

The study of infinite-dimensional representations of locally compact groups, generally acknowledged to have begun with a paper of E. Wigner in 1939 **[W]** * , exploded in the late 1940s and the early 1950s. At that time, there were two major thrusts in the research on infinite-dimensional representations. The first was directed towards results concerning the representation theory of general locally compact groups. This included topics such as induced representations, decomposition of arbitrary unitary representations into irreducible constituents, square-integrable representations, and the Plancherel formula. Much of this work is discussed in Mackey (1963) which gives a good perspective on the early approaches to representation theory. Mackey's article also contains an immense bibliography.

The second major thrust in the late 1940s and early 1950s was in the analysis of the irreducible unitary representations of a particular group or a special class of groups. Outstanding examples of such work are the paper of Bargmann **[B]** on $SL(2, R)$, the book of Gel'fand and Naimark **[G-N]** on the complex, classical

* A bibliography for articles indicated in brackets may be found at the end of the Introduction. References to articles which appear in this volume are given by the author's name followed by the year of publication in the *Bulletin*.

groups, and the paper of Harish-Chandra **[H-C]** on the Plancherel formula for complex, semisimple Lie groups.

The approach of Gel'fand and Naimark to the representation theory of the complex, classical groups was global in nature (that is, they worked directly on the group), while the approaches of Bargmann and Harish-Chandra in their analyses made extensve use of the Lie algebra. The use by Harish-Chandra and others of the Lie algebra of a semisimple Lie group, the universal enveloping algebra, and, in particular, the center of the universal enveloping algebra became known as the "infinitesimal method". Throughout the 1950s, there was considerable work in representation theory using both the global approach and the infinitesimal method. However, as the subject progressed, it became obvious that the infinitesimal method was more powerful. Some indications of this are contained in Harish-Chandra (1955). In this article, Harish-Chandra studies his notion of the character of an admissible representation of a semisimple Lie group. He shows that such a character has a formula on the regular elements analogous to the Weyl character formula. He conjectures that the formula on the regular elements determines the character. He also proves that, for each formula on the regular elements, there are at most a finite number of characters. It is important to note that this implies the result that, given an infinitesimal character, there is only a finite number of associated irreducible characters.

The decade of the 1960s was a dynamic one for representation theory and its applications. Some of the major developments are reflected in the papers of Furstenberg (1965), Kostant (1969), and Wolf (1969). Furstenberg's paper may be regarded as an interesting application of some ideas from representation theory. He discusses the Furstenberg boundary and spherical functions. While representations do not occur explicitly, the concepts play an important role. Kostant's article was extremely influential; it contains many novel techniques, and it was a major victory for the algebraic theory. The paper of Wolf illustrates a variety of analytic and geometric ideas. It foreshadows the work on D-modules and semisimple symmetric spaces which played a major role in representation theory in the 1970s and 1980s. Harish-Chandra's chronicle (1970) of his 1969 Colloquium Lectures, is an absolute tour de force. Here he gives a detailed and lucid exposition of the culmination of his work on harmonic analysis on semisimple Lie groups leading to the Plancherel formula. He also outlines the beginnings of his research on the representation theory of reductive p-adic groups, a field in which he made many deep and incisive contributions in the 1970s.

The role of representation theory in the study of automorphic forms came to center stage during the 1960s. The interplay among representation theory and harmonic analysis, the Selberg trace formula, and the theory of automorphic forms was shaped in the hands of Gel'fand, Langlands, and Harish-Chandra. In fact, this interaction had a powerful impact on Harish-Chandra's theory of the discrete series for semisimple Lie groups. Two of the most important works on this subject are the paper of Langlands **[L]**, in which he presented many of the ideas which constitute the "Langlands' Program", and the book of Jacquet-Langlands **[J-L]**, which develops these ideas for $GL(2)$. The Langlands' Program is still the focus of deep and wide-ranging research which will carry forward well into the next century. The article of Gelbart (1984) gives a clear and useful introduction to the Langlands' Program, and the article of Wallach

(1976) reveals much about the state of the local Selberg trace formula in the early 1970s.

The remaining three papers in this volume present a wide array of ideas and applications arising from representation theory. The paper of Helgason (1977) has two parts. The first is an elegant rendering of the historical foundations of Lie theory. This essay provides a perspective which should be interesting to any mathematician. The second part contains an analysis of representations which act on solution spaces of differential equations and a discussion of the well-known KKMOOT theorem on eigenfunctions of invariant differential operators on symmetric spaces.

In Howe's paper (1980), he presents a convincing demonstration of the importance of the Heisenberg group and its representation theory. The sheer breadth of the applications gives ample evidence of the ubiquity of the Heisenberg group in analysis as well as many other parts of mathematics. Finally, Zimmer (1982) has written a beautiful paper which is full of relationships among representation theory, ergodic theory, and geometry. This paper provides a valuable overview of the author's book **[Z]**.

The first set of References below is for the papers mentioned in this Introduction. The second set of References lists papers related to Representation Theory which appeared in the *Bulletin* but are not published in this volume. The Editors recommend these papers for further reading.

Acknowledgment

The Editors would like to thank Ms. Chris Thivierge of the American Mathematical Society for her help and perseverance in seeing this volume through to publication.

References

[B] V. Bargmann, Irreducible unitary representatios of the Lorentz group, Ann. of Math. (2) **48** (1947), 568–640.

[G-N] I. M. Gel′fand and M. A. Naimark, Unitary representations of the classical groups, Trudy Mat. Inst. Steklov 36 (1950); German transl., Akademie-Verlag, Berlin, 1957.

[H-C] Harish-Chandra, The Plancherel formula for complex semisimple Lie groups, Transl. Amer. Math. Soc. **76** (1954), 485–528.

[J-L] H. Jacquet and R. P. Langlands, Automorphic forms on GL(2), Lecture Notes in Math., vol. 114, Springer-Verlag, Berlin and New York, 1970.

[L] R. P. Langlands, Problems in the theory of automorphic forms, Lectures in Modern Analysis and Applications III (C. T. Taam, ed.), Lecture Notes in Math., vol. 170, Springer-Verlag, Berlin and New York, 1970, pp. 18–61.

[W] E. P. Wigner, On unitary representations of the inhomogeneous Lorentz group, Ann. of Math. (2) **40** (1939), 149–204.

[Z] R. J. Zimmer, Ergodic theory and semisimple groups, Birkhauser, Boston, 1984.

Additional References from the *Bulletin of the American Mathematical Society*

1. C. W. Curtis, Representations of finite groups of Lie type, Bull. Amer. Math. Soc. (N.S.) 1 (1979), 721–757.
2. B. H. Gross, Some applications of Gel′fand pairs to number theory, Bull. Amer. Math. Soc. (N.S.), 24 (1991), 277–301.

3. R. A. Herb, Harish-Chandra and his work, Bull. Amer. Math. Soc. (N.S.), 25 (1991), 1–17.

4. R. E. Howe and Eng-Chye Tan, Homogeneous functions on light cones: The infinitesimal structure of some degenerate principal series representations, Bull. Amer. Math. Soc. (N.S.), 28 (1993), 1–74.

5. J. E. Humphreys, The Steinberg representation, Bull. Amer. Math. Soc. (N.S.), 16 (1987), 247–263.

6. G. W. Mackey, Functions on locally compact groups, Bull. Amer. Math. Soc., 56 (1950), 385–412.

7. ——, Harmonic analysis as the exploitation of symmetry—a historical survey, Bull. Amer. Math. Soc. (N.S.), 3 (1980), 543–698.

ON THE CHARACTERS OF A SEMISIMPLE LIE GROUP

HARISH-CHANDRA

Let G be a connected semisimple Lie group and let Z denote its center. If π is a representation [2c] of G on a Hilbert space $\mathfrak{H}$ we consider the space V consisting of all finite linear combinations of elements of the form

$$\int f(x)\pi(x)\psi dx \qquad (f \in C_c^\infty(G),\ \psi \in \mathfrak{H}),$$

where dx is the Haar measure of G and $C_c^\infty(G)$ is the set of all (complex-valued) functions on G which are everywhere indefinitely differentiable and which vanish outside a compact set. V is called the Gårding subspace of $\mathfrak{H}$. Let R and C be the fields of real and complex numbers respectively and $\mathfrak{g}_0$ the Lie algebra of G. We complexify $\mathfrak{g}_0$ to $\mathfrak{g}$ and denote by $\mathfrak{B}$ the universal enveloping algebra of $\mathfrak{g}$ [2a]. Then there exists a (uniquely determined) representation π_V of $\mathfrak{B}$ on V such that $\pi_V(X)\psi = \lim_{t\to 0} (1/t)\{\pi(\exp tX)\psi - \psi\}$ $(X \in \mathfrak{g}_0,\ \psi \in V,\ t \in R)$. Let $\mathfrak{Z}$ denote the center of $\mathfrak{B}$. We say that π is quasi-simple if there exist homomorphisms η and χ of Z and $\mathfrak{Z}$ respectively into C such that $\pi(\zeta)\phi = \eta(\zeta)\phi$, $\pi_V(z)\psi = \chi(z)\psi$ for all $\zeta \in Z$, $z \in \mathfrak{Z}$, $\phi \in \mathfrak{H}$ and $\psi \in V$. η is then called the central character and χ the infinitesimal character of π. An irreducible unitary representation is automatically quasi-simple [5].

Let A be a bounded linear operator on $\mathfrak{H}$. We say that A is of the trace class or A has a trace if for every complete orthonormal set $(\psi_j)_{j\in J}$ in $\mathfrak{H}$ thc series[1] $\sum_{j\in J} (\psi_j, A\psi_j)$ converges absolutely and its sum is independent of the choice of the complete orthonormal set.[2] We call this sum the trace of A and denote it by Sp A. Now suppose π is quasi-simple and irreducible. Then it can be shown (see [2e]) that for any $f \in C_c^\infty(G)$ the operator $\int f(x)\pi(x)dx$ is of the trace class. If we denote its trace by $T_\pi(f)$ we get a linear function T_π on $C_c^\infty(G)$ which is actually a distribution (see [4; and 2e]). We call this distribution the character of π. Our object is to try to determine T_π.

An address delivered before the New York meeting of the Society on February 25, 1955 by invitation of the Committee to Select Hour Speakers for Eastern Sectional Meetings; received by the editors March 28, 1955.

[1] As usual (ϕ, ψ) denotes the scalar product of the two elements ϕ and ψ in $\mathfrak{H}$.

[2] Actually it can be shown that this independence of the sum follows automatically from the absolute convergence of the series for every orthonormal base.

Let $x \to \mathrm{Ad}(x)$ $(x \in G)$ denote the adjoint representation of G. If λ is an indeterminate and I is the identity mapping of $\mathfrak{g}_0$ we consider the characteristic polynomial $P_x(\lambda) = \det(\lambda I - \mathrm{Ad}(x))$ of $\mathrm{Ad}(x)$. Let l be the highest integer such that $(\lambda - 1)^l$ divides $P_x(\lambda)$ for every $x \in G$. We expand $P_x(\lambda)$ in powers of $(\lambda - 1)$ and denote by $D(x)$ the coefficient of $(\lambda - 1)^l$ in this expansion. Then it is clear that D is an analytic function on G which, in view of our definition of l, cannot be identically zero. The integer l is called the rank of G. Let S denote the set of all $x \in G$ for which $D(x) = 0$. Then S is a closed, nowhere dense subset of G and its complement G' is open and everywhere dense in G. An element $x \in G$ is called singular or regular according as $x \in S$ or $x \in G'$. Since $\mathrm{Ad}(zx) = \mathrm{Ad}(x)$ $(z \in Z)$ it is obvious that $ZS = S$ and $ZG' = G'$. Also since $P_{yxy^{-1}}(\lambda) = P_x(\lambda)$ $(x, y \in G)$, it follows that $D(yxy^{-1}) = D(x)$ and therefore $ySy^{-1} = S$, $yG'y^{-1} = G'$ $(y \in G)$.

When speaking of a real differentiable (or analytic) manifold M let us agree to include the case when M is not connected but the various connected components of M, which are all manifolds in the usual sense, have the same dimension. Under this definition every open subset U of M is again a manifold. Let $C_c^\infty(U)$ denote the set of all complex-valued functions on M which are indefinitely differentiable and which vanish outside some compact subset of U. In particular suppose U is an open subset of G and F is a (complex-valued) function on U. We say that F is locally summable if it is summable on every compact subset of U with respect to the Haar measure of G. T being a distribution on G we say that $T = F$ on U if F is locally summable and

$$T(f) = \int f(x)F(x)dx$$

for all $f \in C_c^\infty(U)$. Our main result may now be stated as follows.

THEOREM 1. *Let π be an irreducible quasi-simple representation of G on $\mathfrak{H}$ and let T_π denote its character. Then there exists an analytic function F_π on G' such that $T_\pi = F_\pi$ on G'.*

Although in general G' is not connected, there always exist a finite number of connected components $G_1, \cdots, G_r$ of G' such that $ZG_i \cap ZG_j = \emptyset$ if $i \neq j$ and $G' = \bigcup_{i=1}^r ZG_i$. Moreover if η_π is the central character of π it is easy to show that $F_\pi(zx) = \eta_\pi(z) F_\pi(x)$ $(z \in Z, x \in G')$. Hence the knowledge of F_π on $G_1 \cup G_2 \cup \cdots \cup G_r$ is sufficient to determine it completely. On the other hand it is possible to give examples in which F_π vanishes everywhere on one of the components G_i without being zero identically on G'. However in case G is either a compact or a complex group, G' is connected and then F_π is completely determined by its restriction on any nonempty open subset of G'.

For any $x \in G$ and $f \in C_c^\infty(G)$ we define the function f^x by $f^x(y) = f(x^{-1}yx)$ $(y \in G)$. Then $f^x \in C_c^\infty(G)$ and $T_\pi(f^x) = T_\pi(f)$ (see [2e]). From this it follows that $F_\pi(xyx^{-1}) = F_\pi(y)$ $(x \in G, y \in G')$. Now let A be a maximal abelian subgroup of G which is not contained in S. (We call such a group a Cartan subgroup of G.) Obviously A is closed in G. Let $A' = A \cap G'$ and $V = \bigcup_{x \in G} xA'x^{-1}$. Then V is an open subset of G' and it is clear from the above remarks that F_π is determined completely on V as soon as we know it on A'. Let $\mathfrak{h}_0$ denote the subalgebra of $\mathfrak{g}_0$ corresponding to A. Then the complexification $\mathfrak{h}$ of $\mathfrak{h}_0$ is a Cartan subalgebra of $\mathfrak{g}$. Let W be the Weyl group (see [2b]) of $\mathfrak{g}$ with respect to $\mathfrak{h}$ so that W is a finite group of nonsingular linear transformations of $\mathfrak{h}$. If Λ is a linear function on $\mathfrak{h}$ and $s \in W$ we define the linear function $s\Lambda$ by the rule $s\Lambda(H) = \Lambda(s^{-1}H)$ $(H \in \mathfrak{h})$. Let $(H_1, \cdots, H_l)$ be a base for $\mathfrak{h}$ over C. A function P on $\mathfrak{h}$ is called a polynomial function if there exists a polynomial $p(x_1, \cdots, x_l)$ in l variables $(x_1, \cdots, x_l)$ with coefficients in C such that $P(H) = p(a_1, \cdots, a_l)$ if $H = a_1H_1 + \cdots + a_lH_l (a_i \in C)$. The degree of p is also called the degree of P. Clearly these definitions are independent of the choice of the base $(H_1, \cdots, H_l)$.

THEOREM 2. *Let π and F_π be as in Theorem 1. Then there exists a linear function Λ on $\mathfrak{h}$ with the following property. For any $a \in A'$ we can choose polynomial functions p_s $(s \in W)$ on $\mathfrak{h}$ such that*

$$F_\pi(a \exp H) = |D(a \exp H)|^{-1/2} \sum_{s \in W} p_s(H) e^{s\Lambda(H)}$$

for all H lying sufficiently near zero in $\mathfrak{h}_0$. Λ is unique up to an operation of W and if N is the number of elements σ in W such that $\Lambda = \sigma\Lambda$, the degree of every p_s is necessarily smaller than N.

In particular if $s\Lambda \neq \Lambda$ except when s is the unit element of W, p_s must all be constants. (It is hardly necessary to point out the resemblance of the above formula to the one given by Weyl [6] for the irreducible characters of a compact semisimple Lie group. It should also be compared with the results of Gelfand and Naimark [7] on the unitary characters of the complex classical groups (see also [2f, p. 511])). Although A' is not necessarily connected, it is possible to select a finite set $B_1, \cdots, B_r$ of its connected components such that $A' = \bigcup_{i=1}^r ZB_i$. Therefore in order to determine F_π on A', it is sufficient to know η_π and the restrictions of F_π on some nonempty open subsets of $B_1, \cdots, B_r$. Hence if Λ is known, Theorem 2 gives us a formula for F_π on A' in terms of a finite number of undetermined constants. On the other hand we shall see presently that Λ is completely determined (up to an operation of W) by the infinitesimal character χ_π of π.

Two Cartan subgroups A_1 and A_2 are said to be conjugate if $A_2 = xA_1x^{-1}$ for some $x \in G$. It is always possible to choose a finite number of distinct Cartan subgroups $A_1, \cdots, A_k$ such that every Cartan subgroup is conjugate to exactly one of these. Then if $A_i' = A_i \cap G'$ and $V_i = \bigcup_{x \in G} xA_i'x^{-1}$, G' is the disjoint union of $V_1, \cdots, V_k$. This shows that if η_π and χ_π are known, F_π is completely determined in terms of a finite number of constants.[3]

Now we come to a brief outline of the proof. Let M be a differentiable manifold, Q a linear mapping of $C_c^\infty(M)$ into itself and x_0 a point in M. We say that Q is a differential operator at x_0 if there exists a coordinate system $(t_1, \cdots, t_m)$ valid on an open neighborhood U of x_0 and indefinitely differentiable functions $g_{i_1 i_2 \cdots i_p}$ on U $(1 \leqq i_1, \cdots, i_p \leqq m,\ 0 \leqq p \leqq q)$ such that if $f \in C_c^\infty(U)$, Qf is zero outside U and

$$Qf = \sum_{0 \leqq p \leqq q} \sum_{1 \leqq i_1, \cdots, i_p \leqq m} g_{i_1 i_2 \cdots i_p} \frac{\partial^p}{\partial t_{i_1} \cdots \partial t_{i_p}} f$$

on U. If Q is a differential operator at every point in M we say simply that it is a differential operator (on M). T being a distribution on M and Q a differential operator we define a distribution $Q'T$ as follows:

$$(Q'T)(f) = T(Qf) \qquad (f \in C_c^\infty(M)).$$

In particular if g is an indefinitely differentiable function on M it defines a differential operator $Q: f \to gf$ $(f \in C_c^\infty(M))$. In this case we write gT to denote $Q'T$ so that $(gT)(f) = T(gf)$. It is clear that the product of two differential operators is again a differential operator and therefore the differential operators form an algebra.

Coming back to G, we note that every $X \in \mathfrak{g}_0$ may be regarded as a differential operator on G as follows:

$$(Xf)(x) = \left\{\frac{d}{dt} f(x \exp tX)\right\}_{t=0} \qquad (f \in C_c^\infty(G),\ x \in G,\ t \in R).$$

Thus it is easy to see that $\mathfrak{B}$ may be identified in a natural way with a subalgebra of the algebra of differential operators on G. Then for any $b \in \mathfrak{B}$ we have a linear transformation b' of the space of distribu-

[3] Actually it is possible to improve Theorems 1 and 2 and show that T_π coincides with an analytic function on an open subset of G which, in general, is larger than G' and therefore has fewer connected components.

tions on G. Since $(b_1b_2)'=b_2b_1$ $(b_1, b_2\in\mathfrak{B})$ the mapping $b\to b'$ is an anti-representation of $\mathfrak{B}$. Let ϕ denote the anti-automorphism of $\mathfrak{B}$ over C which is uniquely determined by the condition that $\phi(X)=-X$ $(X\in\mathfrak{g})$. Then $b\to(\phi(b))'$ $(b\in\mathfrak{B})$ is a representation of $\mathfrak{B}$. T being any distribution on G we now define $bT=(\phi(b))'T$ $(b\in\mathfrak{B})$. Then $(bT)(f)=T(\phi(b)f)$ $(f\in C_c^\infty(G))$. If χ_π is the infinitesimal character of π and $z\in\mathfrak{Z}$, $f\in C_c^\infty(G)$, it is easy to see that

$$\int(\phi(z)f)(x)\pi(x)\psi dx = \pi_0(z)\left(\int f(x)\pi(x)\psi dx\right)$$

$$= \chi_\pi(z)\int f(x)\pi(x)\psi dx$$

for all $\psi\in\mathfrak{H}$. (Here π_0 is the representation of $\mathfrak{B}$ on the Gårding subspace of $\mathfrak{H}$.) From this it follows that $zT_\pi=\chi_\pi(z)T_\pi$ for all $z\in\mathfrak{Z}$. Hence T_π is an eigen-distribution for each differential operator in $\mathfrak{Z}$.

On the other hand let A be a Cartan subgroup of G. Put $A'=A\cap G'$ and $V=\bigcup_{x\in G} xA'x^{-1}$ as before. We regard A' and V as open submanifolds of A and G respectively. Let G^* be the factor space G/A consisting of cosets of the form xA $(x\in G)$. If $h\in A$ and $x^*\in G^*$ we define $h^{x^*}=xhx^{-1}$ where x is any element in the coset x^*. Let dh and dx^* respectively denote the Haar measure on A and the invariant measure on G^*. Then we have the following lemma.

LEMMA 1. *There exists a distribution τ_π on A' with the following property. If $f\in C_c^\infty(V)$, $T_\pi(f)=\tau_\pi(g)$ where g is the function in $C_c^\infty(A')$ given by*

$$g(h) = |D(h)|\int_{G^*} f(h^{x^*})dx^* \qquad (h\in A').$$

Let $\mathfrak{h}_0$ be the Lie algebra of A. Any element $H\in\mathfrak{h}_0$ may be regarded as a differential operator on A so that if $g\in C_c^\infty(A)$,

$$(Hg)(h) = \left\{\frac{d}{dt}g(h\exp tH)\right\}_{t=0} \qquad (h\in A,\ t\in R).$$

Let $\mathfrak{h}$ be the subspace of $\mathfrak{g}$ spanned by $\mathfrak{h}_0$ over C and $\mathfrak{U}$ the subalgebra of $\mathfrak{B}$ generated by $(1, \mathfrak{h})$. Then $\mathfrak{U}$ may be identified in a natural way with a subalgebra of the algebra of differential operators on A. For any distribution τ on A' and $u\in\mathfrak{U}$ we define a distribution $u\tau$ on A' as follows:

$$(u\tau)(g) = \tau(\phi(u)g) \qquad (g\in C_c^\infty(A')).$$

Here ϕ is the automorphism of $\mathfrak{U}$ given by $\phi(H)=-H$ $(H\in\mathfrak{h})$.

For every root α of $\mathfrak{g}$ (with respect to $\mathfrak{h}$), choose an element $X_\alpha \neq 0$ in $\mathfrak{g}$ such that $[H, X_\alpha] = \alpha(H)X_\alpha$ for all $H \in \mathfrak{h}$. We introduce some lexicographic order (see [**2b**]) in the set of all roots and denote by P the set of positive roots under this order. Put $\mathfrak{n} = \sum_{\alpha \in P} CX_\alpha$. Then for every $z \in \mathfrak{Z}$ there exists a unique element $\gamma'(z) \in \mathfrak{U}$ such that $z - \gamma'(z) \in \mathfrak{B}\mathfrak{n}$ (see [**2b**, p. 72]). If $2\rho = \sum_{\alpha \in P} \alpha$ there exists a unique automorphism λ of $\mathfrak{U}$ such that $\lambda(1) = 1$ and $\lambda(H) = H - \rho(H)$ $(H \in \mathfrak{h})$. We put $\gamma(z) = \lambda(\gamma'(z))$ $(z \in \mathfrak{Z})$. Let W be the Weyl group of $\mathfrak{g}$ with respect to $\mathfrak{h}$. It is clear that every $s \in W$ defines an automorphism $u \rightarrow u^s$ of $\mathfrak{U}$ such that $1^s = 1$ and $H^s = sH$ $(H \in \mathfrak{h})$. An element $u \in \mathfrak{U}$ is called an invariant if $u = u^s$ for all $s \in W$. Let J be the subalgebra of $\mathfrak{U}$ consisting of all invariants. Then (see [**2b**, Lemma 38]) the mapping $z \rightarrow \gamma(z)$ $(z \in \mathfrak{Z})$ defines an isomorphism of $\mathfrak{Z}$ onto J. Now let Λ be a linear function on $\mathfrak{h}$. We can extend it uniquely to a homomorphism of $\mathfrak{U}$ into C which takes the value 1 at 1. We agree to denote this extension also by Λ. Then as shown in [**2b**, Theorem 5] we can choose Λ in such a way that $\chi_\pi(z) = \Lambda(\gamma(z))$ for all $z \in \mathfrak{Z}$. Λ is determined up to an operation of W by this condition.

Let $\Delta(h) = |D(h)|^{1/2}$ $(h \in A')$. Then Δ is an analytic function on A' and therefore $\sigma_\pi = \Delta\tau_\pi$ is a well-defined distribution on A'. Now if we transform the differential equation $zT_\pi = \chi_\pi(z)T_\pi$ $(z \in \mathfrak{Z})$ for T_π into a differential equation for σ_π we get the following result which is one of the main steps of our argument.

LEMMA 2. *τ_π being as in Lemma* 1 *put $\sigma_\pi = \Delta\tau_\pi$. Then*

$$\gamma(z)\sigma_\pi = \chi_\pi(z)\sigma_\pi$$

for every $z \in \mathfrak{Z}$.

Now choose a linear function Λ on $\mathfrak{h}$ such that $\chi_\pi(z) = \Lambda(\gamma(z))$ for $z \in \mathfrak{Z}$. Let ζ be an indeterminate and u an element in $\mathfrak{U}$. Since $\mathfrak{U}$ is abelian we can consider the polynomial

$$\prod_{s \in W} (\zeta - u^s).$$

It is clear that every coefficient of this polynomial lies in J and therefore if w is the order of W, there exist uniquely determined elements $z_1(u), \cdots, z_w(u)$ in $\mathfrak{Z}$ such that

$$\prod_{s \in W}(\zeta - u^s) = \zeta^w + \gamma(z_1(u))\zeta^{w-1} + \gamma(z_2(u))\zeta^{w-2} + \cdots + \gamma(z_w(u)).$$

On replacing ζ by u we immediately get the identity

$$u^w + u^{w-1}\gamma(z_1(u)) + u^{w-2}\gamma(z_2(u)) + \cdots + \gamma(z_w(u)) = 0$$

in $\mathfrak{U}$ if we recall that $\mathfrak{U}$ is abelian. Now apply the left side to σ_π and use Lemma 2. Then

$$u^w\sigma_\pi + \chi_\pi(z_1(u))u^{w-1}\sigma_\pi + \cdots + \chi_\pi(z_w(u))\sigma_\pi = 0.$$

But $\chi_\pi(z_j(u)) = \Lambda(\gamma(z_j(u)))$, $1 \leqq j \leqq w$, and since Λ is a homomorphism of $\mathfrak{U}$ into C it is obvious that

$$\prod_{s\in W} (\zeta - \Lambda(u^s)) = \zeta^w + \Lambda(\gamma(z_1(u)))\zeta^{w-1} + \cdots + \Lambda(\gamma(z_w(u))).$$

Therefore the above differential equation for σ_π may be written in the form

$$\prod_{s\in W} (u - \Lambda(u^s))\sigma_\pi = 0.$$

However if $H_1, \cdots, H_l$ is a base for $\mathfrak{h}_0$ over R and $\square = H_1^2 + \cdots + H_l^2$ it is obvious that the differential equation

$$\prod_{s\in W} (\square - \Lambda(\square^s))\sigma_\pi = 0$$

is of the *elliptic* type (see Gårding [**1**]). Hence it follows from the work of Schwartz [**4**, p. 137] and John [**3a**, **3b**] that σ_π must be an analytic function on A'. Now if we take into account Theorem 5 of [**2b**] and the fact that $\prod_{s\in W} (u - \Lambda(u^s))\sigma_\pi = 0$ for every $u \in \mathfrak{U}$, we get Theorem 2 without difficulty.

π being any quasi-simple irreducible representation we denote by T_π, η_π, and χ_π respectively the character, the central character, and the infinitesimal character of π. Also we denote by F_π the analytic function on G' such that $T_\pi = F_\pi$ on G'. Let T be a distribution on G. Since D is an analytic function on G the product D^mT $(m \geqq 0)$ is a well-defined distribution. It is then possible to prove the following result.

LEMMA 3. *There exists an integer $m \geqq 0$ with the following property. Suppose $\pi_1, \cdots, \pi_k$ is a finite set of quasi-simple irreducible representations and $c_1F_{\pi_1} + \cdots + c_kF_{\pi_k} = 0$ $(c_i \in C)$. Then if $T = c_1T_{\pi_1} + \cdots + c_kT_{\pi_k}$, $D^mT = 0$.*

From this lemma one can deduce the following theorem.

THEOREM 3. *Let π_0 be a quasi-simple irreducible representation of G such that F_{π_0} is not identically zero. Then, apart from infinitesimal equivalence* [**2d**, p. 230], *there exist only a finite number of quasi-simple irreducible representations π such that $F_\pi = F_{\pi_0}$.*

Let $\eta \neq 0$ and $\chi \neq 0$ respectively be given homomorphisms of Z and $\mathfrak{Z}$ into C. Let ω denote the set of all quasi-simple irreducible representations π of G such that $\eta_\pi = \eta$ and $\chi_\pi = \chi$. As we have seen, it follows from Theorems 1 and 2 that the functions F_π ($\pi \in \omega$) span a finite-dimensional vector space over C. Hence if ω is not empty, we can choose a finite set of elements $\pi_1, \cdots, \pi_k$ in ω such that $F_{\pi_1}, \cdots, F_{\pi_k}$ form a base for this vector space. Then if π is any representation in ω, $F_\pi = c_1 F_{\pi_1} + \cdots + c_k F_{\pi_k}$ ($c_j \in C$). If one could conclude from this equation that $T_\pi = c_1 T_{\pi_1} + \cdots + c_k T_{\pi_k}$ it would follow (see [2e, Theorem 6]) that π is infinitesimally equivalent to some π_j ($1 \leqq j \leqq k$) and therefore, apart from infinitesimal equivalence, ω has only a finite number of representations. Therefore it is important to consider the following question.

Let $\pi_1, \cdots, \pi_k$ be a finite set of quasi-simple irreducible representations of G such that $c_1 F_{\pi_1} + \cdots + c_k F_{\pi_k} = 0$ ($c_i \in C$). Then is it always true that $c_1 T_{\pi_1} + \cdots + c_k T_{\pi_k} = 0$?

I believe the answer is yes but do not know how to prove it.

References

1. L. Gårding, Math. Scand. vol. 1 (1953) pp. 55–72.
2a. Harish-Chandra, Ann. of Math. vol. 50 (1949) pp. 900–915.
2b. ———, Trans. Amer. Math. Soc. vol. 70 (1951) pp. 28–96.
2c. ———, Proc. Nat. Acad. Sci. U.S.A. vol. 37 (1951) pp. 366–369.
2d. ———, Trans. Amer. Math. Soc. vol. 75 (1953) pp. 185–243.
2e. ———, Trans. Amer. Math. Soc. vol. 76 (1954) pp. 234–253.
2f. ———, Trans. Amer. Math. Soc. vol. 76 (1954) pp. 485–528.
3a. F. John, Comm. Pure Appl. Math. vol. 3 (1950) pp. 273–304.
3b. ———, *Proceedings of the symposium on spectral theory and differential problems*, Stillwater, Okla., 1951, pp. 113–175.
4. L. Schwartz, *Théorie des distributions* I, Paris, Hermann, 1950.
5. I. E. Segal, Proc. Amer. Math. Soc. vol. 3 (1952) pp. 13–15.
6. H. Weyl, Math. Zeit. vol. 24 (1925) pp. 328–395.
7. I. M. Gelfand and M. A. Naimark, Trudi Mat. Inst. Steklova vol. 36 (1950)

Columbia University

INFINITE-DIMENSIONAL GROUP REPRESENTATIONS[1]

GEORGE W. MACKEY

1. Introduction. Let S be a set in which there is given a σ-field of "Borel sets" and let G be a topological group. Let sx be defined for all s in S and all x in G in such a manner that

(a) $sx_1x_2 = (sx_1)x_2$,

(b) $se = s$,

(c) $s, x \to sx$ is a Borel function

from $S \times G$ to S. Under these circumstances we shall say that S is a Borel G space. We shall usually assume that S is *standard* in the sense that it is isomorphic as a Borel space to a Borel subset of a separable complete metric space. Let μ be a Borel measure in S, that is, a σ-finite countably additive measure defined on all Borel subsets of S. If $\mu(Ex) = \mu(E)$ for all E we shall say that μ is invariant. Given an invariant μ in the Borel space S we may form the Hilbert space $\mathfrak{L}^2(S, \mu)$ and observe that for each x in G the operator L_x such that $(L_x(f))(s) = f(sx)$ is a unitary operator. Moreover $x \to L_x$ is a unitary representation of the group G in a sense to be made precise below. When $S = G$, G is locally compact and sx is group multiplication, the measure μ exists and is essentially unique. The unitary representation L in this case is called the *regular representation* of G.

Consider the special case in which G is the compact group K of all rotations in the plane—or equivalently the group obtained from the additive group of the real line by identifying numbers which differ by integral multiples of 2π.

The functions e^{inx} $(n = 0, \pm 1, \pm 2, \cdots)$ form a basis for $\mathfrak{L}^2(K)$ and each member of this basis generates a one-dimensional subspace which is invariant under each L_x. We have, in a sense to be defined below, a decomposition of the regular representation of K into ir-

---

The preparation of this paper was supported in part by the National Science Foundation under grant NSF-G 16428 to the American Mathematical Society; received by the editors May 6, 1963.

[1] Except for the appendix this paper constitutes the text of the "notes" distributed in pamphlet form to the audience at the author's Colloquium Lectures delivered during the 1961 summer meeting of the American Mathematical Society in Stillwater, Oklahoma. These notes were written earlier the same summer while the author was a Walker Ames visiting professor of Mathematics at the University of Washington. In the original version the bibliography was included informally in the text of §11. The supplementary bibliography was written to accompany the appendix.

reducible parts. The fact that this decomposition exists is a simple reformulation of one of the fundamental theorems in the theory of Fourier series. The celebrated Plancherel formula for Fourier transforms provides a similar decomposition for the regular representation of the additive group of the real line. However here one has a continuous decomposition into "infinitesimal" parts.

The theory we are going to describe in these lectures begins with and is to some extent dominated by attempts to answer the following two questions.

(a) To what extent can an arbitrary unitary representation of a group be analyzed into irreducible ones.

(b) Given a group G to what extent can one find all possible irreducible representations of G.

Such answers as are available represent extensions of classical harmonic analysis in various directions and include a large part of what group theory has contributed to analysis. The particular representations defined by invariant measures in Borel G spaces are much more general than they at first appear to be and certain natural generalizations of them come close to being exhaustive. Certainly they are the most important examples occurring naturally in analysis and physics.

When G is finite the theory is purely algebraic in character and (barring classical harmonic analysis whose group theoretical character was clearly recognized only much later) it is with this case that the subject begins. Started by Frobenius in 1900 it was dominated by the names of Frobenius and Schur for the next decade or so. In the mid 1920's it was extended to compact Lie groups by Schur, Weyl and others. The celebrated Peter-Weyl theorem says that the regular representation of a compact Lie group is a direct sum of finite-dimensional irreducible representations. With A. Haar's (1933) discovery of the existence of an invariant measure on any locally compact group it was realized that the Peter-Weyl theorem and much of the rest of the theory applies to arbitrary compact groups. When the compactness hypothesis is discarded there are serious difficulties due to the fact that one may have to deal with continuous decompositions and infinite-dimensional irreducible components. The second difficulty disappears when the group is commutative for then all irreducible representations are one-dimensional and one obtains a close analogue of the theory of the Fourier transform. This theory is based in part on the Pontrjagin-van Kampen duality theory for locally compact groups which was developed in the mid 1930's.

Insofar as unitary representations of compact and commutative groups are concerned the theory was well nigh complete in 1938 and is described in detail in A. Weil's classic treatise. We shall be chiefly concerned with the post World War II theory in which one must deal simultaneously with continuous decompositions and infinite-dimensional irreducible constituents. Although there was an important paper by Wigner in 1939, another by Gelfand and Raikov in 1943 and a few preliminary announcements by Gelfand and Naimark in 1946, this part of the theory did not really get under way until 1947 when papers began to appear at the rate of five to ten a year.

In the fourteen years which have elapsed since this deluge began the theory has developed far too much to be summarized adequately in ten lectures—let alone four. Since we must select we hope we may be excused for emphasizing those parts of the theory with which we have been most directly concerned. Actually we have reason to believe that a fairly large section of the theory will ultimately be fitted into the framework which we shall outline here.

2. **Some basic notions.** We shall deal throughout with topological groups G which are *locally compact* and are separable in the sense that there exists a countable basis for the open sets. By a *unitary representation* L of such a group G we shall mean a mapping $x \rightarrow L_x$ from G to the unitary operators in some separable Hilbert space $\mathcal{H}(L)$ which has the following two properties:

(a) $L_{xy} = L_x L_y$ for all x and y in G.

(b) $x \rightarrow L_x(\phi)$ is a continuous function from G to $\mathcal{H}(L)$ for all ϕ in $\mathcal{H}(L)$.

It is known that property (b) is implied by the ostensibly much weaker property (b′) $x \rightarrow (L_x(\phi), \psi)$ is measurable with respect to Haar measure in G for all ϕ and ψ in $\mathcal{H}(L)$.

By an *intertwining operator* for L and M we shall mean a bounded linear operator T from $\mathcal{H}(L)$ to $\mathcal{H}(M)$ such that $TL_x = M_x T$ for all x in G. The set of all such is a vector space which we denote by $\mathcal{R}(L, M)$. When $L = M$, $\mathcal{R}(L, M) = \mathcal{R}(L, L)$ is a (weakly closed) subalgebra of the algebra of all bounded linear operators on $\mathcal{H}$. It is called the *commuting algebra* of L. If $\mathcal{R}(L, M)$ contains a unitary map U of $\mathcal{H}(L)$ onto $\mathcal{H}(M)$ then $UL_xU^{-1} = M_x$ and L and M are said to be equivalent representations of G. We shall write $L \simeq M$. It is seldom necessary to distinguish between equivalent representations.

If $L^1, L^2, \cdots$ are unitary representations of G we define their *direct sum* $L = L^1 \oplus L^2 \oplus \cdots$ as follows. $\mathcal{H}(L)$ is the direct sum of the Hilbert spaces $\mathcal{H}(L^1), \mathcal{H}(L^2), \cdots$, i.e. the set of all sequences $\phi^1, \phi^2, \cdots$ with $\|\phi^1\|^2 + \|\phi^2\|^2 + \cdots < \infty$, $\phi^j \in \mathcal{H}(L^j)$ and the obvious definition of the Hilbert space operations. $L_x(\phi^1, \phi^2, \cdots) = L_x(\phi^1), L_x(\phi^2), \cdots$.

Let $\mathcal{H}_1$ be a closed subspace of $\mathcal{H}(L)$. If $L_x(\phi) \in \mathcal{H}_1$ for all $x \in G$ and all $\phi \in \mathcal{H}_1$ we say that $\mathcal{H}_1$ is an invariant subspace of $\mathcal{H}(L)$. If $L_x^{\mathcal{H}_1}$ denotes the restriction of L_x to $\mathcal{H}_1$ then $x \rightarrow L_x^{\mathcal{H}_1}$ is a unitary representation of G, which we call the *subrepresentation* defined by $\mathcal{H}_1$. If P is the unique projection whose range is $\mathcal{H}_1$ then $\mathcal{H}_1$ is invariant if and only if $P \in \mathcal{R}(L, L)$ so that the subrepresentations of L correspond one-to-one to the projections in $\mathcal{R}(L, L)$. We shall use L^P as an alternative notation for $L^{\mathcal{H}_1}$. It is easy to see that L is equivalent to $L^P \oplus L^{1-P}$. Thus the following statements about L are equivalent:

(a) L is not equivalent to a (nontrivial) direct sum.

(b) $\mathcal{R}(L, L)$ contains no nontrivial projections.

(c) L has no (nontrivial) subrepresentations.

Using the spectral theorem (b) can be shown to be equivalent to

(b′) $\mathcal{R}(L, L)$ contains only scalar multiples of the identity I.

When any one and hence all of these equivalent statements hold L is said to be *irreducible.*

Let there exist a nonzero member T of $\mathcal{R}(L, M)$. Let N_T denote the set of all ϕ with $T(\phi) = 0$ and let $\overline{W}_T$ denote the closure of the set of all $T(\phi)$. It is immediate that N_T and $\overline{W}_T$ are invariant and it follows from an extension of a classical lemma of Schur that the subrepresentation of L defined by $N_T^{\perp}$ is equivalent to the subrepresentation of M defined by $\overline{W}_T$. Thus $\mathcal{R}(L, M)$ reduces to 0 if and only if no subrepresentation of L is equivalent to any subrepresentation of M. When this is so we say that L and M are *disjoint* and write $L \;\delta\; M$. To pass to the opposite extreme we say that L and M are *quasi equivalent* if no (nonzero) subrepresentation of L is disjoint from M and no (nonzero) subrepresentation of M is disjoint from L. For example if L is irreducible then $L \oplus L$ is *quasi equivalent* but not equivalent to L. Moreover two inequivalent irreducible representations are always disjoint.

It is not hard to show that L^P and L^{1-P} are disjoint if and only if P is in the center of $\mathcal{R}(L, L)$. When this center reduces to multiples of the identity so that it is impossible to decompose L as a direct sum of disjoint parts we say that L is *primary.* When the center is all of $\mathcal{R}(L, L)$ so that $\mathcal{R}(L, L)$ is commutative and every decomposition of L as a direct sum has disjoint parts we say that L is *multiplicity free.*

If $L = L^1 \oplus L^2 \oplus \cdots$ where the L^j are all irreducible then L is primary if and only if $L^i \simeq L^j$ for all i and j and L is multiplicity free if and only if $L^i \simeq L^j$ implies $i = j$.

If $n = 1, 2, \cdots, \aleph_0$ then nL will denote the direct sum $L \oplus L \oplus \cdots$ taken to n terms. If $\aleph_0 L \simeq L$ we shall say that L is infinite. Clearly $\aleph_0 L$ is infinite whenever $L \neq 0$. If no subrepresentation of L is infinite we shall say that L is finite. If L is neither finite nor infinite there exists a unique projection P_f in the center of $\mathcal{R}(L, L)$ such that L^{P_f} is finite and L^{1-P_f} is infinite. Of course $L^{P_f} \mathrel{\delta} L^{1-P_f}$.

Let $L=nL^0$ where L^0 is irreducible. Then L is primary and has an irreducible subrepresentation. Conversely if L is primary and has an irreducible subrepresentation L^0 then $L\simeq nL^0$ for some $n=1, 2, \cdots, \aleph_0$. Any two irreducible subrepresentations L^0 and L^{00} of L are equivalent and $nL^0\simeq n'L^{00}$ implies $n=n'$ and $L^0\simeq L^{00}$. $n=n'$ is called the *multiplicity* of L. A primary representation with an irreducible subrepresentation is said to be of *type* I. There are two further possibilities. It may be that, though L is not irreducible, every proper subrepresentation of L is equivalent to L. Such a representation is necessarily primary and infinite and is said to be primary of *type* III. A primary representation which has a finite subrepresentation but no irreducible subrepresentation is said to be of *type* II. If L is finite, primary and of type II it may always be "cut in half"; that is there exists a primary subrepresentation L^1 such that $L\simeq L^1\oplus L^1$ and L^1 is unique to within equivalence. By iteration $(1/2^k)L$ is defined to within equivalence for all $k=1, 2, \cdots$. Thus we may define λL where λ is any positive real number by writing λ in the form $n + 1/2^{k_1} + 1/2^{k_2} + 1/2^{k_3} + \cdots$ and setting $\lambda L = nL + (1/2^{k_1})L + (1/2^{k_2})L + \cdots$. It can be shown that each λL is finite, primary and quasi equivalent to L and that every representation quasi equivalent to L is either finite and equivalent to a unique λL or else infinite and equivalent to $\aleph_0 L$. It follows that the set of all representation quasi equivalent to a given primary representation has a natural ordering which is either that of the positive integers and $\aleph_0$, the positive reals and $\aleph_0$, or $\aleph_0$ alone.

If $L=L^1\oplus L^2\oplus \cdots$ where the L^j are disjoint and primary then the following statements are easily proved.

(a) Each L^j is of type I if and only if L is quasi equivalent to a multiplicity free representation.

(b) Each L^j is of type III if and only if every representation quasi equivalent to L is equivalent to L.

(c) Each L^j is of type II if and only if L is quasi equivalent to a finite representation and every representation quasi equivalent to L is of the form $M\oplus M$ for some M.

Thus it is natural to make the following definitions. A (not necessarily primary) representation L is of type I if it is equivalent to a multiplicity free representation. It is of type III if every representation quasi equivalent to L is equivalent to L. It is of type II if it is quasi equivalent to a finite representation and every representation quasi equivalent to L is of the form $M \oplus M$ for some M. It can be shown that for any L there exist unique projections P^{I}, P^{II}, and P^{III} in the center of $\mathfrak{R}(L, L)$ such that

(a) $L^{P^j} \circ L^{P^k}$ for $j \neq k$,

(b) $L \simeq L^{P^{I}} \oplus L^{P^{II}} \oplus L^{P^{III}}$,

(c) L^{P^j} is of type j whenever $P^j \neq 0$.

3. **The decomposition of representations.** Let us say that L is discretely decomposable if $L \simeq L^1 \oplus L^2 \oplus \cdots$ where the L^j are irreducible. If M is irreducible then the dimension of $\mathfrak{R}(M, L^1 \oplus L^2 \oplus \cdots)$ is equal to the number of indices k for which $M \simeq L^k$. We shall call this number the multiplicity of M in $L^1 \oplus L^2 \oplus \cdots$. It is clear that L is uniquely determined by the non-negative integers which are the dimensions of the spaces $\mathfrak{R}(M, L)$ whenever L is discretely decomposable. Thus to know all irreducible unitary representations (to within equivalence) is to know all discretely decomposable unitary representations to within equivalence. When G is compact every unitary representation is discretely decomposable but for more general groups discretely decomposable representations are rather special. While it is true that every unitary representation is (in a sense to be defined below) a direct integral of irreducible representations, this decomposition is too nonunique to be useful in the general case in reducing the study of unitary representations to that of irreducible ones. On the other hand there are many important noncompact groups which are known to have only type I representations. For these the direct integral decomposition into irreducibles is essentially unique and there is a sense in which finding the general unitary representation can be reduced to finding the general *irreducible* unitary representation—both to within equivalence of course. When type II and type III representations exist there is a somewhat weaker sense in which finding the most general unitary representation can be reduced to finding the most general *quasi* equivalence class of primary representations. However there is no case known in which one has nontype I representations and can describe (in any reasonable sense) the set of all equivalence classes of irreducibles—let alone the set of all quasi equivalence classes of primary representations.

Let S be a standard Borel space and let μ be a finite measure in S. Let there be given a Hilbert space $\mathcal{H}$ and for each $s \in S$ let L^s be a unitary representation of our group G. Let $L^\bullet$ be a Borel function of s in the sense that $(L^s_x(\phi) \cdot \psi)$ is a complex-valued Borel function on $S \times G$ for all ϕ and ψ in $\mathcal{H}$. We then define a new unitary representation L as follows. $\mathcal{H}(L) = \mathcal{L}^2(S, \mu, \mathcal{H})$; that is, the Hilbert space of all $\mathcal{H}$-valued square summable functions on S. For each $x \in G$ and each

$g \in \mathcal{L}^2(S, \mu, \mathcal{H})$, $L_x(g)(s) = L_x^s(g(s))$. We call L the *direct integral* of the L^s with respect to L^s and denote it by $\int_S L^s d\mu(s)$. [If the L^s are given to us as acting in different Hilbert spaces $\mathcal{H}^s$ or as equivalence classes of representations then we may make the above construction whenever it is possible to map all $\mathcal{H}^s$ of a given dimension onto a single Hilbert space in such a manner that the above condition on L^s is satisfied. It may be shown that the equivalence class of $\int_S L^s d\mu(s)$ does not depend upon which mappings are chosen and that the mappings exist if and only if the Hilbert space "bundle" of all pairs s, ϕ with $\phi \in \mathcal{H}_s$ can be made into a Borel space having certain simple natural properties. In practice there is often a "natural" Borel structure in the bundle of pairs s, ϕ.]

Let ν be a second measure in S having the same null sets as μ. Let ρ be the Radon-Nikodym derivative of ν with respect to μ. Then $f \rightarrow \sqrt{(\rho)}f$ defines a unitary map of $\mathcal{L}^2(S, \nu, \mathcal{H})$ on $\mathcal{L}^2(S, \mu, \mathcal{H})$ which sets up an equivalence between $\int_S L^s d\nu(s)$ and $\int_S L^s d\mu(s)$. In other words in forming direct integrals it is only the family of null sets which counts—not the measure itself. It is even possible to redefine $\int_S L^s d\mu(s)$ in such a way as to avoid choosing a particular μ but we shall not give details here. It is obvious that $\int_{S'} L^s d\mu(s) \simeq \int_S L^s d\mu(S)$ whenever $S' \subseteq S$ and $\mu(S - S') = 0$ and it is true but not obvious that $\int_S L^s d\mu(s) \simeq \int_S M^s d\mu(s)$ if $L^s \simeq M^s$ for almost all s.

For each Borel subset E of S let P_E denote the projection in $\mathcal{L}^2(S, \mu, \mathcal{H})$ which takes f into $\phi_E f$ where ϕ_E is the characteristic function of E. Then $E \rightarrow P_E$ is a *projection-valued* measure; that is a function from Borel sets to projections having the following three properties:

(a) $P_0 = 0$ and $P_S = I$ (I is the identity operator),

(b) $P_{E \cap F} = P_E P_F$,

(c) if $E_i \cap E_j = 0$ for $i \neq j$ then $P_{(E_1 \cup F_2 \cup \cdots)} = P_{E_1} + P_{E_2} + \cdots$.

Moreover each P_E is in $\mathcal{R}(L, L)$. Conversely let L' be any unitary representation of G and let P' be a projection-valued measure defined on the Borel subsets of a standard Borel space S such that P'_E is in $\mathcal{R}(L, L)$ for all E. Then there exists an assignment $s \rightarrow L^s$ of a unitary representation to each s in S and a finite measure μ in S such that $L = \int L^s d\mu(s)$ exists and is equivalent to L'. In addition the unitary mapping setting up the equivalence may be chosen so that it carries P' into the projection-valued measure P associated with the direct integral $\int L^s d\mu(s)$ in the manner described above. The null sets of μ are just the sets E such that $P'_E = 0$ and to within equivalence L^s is uniquely determined for almost all s. Thus a representation L together with a projection-valued measure P such that $P_E L_x = L_x P_E$ for all E and x is essentially the same thing as a representation L together with a particular realization of L as a direct integral.

Actually the space S can be eliminated from the above. Let P^1 and P^2 be projection-valued measures defined on standard Borel spaces S_1 and S_2. Suppose that the range of P^1 is identical with the range of P^2, i.e. that every P^1_E is equal to some P^2_E and vice versa. Then there exists a one-to-one map θ of S_1 on S_2 such that E is a Borel set if and only if $\theta(E)$ is a Borel set and such that $P^1_{\theta(E)} = P^2_E$ for all E. Let $\mathfrak{F}$ be a family of projections in the (separable) Hilbert space $\mathfrak{H}$. Then $\mathfrak{F}$ is the range of a projection-valued measure P defined on a standard Borel space if and only if it has the following properties:

(a) $Q_1Q_2 = Q_2Q_1 \in \mathfrak{F}$ whenever $Q_1 \in \mathfrak{F}$ and $Q_2 \in \mathfrak{F}$.

(b) If $Q \in \mathfrak{F}$ then $1 - Q \in \mathfrak{F}$.

(c) If $Q_1, Q_2, \cdots \in \mathfrak{F}$ and $Q_iQ_j = 0$ for $i \neq j$ then $Q_1 + Q_2 + \cdots \in \mathfrak{F}$.

A family $\mathfrak{F}$ having these three properties is called a *σ Boolean algebra of projections.*

It follows from the above that there is a natural one-to-one correspondence between the representations of a given L as a direct integral and the different σ Boolean algebras of projections in $\mathfrak{R}(L, L)$. Let $\int_S L^s d\mu(s)$ be the essentially unique direct integral decomposition of L associated with the σ Boolean algebra of projections $\mathfrak{F}$ in $\mathfrak{R}(L, L)$. One knows the following about the connections between $\mathfrak{F}$ and the L^s.

(a) Almost every L^s is irreducible if and only if $\mathfrak{F}$ is maximal in $\mathfrak{R}(L, L)$.

(b) Almost every L^s is primary if and only if $\mathfrak{F}$ contains the center of $\mathfrak{R}(L, L)$.

(c) If $\mathfrak{F}$ is equal to the center of $\mathfrak{R}(L, L)$ then there exists a subset N of S such that $\mu(N) = 0$ and such that $L^s \mathrel{\delta} L^{s'}$ if $s \in S - N$ and $s' \in S - N$.

It follows from (a) and Zorn's lemma that every L has a direct integral decomposition into irreducible parts. This decomposition may vary greatly from one maximal $\mathfrak{F}$ to another when L is not of type I. For example there exists a type II primary representation and two direct integral decompositions $L \simeq \int_S L^s d\mu(s)$, $L \simeq \int_{S'} L'^t d\nu(t)$ such that no L^s is equivalent to any L'^t and such that each L^s is irreducible and each L'^t is irreducible.

The decomposition into disjoint primary components defined by taking $\mathfrak{F}$ to be all projections in the center of $\mathfrak{R}(L, L)$ is of course canonical and is called the *central decomposition* of L.

When L is multiplicity free the set of all projections in the center of $\mathfrak{R}(L, L)$ is the set of all projections in $\mathfrak{R}(L, L)$ and hence is maximal. Thus, in this case, the central decomposition of L is into inequivalent irreducible parts. Let $\hat{G}$ denote the set of all equivalence classes of irreducible representations of G. Let L be multiplicity free and let $\int L^s d\mu(s)$ denote the central decomposition chosen so that the L^s are all irreducible. Then the mapping $s \rightarrow L^s$ defines a mapping of S into $\hat{G}$ and takes μ into a measure $\hat{\mu}$ in $\hat{G}$. The null sets of $\hat{\mu}$ are independent of the arbitrary features in $\int_S L^s d\mu(s)$ so that each multiplicity free L defines a unique measure class C_L which in turn determines L to within equivalence. It is possible to define the notion of Borel set in $\hat{G}$ in a natural manner and when this is done the measure class C_L turns out to be a Borel measure class. However if G has nontype I representations there may exist Borel measure classes in $\hat{G}$ which do not arise from any multiplicity free representation. A fairly interesting unsolved problem is that of characterizing intrinsically the measure classes which arise from multiplicity free representations. On the other hand it is known that $\hat{G}$ is a standard Borel space if and only if every representation of G is of type I. When either and hence both of these conditions hold, every Borel measure class in $\hat{G}$ comes from a multiplicity free representation and we have a natural one-to-one correspondence between the measure classes and the equivalence classes of multiplicity free representations. Now the study of type I representations can be reduced (without using direct integral theory) to the study of multiplicity free representations. Hence for type I groups (defined as those having only type I representations) the study of the general unitary representation can be reduced to the study of measure classes in $\hat{G}$—and thus to the determination of $\hat{G}$ and a classical problem in real variable theory.

The "type I to multiplicity free" reduction is a consequence of the following theorem. Let L be any type I representation. Then there exist disjoint multiplicity free representations L^∞, L^1, L^2, $\cdots$ uniquely determined by L to within equivalence such that $L \simeq \aleph_0 L^\infty \oplus L^1 \oplus 2L^2 \oplus \cdots$. Of course some of the L^j may be missing.

Amongst the separable locally compact groups which are known to have only type I representations are the compact groups, the commutative groups, the connected semisimple Lie groups and the connected real algebraic groups. However there are many examples of connected solvable Lie groups which have nontype I representations.

It is known that the regular representation of a discrete group is of type II whenever the subgroup generated by the finite conjugate classes has infinite index. Moreover it is known that a countable restricted direct product of finite groups always has a type II regular representation. No examples are known of type I discrete groups except for finite extensions of commutative groups.

4. Characters, spherical functions and the Plancherel formula. For any Hilbert space let J denote the canonical antilinear mapping of $\mathcal{H}$ on the Banach space adjoint $\overline{\mathcal{H}}$ of $\mathcal{H}$ and if T is a bounded linear operator from $\mathcal{H}_1$ to $\mathcal{H}_2$ let T^0 denote the Banach space adjoint from $\overline{\mathcal{H}}_2$ to $\overline{\mathcal{H}}_1$. Then the ordinary Hilbert space adjoint T^* from $\mathcal{H}_2$ to $\mathcal{H}_1$ will be equal to JT^0J. For each unitary representation L the mapping $x \to L^0_{x^{-1}}$ is a unitary representation which we shall call the *adjoint* of L and denote by $\overline{L}$. The set of all bounded linear operators T from $\overline{\mathcal{H}}_2$ to $\mathcal{H}_1$ such that Trace $(TT^*) < \infty$ is a Hilbert space under the inner product $(T_1 : T_2) = \text{Trace}(T_1T_2^*)$. It is called the Kronecker or tensor product of $\mathcal{H}_1$ with $\mathcal{H}_2$ and denoted by $\mathcal{H}_1 \otimes \mathcal{H}_2$. If A_1 and A_2 are bounded linear operators on $\mathcal{H}_1$ and $\mathcal{H}_2$ respectively, then $T \to A_1TA_2^0$ is a bounded linear operator $A_1 \times A_2$ in $\mathcal{H}_1 \otimes \mathcal{H}_2$. If L^1 and L^2 are unitary representations of G_1 and G_2 respectively, then $x, y \to L^1_x \times L^2_y$ is a unitary representation of $G_1 \times G_2$ called the Kronecker or tensor products $L^1 \times L^2$ of L^1 and L^2. $L^1 \times L^2$ is irreducible whenever L^1 and L^2 are irreducible.

Let $\mathfrak{A}(G)$ denote the set of all set functions of the form $c_1\alpha_1 + \cdots + c_n\alpha_n$ where the α_j are finite Borel measures on G and the c_j are complex constants. $\mathfrak{A}(G)$ is an algebra under *convolution* $\alpha, \beta \to \alpha\beta$ where $\int f(x)d(\alpha\beta)(x) = \int f(xy)d\alpha(x)d\beta(y)$ for all continuous f with compact support. If L is a unitary representation of G and $\alpha \in \mathfrak{A}(G)$ then L_α is defined to be the unique bounded linear operator such that $(L_\alpha(\phi) \cdot \psi) = \int (L_x(\phi) \cdot \psi) d\alpha(x)$ for all ϕ and ψ in $\mathcal{H}(L)$. $\alpha \to L_\alpha$ is a homomorphism of $\mathfrak{A}(G)$ into the ring of all bounded linear operators in $\mathcal{H}(L)$. Moreover $L_{\alpha^*} = L_\alpha^*$ where $\alpha^*(E) = [\alpha(E^{-1})]^-$.[2] If L is finite-dimensional $x \to \text{Trace}(L_x)$ is a complex-valued function $\mathfrak{X}^L$ on G called the *character* of L. L is uniquely determined to within equivalence by its character $\mathfrak{X}^L \cdot \int \mathfrak{X}^L(x) d\alpha(x) = \text{Trace}(L_\alpha)$ and $\mathfrak{X}^L$ determines and is determined by the linear functional $\alpha \to \text{Trace}(L_\alpha)$ on $\mathfrak{A}(L)$. Let $L = n_1L^1 \oplus n_2L^2 \oplus \cdots \oplus n_kL^k$ where the L^j are irreducible and finite-dimensional. Let us define an inner product in $\mathfrak{A}(G)$ by setting $(\alpha : \beta)_{\mathfrak{X}_L} = \text{Trace}(L_{\alpha\beta^*})$. If we identify α with α' whenever $(\alpha - \alpha' : \alpha - \alpha')_{\mathfrak{X}_L} = 0$, $\mathfrak{A}(G)$ becomes a finite-dimensional Hilbert space and right and left translation in $\mathfrak{A}(G)$ define unitary operators in this Hilbert space. Let $V^{\mathfrak{X}_L}$ denote the corresponding unitary

[2] Here and elsewhere the notation [ ]$^-$ is used to avoid putting a bar over a group of symbols. According to the context it means either the conjugate of a complex number or the dual of a Banach space.

representation of $G\times G$. The mapping $f\rightarrow L_f$ is a unitary member of $\mathcal{R}(V^{\mathfrak{X}_L}, L\times\bar{L})$ and hence sets up an equivalence between $V^{\mathfrak{X}_L}$ and a subrepresentation of $L\times\bar{L}$. This representation is easily seen to be the multiplicity free representation $L_1\times\bar{L}_1\oplus L_2\times\bar{L}_2\oplus\cdots\oplus L_k\times\bar{L}_k$ whose restriction to $G\times e$ is quasi equivalent to L. If $\mathfrak{X}_L$ is replaced by $\mathfrak{X}=\lambda_1\mathfrak{X}^{L_1}\oplus\cdots\oplus\lambda_k\mathfrak{X}^{L_k}$ where the λ_j are arbitrary positive real numbers $V^{\mathfrak{X}}$ may be defined just as $V^{\mathfrak{X}_L}$ was and is equivalent to $V^{\mathfrak{X}_L}$. However $f\rightarrow L_f$ is no longer unitary.

The above considerations may be extended to infinite-dimensional representations—but only to some and at the cost of defining a character to be an abstract linear functional defined on a proper subspace of $\mathfrak{A}(G)$ and not necessarily associated with a point function on G. Specifically let us say that the unitary representation L *has a character* if Trace(L_α) exists for some nonzero α in $\mathfrak{A}(G)$. Let $\mathfrak{A}^L(G)$ denote the set of all α for which Trace(L_α) exists and let F_L denote the linear functional on $\mathfrak{A}^L(G):\alpha\rightarrow$Trace$(L_\alpha)$. We note that $\mathfrak{A}^L(G)$ is always an ideal in $\mathfrak{A}(G)$ and in particular is invariant under right and left translation.

We call F_L the *character* of L. Given F_L we define an inner product in $\mathfrak{A}^L(G)$ by setting $(\alpha:\beta)_{F_L}=F_L(\alpha\beta^*)$ and convert $\mathfrak{A}^L(G)$ into a pre-Hilbert space by identifying as above. Left and right translations in the completion of this Hilbert space define a representation of $G\times G$ which we shall denote by V^{F_L}. $f\rightarrow L_f$ is a unitary member of $\mathcal{R}(V^{F_L}, L\times\bar{L})$ and if L is irreducible $f\rightarrow L_f$ sets up an equivalence between V^{F_L} and $L\times\bar{L}$. Thus if the irreducible representation L has a character F_L this character determines L to within equivalence. Indeed L is so determined by the restriction of F_L to any ideal in $\mathfrak{A}^L(G)$ on which it is not identically zero. It can be shown that L cannot have a character in the sense just described unless it is of the form $n_1L^1\oplus n_2L^2\oplus\cdots\oplus n_kL^k\oplus\cdots$ where the L^j are irreducible. When it is, $V^{F_L}\simeq L^1\times\bar{L}^1\oplus\cdots\oplus L^k\times\bar{L}^k\oplus\cdots$ just as in the finite-dimensional case.

The correspondence between certain infinite-dimensional representations and their characters which we have just described can be extended to a much wider class if we give up one-to-oneness and associate representations with characters instead of vice-versa. Let us define a character to be any linear functional F which is defined on a self adjoint ideal $\mathfrak{g}_F$ in $\mathfrak{A}(G)$ and which has the following further properties:

(a) $F(\alpha\beta)=F(\beta\alpha)$ for all α, β in $\mathfrak{g}_F$.

(b) $F(\alpha\alpha^*)\geqq 0$ for all α in $\mathfrak{g}_F$.

(c) F is a Borel function with respect to the topology which $\mathfrak{g}_F$ inherits from the usual norm topology in $\mathfrak{A}(G)$. Then V^F may be defined just as above and can be shown[3] to be a multiplicity free representation of $G\times G$. If G is a type I group there exists a unique measure class in $\hat{G}$ such that $V^F \simeq \int_{\hat{G}} L^s \times \bar{L}^s d\mu(s)$ for any μ in this measure class. The measure class is concentrated in the set of all L's for which F_L is defined on $\mathcal{I}_F$ and there is a *unique member* of it, μ_0 such that $F(\alpha)=\int_{\hat{G}} \mathfrak{X}^{L^s}(\alpha)d\mu_0(s)$ for all α in $\mathcal{I}_F$. We may think of F as being a character of all representations quasi equivalent to $\int L^s d\mu(s)$. Actually there will be many different characters F for a given V^F. They will correspond one-to-one to certain measures in the measure class of μ_0—those for which $\int \mathfrak{X}^{L^s}(\alpha)d\mu(s)$ exists.

Note that though we have lost one-to-oneness for the correspondence between representations and characters we have achieved a certain uniqueness in direct integral decompositions. Any character on a type I group has a *unique* representation as an integral of irreducible characters. Thus whenever a canonical character may be selected for a given representation we have, corresponding to it, a well determined element in the measure class in $\hat{G}$ defined by its direct integral decomposition. For example the regular representation of a unimodular group is that defined by the linear functional $f\mu \rightarrow f(e)$ defined for all continuous f with compact support where μ is Haar measure for the group. The corresponding measure $\hat{\mu}$ in $\hat{G}$ is called the *Plancherel measure* for G and the formula $f(e)=\int_{\hat{G}} \text{Trace}(L^s_{f\mu})d\hat{\mu}(s)$ is called the *Plancherel* formula for G. When G is commutative $\hat{G}$ is itself a group, $\hat{\mu}$ is just Haar measure in $\hat{G}$ and the Plancherel formula for the translate of f by x becomes the Fourier inversion formula. Of course $\hat{\mu}$ depends upon μ which is determined only up to a multiplicative constant.

When G is not a type I group there may exist characters F such that V^F is irreducible but not of the form $L\times\bar{L}$. The restriction of such a V to $G\times e$ will be a type II primary representation. When the regular representation of G is not of type I the character $f\mu\rightarrow f(e)$ can still be decomposed as above but now some of the components will be associated with type II primary representations instead of irreducible ones. Correspondingly the Plancherel measure must extend over type II primary representations and the Plancherel formula adjusted by a suitable redefinition of $\text{Trace}(L^s_{f\mu})$. Similar remarks apply to the decomposition of other characters.

Let L and M be irreducible representations of G having characters

[3] Since this was written it has occurred to the author that it is not obvious that the properties (a), (b) and (c) are strong enough to imply that the relevant Hilbert space is separable. In any event one can simply assume this separability as a separate property (d).

F^L and F^M. Then $L\times M$ has a character whose domain of definition includes all $\alpha\times\beta$ where α and β are in the domains of F^L and F^M respectively. Suppose that this domain includes measures whose supports are in the "diagonal" $\tilde{G}$ of $G\times G$, i.e. the set of all x, y with $x=y$. Then its restriction to these measures defines a character for G. The decomposition of this character defines a unique measure in $\hat{G}$ which we may call the *Plancherel measure* for the decomposition of the Kronecker product of L and M.

Let G be unimodular and let μ be a Haar measure in G. Let L be an irreducible representation of G. Then L is equivalent to a subrepresentation of the regular representation if and only if $x\to(L_x(\phi)\cdot\psi)$ is in $\mathcal{L}^2(G,\mu)$ for all ϕ and ψ in $\mathcal{H}(L)$. Moreover if $(L_x(\phi)\cdot\psi)$ is in $\mathcal{L}^2(G,\mu)$ for some ϕ and ψ with $\phi\neq 0$ and $\psi\neq 0$ then L is equivalent to a subrepresentation of the regular representation. Such representations are said to be *square integrable.* When G is compact, so that $\mu(G)<\infty$ every irreducible L is square integrable and one has the "orthogonality relations."

$$\int (L_x^1(\phi)\cdot\psi)[(L_x^2(\phi')\cdot\psi')d\mu(x)]^- = 0 \text{ if } L^1 \not\simeq L^2$$

and

$$\int (L_x(\phi)\cdot\psi)[(L_x(\phi')\cdot\psi')]^- d\mu(x) = \frac{\mu(G)}{\dim (L)}(\phi\cdot\phi')[(\psi\cdot\psi')]^-$$

for all ϕ, ψ, ϕ' and ψ' in the appropriate Hilbert spaces. Here dim (L) denotes the dimension of $\mathcal{H}(L)$. On the other hand the Plancherel measure $\hat{\mu}$ in $\hat{G}$ assigns the measure dim $(L)/\mu(G)$ to the representation L. Thus we may write the second orthogonality relation in the form

$$\int (L_x(\phi)\cdot\psi)[(L_x(\phi)\cdot\psi)]^- d\mu(x) = \frac{(\phi\cdot\phi)[(\psi\cdot\psi)]^-}{\hat{\mu}(L)}.$$

In this form it makes sense whether or not G is compact and both orthogonality relations can be shown to hold for square integrable representations of arbitrary unimodular groups. It is suggestive to think of $\hat{\mu}(L)$ as the finite ratio of the infinite dimension of $\mathcal{H}(L)$ to the infinite measure of G. In particular it makes sense to talk of the ratio of the dimensions of two infinite-dimensional square integrable representations. Amongst examples of noncompact groups with nontrivial square integrable representations are the group of all 2×2 real matrices of determinant one, certain other noncomplex semisimple Lie groups and certain groups with compact open normal subgroups.

Let K be a compact subgroup of the group G. For each irreducible representation L of G we may consider the primary decomposition of the restriction L^K of L to K. Let $\mathcal{H}_d(L)$ denote the subspace of $\mathcal{H}(L)$ on which L^K is a multiple of the irreducible representation d of K and let P_d^L denote the projection on $\mathcal{H}_d(L)$. Whenever $\mathcal{H}_d(L)$ is finite-dimensional we may form the function $x \rightarrow \text{Trace}(P_d^L L_x)$. It is called the spherical function of L associated with d and denoted by ϕ_d^L. If L is finite-dimensional then $\text{Trace}(L_x) = \sum_d \phi_d^L(x)$. In general we may think of ϕ_d^L as being the d component of the "generalized function" which is the character of L and may think of this character as a formal sum of the ϕ_d^L. If $\mathcal{H}_d(L)$ is finite-dimensional for all L and d we shall say that K is a *large compact* subgroup of G. It is easy to see that if $K_1 \subseteq K_2$ and K_1 is large then K_2 is also large. It is known that the maximal compact subgroup of the quotient of a semisimple Lie group by its center is large and that a compact subgroup K is large whenever there is a commutative normal subgroup N such that $NK = G$. Many of the special properties of semisimple Lie groups hold for all groups which admit a large compact subgroup. For example such a group is always of type I and every irreducible representation has a character. Moreover there exists a single ideal on which all characters are defined. Each irreducible representation L is determined to within equivalence by any nonvanishing ϕ_d^L. For each d the set of all L with $\phi_d^L \neq 0$ and a given value for dim $\mathcal{H}_d(L)$ can be given a natural Hausdorff locally compact topology. Thus though the natural topology which one can introduce into $\hat{G}$ makes it into a non-Hausdorff space—even in very simple cases—this space can be written as a countable union of (overlapping) Hausdorff locally compact subspaces.

Let $\mathfrak{A}^c(G)$ denote the center of $\mathfrak{A}(G)$. If L is an irreducible representation of G then L_α commutes with L_x for all x in G and all α in $\mathfrak{A}^1(G)$ so that L_α must be a scalar multiple of the identity: $L_\alpha = h^L(\alpha)I$. Each h^L is a homomorphism of the commutative Banach algebra $\mathfrak{A}^c(G)$ into the complex numbers. When G is compact the homomorphism h^L determines L to within equivalence. Its restriction to $\mathfrak{L}^1(G)$ is defined by a bounded function which is a constant multiple of $\text{Trace}(L_x)$. For noncompact G this way of defining characters is quite inadequate. For most interesting noncompact groups the conjugate classes are so large that $\mathfrak{A}^c(G)$ is too small for $L \rightarrow h^L$ to be anywhere near to being one-to-one. In fact it is quite often the case that $\mathfrak{A}^c(G)$ contains only zero. On the other hand when G is a connected Lie group one can replace $\mathfrak{A}(G)$ by the algebra $\tilde{\mathfrak{A}}(G)$ of all distributions whose support is the identity of the group. While the center of this algebra is not big enough for h^L to determine L to within equivalence one does have the following result. Let G be a semisimple Lie group with a finite center and let K be a maximal compact subgroup. Let d be an irreducible representation of K. Let h be a homomorphism of the center of $\tilde{\mathfrak{A}}(G)$ onto the complex numbers. Then there are at most a finite number of inequivalent irreducible representations L of G such that $P_d^L \neq 0$ and $h^L = h$. h^L is called the *infinitesimal character* of L.

5. **Induced representations and systems of imprimitivity.** Let H be a closed subgroup of the separable locally compact group G and let L be a unitary representation of H. Let π denote the natural mapping of G onto the space G/H of all H cosets Hx and let $\mathfrak{M}(G/H)$ denote the set of all finite Borel measures α on G/H such that $\alpha(E)=0$ whenever $\pi^{-1}(E)$ has Haar measure zero. We define $\sqrt{\alpha_1\alpha_2}$ to be the measure whose Radon-Nikodym derivative with respect to $\alpha_1+\alpha_2$ is

$$\sqrt{\left(\frac{d\alpha_1}{d(\alpha_1+\alpha_2)}\,\frac{d\alpha_2}{d(\alpha_1+\alpha_2)}\right)}.$$

Let $\mathcal{H}^L$ denote the set of all pairs α, ϕ where α is in $\mathfrak{M}(G/H)$ and ϕ is a Borel function from G to the unit sphere of $\mathcal{H}(L)$ such that $\phi(\xi x) = L_\xi(\phi(x))$ for all ξ in H and all x in G. The set on which $\phi(x) = \phi^1(x)$ is clearly a union of cosets and we identify α, ϕ with α, ϕ^1 whenever $\phi(x)=\phi^1(x)$ almost everywhere with respect to α. Given two pairs α_1, ϕ_1 and α_2, ϕ_2 we define their inner product $(\alpha_1, \phi_1 : \alpha_2, \phi_2)$ by the formula $(\alpha_1, \phi_1 : \alpha_2, \phi_2) = \int_{G/H}(\phi_1(x)\cdot\phi_2(x))d(\sqrt{\alpha_1\alpha_2})(x)$. This formula makes sense because $(\phi_1(x)\cdot\phi_2(x))$ is constant on the cosets Hx. With this inner product $\mathcal{H}^L$ becomes a Hilbert space; that is, there is a unique way of defining linear combinations so that $\mathcal{H}^L$ is a Hilbert space in the usual sense. For each $x \in G$ let U_x^L be the operator taking α, ϕ into α_x, ϕ_x where $\alpha_x(E)=\alpha(Ex)$, $\phi_x(s)=\phi(sx)$. Then U_x^L is a one-to-one inner product preserving map of $\mathcal{H}^L$ on $\mathcal{H}^L$ and hence is a unitary operator. The mapping $x \to U_x^L$ is a unitary representation of G which we call the *representation* of G *induced* by L and denote by U^L. In the special case in which G/H admits an invariant measure μ we may define $\mathcal{H}^L = \mathcal{H}(U^L)$ somewhat more perspicuously as the set of all Borel functions θ from G to $\mathcal{H}(L)$ such that $\theta(\xi x) = L_\xi(\theta(x))$ for all $\xi, x \in H \times G$ and such that $\int_{G/H}(\theta(x)\cdot\theta(x)d\mu(x)) < \infty$. We identify as before and set $(\theta_1 : \theta_2) = \int_{G/H}(\theta_1(x)\cdot\theta_2(x)d\mu(x))$. The natural correspondence between the two Hilbert spaces is that taking θ into the pair α_θ, ϕ_θ where $d\alpha_\theta/d\mu = (\theta(x)\cdot\theta(x))$ and $\phi_\theta(x) = \theta(x)/\sqrt{(\theta(x)\cdot\theta(x))}$.

Amongst the more elementary formal properties of the mapping $L \rightarrow U^L$ are the following:

(a) $U^{\bar{L}} \simeq [U^L]^-$,

(b) $U^{L_1 \times L_2} \simeq U^{L_1} \times U^{L_2}$ if $H_1 \subseteq G_1$, $H_2 \subseteq G_2$ and L_1 and L_2 are unitary representations of H_1 and H_2 respectively,

(c) $U^M \simeq U^L$ if $H \subset H^1 \subset G$ and M is the representation of H^1 induced by L,

(d) $U^{L_1 \oplus L_2} \simeq U^{L_1} \oplus U^{L_2}$ are more generally

$$U^{\int L^\lambda d\mu(\lambda)} \simeq \int U^{L^\lambda} d\mu(\lambda).$$

When L is the one-dimensional identity representation of the subgroup consisting of the identity alone then U^L is easily seen to be the regular representation. It follows from this and the "chain rule," (c) above, that U^L is the regular representation of G whenever L is the regular representation of H.

To a surprisingly large extent the known infinite-dimensional irreducible representations of groups can be thrown into the form U^L where L is a one-dimensional representation of a suitable closed subgroup. For this reason it is useful to have criteria for the irreducibility of induced representations. Irreducibility of L is of course necessary but far from sufficient for the irreducibility of U^L. In the very special case in which there are only a finite number of right H cosets the following necessary and sufficient conditions can be given. If a is an automorphism of G let L^a denote the representation of $a^{-1}(H)$ which carries $a^{-1}(h)$ into L_h and let i_x denote the inner automorphism $y \rightarrow xyx^{-1}$. Then U^L is irreducible if and only if for every x not in H, L^{i_x} and L are disjoint when restricted to $H \cap x^{-1}Hx$. This theorem makes sense no matter how many right H cosets there are but is only known to be true in rather special cases. There are also cases in which it is known to be false and one of the more interesting unsolved problems of the theory is that of finding a suitably modified theorem which is true under reasonably general conditions. When H is normal and a type I group the theorem is true as it stands no matter how big G/H is. It then takes the following simplified form: U^L is irreducible if and only if L^{i_x} and L are inequivalent irreducible representations of H for all x not in H. As an example of the application of this last theorem consider the case in which G is the group of all sense preserving rigid motions of the plane and H is the subgroup of all translations. The theorem tells us, more or less immediately, that U^L is irreducible for every nontrivial irreducible representation L of H. Necessary and sufficient conditions for the irreducibility of U^L are known when L is finite-dimensional and G/H is countable and useful sufficient conditions are known when G is a Lie group and L is finite-dimensional.

Let H_1 and H_2 be closed subgroups of G such that the coset spaces G/H_1 and G/H_2 are finite. The irreducibility theorem described above is an immediate consequence of the following theorem about the dimension of $\mathcal{R}(U^L, U^M)$ where L and M are irreducible representations of H_1 and H_2 respectively. For each x and y in G let $J(x, y, L, M)$ denote the dimension of $\mathcal{R}(L^{i_x}\restriction, M^{i_y}\restriction)$ where $\restriction$ applied to each representation denotes its restriction to their common domain $x^{-1}H_1x\cap y^{-1}H_2y$. Then $J(x, y, L, M)$ depends only upon the double coset $c=H_1xy^{-1}H_2$ to which xy^{-1} belongs so that we may write $J(c, L, M)$. Moreover dim $\mathcal{R}(U^L, U^M)=\sum_{c\in C} J(c, L, M)$ where C is the set of all $H_1:H_2$ double cosets. Applied to the case in which $H_2=G$ this theorem implies the Frobenius reciprocity theorem dim $\mathcal{R}(U^L, M)=$dim $\mathcal{R}(L$, restriction of M to $H_1)$. In addition it implies a criterion for the equivalence of U^L and U^M when L and M are both irreducible. When L and M are both finite-dimensional the theorem about dim $\mathcal{R}(U^L, U^M)$ is an immediate consequence of a theorem representing $U^L\otimes U^M$ as a direct sum over $H_1:H_2$ double cosets on certain other induced representations of G. This theorem of Kronecker products is independent of the hypothesis about the finiteness of G/H_i and holds at least when there are only countably many $H_1:H_2$ double cosets. It actually holds more generally if we are willing to replace direct sums by direct integrals. Moreover it is a consequence of an equally general theorem about the restriction of U^L to H_2. Let C denote the set of all $H_1:H_2$ double cosets and let π denote the mapping taking each $x\in G$ into $H_1\times H_2\in C$. We define a subset E of C to be a Borel set if $\pi^{-1}(E)$ is a Borel set in G. If the points of C are separated by countably many Borel sets we shall say that H_1 and H_2 are regularly related. The unique measure class in C whose null sets are the sets E such that $\pi^{-1}(E)$ is a null set in G will be called the canonical measure class in C. Let H_1 and H_2 be regularly related and let L be a unitary representation of H_1. For each $x\in G$ let ${}_xV$ denote the representation of H_2 induced by the restriction of L^{i_x} to $H_2\cap x^{-1}H_1x$. Then ${}_xV$ is determined to within equivalence by the double coset $c=\pi(x)$ to which x belongs so that we may write ${}_cV$. Moreover U^L restricted to H_2 is equivalent to $\int_C {}_cV d\mu(c)$ where μ is any measure in the canonical measure class. To deduce the Kronecker product theorem from the restriction theorem we recall that $U^L\otimes U^M$ may be regarded as the restriction to the diagonal of $G\times G$ of $U^L\times U^M$

and that $U^L \times U^M \simeq U^{L\times M}$. We leave details to the reader. When U^L and U^M are infinite dimensional the Kronecker product theorem does not have implications about irreducibility because the operators in $\mathcal{H}(U^L) \otimes \mathcal{H}(U^M)$ are only the Hilbert-Schmidt operators. However both the restriction theorem and the Kronecker product theorem are useful tools in dealing with induced representations.

The Frobenius reciprocity theorem, classical for finite groups, was stated above only in the very special case in which G/H is a finite set. Much more general formulations are possible. For example let H and G both be groups whose regular representation are of type I. Let $\int_{\hat{G}} F^s d\zeta(s)$ be quasi equivalent to the regular representations of G and let $\int_{\hat{H}} N^t d\eta(t)$ be quasi equivalent to the regular representation of H where ζ and η are finite measures in $\hat{G}$ and $\hat{H}$ respectively. Then there exists a measure α in $\hat{G}\times\hat{H}$, an α measurable function n from $G\times H$ to the countable ordinals, a function $s\rightarrow\beta_s$ from $\hat{G}$ to measures in $\hat{H}$ and a function $t\rightarrow\gamma_t$ from $\hat{H}$ to measures in $\hat{G}$ such that

(1) $\alpha=\int\beta_s d\zeta(s)=\int\gamma_t dn(t)$,

(2) $U^{N^t}\simeq\int n(s, t) F^s d\gamma_t(s)$ for almost all t,

(3) F^s restricted to H is $\simeq\int n(s, t) N^t d\beta_s(t)$ for almost all s.

In particular if one knows the restriction of F^s for almost all s then one knows U^{N^t} for almost all t and vice versa.

Again let L be a unitary representation of the closed subgroup H of G and for each Borel set E in G/H let P_E denote the operator taking the pair α, ϕ into the pair α_E, ϕ where $\alpha_E(F)=\alpha(E\cap F)$. It is easy to see that each P_E is a projection in $H(U^L)$ and that $E\rightarrow P_E$ is a projection-valued measure. For each x in G and each Borel set in G/H we have $U^L_{x^{-1}}P_E U^L_x = P_{[E]x}$ where $[E]x$ denotes the transform of E by x under the natural action of G on G/H. More generally let $\mathfrak{M}$ be a G space and let V be a unitary representation of G. A projection-valued measure P^1 defined on $\mathfrak{M}$ with each P^1_E an operator in $\mathcal{H}(V)$ will be said to be a *system of imprimitivity* for V *based on* $\mathfrak{M}$ provided that $V_{x^{-1}}P^1_E V_x = P^1{}_{[E]x}$ for all E and x. The system of imprimitivity P for U^L defined above will be called the *canonical system of imprimitivity* for U^L. To within equivalence it is the only system of imprimitivity for U^L based on G/H and together with U determines L uniquely. More specifically we have the following theorem. Let H be any closed subgroup of the separable locally compact group G and let V be any unitary representation of G. Let Q be any system of imprimitivity for V based on the Borel G space G/H. Then there exists a unitary representation L of H, unique to within equivalence, and a unitary mapping W of $\mathcal{H}(U^L)$ on $\mathcal{H}(V)$ such that $WU^L_xW^{-1} = V_x$ for all x in G, and $WP_EW^{-1}=Q_E$ for all Borel subsets of E of

G/H where P is the canonical system of imprimitivity for U^L. Moreover $\mathfrak{R}(L, L)$ is isomorphic to the intersection of $\mathfrak{R}(U^L, U^L)$ with $\mathfrak{R}(P, P)$. This theorem, whose proof is rather long, has many applications. We shall refer to it as the ***imprimitivity theorem.***

6. The relationship between $\hat{N}$ and $\hat{G}$ when N is normal in G. When the group G has a closed normal subgroup N it is to some extent determined by the two "smaller" groups N and G/N and one can reasonably attempt to describe $\hat{G}$ in terms of $\hat{N}$ and $(G/N)^\wedge$. Consider first the special case in which G is isomorphic to $N \times G/N$. The mapping $L, M \rightarrow L \times M$ is one-to-one from $\hat{N} \times (G/N)^\wedge$ into $\hat{G}$ and, for any irreducible representation V of G, its restrictions to $N \times e$ and $e \times G/N$ are primary. However V will be of the form $L \times M$ if and only if these primary restrictions are both of type I. If either N or G/N is a type I group this will always be the case and we have a natural one-to-one mapping of $\hat{N} \times (\hat{G}/N)$ onto $\hat{G}$.

Consider next the case in which N is commutative and G is a "semi-direct product" of N and a second closed subgroup H in the sense that $N \cap H = e$ and $NH = G$. Of course H is isomorphic to G/N. Since N is commutative every irreducible representation of N is of the form $\xi \rightarrow \mathfrak{X}(\xi)I$ where I is the identity operator in a one-dimensional space and $\mathfrak{X}$ is a continuous complex-valued function on N such that $\mathfrak{X}(\xi_1\xi_2) = \mathfrak{X}(\xi_1)\mathfrak{X}(\xi_2)$ and $|\mathfrak{X}(\xi_1)| = |\mathfrak{X}(\xi_2)| = 1$ for all ξ_1 and ξ_2 in N. We shall identify the function $\mathfrak{X}$ with the corresponding representation. For any automorphism α of N there is a unique automorphism α^* of $\hat{N}$ such that $\alpha^*(\mathfrak{X})(\xi) = \mathfrak{X}(\alpha(\xi))$. Let α_h denote the automorphism $\xi \rightarrow h\xi h^{-1}$ defined by each h in H. By the *orbit* $\pi(\mathfrak{X})$ of $\mathfrak{X} \in \hat{N}$ we shall mean the set of all $\alpha_h^*(\mathfrak{X})$ with $h \in H$. By the isotropy group $H_{\mathfrak{X}}$ of $\mathfrak{X}$ we shall mean the subgroup of all $h \in H_{\mathfrak{X}}$ such that $\alpha_h^*(\mathfrak{X}) = \mathfrak{X}$. For each pair $\mathfrak{X}, L$ consisting of a member $\mathfrak{X}$ of $\hat{N}$ and a primary representation L of $H_{\mathfrak{X}}$ let $\mathfrak{X}L$ denote the primary representation of $NH_{\mathfrak{X}}$ which takes ξh into $\mathfrak{X}(\xi)L_h$. Then the induced representation $U^{\mathfrak{X}L}$ is a primary representation of G which is of type j, $j = \mathrm{I}, \mathrm{II}, \mathrm{III}$ if and only if L is of type j. Moreover it is irreducible if and only if L is irreducible. The restriction of $U^{\mathfrak{X}L}$ to N is a multiple of a direct integral of the members of $\pi(\mathfrak{X})$ and every primary representation of G whose restriction to N "lies in $\pi(\mathfrak{X})$" in this sense is of the form $U^{\mathfrak{X}L}$ where L is uniquely determined to within equivalence. If the orbits define a "smooth" equivalence relation in N, in the sense that there exists a Borel set which cuts each orbit just once then *every* primary representation of G is of the form $U^{\mathfrak{X}L}$ for some $\mathfrak{X}$ and some primary representation L of $H_{\mathfrak{X}}$. Thus as $\mathfrak{X}$ varies over a cross section C for the orbits and, for each $\mathfrak{X}$, L varies over the primary representations of $H_{\mathfrak{X}}$, the representations $U^{\mathfrak{X}L}$ vary over all primary representations of G—each occurring just once. In particular if each $H_{\mathfrak{X}}$ is a type I group then G is a type I group. However the orbits need not define a smooth equivalence relation and when they do not there are many irreducible and primary representations in addition to the $U^{\mathfrak{X}L}$. We shall have more to say about these extra representations later but unfortunately nothing like a complete analysis of them is possible. On the other hand the theorem just quoted makes possible a very complete and satisfying description of *all* the irreducible representations of numerous noncompact, noncommutative groups. The proof of this theorem leans heavily on the imprimitivity theorem.

Much of what has just been said may be extended to the case in which N need not be commutative and H need not exist provided that N is a type I group. However we may have to replace the representations L of the groups $H_{\mathfrak{X}}$ by so called *projective* or *ray* representations of certain subgroups of G/N.

A projective unitary representation differs from an ordinary unitary representation in that the requirement that $L_{xy}=L_xL_y$ is replaced by the requirements $L_{xy}=\sigma(x, y)L_xL_y$ and $L_e=I$ where σ is a complex-valued function on $G\times G$. We call σ the *multiplier* of the representation and we call a projective representation whose multiplier is σ a *σ-representation*. It is easy to see that any multiplier σ must be a Borel function from $G\times G$ to the complex numbers of modulus one which satisfies the identity

$$\sigma(x, yz)\sigma(y, z) = \sigma(x, y)\sigma(xy, z)$$

and the normalizing condition $\sigma(e, x)=\sigma(x, e)=1$. Conversely given any function σ with these properties it can be shown that there exist projective representations with multiplier σ. If L is a σ-representation and ρ is a complex-valued Borel function with $\rho(e)=|\rho(x)|=1$ then $x\rightarrow\rho(x)L_x$ is a σ^1-representation where

$$\sigma^1(x, y) = \frac{\rho(x)\rho(y)}{\rho(xy)}\sigma(x, y).$$

In particular $x, y\rightarrow\rho(x)\rho(y)/\rho(xy)$ is always a multiplier. Multipliers of this form (coboundaries) are said to be *trivial* and two multipliers whose quotient is trivial are said to be *similar*. It is clear that the product of two multipliers is a multiplier and that the multipliers form a group under multiplication. The trivial multipliers constitute a subgroup and the corresponding quotient group is called the *multiplier group* of G. It can reduce to the identity but need not do so. The structure of the multiplier group is known for some groups G but much work remains to be done.

For a fixed σ one can develop a theory of the σ-representations of G which is quite analogous to the theory of the ordinary representations of G. In particular practically everything discussed in §§2, 3 and 5 of these lecture notes may be extended to σ-representations. Characters for σ-representations do not seem to have been investigated. On the other hand if σ_1 and σ_2 are nonsimilar multipliers for the same G the theory of the σ_1-representations of G can be as different from the theory of the σ_2-representations of G as the representation theories of two different groups. For example, let $G=H\times\hat{H}$ where H is locally compact and commutative and $\hat{H}$ is its character group. Let $\sigma(h_1, \mathfrak{X}_1; h_2, \mathfrak{X}_2)=\mathfrak{X}_1(h_2)$. Then σ is a nontrivial multiplier for the commutative group G and it can be shown that, to within equivalence, G has just one irreducible σ-representation and that this representation is infinite-dimensional. Commutative groups can also have type II and type III σ-representations.

To return to the main theme of this section let N be a type I normal subgroup of G. For each irreducible unitary representation W of N and each x in G we may form the representation $n\to W_{xnx^{-1}}=W_n^x$. To within equivalence W^x depends only upon the image of x in G/N. Thus $\hat{N}$ is a Borel G/N space in a natural way and we may look at the orbits of the members of $\hat{N}$ under the action of G/N. For each W let H_W denote the group of all x in G such that $W^x\simeq W$. Then $H_W\supseteq N$ and H_W/N is isomorphic to a subgroup of G/N. Just as in the more special case considered above, we may set up a natural one-to-one correspondence between those primary representations of G whose restrictions to N lie in the orbit of W and the primary representations of H_W/N. However there is one important difference. The relevant primary representation of H_W/N may be a σ-representation for some nontrivial σ. Since $W^x\simeq W$ for each $x\in H_W$ there must exist a unitary V_x such that $V_xW_n^xV_x^{-1}=W_n$ for all n in N. When $x\in N$ we may take $V_x=W_x$ but in general V_x is arbitrary up to a multiplicative constant of modulus one and there is no canonical way of choosing this constant. Choosing it arbitrarily (but so that $(V_x(\phi)\cdot\psi)$ is always a Borel function) we get a projective representation $x\to V_x$ of H_W which reduces on N to W. The multiplier for this representation is uniquely determined up to similarity and may be chosen to be of the form $x, y\to\sigma(x^0, y^0)=\sigma^0(x, y)$ where σ is a multiplier for H_W/N and $x\to x^0$ is the canonical mapping of H_W on H_W/N. It is the $(1/\sigma)$-primary representations of H_W/N which correspond to primary representations of G and the one-to-one mapping is that taking L into $U^{V\otimes L^0}$ where L^0 is the $(1/\sigma^0)$-representation $x\to L_{x^0}$ of H_W. Just as before, the induced representations $U^{V\otimes L^0}$ can be shown to include *all* primary representations when the orbits in N define a smooth equivalence relation.

The fact that we may have to consider projective representations of the H_W/N seems at first sight a serious obstacle to using the above described theory inductively to determine the unitary representations of complicated groups in several stages. However it turns out to be possible to carry out the whole discussion starting with projective representations. Thus we can proceed inductively if we consider all possible multipliers for each group concerned.

7. **Ergodic systems of imprimitivity and virtual subgroups.** Let the separable locally compact group G be a semidirect product of the closed subgroups N and H where N is commutative and normal as in the second paragraph of §6. Let V be any unitary representation of G. Let V^0 be the unique (to within equivalence) multiplicity free representation quasi equivalent to the restriction V^N of V to N. By the theory described in §2, V^0 is as associated with a measure class C_V in $\hat{N}$ which determines it to within equivalence. For each Borel subset E of $\hat{N}$ there is a unique measure class $C_V \cap E$ whose null sets are the sets F such that $F \cap E$ is of measure zero with respect to C. Let P_E^V denote the unique projection in the center of $\mathcal{R}(V^N, V^N)$ such that $(V^N)P_E^V$ is quasi equivalent to the representation defined by $C_V \cap E$. Then $E \rightarrow P_E^V$ is a projection-valued measure called the *spectral measure* of V. It is not hard to see that P^V is a system of imprimitivity for V and hence also for the restriction V^H of V to H. Conversely, given a pair consisting of a unitary representation W of H and a system of imprimitivity P for H based on $\hat{N}$, there exists a unique representation V of G such that $V^H = W$ and $P^V = P$. If V is primary then P is *ergodic* in the sense that $P_E = P_{[E]h}$ for all h in H if and only if $P_E = 0$ or $P_{(\hat{N}-E)} = 0$. Conversely if P is ergodic then V is primary if and only if $\mathcal{R}(V^H, V^H) \cap \mathcal{R}(P, P)$ has a trivial center. Thus, to survey all primary unitary representations of G with a given restriction to N is to survey all unitary representations W of H having a fixed ergodic system of imprimitivity P based on N and such that $\mathcal{R}(W, W) \cap \mathcal{R}(P, P)$ has a trivial center.

Let P be a projection-valued measure on $\hat{N}$ which is an ergodic system of imprimitivity for some W. Then either $P_{\pi(\mathfrak{X})} = 1$ for some orbit $\pi(\mathfrak{X})$ or $P_{\pi(\mathfrak{X})} = 0$ for all orbits $\pi(\mathfrak{X})$. In the first case we shall say that P is *transitive* and in the second that P is *strictly ergodic.* When P is transitive it may be regarded as based on $\pi(\mathfrak{X})$ and hence on the coset space $H/H_{\mathfrak{X}}$. By the imprimitivity theorem the corresponding representations of H correspond one-to-one to the primary representations of $H_{\mathfrak{X}}$ and this is how the imprimitivity theorem is used to deduce the analysis of $\hat{G}$ given in §6. Now suppose that P is strictly ergodic and let ϕ be a vector in $\mathcal{H}(P)$ such that $P_E(\phi)=0$ implies $P_E=0$. Then $E \to (P_E(\phi)\cdot\phi)$ is a Borel measure whose null sets are invariant under the action of H. Hence the measure class C to which this measure belongs is invariant. It is *ergodic* in the sense that any Borel set which is invariant under the action of H (modulo sets of measure zero) is either a set of measure zero or the complement of one. It is *strictly ergodic* in the sense that every orbit has measure zero. Conversely it can be shown (see below for further details) that, every ergodic measure class C arises in this way from an ergodic system of imprimitivity P for a unitary representation W of H, that P is strictly ergodic if and only if C is and that P is determined to within equivalence by C and a cardinal number.

It follows from the above that every strictly ergodic invariant measure class in $\hat{N}$ gives rise to a class of "extra" primary (and irreducible) representations of G not covered by the analysis of §6. To find all primary (and irreducible) representations of G when $\hat{N}$ admits strictly ergodic invariant measure classes one must first find all such measure classes and then for each find all associated primary representations. The second of these difficult problems is essentially the same as that of finding all representations W of H with a fixed strictly ergodic system of imprimitivity P and will be considered to some extent below. Next to nothing can be said at present about the first.

Let C be any ergodic invariant measure class in $\hat{N}$. Let $\mathcal{H}_0$ be a separable Hilbert space. Let $\mathcal{H}(C, \mathcal{H}_0)$ denote the set of all pairs α, ϕ where α is a finite Borel measure in $\hat{N}$, ϕ is a Borel function from $\hat{N}$ to the unit sphere of $\mathcal{H}_0$, and $\alpha(E)=0$ whenever E is a C null set. We identify α, ϕ with α, ϕ^1 whenever $\phi(\mathfrak{X})=\phi^1(\mathfrak{X})$ outside of an α null set and make $\mathcal{H}(C, \mathcal{H}_0)$ into a Hilbert space via the inner product

$$(\alpha_1, \phi_1 : \alpha_2, \phi_2) = \int (\phi_1(\mathfrak{X})\cdot\phi_2(\mathfrak{X}))d\sqrt{(\alpha_1\alpha_2)}(\mathfrak{X}).$$

Just as with the definition of $\mathcal{H}(U^L)$ in §6, the action of H on $\hat{N}$ defines a unitary representation W^1 of H and $E \to P_E$ where $P_E(\alpha, \phi) = \alpha_E, \phi$ defines an ergodic system of imprimitivity for $W^1 \cdot P_E = 0$ if and only if E is a C null set. Let Q be any function from $\hat{N}\times H$ to the unitary operators in $\mathcal{H}_0$ which has suitable measurability properties and satisfies the following identities almost everywhere with respect to C.

(*) $Q(\mathfrak{X}, e) \equiv I$,

(**) $Q(\mathfrak{X}, h_1h_2) = Q(\mathfrak{X}, h_1)Q([\mathfrak{X}]h_1, h_2)$.

Let $\tilde{Q}_h$ denote the unitary operator taking α, ϕ into α, $Q_h\phi$ where $Q_h\phi(\mathfrak{X})=Q(\mathfrak{X}, h)(\phi(\mathfrak{X}))$. Then $h\rightarrow\tilde{Q}_hW_h^1=W_h^Q$ is a unitary representation W^Q of H having P as system of imprimitivity. Every unitary representation W of H having P as system of imprimitivity is equivalent so some W^Q via an equivalence which leaves P fixed. Moreover the pair P, W^{Q_1} is equivalent to the pair P, W^{Q_2} if and only if there exists a measurable function $\mathfrak{X}\rightarrow A(\mathfrak{X})$ from $\hat{N}$ to the unitaries in $\mathfrak{H}_0$ such that for each h,

(***) $Q^2(\mathfrak{X}, h) = A(\mathfrak{X})Q^1(\mathfrak{X}, h)A([\mathfrak{X}]h)^{-1}$

for C almost all $\mathfrak{X}$. Finally the commuting algebra intersection $\mathfrak{R}(P, P)\cap\mathfrak{R}(W^Q, W^Q)$ is isomorphic to the algebra $\mathfrak{A}_a$ of all measurable functions $\mathfrak{X}\rightarrow B(\mathfrak{X})$ from N to the unitaries in $\mathfrak{H}_0$ such that for all h,

(***)′ $B([\mathfrak{X}]h)Q(\mathfrak{X}, h) \equiv B(\mathfrak{X})Q(\mathfrak{X}, h)$

for C almost all $\mathfrak{X}$. Of course B_1 and B_2 are to be identified when almost everywhere equal. Finding the most general primary representations of G associated with a given ergodic quasi invariant measure class C is thus equivalent to finding the most general solution Q of (*) and (**) such that $\mathfrak{A}_a$ has a trivial center—identifying two solutions when (***) holds.

When C is concentrated in an orbit $\pi(\mathfrak{X})$ then (**) can be solved and the solutions correlated with the representations of $H_{\otimes}$. In fact this is one way of proving the imprimitivity theorem. On the other hand when C is strictly ergodic we have a well nigh unapproachable problem in noncommutative cohomology theory. One can find explicit solutions of (**)—even all of them in certain rather special cases—but deciding when two are equivalent seems very difficult indeed. Yet there is a much closer analogy with the transitive case than at first meets the eye. There is a sense in which the "generalized orbit" defined by C may be associated with a "virtual" subgroup H_C of H and the equivalence classes of solutions of (*) regarded as representations of this "virtual" subgroup. While no actual subgroup of H is involved the abstract objects H_C behave like subgroups in many ways. One can speak of one virtual subgroup's being contained normally in another, one can consider homomorphisms of one into or onto another, one can form direct products etc. Moreover the theory of the unitary representations of a "virtual" subgroup is very close in structure to the theory of the unitary representations of an actual group. The theory described in §§2 and 3 goes over almost without change though virtual subgroups of type I probably do not exist. One can define characters and Kronecker products and do at least part of the theory of §4. In addition one can induce from one virtual subgroup to a larger one and carry over much if not all of the theory described in §§5 and 6. Indeed parts of this theory take on an improved appearance when virtual subgroups are admitted. In the restriction theorem for induced representations, for example, one can omit the assumption that H_1 and H_2 are regularly related if one is willing to accept representations induced from virtual subgroups of H_2.

The basic idea behind the introduction of the notion of virtual subgroup may be explained as follows. Let S be a standard Borel G space which is *transitive* in the sense that for each s_1 and s_2 in S there exists x in G with $s_1x=s_2$. For each s in S let H_s denote the group of all x in G for which $sx=s$. The groups H_s are mutually conjugate closed subgroups and for each s the coset space G/H_s is equivalent as a G space (in an obvious sense) to the G space S. Since G/H is a transitive standard Borel G space for every closed subgroup H it follows that we have a natural one-to-one correspondence between equivalence classes of transitive standard Borel G spaces on the one hand and conjugate classes of closed subgroups of G on the other. This suggests that notions and theorems about closed subgroups of G should be capable of reformulation so as to be expressed in terms of the corresponding transitive G spaces. It suggests further that—when so reformulated—some of these notions and theorems will apply to G spaces which admit an ergodic invariant measure class but are not strictly transitive and hence not equivalent to coset spaces. When they do we may think of them as being about the nonexistent "virtual" subgroups of G which would define the G spaces if only they were transitive. For example H_1 is continued in some conjugate of H_2 if and only if there is a Borel map θ of G/H_1 on G/H_2 such that $\theta([s]x)=[\theta(s)]x$ for all s in G/H_1 and all x in G. Correspondingly when such a mapping of the ergodic G space S_1 on the ergodic G space S_2 exists we shall say that the virtual subgroup defined by S_1 is a subgroup of the virtual subgroup defined by S_2.

As a less trivial example let us consider the meaning to be attached to the notion of a homomorphism of a virtual subgroup into a group. If ϕ is a continuous homomorphism of the closed subgroup H_1 of G_1 into the group G_2—then the subgroup H_ϕ of $G_1\times G_2$ consisting of all pairs $h, \phi(h)$ with $h\in H_1$ closed, uniquely determines ϕ and is determined to within conjugacy by the $G_1\times G_2$ space $(G_1\times G_2)/H_\phi$. Let S_1 be a Borel subset of G_1 which meets each coset H_1x just once. Then $S_1\times G_2$ meets each H_ϕ coset just once and the natural maps of S_1 on G_1/H_1 and of $S_1\times G_2$ on $(G_1\times G_2)/H_\phi$ are one-to-one. Using these natural maps, S_1 and $S_1\times G_2$ become G_1 and $G_1\times G_2$ spaces equivalent to the coset spaces G/H_1 and $(G_1 \times G_2)/H_\phi$ respectively. Moreover $(s, y)(x, z) = sx, z^{-1}y\pi(s, x)$ where sx is the G_1 space action of x on s, π is a Borel mapping of $S_1 \times G_1$ into G_2 and $z^{-1}y\pi(s, x)$ is the product of three members of the group G_2. If π is any Borel mapping from $S_1 \times G_1$ into G_2 and S_1 is any standard Borel G space then, as is easy to verify $(s, y)(x, z) = sx, z^{-1}y\pi(s, x)$ makes $S_1 \times G_2$ into a Borel $G_1 \times G_2$ space if and only if we have

(†) $\pi(s, e) = e$ for all s in S_1,

(††) $\pi(s, x_1x_2) = \pi(s, x_1)\pi(sx_1, x_2)$

for all s, x_1, x_2 in $S \times G_1 \times G_2$. Moreover if S_1 is transitive then $S_1 \times G_2$ is transitive and the subgroup leaving s_1, e fixed is the set of all h, $\phi(h)$ where $s_1h = s_1$ and ϕ is a Borel (and hence continuous) homomorphism of the subgroup H_{s_1} into G_2. Choosing a different cross section S_1' has the effect of changing π to π' where

(†††) $\pi'(s, x) = a(s)\pi(s, x)a^{-1}(sx)$

and a is a Borel function from S to G_2. Thus the notion of a Borel homomorphism of H_1 into G_2 is equivalent to the notion of an equivalence class of functions satisfying (†) and (††), the equivalence relation being defined by (†††). This latter notion does not explicitly involve H_1 and makes sense for virtual subgroups. Comparing (†), (††), and (†††) with (*), (**), and (***) above we see how the unitary representations of G associated with a given ergodic invariant measure class in N may be regarded as being defined by unitary representations of a virtual subgroup of H.

When $\mathfrak{M}$ is a $G_1 \times G_2$ space each $G_1 \times e$ orbit is carried into another by each element of $e \times G_2$. Thus the set $\mathfrak{M}^{G_1}$ of all $G_1 \times e$ orbits is a G_2 space in a natural way. When $\mathfrak{M}$ is of the form $(G_1 \times G_2)/H_\phi$, where ϕ is a continuous homomorphism of H_1 into G_2, then $\mathfrak{M}^{G_1}$ will be a standard Borel space in a natural way if and only if $\phi(H_1) = H_2$ is closed in G_2. When this is the case G_2/H_2 is equivalent to $\mathfrak{M}^{G_1}$. In other words H_2 will be a homomorphic image of H_1 if and only if G_2/H_2 is $\mathfrak{M}^{G_1}$ for an $\mathfrak{M}$ defined in $G_1/H_1 \times G_2$ by a π satisfying (†) and (††). A definition for the notion of homomorphism of one virtual subgroup onto another virtual subgroup at once suggests itself. We leave the precise formulation to the reader.

Results in the ergodic theory of flows and measure preserving transformations may be regarded as results about virtual subgroups of the additive group of the real line and of the additive group of the integers respectively. When this point of view is taken some known results receive interesting new interpretations. For example Ambrose has shown how to construct a more or less general ergodic flow from a single measure preserving transformation T on a space S and a positive measurable real function f on S. This construction is a special case of the above construction of $\mathfrak{M}^{G_1}$ from S_1 and π. Since G_1 is infinite cyclic π is determined by a single function from S_1 to G_2 and G_2 in this case is the real line. Ambrose's result may thus be looked upon as asserting something about the realizability of virtual subgroups of the real line as homomorphic images of virtual subgroups of the integers. Similar interpretations may be given to Kakutani's notion of induced measures preserving transformations and Anzai's skew product transformations.

Let C_1 be an ergodic invariant measure class in the standard Borel G space S_1. Let S_2 be a second standard Borel G space and let θ be a Borel map from S_1 to S_2 such that $\theta(sx)=\theta(s)x$ for all x and s. Let C_2 be the measure class in S_2 whose null sets E are the sets such that $\theta^{-1}(E)$ is a C_1 null set. Then C_2 is also ergodic and invariant and defines a virtual subgroup of G containing that defined by S_1 and C_1. Now an equivalence class of irreducible representations of the virtual subgroup defined by S_1 and C_1 determines and is determined by an equivalence class of pairs consisting of a unitary representation V of G and a system of imprimitivity P for V based on S_1. Given such a pair let $P^1_E=P_{\theta^{-1}(E)}$ for each Borel set E contained in S_2. Then P^1 is a system of imprimitivity for V based on S_2. The pair P^1, V defines an equivalence class of unitary representation of the virtual subgroup defined by S_2 and C_2 which depends only upon the equivalence class of the pair P, V. In this way we may pass in a canonical way from unitary representations of virtual subgroups to unitary representations of larger virtual subgroups. When the virtual subgroups are actual this passage coincides with that from L to the induced representation U^L described in §5. This generalization of the inducing process may be defined directly in terms of the functions Q. The second form of the definition is the most convenient for many purposes but is rather more complicated.

For further details about virtual subgroups and their unitary representations the reader must be referred to the lecturer's as yet unwritten introductory article on the subject.

8. **Nonunitary representations.** It is natural to attempt to extend the theory we have been describing by allowing the operators L_x to be continuous linear operators in an arbitrary (separable) Banach space or perhaps even in a complete topological vector space. However, very serious obstacles stand in the way of obtaining nearly so complete a theory—even if one sticks to operators in Hilbert space

and drops only the condition that the L_x be unitary. Perhaps the most important of these obstacles is the fact that subrepresentations of nonunitary representations need not have complementary subrepresentations and that correspondingly a nonunitary representation may be reducible without being decomposable as a direct sum. Because of this there is little if any hope of being able to reduce the problem of finding all representations to that of finding all irreducible representations. For example though one knows quite explicitly the finite-dimensional irreducible representations of an n-dimensional vector group one does not know the finite-dimensional representations of such a group when $n>1$. The attempt to determine them all leads to complicated and apparently unsolved problems in the ideal theory of polynomial rings. Moreover for any bounded linear operator A in a Banach space one may form e^{At} for all real t and thus obtain a representation of the real line. It follows that the problem of classifying all representations of the real line includes as a special case that of classifying all bounded linear operators—a problem known to be highly inaccessible even if one considers only operators in Hilbert space. Another difficulty is illustrated by the following. Let C denote the set of all continuous functions with compact support on the separable locally compact group G. This space can be given a tremendous number of different norms which are invariant under translation. The completion of C in any of them yields a Banach space representation of G which is analogous to the regular representation and yet not equivalent to it under the obvious notion of equivalence. If we consider these representations as inequivalent our classification problem is enormously complicated by having appended to it a large chunk of the (apparently hopeless) classification problem for Banach spaces. If we do not we must find a new notion of equivalence which identifies them and is satisfactory in other respects. Such a notion has been proposed (and will be discussed below) but has satisfactory properties only in certain important special cases. This difficulty makes itself felt even for irreducible representations as one can see by replacing the regular representation by an irreducible induced representation. Yet another difficulty appears in the very concept of irreducibility. An important property of irreducible *unitary* representations is that any bounded linear operator in the space of such a representation L may be approximated arbitrarily closely in the strong operator topology by operators of the form $c_1L_{x_1}+ \cdots +c_nL_{x_n}$. However for nonunitary representations it seems to be very difficult if not impossible to prove that this property is implied by the nonexistence of closed invariant subspaces. These irreducible representations which have it are said to be *completely irreducible* and those which do not—if indeed they exist—are left in an unexplored limbo.

In spite of the discouraging facts recounted above there are cogent reasons for giving at least some attention to nonunitary representations. In studying the irreducible unitary representations of certain groups, especially the semisimple Lie groups, one gets a more natural and well rounded theory by including the completely irreducible nonunitary representations. The unitary ones form a rather artificial subset of the whole. Indeed when one studies the representations of semisimple Lie groups via their Lie algebras it is not always easy to see which Lie algebra representations are going to correspond to unitary representations of the group. A more pressing (but closely related) reason for studying nonunitary representations is that there exist irreducible unitary representations which may be thrown into the form U^L where L is a nonunitary one-dimensional representation of a subgroup and (apparently) may not be so expressed for unitary L. Of course the definition of U^L has to be adapted so that it applies to nonunitary L. In addition (as we shall see below) the problem of studying all nondirect sum "extensions" of one irreducible unitary representation by another gives promise of adding interesting new structural features to the "dual object" $\hat{G}$. Finally in so far as one conceives of the theory of infinite-dimensional group representations as being a natural generalization of classical harmonic analysis one hopes ultimately to generalize more than the relatively trivial $\mathcal{L}^2$ theory.

Let L and M be strongly continuous nonunitary representations of the separable locally compact group G in separable Banach spaces $\mathcal{H}(L)$ and $\mathcal{H}(M)$. It has been proposed (by Naimark) to say that L and M are equivalent if there exists a *closed* one-to-one linear operator T from a dense invariant subspace of $\mathcal{H}(L)$ to a dense invariant subspace of $\mathcal{H}(M)$ such that $TL_x(\phi) = M_xT(\phi)$ for all x in G and all ϕ in the domain of T. We shall call this notion *weak equivalence*. It is not hard to show that weak equivalence implies equivalence whenever L and M are unitary. On the other hand it is by no means clear that weak equivalence is even an equivalence relation except in special cases. Suppose that G has a large compact subgroup K in the sense of §4. For each *completely* irreducible representation L of G and each irreducible unitary representation d of K one can define the spherical functions ϕ_d^L more or less as in §4 and it can be shown that L and M are weakly equivalent if and only if $\phi_d^L = \phi_d^M$ for all d. It follows that weak equivalence is an equivalence relation in this case and that it coincides with the "infinitesimal equivalence" of Harish-Chandra when G is a semisimple Lie group with a finite center.

For some purposes it is convenient to reformulate the notion of weak equivalence as follows. Let X be a vector space and let Y be a subspace of the algebraic dual X^* of X which is *total* in the sense that $f(\phi)=0$ for all f in Y implies $\phi=0$. We shall call such a pair a *linear system*. By an isomorphism of the linear system X_1, Y_1 with the linear system X_2, Y_2 we shall mean a one-to-one linear transformation A from X_1 onto X_2 such that A^* is one-to-one from Y_2 onto Y_1. Here A^* is defined by the identity $A^*(f)(\phi)=f(A(\phi))$. By an *automorphism* of X, Y we shall mean an isomorphism of X, Y with itself. By a *linear system representation* L of G we shall mean a homomorphism $x\rightarrow L_x$ of G into the group of automorphisms of a linear system X, Y. If L is a linear system representation of G in X, Y, and X_1 and Y_1 are total subspaces of X and Y invariant under the L_x and L_x^* respectively then restricted to X_1, the L_x define a linear system representation of G in X_1, Y_1. We shall call this second representation a *dense contraction* of the first. Two linear system representations L^1 and L^2 in X^1, Y^1 and X^2, Y^2 respectively will be said to be *equivalent* if there exists an isomorphism A of X^1, Y^1 with X^2, Y^2 such that $AL_x^1=L_x^2A$ for all x in G. Now let L be a Banach space representation. Then L is determined to within (strong) equivalence by the corresponding linear system representation in the linear system $\mathcal{H}(L)$, $[\mathcal{H}(L)]^-$. If T sets up a weak equivalence between L and M then the range and domain of T and T^* define dense contractions of these linear system representations and these dense contractions are equivalent. Conversely if L and M have equivalent dense contractions and A sets up the equivalence then A has a *closed* extension which defines a weak equivalence between L and M. Thus L and M are weakly equivalent if and only if they have equivalent dense contractions. In particular a unitary representation is determined to within equivalence by any dense contraction and, given a linear system representation, we may speak unambiguously of its *unitary expansion* whenever this exists.

Now let L be a Banach space representation of the closed subgroup H of G. Let ${}^0X^L$ denote the vector space of all Borel functions f from G to $\mathcal{H}(L)$ which are bounded on compact sets and satisfy the identity $f(\xi x)=L_\xi f(x)$ for all x in G and all $\xi\in H$. Let ${}^0Y^L$ denote the vector space of all Borel functions g from G to $[\mathcal{H}(L)]^-$ which are bounded on compact sets and satisfy the identity $g(\xi x)=\overline{L}_{\xi^{-1}}(g(x))$ for all x in G and all $\xi\in H$. Then $g(x)(f(x))$ as a function of x is a Borel function on G which is constant on the H cosets and hence defines a Borel function on G/H which is bounded on compact sets. Let ${}^{00}X^L$ denote the set of all pairs consisting of a member f of ${}^0X^L$ and a finite measure α of compact support which is absolutely continuous with respect to the unique invariant measure class in G/H. Let ${}^{00}Y^L$ be similarly defined with respect to ${}^0Y^L$. Given f, α in ${}^{00}X^L$ and g, β in ${}^{00}Y^L$ let $(f, \alpha: g, \beta) = \int_{G/H} g(x)(f(x))d\sqrt{\alpha\beta}$. Let us identify two members f, α and f', α' of ${}^{00}X^L$ if $(f, \alpha: g, \beta) = (f', \alpha': g, \beta)$ for all g, β in X and let us identify two members g, β and g', β' of ${}^{00}Y^L$ correspondingly. Let us denote the sets of equivalence classes by X^L and Y^L respectively. $(f, \alpha: g, \beta)$ of course depends only upon the equivalence classes of the arguments. It can be shown that there is a unique way of defining the linear operations in X^L and Y^L so that X^L, Y^L becomes a linear system in which the linear function defined by g, β is $f, \alpha \rightarrow (f, \alpha: g, \beta)$. X^L, Y^L and $(f, \alpha: g, \beta)$ are invariant under group translation so group translation defines a linear system representation of G in X^L, Y^L. If this linear system representation has a unitary expansion we call it the *unitary representation* of G *induced* by L and denote it by U^L. When L itself is unitary it is easy to see that this definition is equivalent to that given in §5. For any given H one can raise the following rather interesting question: For which nonunitary irreducible representations of H is U^L defined?

Let L and M be Banach space representations of G and let $x \rightarrow A_x$ be any strongly continuous function from G to the bounded linear operators from $\mathcal{H}(M)$ to $\mathcal{H}(L)$ which is a *derivation* in the sense that it satisfies the identity

(*) $A_{xy} = L_x A_y + A_x M_y$

for all x and y in G. For each x let V_x^A be the operator taking ϕ, ψ in $\mathcal{H}(L) \oplus \mathcal{H}(M)$ into $L_x(\phi) + A_x(\psi)$, $M_x(\phi)$. Then $x \rightarrow V_x^A$ will be Banach space representation of G having L as a subrepresentation and M as a quotient representation. Moreover it can be shown rather easily that every representation V having L as a subrepresentation and M as a quotient representation is equivalent to some V^A provided that the invariant subspace defining the subrepresentation L has a (in general nonvariant) complementary closed subspace. A necessary and sufficient condition that there should exist an *invariant* complementary closed subspace so that $V \simeq L \oplus M$ is that A be of the form $x \rightarrow L_x A_0 - A_0 M_x$ where A_0 is some fixed bounded linear operator from $\mathcal{H}(M)$ to $\mathcal{H}(L)$. Let us call such a derivation exact. The set of all continuous derivations is a vector space and the set of all exact derivations is a subspace of it. The dimension of the quotient space is a measure of the number of essentially distinct extensions of L by M that exist. It can of course be zero—indicating that $L \oplus M$ is the only possibility. Let us denote this quotient space by $\mathfrak{D}(L, M)$. It can be different from zero even when L and M are unitary. When G is commutative it is easy to show that $\mathfrak{D}(L, M)=0$ whenever L and M are inequivalent and irreducible. On the other hand if G is commutative, $\hat{G}$ is a commutative Lie group, and L is irreducible and defined by $\mathfrak{X}\in\hat{G}$ then there exists a natural isomorphism of $\mathfrak{D}(L, L)$ with the complexification of the tangent space to G at $\mathfrak{X}$. This suggests the possibility of giving some sort of generalized manifold structure to $\hat{G}$ in the general case whenever it is found that $\mathfrak{D}(L, L)\neq 0$ over a large subset. It would be interesting to know under what circumstances $\mathfrak{D}(L, L)$ has a finite nonzero dimension.

9. **The representations of semisimple Lie groups.** In §§6 and 7 we discussed a method for finding the irreducible unitary representations of particular groups. This method applies only to groups having proper closed normal subgroups and replaces the given problem by the corresponding one for certain "smaller" groups. It leads to complete results only when at some stage all of the smaller groups are commutative with "known" character groups and various other conditions are satisfied. It must be supplemented by other methods whenever one of the smaller groups is *topologically simple* in the sense that it admits no proper closed normal subgroup—unless of course this smaller group is commutative. In this section we shall discuss the extent to which it has been possible to determine the irreducible representations of separable locally compact groups which are topologically simple. Let G be such a group. Since the connected component of the identity is a closed normal subgroup G must be either connected or totally disconnected. The study of nondiscrete totally disconnected simple groups has barely begun and we shall say no more about it here. When G is simple and discrete it almost certainly is not of type I unless it is finite and there seems to be little if any hope of giving a satisfying analysis of the irreducible representations of a nontype I group. For some, but not all, of the known simple finite groups it has been found possible to give a complete list of their irreducible representations—or at least the characters of these representations. However there seems to be little in the way of a general theory and the results are too finite and algebraic in character to fit naturally into this account. We are left with the connected ones and it follows from the solution to Hilbert's fifth problem that a connected topologically simple locally compact group is in fact a simple Lie group.

The simple Lie groups are all known, and extensive (though still incomplete) work has been done on the problem of determining their irreducible representations, the characters of these and the Plancherel measure in $\hat{G}$. Much of this work can be formulated so that it applies to general semisimple Lie groups, i.e. groups which are locally isomorphic to direct products of connected simple Lie groups. For simplicity of exposition we shall consider only those semisimple Lie groups which have finite centers and no compact components. The irreducible representations of compact simple Lie groups are all finite-dimensional and their complete determination is classical. Basic to the general formulation is the Iwasawa structure theorem which may be stated as follows in our special case. Let G be a semisimple Lie group with a finite center and no compact components. Then G has closed connected subgroups N, D and K having the following properties:

(1) N is nilpotent, D is a vector group and K is compact.

(2) The mapping $n, d, k \rightarrow ndk$ from $N \times D \times K$ to G is a homeomorphism onto.

(3) If A is the centralizer of D in K then $S = NDA$ is a semidirect product of N and DA.

(4) K is contained in no properly larger compact subgroup of G. As an example of the Iwasawa structure theorem it is illuminating to consider the group $SL(n, C)$ of all $n \times n$ complex matrices of determinant one. Here we may choose N, D and K as follows: N is the group of all matrices in $SL(n, C)$ which are one on the main diagonal and zero above it. D is the group of all matrices in $SL(n, C)$ which are real on the main diagonal and zero off of it. K is the group of all unitary matrices in $SL(n, C)$. With these choices of N, D and K, A comes out to be the group of all matrices in $SL(n, C)$ which are zero off of the main diagonal and of absolute value one on it and S comes out to be the group of all matrices in $SL(n, C)$ which are zero above the main diagonal.

Returning to the general case let W be the normalizer of D in K. Then $A \subseteq W$ and W/A defines a finite group of automorphisms of DA and hence of $\hat{D} \times \hat{A}$. Let $(\hat{D} \times \hat{A})'$ denote the set of all members of $\hat{D} \times \hat{A}$ whose transforms under W/A are all distinct. For each $\mathfrak{X} \in \hat{D} \times \hat{A}$ let $\mathfrak{X}^0$ denote the one-dimensional representation of S, $nda \rightarrow da \rightarrow \mathfrak{X}(da)I$. It can be shown that the induced unitary representation $U^{\mathfrak{X}^0}$ of G is irreducible whenever $\mathfrak{X} \in (\hat{D} \times \hat{A})'$. It has been conjectured that $U^{\mathfrak{X}^0}$ is irreducible for all $\mathfrak{X}$ and this has been proved for the classical complex groups and for the group of all $n \times n$ real matrices of determinant one. If $\mathfrak{X}_1$ and $\mathfrak{X}_2$ are in $(\hat{D} \times \hat{A})'$ then the irreducible representations $U^{\mathfrak{X}_1^0}$ and $U^{\mathfrak{X}_2^0}$ are equivalent if and only if $\mathfrak{X}_1$ and $\mathfrak{X}_2$ belong to the same W/A orbit. Let C be a cross section for the orbits in $(D\times A)'$ under W/A. Then $\mathfrak{X}\rightarrow U^{\mathfrak{X}^0}$ defines a one-to-one mapping θ of C into $\hat{G}$. In the important special case in which G is a complex analytic Lie group it can be shown that $\hat{G}-\theta(C)$ is of measure zero with respect to the Plancherel measure $\hat{\mu}$. Thus in this case there is a sense in which the $U^{\mathfrak{X}^0}$ comprise "almost all" unitary representations of G. Moreover the Plancherel measure can be described without finding any further irreducible representations of G. It can be shown that the "transformed measure" $\theta(\hat{\mu})$ such that $\theta(\hat{\mu}(E))=\hat{\mu}(\theta(E))$ is absolutely continuous with respect to the restriction of Haar measure in $\hat{D}\times\hat{A}$ to C and the Radon-Nikodym derivative has been explicitly computed.

Even when G is complex analytic the $U^{\mathfrak{X}^0}$ for $\mathfrak{X}\in\hat{D}\times\hat{A}$ are known *not* to exhaust the irreducible representations of G. Further irreducible representations of G may be constructed in two ways. If $\mathfrak{X}\notin(\hat{D}\times\hat{A})'$ it will be possible to extend $\mathfrak{X}^0$ from S to a larger subgroup and consider the representations of G induced by this extended one dimensional representation. For example let G be the group of all $n\times n$ unimodular complex matrics and let $G_{r,n-r}$ denote the subgroup of all matrices which have zeros in the upper right submatrix of r rows and $n-r$ columns. Let $G'_{r,n-r}$ denote the subgroup of $G_{r,n-r}$ having zeros in the lower left submatrix of r columns and $n-r$ rows. For each element of $G'_{r,n-r}$ we may compute the determinant of the $r\times r$ matrix in the upper left hand corner and obtain in this way a homomorphism of $G'_{r,n-r}$ and hence of $G_{r,n-r}$ into the multiplicative group of nonzero complex numbers. Every one-dimensional unitary representation of this last group defines one of $G_{r,n-r}$ which is not in $(\hat{D}\times\hat{A})'$ when restricted to $D\times A$. These unitary representations of $G_{r,n-r}$ induce unitary representations of G which can be shown to be irreducible. They belong to what Gelfand and Naimark call the *degenerate series*. There are analogous degenerate series associated with partitions of n into more than two summands. Returning to the general case again let $\mathfrak{X}$ be a homomorphism of $D\times A$ into the multiplicative group of all nonzero complex numbers such that $|\mathfrak{X}(da)|\neq 1$. Let $\mathfrak{X}^0$ be the corresponding *non*unitary one-dimensional representation of S. Then, as explained in §8, $U^{\mathfrak{X}^0}$ might exist as the unique unitary expansion of a certain linear system representation. Moreover $U^{\mathfrak{X}^0}$ has been shown to be irreducible, when it exists, under conditions analogous to those described above. For the classical complex groups sufficient conditions for the existence of $U^{\mathfrak{X}^0}$ are known and the corresponding irreducible representations of G constitute what Gelfand and Naimark call the supplementary series. Of course one can form $U^{\mathfrak{X}^0}$ also when $\mathfrak{X}^0$ is a nonunitary one-dimensional representation of a subgroup containing S. The representations obtained in this way are said to belong to the supplementary degenerate series. Whether the principal, supplementary, degenerate, and supplementary degenerate series exhaust the irreducible representations of the complex semisimple Lie groups is an open question. Naimark and Berezin have independent papers purporting to show that the answer is yes for the classical complex groups but Berezin's argument seems to have a serious obscurity or two if not an actual gap and the completion of Neumark's argument has yet to be published. For the special case of the group of all 2×2 complex matrices of determinant one there is no degenerate series and an affirmative answer has been established.

When G is not a complex analytic group much less information is available. In particular $\hat{G} - \theta(C)$ is not of $\hat{\mu}$ measure zero and the representations in $\theta(C)$ are not sufficient to define $\hat{\mu}$ and the Plancherel formula. This fact seems to be closely related to the fact that it is only for the complex groups that almost every group element is conjugate to a member of the maximal commutative subgroup DA. For more general semisimple Lie groups one has a finite number (greater than one) of mutually nonconjugate maximal commutative subgroups and each one seems to have associated with it a series of irreducible unitary representations whose $\hat{\mu}$ measure is not zero. However the situation has been completely elucidated only for the group $SL(n, R)$ of all $n \times n$ real matrices of determinant one and even here one does not know *all* irreducible unitary representations—only enough so that the rest have Plancherel measure zero. It is only for the group $SL(2, R)$ that *all* irreducible representations are known. This group has to within conjugacy *one* maximal commutative subgroup other than $DA = D$. This is the compact group K of all matrices of the form

$$\begin{Vmatrix} \cos\theta & \sin\theta \\ -\sin\theta & \cos\theta \end{Vmatrix}.$$

For each irreducible representation

$$\begin{Vmatrix} \cos\theta & \sin\theta \\ -\sin\theta & \cos\theta \end{Vmatrix} \to e^{in\theta}$$

of this group we have an irreducible representation V^n of $G = SL(2, R)$ defined by an analogue of the inducing process in which Borel functions are replaced by complex analytic functions. The coset space G/K is in a natural way a complex analytic manifold analytically equivalent to the upper half-plane. The representations V^n and the members of the principal and supplementary series exhaust the irreducible representations of G. For $\mathrm{SL}(n, R)$ a typical maximal commutative subgroup is defined by a positive integer k with $2k \leqq n$ and consists of all matrices in $\mathrm{SL}(n, R)$ such that each submatrix

$$\begin{Vmatrix} a_{2j,2j} & a_{2j,2j+1} \\ a_{2j+1,2j} & a_{2j+1,2j+1} \end{Vmatrix}$$

with $j \leqq k$ is of the form

$$\lambda_j \begin{Vmatrix} \cos\theta_j & \sin\theta_j \\ -\sin\theta_j & \cos\theta_j \end{Vmatrix}$$

with $\lambda_j > 0$ and all other nondiagonal elements are zero. Let us denote this commutative subgroup by D_k. Let G_k denote the subgroup of all matrices in $\mathrm{SL}(n, R)$ whose elements are zero except on the main diagonal and in the submatrices

$$\begin{Vmatrix} a_{2j,2j} & a_{2j,2j+1} \\ a_{2j+1,2j} & a_{2j+1,2j+1} \end{Vmatrix}, \quad \text{with} \quad j \leqq k.$$

Then G_k is isomorphic to the direct product of k replicas of $\mathrm{SL}(2, R)$, $n-k-1$ replicas of the multiplicative group of the positive real numbers and $n-1-2k$ two element groups. Let us denote the product of the noncommutative factors by G_k' and the product of the commutative factors by A_k. Then $D_k = (G_k' \cap D_k)A_k$ and the general member of D_k is defined by a k tuple $n_1, \cdots, n_k$ of integers and a character $\mathfrak{X}$ of $\hat{A}_k$. $n_1, \cdots, n_k$ and $\mathfrak{X}$ may in turn be associated with the irreducible representation $V^{n_1} \times V^{n_2} \times \cdots \times V^{n_k} \times \mathfrak{X}$ of G_k. Now let N_k be the group of all matrices in $\mathrm{SL}(n, R)$ which are one on the main diagonal, zero above it, and have $a_{2j+1,2j} = 0$ for $j \leqq k$. Then $N_k G_k$ is a semidirect product of N_k and G_k and the representation $V^{n_1} \times \cdots \times V^{n_k} \times \mathfrak{X}$ of G_k defines one of $N_k G_k$ which is the identity on N_k. We shall denote it by $L^{n_1, n_2, \cdots, n_k, \mathfrak{X}}$. The representations

$$U^{L^{n_1, n_2, \cdots, n_k, \mathfrak{X}}}$$

of $\mathrm{SL}(n, R)$ induced by $L^{n_1, n_2, \cdots, n_k, \mathfrak{X}}$ have been shown to be irreducible for all $n_1, \cdots, n_k, \mathfrak{X}$. In the aggregate they are of nonzero Plancherel measure and constitute the part of the principal series associated with D_k.

In the general case, Harish-Chandra has indicated a method for reducing the problem of finding a series of irreducible representations for each maximal abelian subgroup A to the case in which A is compact. Moreover he has solved the problem for compact A when the maximal compact subgroup K has a one-dimensional center. In this case one obtains a discrete series of square integrable representations quite analogous to the representations V^n of SL(2, R).

There are a number of things that one can say in general about the representations of semisimple Lie groups. The subgroup K of the Iwasawa decomposition is always large in the sense of §4 so that one can define spherical functions and one knows that all representations are of type I. Moreover all irreducible representations have characters in the sense of §4 and these characters may be described by Schwartzian distributions on the group. In many concrete cases these distributions are actually functions and in all cases it has been shown that the distributions reduce to functions—in fact analytic functions—on a dense open subgroup. In addition to the distribution characters one has the infinitesimal characters defined at the end of §4. There is a natural mapping of the complex linear functionals on a Cartan subalgebra of the Lie algebra of the group onto the maximal ideals of the center of the infinitesimal group algebra and hence onto the "potential" infinitesimal characters. One knows which of these "potential" infinitesimal characters can actually arise from Banach space representations and those that arise from unitary representations in a Hilbert space form a small subset. In the case of the group of all 2×2 complex matrices of determinant one the irreducible Banach space representations have all been determined—of course only to within weak equivalence.

10. **Applications to quantum mechanics.** Let $\mathfrak{M}$ denote the "configuration space" of a classical mechanical system. For example, if the classical system describes the motion of n particles subject to no constraints $\mathfrak{M}$ will be an open subset of a $3n$-dimensional vector space. We shall suppose for the time being that $\mathfrak{M}$ is an open subset of a finite-dimensional vector space V so that the *phase space* of the system may be taken to be the direct product $\mathfrak{M}\times V^*$ where V^* is the algebraic dual of V. An *observable* in classical mechanics is a real valued Borel function on $\mathfrak{M}\times V^*$. In order to facilitate the passage to quantum mechanics it will be convenient to call a point in $\mathfrak{M}\times V^*$ a *pure* state and use the word state to denote a probability measure in $\mathfrak{M}\times V^*$. Here of course we identify a point in $\mathfrak{M}\times V^*$ with the probability measure concentrated in that point. The *dynamical group* of the system is a homomorphism $t\rightarrow U_t$ of the additive group of the real line into the group of all one-to-one differentiable maps of $\mathfrak{M}\times V^*$ onto itself. It is usually defined by the vector field in $\mathfrak{M}\times V^*$ which is its "infinitesimal generator" and is restricted by further requirements which we shall not describe here. If A is an observable, q, p

a pure state and t a real number, then $A(q, p)$ is the value of the observable A in the pure state q, p and $U_t(q, p)$ is the (pure) state of the system t time units after it was in the pure state q, p. If A is an observable, α is a state and E is a Borel subset of the real line, then $\alpha(A^{-1}(E))$ is the *probability* that a measurement of A will lead to a value in E when the system is in the state α. In classical statistical mechanics one studies states rather than pure state. Clearly the dynamical group U induces a one parameter group of transformations of the space of all states onto itself. The states form a convex set in a natural way and the pure states are just the extreme points of this convex set.

Let $\mathcal{L}_c$ denote the σ-Boolean algebra of all Borel subsets of $\mathfrak{M}\times V^*$. Then the observable A is completely determined by the mapping $E\to A^{-1}(E)$ of the σ-Boolean algebra of all Borel subsets of the real line into $\mathcal{L}_c$, the state α *is* the mapping $F\to\alpha(F)$ from $\mathcal{L}_c$ to the unit interval and the dynamical group $t\to U_t$ is completely determined by the homomorphism which it defines of the real line into the group of all automorphisms of $\mathcal{L}_c$. Quantum mechanics differs from classical mechanics in that the σ-Boolean algebra $\mathcal{L}_c$ is replaced by the complete lattice $\mathcal{L}_q$ of all projections in a complete separable infinite-dimensional Hilbert space $\mathcal{H}$. The observables correspond one-to-one to the projection-valued measures on the line (see §3) and the states to the *probability measures on the projections*, that is to the functions α from the projections to the unit interval such that $\alpha(P_1+P_2+\cdots)=\alpha(P_1)+\alpha(P_2)+\cdots$ whenever $P_iP_j=P_jP_i=0$ and such that $\alpha(0)=0$, $\alpha(I)=1$. If $E\to Q_E$ "is" an observable and α "is" a state then $\alpha(Q_E)$ is the probability that a measurement of the observable will lead to a value in E when the system is in the state α. For each unit vector ϕ in $\mathcal{H}$, $P\to(P(\phi), \phi)$ is a probability measure on the projections which we shall denote by α_ϕ. It has been shown that every probability measure on the projections may be put into the form $\gamma_1\alpha_{\phi_1}+\gamma_2\alpha_{\phi_2}+\cdots$ where the ϕ_j are mutually orthogonal, the γ_j are positive and $\gamma_1+\gamma_2+\cdots=1$. The pure states, that is the extreme points in the convex set of all states, are just the α_ϕ. Via the spectral theorem there is a natural one-to-one correspondence between self-adjoint operators and projection-valued measures on the line—hence a natural one-to-one correspondence between observables and self-adjoint operators in $\mathcal{H}$. If A is a selfadjoint operator and ϕ a unit vector let $E\to Q_E^A$ denote the projection-valued measure corresponding to A. Then in the pure state defined by ϕ the observable corresponding to A has the probability distribution $E\to(Q_E^A(\phi), \phi)$. This is the basic statistical assertion of quantum mechanics as formulated by von Neumann.

The dynamical group in quantum mechanics is a continuous homomorphism $t\rightarrow U_t$ of the additive group of the real line into the (suitably topologized) group of all automorphisms of $\mathcal{L}_q$. In general an automorphism of $\mathcal{L}_q$ will be defined by a unitary or antiunitary operator in $\mathcal{H}$ which is unique to within multiplication by a complex constant of modulus one. It can be shown that the U_t may all be defined by unitary operators in $\mathcal{H}$ and that the ambiguous constants may be chosen so that U_t is defined by the unitary operator W_t where $t\rightarrow W_t$ is a unitary representation of the real line. We shall call W the unitary dynamical group of the system. It is uniquely determined by U to within multiplication by e^{iat} where a is a real constant. Using Stone's theorem we may write $W_t=e^{-iHt}$ where H is a selfadjoint operator. When units are suitably chosen (so that Planck's constant is 2π) the observable associated with H is the quantum mechanical analogue of the total energy. H is called the Hamiltonian of the system. Let ϕ be a unit vector defining the state of the system at some instant of time. Then t units of time later, the state is described by $W_t(\phi) = e^{-iHt}(\phi)$ and when the derivatives exist $d(W_t(\phi))/dt = -iHW_t(\phi)$. This last equation is Schrödinger's equation in abstract form.

Since classical mechanical observables are in a sense "limiting cases" of quantum mechanical observables no one-to-one correspondence is to be expected. On the other hand those classical observables corresponding to functions A on $\mathfrak{M}\times V^*$ such that $A(q, p)$ is independent of p, i.e. the *configuration observables*, have unique quantum counterparts and the corresponding selfadjoint operators commute. Let P_F be the operator corresponding to the function on $\mathfrak{M}$ which is 1 on the Borel subset F of $\mathfrak{M}$ and zero outside it. Then $F\rightarrow P_F$ is a projection-valued measure defined on $\mathfrak{M}$ and for any Borel function A on $\mathfrak{M}$ the corresponding quantum observable is that defined by the projection-valued measure on the line: $E\rightarrow P_{A^{-1}(E)}$. Thus the configuration observables in quantum mechanics are completely specified by giving a projection-valued measure P on the classical configuration space $\mathfrak{M}$. Moreover in so far as there are no observables except those expressible in terms of the configuration observables and their time derivatives the entire system is determined by giving the projection-valued measure P and the unitary dynamical group W.

Suppose now that our system is symmetric under the action of a separable locally compact group G and that there is a closed subgroup G_0 of G such that $\mathfrak{M}$ is a G_0 space in a natural way (see §1). The first supposition means that there is a natural homomorphism $\beta\rightarrow T_\beta$ of G into the group of automorphisms of $\mathcal{L}_q$ and assuming (for simplicity) that G is connected we may realize each T_β by a unitary S_β in our Hilbert space $\mathfrak{H}$. The S_β may be chosen so that $\beta \to S_\beta$ is a σ-representation of G for some (not necessarily trivial) σ. Moreover it follows from general invariance principles that the P_E and the S_β must be related as follows: $S_\beta P_E S_\beta^{-1} = P_{[E]\beta}$ for all β in G_0 and all $E \subseteq \mathfrak{M}$. In other words P must be a system of imprimitivity for the restriction S^0 of S to G_0. This fact has many interesting implications. In the appropriate context and slightly strengthened it is a restatement of the celebrated Heisenberg commutation relations. In what sense this is so will appear below.

For each continuous homomorphism $x \to \beta_x$ of the real line into G, $x \to S_{\beta_x}$ is a σ-representation of the real line which becomes an ordinary unitary representation on multiplying by a suitable complex-valued function of x. By Stone's theorem this unitary representation may be put into the form $x \to e^{iAx}$ where A is a selfadjoint operator. When the β_x all belong to G_0, the observable corresponding to A is called the *momentum observable* associated with the corresponding one parameter group of transformations of $\mathfrak{M}$. It is unique up to an additive constant. Modulo this additive constant, the momentum observables in quantum mechanics correspond one-to-one to the so called generalized momenta in classical mechanics. If G_0 contains the group of all translations in physical space then the observable defined by the one parameter subgroup of all translations in a particular direction is called the *total linear momentum* in that direction. If G_0 contains the group of all rotations about a particular axis in space then the observable defined by the corresponding one parameter subgroup is called the *total angular momentum* about that axis.

Suppose now that our system is a one particle system in the sense that $\mathfrak{M}$ is three-dimensional Euclidean space and that G_0 is the group of all rigid motions in $\mathfrak{M}$. Then the projection valued measure P on $\mathfrak{M}$ is a transitive system of imprimitivity for S^0 and we may apply the imprimitivity theorem of §5 to obtain rather detailed information about the form of P and S and hence of the configuration and momentum observables. The subgroup K leaving the origin 0 in $\mathfrak{M}$ fixed is the three-dimensional rotation group and the pair P, S^0 is determined to within equivalence by a projective representation L of K. Using the fact that G_0 is a semidirect product of $\mathfrak{M}$ and K we can rewrite the conclusion of the imprimitivity theorem as follows: $\mathfrak{H}$ may be realized as the space of all functions ϕ from $\mathfrak{M}$ to $\mathfrak{H}(L)$ which are square integrable with respect to Lebesgue measure μ in $\mathfrak{M}$ in such a manner that

(a) P_E is multiplication by the characteristic function of E and, more generally, for any function A on $\mathfrak{M}$ the corresponding quantum mechanical observable is that associated with the operator $\psi\rightarrow A\psi$.

(b) If β is a translation in G_0 then $S^0_\beta(\psi)(q)=\psi(q\beta)$. If $\beta\in K$ then $S^0_\beta\psi(q)=L_\beta\psi(q\beta)$.

From (b) we deduce at once that the selfadjoint operator defining the x component of linear momentum is $\psi\rightarrow i\partial\psi/\partial x$ and similarly for y and z. (We are of course assuming that Planck's constant is 2π.) If we add the assumption that there are no observables except those which can be obtained as weak limits of the configuration and momentum observables we may conclude that L is irreducible. As is well known there is to within similarity just one multiplier σ_0 for K other than the identity. Moreover there is to within equivalence just one irreducible projective representation of K of each finite dimension. Those of odd dimension are ordinary representations and those of even dimension are σ_0-representations. It is customary to denote the irreducible $2j+1$ dimensional projective representation of K by D_j where $j=0, 1/2, 1, 3/2, 2, \cdots$. The integer or half integer j is an important invariant of the one particle system. It is called the *spin* of the particle. Each angular momentum operator can be written naturally in the form $\Omega+\Omega'$ where Ω is independent of j and Ω' has $2j+1$ proper values varying from $-j$ to j in steps of 1. Ω' is called the *spin angular momentum.*

To obtain information about the form of the unitary dynamical group $t\rightarrow e^{-itH}$ we must make further assumptions. Assuming that the spin is zero and that H commutes with the S^0_β it is easy to deduce that H must be of the form $\theta(H_0)$ where θ is a Borel function and H_0 is the Laplacian. Similar but more complicated conclusions can be drawn for particles of higher spin. The possibilities for θ can be sharply reduced by taking into account the nonabsolute character of space. This is done by imbedding G_0 in the group G of all automorphisms of space-time and demanding (a) that S^0 be the restriction to G_0 of some projective representation S of G and (b) that W be the restriction of S to the group of translations in time. When G is the Galilean group we find that θ must be multiplication by a negative real number which turns out to be $-1/2$ times the "mass" of the particle. In particular we recover the classical free particle Schrödinger equation. More generally we may forget about θ and analyze the possibilities for relativistic free particles of arbitrary spin by studying the possible extensions of S^0 from G_0 to G as above. When G is the inhomogeneous Lorentz group we find that there is a one-parameter family of possibilities for each possible spin and that this parameter may be identified as above with the mass of the particle. The same is true when G is the Galilean group. One obtains the usual Schrödinger equation for a particle in a potential field by assuming that the velocity observables are the same as they are for a free Galilean particle of spin zero.

Now let us change our assumption on $\mathfrak{M}$ by supposing that it is an arbitrary finite-dimensional vector space. Let G_0 be the group of all translations in $\mathfrak{M}$ and suppose that S^0 is an ordinary representation of G_0. Each vector ϕ in $\mathfrak{M}$ defines a one-parameter subgroup of G_0 and hence a monentum observable with selfadjoint operator B_ϕ. Each dual vector f in $\mathfrak{M}^*$ is a Borel function on $\mathfrak{M}$ and hence defines a configuration observable with operator A_f. The fact that the projection-valued measure on $\mathfrak{M}$ defining the A_f is a system of imprimitivity for S^0 translates at once into the identity $e^{iA_f}e^{iB_\phi} = e^{iB_\phi}e^{iA_f}e^{if(\phi)}$. If we let $\phi_1, \phi_2, \cdots, \phi_n$ be a basis for $\mathfrak{M}$ and let $f_1, f_2, \cdots, f_n$ be the dual basis for $\mathfrak{M}^*$ we get the Heisenberg commutation relations in integrated or Weyl form. To within equivalence these relations have a unique irreducible solution. This theorem, first proved by Stone and von Neumann, is an immediate consequence of the spectral theorem and the imprimitivity theorem. Indeed, modulo the spectral theorem, it is equivalent to the special case of the imprimitivity theorem in which the group is a vector group and acts on itself by way of the group operation.

The interaction of matter with the electromagnetic field is handled in quantum mechanics by regarding the electromagnetic field as a classical mechanical system with an infinite number of degrees of freedom, i.e. having infinite-dimensional configuration and phase spaces. The corresponding quantum system turns out to be equivalent to that for a system of an "indefinite number" of "indistinguishable particles." These particles are identified with the photons of the old quantum theory and in this way the wave-particle paradox is neatly resolved. There is reason to believe that all of the elementary particles of physics—electrons, mesons, etc.—can be best understood as related to some "field" as photons are related to the electromagnetic field. On the other hand the problem of "quantizing a field," that is finding the quantum system corresponding to a classical mechanical system with an infinite-dimensional phase space of a certain kind, is not nearly so well understood as the problem of "quantizing" a system with a finite number of degrees of freedom. When one attempts to adapt the procedures described above one meets all sorts of difficulties and ambiguities corresponding to the complications one encounters on passing from finite-dimensional linear algebra to linear algebra in function spaces and from ordinary to partial differential equations.

It is usual to start by assuming that the Heisenberg commutation relations hold (or in dealing with particles of half integral spin that the so called anticommutation relations hold). The Heisenberg commutation relations are easily extended to the infinite-dimensional case—once one has decided what function space to choose for $\mathfrak{M}$ and what subspace of the algebra dual to use for $\mathfrak{M}^*$. However a serious obstacle immediately arises in that these relations now have a continuous infinity of inequivalent irreducible solutions. Moreover the solution which it is appropriate to use in any given problem is not only hard to determine but seems to be sensitive to small changes in the parameters of that problem. For this reason considerable attention has been given to the problem of determining all solutions of these relations. Unfortunately this problem differs from that presented by the finite-dimensional case in that a transitive system of imprimitivity is replaced by a strictly ergodic one—and one is brought face to face with the difficulties discussed in §7. Since a vector space acts transitively on itself one might wonder where the strict ergodicity comes from. The answer is that one cannot allow G_0 to consist of all translations in $\mathfrak{M}$ without reaching a contradiction—due essentially to the fact that an infinite-dimensional vector space cannot have an invariant measure class. One must either weaken the requirement that P_E be a projection-valued measure defined on all Borel subsets of $\mathfrak{M}$ or else replace G_0 by a proper subgroup of the translation in $\mathfrak{M}$. These alternatives are more or less equivalent and both lead to strict ergodicity.

Many other difficulties present themselves and the problem of extending quantum mechanics to infinite-dimensional systems must be considered largely unsolved. Of course the physicists have their heuristic formulations—and these work surprisingly well in certain contexts, at least when certain semi ad hoc "renormalizations" are made to eliminate annoying divergent integrals. However even they are not satisfied with the present form of the theory and a mathematician naturally would like an infinite-dimensional theory as complete and satisfying as the existent finite-dimensional one. Perhaps one can hope that the theory of infinite-dimensional group representations will be instrumental in producing one.

11. Historical and bibliographical remarks. The purpose of the following is to orient the reader in the literature of the material covered in the preceding sections. No attempt has been made to list *all* relevant papers and it is quite probable that several important ones have been omitted through oversight. We apologize in advance to their authors.

The material on primary representations and their classification into types I, II, and III is essentially a translation into the language of group representations of the fundamental work of Murray and von Neumann [91] on the dimension theory of operator rings. The "global" theory in which the notion of type is applied to nonprimary representations is correspondingly a translation of work of Dixmier [9] and Kaplansky [68]. This point of view toward the Murray-von Neumann theory was first published by Mackey [80]. Direct integral decompositions were invented by von Neumann in 1938, but his results were not published until 1949 [94]. They were first systematically applied to general group representations by Mautner [83], [84]. However direct integral decompositions were used in special cases by Gelfand and Naimark in 1947 [30] and Kolmogoroff and Adelson-Velskii may be given credit for having anticipated at least a part of the general theory. Various authors have given alternative treatments including Godement [40], Segal [97] and Mackey [79].

The study of the problem of classifying groups according to whether they have only type I representations or not was begun by Mautner in the two papers cited above and continued in later papers [85], [87]. Kaplansky [70] simplified and extended some of Mautner's work. The fact that semisimple Lie groups have only type I representations is due to Harish-Chandra [50]. For the case of finite center, substantial simplifications have been achieved by Godement [44] and Stinespring [100] using a device due to Kaplansky.

The theorem connecting measure classes in $\hat{G}$ and multiplicity free representations of G generalizes the classical Hahn-Hellinger theory for selfadjoint operators in Hilbert space. This generalization is due to Mackey [81] who at the same time defined the Borel structure in $\hat{G}$. The theorem was proved under the dual assumption that $\hat{G}$ has a metrically standard Borel structure and G has only type I representations. In a recent paper [35] Glimm proved that these two assumptions are equivalent, that they imply that $\hat{G}$ is standard and that they are equivalent to a number of other important regularity properties of G. Certain extensions and refinements (of the results in the cited Transactions paper) have been found by Ernest [21] and Guichardet [46]. Ernest considers measure classes in the space of all quasi equivalence classes of primary representations and his results have been improved in an unpublished work of Dixmier. We remark that the notions and results of §§2 and 3 apply equally well to the * representations of * algebras and that much of the quoted literature deals with the algebra formulation.

The introduction of the Borel structure in $\hat{G}$ was inspired by the fact that the natural topology in $\hat{G}$ is seldom Hausdorff—even for quite well-behaved groups—and the fact that only the Borel sets are needed for direct integral theory. On the other hand, the topology is not as bad as it appears to be and there is a growing literature devoted to its study. In addition to a fundamental paper published by Kaplansky in 1951 [**69**], there are recent contributions due to Fell [**22**], [**23**] and Dixmier [**19**], [**20**]. The paper of Glimm cited above is also relevant here.

The fact that every separable locally compact group G has a sufficient number of irreducible unitary representations may be proved by applying the direct integral decomposition theorem to the regular representation of G. However it was first proved (for not necessarily separable groups) by applying the Krein-Milman theorem to the convex set of all "positive definite functions" on the group. This was done by Gelfand and Raikov [**33**] and a general theory of positive definite functions and their connections with representations was later published by Godement [**38**]. In dealing with representations of * algebras instead of groups the positive definite functions are replaced by the positive linear functionals or "states" as they are often called. The theory of states for operator algebras has had an extensive development which we shall not attempt to summarize here. We shall only mention a paper of Kadison [**67**] in which representations of general operator algebras are described by ideals of null sets in the "pure state space" of the algebra in partial analogy with Mackey's generalization of the Hahn-Hellinger theory.

The general theory of characters as linear functionals on ideals in group algebras described in §4 is a mild reworking of ideas of Godement [**43**], [**44**]. An earlier and less widely applicable theory is also due to Godement [**41**]. Godement's two 1954 papers are closely connected with work of Segal on "noncommutative integration" [**98**] on the one hand and work of Dixmier on "quasiunitary algebras" [**10**] on the other. Godement did not complete the program begun in his 1954 Annals papers but it has been recently resumed by Guichardet [**47**]. The abstract Plancherel theorem (for the regular representation) was first found by Segal [**95**], [**96**] and Mautner [**84**].

Orthogonality relations for infinite-dimensional irreducible representations were first found by Bargmann [**1**] in connection with his studies of SL(2, R). His results were generalized and thrown into abstract form by Godement [**36**], [**37**]. The connection between the orthogonality relations and the Plancherel measure was established for semisimple Lie groups by Harish-Chandra [**58**]. That this connection holds for arbitrary unimodular separable locally compact groups is an unpublished result of Mackey.

The spherical functions ϕ_d^L were introduced by Gelfand [26] in the special case in which d is the one-dimensional identity and G is a Lie group. For d general and G a semisimple Lie group they were first defined by Harish-Chandra [49]. Returning to the case in which d is the one-dimensional identity, some of Gelfand's ideas were generalized to arbitrary locally compact groups by Mautner [86]. A study of the general case was made by Godement [42]. Godement's paper makes quite clear and explicit the role played by large compact subgroups in the theory of semisimple Lie groups and is an excellent introduction to the work of Harish-Chandra. Gelfand's ideas are developed more fully in a joint work with Berezin [5]. The few remarks made in §4 are far from doing justice to the extensive theory involved here. The interested reader is referred to the above cited literature for further details.

Infinitesimal characters were introduced by Harish-Chandra [48] and Gelfand [25] but not in terms of the algebra $\tilde{\mathfrak{a}}(G)$ as described in §4. That $\tilde{\mathfrak{a}}(G)$ is isomorphic to the universal enveloping algebra of Harish-Chandra is quoted in Godement's paper [42] as an unpublished result of L. Schwartz. The theorem stated at the end of §4 is also due to Harish-Chandra [50].

The imprimitivity theorem and the definition and properties of induced representations will be found in various papers of Mackey [77], [78], [79], [80]. For finite groups the notion of induced representation is due to Frobenius. This definition and the Frobenius reciprocity theorem were extended to compact groups by Weil in his book. Mackey's definition in the general case was in part inspired by a construction in a note of Godement [39]. The notion of induced representation—at least when the inducing representation is one-dimensional—is implicit in the early work of Bargmann [1] and Gelfand and Naimark [30] on the 3- and 4-dimensional Lorentz group. The first analysis of a semidirect product in the infinite case was Wigner's study of the inhomogeneous Lorentz group [103]. The second was Gelfand and Naimark's determination [31] of the irreducible representations of the group of projective transformations of the real line. The general theory of semidirect products which follows from the imprimitivity theorem is described in [77] and [79]. The extension of the semidirect product theory to the case in which N is an arbitrary closed type I normal subgroup of G and in which one deals with projective unitary representations is due to Mackey [82]. Projective representations of the inhomogeneous Lorentz group are considered by Wigner [**103**] and multipliers for projective representations of Lie groups are studied systematically by Bargmann [**2**].

A considerable development of the theory of induced representations has been given by Bruhat in his thesis [**7**]. He restricts himself to Lie groups but compensates for this by considering representations in topological vector spaces. Using the distributions of L. Schwartz he is able to obtain useful sufficient conditions for the irreducibility of induced representations. Quite recently [**8**] he has been able to extend many of his results to arbitrary locally compact groups by way of an extension of the notion of distribution and has applied them to demonstrate the irreducibility of certain representations of totally disconnected analogues of the classical groups. Blattner [**6**] has improved one of the results in Bruhat's 1956 paper and made several other contributions to the theory. Loomis has given a new proof of the imprimitivity theorem which uses positive definite functions and avoids separability assumptions [**76**].

There is some material on ergodic systems of imprimitivity in [**82**] but most of §7 is to be regarded as a preliminary announcement of work in progress by Mackey.

At an early stage (1951) of his extensive development of the representation theory of semisimple Lie groups (see below), Harish-Chandra decided, at the suggestion of Chevalley, not to restrict himself to unitary representations. The paper of Godement on spherical function cited above is concerned throughout with Banach space representations and introduces the notion of complete irreducibility. Naimark gave his definition of equivalence for Banach space representations in [**92**] which at the same time announced a solution of the problem of determining (to within equivalence) all completely irreducible Banach space representations of the Lorentz group. Details are published in his book [**93**]. The paper of Berezin mentioned in §9 [**4**] deals with the problem of extending Naimark's results to all the complex classical groups. The papers of Bruhat cited above deal with nonunitary representations and so do two recent notes of Zhelobenko [**104**], [**105**]. Zhelobenko announces some preliminary but provocative results on the classification of reducible nonunitary representations.

In §8 the statements about linear system representations, the definition of U^L when L is nonunitary and the definition of $\mathfrak{D}(L, M)$ are unpublished remarks of Mackey. The definition of $\mathfrak{D}(L, M)$ was in part inspired by a note of Zhelobenko.

The problem of doing non $\mathfrak{L}^2$ harmonic analysis on noncompact noncommutative groups has barely begun but the important special case of the 2×2 real unimodular group has been extensively studied in three long papers by Mautner and Ehrenpreis [**88**], [**89**], [**90**]. Kunze and Stein [**75**] have done further work on this group and Kunze [**74**] has proved a rather general theorem about $\mathcal{L}^p$ Fourier transforms. A note of Gelfand [**27**] is also relevant here.

The unitary representation theory of semisimple Lie groups began with studies of the 2×2 real and complex unimodular groups by Bargmann [**1**] on the one hand and Gelfand and Naimark on the other. Bargmann's paper [**1**] contained a detailed study of the real case and some indications concerning the complex one. Gelfand and Naimark [**30**] treated the complex case only. Later they wrote a series of papers extending their results to all of the classical complex groups. This work was published in book form [**32**] by the Steklov Institute in 1950. Another approach to the Plancherel formula for these groups will be found in a paper by Gelfand and Graev [**28**]. The cited results on the $n\times n$ real unimodular group with $n>2$ are due to Gelfand and Graev [**29**]. As with all the results described in §9 the lecturer has taken liberties with the original formulations in order to bring out the connection with the theory of induced representations. Graev has undertaken further studies of real simple Lie groups on his own. Most of his results [**45**] concern the group of all $n\times n$ complex non-singular matrices which leave invariant an indefinite Hermitian form.

The study of the infinite-dimensional irreducible representations of *abstract* semisimple Lie groups was begun by Harish-Chandra and has led to a long series of articles [**48**], [**49**], $\cdots$, [**66**]. Insofar as possible Harish-Chandra works with the Lie algebra of the group rather than the group itself. Many of his results are theorems about Lie algebras and a large part of his 1953 Transactions paper is devoted to developing the theory necessary to transform results about infinite-dimensional representations of Lie algebras into results about infinite-dimensional representations of Lie groups. The general results cited in §9 about infinitesimal and distribution characters and about the Plancherel formula for complex groups are due to Harish-Chandra and are proved in the Transactions papers. The American Journal papers are devoted to various aspects of the still unsolved problem of finding almost all irreducible representations and the Plancherel measure for the general real semisimple Lie group.

A systematic study of the unitary representations of simply connected nilpotent Lie groups has been undertaken by Dixmier [**11**], [**12**], [**14**], [**15**], [**16**], [**17**], [**18**]. Not only are all such groups of type I but their irreducible representations have distribution characters. Moreover barring a certain exceptional set of Plancherel measure zero the irreducible representations are determined by their infinitesimal characters. Various particular cases are examined in detail. Further contributions to the theory of nilpotent groups have been made by Kirillov [71], [72, [73], Takenouchi [101], and Fell [24].

The ideas and viewpoints of §10 have occurred in various forms with varying degrees of rigor to a number of people and it would be difficult indeed to attach them to their "true" inventors. We shall content ourselves with two more or less indisputable attributions and a few references to papers where further details may be found. We remark that the purpose of §10 is less to give a survey of known results than to give some idea of what quantum mechanics is about and how infinite-dimensional group representations may be applied to it.

The result concerning the form of the general probability measure on the projections is a theorem of Gleason [34]. The relationship between the irreducible unitary representations of the inhomogeneous Lorentz group and the possible elementary particles was first pointed out by Bargmann and Wigner [3].

A treatment of elementary quantum mechanics from the point of view of §10 will be found in Harvard lecture notes of Mackey.[4]

The Centre National de la Recherche Scientifique in France has issued a volume containing the proceedings of a conference held in Lille in 1957 and entitled *Les problèmes mathématiques de la théorie quantique des champs*. This volume contains a long survey article by Segal [99] and another by Wightman [102]. These articles together form an excellent introduction to the various attempts to put the quantum theory of fields on a rigorous basis and refer to most of the relevant literature up to that time.

Appendix

(Added June 1963)

In this appendix I shall mention briefly a number of recent developments but shall make no attempt to be exhaustive. Some of these could have been included in the main text. Others had not been published in the summer of 1961 when the main text was written.

A good deal of recent work has been devoted to the problem of decomposing unitary representations of the form U^L (see §5) where L is a finite-dimensional unitary representation of the closed subgroup H of G and G is a semisimple Lie group with a finite center.

[4] To be published in October by W. A. Benjamin Inc., New York.

In most of this work H is either a discrete group such that G/H has a finite invariant measure or the subgroup of fixed elements for some involutary automorphism of G. For the case in which L is the one-dimensional identity and H is a maximal compact subgroup the theory of U^L is very closely related to the theory of spherical functions. Harish-Chandra in **[130]** and **[131]** has worked toward finding a Plancherel formula for the decomposition in this case and further results have been announced by Gindinkin and Karpelevich in **[127]**.

In **[121]**, **[122]** and **[123]** Gelfand and Graev describe a general method for attacking the problem of decomposing U^L when L is the one-dimensional identity and apply it to a number of examples. The method may be formulated quite generally but works effectively only when the subgroup N (the horospherical subgroup) is properly chosen. If $S=G/H$ and N is any closed subgroup of G then a horosphere (with respect to N) is a subset of S of the form $sxNx^{-1}$ where s and x are arbitrary fixed elements of S and G respectively. G acts in a natural way on the space S^N of all horospheres; though not in general transitively. Each horosphere may be equipped with a measure in a natural way as may the space S^N. If f is a suitably restricted function on S we may integrate it over each horosphere and thus obtain a function f' on S^N. The mapping $f\rightarrow f'$ intertwines U^L with the representation defined in $\mathcal{L}^2(S^N)$ by the action of G (for simplicity we consider only the special case in which L is the identity). When N is suitably chosen it is usually quite easy to analyze this second representation. Thus the analysis of U^L reduces to the analysis of the "Radon transform" $f\rightarrow f'$ and its inverse. This is the problem in integral geometry to which the titles of **[121]** and **[122]** refer. This problem is discussed at length in **[124]**.

The case in which H is discrete and G/H has finite invariant measure is of special interest because of its connection with problems in number theory and the theory of automorphic functions. When G/H is compact it is not hard to show **[125]** that U^L is a discrete direct sum of irreducible representations each of which occurs with finite multiplicity. When G/H is not compact U^L may not be a discrete direct sum of irreducibles but Harish-Chandra has shown **[132]** that such direct summands as do exist all occur with finite multiplicity—at least when L is the one-dimensional identity. His argument requires that H satisfy certain auxiliary conditions too complicated to state here. Gelfand and Pyatetzki-Shapiro **[126]** give a somewhat different theorem for noncompact G/H. Assuming that L is the one-dimensional identity and that H satisfies auxiliary conditions related to those of Harish-Chandra they single out a closed invariant subspace $\mathcal{H}_0$ of $\mathfrak{L}^2(G/H)$ (defined by means of compact horospheres) and sketch a proof of the fact that this space is a discrete direct sum of irreducible invariant subspaces. When G/H is compact this subspace coincides with the whole space.

In the theory of G/H, when H is discrete, a central role is played by a noncommutative generalization of the Poisson summation formula known as the Selberg trace formula. For the moment let G be a finite group and let H be any subgroup of G. Then U^L is finite-dimensional and hence has a character $\mathfrak{X}$: $\mathfrak{X}(x) = \text{Trace}(U_x^L)$. Let $\mathfrak{X}(x) \equiv n_1\mathfrak{X}_1(x) + n_2\mathfrak{X}_2(x) + \cdots + n_k\mathfrak{X}_k(x)$ where the n_j are positive integers and the $\mathfrak{X}_j$ are the characters of the irreducible constituents of U^L. Let $\mathfrak{X}^0$ be the character of L and let $\mathfrak{X}^{00}$ be the function on G which coincides with $\mathfrak{X}^0$ in H and is zero outside of H. By a well-known elementary formula we have $\mathfrak{X}(x) = (1/o(H)) \sum_{y \in G} \mathfrak{X}^{00}(yxy^{-1})$. The Selberg trace formula for finite G is obtained by equating the two expressions for $\mathfrak{X}$, multiplying by an arbitrary complex-valued function f on G, summing over G and rearranging the left side. It is

$$(1/o(H)) \sum_{y \in G} \sum_{x \in H} f(y^{-1}xy)\, \mathfrak{X}^0(x) = \sum_{j=1}^{k} n_j \left(\sum_{x \in G} f(x)\, \mathfrak{X}_j(x) \right).$$

Note that the inner sum on the left-hand side, as a function of y, is a constant on the right H cosets. Thus the left side is really a sum over G/H. Once this has been observed it is clear that the trace formula may be stated (though not necessarily proved) when H is any discrete subgroup of a separable locally compact group G such that (a) G/H has a finite invariant measure μ and (b) U^L is a discrete direct sum of irreducible representations each of which has a character and occurs with finite multiplicity. One need only replace $\sum_{x \in G} f(x)\mathfrak{X}_j(x)$ by $\int_G f(x)\mathfrak{X}_j(x)dx$ and $(1/o(H)) \sum_{y \in G}$ by an integral with respect to μ. Proofs under various hypotheses have been given in [**125**], [**146**] and [**147**]. In [**146**] and [**147**] it is assumed that G has a large compact subgroup (cf. §4) and only those functions f are considered which are constant on the right K cosets. Under these circumstances the integrals $\int f(x)\mathfrak{X}_j(x)$ may be expressed in terms of spherical functions and one need make no assumption about the existence of characters for the irreducible constituents of U^L. In [**125**] attention is restricted to the case in which L is the one-dimensional identity.

Answering two questions raised by Kaplansky, Mackey [**141**] has shown that a nontype I group can have a type I regular representation and Thoma [**148**] has shown that a discrete group G cannot be of type I unless it has a commutative normal subgroup N such that

G/N is finite. Dixmier [111] has shown that the universal covering group of a Lie group need not be of type I even if the group itself is of type I.

The details of Ernest's paper [21] have been published [118] and the "unpublished" supplementary remarks of Dixmier referred to in the main text will be found in [115]. Closely related results will be found in [149]. Further results about the Borel structure in $\hat{G}$ and related spaces will be found in the thesis of Guichardet [129] and in papers of Effros [117] and Naimark [145]. Guichardet considers the Borel structure in the space of all "normal" irreducible representations showing it to be analytic and identical with that derived from a certain natural topology in this space. He also gives an example showing that a group with a complete system of finite-dimensional irreducible representations need not have a smooth dual. Effros is concerned with the Borel structure in the space of primitive ideals of a C^* algebra.

The work of Dixmier and Fell on the topology of G has been carried further in [112], [113], [116], [119], and [120]. Fell in particular has studied the continuity properties of the inducing process and Glimm [128] has introduced and studied a topology in the space of all irreducible pairs P, L where P is a system of imprimitivity for the representation L of G. The G space on which P is defined is fixed and given in advance. Results close to those in [119] will be found in [150].

Guichardet's reworking of Godement's character theory as announced in [47] is carried out in detail in the first chapter of [129].

Various contributions to the general theory of induced representations have been made by Blattner [106], [107], [108], Kleppner [136], [137] and Moore [144]. [106] concerns an abstract formulation of the analogue of the inducing process in which Borel functions are replaced by functions which are "partially analytic." Kunze [139] and Kobayashi [138] have proved theorems about the irreducibility of such representations in which the inducing representation is one-dimensional and the analyticity is complete. In [144] Moore considers a variant of the notion of induced representation in which, roughly speaking, summability replaces square summability. The induced representations defined in this way are not Hilbert space representations but yield a rather elegant version of the Frobenius reciprocity theorem in which one has a linear *isometry* between the relevant spaces of intertwining operators. This work of Moore has been extended and developed further by Kleppner in [137]. These generalizations of the Frobenius reciprocity theorem go in quite a different direction from those in [80] since an irreducible representation L admits a nonzero intertwining operator with a reducible representation M only if M has *discrete* irreducible components.

The generalized distributions studied by Bruhat in [**8**] were introduced and studied independently by Maurin [**142**] and G. I. Kac [**134**]. Kunze and Stein have continued their work on nonunitary representations in [**140**].

A start on the problem of determining the irreducible representations of totally disconnected simple groups is made in [**8**], [**109**], [**110**], and [**143**]. Mautner in [**143**] develops a theory of spherical functions for 2×2 unimodular matrix groups over totally disconnected fields using the subgroup of matrices with integer coefficients as compact subgroup. Among other things he obtains the Plancherel formula for the coset space. Bruhat in [**109**] studies the $n\times n$ case and attempts to prove that the compact group in question is large (§4). He is only partially successful and the problem of deciding whether the totally disconnected analogues of the classical groups are type I groups or not remains a challenging open question. In [**110**] the results of [**109**] are extended to totally disconnected analogues of the unitary symplectic and orthogonal groups.

Dixmier in [**114**] gives a complete determination of the irreducible representations of the DeSitter group—correcting and rigorizing earlier formal work of L. H. Thomas and T. D. Newton. He indicates that analogous arguments can be used to deal with all semisimple Lie groups. He also investigates the integrability and square integrability of the representations and in this particular case verifies a conjecture of Harish-Chandra.

Harish-Chandra in [**133**] announces some results (too involved to quote here) allowing him to associate square integrable representations with every character of every maximal abelian subgroup of a semisimple Lie group with finite center. This accomplishes an essential step in his program of determining the Plancherel formula for general semisimple Lie groups.

The details of the work of Kirillov on nilpotent Lie groups appear in [**135**]. He shows among other things that the equivalence classes of irreducible unitary representations of a simply connected nilpotent Lie group correspond in a natural one-to-one manner to the orbits of the canonical action of the group on the dual of its Lie algebra. This correspondence has elegant properties with respect to the formation of tensor products and the restriction of representations to subgroups.

BIBLIOGRAPHY

1. V. Bargmann, *Irreducible unitary representations of the Lorentz group*, Ann. of Math. (2) **48** (1947), 568–640.

2. ———, *On unitary ray representations of continuous groups*, Ann. of Math. (2) **59** (1954), 1–46.

3. V. Bargmann and E. P. Wigner, *Group theoretical discussion of relativistic wave equations*, Proc. Nat. Acad. Sci. U.S.A. **34** (1948), 211–223.

4. F. A. Berezin, *Laplace operators on semi-simple Lie groups*, Trudy Moskov. Math. Obšč. **6** (1957), 371–463. (Russian)

5. F. A. Berezin and I. M. Gelfand, *Some remarks on the theory of spherical functions on symmetric Riemannian manifolds*, Trudy Moskov Mat. Obšč. **5** (1956), 311–351. (Russian)

6. R. J. Blattner, *On induced representations*, Amer. J. Math. **83** (1961), 79–98.

7. F. Bruhat, *Sur les représentations induites des groups de Lie*, Bull. Soc. Math. France **84** (1956), 97–205.

8. ———, *Distributions sur un groupe localement compact et applications à l'étude des représentations des groupes p-adiques*, Bull. Soc. Math. France **89** (1961), 43–75.

9. J. Dixmier, *Sur la réduction des anneaux d'opérateurs*, Ann. Sci. École. Norm. Sup (3) **68** (1951), 185–202.

10. ———, *Algèbres quasi-unitaires*, Comment. Math. Helv. **26** (1952), 275–322.

11. ———, *On unitary representations of nilpotent Lie groups*, Proc. Nat. Acad. Sci. U.S.A. **43** (1957), 985–986.

12. ———, *Sur les représentations unitaires des groupes de Lie nilpotents*. II, Bull. Soc. Math. France **85** (1957), 325–388.

13. ———, *Sur les représentations unitaires des groupes de Lie algébriques*, Ann. Inst. Fourier (Grenoble) **7** (1957), 315–328.

14. ———, *Sur les représentations unitaires des groupes de Lie nilpotents*. III, Canad. J. Math. **10** (1958), 321–348.

15. ———, *Sur les représentations unitaires des groupes de Lie nilpotents*. I, Amer. J. Math. **81** (1959), 160–170.

16. ———, *Sur les représentations unitaires des groupes de Lie nilpotents*. IV, Canad. J. Math. **11** (1959), 321–344.

17. ———, *Sur les représentations unitaires des groupes de Lie nilpotents*. V, Bull. Soc. Math. France **87** (1959), 65–79.

18. ———, *Sur les représentations unitaires des groupes de Lie nilpotents*. VI, Canad. J. Math. **12** (1960), 324–352.

19. ———, *Sur les C* algèbres*, Bull. Soc. Math. France **88** (1960), 95–112.

20. ———, *Sur les structures boréliennes du spectre d'une C*-algèbre*, Inst. Hautes Etudes Sci. Publ. Math. **6** (1960), 5–11.

21. J. Ernest, *A decomposition theory for unitary representations of locally compact groups*, Bull. Amer. Soc. Math. **67** (1961), 385–388.

22. J. M. G. Fell, *C* algebras with smooth dual*, Illinois J. Math. **4** (1960), 221–230.

23. ———, *The dual spaces of C* algebras*, Trans. Amer. Math. Soc. **94** (1960), 365–403.

24. ———, *A new proof that nilpotent groups are* CCR, Proc. Amer. Math. Soc. **13** (1962), 93–99.

25. I. M. Gelfand, *The center of an infinitesimal group ring*, Mat. Sb. (N.S.) **26** (68) (1950), 103–112. (Russian)

26. ———, *Spherical functions in symmetric Riemann spaces*, Dokl. Akad. Nauk. SSSR **70** (1956), 5–8. (Russian)

27. ———, *The structure of a ring of rapidly decreasing functions on a Lie group*, Dokl.. Akad. Nauk. SSSR **124** (1959), 19–21. (Russian)

28. I. M. Gelfand and M. I. Graev, *Analogue of the plancherel formula for the classical groups*, Trudy Moskov. Mat. Obšč. **4** (1955), 375–404. (Russian)

29. ———, *Unitary representations of the real unimodular group (principal non-degenerate series)*, Izvestiya Akad. Nauk. SSSR **17** (1953), 189–248. (Russian)

30. I. M. Gelfand and M. A. Naimark, *Unitary representations of the Lorentz group*, Izvestiya Akad. Nauk SSSR **11** (1947), 411–504. (Russian)

31. ———, *Unitary representations of the group of linear transformations of the straight line*, Dokl. Akad. Nauk. SSSR **55** (1947), 567–570. (Russian)

32. ———, *Unitary representations of the classical groups*, Trudy Mat. Inst. Steklov **36** (1950), 288. (Russian)

33. I. M. Gelfand and D. A. Raikov, *Irreducible unitary representations of arbitrary locally bicompact groups*, Mat. Sb. (N.S.) **13** (**55**) (1943), 301–316. (Russian)

34. A. M. Gleason, *Measures on the closed subspaces of a Hilbert space*, J. Math. Mech. **6** (1957), 885–894.

35. J. Glimm, *Type I C^* algebras*, Ann. of Math. (2) **73** (1961), 572–612.

36. R. Godement, *Sur les relations d'orthogonalité de V. Bargmann*. I. *Résultats préliminaires*, C. R. Acad. Sci. Paris **225** (1947), 521–523.

37. ———, *Sur les relations d'orthogonalité de V. Bargmann*. II. *Démonstration générale*, C. R. Acad. Sci. Paris **225** (1947), 657–659.

38. ———, *Les fonctions de type positif et la théorie des groupes*, Trans. Amer. Math. Soc. **63** (1948), 1–84.

39. ———, *Sur la transformation de fourier dans les groupes discrets*, C. R. Acad. Sci. Paris **228** (1949), 627–628.

40. ———, *Sur la théorie des représentations unitaires*, Ann. of Math. (2) **53** (1951), 68–124.

41. ———, *Mémoire sur la théorie des caractères dans les groupes localement compacts unimodulaires*, J. Math. Pures Appl. **30** (1951), 1–110.

42. ———, *A theory of spherical functions*. I, Trans. Amer. Math. Soc. **73** (1952), 496–556.

43. ———, *Théorie des caractères*. I. *Algèbres unitaires*, Ann. of Math. (2) **59** (1954), 47–62.

44. ———, *Théorie des caractères*. II. *Définition et propriétés générales des caractères*, Ann. of Math. (2) **59** (1954), 63–85.

45. M. I. Graev, *Unitary representation of real simple Lie groups*, Trudy Moskov. Mat. Obšč. **7** (1958), 335–389. (Russian)

46. A. Guichardet, *Sur un problème posé par G. W. Mackey*, C. R. Acad. Sci. Paris **250** (1960), 962–963.

47. ———, *Sur les caractères des algèbres de Banach à involution*, C. R. Acad. Sci. Paris **252**(1961), 2800–2862.

48. Harish-Chandra, *On some applications of the universal enveloping algebra of a semi-simple Lie algebra*, Trans. Amer. Math. Soc. **70** (1951), 28–96.

49. ———, *Representations of semi-simple Lie groups*. II, Proc. Nat. Acad. Sci. U.S.A. **37** (1951), 170–173.

50. ———, *Representations of a semi-simple Lie group on a Banach space*. I, Trans. Amer. Math. Soc. **75** (1953), 185–243.

51. ———, *Plancherel formula for complex semi-simple Lie groups*, Proc. Nat. Acad. Sci. U.S.A. **37** (1951), 813–818.

52. ———, *Representations of semi-simple Lie groups*. II, Trans. Amer. Math. Soc. **76** (1954), 26–65.

53. ———. *Representations of semi-simple Lie groups*. III, Trans. Amer. Math. Soc. **76** (1954), 234–253.

54. ———, *The Plancherel formula for complex semi-simple Lie groups*, Trans. Amer. Math. Soc. **76** (1954), 485–528.

55. ———, *On the characters of a semi-simple Lie group*, Bull. Amer. Math. Soc. **61** (1955), 389–396.

56. ———, *Representations of semi-simple Lie groups*. IV, Amer. J. Math. **77** (1955), 743–777.

57. ———, *Representations of semi-simple Lie groups*, V, Amer. J. Math. **78** (1956), 1–41.

58. ———, *Representations of semi-simple Lie groups*, VI, Amer. J. Math. **78** (1956), 564–628.

59. ———, *The characters of semi-simple Lie groups*, Trans. Amer. Math. Soc. **83** (1956), 98–163.

60. ———, *On a lemma of F. Bruhat*, J. Math. Pures Appl. (9) **35** (1956), 203–210.

61. ———, *Differential operators on a semi-simple Lie algebra*, Amer. J. Math. **79** (1957), 87–120.

62. ———, *Fourier transforms on a semi-simple Lie algebra*. I, Amer. J. Math. **79** (1957), 193–257.

63. ———, *Fourier transforms on a semi-simple Lie algebra*, II, Amer. J. Math. **79** (1957), 653–686.

64. ———, *A formula for semi-simple Lie groups*. Amer. J. Math. **79** (1957), 733–760.

65. ———, *Spherical functions on a semi-simple Lie group*. I, Amer. J. Math. **80** (1958), 241–310.

66. ———, *Spherical functions on a semi-simple Lie group*. II, Amer. J. Math. **80** (1958), 553–613.

67. R. V. Kadison, *Unitary invariants for representations of operator algebras*, Ann. of Math. (2) **66** (1957), 304–379.

68. I. Kaplansky, *Projections in Banach algebras*, Ann. of Math. (2) **53** (1951), 235–249.

69. ———, *The structure of certain operator algebras*, Trans. Amer. Math. Soc. **70** (1951), 219–255.

70. ———, *Group algebras in the large*, Tôhoku Math. J. **3** (1951), 249–256.

71. A. A. Kirillov, *Induced representations of nilpotent Lie groups*, Dokl. Akad. Nauk SSSR **128** (1959), 886–889. (Russian)

72. ———, *On unitary representations of nilpotent Lie groups*, Dokl. Akad. Nauk SSSR **130** (1960), 966–968. (Russian)

73. ———, *Unitary representations of nilpotent Lie groups*, Dokl. Akad. Nauk SSSR **138** (1961), 283–284. (Russian)

74. R. A. Kunze, *L_p Fourier transforms on locally compact unimodular groups*, Trans. Amer. Math. Soc. **89** (1958), 519–540.

75. R. A. Kunze and E. M. Stein, *Uniformly bounded representations and harmonic analysis on the 2×2 real unimodular group*, Amer. J. Math. **82** (1960), 1–62.

76. L. H. Loomis, *Positive definite functions and induced representations*, Duke Math. J. **27** (1960), 569–580.

77. G. W. Mackey, *Imprimitivity for representations of locally compact groups*. I, Proc. Nat. Acad. Sci. U.S.A. **35** (1949), 537–545.

78. ———, *On induced representations of groups*, Amer. J. Math. **73** (1951), 576–592.

79. ———, *Induced representations of locally compact groups*. I, Ann. of Math. (2) **55** (1952), 101–139.

80. ———, *Induced representations of locally compact groups*. II, Ann. of Math. (2) **58** (1953), 193–221.

81. ———, *Borel structure in groups and their duals*, Trans. Amer. Math. Soc. **85** (1957), 134–165.

82. ———, *Unitary representations of group extensions*. I, Acta Math. 99 (1958), 265–311.

83. F. I. Mautner, *Unitary representations of locally compact groups*. I, Ann. of Math. (2) **51** (1950), 1–25.

84. ———, *Unitary representations of locally compact groups*. II, Ann. of Math. (2) **52** (1950), 528–556.

85. ———, *The structure of the regular representation of certain discrete groups*, Duke Math. J. **17** (1950), 437–441.

86. ———, *Fourier analysis and symmetric spaces*, Proc. Nat. Acad. Sci. U.S.A. **37** (1951), 529–533.

87. ———, *The regular representation of a restricted direct product of finite groups*, Trans. Amer. Math. Soc. **70** (1951), 531–548.

88. F. I. Mautner and L. Ehrenpreis, *Some properties of the Fourier transform on semi-simple Lie groups*. I, Ann. of Math. (2) **61** (1955), 406–439.

89. ———, *Some properties of the Fourier transform on semi-simple Lie groups*. II, Trans. Amer. Math. Soc. **84** (1957), 1–55.

90. ———, *Some properties of the Fourier transform on semi-simple Lie groups*. III, Trans. Amer. Math. Soc. **90** (1959), 431–484.

91. F. J. Murray and J. von Neumann, *On rings of operators*, Ann. of Math (2) **37** (1936), 116–229.

92. M. A. Naimark, *On irreducible linear representations of the proper Lorentz group*, Dokl. Akad. Nauk SSSR **97** (1954), 969–972. (Russian)

93. ———, *Linear representations of the Lorentz group*, 376 pp. Fizmatgiz, Moscow, 1958. (Russian)

94. J. von Neumann, *On rings of operators. Reduction theory*, Ann. of Math (2) **50** (1949), 401–485.

95. I. E. Segal, *The two-sided regular representation of a unimodular locally compact group*, Ann. of Math. (2) **51** (1950), 293–298.

96. ———, *An extension of Plancherel's formula to separable unimodular groups*, Ann. of Math. (2) **52** (1950), 272–292.

97. ———, *Decomposition of operator algebras*. I and II, Mem. Amer. Math. Soc. No. 9 (1952), 67+66 pp.

98. ———, *A non-commutative extension of abstract integration*, Ann. of Math (2) **57** (1953), 401–457.

99. ———, *Caractérisation mathématique des observables en théorie quantique des champs et ses conséquences pour la structure des particules libres*, Report of Lille conference on quantum fields, C.N.R.S., Paris, 1959, pp. 57–163.

100. W. F. Stinespring, *A semi-simple matrix group is of type* I, Proc. Amer. Math. Soc. **9** (1958), 965–967.

101. O. Takenouchi, *Sur la facteur représentation d'un group de Lie résoluble de type* E, Math. J. Okayama Univ. **7** (1957), 151–161.

102. A. S. Wightman, *Quelques problèmes mathématiques de la théorie quantique relativiste*, Report of Lille conference on quantum fields, C.N.R.S., Paris, 1959, pp. 1–35.

103. E. Wigner, *On unitary representations of the inhomogeneous Lorentz group*, Ann. of Math. (2) **40** (1939), 149–204.

104. D. P. Zhelobenko, *A description of a certain class of Lorentz group representations*, Dokl. Akad. Nauk SSSR **125** (1958), 586–589. (Russian)

105. ———, *Linear representations of the Lorentz group*, Dokl. Akad. Nauk SSSR **126** (1959), 482–485. (Russian)

BIBLIOGRAPHY (SUPPLEMENT)

106. R. J. Blattner, *On induced representations. II. Infinitesimal induction*, Amer. J. Math. **83** (1961), 499–512.

107. ———, *On a theorem of G. W. Mackey*, Bull. Amer. Math. Soc. **68** (1962), 585–587.

108. ———, *A theorem on induced representations*, Proc. Amer. Math. Soc. **13** (1962), 881–884.

109. F. Bruhat, *Sur les représentations de groupes classiques p-adiques*. I, Amer. J. Math. **83** (1961), 321–338.

110. ———, *Sur les représentations de groupes classiques p-adiques*. II, Amer. J. Math. **83** (1961), 343–368.

111. J. Dixmier, *Sur le revêtment universel d'un groupe de Lie de type* I, C. R. Acad. Sci. Paris 252 (1961), 2805–2806.

112. ———, *Points séparés dans le spectre d'une C*-algèbre*, Acta Sci. Math. Szeged 22 (1961), 115–128.

113. ———, *Points isolés dans le dual d'un groupe localement compact*, Bull. Soc. Math. France 85 (1961), 91–96.

114. ———, *Représentations intégrables du groupe de deSitter*, Bull. Soc. Math. France **89** (1961), 9–41.

115. ———, *Dual et quasi dual d'une algèbre de Banach involutive*, Trans. Amer. Math. Soc. **107** (1963), 278–283.

116. J. Dixmier et P. Bernat, *Sur le dual d'un groupe de Lie*, C. R. Acad. Sci. Paris **250** (1960), 1778–1779.

117. E. Effros, *A decomposition theory for representations of C*-algebras*, Trans. Amer. Math. Soc. **107** (1963), 83–106.

118. J. Ernest, *A decomposition theory for unitary representations of locally compact groups*, Trans. Amer. Math. Soc. **104** (1962), 252–277.

119. J. M. G. Fell, *The structure of algebras of operator fields*, Acta. Math. **106** (1961), 233–280.

120. ———, *Weak containment and induced representations of groups*, Canad. J. Math. **14** (1962), 237–268.

121. I. M. Gelfand, *Integral geometry and its relation to the theory of representations*, Uspehi Mat. Nauk **15** (1960), 155–164. (Russian)

122. I. M. Gelfand and M. I. Graev, *Geometry of homogeneous spaces, representations of groups in homogeneous spaces and related questions of integral geometry*, Trudy Moskov Mat. Obšč. **8** (1959), 321–390. (Russian)

123. ———, *Resolution of Lorentz group representations into irreducible representations on spaces of functions defined on symmetrical spaces*, Dokl. Akad. Nauk SSSR **127** (1959), 250–253. (Russian)

124. I. M. Gelfand and N. Y. Vilenkin, *Integral geometry and connections with questions in the theory of representations*, Moscow, 656 pp., Fizmatgiz, 1962. (Russian)

125. I. M. Gelfand and I. Pyatetzki-Shapiro, *Theory of representations and theory of automorphic functions*, Uspehi Mat. Nauk **14** (1959), 171–194. (Russian)

126. ———, *Unitary representations in homogeneous spaces with discrete stationary subgroups*, Dokl. Akad. Nauk SSSR **147** (1962), 17–20. (Russian)

127. S. G. Gindikin and F. I. Karpelevic, *Plancherel measure of Riemannian*

symmetric spaces of non positive curvature, Dokl. Akad. Nauk **145** (1962), 252–255. (Russian)

128. J. Glimm, *Families of induced representations*, Pacific J. Math. **12** (1962), 885–911.

129. A. Guichardet, *Caractères des algèbres de Banach involutives*, Ann. Inst. Fourier (Grenoble) **13** (1962), 1–81.

130. Harish-Chandra, *Spherical functions on a semisimple Lie group*. I, Amer. J. Math. **80** (1958), 241–310.

131. ———, *Spherical functions on a semisimple Lie group*. II, Amer. J. Math. **80** (1958), 553–613.

132. ———, *Automorphic forms on a semisimple Lie group*, Proc. Nat. Acad. Sci. U.S.A. **45** (1959), 570–573.

133. ———, *Invariant eigendistributions on semisimple Lie groups*, Bull. Amer. Math. Soc. **69** (1963), 117–123.

134. G. I. Kac, *Generalized functions on a locally compact group and decompositions of unitary representations*, Trudy Moskov Mat. Obšč. **10** (1961), 3–40. (Russian)

135. A. A. Kirillov, *Unitary representations of nilpotent Lie groups*, Uspehi Mat. Nauk **106** (1962), 57–110. (Russian)

136. A. Kleppner, *The structure of some induced representations*, Duke Math. J. **29** (1962), 555–572.

137. ———, *Intertwining forms for summable induced representations*, (to appear).

138. S. Kobayashi, *On automorphism groups of homogeneous complex manifolds*, Proc. Amer. Math. Soc. **12** (1961), 359–361.

139. R. A. Kunze, *On the irreducibility of certain multiplier representations*, Bull. Amer. Math. Soc. **68** (1962), 93–94.

140. R. A. Kunze and E. M. Stein, *Uniformly bounded representations*. II. *Analytic continuation of the principal series of representations of the $n\times n$ complex unimodular group*, Amer. J. Math **83** (1961), 723–786.

141. G. W. Mackey, *Induced representations and normal subgroups*, Proc. Internat. Sympos. Linear Spaces, Jerusalem pp. 319–326, Pergamon, Oxford, 1960.

142. K. Maurin, *Distributionen auf Yamabe-Gruppen. Harmonische Analyse auf einer Abelschen l.k. Gruppe*, Bull. Acad. Polon. Sci. Sér. Sci. Math. Astronom. Phys. **9** (1961), 845–850.

143. F. I. Mautner, *Spherical functions over p-adic fields*, Amer. J. Math. **80** (1958), 441–457.

144. C. C. Moore, *On the Frobenius reciprocity theorem for locally compact groups*, Pacific J. Math. **12** (1962), 359–365.

145. M. A. Naimark, *Decomposition into factor representations of unitary representations of locally compact groups*, Sibirsk. Mat. Ž. **2** (1961), 89–99. (Russian)

146. A. Selberg, *Harmonic analysis and discontinuous groups in weakly symmetric Riemannian spaces with applications to Dirichlet series*, J. Indian Math. Soc. **20** (1956), 47–87.

147. T. Tamagawa, *On Selberg's trace formula*, J. Fac. Sci. Univ. Tokyo Sect. I **8** (1960), 363–386.

148. E. Thoma, *Über unitare Darstellungen abzahlbarer, diskreter Gruppen*, (to appear).

149. M. Takesaki, *On some representations of C* algebras*, Tôhoku Math. J. **15** (1963), 79–95.

150. J. Tomiyama, *Topological representation of C* algebras*, Tôhoku Math. J. **14** (1962), 187–204.

HARVARD UNIVERSITY

TRANSLATION-INVARIANT CONES OF FUNCTIONS ON SEMI-SIMPLE LIE GROUPS

HARRY FURSTENBERG[1]

Introduction. Originally, the phrase harmonic analysis had a function-theoretic meaning, referring to the decomposition of a function into exponentials. In the current interpretation, particularly in connection with noncommutative groups, the term refers not to functions but to representations, and harmonic analysis is regarded as part of the theory of group representations. This shift in interpretation was motivated by the L^2-theory for compact groups. The decomposition of the L^2-space of a noncommutative compact group analogous to the Fourier decomposition for the circle involves multidimensional subspaces, and, as a result, there is no longer a canonical choice of a basis for the L^2-space analogous to the set $\{e^{in\theta}\}$ for the circle. The subspaces, on the other hand, are canonically determined, and correspond to the various irreducible representations of the group. It therefore became natural to regard irreducible representations as the basic building blocks of the theory in the place of the exponential functions.

Our purpose here is to call attention to some examples in the theory of semi-simple noncompact Lie groups where the classical setup prevails. That is, we shall find a class of functions on these groups which appear to play a role similar to that played by the exponentials for the circle or the real line. In terms of these functions, a form of spectral synthesis will be valid. Namely, for certain translation-invariant classes of functions on the group, we shall find that each function of such a class admits a unique representation as a generalized linear combination of the "exponentials" in that class. Admittedly, this result corresponds to a relatively easy case of spectral synthesis for the line. However, it is hoped that by pursuing this analogy further, other fruitful applications may be found.

The prototype of the theorem we shall prove is a theorem of Choquet and Deny [3] for $\boldsymbol{R}^n$ (or any locally compact commutative group). Let μ be a positive bounded Borel measure on $\boldsymbol{R}^n$ that does

An address delivered before the New York meeting of the Society on April 21, 1964 under the title *Aspects of harmonic analysis on Lie groups*, by invitation of the Committee to Select Hour Speakers for Eastern Sectional Meetings; received by the editors September 20, 1964.

[1] Work supported in part by Army Research Office, Project in Probability, Princeton University and the Air Force Office of Scientific Research.

not have its support in a proper subgroup of $\boldsymbol{R}^n$. Form the cone V of (unbounded) positive measures ν on $\boldsymbol{R}^n$ satisfying $\mu * \nu = \lambda\nu$, where $\lambda > 0$ is fixed. A general result of Choquet and others to which we shall refer more explicitly later ensures that V is spanned by its extremal rays. Moreover, choosing appropriately a single element from each extremal ray, there will be a unique integral representation of each member of V in terms of these extremals. The question arises as to the nature of the extremal measures of V. The answer given by Choquet and Deny is that the extremal measures have the form $f(x)dx$ where $f(x)$ is proportional to an exponential: $f(x) = ce^{u \cdot x}$. In other words, the measures of V all correspond to functions (in fact, convex functions), and these functions are linear combinations of the exponentials in V: $F \in V \Leftrightarrow F(x) = \int e^{u \cdot x}\, d\omega(u)$ where ω is a measure concentrated in the set of u for which $\int e^{u \cdot x}\, d\mu(x) = \lambda$.

It is worthwhile sketching a proof of the fact that the extremals of V correspond to exponentials. The cone V is invariant under translation. Now the equation $\nu = \lambda^{-1}\mu * \nu$ implies that ν is a linear combination of its translates. This means that ν cannot be an extremal of V unless all its translates by points in the support of μ are proportional to ν. If the support of μ generates $\boldsymbol{R}^n$ as a group, this implies that all translates of ν are proportional to ν. One sees easily that ν must then be given by an exponential.

For a noncommutative group the convolution $\mu * \nu$ may still be defined and we can form the cone V of solutions to $\mu * \nu = \lambda\nu$ as before. This time, however, V is invariant under translation on the *right*: $\nu \to \nu * g$ (this will presently be defined more precisely) and the equation $\nu = \lambda^{-1}\mu * \nu$ implies that ν is a linear combination of *left* translates. The above argument, therefore, breaks down, and one can inquire what class of functions can occur as the extremals of V in the general case. For a compact group G this question is easily answered, because V will be empty unless $\lambda = \mu(G)$ and ν must be proportional to the Haar measure of G. But for noncompact groups the question is more interesting and for the most part unsolved. We shall solve this problem only for semi-simple Lie groups, and even here only under additional hypotheses regarding μ.

1. **Preliminaries.** If G is a topological group, by a *G-space* is meant a space X and a continuous map $(g, x) \to gx$ of $G \times X \to X$ satisfying $(g_1 g_2)x = g_1(g_2 x)$ and $ex = x$ for e the identity of G. X is a *homogeneous* G-space if G is transitive on X; that is for $x, y \in X$, there exists $g \in G$ with $gx = y$. X may then be identified with the coset space G/H, where, for some $x_0 \in X$, H is the subgroup that leaves x_0 fixed.

If X is a G-space, μ a Borel measure on G and ν a Borel measure on X, then the convolution $\mu * \nu$ may be defined by

$$\int_X f(y)\, d\mu * \nu(y) = \int_X \int_G f(gx)\, d\mu(g) d\nu(x),$$

providing this integral is finite for continuous functions f of compact support. In particular, the convolution of two bounded measures on G is defined. We shall interchangeably denote the unit mass concentrated at a point x of a space X by δ_x or by x itself. The convolution $\delta_g * \mu$ for $g \in G$ and μ a measure on a G-space (or on G itself) will sometimes be denoted $g\mu$; similarly μx is the same as $\mu * \delta_x$ where μ is a measure on G and x belongs to the G-space X. Notice that $\delta_g * \delta_x = \delta_{gx}$, so that our notation for convolution is consistent with the notation for multiplication. It is frequently useful to regard convolution as a linear extension of multiplication so that

$$\mu * \nu = \int_G g\nu\, d\mu(g) = \int_X \mu x\, d\nu(x) = \int_G \int_X gx\, d\mu(g) d\nu(x).$$

The space of unbounded regular measures on a locally compact space may be identified with the dual of the space of continuous functions with compact support on the space. When we speak of the weak topology on measures, we have in mind the weak* topology of this dual space. In addition, it will often be convenient to denote an integral $\int f(x)\, d\mu(x)$ simply as $\mu(f)$.

On a locally compact group G a locally integrable function $f(g)$ determines a measure ω on G by $d\omega(g) = f(g)dg$, dg denoting the left invariant Haar measure on G. We shall often find it convenient to identify the function f with the measure ω. Thus the convolution of two functions $f_1 * f_2$, or of a function with a measure, $f * \mu$ or $\mu * f$, is defined as the function corresponding to the measure which is the convolution of the two measures involved.

On a group G we use the notation R_γ to denote the operator $R_\gamma f(g) - f(g\gamma)$. It might be noticed that, as a measure, $R_\gamma f$ corresponds to the convolution $f * \delta_{\gamma^{-1}}$. In our discussion we shall be concerned primarily with translates on the right: $f \to R_\gamma f$. As a result we shall always understand, by the *translate* of a function f, a right translate $R_\gamma f$.

If X and Y are two G-spaces and j is a map of X into Y, we shall say that j is *equivariant* if $j(gx) = gj(x)$, for all $x \in X$, $g \in G$.

By a *semi-simple group* G we shall mean one which is noncompact and has a finite center. Such a group admits an *Iwasawa decomposi-*

tion $G=K\cdot A\cdot N$. Here K is any maximal compact subgroup of G, A is a vector group and N is nilpotent, $A\cdot N$ is a solvable group whose commutator subgroup is N. The decomposition is not direct but each $g\in G$ admits a unique representation $g=kan$. The letter K will always refer to a fixed but arbitrary maximal compact subgroup of G. m_K denotes the normalized Haar measure on K. In general, if μ is a positive measure on a space X and $\mu(X)=1$, we refer to μ as a *probability* measure.

2. **Multipliers and multiplier functions.** To generalize the notion of an exponential function to an arbitrary group we may proceed as follows. The exponentials are characterized by the property that they have unit value at the identity and any translate is proportional to the function itself. On a semi-simple group only the constant function has this property. (Such a function defines a homomorphism from the group to the complex numbers and the homomorphic image of a semi-simple group is semi-simple.) So we modify this condition by supposing instead that we have a family of functions $s(\cdot,\xi)$, where ξ ranges over an index set X, with the property that $s(e,\xi)=1$ and that a (right) translate of each of the $s(\cdot,\xi)$ is proportional to some other of the $s(\cdot,\xi)$. We may take X to be the family of functions occurring here with the identification of proportional functions. M is then a G-space, and we may write $R_\gamma s(\cdot,\xi)\sim s(\cdot,\gamma\xi)$. More precisely, we have

$$R_\gamma s(g,\xi) = s(g\gamma,\xi) = h_\xi(\gamma)s(g,\gamma\xi).$$

Setting $g=e$, we have $s(\gamma,\xi)=h_\xi(\gamma)$ or

$$s(g\gamma,\xi) = s(g,\gamma\xi)s(\gamma,\xi). \tag{2.1}$$

A function satisfying (2.1) is usually referred to as a *multiplier*. As a function from G to the module of functions on M it is referred to as a *crossed homomorphism* or a 1-*cocycle*. Our usage will be more restrictive and we take the following as our definition.

DEFINITION 2.1. Let X be a G-space. A continuous function s from $G\times X$ to the positive reals is called a *multiplier* if it satisfies (2.1). The group of all such functions is denoted $M(X)$ (where the group G is assumed fixed). If $s\in M(X)$ and $\xi\in X$ then the function $s(\cdot,\xi)$ is called a *multiplier function*. The set of all multiplier functions corresponding to multipliers in $M(X)$ is denoted $E(X)$.

If X consists of a single point, then (2.1) defines a positive character (homomorphism into $(0,\infty)$) on G. Thus the positive exponentials are the multiplier functions on R^n corresponding to a trivial X. As we have remarked, for G semi-simple, if X is trivial, then $E(X)$ consists of the function 1. On the other hand, if X is taken too large, $E(X)$ is again uninteresting because it contains all positive continuous functions f with $f(e)=1$. For example, if we take $X=G$, and if f satisfies $f(e)=1$, define $s\in M(G)$ by $s(g_1, g_2)=f(g_1g_2)/f(g_2)$. It is easily seen that s satisfies (2.1) and $f(g)=s(g, e)$. What we shall find is that in the case of a semi-simple G, there is a choice of X for which the functions of $E(X)$ play the role of the exponentials. In particular, these functions will turn out to be the extremals in our version of the Choquet-Deny theorem.

If $p(\xi)$ is a positive continuous functions on the G-space X, we may form a multiplier in $M(X)$ by setting

$$s(g, \xi) = p(g\xi)/p(\xi). \tag{2.2}$$

We call these multipliers *trivial.* (In the terminology of cohomology theory, these would be coboundaries.) We shall call two multipliers *equivalent* if their quotient is trivial. The (cohomology) group of equivalence classes will be denoted $H(X)$.

Similar considerations arise in the representation theory of G. If X is a G-space we may define a representation of G on some linear space of functions on X by setting $U_gf(\xi)=s(g^{-1}, \xi)f(g^{-1}\xi)$. The condition that U_g be a representation is just the multiplier equation (2.1) for s. Moreover, if U_g and V_g arise from equivalent multipliers, then U_g and V_g are equivalent representations. It will not be surprising that the space X that we will study has also been studied in connection with the irreducible representations for semi-simple Lie groups [1].

Let G be semi-simple and K a maximal compact subgroup, and let X be a G-space.

DEFINITION 2.2. σ is a *K-multiplier* if $\sigma\in M(X)$ and $\sigma(k, \xi)=1$ for all $k\in K$, $\xi\in X$. The set of K-multipliers forms a group $M_K(X)$, and the corresponding set of multiplier functions is denoted $E_K(X)$.

LEMMA 2.1. *Let X be a homogeneous G-space and suppose that the subgroup K is already transitive on X. Then the natural map of $M_K(X)$ into $H(X)$ is an isomorphism onto.*

PROOF. To show that this map is 1-1 suppose $\sigma\in M_K(X)$ and is trivial: $\sigma(g\xi)$, $=p(g\xi)/p(\xi)$. Then $p(k\xi)=p(\xi)$ and since K is transitive, p is constant so that $\sigma\equiv 1$. To show the map is onto we associate to every multiplier an equivalent K-multiplier. For $s\in M(X)$, set

$$(2.3)\qquad \sigma(g, \xi) = \frac{\int_K s(kg, \xi)\, dk}{\int_K s(k, \xi)\, dk} = \frac{\int_K s(k, g\xi)\, dk}{\int_K s(k, \xi)\, dk}\, s(g, \xi),$$

from which we see that $\sigma \sim s$ and $\sigma(k, \xi) = 1$.

From this we can see that, in many cases, although the group $M(X)$ is infinite-dimensional, the group $H(X)$ is finite-dimensional. For if we let $X = G/L$, where L is a subgroup of G, then K is transitive on X if and only if $KL = G$. If $\sigma \in M_K$ and $\xi_0 \in G/L$ corresponds to the coset L, then $\sigma(g, \xi) = \sigma(g, k\xi_0) = \sigma(gk, \xi_0)$ for some $k \in K$. So σ is determined by the function $\sigma(g, \xi_0)$. Since $g = k'l$ for some $k' \in K$, $l \in L$, σ is determined by the restriction of $\sigma(\cdot, \xi_0)$ to L. But this is a positive character by (2.1), since $l\xi_0 = \xi_0$. In particular, if L has finitely many connected components then its (positive) character group is a subgroup of a vector group. It follows that $M_K(X)$ and, by Lemma 2.1, also $H(X)$, is isomorphic to a subgroup of a vector group.

In certain cases we can assert that $H(X)$ is trivial.

LEMMA 2.2. *If L is compact then* $H(G/L) = 1$.

PROOF. Let $s \in M(G/L)$ and let $\xi_0 \in G/L$ correspond to the coset L. Then $s(l, \xi_0)$ is a positive character on L. But, since L is compact, $s(l, \xi_0) = 1$. It follows that $s(gl, \xi_0) = s(g, l\xi_0)s(l, \xi_0) = s(g, \xi_0)$, so that $s(g, \xi_0) = p(g\xi_0)$ for some positive continuous function p on G/L. Hence $s(g, g'\xi_0) = s(gg', \xi_0)/s(g', \xi_0) = p(gg'\xi_0)/p(g'\xi_0)$ or $s(g, \xi) = p(g\xi)/p(\xi)$.

As a final remark we point out that if j is an equivariant map of one G-space X into another G-space Y, then there is induced a map $M(Y) \to M(X)$. Namely we assign to each $s \in M(Y)$ the multiplier s^* defined by $s^*(g, x) = s(g, j(x))$. s^* satisfies (2.1) if s does. If K is transitive on X, then it is possible to show that the map of $M(Y) \to M(X)$ induces an isomorphism of $H(Y)$ into $H(X)$.

3. Examples. In this section we shall recall some familiar integral representation formulae which we shall see express a function as a linear combination of multiplier functions. In each case the function being represented satisfies a convolution equation or a family of such equations, and the formula will turn out to be an instance of the generalized Choquet-Deny representation.

Let G be the connected group of rigid motions of the plane $\boldsymbol{R}^2$. G is the semidirect product $K \cdot \boldsymbol{R}^2$ where K is the subgroup of notations and $\boldsymbol{R}^2$ the subgroup of pure translations. $\boldsymbol{R}^2$ is a normal subgroup of G and $G/\boldsymbol{R}^2$ may be identified with K. We shall compute $M_K(G/\boldsymbol{R}^2)$.

(Note that by Lemma 2.2, $H(G/K)$ is trivial, so that all multipliers of $M(G/K)$ are trivial.) Identifying $\boldsymbol{R}^2$ with the complex plane, we may denote an element $g\in G$ by the pair $(e^{i\theta}, w)$ if $g(z)=e^{i\theta}z+w$. The action of G on $K=G/\boldsymbol{R}^2$ is given by $(e^{i\theta}, w)\zeta=e^{i\theta}\zeta$ where $|\zeta|=1$. We then find that if $\sigma\in M_K(G/\boldsymbol{R}^2)$,

$$\sigma(e^{i\theta}, w; \zeta) = \sigma(1, e^{-i\theta}\zeta^{-1}w; 1) = e^{\Lambda(e^{-i\theta}\zeta^{-1}w)},$$

where Λ denotes a homomorphism of $\boldsymbol{C}\rightarrow\boldsymbol{R}$. This means $\Lambda(z)=\Re z\tau$ for some complex τ. The general form of a multiplier in $M_K(G/\boldsymbol{R}^2)$ is therefore

$$\sigma_\tau(e^{i\theta}, w; \zeta) = e^{\Re e^{-i\theta}\zeta^{-1}w\tau}. \tag{3.1}$$

Consider now the zero-order Bessel function of an imaginary argument: $f(r)=J_0(ir)$. The function on $\boldsymbol{R}^2$ given by $F(z)=f(|z|)$ is an eigenfunction of the Laplacian $\Delta=D_x^2+D_y^2$ on the plane. This implies that it is an eigenfunction of the mean-value operators:

$$T_\rho F(z) = \frac{1}{2\pi}\int_0^{2\pi} F(z+\rho e^{i\theta})\,d\theta.$$

We now lift the function $F(z)$ to the group G by setting $\Phi(e^{i\theta}, w) = F(w)$. If we write the equations $T_\rho F=\lambda_\rho F$ in terms of Φ we find that Φ satisfies a family of convolution equations on the group. According to the theorem to be proven it follows from this that Φ can be expressed as an integral over a set of multiplier functions in $E(G/\boldsymbol{R}^2)$. Moreover, since $\Phi(kg)=\Phi(g)$ for $k\in K$, these multiplier functions will be in $E_K(G/\boldsymbol{R}^2)$, and so are given by (3.1). This is indeed the case, for by the Bessel formula

$$f(r) = \frac{1}{2\pi}\int_0^{2\pi} e^{-r\cos\omega}\,d\omega = \int e^{-\Re r\zeta^{-1}}\,dm(\zeta),$$

where $dm(\zeta)$ refers to normalized Lebesgue measure on the unit circle. It follows that

$$F(z) = \int e^{-\Re z\zeta^{-1}}\,dm(\zeta),$$

$$\Phi(e^{i\theta}, w) = \int e^{-\Re w\zeta^{-1}}\,dm(\zeta) = \int e^{-\Re we^{-i\theta}\zeta^{-1}}\,dm(\zeta)$$

$$= \int \sigma_1(e^{i\theta}, w; \zeta)\,dm(\zeta).$$

Bessel's formula is a special case of a formula for spherical functions on any symmetric space of noncompact type. (See [8, Chapter X] for details.) Let us consider the case of a space of constant negative curvature. The space may then be taken as G/K where G is semisimple. A spherical function $F(x)$ on G/K is defined by the conditions: $F(kx)=F(x)$ for all $k\in K$, $F(x_0)=1$ where x_0 corresponds to the coset K, and F is an eigenfunction of all differential operators on G/K which commute with the action of G on G/K. If $F(x)$ is lifted to G we obtain a function $\phi(g)=F(gK)$ which, as may be shown, is characterized by the equation:

$$\int_K \phi(g_1kg_2)\,dk = \phi(g_1)\phi(g_2), \qquad g_1, g_2 \in G. \tag{3.2}$$

We notice that ϕ satisfies a family of convolution equations. In fact, (3.2) may be rewritten (see §1) as

$$m_K * g_1^{-1} * \phi = \phi(g_1)\phi. \tag{3.3}$$

The integral representation formula for these functions is due to Harish-Chandra [8, Chapter X, Theorem 6.16]. We define a function H from G to A, where A is the vector part of the Iwasawa decomposition of G, by $H(kan)=a$. Then, for every spherical function ϕ on G, there is a homomorphism $\Lambda: A\to \mathbf{C}$ with

$$\phi(g) = \int_K e^{\Lambda(H(gk))}\,dk. \tag{3.4}$$

When ϕ is real and positive, Λ is real-valued. We shall show that in this case, the integrand of (3.4) represents a multiplier in $M_K(G/AN)$, where KAN is the Iwasawa decomposition of G. Namely, set $\sigma(g, kAN)=e^{\Lambda(H(gk))}$. Since $H(k')$ is the identity of A if $k'\in K$, it follows that $\sigma(k', kAN)=1$. We next check the multiplier equation. Suppose $\xi=kAN$ for $k\in K$, and $g_1, g_2\in G$ with

$$g_2k = k_2a_2n_2, \qquad g_1k_2 = k_1a_1n_1.$$

Then $g_1g_2k=g_1k_2a_2n_2=k_1a_1n_1a_2n_2=k_1a_1a_2n_1'n_2$ since N is normal in AN. Then

$$\sigma(g_1g_2, \xi) = e^{\Lambda(H(g_1g_2k))} = e^{\Lambda(a_1)+\Lambda(a_2)},$$

$$\sigma(g_1, g_2\xi) = \sigma(g_1, k_2AN) = e^{\Lambda(H(g_1k_2))} = e^{\Lambda(a_1)},$$

$$\sigma(g_2, \xi) = \sigma(g_2, kAN) = e^{\Lambda(H(g_2k))} = e^{\Lambda(a_2)},$$

so that $\sigma(g_1g_2, \xi)=\sigma(g_1, g_2\xi)\sigma(g_2, \xi)$. (3.4) may therefore be rewritten

$$\phi(g) = \int_K \sigma(g, kAN)\, dk = \int_{G/AN} \sigma(g, \xi)\, dm(\xi). \tag{3.5}$$

Thus the Harish-Chandra formula in the case of positive spherical functions is a special case of the generalized Choquet-Deny representation. It should be pointed out that our methods fail in the case of complex-valued spherical functions although the formula (3.4) is still valid.

Another example of a spectral decomposition for functions on a semi-simple group is given by the Poisson formula for harmonic functions in the unit disc. Let G denote the connected group of conformal maps of the unit disc $D=\{z: |z|<1\}$ onto itself. A function $h(z)$ on the disc determines a function $f(g)=h(g^{-1}(0))$ on G. Let K again denote the group of rotations about the origin. The mean-value property for harmonic functions says that $h(0)=\int_K h(kz)\, dk$ for any $z\in D$. If h is harmonic, so is the function obtained by composing h with the transformation g^{-1}, $g\in G$. This gives

$$h(g^{-1}(0)) = \int_K h(g^{-1}kz)\, dk. \tag{3.6}$$

If we write $z=\gamma^{-1}(0)$, (3.6) becomes

$$f(g) = \int_K f(\gamma k g)\, dk, \qquad g, \gamma \in G. \tag{3.7}$$

Thus the functions $f(g)$ which correspond to harmonic functions on D satisfy a family of convolution equations:

$$m_K * \gamma^{-1} * f = f. \tag{3.8}$$

Assume that $h\geqq 0$. The Poisson formula gives

$$h(z) = \int_0^{2\pi} P(z, \theta)\, d\nu(\theta),$$

where P denotes the Poisson kernel, and ν is a positive measure. Now one may verify that if $z=g^{-1}(0)$ then

$$P(z, \theta) = |g'(e^{i\theta})|$$

or

$$f(g) = \int_0^{2\pi} |g'(e^{i\theta})|\, d\nu(\theta). \tag{3.9}$$

We recognize by the formula for differentiating a composite function that the integrand in (3.9) is a multiplier in $M_K(\partial D)$ where ∂D is the unit circle.

The Poisson formula for harmonic functions also generalizes to arbitrary noncompact symmetric spaces. Moreover, a similar representation is valid for any positive eigenfunction of the Laplace-Beltrami operator on a symmetric space. This was shown by Dynkin [5] for the symmetric space $\mathrm{SL}(n, \mathbf{C})/\mathrm{SU}(n)$. Once again it may be shown that the extremal solutions occurring in Dynkin's representation are multiplier functions for a certain compact G-space which we will call $B(G)$ and regarding which we shall presently be more specific.

Another generalization of the Poisson formula for harmonic functions was given in [6]. For our present purposes it will be convenient to deviate somewhat from the notation used in [6]. Let μ be a probability measure on G which is absolutely continuous and whose support contains a neighborhood of the identity of G. Consider the *bounded* solutions to

$$f(g) = \int_G f(g'g)\, d\mu(g'). \tag{3.10}$$

This is a generalization of (3.7) which was satisfied by ordinary harmonic functions. It is shown that, given μ, there is determined a probability measure ν on the G-space $B(G)$ referred to in the preceding paragraph, with the property that every bounded solution of (3.10) is given by

$$f(g) = \int_{B(G)} f(g^{-1}x)\, d\nu(x), \tag{3.11}$$

where f is a bounded measurable function on $B(G)$. The measure ν satisfies $\mu * \nu = \nu$ and from this one can infer that the transforms $g\nu$ are absolutely continuous with respect to ν, and that $(dg\nu/d\nu)(x)$ is positive and continuous on $G \times B(G)$. (3.11) may therefore be written

$$f(g) = \int_{B(G)} \frac{dg^{-1}\nu}{d\nu}(x) f(x)\, d\nu(x). \tag{3.12}$$

We claim that $s(g, x) = (dg^{-1}\nu/d\nu)(x)$ is a multiplier in $M(B(G))$. By the definition of the Radon-Nikodym derivative it is easily seen that

$$\frac{dg^{-1}\nu_1}{dg^{-1}\nu_2}(x) = \frac{d\nu_1}{d\nu_2}(gx). \tag{3.13}$$

So

$$\frac{dg_2^{-1}g_1^{-1}\nu}{d\nu}(x) = \frac{dg_2^{-1}g_1^{-1}\nu}{dg_2^{-1}\nu}(x)\frac{dg_2^{-1}\nu}{d\nu}(x)$$
$$= \frac{dg_1^{-1}\nu}{d\nu}(g_2x)\frac{dg_2^{-1}\nu}{d\nu}(x),$$

which is the multiplier equation for s. Thus the measure μ on G determines a multiplier $s\in M(B(G))$ and the *bounded* solutions of (3.10) are integrals over multiplier functions corresponding to s:

$$f(g) = \int_{B(G)} s(g, x)f(x)\, d\nu(x).$$

We note that this is the first example in which a multiplier is used which is not a K-multiplier. Naturally, a function cannot be represented in terms of K-multiplier functions unless it satisfies $f(kg) = f(g)$ for $k\in K$.

It should be mentioned that in case the support of μ does not contain a neighborhood of the identity a similar representation is valid, but the relevant space may not be $B(G)$ but rather one of finitely many covering spaces of $B(G)$.

The space $B(G)$ which occurs in these formulas may be defined as follows. If $G=KAN$ is an Iwasawa decomposition of G we let T denote the normalizer of AN, that is the set of all $g\in G$ with $gANg^{-1} \subset AN$. T is also obtained by adjoining to AN the centralizer K_0 of A in K: $T=K_0AN$. $B(G)$ is then defined as G/T. T is not uniquely determined, but it is unique up to conjugacy. As a result, the various versions of G/T are all isomorphic as G-spaces. An alternative definition of $B(G)$ occurs in [6]. We define a *boundary* of a Lie group G to be a compact G-space with the property that for any probability measure π on the space, some sequence $g_n\pi$, $g_n\in G$, converges to a point measure. All boundaries of G are equivariant images of one of them, and this one is $B(G)$. The equivalence of the two definitions was proven in [10].

The space $B(G)$ is an equivariant image of the G-space G/AN that occurs in (3.5). It may be shown that, in fact, the multiplier σ in formula (3.5) which belongs to $M_K(G/AN)$ actually is the image of a multiplier in $M_K(B(G))$ (see end of §2) so that the integral in (3.5) may also be taken over $B(G)$. Thus in each of the cases studied, the integral representation involves multipliers in $M(B(G))$.

4. Tychonoff groups. The significance of the subgroup T may be attributed to the fact that it possesses a certain "fixed-point" property which we will elucidate here. We consider a subcone V of a linear topological space S. We say that V has a *compact base* if there exists a continuous linear functional L on S such that $L(x)>0$ for $x\in V$ except for $x=0$, and such that $V_1=\{x\in V: L(x)=1\}$ is compact. If τ is a continuous linear transformation of S taking V into itself then τ has a fixed (invariant) ray in V. For, by the Schauder-Tychonoff fixed-point theorem [4], the transformation

$$x \longrightarrow \frac{\tau x}{L(\tau x)}$$

which is continuous and takes $V_1 \to V_1$ must have a fixed point in V_1. The ray through this fixed point is clearly left fixed by τ. This result may be extended to a one-parameter group of transformations $\tau(t)$ of S which take V into itself, and there will exist a ray in V invariant under the entire group.

DEFINITION 4.1. A Lie group G is a *Tychonoff* group if, whenever G acts continuously by linear transformations on a locally convex, linear, topological space S, taking a cone V with compact base into itself, then G has a fixed ray in V.

A Tychonoff group also has the following fixed-point property: If it acts by affine transformations on a compact convex set, it has a fixed point in the set. (An affine transformation is one that preserves the convex structure: $\tau(\alpha x+(1-\alpha)y)=\alpha\tau x+(1-\alpha)\tau y$, $0\leqq\alpha\leqq 1$.) The converse, however, is not true; a group may have the latter fixed-point property without being a Tychonoff group. For example, every solvable group has the fixed-point property relative to compact convex sets. (See [6].) On the other hand, the group of matrices

$$\left\{\begin{pmatrix} x & 0 \\ 0 & y\end{pmatrix}, \begin{pmatrix} 0 & u \\ v & 0\end{pmatrix} : x, y, u, v > 0\right\}$$

is solvable, maps the first quadrant of $\boldsymbol{R}^2$ into itself, but has no fixed ray. An example of a connected solvable group which is not a Tychonoff group is the connected group of rigid motions of the plane. It will develop that if this group were a Tychonoff group, there would be no nonconstant positive eigenfunctions of the Laplacian in the plane.

The following lemmas will be useful in establishing that certain groups are Tychonoff groups.

LEMMA 4.1. *Let A be a linear transformation on $\boldsymbol{R}^r$ with strictly positive eigenvalues. If for some vector x, the set of vectors $\{A^n x: -\infty<n<\infty\}$ is bounded, then $Ax=x$.*

PROOF. Since A satisfies its characteristic equation, it follows that all the vectors $A^n x$ are contained in the subspace H spanned by $\{x, Ax, \cdots, A^{r-1}x\}$. Clearly each vector $y \in H$ has the property that the set $\{A^n y\}$ is bounded. So the restriction of A to H is an operator with all its powers, positive and negative, bounded. Therefore all its eigenvalues are on the unit circle. But they must be positive and so the restriction of A to H has only the eigenvalue 1. We may therefore express this restriction as $I+N$, where I is the identity on H and N is nilpotent on H. Let i be the least integer with $N^i=0$. If $i \geqq 2$, form $(I+N)^n N^{i-2} = N^{i-2} + nN^{i-1}$. This must be bounded as $n \to \infty$ which implies $N^{i-1}=0$ and this is a contradiction. It follows that $N=0$, or the restriction of A to H is the identity. Hence $Ax=x$.

LEMMA 4.2. *Let G, G_1 be connected Lie groups with G_1 a normal subgroup of G and let $\mathfrak{G}$, $\mathfrak{G}_1$ denote their respective Lie algebras. Suppose*
(a) *G_1 is a Tychonoff group,*
(b) *G/G_1 is 1-dimensional,*
(c) *the adjoint representation of $\mathfrak{G}$ restricted to $\mathfrak{G}_1$ has matrices with only real eigenvalues. Then G is a Tychonoff group.*

PROOF. We suppose that G operates on a linear space S, that V is a cone in S with compact base, and that L is a continuous linear functional on S, positive on V, and such that $\{x: L(x)=1\}$ intersects V in a compact set. Since G_1 is a Tychonoff group, G_1 leaves fixed at least one ray of V. If x is a vector on such a ray there is determined a positive character λ on G_1 defined by $g_1x=\lambda(g_1)x$. Also, given a character λ, there will exist a cone (possibly empty) $V_\lambda \subset V$ of vectors $y \in V$ for which $g_1y=\lambda(g_1)y$. For $g \in G$, $g^{-1}G_1g \subset G_1$ and we may define the transform λ^g of the character λ by $\lambda^g(g_1)=\lambda(g^{-1}g_1g)$. If $x \in V_\lambda$,

$$\lambda^g(g_1)gx = \lambda(g^{-1}g_1g)gx = gg^{-1}g_1gx = g_1gx, \tag{4.1}$$

from which it follows that $gV_\lambda \subset V_{\lambda^g}$. Another consequence of (4.1) is that if V_λ is nonempty, then $\lambda^g(g_1)$ is bounded as a function of g. For we have

$$\lambda^g(g_1)L(gx) = L(g_1gx),$$
$$\lambda^g(g_1) = L(g_1gx)/L(gx),$$

and $L(g_1y)/L(y) \leqq \max L(g_1z)$, where z ranges over the compact set of vectors in V for which $L(z)=1$.

If G_1' denotes the commutator subgroup of G_1, then the group of positive characters on G_1 may be identified with the group of positive characters on G_1/G_1'. Since the latter group is abelian, it is the image under the exponential map of its Lie algebra $\mathfrak{G}_1/[\mathfrak{G}_1, \mathfrak{G}_1]$. Thus the positive character group of G_1 may be identified with a subspace W of the dual space $(\mathfrak{G}_1/[\mathfrak{G}_1, \mathfrak{G}_1])^*$. Now it may be checked that the automorphism $\lambda\rightarrow\lambda^g$ of the positive character group of G_1 corresponds to the operation $X\rightarrow\text{Ad } g(X)$ on $\mathfrak{G}_1$. Since ad $\mathfrak{G}$ restricted to $\mathfrak{G}_1$ has all real eigenvalues, its image under the exponential map, namely Ad G, has all its eigenvalues positive. This remains true when we pass to the quotient space $\mathfrak{G}_1/[\mathfrak{G}_1, \mathfrak{G}_1]$ of $\mathfrak{G}_1$, and to its dual, and to the subspace W of the dual.

We see then that the set of positive characters on G_1 forms a linear space on which we have a representation of G given by $\lambda\rightarrow\lambda^g$, and the eigenvalues of the transformations of this representation are all positive.

We know that for some λ_0, V_{λ_0} is nonempty. This means by our previous observation that $\lambda_0^g(g_1)$ is bounded as a function of $g\in G$. But since the map $\lambda\rightarrow\lambda^g$ has positive eigenvalues, we see by Lemma 4.1 that $\lambda_0^g=\lambda_0$. If we also recall that $gV_{\lambda_0}\subset V_{\lambda_0^g}$, we find that G takes V_{λ_0} into itself. We shall now show that some ray of V_{λ_0} is left fixed by all of G.

Let V'_{λ_0} denote the set of vectors $x\in V_{\lambda_0}$ with $L(x)=1$. If for each $g\in G$ we define g^* by

$$g^*(x) = gx/L(gx)$$

then each g^* takes V'_{λ_0} into itself. $g\rightarrow g^*$ is a homomorphism of G into a group of continuous transformations of V'_{λ_0}. However, G_1^* leaves V'_{λ_0} pointwise fixed since each g_1x is proportional to x for $x\in V'_{\lambda_0}$. Thus the action of G^* on V'_{λ_0} is that of a one-parameter group. As a result G^* has a fixed point and this means that G has a fixed ray in V_{λ_0}. This completes the proof of the lemma.

COROLLARY. *A connected subgroup of the group of upper triangular real $r\times r$ matrices is a Tychonoff group.*

PROOF. Such a group is solvable and we can find a chain $G=G_0\supset G_1\supset G_2\cdots\supset G_n=e$ with G_i/G_{i+1} 1-dimensional and G_{i+1} normal in G_i. Let Q denote the group of upper triangular matrices with positive entries on the diagonal. Since G is connected, $G\subset Q$. If $\mathfrak{Q}$ is the Lie algebra of Q, then $\mathfrak{Q}$ consists of real triangular matrices. For $q\in\mathfrak{Q}$, the eigenvalues of ad q are differences of diagonal entries of q, and so they are real. This will be true for any subalgebra of $\mathfrak{Q}$ and so by Lemma 4.2 we may proceed inductively to prove that G is a Tychonoff group.

COROLLARY. *If G is a semi-simple Lie group with an Iwasawa decomposition $G=KAN$, then AN is a Tychonoff group.*

PROOF. According to [11, Exposé 11, p. 14], AN may be represented (faithfully) by upper triangular matrices.

THEOREM 4.1. *If G is a semi-simple Lie group, the subgroup T defined at the end of* §3 *is a Tychonoff group.*

PROOF. We recall that $T=K_0AN$ where K_0 is the centralizer of A in K. Now since N is the commutator subgroup of AN, every positive character λ on AN is identically 1 on N, and is determined by its values on A. If $k_0\in K_0$, then $\lambda^{k_0}(a)=\lambda(k_0^{-1}ak_0)=\lambda(a)$ for $a\in A$, so that $\lambda^{k_0}=\lambda$. Hence $k_0V_\lambda\subset V_{\lambda^{k_0}}=V_\lambda$, where V_λ denotes the subset of x in a cone V (with compact base) with $gx=\lambda(g)x$ for $g\in AN$. Since AN is a Tychonoff group, V_λ will be nonempty for some λ. Let $x_0\in V_\lambda$ and form

$$x = \int_{K_0} k_0x_0\, dk_0.$$

Since $k_0V_\lambda\subset V_\lambda$, $x\in V_\lambda$, and so AN leaves fixed the ray to which x belongs. But clearly K_0 leaves x fixed so that the ray to which x belongs is left fixed by all of T.

By [6, Theorem 1.10] it follows that the subgroup T is maximal with respect to the property of being a Tychonoff subgroup of G.

5. **Spherical and semi-spherical functions.** Before obtaining our generalization of the Choquet-Deny theorem, we shall discuss a special case connected with a class of functions which we shall call *semi-spherical.* Throughout this section, G is a semi-simple group, K a fixed maximal compact subgroup, T the Tychonoff subgroup defined in §3 and $B(G)=G/T$. If $\gamma\in G$, let A_γ denote the operator

$$A_\gamma f(g) = \int_K f(\gamma kg)\, dk = m_K * \gamma^{-1} * f(g). \tag{5.1}$$

DEFINITION 5.1. A positive function f on G is *semi-spherical* if it is an eigenfunction of each A_γ, $\gamma\in G$.

The reason for our terminology is that a spherical function fulfills this condition, as we see from equation (3.2). But whereas a spherical function is bi-invariant under K: $\phi(k_1gk_2)=\phi(g)$, a semi-spherical function need only satisfy $f(kg)=f(g)$. (In fact, it is not difficult to show that a semi-spherical function which is bi-invariant under K is actually spherical.)

If f is semi-spherical, there is defined a function $\phi(\gamma)$ on G by $A_\gamma f=\phi(\gamma)f$. We notice that $\phi(\gamma)$ is spherical:

$$\int_K \int_K f(g_1 k g_2 k' g_3)\, dk dk' = \int_K A_{g_1} f(g_2 k' g_3)\, dk' \tag{5.2}$$
$$= \phi(g_1) A_{g_2} f(g_3) = \phi(g_1)\phi(g_2) f(g_3).$$

But the first integral of (5.2) is also

$$\int_K A_{g_1 k g_2} f(g_3)\, dk = \left(\int_K \phi(g_1 k g_2)\, dk \right) f(g_3)$$

and so

$$\int_K \phi(g_1 k g_2)\, dk = \phi(g_1)\phi(g_2),$$

which is the defining equation of a spherical function.

DEFINITION 5.2. We say a semi-spherical function f *belongs to* the spherical function ϕ if

$$\int_K f(\gamma k g)\, dk = \phi(\gamma) f(g); \tag{5.3}$$

the cone of semi-spherical functions belonging to ϕ will be denoted V_ϕ.

V_ϕ is never empty since $\phi \in V_\phi$. Notice also that V_ϕ is closed under right translation, $R_\gamma V_\phi \subset V_\phi$, since the translates R_γ commute with $A_{\gamma'}$.

LEMMA 5.1. *If $\sigma \in M_K(X)$ where X is a G-space such that K is transitive on X, then the multiplier functions $\sigma(\cdot, x)$ are semi-spherical and belong to the spherical function*

$$\phi(g) = \int_X \sigma(g, x)\, dm(x). \tag{5.4}$$

Here m denotes the unique K-invariant probability measure on X. (m is unique because K is transitive on X.)

PROOF.

$$\int \sigma(\gamma k g, x)\, dk = \left(\int \sigma(\gamma k, gx)\, dk \right) \sigma(g, x)$$
$$= \left(\int \sigma(\gamma, kgx)\, dk \right) \sigma(g, x)$$
$$= \left(\int \sigma(\gamma, y)\, dm_K * gx(y) \right) \sigma(g, x).$$

Now the measure $m_K * gx$ is a K-invariant probability measure on X and so $m_K * gx = m$. Thus

$$A_\gamma \sigma(\cdot, x) = \phi(\gamma)\sigma(\cdot, x)$$

which proves simultaneously that σ is semi-spherical, that it belongs to ϕ, and that the function ϕ defined by (5.4) is a spherical function.

Notice that this yields the easier half of the Harish-Chandra representation formula; namely, that the function defined by (3.4) is a spherical function. For we have shown that the integral in (3.4) is of the form (5.4) for the space $X = G/AN$.

We will show that the multiplier functions of Lemma 5.1 generate the set of *all* semi-spherical functions. More precisely, we will show that we may take $X = B(G)$ and each V_ϕ is spanned by the functions in $V_\phi \cap E_K(B(G))$.

To begin with we examine how the multiplier functions of $E_K(B(G))$ are expressed as measures. Suppose χ is a positive character on the group T, and form the measure m_T^χ on T defined by

$$dm_T^\chi(t) = \chi(t^{-1})dt, \tag{5.5}$$

dt denoting left invariant Haar measure on T. From (5.5) we obtain

$$tm_T^\chi = \chi(t)m_T^\chi \tag{5.6}$$

for $t \in T$. Inasmuch as right-invariant Haar measure on T is related to the left-invariant Haar measure by multiplication by a positive character, it follows that the measure m_T^χ is also related to right-invariant Haar measure in a manner similar to (5.5). As a result we may also write

$$m_T^\chi t = \chi'(t)m_T^\chi, \tag{5.7}$$

where χ' is some other positive character in T.

Now form the measure $m_K * m_T^\chi = \nu$. If $g = kt$, with $k \in K$, $t \in T$, then $m_K * g * m_T^\chi = \chi(t)m_K * m_T^\chi = \chi(t)\nu$. Since $KT = G$, every g has such a decomposition. Viewing an arbitrary measure on G as a linear combination of point measures we find that for any bounded measure ω,

$$m_K * \omega * m_T^\chi = c\nu \tag{5.8}$$

for some constant c. From this it follows that ν is absolutely continuous, since ω could be chosen absolutely continuous. Let f be the density of ν: $d\nu(g) = f(g)dg$. Then

$$f(kg) = f(g), \qquad f(gt) = \chi'(t^{-1})f(g) \tag{5.9}$$

by (5.7). We may therefore define a multiplier by

$$\sigma(g, \gamma T) = f(g\gamma)/f(\gamma)$$

since if γ is replaced by γt, both numerator and denominator are multiplied by $\chi'(t^{-1})$. It is, moreover, clear that σ is a multiplier and is in $M_K(B(G))$. Thus the measure $m_K * m_T^\chi$, as a function, is proportional to the multiplier function $\sigma(g, T)$.

We shall need the following lemma whose proof we leave to the Appendix.

LEMMA 5.2. *If π is a measure on G satisfying $t\pi = \chi(t)\pi$ for $t \in T$, and χ a positive character on T, then $\pi = m_T^\chi * \mu$ for some positive measure μ on G.*

We now turn to the proof of the fact that the K-multiplier functions for $B(G)$ span the cones V_ϕ. Suppose $h \in V_\phi$. We then have

$$m_K * \gamma^{-1} * h = \phi(\gamma)h. \tag{5.10}$$

Form the cone V' of all positive measures π with the property that $m_K * \gamma^{-1} * \pi$ is proportional to h for all $\gamma \in G$. Then $h \in V'$ and also V' is closed under translation from the left $\pi \to g\pi$. We claim that V' is a cone with compact base in the weak topology on measures. If ψ is some positive function with compact support on G with

$$\int \psi(g)h(g)dg > 0,$$

define a linear functional L by

$$L(\pi) = m_K * \pi(\psi) \Big/ \int \psi(g)h(g)\, dg.$$

L is continuous and the intersection V_1' of V' with the set of π such that $L(\pi) = 1$ consists of those π in V' for which $m_K * \pi = h$. We claim this set is compact. Since it is closed, it suffices to show that for each compact set $J \subset G$, $\pi(J)$ is bounded as π ranges over V_1'. But

$$m_K * \pi(KJ) = \int_K \int_G C_{KJ}(kg)\, dk d\pi(g),$$

with C_{KJ} the indicator function of KJ. But $C_{KJ}(kg) \geqq C_K(k)C_J(g)$ so

$$\int_{KJ} h(g)\, dg \geqq m_K(K)\pi(J) = \pi(J),$$

which gives the desired result. So V' is a cone with compact base.

Since G and therefore T operates on V' on the left, and since T is a Tychonoff group (Theorem 4.1), there is a measure $\pi \in V'$ with $t\pi = \chi(t)\pi$ for all $t \in T$ and some positive character χ. By Lemma 5.2, $\pi = m_T^\chi * \mu$ for some measure μ on G. Thus for some χ, the measure $m_T^\chi * \mu$ is in V'. This means that $m_K * \gamma^{-1} * m_T^\chi * \mu$ is proportional to h for every γ and, in particular, $m_K * m_T^\chi * \mu$ is proportional to h. But $m_K * m_T^\chi \leftrightarrow \sigma(\cdot, x_0)$ for $\sigma \in M_K(B(G))$ and for x_0 the point of $B(G)$ corresponding to the coset T. $m_K * m_T^\chi * \mu$ is a linear combination of right translates of $m_K * m_T^\chi$ and so it follows that

$$
\begin{aligned}
h(g) &= \int_G \sigma(g\gamma, x_0)\, d\mu(\gamma) = \int_G \sigma(g, \gamma x_0)\sigma(\gamma, x_0)\, d\mu(\gamma) \\
&= \int_{B(G)} \sigma(g, x)\, d\mu^*(x)
\end{aligned}
\tag{5.11}
$$

for some positive measure μ^* on $B(G)$. Now by Lemma 5.1, $\sigma(\cdot, x)$ belongs to some spherical function ϕ' defined by

$$\phi'(g) = \int_{B(G)} \sigma(g, x)\, dm(x),$$

with m, as usual, denoting the unique K-invariant probability measure on $B(G)$. But then the function $h(g)$ satisfying (5.11) must belong to the same ϕ'. Our assumption, however, was that $h \in V_\phi$. We must, therefore, have $\phi' = \phi$. We have thereby proven

THEOREM 5.1. *Let ϕ be a positive spherical function and V_ϕ the cone of semi-spherical functions belonging to ϕ. For each $h \in V_\phi$ there is a multiplier $\sigma \in M_K(B(G))$ with $\sigma(\cdot, x) \in V_\phi$ for each $x \in B(G)$ and such that*

$$h(g) = \int_{B(G)} \sigma(g, x)\, d\mu^*(x) \tag{5.12}$$

for some positive measure μ^* on $B(G)$. The multiplier σ which occurs in (5.12) *must satisfy*

$$\phi(g) = \int_{B(G)} \sigma(g, x)\, dm(x), \tag{5.13}$$

where m is the unique K-invariant probability measure on $B(G)$.

It should be noticed that we have incidentally proven that every positive spherical function ϕ admits a representation (5.13), which is essentially the Harish-Chandra formula.

If the K-multiplier functions $\sigma(\cdot, x)$ belong to V_ϕ we shall simply say that σ belongs to ϕ. Thus every spherical function has a K-multiplier belonging to it. In fact there is always a finite set of K-multipliers belonging to each spherical function, the cardinality of this set being, in general, equal to the order of the Weyl group of G. (See [8, Chapter X].) It will develop that even though a number of distinct K-multipliers (over $B(G)$) belong to ϕ, only one of them is needed to span V_ϕ.

6. **The Choquet-Deny representation.** The considerations of this section are based on [3]. We are interested in determining when the points of a convex cone can be expressed as resultants of measures placed on the extremal points of the cone. We call a point x of a cone V an extremal point if it lies on an extremal ray, namely, if $y \in V$ and $x-y \in V$ imply $y=\lambda x$ for $\lambda \leqq 1$. We shall be concerned with cones of positive (unbounded) measures on a separable locally compact space X. We regard the space of such measures as the dual to the space C^0 of continuous functions of compact support on X and we endow it with the corresponding weak topology. Suppose V is a closed cone of positive measures on X and suppose μ is a measure on V. We form the *resultant* $\bar{\mu}$ by setting

$$\bar{\mu}(\psi) = \int \pi(\psi)\, d\mu(\pi), \tag{6.1}$$

where ψ is a function of compact support, and the integral on the right is assumed to exist. When it does, the resultant $\bar{\mu}$ is again a measure in V. First of all, it is a positive measure because (6.1) defines a positive linear functional on C^0. Also $\bar{\mu} \in V$ for if $\bar{\mu} \notin V$ there would exist, by the Hahn-Banach theorem, a function $\psi \in C^0$ with $\bar{\mu}(\psi) < 0$ and $\pi(\psi) \geqq 0$ for all $\pi \in V$. This is impossible by (6.1).

In general a cone V is a *lattice* if whenever $x_1, x_2 \in V$, there exists an infimum $y = \inf(x_1, x_2)$ with respect to the ordering of V. That is, there exists $y \in V$ with $x_1 - y$, $x_2 - y \in V$, and if, for some other $y' \in V$, $x_1 - y'$, $x_2 - y' \in V$, then $y - y' \in V$.

The result we shall need is the following:

THEOREM 6.1. *Let V be a weakly closed cone of positive measures on a separable, locally compact space X and let $\mathcal{E}$ denote the extremal rays of V. Suppose there is a positive function of compact support ψ with $\pi(\psi) > 0$ for all $\pi \in V$, $\pi \neq 0$, and let $V_1 = \{\pi : \pi(\psi) = 1\}$. Then for each $\pi \in V$ there exists a measure μ on a Borel subset of $\mathcal{E} \cap V_1$ such that π is the resultant $\bar{\mu}$. If, moreover, V is a lattice, then the measure μ is unique.*

Clearly if we wish to have a unique representation in a certain set, then this set can intersect each ray in at most one point. That is the reason for restricting the representation to V_1.

The proof of Theorem 6.1 is based on the corresponding result for compact, convex sets. If W is a compact, convex set it is possible to define the resultant $\bar{\mu}$ of a probability measure μ on W, and $\bar{\mu}$ will be a point of W. We then have [2], [9]

THEOREM A. *If W is metrizable then the set of extremals of W is a Borel set and every point of W is the resultant of some measure on the set of extremals of W.*

A convex set W may be taken to be the base of a convex cone. For example, if W is in the linear space S we may form the cone $\tilde{W}$ in $\boldsymbol{R}\times S$ by setting $\tilde{W}=\{(t, tw): t>0, w\in W\}$. We say that W is a *simplex* if $\tilde{W}$ is a lattice.

THEOREM B [2], [9]. *If W is metrizable and is a simplex, then each point of W is the resultant of a unique probability measure on the set of extremals of W.*

We apply these theorems to the cone V of Theorem 6.1 by showing that each point of V lies in some compact convex subset $W\subset V$. Let $\pi\in V$. By the separability of X there will be some continuous function h everywhere positive in X and such that $\pi(h)<\infty$. Suppose $\pi(h)=1$. Let W be the set of $\pi'\in V$ with $\pi'(h)\leqq 1$. This set is closed in V and moreover, since $h(x)>0$ everywhere, so that on every compact set $h(x)$ is bounded from below, the measures π' in W are bounded in every compact subset of X. Thus W is a compact convex metrizable set. Let us show that each extremal of W with the exception of the point 0 is an extremal of V. If π' is an extremal of W and $\pi'=\pi_1'+\pi_2'$ with $\pi_i'\in V$, then $\pi_i'(h)>0$ unless $\pi_i'=0$. So

$$\pi' = \pi_1'(h)\,\frac{\pi_1'}{\pi_1'(h)} + \pi_2'(h)\,\frac{\pi_2'}{\pi_2'(h)},$$

which expresses π' as a convex combination of measures in W since $\pi_1'(h)+\pi_2'(h)=1$. It follows that π_1' and π_2' are proportional to π'. Now by Theorem A, π is the resultant of a measure on the extremal points of W:

$$\pi = \int \pi'\, d\mu(\pi') = \int \frac{\pi'}{\pi'(\psi)}\,\pi'(\psi)\, d\mu(\pi')$$

and this expresses π as the resultant of a measure on $\mathcal{E}\cap V_1$.

We observe that conversely, if π is given as the resultant of a measure on $\mathcal{E}\cap V_1$, then this measure corresponds to one on the extremals of W. Namely, if

$$\pi = \int \pi' \, d\mu(\pi'),$$

then $\int\pi'(h)\,d\mu(\pi')=1$ and $\pi'(h)<\infty$ for almost all π' with respect to μ. We can then write

$$\pi = \int \frac{\pi'}{\pi'(h)}\,\pi'(h)\,d\mu(\pi')$$

and $\pi'(h)d\mu(\pi')$ is a probability measure, and the measures $\pi'/\pi'(h)$ are extremals of W if the π' are extremals of V. As a result of this observation it follows that to prove the uniqueness portion of Theorem 6.1 it suffices, in view of Theorem B, to show that if V is a lattice then W is a simplex. Equivalently we must show that $\tilde{W}$ is a lattice. This follows readily. If $x_1=(t,\,t\pi_1)$, $x_2=(s,\,s\pi_2)$ are two points of $\tilde{W}$ where $\pi_1,\,\pi_2\in W$, then if $t<s$, we set

$$y = \inf_{\tilde{W}}(x_1, x_2) = \left(t,\, t\inf_V\left(\pi_1, \frac{s}{t}\,\pi_2\right)\right).$$

Notice that $y\in\tilde{W}$ inasmuch as $\inf_V(\pi_1,\,s\pi_2/t)$ is in W (being bounded by $\pi_1\in W$). If we rewrite y as

$$(t,\,\inf_V(t\pi_1,\,s\pi_2)),$$

it becomes clear that $x_1-y,\ x_2-y\in\tilde{W}$ and if $x_1-y',\ x_2-y'\in\tilde{W}$ then $y-y'\in\tilde{W}$.

The uniqueness of the representation in Theorem 6.1 for the case that V is a lattice may also be formulated as follows:

COROLLARY. *If V is a cone of measures as in Theorem* 6.1 *and V is a lattice, then if ν is a signed measure on $\mathcal{E}\cap V_1$ such that $\int\pi'\,d\nu(\pi')\in V$, then $\nu\geqq 0$.*

PROOF. Write $\nu=\nu_1-\nu_2$ with $\nu_i\geqq 0$ and suppose that $\int\pi'\,d\nu(\pi')=\int\pi'\,d\nu_3(\pi')$ with $\nu_3\geqq 0$. Then $\bar{\nu}_1=\bar{\nu}_2+\bar{\nu}_3$ and, by uniqueness, $\nu_1=\nu_2+\nu_3$. Then $\nu=\nu_3\geqq 0$.

Theorem 6.1 is used to prove the following result which we refer to as the Choquet-Deny representation.

THEOREM 6.2. *Let G be a separable locally compact group and suppose that $\{\mu_\alpha,\,\alpha\in\Gamma\}$ is a commuting family of positive measures with com-*

pact supports on G $(\mu_\alpha * \mu_\beta = \mu_\beta * \mu_\alpha,\ \alpha,\ \beta \in \Gamma)$. *Let* V *denote the set of positive measures* π *on* G *satisfying* $\mu_\alpha * \pi = \pi$ *for all* $\alpha \in \Gamma$. *Assume moreover that there exists a continuous function of compact support* $\psi \geqq 0$ *on* G *with* $\pi(\psi) > 0$ *for all* $\pi \in V$, $\pi \neq 0$. *If* $\mathcal{E}$ *denotes the set of extremals of* V *and* $V_1 = \{\pi : \pi(\psi) = 1\}$, *then, for each* $\pi \in V$ *there exists a unique Borel measure* ν_π *on* $\mathcal{E} \cap V_1$ *with* $\pi = \bar{\nu}_\pi$.

PROOF. Since the measures μ_α have compact support it follows that V is closed. Thus the existence of a measure ν_π is guaranteed by the first part of Theorem 6.1. To prove uniqueness we must show that V is a lattice. Suppose then that $\pi_1, \pi_2 \in V$. Denote by π' the infimum of π_1 and π_2 in the lattice of all positive measures. π' may be defined by

$$\frac{d\pi'}{d(\pi_1 + \pi_2)}(g) = \min\left(\frac{d\pi_1}{d(\pi_1 + \pi_2)}(g), \frac{d\pi_2}{d(\pi_1 + \pi_2)}(g)\right).$$

Let S denote the convolution semigroup of measures on G generated by the μ_α, $\alpha \in \Gamma$. We shall consider the family $S * \pi'$ of measures $\mu * \pi'$, $\mu \in S$. We have

$$\mu_\alpha * \pi' \leqq \mu_\alpha * \pi_i = \pi_i, \qquad i = 1, 2,$$

so that $\mu_\alpha * \pi' \leqq \pi'$. Hence $\mu * \pi' \leqq \pi'$ for any $\mu \in S$. By the commutativity of S we also have

$$\mu' * (\mu * \pi') \leqq \mu * \pi' \tag{6.2}$$

for $\mu, \mu' \in S$.

Choose a dense subset $\{\mu_n * \pi'\}$ in $S * \pi'$. This may be done because G is separable. Let

$$\pi = \lim_{n \to \infty} \mu_1 * \cdots * \mu_n * \pi'.$$

This limit exists because the sequence of measures is decreasing by (6.2). Clearly $\pi \leqq \mu * \pi'$ for any $\mu \in S$. Also $\mu_\alpha * \pi \leqq \pi$. But $\mu_\alpha * \pi$ is a limit of measures in $S * \pi'$ and so $\pi \leqq \mu_\alpha * \pi$. Hence $\pi = \mu_\alpha * \pi$ and $\pi \in V$. Since $\pi \leqq \pi'$ we have $\pi_1 - \pi \in V$ and $\pi_2 - \pi \in V$. Now suppose $\pi_1 - \tilde{\pi}$, $\pi_2 - \tilde{\pi} \in V$ for some $\tilde{\pi} \in V$. Then $\tilde{\pi} \leqq \pi'$ and $\mu * \tilde{\pi} \leqq \mu * \pi'$ for all $\mu \in S$. But if $\tilde{\pi} \in V$ then $\mu * \tilde{\pi} = \tilde{\pi}$, so $\tilde{\pi} \leqq \mu * \pi'$ for $\mu \in S$, and so $\tilde{\pi} \leqq \pi$ or $\pi - \tilde{\pi} \in V$. Therefore $\pi = \inf(\pi_1, \pi_2)$ with respect to V. This proves the theorem.

7. Application to semi-spherical functions. We apply the foregoing to obtain more information regarding the representation of semi-spherical functions in terms of multiplier functions. In particular we shall see that for each spherical function ϕ $(\phi > 0)$ there is a single

K-multiplier $\sigma_\phi \in M_K(B(G))$ such that V_ϕ is spanned by $\sigma_\phi(\cdot, x)$. G is again a semi-simple Lie group.

LEMMA 7.1 [7], [8]. *The measures $m_K * g * m_K$ for $g \in G$ commute with one another.*

PROOF. We use the fact that there is a map $g \to g'$ of G into itself with the following properties: (a) $(g')' = g$, (b) $(g_1 g_2)' = g_2' g_1'$, (c) $k' = k^{-1}$ for $k \in K$, (d) if $P = \{g : g' = g\}$ then $KP = PK = G$. (In case $G = \mathrm{SL}(n, R)$, g' denotes transposition.)

We extend the operation $g \to g'$ to measures $\mu \to \mu'$. We then have $m_K' = m_K$ and $(m_K * p * m_K)' = m_K * p * m_K$ for $p \in P$. Now every $g \in G$ may be expressed as $g = kp$ for $k \in K$, $p \in P$ and $m_K * g * m_K = m_K * p * m_K$. It follows that $(m_K * g * m_K)' = m_K * g * m_K$. From this it follows that generally $(m_K * \mu * m_K)' = m_K * \mu * m_K$. So $m_K * g_1 * m_K * m_K * g_2 * m_K$ is invariant under $(\ )'$, but, on the other hand, using (b) it must equal $(m_K * g_2 * m_K)' * (m_K * g_1 * m_K)' = m_K * g_2 * m_K * m_K * g_1 * m_K$. This proves the lemma.

Now the functions in V_ϕ are the solutions $\geqq 0$ of

$$m_K * \gamma^{-1} * f = \phi(\gamma) f. \tag{7.1}$$

The measure solutions of (7.1) are actually functions because f is proportional to $m_K * \mu * f$ for any μ, in particular, an absolutely continuous one. Since $m_K * f = f$ we may rewrite (7.1) as

$$\phi(\gamma)^{-1} m_K * \gamma^{-1} * m_K * f = f. \tag{7.2}$$

Since the measures $m_K * \gamma^{-1} * m_K$ are mutually commutative and of compact support, the cone V_ϕ of solutions to (7.2) is one to which Theorem 6.2 applies. Moreover, a function $f \in V_\phi$ must be everywhere positive unless $f \equiv 0$, since by

$$\int_K f(\gamma k g)\, dk = \phi(\gamma) f(g); \tag{7.3}$$

if $f(g) = 0$, since f is continuous, f must vanish everywhere. Hence for any function $\psi \geqq 0$ of nonempty compact support, $\int f(g)\psi(g)\, dg > 0$. Choose ψ so that $\psi(gk) = \psi(g)$ for $k \in K$ and $\int \psi(g)\phi(g)\, dg = 1$. Then

$$\begin{aligned}\int_G f(g)\psi(g)\, dg &= \int_K \int_G f(gk)\psi(g)\, dg dk \\ &= \int_G \phi(g)\psi(g)\, dg\ f(e) = f(e).\end{aligned}$$

The set V_1 of Theorem 6.2 is therefore the set of functions with unit value at the identity. We refer to $\mathcal{E}\cap V_1$ as the *normalized* extremals. Theorem 6.2 then asserts that each semi-spherical function in V_ϕ has a unique representation in terms of normalized extremals of V_ϕ. Now by Theorem 5.1, every function in V_ϕ is an integral over multiplier functions $\sigma(\cdot, x)$ in V_ϕ (equation (5.12)). It follows that an extremal of V_ϕ must already be proportional to a multiplier function. Since the multiplier functions satisfy $\sigma(e, x)=1$, we see that the normalized extremals of V_ϕ are multiplier functions.

We next show that only one K-multiplier $\sigma_\phi \in M_K(B(G))$ can be such that the functions $\sigma_\phi(\cdot, x)$ are extremals of V_ϕ. Suppose that $\sigma_1(\cdot, x)$ and $\sigma_2(\cdot, x)$ are all extremals in V_ϕ. (Note that if $\sigma_1(\cdot, x_1)$ is an extremal so is $R_\gamma\sigma_1(\cdot, x_1)=\sigma_1(\cdot, \gamma x_1)\sigma_1(\gamma, x_1)$ so that all $\sigma_1(\cdot, x)$ are extremals if one of them is.) Now if $\sigma_1(\cdot, x)\in V_\phi$ then $\int_{B(G)}\sigma_1(g, x)\,dm(x) =\phi(g)$. The same holds for σ_2. But $\phi\in V_\phi$ has a unique representation in terms of its extremals (Theorem 6.2) and so the set of functions $\sigma_1(\cdot, x)$ must be identical with the set $\sigma_2(\cdot, x)$. Now both these sets of functions B_1 and B_2 form G-spaces with the operations $h\rightarrow R_g h/h(g)$. Moreover it is easy to see that $x\rightarrow\sigma_i(\cdot, x)$ then gives an equivariant map of $B(G)$ onto B_i, $i=1, 2$. But the spaces B_i are the same, so if $\sigma_1\neq\sigma_2$, we obtain two distinct equivariant maps of $B(G)$ onto $B_1=B_2$.

LEMMA 7.2. *If j_1: $B(G)\rightarrow B$, j_2: $B(G)\rightarrow B$ are two equivariant maps of $B(G)$ onto the same space B, then $j_1=j_2$.*

PROOF. It suffices to show that for some point $x_0\in B(G)$, $j_1(x_0) =j_2(x_0)$. For any $x\in B(G)$ form the measure $\pi=\frac{1}{2}\{\delta_{j_1(x)}+\delta_{j_2(x)}\}$ on B. Since B is a boundary (end of §3), there exists a sequence $\{g_n\}\subset G$ with $g_n\pi$ converging to a point measure. But extracting a subsequence we will have $g_{n_k}x\rightarrow x_0$ for some $x_0\in B(G)$ and so $g_{n_k}\pi\rightarrow\frac{1}{2}\{\delta_{j_1(x_0)}+\delta_{j_2(x_0)}\}$. If this is a point measure we must have $j_1(x_0)=j_2(x_0)$. Hence $j_1=j_2$.

It follows by this lemma that the maps $x\rightarrow\sigma_i(\cdot, x)$ must be the same for $i=1, 2$, and so $\sigma_1=\sigma_2$. We have thereby proved

THEOREM 7.1. *To each positive spherical function ϕ on G there is a multiplier $\sigma_\phi\in M_K(B(G))$ such that every positive solution to*

$$\int_K f(\gamma k g)\,dk = \phi(\gamma)f(g)$$

is given by

$$f(g) = \int_{B(G)} \sigma_\phi(g, x)\,d\nu(x), \tag{7.4}$$

where ν is a positive measure on $B(G)$.

For, the function $f \in V_\phi$ has a representation in terms of normalized extremals of V_ϕ. We have seen that these extremals must be multiplier functions and, moreover, only one multiplier belonging to ϕ can give rise to extremals. We denote that multiplier by σ_ϕ.

There is still one improvement that may be made on Theorem 7.1. We know that a measure on the set of extremals of V_ϕ is uniquely determined by its resultant, but we cannot as yet identify $B(G)$ with the set of normalized extremals. Namely the map $x \to \sigma_\phi(\cdot, x)$ may not be 1-1. We shall see later that this map is 1-1, so that the measure ν in (7.4) is uniquely determined by the function f.

According to Lemma 5.1, every multiplier function in $E_K(B(G))$ is semi-spherical and belongs to some V_ϕ. We may then apply Theorem 7.1 to obtain an integral representation of any multiplier function with respect to $B(G)$ in terms of some σ_ϕ. Even in the simplest cases this leads to nontrivial relationships.

Let us illustrate this for the case of the group G of analytic automorphisms of the unit disc. In the Iwasawa decomposition of this group G, the vector group is 1-dimensional. It follows that $M_K(B(G))$ is 1-dimensional ($\sigma(g, \gamma T)$ is determined by the positive character $\sigma(t, T)$ which is determined by $\sigma(a, T)$). For this group, $B(G)$ is the boundary of the unit disc, $B(G) = \{e^{i\theta}\}$. We have seen in §3 that $|g'(e^{i\theta})|$ is a K-multiplier with respect to $B(G)$ and so it follows that all the K-multipliers in $M_K(B(G))$ are given by

$$\sigma_s(g, e^{i\theta}) = |g'(e^{i\theta})|^s, \qquad -\infty < s < \infty. \tag{7.5}$$

If we denote by m normalized Lebesgue measure on the unit circle, then (7.5) can also be written as

$$\sigma_s(g, e^{i\theta}) = \left[\frac{dg^{-1}m}{dm}(e^{i\theta})\right]^s. \tag{7.6}$$

We recall from §3 that this expression always represents a multiplier. The spherical functions on G and therefore given by

$$\phi_s(g) = \int \sigma_s(g, e^{i\theta})\, dm(e^{i\theta}).$$

Now, in general, if G is a semi-simple group and $B(G)$ is its associated G-space we can form a 1-parameter family of spherical functions by taking

$$\phi_s(g) = \int_{B(G)} \left[\frac{dg^{-1}m}{dm}(x)\right]^s dm(x), \qquad -\infty < s < \infty.$$

We then have

$$
\begin{aligned}
\phi_{1-s}(g) &= \int_{B(G)} \left[\frac{dg^{-1}m}{dm}(x)\right]^{-s} \frac{dg^{-1}m}{dm}(x)\, dm(x) \\
&= \int_{B(G)} \left[\frac{dg^{-1}m}{dm}(x)\right]^{-s} dg^{-1}m(x) \\
&= \int_{B(G)} \left[\frac{dg^{-1}m}{dm}(g^{-1}x)\right]^{-s} dm(x) \qquad (7.7)\\
&= \int_{B(G)} \left[\frac{dm}{dgm}(x)\right]^{-s} dm(x) \\
&= \int_{B(G)} \left[\frac{dgm}{dm}(x)\right]^{s} dm(x) = \phi_s(g^{-1}).
\end{aligned}
$$

In our case it can also be verified that $\phi_s(g)=\phi_s(g^{-1})$. (Every $g\in G$ can be written $g=k_1hk_2$ where k_1 and k_2 are rotations and

$$h(z) = \frac{z-\tau}{1-\tau z}, \qquad -1<\tau<1.$$

But $h=h^{-1}$, so $\phi_s(g)=\phi_s(h)=\phi_s(h^{-1})=\phi_s(g^{-1})$.) For this group it follows that σ_s and σ_{1-s} belong to the same spherical function. As a result, for each s we must be able to express $\sigma_s(\cdot, e^{i\theta})$ in terms of $\sigma_{1-s}(\cdot, e^{i\theta})$, or vice versa. Now it can be seen that $\sigma_{1/2+t}$ grows more rapidly than $\sigma_{1/2-t}$ and, as a result, the latter must be expressed in terms of the former. Granting that such an expression exists, it is not difficult to determine what it must be. If we write $\sigma_{1-s}(g, 1) =\int\sigma_s(g, e^{i\theta})\, d\nu(\theta)$, and use the multiplier property for σ_{1-s} and σ_s and the uniqueness of the decomposition in terms of $\sigma_s(\cdot, x)$—the extremal of V_{ϕ_s}—we conclude that

$$\frac{dt\nu}{d\nu}(e^{i\theta}) = \sigma_{1-s}(t, 1)\sigma_s(t^{-1}, e^{i\theta})$$

for $t\in G$ and $t1=1$. This determines the measure ν. If we write

$$\sigma_s(g, e^{i\theta}) = \left(\frac{1-r^2}{1-2r\cos(\theta-\phi)+r^2}\right)^s = P(r, \theta-\phi)^s,$$

the relationship in question is

$$
\begin{aligned}
P(r, \theta)^{1-s} &= \gamma_s \int_0^{2\pi} P(r, \theta-\phi)^s (1-\cos\phi)^{s-1}\, d\phi, \qquad (7.8)\\
\gamma_s &= \left[\int_0^{2\pi} (1-\cos\phi)^{s-1}\, d\phi\right]^{-1}.
\end{aligned}
$$

(7.8) is proved by these means to be true for $s>1/2$. Since both sides of (7.8) are analytic for $\Re s>1/2$, the same formula holds for these complex values as well. Apparently a limiting form of (7.8) is still valid for $\Re s=1/2$. In this case the kernel $(1-\cos\phi)^{1/2+it}$ is singular, but with the proper interpretation it defines a unitary operator on $L^2(0, 2\pi)$. Equation (7.8) then shows that this operator is the intertwining operator relating the equivalent unitary representations obtained from the complex multipliers $\sigma_{1/2+it}$ and $\sigma_{1/2-it}$ (see §2).

8. **The general convolution equation.** We now come to our main result which deals with solutions to the general convolution equation

$$\int f(g'g)\,d\mu(g') = \lambda f(g), \qquad \lambda > 0. \tag{8.1}$$

Unfortunately, we shall not be able to handle the case of an arbitrary positive measure μ. We shall have to assume, to begin with, that μ is absolutely continuous and we shall also make a special assumption regarding the support of μ. Specifically, we shall take $d\mu(g)=\psi(g)dg$ where $\psi(g)$ is a bounded measurable function of compact support Δ, and assume that Δ has the property that some power Δ^n contains a neighborhood of the identity. These hypotheses regarding ψ will be assumed in force throughout our discussion. Our equation now takes the form

$$\int f(g'g)\psi(g')\,dg' = \lambda f(g) \tag{8.2}$$

and we denote by $V_\lambda(\psi)$ the cone of non-negative solutions to (8.2). We shall show that $V_\lambda(\psi)$ is spanned by its extremals and that the extremals are multiplier functions. Note that unlike equation (5.3), the solutions to (8.2) need not satisfy $f(kg)=f(g)$. As a result, the multiplier functions that we deal with now are no longer K-multiplier functions. As usual, $V_\lambda(\psi)$ is translation invariant, and so, for a multiplier $s\in M(B(G))$, if $s(\cdot, x_1)\in V_\lambda(\psi)$ for some $x_1\in B(G)$, then $s(\cdot, x)\in V_\lambda(\psi)$ for all $x\in B(G)$.

Lemma 8.1. *In each equivalence (cohomology) class of $H(B(G))$ there is at most one multiplier s with $s(\cdot, x)\in V_\lambda(\psi)$.*

Proof. Let us denote by $\psi^{(n)}$ the n-fold convolution of ψ with itself. Notice that under our hypotheses, $\psi^{(n)}$ will be continuous of compact support containing a neighborhood of the identity for some $n\geqq 2$. (8.2) implies that

$$\int f(g'g)\psi^{(n)}(g')\,dg' = \lambda^n f(g). \tag{8.3}$$

If $s(\cdot, x)$ satisfies (8.2), then

$$\int s(g, x)\psi^{(n)}(g)\,dg = \lambda^n.$$

If $s' \sim s$ then $s'(g, x) = (p(gx)/p(x))s(g, x)$, so that if $s'(\cdot, x) \in V_\lambda(\psi)$,

$$\int p(gx)s(g, x)\psi^{(n)}(g)\,dg = \lambda^n p(x).$$

Then

$$p(x) = \frac{\int p(gx)s(g, x)\psi^{(n)}(g)\,dg}{\int s(g, x)\psi^{(n)}(g)\,dg}. \tag{8.4}$$

If $\Gamma \subset B(G)$ denotes the set where $p(x)$ attains its maximum, and $\psi^{(n)}(g) > 0$, then $g\Gamma \subset \Gamma$, according to (8.4). Since this set of g comprises a neighborhood of the identity, $G\Gamma \subset \Gamma$ so that $\Gamma = B(G)$. Hence p is constant and $s' = s$. This proves the lemma.

If the function f on G corresponds to the measure π, then equation (8.1) may be rewritten $\tilde{\mu} * \pi = \lambda\pi$. Here $\tilde{\mu}$ is the measure corresponding to the functions $\psi(g^{-1})$. Because $\tilde{\mu}$ is absolutely continuous, π must also be, so the set of positive solutions to $\tilde{\mu} * \pi = \lambda\pi$ is $V_\lambda(\psi)$. We also note that the non-negative solutions to (8.2) are all continuous and everywhere positive. This follows from (8.3) since $\psi^{(n)}$ is continuous, and if $f(g) = 0$ then $f(g'g) = 0$ for g' in a neighborhood of the identity of G.

We may now apply Theorem 6.2 to the cone $V_\lambda(\psi)$. For, to begin with, this cone coincides with the cone of measures satisfying $\tilde{\mu} * \pi = \lambda\pi$. Secondly, $\int f(g)\psi(g)\,dg = \lambda f(e) > 0$ for every $f \in V_\lambda(\psi)$. So if $\mathcal{E}$ denotes the set of extremals of $V_\lambda(\psi)$ and $V'_\lambda(\psi)$ denotes the set of functions in $V_\lambda(\psi)$ with $f(e) = 1$ we have

LEMMA 8.2. *For every function $f \in V_\lambda(\psi)$ there is a unique measure ν on $V'_\lambda(\psi) \cap \mathcal{E}$ with*

$$f(g) = \int_{V'_\lambda(\psi)\cap\mathcal{E}} f_\xi(g)\,d\nu(\xi). \tag{8.5}$$

The deeper part of our analysis concerns the identification of the extremal functions in $V_X'(\psi)$ with multiplier functions. We shall first prove

LEMMA 8.3. *For each extremal $f\in V_X'(\psi)\cap\mathcal{E}$, there is a G-space X on which K is transitive, and a multiplier $s\in M(X)$, such that $f(g)=s(g,\xi)$ for some $\xi\in X$.*

PROOF. To begin with, let us notice that the space $V_X'(\psi)\cap\mathcal{E}$ is a G-space. For if $f(g)$ is an extremal of $V_X'(\psi)$, so is the function $f(g\gamma)/f(\gamma)$. Thus G operates on $V_X'(\psi)\cap\mathcal{E}$ by right translation (and normalization). Let us form the subset X of extremals $f(g\gamma)/f(\gamma)$, $\gamma\in G$. If $L=\{\gamma: f(g\gamma)=c(\gamma)f(g)\}$, where $c(\gamma)$ is some arbitrary constant, then $X=G/L$. We can define a function s on $G\times X$ by

$$s(g,\gamma L) = f(g\gamma)/f(\gamma). \tag{8.6}$$

s is well defined since if γ is replaced by γl, $l\in L$, both numerator and denominator in (8.6) are multiplied by $c(l)$. Moreover, s is a multiplier in $M(X)$. To prove the lemma, it is merely necessary to show that K is transitive on X. (This will, incidentally, imply that X is compact, a fact which is not otherwise obvious.)

We shall use the following notation. If π_1 and π_2 are two measures we shall write $\pi_1\succ\pi_2$ if, for some $\epsilon>0$, $\pi_1\geqq\epsilon\pi_2$. We call a measure π *radial* if it satisfies $k_1*\pi*k_2=\pi$ for $k_1, k_2\in K$.

If ρ is any continuous function on G with compact support, there is an n with $\tilde{\psi}^{(n)}\succ\rho$. $\tilde{\psi}$ is defined by $\tilde{\psi}(g)=\psi(g^{-1})$. Choose ρ radial and positive in a neighborhood of the identity. Then

$$\tilde{\psi}^{(n)}*f=\lambda^n f,$$

and so $\rho*f\prec f$. If we now choose $\rho>1$ on the support of $\tilde{\psi}$, we will also have $\tilde{\psi}*f\prec\rho*f$. Hence

$$f\prec\rho*f\prec f. \tag{8.7}$$

Since ρ is radial, $m_K*\rho=\rho*m_K=\rho$. By Lemma 7.1,

$$\rho*f*m_K=\rho*m_K*f*m_K=m_K*f*m_K*\rho=m_K*f*\rho. \tag{8.8}$$

If ρ' is another radial function whose support contains a neighborhood of the identity, and $\rho'\geqq 0$, we will have $\rho'*\rho\succ\rho$. So

$$\rho*f*\rho'=\rho*m_K*f*\rho'=m_K*f*\rho'*\rho\succ m_K*f*\rho, \tag{8.9}$$

or $\rho*f*\rho'\succ\rho*f*m_K$. By (8.7),

$$f*\rho'\succ f*m_K. \tag{8.10}$$

Returning to the functional notation we have

$$\int_G f(gg'^{-1})\rho'(g')\,dg' \succ \int_K f(gk)\,dk.$$

Setting $f(g) = s(g, \xi_0)$ (see (8.6)), we can write

$$\int_G s(g, g'^{-1}\xi_0)s(g'^{-1}, \xi_0)\rho'(g')\,dg' \succ \int_K s(g, k\xi_0)s(k, \xi_0)\,dk$$

or

$$\int_G s(g, g'^{-1}\xi_0)\rho'(g')\,dg' \succ \int_K s(g, k\xi_0)\,dk$$

which may be written

$$\int_X s(g, \xi)\,d\omega_1(\xi) \succ \int_X s(g, \xi)\,d\omega_2(\xi). \tag{8.11}$$

Now the functions $s(g, \xi)$ are extremals and $V_\lambda(\psi)$ is a lattice. By the corollary to Theorem 6.1, (8.11) implies that $\omega_1 \succ \omega_2$. Now X is a manifold as the quotient of a Lie group by a closed subgroup, and the measure ω_1, as image under the natural map of $G \to G/L$ of an absolutely continuous measure, is itself absolutely continuous on X. (This means that the restriction to a coordinate neighborhood is absolutely continuous with respect to Lebesgue measure induced on that coordinate neighborhood.) On the other hand, $\omega_2 = m_K * \xi_0$ is concentrated on the submanifold $K\xi_0 \subset X$. A measure on a submanifold of X can be absolutely continuous on X only if the submanifold is open. But $K\xi_0$ is closed and X is connected. It follows that $K\xi_0 = X$ which is what was needed to prove the lemma.

To prove that the space X occurring in Lemma 8.3 is $B(G)$ or an equivariant image of $B(G)$, we use the next lemma. We shall continue to use the notation $\pi_1 \succ \pi_2$ introduced in the proof of Lemma 8.3.

LEMMA 8.4. *Let Y and Z be two homogeneous K-spaces and suppose there is a map from Y to the space of probability measures on Z, $y \to \theta_y$, such that $\theta_{ky} = k\theta_y$ for $k \in K$, and whenever μ_1, μ_2 are two positive measures on K such that $\mu_1 * \theta_y \succ \mu_2 * \theta_y$, we have $\mu_1 * y \succ \mu_2 * y$. Then there is a map $j: Z \to Y$, equivariant with respect to K, and such that $j^{-1}(y)$ coincides with the support of θ_y for each $y \in Y$.*

PROOF. We set $Y = K/K_1$, $Z = K/K_2$. If m_{K_1} denotes the Haar measure on K_1, then Y may be identified with the space of measures $\{km_{K_1}: k \in K\}$ on K. Similarly, the space of measures on Z can be identified with the space of measures $\{\nu * m_{K_2}: \nu \text{ a measure on } K\}$. Namely, if ω is a measure on Z, set

$$\tilde{\omega}(F) = \int_{K/K_2}\left[\int_{K_2} F(kk')\,dk'\right] d\omega(kK_2),$$

so that $\tilde{\omega}$ is a measure on K. Then $\tilde{\omega}k=\tilde{\omega}$ for $k\in K_2$, and so $\tilde{\omega}=\tilde{\omega} * m_{K_2}$. Since $\tilde{\omega}\to\omega$ under the natural map of $K\to Z$, the correspondence $\tilde{\omega}\leftrightarrow\omega$ is 1-1.

Let $y_0\in Y$ correspond to the coset K_1 or to the measure m_{K_1}. If θ_{y_0} corresponds to $\nu * m_{K_2}$, then, since K_1 leaves y_0 fixed, it must leave θ_{y_0} fixed, so that $\theta_{y_0}=m_{K_1} * \theta_{y_0}$, and θ_{y_0} corresponds to $m_{K_1} * \nu * m_{K_2}$. The map $y\to\theta_y$ is then given by

$$km_{K_1} \to km_{K_1} * \nu * m_{K_2}. \tag{8.12}$$

The hypothesis of the lemma may then be rewritten: $\mu_1 * m_{K_1} * \nu * m_{K_2} \succ \mu_2 * m_{K_1} * \nu * m_{K_2}$ implies $\mu_1 * m_{K_1} \succ \mu_2 * m_{K_1}$. We can make this more precise. Suppose, in fact, that $\mu_1 * m_{K_1} * \nu * m_{K_2} \geqq \mu_2 * m_{K_1} * \nu * m_{K_2}$. Then for some $C>0$, $\mu_1 * m_{K_1} \geqq C\mu_2 * m_{K_1}$. Take the largest possible C for which this is true and suppose $C<1$. Then

$$(\mu_1 * m_K - C\mu_2 * m_{K_1}) * m_{K_1} * \nu * m_{K_2} \geqq (1-C)\mu_2 * m_{K_1} * \nu * m_{K_2}.$$

Now $\mu_1 * m_K - C\mu_2 * m_{K_1}$ is a positive measure, so we can apply the condition in the lemma again. We get, for some $\epsilon>0$,

$$(\mu_1 * m_{K_1} - C\mu_2 * m_{K_1}) * m_{K_1} \geqq \epsilon\mu_2 * m_{K_1},$$

or

$$\mu_1 * m_{K_1} \geqq (C+\epsilon)\mu_2 * m_{K_1}.$$

This contradicts the choice of C. It follows that $C\geqq 1$. What we have shown is that, for a signed measure μ on K,

$$\mu * m_{K_1} * \nu * m_{K_2} \geqq 0 \quad \text{implies} \quad \mu * m_{K_1} \geqq 0. \tag{8.13}$$

Suppose, more generally, that we have two positive measures ω_1 and ω_2 on K with the property that $\mu * \omega_1\geqq 0$ implies $\mu * \omega_2\geqq 0$. We claim that we must have $\omega_2=\omega_1 * \omega$ for some positive measure ω on K. The proof is based on the Hahn-Banach theorem. Namely, form the cone of all $\omega_1 * \omega$, $\omega\geqq 0$. This cone is weakly closed. If ω_2 did not belong to this cone, it could be separated from it by a continuous linear functional. That is, there would exist a continuous function h with $\omega_1 * \omega(h)\geqq 0$ for all $\omega\geqq 0$ on K, but $\omega_2(h)<0$. Now the first inequality implies that $\tilde{h} * \omega_1\geqq 0$ ($\tilde{h}(g)=h(g^{-1})$). Hence $h * \omega_2\geqq 0$ which in turn implies $\omega_2(h)\geqq 0$. So ω_2 must belong to the cone in question, or $\omega_2=\omega_1 * \omega$.

By (8.13) we therefore have $m_{K_1}=m_{K_1} * \nu * m_{K_2} * \omega$ for a positive measure ω on K. Let P be the support of ν and Q the support of ω.

The support of a convolution of positive measures is the product of the supports, so $PK_2Q \subset K_1$. This gives $PK_2P^{-1} \subset PK_2QQ^{-1}K_2P^{-1} \subset K_1$.

To define the map j it is merely necessary to verify that every point of Z belongs to the support of some θ_y and that the supports of θ_{y_1} and θ_{y_2} do not overlap unless $y_1 = y_2$. As to the first question, we notice that since some point belongs to the support of some θ_y, and Z is homogeneous, every point belongs to the support of some θ_y. For the second question we observe that the support of θ_{kK_1} is the image in Z of kK_1PK_2 by (8.12). So if

$$k'K_1PK_2 \cap k''K_1PK_2 \neq \emptyset,$$

then $k'^{-1}k'' \in K_1PK_2P^{-1}K_1^{-1} \subset K_1$ and $k'K_1 = k''K_1$. This completes the proof of the lemma.

We return to the extremal f in $V_X'(\psi)$ and the space X which it generates. We have $f = s(\cdot, \xi_0)$, $s \in M(X)$, and we have shown that K is transitive on X. At this point we invoke the theory of semi-spherical functions. By Lemma 2.1, every multiplier in $M(X)$ is equivalent to a K-multiplier:

$$s(g, \xi) = \frac{p(g\xi)}{p(\xi)}\, \sigma(g, \xi), \qquad \sigma \in M_K(X), \tag{8.14}$$

where $p(\xi)$ is continuous and positive in X. By Lemma 5.1, the functions $\sigma(\cdot, \xi)$ are semi-spherical and belong to the spherical function

$$\phi(g) = \int_X \sigma(g, \xi)\, dm(\xi),$$

where m is the unique K-invariant measure on X. By Theorem 7.1, the semi-spherical functions belonging to ϕ are all generated by some $\sigma_\phi \in M_K(B(G))$:

$$\sigma(g, \xi) = \int_{B(G)} \sigma_\phi(g, x)\, d\theta_\xi'(x),$$

for some measure θ' on $B(G)$. Let B' be the equivariant image of $B(G)$ obtained by identifying $x_1, x_2 \in B(G)$ if $\sigma_\phi(g, x_1) = \sigma_\phi(g, x_2)$. (One verifies readily that this is an equivalence relationship.) σ_ϕ defines a multiplier σ_ϕ' in $M_K(B')$ and we may write

$$\sigma(g, \xi) = \int_{B'} \sigma_\phi'(g, y)\, d\theta_\xi(y). \tag{8.15}$$

Because the $\sigma_\phi'(\cdot, y)$, $y \in B'$, are all distinct, the points of B' correspond to the normalized extremals of V_ϕ. In particular, the measures θ_ξ in (8.15) are uniquely determined. Since $\sigma(g, k\xi)=\sigma(gk, \xi)$,we conclude from (8.15) that $k\theta_\xi=\theta_{k\xi}$. We now show that the map $\xi\to\theta_\xi$ from X to the probability measures on B' satisfies the condition of Lemma 8.4. Suppose then that μ_1, μ_2 are positive measures on K with $\mu_1 * \theta_\xi \succ \mu_2 * \theta_\xi$ for some $\xi\in X$. Note that by (8.14), $s\succ\sigma\succ s$. As functions of g we then have

$$\int_K s(g, k\xi)\, d\mu_1(k) \succ \int_K \sigma(g, k\xi)\, d\mu_1(k)$$

$$= \int_K \sigma(gk, \xi)\, d\mu_1(k) = \int_{B'}\int_K \sigma'_\phi (gk, y)\, d\mu_1(k) d\theta_\xi(y)$$

$$= \int_{B'}\int_K \sigma'_\phi (g, ky)\, d\mu_1(k) d\theta_\xi(y) = \int_{B'} \sigma'_\phi (g, z)\, d\mu_1 * \theta_\xi(z)$$

$$\succ \int_{B'} \sigma'_\phi (g, z)\, d\mu_2 * \theta_\xi(z) = \int_{B'}\int_K \sigma'_\phi (g, ky)\, d\mu_2(k) d\theta_\xi(y)$$

$$= \int_{B'}\int_K \sigma'_\phi (gk, y)\, d\mu_2(k) d\theta_\xi(y) = \int_K \sigma(gk, \xi)\, d\mu_2(k)$$

$$= \int_K \sigma(g, k\xi)\, d\mu_2(k) \succ \int s(g, k\xi)\, d\mu_2(k).$$

Now the extreme ends of this series of inequalities are integrals over $s(g, \eta)$, $\eta\in X$, and these are extremals in $V'_\lambda(\psi)$. By the corollary to Theorem 6.1, since $V_\lambda(\psi)$ is a lattice, we must have $\mu_1 * \xi \succ \mu_2 * \xi$. The conditions of Lemma 8.4 are thereby verified and we can conclude that there is a map j: $B'\to X$ such that θ_ξ has its support equal to $j^{-1}(\xi)$.

Finally we must show that j is equivariant as a map of G-spaces: $j(gy)=gj(y)$. By (8.15),

$$\begin{aligned} \sigma(g\gamma, \xi) &= \sigma(g, \gamma\xi)\sigma(\gamma, \xi) \\ &= \sigma(\gamma, \xi) \int_{B'} \sigma'_\phi (g, z)\, d\theta_{\gamma\xi}(z). \end{aligned} \tag{8.16}$$

On the other hand,

$$\begin{aligned} \sigma(g\gamma, \xi) &= \int_{B'} \sigma'_\phi (g\gamma, y)\, d\theta_\xi(y) = \int_{B'} \sigma'_\phi (g, \gamma g)\sigma'_\phi (\gamma, y)\, d\theta_\xi(g) \\ &= \int \sigma'_\phi (g, z)\sigma'_\phi (\gamma, \gamma^{-1}z)\, d\gamma\theta_\xi(z). \end{aligned} \tag{8.17}$$

By the uniqueness of the Choquet-Deny representation in V_ϕ we must have

$$\sigma(\gamma, \xi)d\theta_{\gamma\xi}(z) = \sigma_\phi'(\gamma, \gamma' z)d\gamma\theta_\xi(z).$$

This implies that

$$(8.18) \qquad \gamma\theta_\xi \prec \theta_{\gamma\xi}$$

for any $\gamma \in G$. Suppose now that $j(y) = \xi$. Then $y \in \text{support}(\theta_\xi)$ and $gy \in \text{support}(g\theta_\xi)$. By (8.18) this implies that $gy \in \text{support}(\theta_{g\xi})$ or $j(g\xi) = gy$. This proves that j is an equivariant map of B' onto X. Now B', in turn, is an equivariant image of $B(G)$. But the map $B(G) \to X$ induces an injection of $M(X) \to M(B(G))$ and so the multiplier $s \in M(X)$ corresponds to a multiplier $s^* \in M(B(G))$. We have proved

THEOREM 8.1. *The cone $V_\lambda(\psi)$ of non-negative solutions to*

$$\int f(g'g)\psi(g')\, dg' = \lambda f(g)$$

admits a Choquet-Deny representation in terms of its normalized extremals. Each normalized extremal has the form $s(\cdot, x)$, where $s \in M(B(G))$ and $x \in B(G)$. Moreover, if $\eta \in H(B(G))$, there is at most one multiplier $s_\eta \in \eta$ such that $s_\eta(\cdot, x) \in V_\lambda(\psi)$.

9. **Basic multipliers.** Theorem 8.1 tells us that the extremals of $V_\lambda(\psi)$ are proportional to multiplier functions in $E(B(G))$. However, not all the multiplier functions in $V_\lambda(\psi)$ need be extremals. For example, taking $\phi \equiv 1$ on G, the multipliers $(dg^{-1}m/dm)(x) = \sigma_1(g, x)$ and $\sigma_0(g, x) \equiv 1$ in $M_K(B(G))$ both belong to V_ϕ. However,

$$1 = \int_{B(G)} \frac{dg^{-1}m}{dm}(x)\, dm(x),$$

so that σ_0 cannot be an extremal of V_ϕ. In this section we shall introduce a class of multipliers, the *basic* multipliers, such that the normalized extremals of $V_\lambda(\psi)$ coincide exactly with the set of basic multiplier functions in $V_\lambda(\psi)$. In the course of this discussion we shall resolve another question. Namely we shall show that the space X of the preceding section is not just an image of $B(G)$ but coincides with $B(G)$. In addition we will find that the space $B' = B(G)$ so that the space $B(G)$ coincides with the set of extremals in each V_ϕ.

DEFINITION 9.1. $s \in M(B(G))$ is *basic* if, for a signed measure ω on $B(G)$,

$$\int s(g, x)\, d\omega(x) \geqq 0 \text{ for all } g \quad \text{implies} \quad \omega \geqq 0.$$

We denote by $M_b(B(G))$ the set of all basic multipliers, and by $E_b(B(G))$ the set of all multiplier functions coming from basic multipliers.

In particular, $\int s(g, x)\, d\omega(x) = 0$ implies $\omega = 0$ for s basic, and so the functions $s(\cdot, x)$, $x \in B(G)$, must all be distinct. It follows that a multiplier on $B(G)$ induced by a multiplier in an equivariant image B' of $B(G)$ cannot be basic unless $B' = B(G)$.

LEMMA 9.1. *$s \in M(B(G))$ is basic if and only if*

$$\int s(g, x)\, d\omega_1(x) \succ \int s(g, x)\, d\omega_2(x) \tag{9.1}$$

implies $\omega_1 \succ \omega_2$ for any two positive measures ω_1, ω_2 on $B(G)$.

The proof is similar to the proof of (8.13) in Lemma 8.4.

LEMMA 9.2. *If s_1 and s_2 are equivalent multipliers then s_1 is basic if and only if s_2 is basic.*

PROOF. For, if s_1 and s_2 are equivalent, $s_1 \succ s_2 \succ s_1$, so that (9.1) for s_1 is equivalent to the same assertion for s_2.

Thus the condition of being basic depends only on the equivalence class of s and we may introduce the set $H_b(B(G))$ of basic (cohomology) classes.

LEMMA 9.3. *s is basic if and only if the cone of functions on $B(G)$ spanned by the functions $s(g, \cdot)$ is dense in the cone of all positive continuous functions on $B(G)$.*

That s is basic under the stated condition is immediate. The converse is a consequence of the Hahn-Banach theorem.

According to Lemma 9.2, s is basic if and only if the K-multiplier equivalent to s is basic. We will now show that the basic K-multipliers are just the σ_ϕ which we saw spanned the cones V_ϕ.

THEOREM 9.1. *$\sigma \in M_K(B(G))$ is basic if and only if $\sigma = \sigma_\phi$ for some spherical function ϕ.*

PROOF. We shall first show that if σ is basic it coincides with some σ_ϕ. This is the easier of the two directions. Suppose then that σ is basic. In any case $\sigma \in V_\phi$ for some spherical function ϕ by Lemma 5.1. Then, as in the proof of Theorem 8.1,

$$\sigma(g, x) = \int_{B'} \sigma'_\phi(g, y)\, d\theta_x(y). \tag{9.2}$$

Now suppose that $\mu_1 * \theta_\xi \succ \mu_2 * \theta_\xi$ where μ_1, μ_2 are measures on K. Then

$$\begin{aligned}
&\int_B \sigma(g, y)\, d\mu_i * x(y) \\
&= \int_K \sigma(g, kx)\, d\mu_i(k) = \int_K \sigma(gk, x)\, d\mu_i(k) \\
&= \int_K \int_{B'} \sigma'_\phi(gk, z)\, d\theta_x(z) d\mu_i(k) = \int_K \int_{B'} \sigma'_\phi(g, kz)\, d\theta_x(z) d\mu_i(k) \\
&= \int_{B'} \sigma'_\phi(g, u)\, d\mu_i * \theta_x(u),
\end{aligned}$$

for $i = 1, 2$. Since $\mu_1 * \theta_x \succ \mu_2 * \theta_x$ it follows that

$$\int \sigma(g, y)\, d\mu_1 * x(y) \succ \int \sigma(g, y)\, d\mu_2 * x(y).$$

By Lemma 9.1, since σ is basic, this implies $\mu_1 * x \succ \mu_2 * x$. We see that the map $x \to \theta_x$ from $B(G)$ to B' fulfills the conditions of Lemma 8.4. This means that there is an equivariant map (see the proof of Theorem 8.1) $j: B' \to B(G)$ such that θ_x has its support on $j^{-1}(x)$. But B' is itself an equivariant image of $B(G)$—$j': B(G) \to B'$, so that jj' is an equivariant map of $B(G)$ onto itself. Now, by Lemma 7.2, the identity is the only equivariant map of $B(G)$ onto itself. So jj' is the identity. Since j' is onto, j is an isomorphism. Hence $B' \approx B(G)$ and j is an equivariant map of $B(G)$ onto itself; hence the identity. Since θ_x has its support on $j^{-1}(x)$, it follows that $\theta_x = \delta_x$ and $\sigma(g, x) = \sigma'_\phi(g, x) = \sigma_\phi(g, x)$. This proves that if σ is basic, then $\sigma = \sigma_\phi$.

Now consider the K-multiplier σ_ϕ for some spherical ϕ. We wish to prove that σ_ϕ is basic. We must first establish that σ_ϕ separates points in $B(G)$. To do this introduce the space B' as before: B' is obtained from $B(G)$ by identifying $x_1, x_2 \in B(G)$ if $\sigma_\phi(g, x_1) = \sigma_\phi(g, x_2)$ for all $g \in G$. Let j denote the ensuing natural map of $B(G) \to B'$. We then define σ'_ϕ as before by $\sigma'_\phi(g, j(x)) = \sigma_\phi(g, x)$.

For each $\xi \in B'$, let $K_\xi \subset K$ denote the subgroup of K that leaves ξ fixed. K_ξ takes $j^{-1}(\xi)$ into itself and is transitive on $j^{-1}(\xi)$ since K is transitive on $B(G)$. So there is a unique probability measure θ_ξ on $j^{-1}(\xi)$ invariant under K_ξ. We see readily that $k\theta_\xi = \theta_{k\xi}$ for $k \in K$. If

$\gamma \in G$ and $\gamma\xi = \xi$, then γ determines a differentiable homeomorphism of $j^{-1}(\xi)$ with itself and so $\gamma\theta_\xi$ is absolutely continuous with respect to θ_ξ. Since every $g \in G$ is a product $k\gamma$ with $k \in K$, $\gamma\xi = \xi$ (since $K\xi = B'$), it follows that $g\theta_\xi = k\gamma\theta_\xi$ is absolutely continuous with respect to $k\theta_\xi = \theta_{k\xi} = \theta_{k\gamma\xi} = \theta_{g\xi}$. Then $g^{-1}\theta_{g\xi}$ is absolutely continuous with respect to θ_ξ and the expression

$$\sigma(g, x) = \frac{dg^{-1}\theta_{g\xi}}{d\theta_\xi}(x)\sigma_\phi'(g, \xi) \tag{9.3}$$

is well defined for x in the support of θ_ξ, that is, for $j(x) = \xi$. We claim that with this definition of ξ, $\sigma(g, x)$ defines a multiplier in $M_K(B(G))$. $\sigma(k, x) = 1$ since $k^{-1}\theta_{k\xi} = \theta_\xi$ and $\sigma_\phi'(k, \xi) = 1$. What remains to be verified is the multiplier identity. Letting $\xi = j(x)$ we have, making use of (3.13):

$$\begin{aligned}
\sigma(g_1g_2, x) &= \frac{dg_2^{-1}g_1^{-1}\theta_{g_1g_2\xi}}{d\theta_\xi}(x)\sigma_\phi'(g_1g_2, \xi) \\
&= \frac{dg_1^{-1}\theta_{g_1g_2\xi}}{dg_2\theta_\xi}(g_2x)\sigma_\phi'(g_1, g_2\xi)\sigma_\phi'(g_2, \xi) \\
&= \frac{dg_1^{-1}\theta_{g_1g_2\xi}}{d\theta_{g_2\xi}}(g_2x)\frac{d\theta_{g_2\xi}}{dg_2\theta_\xi}(g_2x)\sigma_\phi'(g_1, g_2\xi)\sigma_\phi'(g_2, \xi) \\
&= \sigma(g_1, g_2x)\frac{dg_2^{-1}\theta_{g_2\xi}}{d\theta_\xi}(x)\sigma_\phi'(g_2, \xi) \\
&= \sigma(g_1, g_2x)\sigma(g_2, x).
\end{aligned}$$

Here we have used the fact that $j(g_2x) = g_2\xi$ if $j(x) = \xi$.

We next verify that this multiplier σ also belongs to ϕ. It suffices to show that

$$\int_{B(G)} \sigma(g, x)\, dm(x) = \phi(g). \tag{9.4}$$

But

$$\begin{aligned}
\int_{B(G)} \sigma(g, x)\, dm(x) &= \int_K \sigma(g, kx_0)\, dk \\
&= \int_K \int_{K'} \sigma(g, kk'x_0)\, dk dk'.
\end{aligned}$$

Here x_0 is any point of $B(G)$ and K' is any subgroup of K. If we let

$\xi_0=j(x_0)$ and set $K'=K_{\xi_0}$, then $k'x_0$ has the distribution θ_{ξ_0} for k' uniformly distributed in K'. So

$$\int_{B(G)}\sigma(g,x)\,dm(x)=\int_K\int_{B(G)}\sigma(g,ky)\,d\theta_{\xi_0}(y)dk$$
$$(9.5)\qquad =\int_K\int_{B(G)}\sigma(gk,y)\,d\theta_{\xi_0}(y)dk.$$

Now

$$\int_{B(G)}\sigma(g,y)\,d\theta_{\xi_0}(y)=\int_{j^{-1}(\xi_0)}\sigma(g,y)\,d\theta_{\xi_0}(y)$$
$$(9.6)\qquad =\int_{j^{-1}(\xi_0)}\frac{dg^{-1}\theta_{g\xi_0}}{d\theta_{\xi_0}}(y)\sigma'_\phi(g,\xi_0)\,d\theta_{\xi_0}(y)$$
$$=\sigma'_\phi(g,\xi_0)\int dg^{-1}\theta_{g\xi_0}(y)=\sigma'_\phi(g,\xi_0).$$

By (9.5),

$$\int_{B(G)}\sigma(g,x)\,dm(x)=\int_K\sigma'_\phi(gk,\xi_0)\,dk=\int_K\sigma_\phi(gk,x_0)\,dk$$
$$=\int_K\sigma_\phi(g,kx_0)\,dk=\int_{B(G)}\sigma_\phi(g,x)\,dm(x)=\phi(g).$$

This proves that σ belongs to ϕ. But (9.6) also shows that $\sigma_\phi(g,x_0)=\sigma'_\phi(g,\xi_0)$ is expressed as a linear combination of functions in V_ϕ. Since the σ_ϕ are extremals this can only take place if all the $\sigma(g,y)$ for $y\in j^{-1}(\xi_0)$ coincide with $\sigma_\phi(g,x_0)$. In other words, $dg^{-1}\theta_{g\xi_0}/d\theta_{\xi_0}$ must be identically 1, or, equivalently, the measure $g^{-1}\theta_{g\xi_0}$ coincides with θ_{ξ_0}. The same argument implies that for all $g\in G$, $\xi\in B'$, $g\theta_\xi=\theta_{g\xi}$. But $B(G)$ is a boundary (see §3) so that some $g_n\theta_\xi$ tends to a point measure. Setting $g_n=k_n\gamma_n$, where $\gamma_n\xi=\xi$ and $k_n\in K$, we must have $\gamma_n\theta_\xi=\theta_{\gamma_n\xi}=\theta_\xi$ tending to a point measure. This means that θ_ξ is a point measure and so j is 1-1. We have thereby proved that $\sigma_\phi(g,x)$ separates points in $B(G)$.

It is now an easy matter to show that σ_ϕ is basic. For, since the functions $\sigma_\phi(\cdot,x)$ are distinct for distinct $x\in B(G)$, it follows that $B(G)$ coincides with the normalized extremals of V_ϕ. Now suppose $\int_{B(G)}\sigma_\phi(g,x)\,d\omega(x)\geqq 0$ for some signed measure ω on $B(G)$. From the fact that V_ϕ is a lattice, by the corollary to Theorem 6.1, this inequality implies that $\omega \geqq 0$. This, however, was just the condition for σ_ϕ to be basic. This completes the proof of the theorem.

COROLLARY. *A multiplier s is basic if and only if it is equivalent to σ_ϕ for some spherical function ϕ.*

Very little is known regarding the subset $H_b(B(G))$ of the linear space $H(B(G))$. For example, is it a semigroup, and is it closed under the operations $\sigma \rightarrow \sigma^r$, $r \geqq 1$?

10. **Application to the Choquet-Deny representation.** We shall apply the results of §9 to obtain more precise information regarding the extremals of $V_\lambda(\psi)$.

THEOREM 10.1. *If $s(\cdot, x)$ is an extremal of $V_\lambda(\psi)$, then the multiplier s is basic.*

PROOF. As in the proof of Theorem 9.1, we form the space B' which is obtained from $B(G)$ by identifying $x_1, x_2 \in B(G)$ if $s(g, x_1) = s(g, x_2)$. We have an equivariant map $j: B(G) \rightarrow B'$, and $s(g, x) = s'(g, j(x))$, where $s' \in M(B')$. Again, it will suffice to prove that $B' = B(G)$. For, $V_\lambda(\psi)$ is a lattice, and so by the corollary to Theorem 6.1, if $\int s'(g, \xi)\, d\omega(\xi) \geqq 0$, then $\omega \geqq 0$.

Let σ' be the K-multiplier equivalent to s' on B'. The functions $\sigma'(\cdot, \xi)$ are semi-spherical and belong to V_ϕ for some ϕ. So, by Theorem 7.1,

$$\sigma'(g, \xi) = \int_{B(G)} \sigma_\phi(g, x)\, d\theta_\xi(x), \tag{10.1}$$

where σ_ϕ is the basic multiplier belonging to ϕ. Because σ_ϕ is basic, the measures θ_ξ in (10.1) are uniquely determined. Now use the multiplier equation:

$$\begin{aligned}\sigma'(\gamma g, \xi) &= \int \sigma_\phi(\gamma g, x)\, d\theta_\xi(x) \\ &= \int \sigma_\phi(\gamma, gx)\sigma_\phi(g, x)\, d\theta_\xi(x) = \int \sigma_\phi(\gamma, x)\sigma_\phi(g, g^{-1}x)\, dg\theta_\xi(x).\end{aligned}$$

But $\sigma'(\gamma g, \xi)$ is also equal to

$$\sigma'(\gamma, g\xi)\sigma'(g, \xi) = \int \sigma_\phi(\gamma, x)\sigma'(g, \xi)\, d\theta_{g\xi}(x).$$

Comparing the two expressions as functions of γ, we have

$$\sigma_\phi(g, g^{-1}x)\, dg\theta_\xi(x) = \sigma'(g, \xi)\, d\theta_{g\xi}(x). \tag{10.2}$$

Now we can write

$$s'(g, \xi) = \frac{p'(g\xi)}{p(\xi)} \sigma'(g, \xi),$$

where p' is a positive continuous function on B'. Since $s'(\cdot, \xi) \in V_\lambda(\psi)$,

$$\int p'(g\xi)\sigma'(g, \xi)\psi(g)\, dg = \lambda p'(\xi). \tag{10.3}$$

Form the convex set Q of all functions $p(x)$, continuous and nonnegative on $B(G)$, and satisfying

$$\int p(x)\, d\theta_\xi(x) = p'(\xi). \tag{10.4}$$

If we recall the proof of Theorem 8.1 we see that the maps $\xi \to \theta_\xi$ of (8.15) and (10.1) are the same. As a result θ_ξ has its support on $j^{-1}(\xi)$. It follows that Q is nonempty since the function $p = p' \circ j^{-1}$ will belong to Q.

Define a linear transformation T on the space of continuous functions on $B(G)$ by

$$Tp(x) = \lambda^{-1} \int p(gx)\sigma_\phi(g, x)\psi(g)\, dg.$$

Then

$$\lambda^{-1} \int\int p(gx)\sigma_\phi(g, x)\psi(g)\, dg d\theta_\xi(x) = \lambda^{-1} \int\int p(x)\sigma_\phi(g, g^{-1}x)\, dg\theta_\xi(x)\psi(g)\, dg.$$

By (10.2), we have

$$\int Tp(x)\, d\theta_\xi(x) = \lambda^{-1} \int\int p(x)\sigma'(g, \xi)\, d\theta_{g\xi}(x)\psi(g)\, dg. \tag{10.5}$$

If $p \in Q$, this equals

$$\lambda^{-1} \int p'(g\xi)\sigma'(g, \xi)\psi(g)\, dg = p'(\xi).$$

It follows that T takes Q into itself.

The functions in TQ form an equicontinuous family. For

$$
\begin{aligned}
&| Tp(kx) - Tp(x) | \\
&\qquad \leq \lambda^{-1} \int p(gx)\sigma_\phi(g, x)(| \psi(k^{-1}g) - \psi(g) |)\, dg. \\
(10.6) &\qquad = \lambda^{-1} \int\int p(k_1gx)\sigma_\phi(g, x)(| \psi(k^{-1}k_1g) - \psi(k_1g) |)\, dg dk_1 \\
&\qquad \leq A \int p(y)\, dm(y) \cdot \max_g | \psi(k^{-1}g) - \psi(g) |,
\end{aligned}
$$

where $A = \lambda^{-1} \max \sigma_\phi(g, x)$ and the maximum is taken over all $x \in B(G)$ and all g for which some $k'g$ is in the support of ψ, $k' \in K$. Now, by (10.4),

$$\int p(y)\, dm(y) = \int p'(\xi)\, dm(\xi)$$

so that the right-hand side of (10.6) goes to 0 as $kx \rightarrow x$ independently of $p \in Q$.

As a result, $\overline{TQ}$ is a compact convex set and T maps $\overline{TQ}$ into itself. It follows that T has a fixed point in Q; that is, there exists a function $p(x)$ satisfying (10.4) and

$$\int\int p(gx)\sigma_\phi(g, x)\psi(g)\, dg = \lambda p(x). \tag{10.7}$$

If we set $f_x(g) = p(gx)\sigma_\phi(g, x)$, we then find that $f_x \in V_\lambda(\psi)$. Now

$$\int f_x(g)\, d\theta_\xi(x) = \int p(gx)\sigma_\phi(g, x)\, d\theta_\xi(x) = \int p(x)\sigma_\phi(g, g^{-1}x)\, dg\theta_\xi(x).$$

By (10.2), this gives

$$\int f_x(g)\, d\theta_\xi(x) = \int p(x)\sigma'(g, \xi)\, d\theta_{g\xi}(x) = \sigma'(g, \xi)p'(g\xi) = p'(\xi)s'(g, \xi),$$

since $p \in Q$. This expresses $s'(\cdot, \xi)$ as a linear combination of the functions f_x which are in $V_\lambda(\psi)$. However, the $s'(\cdot, \xi)$ were taken to be extremals of $V_\lambda(\psi)$. It must therefore be the case that all the f_x with x in the support of θ_ξ are proportional. If j were not 1-1, that is, $j(x_1) = j(x_2)$ for some pair $x_1, x_2 \in B(G)$, then, since f_{x_1} and f_{x_2} are proportional,

$$\frac{p(gx_1)}{p(x_1)} \sigma_\phi(g, x_1) = \frac{p(gx_2)}{p(x_2)} \sigma_\phi(g, x_2). \tag{10.8}$$

Replace g by kg and integrate both sides of (10.8) over K. This gives

$$\frac{\sigma_\phi(g, x_1)}{p(x_1)} = \frac{\sigma_\phi(g, x_2)}{p(x_2)} .$$

Setting $g=e$ we find $p(x_1)=p(x_2)$ and $\sigma_\phi(\cdot, x_1)=\sigma_\phi(\cdot, x_2)$. Since σ_ϕ is basic this cannot happen unless $x_1=x_2$. This proves that j is 1-1 and hence that $B'=B(G)$. This completes the proof of the theorem.

As a result, the statement of Theorem 8.1 may be made more precise. That is, the normalized extremals of $V_\lambda(\psi)$, which are already known to be multiplier functions, are now known to be basic multiplier functions. We now give a more precise form to the Choquet-Deny representation for $V_\lambda(\psi)$. For each cohomology class $\eta \in H(B(G))$, we denote by σ_η the unique K-multiplier in that class. σ_η belongs to a spherical function which we shall denote ϕ_η.

LEMMA 10.1. *For each cohomology class $\eta \in H(B(G))$, there exists a uniquely determined positive constant $\Lambda(\eta)$ and a positive continuous function $p_\eta(x)$ on $B(G)$, unique up to a constant multiple, such that*

$$\int p_\eta(gx)\sigma_\eta(g, x)\psi(g)\, dg = \Lambda(\eta)p_\eta(x). \tag{10.9}$$

PROOF. Let Q denote the convex set of non-negative functions $p(x)$ on $B(G)$ satisfying $\int p(x)\, dm(x)=1$. Let T denote the transformation of Q into itself defined by

$$Tp(x) = c(p) \int p(gx)\sigma_\eta(g, x)\psi(g)\, dg,$$

where $c(p)$ is a constant chosen so that $Tp \in Q$. One verifies that the functions of TQ form an equicontinuous family so that $\overline{TQ}$ is a compact convex set. If p_η is the fixed point of T in $\overline{TQ}$ then there is a constant $\Lambda(\eta)$ so that (10.9) is valid.

The uniqueness of $\Lambda(\eta)$ follows from

$$\frac{\int p(gx)\sigma_\eta(g, x)\psi(g)\, dg}{\int p'(gx)\sigma_\eta(g, x)\psi(g)\, dg} = \frac{\lambda}{\lambda'} \frac{p(x)}{p'(x)} . \tag{10.10}$$

If $\lambda/\lambda' < 1$ we choose x so that p/p' attains its minimum at x; then (10.10) becomes contradictory. So $\lambda \geqq \lambda'$ and similarly $\lambda' \leqq \lambda$. Thus there is a unique eigenvalue for which (10.9) can have a solution. Now if p_η satisfies (10.9), then the multiplier functions

$$\frac{p_\eta(gx)}{p_\eta(x)}\sigma_\eta(g, x)$$

belong to $V_{\Lambda(\eta)}(\psi)$, and, by Lemma 8.1, p_η is thereby uniquely determined. This completes the proof of the lemma.

If we compare the multiplier functions on G with the exponentials e^{at} on the real line, then we see that $\Lambda(\eta)$ corresponds to the Laplace transform of the functions $\psi(g)$.

We can now formulate the Choquet-Deny representation as follows:

THEOREM 10.2. *Restrict the function $\Lambda(\eta)$ to the subset $H_b(B(G)) \subset H(B(G))$, so that for each $\lambda > 0$, $\Lambda^{-1}(\lambda) \subset H_b(B(G))$. Then, for each function $f \in V_\lambda(\psi)$, there exists a measure ω on $\Lambda^{-1}(\lambda) \times B(G)$ such that*

$$f(g) = \int_{\Lambda^{-1}(\lambda)\times B(G)} p_\eta(gx)\sigma_\eta(g, x)\, d\omega(\eta, x). \tag{10.11}$$

In considering $\Lambda^{-1}(\lambda) \times B(G)$ as a measure space, we have imbedded it in $V_\lambda(\psi)$ and taken the induced Borel structure. We do not as yet know enough about the function $\Lambda(\eta)$ to claim that $\Lambda^{-1}(\lambda)$ is a Borel subset of $H(B(G))$.

What has been gained by our analysis as regards the integral equation

$$\int_G f(g'g)\psi(g')\, dg' = \lambda f(g)$$

is this. We have replaced a single equation for a function on a noncompact space and which, moreover, does not have a unique solution, by a family of equations (10.9) for a function $\Lambda(\eta)$ and a family of functions $p_\eta(x)$ on the compact space $B(G)$ which now have a unique solutions.

Theorem 10.2 still leaves open the uniqueness of the measure ω. Although we know that the representation in terms of normalized extremals of $V_\lambda(\psi)$ is unique, we have not shown that every multiplier function corresponding to a point in $\Lambda^{-1}(\lambda) \times B(G)$ is an extremal. This we shall do in the next theorem.

THEOREM 10.3. *The normalized extremals of $V_\lambda(\psi)$ consist exactly of those multiplier functions $s(\cdot, x)$, $x \in B(G)$, such that s is basic, and $s(\cdot, x)$ belongs to $V_\lambda(\psi)$.*

PROOF. We have already shown in Theorem 10.1 that if $s(\cdot, x)$ is extremal in $V_\lambda(\psi)$ then s is basic. We now show the converse. Suppose then that s is basic and $s(\cdot, x) \in V_\lambda(\psi)$. By Theorem 10.2 there is some measure ω with

$$s(g, x_0) = \int_{\Lambda^{-1}(\lambda)\times B(G)} p_\eta(gx)\sigma_\eta(g, x)\, d\omega(\eta, x). \tag{10.12}$$

Here we may assume that the multiplier functions

$$s_\eta(g, x) = \frac{p_\eta(gx)}{p_\eta(x)} \sigma_\eta(g, x) \tag{10.13}$$

are extremals of $V_\lambda(\psi)$ for those η occurring in this representation. If $s \sim \sigma$ where $\sigma \in M_K(B(G))$, then, by (2.3),

$$\int_K s(kg, x)\, dk = \left(\int_K s(k, x)\, dk\right)\sigma(g, x) = S(x)\sigma(g, x).$$

By (10.12)

$$S(x_0)\sigma(g, x_0) = \int_{\Lambda^{-1}(\lambda)\times B(G)} \left(\int p_\eta(y)\, dm(y)\right)\sigma_\eta(g, x)\, d\omega(\eta, x)$$

or

$$\sigma(g, x_0) = \int_{\Lambda^{-1}(\lambda)\times B(G)} \sigma_\eta(g, x)\, d\omega'(\eta, x). \tag{10.14}$$

Setting $g = e$, we see that ω' is a probability measure. Replacing g by gk and integrating both sides of (10.14) over K we find

$$\phi(g) = \int_{\Lambda^{-1}(\lambda)} \phi_\eta(g)\, d\omega''(\eta).$$

Here ϕ is the spherical function to which σ belongs. Now replace g by gkg and integrate over K. We have, by (3.2),

$$\begin{aligned} \phi(g)^2 &= \int_{\Lambda^{-1}(\lambda)} \phi_\eta(g)^2\, d\omega''(\eta) \\ &= \left\{\int_{\Lambda^{-1}(\lambda)} \phi_\eta(g)\, d\omega''(\eta)\right\}^2. \end{aligned} \tag{10.15}$$

Now ω'' is again a probability measure, so (10.15) shows that ϕ_η is essentially constant as a function of η. This means that ω'' is concentrated on classes η with the property that $\phi_\eta = \phi$. All we need, however, is the fact that for some η, $\phi_\eta = \phi$. We have s basic and $\sigma \sim s$. It follows that σ is basic; so $\sigma = \sigma_{\phi'}$ for some spherical function ϕ', by

Theorem 9.1. Since $\sigma_{\phi'}$ belongs to ϕ' and σ belongs to ϕ, we must have $\phi'=\phi$ and so $\sigma=\sigma_\phi$. Since the η occurring in (10.12) are basic classes, σ_η is basic and so $\sigma_\eta=\sigma_{\phi_\eta}=\sigma_\phi=\sigma$. As a result, $s\sim\sigma_\eta$. But $\sigma_\eta\sim s_\eta$ ((10.13)), so $s\sim s_\eta$. Now the multiplier functions $s(\cdot, x)$ and $s_\eta(\cdot, x)$ are in $V_\lambda(\psi)$ and, by Lemma 8.1, since $s\sim s_\eta$, we must have $s=s_\eta$. Since the functions $s_\eta(\cdot, x)$ are extremal, it follows that the functions $s(\cdot, x)$ are extremal and this proves our theorem.

This theorem implies that the normalized extremals of $V_\lambda(\psi)$ correspond exactly to the points of $\Lambda^{-1}(\lambda)\times B(G)$, the correspondence being given by

$$(\eta, x) \to \frac{p_\eta(gx)}{p_\eta(x)}\sigma_\eta(g, x).$$

Since the representation of elements in $V_\lambda(\psi)$ in terms of measures on the normalized extremals is unique, we have:

COROLLARY. *For each function $f\in V_\lambda(\psi)$, the measure ω in* (10.11) *is unique.*

The eigenvalue function $\Lambda(\eta)$ was defined originally for all $\eta\in H(B(G))$. It is not an arbitrary function; in particular, it is determined in all of $H(B(G))$ by its values in $H_b(B(G))$. This follows from the fact that the Choquet-Deny representation depends only on the restriction of $\Lambda(\eta)$ to $H_b(B(G))$. However, we can make a more precise statement.

DEFINITION 10.1. We shall say that two classes η_1 and η_2 in $H(B(G))$ are *related* if $\phi_{\eta_1}=\phi_{\eta_2}$.

Thus every class is related to one and only one basic class.

THEOREM 10.4. *If η_1 and η_2 are related, then* $\Lambda(\eta_1)=\Lambda(\eta_2)$.

PROOF. Suppose s is a multiplier in the class η_1 such that $s(\cdot, x)\in V_\lambda(\psi)$. Then $\lambda=\Lambda(\eta_1)$. By Theorem 10.2,

$$s(g, x_0) = \int_{\Lambda^{-1}(\lambda)\times B(G)} p_\eta(gx)\sigma_\eta(g, x)\, d\omega(\eta, x). \tag{10.16}$$

Replacing g by k_1gk_2 and integrating both sides of (10.16) over $K\times K$ we obtain, as in the proof of Theorem 10.3,

$$\phi_{\eta_1}(g) = \int_{\Lambda^{-1}(\lambda)} \phi_\eta(g)\, d\omega''(\eta). \tag{10.17}$$

As in the proof of Theorem 10.3, it follows that η is constant in the integration (10.17), and $\phi_{\eta_1}=\phi_\eta$. Since $\eta\in\Lambda^{-1}(\lambda)$, we have $\lambda=\Lambda(\eta_1)=\Lambda(\eta)$. Now η is the uniquely determined basic class related to η_1 since $\phi_{\eta_1}=\phi_\eta$ and η is basic. The same reasoning then implies $\Lambda(\eta_2)=\Lambda(\eta)$, so $\Lambda(\eta_1)=\Lambda(\eta_2)$. This completes the proof of the theorem.

11. Irreducible multipliers. If $s\in M(B(G))$, let $V(s)$ denote the closed cone of functions on G generated by $s(\cdot, x)$, $x\in B(G)$. Here closure will mean with respect to the topology of pointwise convergence on G. $V(s)$ is identical with the functions

$$f_\omega(g) = \int_{B(G)} s(g, x)\, d\omega(x),$$

where ω ranges over the positive measures on $B(G)$. That this family is closed follows from two facts. First, if $f_\omega(e)$ is bounded, then the measures ω are bounded and lie in a compact set. Secondly, the map $\omega\to f_\omega$ is continuous with respect to the weak topology on positive measures.

The cones $V(s)$ are obviously closed under (right) translation. We may inquire for which s the cone $V(s)$ will be *irreducible*, that is, have no proper closed, translation-invariant subcone. For example, if ν is a smooth measure on $B(G)$, the multiplier $s(g, x)=(dg^{-1}\nu/d\nu)(x)$ is not irreducible. For $\int s(g, x)\, d\nu(x)=\nu(B(G))$ is constant, and if $V(s)$ contains a constant it cannot be irreducible unless it reduces to the cone of positive constants.

DEFINITION 11.1. We say $s\in M(B(G))$ is *irreducible* if the cone $V(s)$ contains no proper closed translation-invariant subcone.

Let m denote the K-invariant probability measure on $B(G)$. We shall use σ_0 to denote the K-multiplier defined by

$$\sigma_0(g, x) = \frac{dg^{-1}m}{dm}(x). \tag{11.1}$$

LEMMA 11.1. *If s^* is a basic multiplier, then $s=\sigma_0/s^*$ is irreducible.*

PROOF. A typical function in $V(s)$ is $f(g)=\int s(g, x)\, d\omega(x)$. To prove that s is irreducible, we must show that given any $f\in V(s)$, the functions $s(\cdot, x)$ can be retrieved from f by taking limits of positive linear combinations of translates. We first form

$$F(g) = \int_K f(gk)\, dk = \int_K \int_{B(G)} s(g, kx)s(k, x)\, d\omega(x)\, dk. \tag{11.2}$$

Now, fixing $x_0\in B(G)$, let $x=k'x_0$. Then

$$\int_K s(g, kx)s(k, x)\,dk = \frac{1}{s(k', x_0)}\int_K s(g, kk'x_0)s(kk', x_0)\,dk$$

$$(11.3)\qquad = \frac{1}{s(k', x_0)}\int_K s(g, kx_0)s(k, x_0)\,dk = \frac{1}{s(k', x_0)}F_0(g).$$

From (11.2) and (11.3) we see that $F(g)$ is proportional to $F_0(g)$, where the latter function is independent of ω, and thus independent of the function f in $V(s)$. It follows that the function $F_0(g)$ belongs to every closed translation-invariant subcone of $V(s)$. $F_0(g)$ may be re-written as

$$(11.4)\qquad F_0(g) = \int_{B(G)} s(g, x)h(x)\,dm(x),$$

where $h(x)$ is a positive continuous function on $B(G)$.

Recall Lemma 9.3 according to which s^* is basic only if the functions of the form $\int s^*(g, x)\,d\mu(g)$, $\mu \geqq 0$, are dense in the positive continuous functions on $B(G)$. In particular, if U is any neighborhood of a point $x_0 \in B(G)$ and $\epsilon > 0$, there exists a measure μ on G with

$$(11.5)\qquad \begin{gathered} N(x) = \int s^*(g, x)\,d\mu(g) < \epsilon \quad \text{for} \quad x \notin U, \\ \int_{B(G)} N(x)\,dm(x) = 1. \end{gathered}$$

Then

$$\begin{aligned} F_0(g\gamma^{-1}) &= \int_{B(G)} s(g, \gamma^{-1}x)s(\gamma^{-1}, x)h(x)\,dm(x) \\ &= \int_{B(G)} s(g, x)s(\gamma^{-1}, \gamma x)h(\gamma x)\,d\gamma^{-1}m(x) \\ &= \int_{B(G)} s(g, x)s(\gamma, x)^{-1}h(\gamma x)\sigma_0(\gamma, x)\,dm(x) \\ &= \int_{B(G)} s(g, x)s^*(\gamma, x)h(\gamma x)\,dm(x). \end{aligned}$$

So

$$\int_G F_0(g\gamma^{-1})\,d\mu(\gamma) = \int_{B(G)} s(g, x)\left[\int_G s^*(\gamma, x)h(\gamma x)\,d\mu(\gamma)\right] dm(x).$$

Since $\int_G s^*(\gamma, x)\, d\mu(\gamma)$ approximates a δ-function in x with respect to the measure m, and since $h(x)$ is bounded from above and below, it follows that

$$\int_G s^*(\gamma, x)h(\gamma x)\, d\mu(\gamma)$$

approximates a multiple of the same δ-function. More precisely, we see from (11.5) that

$$\left| \frac{\int F_0(g\gamma^{-1})\, d\mu(\gamma)}{\int F_0(\gamma^{-1})\, d\mu(\gamma)} - s(g, x_0) \right| \leqq \max_{x\in U} C \left| s(g, x) - s(g, x_0) \right|,$$

where C is a constant depending upon $h(x)$. Consequently, with an appropriate choice of measures μ_n,

$$\frac{\int F_0(g\gamma^{-1})\, d\mu_n(\gamma)}{\int F_0(\gamma^{-1})\, d\mu_n(\gamma)} \to s(g, x_0), \quad \text{as } n \to \infty,$$

for each g. It follows that $s(\cdot, x_0)$ belongs to every closed translation-invariant cone containing $F_0(g)$ and so $V(s)$ is irreducible. This completes the proof of the lemma.

We turn now to K-multipliers.

LEMMA 11.2. *For each spherical function ϕ there is one and only one K-multiplier belonging to ϕ which is irreducible.*

PROOF. That at least one K-multiplier belonging to ϕ is irreducible is easily established. If $\phi(g)$ is spherical, then $\tilde{\phi}(g) = \phi(g^{-1})$ is also spherical. Suppose σ^* is the basic multiplier belonging to $\tilde{\phi}$, and form $\sigma = \sigma_0/\sigma^*$. Then

$$\begin{aligned}\int_{B(G)} \sigma(g, x)\, dm(x) &= \int_{B(G)} \sigma(g, g^{-1}x) \frac{dgm}{dm}(x) dm(x) \\ &= \int_{B(G)} \frac{1}{\sigma(g^{-1}, x)} \sigma_0(g^{-1}, x)\, dm(x) \\ &= \int_{B(G)} \sigma^*(g^{-1}, x)\, dm(x) \\ &= \tilde{\phi}(g^{-1}) = \phi(g).\end{aligned}$$

So σ belongs to ϕ and is irreducible by the foregoing lemma.

Now suppose that σ_1 and σ_2 are related K-multipliers (Definition 10.1) which are both irreducible. Since $\phi(g)=\int_{B(G)}\sigma_i(g, x)\, dm(x)$, we have $\phi\in V(\sigma_1)\cap V(\sigma_2)$. If both cones are irreducible, we must have $V(\sigma_1)=V(\sigma_2)$. In particular, we have

$$\tag{11.6} \begin{aligned} \sigma_2(g, x_0) &= \int \sigma_1(g, x)\, d\omega_1(x), \\ \sigma_1(g, x_0) &= \int \sigma_2(g, x)\, d\omega_2(x) \end{aligned}$$

for two positive measures ω_1, ω_2 on $B(G)$. One can find measures $\mu_1, \mu_2 \geqq 0$ on K with $\omega_i=\mu_i * x_0$, and (11.6) becomes

$$\sigma_2(g, x_0) = \int_K \sigma_1(gk, x_0)\, d\mu_1(k),$$

$$\sigma_1(g, x_0) = \int_K \sigma_2(gk, x_0)\, d\mu_2(k).$$

Combining these two:

$$\tag{11.7} \sigma_1(g, x_0) = \int_K \sigma_1(gk, x_0)\, d\mu_2 * \mu_1(k).$$

Let $\mu^{(n)}$ denote the n-fold convolution of $\mu_2 * \mu_1$ with itself. It is not hard to see that $n^{-1}[\mu^{(1)}+ \cdots +\mu^{(n)}]$ converges, as $n\to\infty$, to the Haar measure of the least subgroup $K_1\subset K$ which contains the support of $\mu_2 * \mu_1$ [6, pp. 343–344]. Since (11.7) would be valid with $\mu^{(n)}$ replacing $\mu_2 * \mu_1$, it follows that one may write

$$\sigma_1(g, x_0) = \int_{K_1} \sigma_1(gk, x_0)\, dk.$$

Hence, $\sigma_1(gk, x_0)=\sigma_1(g, x_0)$ for every $k\in K_1$. If P_2 and P_1 denote, respectively, the supports of μ_2 and μ_1, then $P_2P_1\subset K_1$. In particular, P_1 is contained in some left coset k_1K_1. Then

$$\sigma_2(g, x_0) = \int_{K_1} \sigma_1(gk_1k, x_0)\, dk_1^{-1}\mu_1(k).$$

But the support of $k_1^{-1}\mu_1$ is in K_1 and the $k\in K_1$ leave σ_1 invariant. So

$$\sigma_2(g, x_0) = \sigma_1(gk_1, x_0) = \sigma_1(g, k_1x_0) = \sigma_1(g, x_1).$$

But we have seen in §7, using Lemma 7.2, that this implies $\sigma_1=\sigma_2$.

This completes the proof of the lemma.

We can now prove the main result of this section.

THEOREM 11.1. *A multiplier s is irreducible if and only if σ_0/s is basic.*

PROOF. The implication in one direction is given in Lemma 11.1. What remains to be shown is that if s is irreducible, then σ_0/s is basic. We claim that if s is irreducible and σ is the K-multiplier equivalent to s, then σ is irreducible. To see this, consider the operator with domain $V(s)$ defined by

$$Tf(g) = \int_K f(kg)\, dk = m_K * f(g).$$

Since T commutes with right translation, it follows that $TV(s)$ is a translation-invariant cone. Moreover,

$$Ts(g, x) = \int_K s(kg, x)\, dk = \sigma(g, x)\left[\int s(k, x)\, dk\right]$$

so that $\sigma(\cdot, x)\in TV(s)$. It is also easy to see that $TV(s)$ is closed, so it follows that $TV(s)\supset V(\sigma)$. Now, since $V(s)$ is irreducible, it follows that $TV(s)$ is irreducible, which implies that $TV(s)=V(\sigma)$, and that $V(\sigma)$ is irreducible, whence σ is irreducible. Let σ belong to ϕ. By the foregoing lemma, σ is the only irreducible multiplier belonging to ϕ. On the other hand, we know that if σ^* is the basic multiplier belonging to $\phi(g^{-1})$, then σ_0/σ^* is irreducible and belongs to $\phi(g)$. So we must have $\sigma=\sigma_0/\sigma^*$. In other words, if σ is irreducible, σ_0/σ is basic. But then σ_0/s is basic, for $\sigma\sim s$ implies $\sigma_0/\sigma\sim\sigma_0/s$. This completes the proof of the theorem.

COROLLARY. *If s_1 and s_2 are equivalent multipliers in $M(B(G))$, then s_1 is irreducible if and only if s_2 is irreducible.*

If we consider the exponentials in the real line we notice that they obviously have the irreducibility property—the translation-invariant cones they generate are irreducible (in fact, 1-dimensional). Thus, for the real line, the properties of irreducibility and forming the extremals in the Choquet-Deny representation are shared by the same class of functions. In the case of semi-simple groups these properties split: one characterizes the irreducible multiplier functions and the other the basic multiplier functions. (It is probably true that the only multipliers which are simultaneously basic and irreducible are those equivalent to $\sigma_0^{1/2}$.)

12. **Irreducible cones.** We have mentioned the fact that the positive exponentials on the group of reals generate irreducible translation-invariant cones. In fact this property characterizes the exponentials. More precisely

THEOREM 12.1. *If V is a cone of non-negative continuous functions on the reals, closed under translation and closed in the topology of pointwise convergence, and if V is irreducible (that is, no proper subcone of V has the same properties), then V consists of the positive multiples of some exponential $f(t)=e^{at}$, $-\infty<a<\infty$.*

We will not give the proof of this inasmuch as it resembles that of the analogous result for semi-simple groups which we will give. Let us, however, sketch an alternative proof. If we know that V has a compact base, then, since the reals form a Tychonoff group and act on V by translation, V must have a fixed ray, i.e., a 1-dimensional translation-invariant subcone. This must be all of V and, as is easily seen, this implies that V consists of multiples of an exponential. The problem is, therefore, to show that V has a compact base. This may be done by showing that if V is irreducible, the functions in V are convex, secondly, that, in fact, their logarithms are convex, and, finally, that their logarithms satisfy a uniform Lipschitz condition.

For an arbitrary topological group one can attempt to characterize the irreducible (right) translation-invariant cones. In the semi-simple case, this is most easily done if we assume further that the functions in the cone are invariant under left translation by elements in a maximal compact subgroup. Note that if a cone is irreducible (with respect to right translations) and if one function has this property, then so do all functions in the cone.

THEOREM 12.2. *Let V be a cone of continuous non-negative functions on G closed under (right) translation and closed in the topology of pointwise convergence. Assume V is irreducible, that is, that no proper subcone of V has these properties. If, furthermore, there is a function $f\in V$ with $f(kg)=f(g)$ for all $k\in K$, then V coincides with $V(\sigma)$ for some irreducible K-multiplier σ.*

We remark that since the functions of V may be thought of as functions on $K\backslash G$, the question is that of characterizing cones of non-negative functions on a symmetric space, invariant under the isometries of the space, and irreducible with respect to this property.

PROOF. We convert the problem into one involving cones of measures. Identify V with a subcone of the cone of positive (unbounded) measures on G and let $\tilde{V}$ denote the closure of V in this space. Suppose ω is a measure on G corresponding to a continuous function of compact support. Then $V * \omega \subset V$ by the translation invariance of V, and, since $\pi_n \to \pi$ weakly implies $\pi_n * \omega \to \pi * \omega$ pointwise as functions, it follows that $\tilde{V} * \omega \subset V$. Suppose that $\tilde{V}$ is not irreducible, and suppose that $W \subset \tilde{V}$ is also invariant and weakly closed. Then $W * \omega \subset W \cap V$ so, then, $W \cap V$ is nonempty. By the irreducibility of V, $W \cap V = V$ or $W \supset V$ and, since W is weakly closed, $W = \tilde{V}$. Hence $\tilde{V}$ is irreducible.

Let $\tilde{V}_K$ denote the set of measures $\pi \in \tilde{V}$ such that $\pi * k = \pi$ for each $k \in K$. Under the hypotheses of the theorem, it is true, for every $\pi \in V$, that $k * \pi = \pi$. So the measures in V_K have the form $m_K * \pi' * m_K$. By Lemma 7.1, measures of this form are commutative when the product is defined.

Notice that V_K is invariant under the operations $\pi \to \pi * \gamma * m_K$, $\gamma \in G$. We claim that V_K is minimal with respect to this property. Let $W \subset \tilde{V}_K$ be closed and invariant under $\pi \to \pi * \gamma * m_K$ for all $\gamma \in G$. Form $W' = \{\pi \in \tilde{V} : \pi * g * m_K \in W \text{ for all } g \in G\}$. Clearly W' is right-invariant, weakly closed, and $W' \supset W$ so it is not empty. So $W' = \tilde{V}$, or $\tilde{V} * m_K = \tilde{V}_K \subset W$. Thus $W = V_K$.

Let ω be a measure of compact support on G and form $\tilde{V}_K * \omega * m_K$. This cone is closed, for, if $\pi_n * \omega * m_K$ converges, a subsequence of $\{\pi_n\}$ must converge weakly $(\pi * \omega * m_K(\Delta_1\Delta_2) \geqq \pi(\Delta_1)\omega * m_K(\Delta_2))$. We also have $\tilde{V}_K * \omega * m_K * \gamma * m_K = \tilde{V}_K * \gamma * m_K * \omega * m_K \subset \tilde{V}_K * \omega * m_K$. Also $\tilde{V}_K * \omega * m_K \subset \tilde{V} * m_K = \tilde{V}_K$, so that $\tilde{V}_K * \omega * m_K$ is a subcone of V_K invariant under $\pi \to \pi * \gamma * m_K$. So $\tilde{V}_K * \omega * m_K = \tilde{V}_K$.

Now V_K is a weakly closed cone of positive measures on G, so, by Theorem 6.1, it is spanned by its extremals. Let π_0 be an extremal of $\tilde{V}_K$. By what has just been shown $\pi_0 = \pi_\omega * \omega * m_K$ for some measure $\pi_\omega \in \tilde{V}_K$. But this expresses π_0 as a linear combination of measures $\pi_\omega * g * m_K \in \tilde{V}_K$. Since π_0 is extremal, these must be proportional for almost all g in the support of ω. By continuity, this will be true for all g in the support of ω. Moreover, if the identity of G is in the support of ω, then π_0 will itself be proportional to π_ω. We then find that $\pi_0 * g * m_K$ is proportional to π_0 for all g in the support of ω. Since ω had arbitrarily large support, we must have

$$\pi_0 * g^{-1} * m_K = \phi(g)\pi_0. \tag{12.1}$$

Since $\pi_0 = m_K * \pi_0 * m_K$, we have

$$\pi_0 * g * m_K = m_K * \pi_0 * m_K * g * m_K = m_K * g * m_K * \pi_0 * m_K = m_K * g * \pi_0,$$

so that

$$m_K * g^{-1} * \pi_0 = \phi(g)\pi_0. \tag{12.2}$$

This, however, is the defining equation (7.1) for semi-spherical functions. That is, ϕ must be spherical and π_0 corresponds to a function $\Phi \in V_\phi$. So

$$\Phi(g) = \int_{B(G)} \sigma_\phi(g, x)\, d\nu(x)$$

for some $\nu \geqq 0$ on $B(G)$. But $\Phi(gk) = \Phi(g)$ since $\pi_0 \in \tilde{V}_k$, so that

$$\Phi(g) = \int_{B(G)} \int_K \sigma_\phi(g, kx)\, dk d\nu(x)$$

$$= \int_{B(G)} \sigma_\phi(g, x)\, dm(x) = \phi(g).$$

We have proven that if V is irreducible, some spherical function $\phi \in \tilde{V}$. Now $\tilde{V} * \omega \subset V$ if ω corresponds to a continuous function with compact support, and $\phi * \omega$ is proportional to ϕ if ϕ is spherical and ω is radial. Hence, $\phi \in V$. Now let σ be the (unique) irreducible K-multiplier belonging to ϕ (Lemma 11.2). Then $V(\sigma)$ contains ϕ and, hence, $V(\sigma) \cap V$ is nonempty. It follows that $V = V(\sigma)$. This completes the proof of the theorem.

REMARK 1. The result would not have been altered if the topology on the cone V had been taken to be that of uniform convergence on compact subsets.

REMARK 2. Whenever a group is represented by automorphisms of some structure, we may speak of an irreducible representation when the structure is minimal with respect to invariance under the action of the group. Traditionally, the structure is assumed to be that of a (topological) linear space; in this section we have taken it to be that of a closed cone. Another possibility would be to let the structure be that of a compact convex set. For example, every bounded measurable function on G generates a translation-invariant compact convex set in $L^\infty(G)$ (with the weak* topology), and we may inquire when this gives an irreducible representation of G. Using [6] it is easy to show that, up to isomorphism, the irreducible representations of a semi-simple G on compact convex sets are just its representations on the space of probability measures on one of its boundary spaces (the equivariant images of $B(G)$). With this it may be shown that the irreducible weak* compact convex sets in $L^\infty(G)$ are just those which are spanned by a function $f(g)$ satisfying $f(gk) = f(g)$ for all $k \in K$ and which, as a function on the symmetric space G/K, is harmonic.

APPENDIX. LEMMA 5.2. *If π is a measure on G satisfying $t\pi=\chi(t)\pi$ for $t\in T$, and χ a positive character on T, then $\pi=m_T^\chi * \mu$ for some positive measure μ on G.*

PROOF. Form the cone V of measures π on G satisfying $t\pi=\chi(t)\pi$, $\pi\geqq 0$. Let $h(g)$ be a non-negative continuous function on G with compact support and such that $h(k)>0$ for $k\in K$. If, for $\pi\in V$, $\pi(h)=0$, then $\int h(tg)\,d\pi(g)=0$ for all $t\in T$. But the interiors of the supports of the functions $h(tg)$ cover G since $TK=G$. This would then imply that $\pi=0$, so we see that $\pi(h)>0$ for all $\pi\in V$, $\pi\neq 0$. We can now apply Theorem 6.1 to the weakly closed cone V, and we find that each $\pi\in V$ has an integral representation in terms of normalized extremals of V. We see then that the lemma will be proved once we establish that the extremals of V are the measures $m_T^\chi * \gamma$, $\gamma\in G$, up to a multiplicative constant.

Let $\pi\in V$ and form the measure π_Δ given by

$$\pi_\Delta(f) = \int f(g)C_\Delta(Tg)\,d\pi(g),$$

where Δ is a set in $T\backslash G$ and C_Δ denotes its characteristic function. Now it is readily seen that π_Δ again belongs to V. So if π is extremal in V it must be concentrated on a single coset $T\gamma$. Then $\pi * \gamma^{-1}=\pi_0$ is concentrated on T and satisfies $t\pi_0=\chi(t)\pi_0$. Define the measure π_1 on T by

$$d\pi_1(t) = \chi(t)\,d\pi_0(t).$$

If $\tau\in T$, then

$$\frac{d\tau\pi_1}{d\tau\pi_0}(t) = \frac{d\pi_1}{d\pi_0}(\tau^{-1}t) = \chi(\tau^{-1}t).$$

But $(d\tau\pi_0/d\pi_0)(t)=\chi(\tau)$, so

$$\frac{d\tau\pi_1}{d\pi_0}(t) = \frac{d\tau\pi_1}{d\tau\pi_0}(t)\frac{d\tau\pi_0}{d\pi_0}(t) = \chi(\tau^{-1}t)\chi(\tau) = \chi(t) = \frac{d\pi_1}{d\pi_0}(t).$$

Hence $\tau\pi_1=\pi_1$ for all $\tau\in T$. So π_1 is proportional to Haar measure on T and

$$d\pi_0(t) = c\chi(t^{-1})\,dt,$$

and $\pi_0=cm_T^\chi$. Since every extremal π of V has the form $\pi_0 * \gamma^{-1}$ for some $\gamma\in G$, our lemma is proven.

Bibliography

1. F. Bruhat, *Sur les représentations induites des groupes de Lie*, Bull. Soc. Math. France **84** (1956), 97–205.

2. G. Choquet, *Le théorème de représentation intégrale dans les ensembles convexes compacts*, Ann. Inst. Fourier (Grenoble) **10** (1960), 333–344.

3. G. Choquet et J. Deny, *Sur l'équation de convolution* $\mu=\mu * \sigma$, C. R. Acad. Sci. Paris **250** (1960), 799–801.

4. N. Dunford and J. T. Schwartz, *Linear operators*, Interscience, New York, 1958.

5. E. B. Dynkin, *Non-negative eigenfunctions of the Laplace-Beltrami operator and Brownian motion on certain symmetric spaces*, Dokl. Akad. Nauk SSSR **141** (1961), 288–291 (Russian); translated as Soviet Math. Dokl. 2 (1961), 1433–1436.

6. H. Furstenberg, *A Poisson formula for semi-simple Lie groups*, Ann. of Math. (2) **77** (1963), 335–386.

7. I. Gel'fand, *Spherical functions on symmetric Riemann spaces*, Dokl. Akad. Nauk SSSR **70** (1950), 5–8. (Russian)

8. S. Helgason, *Differential geometry and symmetric spaces*, Academic Press, New York, 1962.

9. L. Loomis, *Unique direct integral decompositions on convex sets*, Amer. J. Math. **84** (1962), 509–526.

10. C. C. Moore, *Compactifications of symmetric spaces*, Amer. J. Math. **86** (1964), 201–218.

11. Séminaire Sophus Lie, *Théorie des algèbres de Lie*, École Normale Supérieure 1954/1955.

University of Minnesota

ON THE EXISTENCE AND IRREDUCIBILITY OF CERTAIN SERIES OF REPRESENTATIONS[1]

BY BERTRAM KOSTANT

1. **Introduction.** 1. By the principal series one means here the unitary representations of a semisimple Lie group G arising from induction to G by characters on MAN corresponding to characters on A. Although long conjectured to be irreducible, this family of representations has been shown to be irreducible only for special groups. For example see [**9**] for complex G and see [**3**] for the group $\mathrm{Sl}(n, \boldsymbol{R})$. In the general case (all G) irreducibility has been proved by Bruhat [**1**] using analytic methods, only however, for the "regular" characters on A. A proof of the irreducibility of all the elements of the principal series is but one application of certain algebraic results, stated here, on modules of the universal enveloping algebra U of the Lie algebra $\mathfrak{g}$ of G.

A second application is the proof of the existence and irreducibility of the complementary series for all semisimple Lie groups generalizing in a natural way the case of $\mathrm{Sl}(2, \boldsymbol{R})$. It is shown also that if $\dim A = 1$ (split rank 1 case) then except for possibly the trivial (one dimensional) representation the most general irreducible unitary representation of G admitting a fixed vector for K ($\mathrm{Ad}_{\mathfrak{g}}K$ is the maximal compact subgroup of $\mathrm{Ad}_{\mathfrak{g}}G$) belongs either to the principal or complementary series.

1.2. If $\mathfrak{a}'_c$ is the complex dual to the Lie algebra a of A then any $\lambda \in \mathfrak{a}'_c$ defines a one dimensional representation $b \rightarrow b^{\lambda}$ of $B = MAN$. If X^{λ} is the space of all analytic K-finite functions f on G such that $f(ab) = b^{-\lambda}f(a)$ where $a \in G$, $b \in B$ then X^{λ} is in a natural way a U-module. The results above are mainly applications of a theorem (Theorem 2) giving a necessary and sufficient condition on λ for X^{λ} to be an irreducible (in the usual algebraic sense) U-module. In particular there arises, in a natural way, a region in $\mathfrak{a}'_c$ which we call the critical strip (CS) for which X^{λ} is always U-irreducible. The critical strip contains all of the λ corresponding to the principal series and its closure contains all the λ corresponding to the complementary series.

---

[1] By invitation the author addressed the annual meeting of the American Mathematical Society in Cincinnati on January 22, 1962 on the topic *A survey of Lie group representations*. The present paper is partially an outgrowth of some of the ideas on multiplicities of representations stated during that talk; received by the editors March 3, 1969.

1.3. Now X^λ contains a unique function, 1_λ, which is the constant 1 on K. In the general case X^λ is not irreducible and in fact the cyclic U-submodule $X_e^\lambda = U \cdot 1_\lambda$ is not equal to X^λ. It is proved, however, that if C is the complex Weyl chamber, suitably translated, then X_e^λ is U-irreducible for all $\lambda \in -C$ and that every irreducible U-module Z such that $Z^{\mathfrak{k}} \neq 0$ ($\mathfrak{k}$ is the Lie algebra of K) is equivalent to X_e^λ for some $\lambda \in -C$. On the other hand, if $\lambda \in C$ one always has $X_e^\lambda = X^\lambda$ and that every U-module Y cyclically generated by an element in $Y^{\mathfrak{k}}$, where dim $Y^{\mathfrak{k}} = 1$, is equivalent to a quotient of X^λ for some $\lambda \in C$.

The completeness result on the unitary representations of G in the split rank 1 case is a consequence of another theorem which asserts that in the split rank 1 case if G is chosen as defined in §4.1 the multiplicity of any irreducible representation of K in X^λ, for any $\lambda \in \mathfrak{a}'_{\mathbf{C}}$, is at most one.

2. **The principal series.** 1. Let G be a Lie group, not necessarily connected. Let $\mathfrak{g}$ be its Lie algebra, so that $\mathfrak{g}$ is a Lie algebra over $\boldsymbol{R}$. Assume that $\mathfrak{g}$ is semisimple and assume that for any $a \in G$ the operator Ad a on the complexification of $\mathfrak{g}_{\mathbf{C}}$ of $\mathfrak{g}$, defined by conjugation of G by a, lies in the adjoint group of $\mathfrak{g}_{\mathbf{C}}$.

REMARK 1. The second assumption is of course satisfied if G is connected. However it is also satisfied if, for example, G is the set of rational points of a complex connected algebraic group defined over $\boldsymbol{R}$.

We recall that a subalgebra $\mathfrak{b} \subseteq \mathfrak{g}$ is called parabolic in case the complexification of $\mathfrak{b}$ contains a maximal solvable subalgebra (a Borel subalgebra) of $\mathfrak{g}_{\mathbf{C}}$.

Let $\mathfrak{b} \subseteq \mathfrak{g}$ be a minimal parabolic subalgebra of $\mathfrak{g}$ (all such are conjugate) and let $\mathfrak{n}$ be the nil-radical of $\mathfrak{b}$. Then $\mathfrak{b}$ can be written as a semidirect sum $\mathfrak{b} = \mathfrak{h} + \mathfrak{n}$ where the subalgebra $\mathfrak{h} \subseteq \mathfrak{b}$ is unique up to conjugacy.

An element $x \in \mathfrak{g}$ is called real semisimple in case ad x is diagonizable on $\mathfrak{g}$ (and hence real eigenvalues).

Now let $\mathfrak{a}$ be the set of all elements in the center of $\mathfrak{h}$ which are real semisimple. Then one knows that $\mathfrak{a}$ is an abelian subalgebra of $\mathfrak{g}$ maximal with the property that all of its elements are real semisimple. Furthermore if $A \subseteq G$ is the corresponding subgroup then A is simply connected and hence $A \cong \boldsymbol{R}^r$ where $r = \dim \mathfrak{a}$.

Now let G^A be the centralizer of A in G and let $N \subseteq G$ be the subgroup corresponding to $\mathfrak{n}$. Then one knows that G^A normalizes N and $G^A \cap N = (e)$. Let $B = G^A N$. Then the Lie algebra of B is $\mathfrak{b}$. Moreover there exists a closed subgroup $M \subseteq G^A$ such that G^A is a direct product

$G^A = MA$. Moreover if D (= Ker Ad) is the centralizer of G_e in G (G_e is the identity component of G) then D is discrete, normal and $\mathfrak{a}$ can be chosen so that $D \subseteq M$. With this additional property M is unique and we shall assume that M is so chosen. One thus has $B = MAN$.

Now let $\mathfrak{a}'_{\mathbf{C}} = \mathrm{Hom}_{\mathbf{R}}(\mathfrak{a}, \mathbf{C})$ be the complex dual to $\mathfrak{a}$. Any $\lambda \in \mathfrak{a}'_{\mathbf{C}}$ defines a character $b \to b^\lambda$ on B as follows: For $b \in MN$ put $b^\lambda = 1$. If $b \in A$ then one may uniquely write $b = \exp x$, for $x \in \mathfrak{a}$. Put $b^\lambda = \exp \langle \lambda, x \rangle$. One obtains in this way all complex characters on B which are trivial on M.

2.2. We will be concerned with the representations of G induced by the one dimensional representations of B given by $b \to b^\lambda$, $\lambda \in \mathfrak{a}'_{\mathbf{C}}$. Let $\lambda \in \mathfrak{a}'_{\mathbf{C}}$ and let $X^{(\lambda)}$ be the space of all measurable functions f on G such that $f(ab) = b^{-\lambda} f(a)$ for all $a \in G$, $b \in B$. Then $X^{(\lambda)}$ is a G-module where if $a \in G$ and $f \in X^{(\lambda)}$ then $a \cdot f \in X^{(\lambda)}$ is given by $(a \cdot f)(g) = f(a^{-1} g)$ for all $g \in G$.

Now for certain λ one obtains a unitary representation: Indeed we first observe that G/D is a centerless Lie group with at most a finite number of connected components. As such it contains a maximal compact subgroup K_0, unique up to conjugacy. Let $K \subseteq G$ be the unique subgroup of G containing D and such that $K/D = K_0$. Let $\mathfrak{k} \subseteq \mathfrak{g}$ be the Lie algebra of K so that $\mathfrak{k}$ corresponds to a maximal compact subgroup of the adjoint group of $\mathfrak{g}$. We can choose K_0, and do so, so that the Cartan involution of $\mathfrak{g}$ defined by $\mathfrak{k}$ leaves $\mathfrak{h}$, the fixed reductive part of $\mathfrak{b}$, stable.

Now let $L = K \cap M$. Then the K-homogeneous space K/L is connected compact and has a unique K-invariant measure ω of total volume 1. Now let $\mathfrak{a}' \subseteq \mathfrak{a}'_{\mathbf{C}}$ be the real dual to $\mathfrak{a}$ and let $\rho \in \mathfrak{a}'$ be defined by $\langle \rho, x \rangle = \frac{1}{2}\, \mathrm{tr}\, \mathrm{ad}\, x \mid \mathfrak{n}$ for all $x \in \mathfrak{a}$.

Now if $\lambda \in \mathfrak{a}'_{\mathbf{C}}$ is of the form $\lambda = \rho + i\nu$ where $\nu \in \mathfrak{a}'$, let $\mathcal{H}_\nu$ be the Hilbert space of all $f \in X^{(\lambda)}$ such that

$$\int_{K/L} |f|^2 \omega < \infty.$$

The integral is well defined since $f \mid K$ is constant on the left cosets of L in M. Furthermore $\mathcal{H}_\nu$ is stable under the action of G defining a representation s_ν of G on $\mathcal{H}_\nu$ which is in fact unitary.

The family or series of unitary representations s_ν obtained as ν runs through $\mathfrak{a}'$ is called the principal series of class zero or simply, in our terminology, principal series.

It has been long conjectured that all members of the principal series are irreducible. For certain special groups (e.g. some classical groups) this has been proved. Using analytic methods (e.g. the Schwartz-Kernel theorem) Bruhat has proved irreducibility for the case where ν is a regular element of $\mathfrak{a}'$. If G is complex, Parthasarathy, Rao and Varadarajan [9] have proved irreducibility for all ν. Based on a similar approach we can prove

THEOREM 1. *For all G every member of the principal series s_ν is an irreducible unitary representation. Moreover s_ν is equivalent to s_{ν_1} if and only if ν and ν_1 are conjugate under the Weyl group.*

3. **The U-module and the critical strip.** 1. Theorem 1 is an application of the much more complete result which follows: The method used in the proof is purely algebraic and is modeled after [9] in the sense that it involves a detailed examination of a family of polynomial valued matrices. The result rests upon [6], [7] in the same sense that [9] rests upon [5].

Now for any $\lambda \in \mathfrak{a}'_C$ let X^λ be the subspace of all $f \in X^{(\lambda)}$ which are K-finite, i.e., $K \cdot f$ spans a finite dimensional space. The elements of X^λ are analytic functions (real sense) in G and although X^λ is no longer in general a G-module it is, however, a U-module where U is the universal enveloping algebra of $\mathfrak{g}$ over $\mathbf{C}$. Indeed if $f \in X^\lambda$ then $x \cdot f \in X^\lambda$ where $x \in \mathfrak{g}$ and $x \cdot f$ is the function on G given by

$$x \cdot f(g) = \frac{d}{dt} f(\exp -tx \cdot g) \Big|_{t=0}.$$

The next result gives the set of all $\lambda \in \mathfrak{a}'_C$ such that X^λ is an irreducible (in the usual algebraic sense) U-module.

Let $\Lambda \subseteq \mathfrak{a}'$ be the set of (sometimes referred to as restricted) roots for the action of $\mathfrak{a}$ on $\mathfrak{g}$ and for any $\phi \in \Lambda$ let $\mathfrak{g}^\phi \subseteq \mathfrak{g}$ be the corresponding root space. A lexicographical ordering may be introduced into $\mathfrak{a}'$ so that if Λ_+ is the set of positive roots then

$$\mathfrak{n} = \sum_{\phi \in \Lambda_+} \mathfrak{g}^\phi.$$

Now for any $\phi \in \Lambda$ the space $[\mathfrak{g}^\phi, \mathfrak{g}^{-\phi}] \cap \mathfrak{a}$ is one dimensional and is spanned by a unique element w_ϕ such that $\langle \phi, w_\phi \rangle = 1$. If $\phi \in \Lambda$ then $\phi/2$ or 2ϕ may be a root. Let Λ^1 be the set of all $\phi \in \Lambda$ such that $\phi/2$ is not a root (the set Λ^1 may be characterized as the set of roots which may be embedded in some simple system) and let Λ^1_+ be the positive elements in Λ^1. Of course if $\phi \in \Lambda^1_+$ then 2ϕ may be a root.

For any $\phi \in \Lambda^1_+$ let $T_\phi \subset \boldsymbol{R}$ be the open interval, symmetric about the origin, given by $T_\phi = \{t \in \mathbf{R} \mid |t| < \dim g^\phi/2$ if 2ϕ is not a root and $|t| < \dim g^\phi/2+1$ if 2ϕ is a root$\}$.

We now define a region in $\mathfrak{a}'_{\mathbf{C}}$, symmetric around the point ρ, which arises in a natural way from our theory and which plays a special role in the representation theory. We will call the region the critical strip, CS, and it is defined by

$$\mathrm{CS} = \{\lambda \in \mathfrak{a}'_{\mathbf{C}} \mid \mathrm{Re}\langle \lambda - \rho, w_\phi\rangle \in T_\phi \quad \text{for all } \phi \in \Lambda^1_+\}.$$

For $\mathrm{Sl}(2, \mathbf{R})$ it reduces to the strip of all $z\phi$, $z \in \mathbf{C}$, $(\phi) = \Lambda^1_+$, such that $0 < \mathrm{Re}\ z < 1$.

Now for any root $\phi \in \Lambda^1_+$ let m_ϕ be the integer or half-integer given by $m_\phi = (\dim \mathfrak{g}^\phi/2) + \dim \mathfrak{g}^{2\phi}$ and let $n_\phi = 1$ if 2ϕ is not a root and $n_\phi = 2$ if 2ϕ is a root.

It is somewhat easier to describe the set of all $\lambda \in \mathfrak{a}'_{\mathbf{C}}$ such that X^λ is not U-irreducible.

THEOREM 2. *Let $\lambda \in \mathfrak{a}'_{\mathbf{C}}$ be arbitrary then the U-module X^λ is not irreducible if and only if there exists a root $\phi \in \Lambda^1_+$ such that*

(1) $\langle \lambda - \rho, w_\phi\rangle \notin T_\phi$ *and*

(2) $\langle \lambda - \rho, w_\phi\rangle + m_\phi = 0 \mod n_\phi \mathbf{Z}$.

In particular X^λ is U-irreducible whenever λ lies in the critical strip CS.

3.2. A U-module X will be said to be admissible in case $X^{\mathfrak{k}}$, the annihilator of $\mathfrak{k}$ in X, is one dimensional. Let $\mathfrak{C}$ be the set of U equivalence classes $\{X\}$ of all admissible U-modules X.

Now for any $u \in U$ one knows there exists a unique element $p^u \in U(\mathfrak{a})$, (the enveloping algebra of $\mathfrak{a}$) such that $u - p^u \in \mathfrak{n}U + U\mathfrak{k}$. Also if $U^{\mathfrak{k}}$ is the centralizer of $\mathfrak{k}$ in U then the map $u \to p^u$ is a homomorphism on $U^{\mathfrak{k}}$. Now regarding $U(\mathfrak{a})$ as the polynomial ring on $\mathfrak{a}'_{\mathbf{C}}$ let $\mathfrak{C}_\lambda \subseteq \mathfrak{C}$, for any $\lambda \in \mathfrak{a}'_{\mathbf{C}}$, be the set of all $\{X\} \in \mathfrak{C}$ such that $u \cdot \xi = p^u(\lambda)\xi$ for all $u \in U^{\mathfrak{k}}$ where $X^{\mathfrak{k}} = (\xi)$.

Now if W is the Weyl group operating in $\mathfrak{a}'_{\mathbf{C}}$ let $\tilde{W}$ be the group of affine transformations in $\mathfrak{a}'_{\mathbf{C}}$ of the form $\lambda \to \tilde{\sigma}\lambda = \sigma(\lambda - \rho) + \rho$ where $\sigma \in W$. One has

PROPOSITION 1. *$\mathfrak{C}$ is a union of $\mathfrak{C}_\lambda$ over all $\lambda \in \mathfrak{a}'_{\mathbf{C}}$. Moreover if λ, $\nu \in \mathfrak{a}'_{\mathbf{C}}$ then $\mathfrak{C}_\lambda$ and $\mathfrak{C}_\nu$ are either disjoint or equal and they are equal if and only if λ and ν are $\tilde{W}$ conjugate.*

REMARK 2. One readily shows that for all $\lambda \in \mathfrak{a}'_{\mathbf{C}}$ one has $\{X^\lambda\} \in \mathfrak{C}_\lambda$.

3.3. With regard to irreducibility one has

PROPOSITION 2. *For any $\lambda \in \mathfrak{a}'_{\mathbf{C}}$ there exist a unique element $\{Z^\lambda\} \in \mathfrak{C}_\lambda$*

such that Z^λ is U-irreducible. Moreover any irreducible U-module Z such that $Z^{\mathfrak{k}} \neq 0$ is admissible and hence is equivalent to some Z^λ where λ is unique up to $\tilde{W}$ conjugacy in $\mathfrak{a}'_C$.

For any $\lambda \in \mathfrak{a}'_C$ Z^λ may be explicitly given by putting $Z^\lambda = U/L_\lambda$ where L_λ is the left ideal in U defined by

$$L_\lambda = \{u \in U \mid p^{(vu)_0}(\lambda) = 0 \quad \text{for all } v \in U\}.$$

Here $w_0 \in U^k$, for any $w \in U$, is the component of w in $U^{\mathfrak{k}}$ corresponding to the direct sum $U = [k, U] + U^k$.

REMARK 3. Note that by Theorem 2 one has that $\{X^\lambda\} = \{Z^\lambda\}$ if and only if λ does not satisfy the condition of Theorem 2. In particular the equality holds if $\lambda \in CS$.

3.4. Now let $u \to u^s$ be the conjugate-linear, antilinear $((uv)^s = v^s u^s)$ involution of U satisfying $x^s = -x$ for $x \in \mathfrak{g}$. A Hermitian bilinear form $((\xi, \eta) = \overline{(\eta, \xi)}, c(\xi, \eta) = (c\xi, \eta) = (\xi, \bar{c}\eta))$ on a U-module X will be said to be U-invariant in case $(u\xi, \eta) = (\xi, u^s\eta)$ for all $u \in U$, $\xi, \eta \in X$. If $\lambda \in \mathfrak{a}'_C$ its conjugate $\bar{\lambda} \in \mathfrak{a}'_C$ is defined relative to the real form $\mathfrak{a}'$ of $\mathfrak{a}'_C$.

PROPOSITION 3. *Let $\lambda \in \mathfrak{a}'_C$. Then there exists a nontrivial U-invariant Hermitian bilinear form on Z^λ if and only if λ and $-\bar{\lambda} + 2\rho$ are $\tilde{W}$ conjugate. Moreover in this case the form is unique up to a real scalar.*

If λ and $-\bar{\lambda} + 2\rho$ are $\tilde{W}$ conjugate then the U-invariant Hermitian form may be given explicitly by

$$(u \cdot \xi_\lambda, v \cdot \xi_\lambda) = p^{(v^s u)_0}(\lambda)$$

where $u, v \in U$ and $\xi_\lambda \in (Z^\lambda)^k$ is the image of 1 in $Z^\lambda = U/L_\lambda$. We denote the form by B_λ and we will be concerned as to when it is positive definite so that Nelson's theorem, [8], can be applied.

4. **The polynomial valued matrices P^γ, Q^γ and R^γ.** 1. Now let G_θ be the set of all elements of the adjoint group of $\mathfrak{g}_C$ leaving $\mathfrak{g}$ stable. Note that the considerations above may be applied to G_θ instead of G. We let K_θ and M_θ be the subgroups of G_θ corresponding to the subgroups K and M of G. Also let Γ be the set of all equivalence classes $\{V\}$ of all irreducible (finite dimensional, continuous) K_θ-modules V such that $V^{M_\theta} \neq 0$.

For any $\gamma \in \Gamma$ fix a K_θ-module V_γ in the class γ and let $l(\gamma) = \dim V_\gamma^{M_\theta}$.

Also let $H \subseteq U$ be the subspace spanned by all powers x^j, $j = 0, 1, \cdots$, for all nilpotent elements x in the complexification $\mathfrak{p}_C$ of $\mathfrak{p}$. Then if $\gamma \in \Gamma$ is arbitrary one knows from [6] that $\dim E_\gamma = l(\gamma)$ where $E_\gamma = \operatorname{Hom}_{K_\theta}(V_\gamma, H)$.

Now if v_i, $i=1, 2, \cdots, l(\gamma)$, is a basis of $V_\gamma^{M_\theta}$ and ϵ_j, $j=1, 2, \cdots, l(\gamma)$, is a basis of E_γ we define an $l(\gamma)\times l(\gamma)$ matrix P^γ with coefficients in $U(a)$ by defining $P_{ij}^\gamma = p^{\epsilon_j(v_i)}$. It is clear that a change of basis of E_γ and $V_\gamma^{M_\theta}$ only changes P^γ into a matrix of the form $AP^\gamma B$ where A and B are nonsingular matrices with coefficients in $\mathbf{C}$. In particular $p^\gamma = \det P^\gamma$ is an element of $U(\mathfrak{a})$ which, up to a scalar multiple, is independent of the choice of the bases. Also for any $\lambda \in \mathfrak{a}_{\mathbf{C}}$ the rank of the scalar matrix $P^\gamma(\lambda)$ is also independent of the basis.

Now if $\lambda \in \mathfrak{a}_{\mathbf{C}}'$ then the induced module X^λ is not always U-irreducible (see Theorem 2). In fact if 1_λ is the unique element of X^λ which as a function on G equals the constant 1 on K (and hence is a generator of $(X^\lambda)^{\mathfrak{k}}$) then in general the cyclic submodule $U\cdot 1_\lambda = X_e^\lambda \subseteq X^\lambda$ is not equal to X^λ. A measure of its failure to equal X^λ is given in the next proposition and is expressed in terms of the matrices P^γ.

If $\gamma \in \Gamma$ and X_γ^λ is the finite dimensional space of all $f\in X^\lambda$ which transforms under K_θ according to γ then it is obvious that X^λ is a direct sum of X_γ^λ over all γ. The same is true of X_e^λ. In X^λ the multiplicity of γ is obviously $l(\gamma)$. In X_e^λ, however, the question of multiplicity is less trivial.

PROPOSITION 4. *For any $\lambda \in \mathfrak{a}_{\mathbf{C}}'$ and $\gamma\in\Gamma$ the multiplicity of γ in the U-cyclic submodule $U\cdot 1_\lambda = X_e^\lambda$ of the induced module X^λ is exactly the rank of the matrix $P^\gamma(\lambda)$.*

In particular one has

COROLLARY TO PROPOSITION 4. *For any $\lambda\in\mathfrak{a}_{\mathbf{C}}'$ one has $X^\lambda = U\cdot 1_\lambda$ if and only if $p^\gamma(\lambda)\neq 0$ for all $\gamma\in\Gamma$.*

4.2. Now for any $\gamma\in\Gamma$ one defines another $l(\gamma)\times l(\gamma)$ matrix Q^γ with coefficients in $U(\mathfrak{a})$ by the relation $Q^\gamma(\lambda) = P^\gamma(-\bar\lambda+2\rho)^*$ for all $\lambda\in a_{\mathbf{C}}'$. Here * denotes the Hermitian adjoint of the matrix.

Now finally let R^γ be the $l(\gamma)\times l(\gamma)$ matrix with coefficients in $U(\mathfrak{a})$ defined by $R^\gamma = Q^\gamma P^\gamma$. Now whereas the P^γ describe the structure of U-module X_e^λ the R^γ describe the structure of the irreducible U modules Z^λ. Moreover it determines the signature of the U-invariant Hermitian structure on Z^λ when such a structure exists (that is when λ and $-\bar\lambda+2\rho$ are $\tilde W$ conjugate). Also R^γ is invariant under the action of $\tilde W$.

The U-modules Z^λ are in a natural way K_θ-modules since the ideals $L_\lambda\subseteq U$ are stable under the adjoint action of K_θ on U. It follows that Z^λ is a direct sum of the finite dimensional subspaces Z_γ^λ, $\gamma\in\Gamma$, of all elements in Z^λ which transform under K_θ according to the representation class γ.

PROPOSITION 5. *For any* $\lambda \in \mathfrak{a}'_C$, $\gamma \in \Gamma$, *the multiplicity of* γ *in the irreducible U-module* Z^λ *is exactly the rank of the matrix* $R^\gamma(\lambda)$. *Moreover for any* $\sigma \in W$ *one has* $R^\gamma(\lambda) = R^\gamma(\tilde{\sigma}\lambda)$.

Finally if λ *and* $-\bar{\lambda}+2\rho$ *are* $\tilde{W}$ *conjugate (so that the Hermitian bilinear form* B_λ *on* Z^λ *(see* §3.4) *is defined) then* $R^\gamma(\lambda)$ *is a Hermitian matrix. Furthermore* $R^\gamma(\lambda)$ *is positive semidefinite if and only if* $B_\lambda | Z^\lambda_\gamma$ *is positive definite. In particular then (since the* Z^λ_γ, *over all* $\gamma \in \Gamma$, *are in any case* B_λ *orthogonal)* B_λ *defines a pre-Hilbert space structure on* Z^λ *if and only if* $R^\gamma(\lambda)$ *is positive semidefinite for all* $\gamma \in \Gamma$.

Now for any $\gamma \in \Gamma$ let $r^\gamma = \det R^\gamma$. One thus has that $r^\gamma(\lambda) \neq 0$ for all γ if and only if $(Z^\lambda) = (X^\lambda)$. That is, one has

COROLLARY TO PROPOSITION 5. *For any* $\lambda \in \mathfrak{a}'_C$ *the induced U-module* X^λ *is U-irreducible if and only if* $r^\gamma(\lambda) \neq 0$ *for all* $\gamma \in \Gamma$.

Let $\gamma \in \Gamma$ be arbitrary and put $q^\gamma = \det Q^\gamma$ so that $r^\gamma = q^\gamma p^\gamma$. But since $Q^\gamma(\lambda) = P^\gamma(-\bar{\lambda}+2\rho)^*$ one has that

$$r^\gamma(\lambda) = \overline{p^\gamma}(-\lambda + 2\rho)p^\gamma(\lambda)$$

where conjugation in $U(\mathfrak{a})$ is the conjugate linear automorphism of $U(\mathfrak{a})$ which is the identity on $\mathfrak{a}$. The determination of the r^γ thus depends upon the knowledge of the p^γ. It is the latter that we can explicitly give. Practically all the results announced here follow from the knowledge of the p^γ. However, one shows that the determination of the p^γ reduces to the split rank 1 case.

5. The split rank one case and 2-transitivity on a unit sphere. 1. In this section we shall assume that dim $\mathfrak{a} = 1$. We first need an explicit description of Γ. We find that the elements of Γ are parameterized by a pair of integers $i(\gamma)$ and $j(\gamma)$ in a manner now to be described.

Let $\mathfrak{s}_C \subseteq \mathfrak{g}_C$ be a normal principal TDS in the sense of [7] which is stable under conjugation over $\mathfrak{g}$, and let $\mathfrak{s} = \mathfrak{g} \cap \mathfrak{s}_C$. Let S_θ be the intersection of G_θ with the subgroup of the adjoint group of $\mathfrak{g}_C$ corresponding to $\mathfrak{s}_C$. The group S_θ is centerless and is never connected. In fact if $\mathfrak{t}$ is the one dimensional algebra given by $\mathfrak{t} = \mathfrak{s} \cap \mathfrak{k}$ and T_θ is the subgroup of S_θ normalizing $\mathfrak{t}$ then T_θ is isomorphic to the orthogonal group $O(2, \boldsymbol{R})$.

Now let $b \in T_\theta$ be a fixed element not in the identity component of

T_θ. Then $b^2=1$. The subspace $[\mathfrak{t}, \mathfrak{p}]$ is stable under b and we let $\mathfrak{d}$ be the $+1$ eigenspace of b in $[\mathfrak{t}, \mathfrak{p}]$.

REMARK 4. We shall not use the fact but it can be shown that the only possible dimensions of $\mathfrak{d}$ are 1, 2, 4 or 8. Moreover there is a natural structure of a division algebra on $\mathfrak{d}$ with the identity lying in the one dimensional space $\mathfrak{d}\cap\mathfrak{s}$. Of course the division algebras in the four cases are

(1) the real numbers,
(2) the complex numbers,
(3) the quaterions and
(4) the Cayley numbers.

They occur, respectively, in the case when G is a suitable real form of

(1) the complex orthogonal group,
(2) the complex special unitary group,
(3) the complex symplectic group and
(4) F_4.

One also can show that dim $\mathfrak{d}=\dim \mathfrak{g}^{2\alpha}+1$ where $(\alpha)=\Lambda_+^1$ which is of course consistent with the known fact that the dim $\mathfrak{g}^{2\phi}$ for any root ϕ takes on only the values 0, 1, 3, 7.

5.2. Now let SO($\mathfrak{d}$) and O($\mathfrak{d}$) be, respectively, the special orthogonal group and orthogonal groups operating on $\mathfrak{d}$ relative to the restriction of the Killing form to $\mathfrak{d}$. Consider the (complex) symmetric algebra $\mathcal{S}(\mathfrak{d})=\mathcal{S}$ over $\mathfrak{d}$ and its homogeneous components $\mathcal{S}_k$.

Of course the action of O($\mathfrak{d}$) on $\mathfrak{d}$ extends to a group of automorphisms of $\mathcal{S}$. The homogeneous irreducible components of O($\mathfrak{d}$) on $\mathcal{S}$ are of course classical and well known. They are described as follows: Let q be the nonzero (quadratic) invariant of O($\mathfrak{d}$) in $\mathcal{S}_2$ and for any j let $L_j\subseteq\mathcal{S}_j$ be the space of harmonic elements in $\mathcal{S}_j$ (harmonic polynomials when $\mathcal{S}$ is regarded as the polynomial ring on the dual $\mathfrak{d}'$).

Of course $L_j=0$ if $j\geqq 2$ when dim $\mathfrak{d}=1$. In what follows i and j take all nonnegative integral values except when dim $\mathfrak{d}=1$ (orthogonal case) where j is restricted to 0 and 1. With such a restriction the most general homogeneous irreducible component of O($\mathfrak{d}$) on $\mathcal{S}$ is uniquely of the form q^iL_j. For SO($\mathfrak{d}$) the same statement is true except that when dim $\mathfrak{d}=2$ (unitary case) the two dimensional space L_j, $j>0$, is uniquely a direct sum of two irreducible components L_j^+ and L_j^- and hence the most general homogeneous irreducible component of SO($\mathfrak{d}$) on $\mathcal{S}$ is uniquely of the form q^iL_0, $q^iL_j^+$ or $q^iL_j^-$.

Now let J be the centralizer of T_θ in K_θ. Clearly $\mathfrak{d}$ is stable under J inducing a map $J\rightarrow$O($\mathfrak{d}$) and hence $\mathcal{S}$ is naturally a J-module.

A crucial fact is given in the second paragraph of

THEOREM 3. *The group J is connected and hence its action on $\mathfrak{d}$ induces a map $J \to \mathrm{SO}(\mathfrak{d})$. Moreover, if* dim $\mathfrak{d}=1, 2$ *this map is surjective and if* dim $d>2$ *then J is 2-transitive on the unit sphere of $\mathfrak{d}$ (e.g. if* dim $d=8$ *then J operates as* Spin 7).

Furthermore (as a consequence of the 2-transitivity) in all cases the homogeneous irreducible components of J on $\mathcal{S}$ are exactly the same as the homogeneous irreducible components of $\mathrm{SO}(\mathfrak{d})$ *on $\mathcal{S}$. That is, the $q^iL_j(q^iL_j^+, q^iL_j^-$ and q^iL_0 if* dim $\mathfrak{d}=2$) *are all irreducible and, when of the same degree, are inequivalent under the action of J.*

5.3. Now in it there exists a unique, up to sign, element z such that the eigenvalues of ad $z|\mathfrak{s}_C$ are 0, ± 1. If V is any finite dimensional K_θ module it is of course a $\mathfrak{k}_C$ module and it follows easily that the eigenvalues of z in V are integers.

The structure of Γ arises from

THEOREM 4. *Let $\gamma \in \Gamma$ be arbitrary and let k be the highest eigenvalue of z operating in V_γ and let $V_\gamma(z)$ be the corresponding eigenspace. Then $V_\gamma(z)$ is stable and irreducible under the action of J. Moreover k is a nonnegative integer and there exists a unique J-irreducible component $Y_\gamma \subseteq \mathcal{S}_k$ which is equivalent to the J-module $V_\gamma(z)$. Furthermore the correspondence $\gamma \to Y_\gamma$ thus set up between Γ and the set of all homogeneous irreducible components of J on $\mathcal{S}$ is a bijection.*

5.4. Now if $\gamma \in \Gamma$ then Y_γ by Theorem 4 is necessarily in any case a subspace of q^iL_j for a unique i and j. We put $i=i(\gamma)$ and $j=j(\gamma)$ defining the two parameters mentioned in the beginning of §5. Note that $k=2i(\gamma)+j(\gamma)$ where $Y_\gamma \subseteq \mathcal{S}_k$.

Now Λ_+^1 has only one element α and hence there is a unique element $w \in \mathfrak{a}$ such that $\langle \alpha, w \rangle = 1$. Thus $U(\mathfrak{a})$ is the polynomial ring in w and hence if $p \in U(\mathfrak{a})$ we may write $p = c_0 \prod_{i=1}^n (w+c_i)$ where $c_j \in \mathbf{C}$.

We can now state

THEOREM 5. *Let $\gamma \in \Gamma$ and let $i=i(\gamma), j=j(\gamma)$. Then up to a nonzero scalar multiple*

$$p^\gamma = [w(w+2) \cdots (w+2(i+j)-2)] \cdot [(w-s+1)(w-s+3) \cdots (w-s+2i-1)]$$

where $s = \dim \mathfrak{g}^{2\alpha}$ ($=\dim \mathfrak{d} - 1$ *in the notation of* §5.1).

Any element λ in $\mathfrak{a}_C'$ is of the form $c\alpha$ where $c \in \mathbf{C}$ and any $\lambda \in \mathfrak{a}'$ (the real elements) is of the form $t\alpha$ where $t \in \mathbf{R}$. The ordering in $\mathfrak{a}'$ is of course consistent with the natural ordering $\mathbf{R}$. An element $\lambda \in \mathfrak{a}_C'$ is called a root of p^γ if $p^\gamma(\lambda)=0$.

REMARK 5. Note that by Theorem 5 all the roots of p^γ are not only real but of the form $n\alpha$ where n is an integer. Furthermore n is in fact an even integer if $s\neq 0$ since $s\neq 0$ implies s is odd (it is restricted to 1, 3 or 7).

Next note that the maximal root of p^γ is 0 when $s=0$ and $s-1$ when $s>0$. Since $\rho=(s+\dim \mathfrak{g}^\alpha/2)\alpha$ it follows that ρ is larger than the largest root and in fact between the two there is an interval of "width" $\dim \mathfrak{g}^\alpha/2$ if $s=0$ and $\dim \mathfrak{g}^\alpha/2+1$ if $s>0$.

Now $q^\gamma(\lambda)=\overline{p^\gamma}(-\lambda+2\rho)$. But since p^γ is real it follows that q^γ is obtained by simply reflecting p^γ through ρ.

COROLLARY TO THEOREM 5. *Using the notation of Theorem* 5, *if* p^γ *is normalized so that the scalar multiple is* 1 *one has, where* $m=\dim \mathfrak{g}^\alpha/2+s$,

$$\begin{aligned} q^\gamma = (-1)^{2i+j}[(w-2m)(w-2m-2)\cdots(w-2m-2(i+j)+2)] \\ \cdot[(w-2m+s-1)(w-2m+s-3)\cdots \\ \cdot(w-2m+s-2i+1)]. \end{aligned}$$

REMARK 6. Since $r^\gamma=q^\gamma p^\gamma$ one obtains an explicit expression for r^γ and like p^γ all the roots of r^γ are real and of the form $n\alpha$ where n is an integer and if $s>0$, n is necessarily an even integer.

One notes also that independent of γ, r^γ has no root of the form $\rho+c\alpha$ where $|\mathrm{Re}\, c|<\dim \mathfrak{g}^\alpha/2$ if $s=0$ and $|\mathrm{Re}\, c|<\dim \mathfrak{g}^\alpha/2+1$ if $s>0$. This set is what we have called the critical strip.

5.5. The situation in the case at hand ($\dim \mathfrak{a}=1$) is even nicer than indicated by the last two results. Not only do we know p^γ, q^γ and r^γ, but also, P^γ, Q^γ and R^γ. This is a consequence of

THEOREM 6. *In case* $\dim \mathfrak{a}=1$ (*our present assumption*) *one has* $l(\gamma)=1$ *for all* γ. *Thus* $P^\gamma=p^\gamma$, $Q^\gamma=q^\gamma$ *and* $R^\gamma=r^\gamma$.

The proof of Theorem 6 comes out in the course of proving Theorem 4.

6. The determination of p^γ, q^γ and r^γ in the general case. 1. We now return to the general case, that is, no assumptions are made about $\dim \mathfrak{a}$.

Let F be the group (of order 2^r, $r=\dim \mathfrak{a}$) of all elements of order 2 in that subgroup of the adjoint group of g_C whose Lie algebra is $\mathfrak{a}_C$. One has $F\subseteq K_\theta$.

Now for any $\phi\in\Lambda^1_+$ let $\mathfrak{g}_\phi$ be the commutator subalgebra of the reductive subalgebra $\mathfrak{a}+\mathfrak{m}+\sum_{j=-2}^{2}\mathfrak{g}^{j\phi}$. Let G_ϕ be the subgroup of G_θ generated by F and the connected subgroup (normalized by F) of

G_θ whose Lie algebra is g_ϕ. The Lie algebra of G_ϕ is again $\mathfrak{g}_\phi$ and we may apply all the preceding results to G_ϕ instead of G. Here the subalgebras of $\mathfrak{g}_\phi$ corresponding to $\mathfrak{k}$, $\mathfrak{a}$ and $\mathfrak{n}$ are just the intersection of these algebras with $\mathfrak{g}_\phi$ and are denoted by $\mathfrak{k}_\phi$, $\mathfrak{a}_\phi$ and $\mathfrak{n}_\phi$. The subgroups corresponding to K and M are denoted by K_ϕ and M_ϕ. Furthermore this is the split rank 1 case since $\dim \mathfrak{a}_\phi = 1$ and hence also the results of §5 apply where the role of w is played by the unique element $w_\phi \in \mathfrak{a}_\phi \subseteq \mathfrak{a}$ such that $\langle \phi, w_\phi \rangle = 1$.

Now let $K_{\phi,\theta}$, $M_{\phi,\theta}$ and Γ_ϕ play the roles for G_ϕ what K_θ, M_θ and Γ play for G. Now K_ϕ operates on $\mathfrak{g}_C$ and its restriction to $\mathfrak{g}_{\phi,C}$ induces a homomorphism $K_\phi \to K_{\phi,\theta}$. However, since $F \subseteq K_\phi$ this map is a surjection. Moreover the kernel of this map is central in K_ϕ and hence is a normal subgroup of M_ϕ. But since the latter maps onto $M_{\phi,\theta}$ it follows that we can regard Γ_ϕ as the set of all equivalence classes of (finite dimensional, continuous) irreducible K_ϕ modules V such that $V^{M_\phi} = 0$. Also by Theorem 6 one has that $\dim V_\delta^{M_\phi} = 1$ for all $\delta \in \Gamma_\phi$.

Now for any $\delta \in \Gamma_\phi$ the integral functions $i(\delta)$ and $j(\delta)$ are defined (as in §5) and hence p^δ is given by Theorem 5 with w_ϕ substituted for w.

It is necessary to consider a slight shift in p^δ (since ϕ is not in general a simple root in Λ_+^1). Let d_ϕ be the integral or half-integral scalar given by $d_\phi = m_\phi - \langle \rho, w_\phi \rangle$ where $m_\phi = \dim \mathfrak{g}^\phi/2 + \dim \mathfrak{g}^{2\phi}$ ($d_\phi = 0$ if ϕ is simple) and let $p^{(\delta)}$ be the polynomial in w_ϕ obtained from p^δ by substituting $w_\phi + d_\phi$ for w_ϕ. Thus if $i = i(\delta)$, $j = j(\delta)$ one has explicitly

$$\begin{aligned} p^{(\delta)} = {} & [(w_\phi + d_\phi)(w_\phi + d_\phi + 2) \cdots (w_\phi + d_\phi + 2(i+j) - 2)] \\ & \cdot [(w_\phi + d_\phi - s_\phi + 1)(w_\phi + d_\phi - s_\phi + 3) \cdots \\ & \qquad \cdot (w_\phi + d_\phi - s_\phi + 2i - 1)] \end{aligned}$$

where $s_\phi = \dim \mathfrak{g}^{2\phi}$. Also $q^{(\delta)}$ is obtained by substituting $-w_\phi + 2\langle \rho, w_\phi \rangle$ for w_ϕ in p^δ and one puts $r^{(\delta)} = q^{(\delta)} p^{(\delta)}$.

6.2. Now let $\gamma \in \Gamma$ be arbitrary we wish to determine p^γ, q^γ and $r^\gamma \in U(\mathfrak{a})$ explicitly. Now let $\phi \in \Lambda_+^1$ and since $K_\phi \subseteq K_\theta$ we can consider the K_ϕ-submodule $V_\gamma(\phi)$ of V_γ generated by $V_\gamma^{M_\theta}$. Now since $M_\phi \subseteq M_\theta$ it follows from Theorem 6 that in a complete reduction of $V_\gamma(\phi)$ into irreducible K_ϕ components $V_\gamma^i(\phi)$ there are exactly $l(\gamma)$ components and that each $V_\gamma^i(\phi)$ belongs to a class $\delta^i(\phi) \in \Gamma_\phi$. Thus γ defines elements $\delta^i(\phi) \in \Gamma_\phi$, $i = 1, 2, \cdots, l(\gamma)$. We now state

THEOREM 7. *For any* $\gamma \in \Gamma$ *one has, up to a nonzero scalar multiple,*

$$p^{\gamma} = \prod_{\phi \in \Lambda^1_+} \prod_{i=1}^{l(\gamma)} p^{(\delta^i(\phi))}.$$

Similarly q^{γ} *and* r^{γ} *are respectively the products of* $q^{(\delta^i(\phi))}$ *and* $r^{(\delta^i(\phi))}$ *over all* $\phi \in \Lambda^1_+$ *and* $i = 1, 2, \cdots, l(\gamma)$.

REMARK 7. From the expression given for $p^{(\delta)}$, $q^{(\delta)}$ and $r^{(\delta)}$ one notes that all the prime factors of p^{γ}, q^{γ} and r^{γ} are polynomials of the first degree.

6.3. Theorems 1 and 2 are consequences of Theorem 7. However, there are other conclusions. Let C (complex Weyl chamber) be the set of all $\lambda \in \mathfrak{a}'_c$ such that Re $\langle \lambda - \rho, w_{\phi} \rangle \geqq 0$ for all $\phi \in \Lambda^1_+$. Every $\lambda \in \mathfrak{a}'_c$ is $\tilde{W}$-conjugate to an element in C or its negative $-C$. As a consequence of the explicit expressions of p^{γ} and q^{γ} one has

COROLLARY TO THEOREM 7. *If* $\lambda \in C$ *then* $p^{\gamma}(\lambda) \neq 0$ *for every* $\gamma \in \Gamma$ *and hence, for all,* γ $P^{\gamma}(\lambda)$ *is nonsingular.*

On the other hand if $\lambda \in -C$ *then* $q^{\gamma}(\lambda) \neq 0$ *for all* $\gamma \in \Gamma$ *so that* $Q^{\gamma}(\lambda)$ *is nonsingular and hence*

$$\text{rank } R^{\gamma}(\lambda) = \text{rank } P^{\gamma}(\lambda).$$

Now whereas the irreducible U-module Z^{λ} depends only on the $\tilde{W}$-orbit of λ this is not true of the induced module X^{λ} or of the cyclic submodule $X^{\lambda}_e = U \cdot 1_{\lambda}$. As a consequence of the previous corollary and Propositions 4 and 5 one has

THEOREM 8. *For any* $\lambda \in C$ *one has* $X^{\lambda}_e = X^{\lambda}$ *so that the induced module* X^{λ} *is* U*-cyclic. Moreover if* Y *is any* U*-module cyclically generated by* $Y^{\mathfrak{k}}$, *where* dim $Y^{\mathfrak{k}} = 1$, *then* Y *is equivalent to a quotient of* X^{λ} *for some* $\lambda \in C$.

For any $\lambda \in -C$ *the cyclic submodule* $X^{\lambda}_e \subseteq X^{\lambda}$ *is* U*-irreducible and one obtains, up to equivalence, all irreducible* U*-modules* Z, *such that* $Z^{\mathfrak{k}} \neq 0$, *in this way.*

7. **Unitary representations and the complementary series.** 1. Now let $\mathfrak{a}^*$ be the set of all $\lambda \in \mathfrak{a}'_c$ such that

(1) λ and $-\bar{\lambda} + 2\rho$ are $\tilde{W}$-conjugate (so that the U-invariant Hermitian bilinear form B_{λ} exists on Z^{λ}) and

(2) B_{λ} is positive definite on Z^{λ}, (therefore inducing a pre-Hilbert structure on Z^{λ}).

Now let $\lambda \in \mathfrak{a}^*$ and let $\mathcal{H}^{\lambda}$ be the Hilbert space completion of Z^{λ}. It is easy to see that Nelson's theorem (see [8]) applies defining a unique unitary representation π^λ of G in $\mathcal{H}^\lambda$ whose differential induces the given U-module structure on Z^λ.

Irreducibility follows easily. Also completeness follows using well-known results of Harish-Chandra, [2], relating the action of $\mathfrak{g}$ with that of G. That is, one has

PROPOSITION 6. *For any $\lambda \in a^*$, π^λ is an irreducible unitary representation of G. Moreover π^λ is equivalent to π^ν if and only if λ and ν are $\tilde{W}$-equivalent. Furthermore for any $\lambda \in a^*$ one has $(H^\lambda)^K \neq 0$ and, up to equivalence, one thus obtains all irreducible unitary representations of G on a Hilbert space H such that $H^K \neq 0$.*

7.2. The obvious question is: what is the subset $\mathfrak{a}^* \subseteq \mathfrak{a}'_c$? Condition (1) for an element $\lambda \in \mathfrak{a}'_c$ to lie in $\mathfrak{a}^*$ is easy enough to deal with. If $\mu = \lambda - \rho$ this is the condition that μ and $-\bar{\mu}$ are W-conjugate. That is, writing $\mu = \mu_1 + i\mu_2$ where $\mu_1, \mu_2 \in \mathfrak{a}'$ then condition (1) is that there should exist $\sigma \in W$ such that $\sigma\mu_1 = -\mu_1$ and $\sigma\mu_2 = \mu_2$.

If condition (1) is satisfied for $\lambda \in \mathfrak{a}'_c$ then as we noted above the matrix $R^\gamma(\lambda) = P^\gamma(-\bar{\lambda} + 2\rho)^* P^\gamma(\lambda)$ is Hermitian for all $\gamma \in \Gamma$ (see Proposition 5). By Proposition 5 condition (2) is that the Hermitian matrix $R^\gamma(\lambda)$ should be positive semidefinite for all $\gamma \in \Gamma$.

But now if $\lambda = \rho$ then $R^\gamma(\rho) = P^\gamma(\rho)^* P^\gamma(\rho)$ and hence $\rho \in \mathfrak{a}^*$. On the other hand, if $\lambda \in \mathfrak{a}'_c$ only satisfies condition (1) then any element on the line segment $\lambda(t) = t\lambda + (1-t)\rho$ also satisfies condition (1) and hence $R^\gamma(\lambda(t))$ is a continuous curve of Hermitian matrices. But now $\rho \in \mathrm{CS}$, the critical strip, and if $\lambda \in \overline{\mathrm{CS}}$ (the closure of CS) then $\lambda(t) \in \mathrm{CS}$ for all $0 \leqq t < 1$ and hence, for such t, $R^\gamma(\gamma(t))$ is a curve of nonsingular Hermitian matrices. But $R^\gamma(\gamma(0))$ is positive definite and since the eigenvalues cannot change sign along $\lambda(t)$ for $0 \leqq t < 1$ it follows that $R^\gamma(\lambda)$ is positive semidefinite. Thus for elements $\lambda \in \overline{\mathrm{CS}}$ condition (1) is necessary and sufficient for $\lambda \in \mathfrak{a}^*$. That is, we have proved

THEOREM 9. *Let λ be an arbitrary element in the closure of the critical strip (see §3). Then the irreducible U-module Z^λ corresponds to a (necessarily irreducible) unitary representation of G if and only if there exists σ as in the Weyl group such that $\sigma\mu_1 = -\mu_1$ and $\sigma\mu_2 = \mu_2$ where $\mu_1, \mu_2 \in \mathfrak{a}'$ and $\lambda - \rho = \mu_1 + i\mu_2$.*

REMARK 8. To clarify the word "corresponds" one recalls that to any unitary representation of G on a Hilbert space $\mathcal{H}$ there is induced a U-module structure on the space $\mathcal{H}^\infty$ of all C^∞ elements in $\mathcal{H}$. We will say that a U-module X corresponds to the given unitary representation of G in case X is U-equivalent to a U-submodule Y of $\mathcal{H}^\infty$

where Y is dense in $\mathcal{H}$. However, if Z^λ corresponds to the unitary G module $\mathcal{H}$ it follows easily from [2] that

(1) the representation of G is irreducible,

(2) all the elements of Y are well behaved and

(3) Y is the set of all K-finite vectors in $\mathcal{H}$.

7.3. Now among all the elements of $\overline{\text{CS}}$ which satisfy (condition (1) of §7.2) the condition of Theorem 9 there are two extreme types. That is, if $\lambda \in \overline{\text{CS}}$ and $\lambda - \rho = \mu = \mu_1 + i\mu_2$ then

(a) if μ is pure imaginary, i.e. if $\mu_1 = 0$, λ automatically satisfies condition (1) and hence $\lambda = \rho + i\mu_2 \in \mathfrak{a}^*$. Here, however, comparing with §2, π^λ is just an element of the principal series (π^λ is equivalent to s_{μ_2}) and one obtains all the elements of the principal series this way.

Next

(b) if μ is real, i.e. $\mu_2 = 0$, then condition (1) of §7.2 reduces to the condition that μ and $-\mu$ are W-conjugate.

We will call the family of unitary representations π^λ, where this condition holds, the complementary series since it is a natural generalization of such a designated series when $G = \mathrm{Sl}(2, \boldsymbol{R})$. In summary let CI (the critical interval) be the set of all $\lambda \in \mathfrak{a}'$ such that $\langle \lambda - \rho, w_\phi \rangle \in T_\phi$ for all $\phi \in \Lambda_+^1$ where $T_\phi = \{t \in \boldsymbol{R} \mid |t| < \dim \mathfrak{g}^\phi/2$ if $2\phi \in \Lambda$ and $|t| < \dim \mathfrak{g}^\phi/2 + 1$ if $2\phi \in \Lambda\}$.

Corollary to Theorem 9. *Let $\lambda \in \overline{CI}$ then the irreducible U-module Z^λ corresponds (see Remark 7) to a unitary representation of G (necessarily π^λ) if and only if μ and $-\mu$ are W-conjugate where $\mu = \lambda - \rho$.*

The family of unitary representations π^λ for such λ is called the complementary series.

Remark 9. Note that if $-1 \in W$ then all the elements of $\overline{\text{CI}}$ satisfy the condition of the corollary above.

In general we have not as yet been able to determine all the elements of $\mathfrak{a}^*$ lying outside of $\overline{\text{CS}}$.

For the split rank 1 case, however, $\mathfrak{a}^*$ can be completely determined giving therefore in that case the set of all irreducible unitary representations of G with a nonzero K-fixed vector. The point is that if $\dim \mathfrak{a} = 1$ then by Theorem 6, $R^\gamma = r^\gamma$ and using Remark 5 it is easy to determine the set of all $\lambda \in \mathfrak{a}'_C$ such that $r^\gamma(\lambda) \geqq 0$ for all $\gamma \in \Gamma$. It turns out that except for $\lambda = 0$ or 2ρ the most general such λ lies in $\overline{\text{CS}}$. But λ satisfies condition (1) of §7.3 if and only if $\lambda - \rho$ is either real or pure imaginary. This proves

Theorem 10. *Assume* $\dim \mathfrak{a} = 1$. *Then except for possibly the identity representation (which may be a member of the complementary series)*

every irreducible unitary representation of G with a nonzero K-fixed vector is either a member of the principal series or the complementary series. Furthermore (since $-1\in W$) the complementary series is defined for all $\lambda\in\overline{CI}$.

REMARK 10. Note that Theorem 10 points up a single irreducible unitary representation, distinguished in the following sense: The complement of CI in $\overline{\mathrm{CI}}$ consists of 2 points which are conjugate under $\tilde{W}$ and hence this orbit defines a single irreducible unitary representation π^λ of G. In case dim $\mathfrak{g}^{2\alpha}=0$, 1 (the orthogonal and unitary case), where $(\alpha)=\Lambda_+^1$, this representation is nothing more than the trivial one. However, if dim $\mathfrak{g}^{2\alpha}=3$, 7 (the symplectic and F_4 cases) this representation is not trivial and it is distinguished among all nontrivial irreducible unitary representations of G with a K-fixed vector in that it is the only one such that the subspace of K-finite vectors, Z^λ, is not U-equivalent to the induced module X^λ, i.e. X^λ is not U-irreducible.

One notes also that in the latter case and only in this case (dim $g^{2\alpha}=3$, 7) the identity representation is not a member of the complementary series and, hence, is in fact isolated from all other unitary representations. This should perhaps be related with recent results of Kajdan. See [4].

REFERENCES

1. François Bruhat, *Sur les representation induites des groupes de Lie*, Bull. Soc. Math. France **84** (1956), 97–205.

2. Harish-Chandra, *Representations of a semi-simple Lie group on a Banach Space.* I, Trans. Amer. Math. Soc. **75** (1953), 185–243.

3. I. M. Gel'fand and M. I. Graev, *Unitary representations of the real unimodular group (principal nondegenerate series)*, Izv. Akad. Nauk SSSR Ser. Mat. **17** (1953), 189–248; English transl., Amer. Math. Soc. Transl. (2) **2** (1956), 147–205.

4. D. A. Kajdan, *Sur les relations entre l'espace dual d'un groupe et la structure de ses sous-groupes fermês*, Funkcional. Anal. i Priložen. **1** (1967), 71–74.

5. Bertram Kostant, *Lie group representations on a polynomial ring*, Amer. J. Math. **85** (1963), 327–404.

6. Bertram Kostant and Stephen Rallis, *On representation associated with symmetric spaces*, Bull. Amer. Math. Soc. **75** (1969), 884–888.

7. ———, *On orbits associated with symmetric spaces*, Bull Amer. Math. Soc. **75** (1969), 879–883.

8. Edward Nelson, *Analytic vectors*, Ann. of Math. (2) **70** (1959), 572–615.

9. K. R. Parthasarathy, R. Ranga Rao and V. S. Varadarajan, *Representations of complex semi-simple Lie groups and Lie algebras*, Ann. of Math. (2) **85** (1967), 383–429.

MASSACHUSETTS INSTITUTE OF TECHNOLOGY, CAMBRIDGE, MASSACHUSETTS 02139

THE ACTION OF A REAL SEMISIMPLE GROUP ON A COMPLEX FLAG MANIFOLD. I: ORBIT STRUCTURE AND HOLOMORPHIC ARC COMPONENTS[1]

BY JOSEPH A. WOLF

Table of Contents

1. Introduction. This paper describes a topic that is of some interest in Lie groups and in differential and algebraic geometry. The topic shows good promise of being the "correct" context for explicit realization of those series of irreducible unitary representations of semisimple Lie groups that come into the Plancherel formula, so it probably is also of interest in harmonic analysis. We start with an example.

Let X be the Riemann sphere, viewed as $\mathbf{C}\cup\{\infty\}$ via stereographic projection. Then the group G of all holomorphic automorphisms of X consists of the linear fractional transformations

$$\pm\begin{pmatrix} a & c \\ b & d\end{pmatrix} : z \to \frac{az+b}{cz+d}, \qquad \det\begin{pmatrix} a & c \\ b & d\end{pmatrix} = 1;$$

Detailed version of an address delivered at the Riverside meeting of the American Mathematical Society on November 16, 1968, by invitation of the Committee to Select Hour Speakers for Far Western Sectional Meetings; received by the editors May 21, 1969.

[1] Research partially supported by N.S.F. Grants GP-2439, GP-5798 and GP-8008, and by an Alfred P. Sloan Research Fellowship.

that expresses G as a complex matrix group $\mathbf{SL}(2, \mathbf{C})/\{\pm I\}$. Let $G_0 = \mathbf{SL}(2, \mathbf{R})/\{\pm I\}$, subgroup of elements represented by real matrices, so G is the complexification of G_0. The complex group G is transitive on X but the real group G_0 is not. In fact X is the union of two open G_0-orbits

$$G_0(\sqrt{1}) = \{z \in \mathbf{C}: \operatorname{Im} z > 0\} \quad \text{and} \quad G_0(-\sqrt{-1}) = \{z \in \mathbf{C}: \operatorname{Im} z < 0\}$$

and the closed G_0-orbit

$$G_0(0) = \{z \in \mathbf{C}: \operatorname{Im} z = 0\} \cup \{\infty\}.$$

The Cayley transform

$$c = \pm \frac{1}{\sqrt{2}} \begin{pmatrix} 1 & \sqrt{-1} \\ \sqrt{-1} & 1 \end{pmatrix} \in G$$

of X relates these orbits, so they are

$$G_0(x_0) \text{ and } G_0(c^2x_0) \text{ open,} \qquad G_0(cx_0) \text{ closed,} \qquad x_0 = \sqrt{-1}.$$

The series of irreducible unitary representations of G_0 are the "discrete series," the "principal series" and the "complementary series." Discrete series representations are the square integrable ones; they can be realized on the spaces of L^2-holomorphic sections of G_0-homogeneous holomorphic line bundles over the open orbits. Principal series representation can be realized on L^2 (relative to invariant measure for the rotation subgroup of G_0) sections of certain G_0-homogeneous complex line bundles over the closed orbit. The complementary series does not contribute to Plancherel measure.

In general we start with a complex manifold $X = G/P$, where G is a complex semisimple Lie group and P is a complex subgroup of the sort called *parabolic*. The latter means any of the following conditions, which are equivalent:

$X = G/P$ is compact,

X is a compact simply connected kaehler manifold,

the complex manifold X is a projective variety,

X is a closed G-orbit in a projective representation.

Then X is a *complex flag manifold*. Let G_0 be a real form of G, i.e. a real Lie group whose complexification is G. Then G_0 acts on X as a subgroup of G. It turns out that there are only a finite number of

G_0-orbits on X and that there is just one closed orbit. The decomposition of X into G_0-orbits, and especially the structure of the open orbits, is studied in some detail in Chapter I.

We then take an arbitrary orbit $G_0(x)\subset X$ and partition it into "maximal complex analytic pieces" that we call its *holomorphic arc components.* If $z\in G_0(x)$ then $S_{[z]}$ denotes the holomorphic arc component of $G_0(x)$ through z and $N_{[z],0}$ denotes the identity component of the Lie subgroup $\{g\in G_0: gS_{[z]}=S_{[z]}\}$; then $S_{[z]}=N_{[z],0}(z)$ so it is a real submanifold of X. The best situation is that in which some (hence every) $S_{[z]}\subset G_0(x)$ carries a positive $N_{[z],0}$-invariant Radon measure; then we say that $G_0(x)$ is *measurable.* If $G_0(x)$ is measurable and $z\in G_0(x)$ then $S_{[z]}$ is a complex submanifold of X, in fact is an open $N_{[z],0}$-orbit on the complex flag submanifold $N^C_{[z],0}(z)$ of X, and its invariant measure is the volume element of an $N_{[z],0}$-invariant, possibly indefinite signature, kaehler metric. For example, the closed G_0-orbit on X is always measurable, and we have a method for deciding measurability of the open G_0-orbits. Holomorphic arc components and global conditions on them such as measurability of the orbit, are studied in Chapter II.

In Chapter III we work out the hermitian symmetric case in complete detail, extending earlier work of Korányi-Wolf [**2**], Wolf-Korányi [**15**] and Takeuchi [**9**]. This works well because every orbit is measurable. The Riemann sphere, described above, is the simplest of the hermitian symmetric cases; in fact its extreme simplicity can be misleading, so we look at the example of a general complex Grassmann manifold.

Let X be the complex Grassmann manifold consisting of the k-planes through the origin in $\mathbf{C}^n$. If $\{v_1, \cdots, v_k\}\subset\mathbf{C}^n$ is linearly independent then $v_1\wedge\cdots\wedge v_k\in X$ denotes its span. The complex general linear group acts on X by $g: v_1\wedge\cdots\wedge v_k\to g(v_1)\wedge\cdots\wedge g(v_k)$; there the scalars act trivially so the resulting group of transformations is the complex Lie group

$$G = \mathbf{GL}(n, \mathbf{C})/\{aI: a\neq 0\} = \mathbf{SL}(n, \mathbf{C})/\{e^{2\pi i r/n}I\}$$

of complex dimension n^2-1. Let $\{e_1, \cdots, e_n\}$ be a fixed basis of $\mathbf{C}^n$ and $x_0=e_1\wedge\cdots\wedge e_k\in X$ our base point. Then the isotropy subgroup of G at x_0 is the complex subgroup

$$P = \{g \in \mathbf{SL}(n, \mathbf{C}): gx_0 = x_0\}/\{e^{2\pi i r/n}I\}$$

of complex dimension $k^2+(n-k)^2+nk-1$. That exhibits X as a complex flag manifold[2] G/P of complex dimension $k(n-k)$.

Consider the hermitian form $\langle \sum_1^n z^i e_i, \sum_1^n w^j e_j \rangle = -\sum_1^k z^i \bar{w}^i + \sum_{k+1}^n z^i \bar{w}^i$ of signature $(k, n-k)$ on $\mathbf{C}^n$. The subgroup of $\mathbf{GL}(n, \mathbf{C})$ preserving it is the indefinite unitary group $\mathbf{U}^k(n)$. The latter acts on X as the real group

$$G_0 = \mathbf{U}^k(n)/\{e^{2\pi i\theta}I : \theta \text{ real}\} = \mathbf{SU}^k(n)/\{e^{2\pi i r/n}I\}$$

whose complexification is G. If $x \in X$ we associate a triple (a, b, c) where $\langle\ ,\ \rangle$ has rank $a+b$ on x, with a negative squares and b positive ones, and $a+b+c=k$. The only restrictions are

$$a+b+c=k, \quad 0 \leqq a \leqq k, \quad 0 \leqq b \leqq n-k, \quad 0 \leqq c \leqq \min(k, n-k).$$

If $y \in G_0(x)$ it gives the same triple; if $y \in X$ gives the same triple then Witt's Theorem says $y \in G_0(x)$. Thus there are precisely $\frac{1}{2}(l+1)(l+2)$ G_0-orbits on X, $l=\min(k, n-k)$, given by

$$G_0(c_1c_2 \cdots c_s c_{s+1}^2 c_{s+2}^2 \cdots c_t^2 x_0), \qquad 0 \leqq s \leqq t \leqq l = \min(k, n-k),$$

where c_i are the partial Cayley transforms defined by

$$c_i(e_i) = \frac{1}{\sqrt{2}}(e_i + \sqrt{-1}\,e_{k+i}), \quad c_i(e_{k+i}) = \frac{1}{\sqrt{2}}(\sqrt{-1}\,e_i + e_{k+i}),$$

$$c_i(e_j) = e_j \qquad \text{if } i \neq j \neq k+i.$$

For $(\prod_{1\leqq i\leqq s} c_i)(\prod_{s<j\leqq t} c_j^2)x_0$ has "signature" $(k-t, t-s, s)$. A glance at determinants shows that there are

$$l+1 \text{ open orbits } G_0(c_1^2c_2^2 \cdots c_t^2 x_0), \qquad 0 \leqq t \leqq l,$$
$$\text{just one closed orbit } G_0(c_1c_2 \cdots c_l x_0).$$

More precisely, the closure of $G_0(c_1c_2 \cdots c_s c_{s+1}^2 c_{s+2}^2 \cdots c_t^2 x_0)$ is the union of all $G_0(c_1c_2 \cdots c_u c_{u+1}^2 c_{u+2}^2 \cdots c_v^2 x_0)$ such that $v-u \leqq t-s$ and $t \leqq v$. Thus, for example, in the case $1=k<n$ of complex projective $(n-1)$-space (of which the Riemann sphere is the case $n=2$) there are

[2] Alternatively one can embed X in complex projective space of dimension $\binom{n}{k}-1$ by Plücker coordinates as follows. If $x \in X$ has basis $\{v_1, \cdots, v_k\}$, say $v_i = \sum_j a_k^i e_j$, then x goes to the point with homogeneous coordinates given by the determinants of the $\binom{n}{k}$ matrices $(a^{j_1}, a^{j_2}, \cdots, a^{j_k})$, $1 \leqq j_1 < \cdots < j_k \leqq n$, where $a^j = {}^t(a_1^i, \cdots, a_k^i)$. The same embedding is achieved as the closed orbit of the projective representation of $\mathbf{SL}(n, \mathbf{C})$ associated to the kth alternation $\Lambda^k(\alpha)$ of the usual (vector) representation α of degree n.

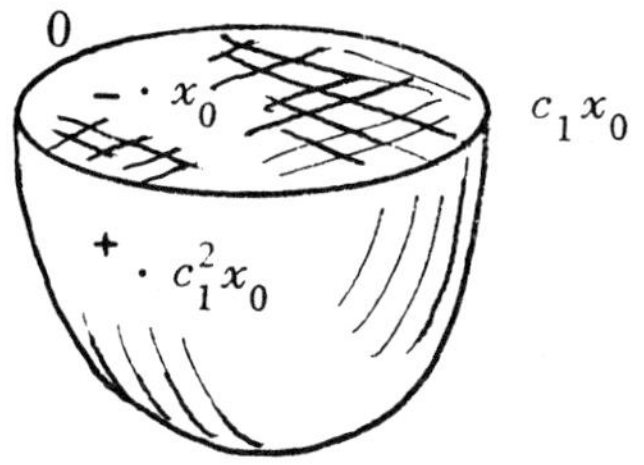

$k=1$, $n \geqq 2$

$-$: open: negative definite lines in $\mathbf{C}^n$
$+$: open: positive definite lines in $\mathbf{C}^n$
0: closed: isotropic lines in $\mathbf{C}^n$

3 orbits, and in the case $2=k \leqq n/2$ there are 6 orbits, whose boundary structure is illustrated in the figures.[3]

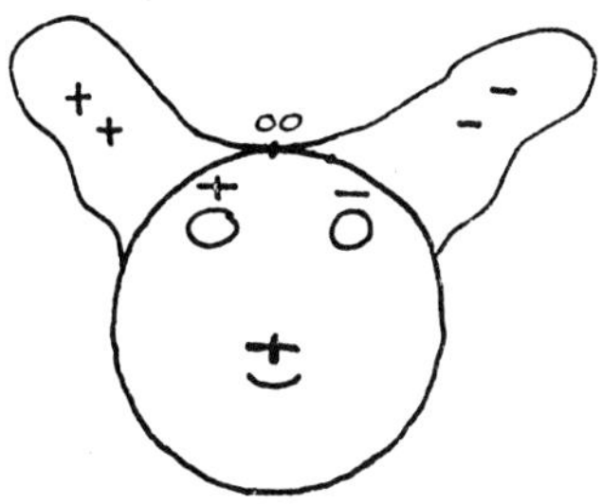

$k=2$, $n \geqq 4$

$--$: $G_0(x_0)$ open: negative definite planes
$+-$: $G_0(c_1^2 x_0)$ open: indefinite nondegenerate planes
$++$: $G_0(c_1^2 c_2^2 x_0)$ open: positive definite planes
$+0$: $G_0(c_1 c_2^2 x_0)$ intermediate: degenerate positive semidefinite planes
-0: $G_0(c_1 x_0)$ intermediate: degenerate negative semidefinite planes
00: $G_0(c_1 c_2 x_0)$ closed: totally isotropic planes

Back with the general complex grassmannian, we note that the open orbit $G_0(c_1^2 c_2^2 \cdots c_t^2 x_0)$ consisting of the k-planes of signature $(k-t, t)$ has coset space representation $\mathbf{U}^k(n)/\mathbf{U}^t(k) \times \mathbf{U}^{k-t}(n-k))$; so it has a G_0-invariant indefinite-kaehler metric of signature $(t(n-2k), k(n-k)-t(n-2k))$ whose volume element will prove useful. Indefiniteness of the metric is related to the existence of compact subvarieties. G_0 has maximal compact subgroup

$$K = \{\mathbf{U}(k) \times \mathbf{U}(n-k)\}/\{e^{2\pi i\theta}I : \theta \text{ real}\}$$

from the maximal compact subgroup $\mathbf{U}(k) \times \mathbf{U}(n-k) \subset \mathbf{U}^k(n)$. Now

$$K(c_1^2 c_2^2 \cdots c_t^2 x_0) = \{\mathbf{U}(k)/\mathbf{U}(t) \times \mathbf{U}(k-t)\} \times \{\mathbf{U}(n-k)/\mathbf{U}(k-t) \times \mathbf{U}(n-2k+t)\}$$

which is the direct product of the complex grassmannians of t-planes in $\mathbf{C}^k$ and $(k-t)$-planes in $\mathbf{C}^{n-k}$. Moreover the maximal compact sub-

[3] The figure for $k=2$ is due to A. Korányi.

varieties of the open orbits are the

$$g\cdot K(c_1^2c_2^2 \cdots c_t^2x_0) \subset G_0(c_1^2c_2^2 \cdots c_t^2x_0), \qquad g \in G_0.$$

Furthermore there exist holomorphic fibrations

$$\beta_t\colon G_0(c_1^2c_2^2 \cdots c_t^2x_0) \to K(c_1^2c_2^2 \cdots c_t^2x_0)$$

given as follows. First notice that the intended base space of β_t consists of all $v_1\wedge \cdots \wedge v_{k-t}\wedge w_1 \wedge \cdots \wedge w_t$ where $v_1\wedge \cdots \wedge v_{k-t}$ is a subspace of $e_1\wedge \cdots \wedge e_k$ and $w_1\wedge \cdots \wedge w_t$ is a subspace of $e_{k+1}\wedge \cdots \wedge e_n$. If $x\in X$ has signature $(k-t, t)$, its orthogonal projection to $e_1\wedge \cdots \wedge e_k$ is a subspace $v_1\wedge \cdots \wedge v_{k-t}$, and its orthogonal projection to $e_{k+1}\wedge \cdots \wedge e_n$ is some $w_1\wedge \cdots \wedge w_t$; by definition $\beta_t(x) = v_1\wedge \cdots \wedge v_{k-t}\wedge w_1\wedge \cdots \wedge w_t$. It is not too hard to check that this defines a holomorphic fibre bundle with total space and base as mentioned, whose fibre over $k(c_1^2c_2^2 \cdots c_t^2x_0)$, $k\in K$, is naturally isomorphic to the product of open orbits in smaller grassmannians,

negative definite t-planes in a $\mathbf{C}^{2t}$ of signature (t, t) and
negative definite $(k-t)$-planes in a $\mathbf{C}^{n-2t}$ of signature $(k-t, n-k-t)$.

One can use the maximal compact subvarieties to check directly that the complex manifold $G_0(c_1^2c_2^2 \cdots c_t^2x_0)$ has nonconstant holomorphic functions if, and only if, either $t=0$ or $t=k=n/2$. Thus the closest "approximation" to holomorphic functions on $G_0(c_1^2c_2^2 \cdots c_t^2x_0)$ is that of holomorphic sections of holomorphic line bundles $\beta_t^*\mathcal{L}$ where $\mathcal{L}\to K(c_1^2c_2^2 \cdots c_t^2x_0)$ is a holomorphic line bundle with nonzero positive semidefinite Chern class. Look back to the case $k=1$ of complex projective $(n-1)$-space. There $G_0(x_0)$ is the unit ball in $\mathbf{C}^{n-1}$, which carries holomorphic functions, and β_0 is fibration over the single point x_0. The other open orbit $G_0(c_1^2x_0)$ is the exterior of that unit ball with a complex projective $(n-2)$-space $K(c_1^2x_0)$ attached; if $n>2$ it carries no nonconstant holomorphic functions; β_1 projects it to the polar hyperplane $K(c_1^2x_0)$ of x_0, and eligible bundles $\beta_1^*\mathcal{L}$ are the $\beta_1^*\mathcal{H}^q$, $q\geqq 1$, where $\mathcal{H}\to K(c_1^2x_0)$ is the hyperplane bundle over that projective space.

We now cut an arbitrary G_0-orbit on X into complex submanifolds of X. If $g\in G$ (resp. $g\in G_0$) then $\tilde{g}\in \mathbf{SL}(n, \mathbf{C})$ (resp. $\tilde{g}\in \mathbf{SU}^k(n)$) denotes a representing linear transformation of $\mathbf{C}^n$. Now define, for $0\leqq s\leqq l=\min(k, n-k)$,

$$G_{[s]} = \{g\in G\colon \tilde{g}e_i = e_i \text{ and } \tilde{g}e_{k+i} = e_{k+i} \quad \text{for } 1\leqq i\leqq s\},$$

$$G_{[s],0} = \{g\in G_0\colon \tilde{g}e_i = e_i \text{ and } \tilde{g}e_{k+i} = e_{k+i} \quad \text{for } 1\leqq i\leqq s\}.$$

Then

$$G_{[s]}(c_1c_2 \cdots c_s x_0) = G_{[s]}(c_1c_2 \cdots c_s c_{s+1}^2 c_{s+2}^2 \cdots c_t^2 x_0)$$

for $s \leqq t \leqq l$, and that manifold is the sub-grassmannian of X consisting of all $(k-s)$-planes in the $\mathbf{C}^{n-2s}$ with basis $\{e_{s+1}, \cdots, e_k; e_{k+s+1}, \cdots, e_n\}$. Furthermore

$$G_{[s],0}(c_1c_2 \cdots c_s c_{s+1}^2 c_{s+2}^2 \cdots c_t^2 x_0) \text{ is open in } G_{[s]}(c_1c_2 \cdots c_s x_0),$$

and thus is a complex submanifold of X with an invariant indefinite-kaehler metric of signature

$$((t-s)(n-2s-2k), \qquad k(n-2s-k) - (t-s)(n-2s-2k)).$$

Now the point is that $G_0(c_1c_2 \cdots c_s c_{s+1}^2 c_{s+2}^2 \cdots c_t^2 x_0)$ is a union of submanifolds $k \cdot G_{[s],0}(c_1c_2 \cdots c_s c_{s+1}^2 c_{s+2}^2 \cdots c_t^2 x_0)$ with $k \in K$, that any two such submanifolds are equal or disjoint, and that any holomorphic map of the unit disc into X with image in

$$G_0(c_1c_2 \cdots c_s c_{s+1}^2 c_{s+2}^2 \cdots c_t^2 x_0)$$

must have image in one of those submanifolds. Thus every G_0-orbit is decomposed into a disjoint union of indefinite-kaehler submanifolds of X that are maximal among complex submanifolds of X contained in the orbit. They are the *holomorphic arc components* of the orbit. It now follows that every G_0-orbit on X is *measurable.*

Let $y = c_1c_2 \cdots c_s c_{s+1}^2 c_{s+2}^2 \cdots c_t^2 x_0$. Now every holomorphic arc component of the arbitrary G_0-orbit $G_0(y) \subset X$ is an open orbit $kG_{[s],0}k^{-1}(ky)$, $k \in K$, on the smaller grassmannian $kG_{[s]}(y)$. Thus it has a holomorphic fibration

$$\beta_{(k)}: kG_{[s],0}(y) \to k(K \cap G_{[s],0})(y)$$

over its maximal compact subvariety. It turns out that the $\beta_{(k)}$ fit together in the sense that

$$\text{if} \quad k_1 G_{[s],0}(y) = k_2 G_{[s],0}(y) \quad \text{then} \quad \beta_{(k_1)} = \beta_{(k_2)}.$$

The result is a K-equivariant fibration

$$\beta: G_0(y) \to K(y)$$

whose fibres coincide with those of its restrictions $\beta_{(k)}$, and which thus projects every holomorphic arc component holomorphically to its maximal compact subvariety. Thus one obtains "generalized Siegel domain" realizations of $G_0(y)$ by fitting together the standard Siegel domain realizations of the β-fibres.

Finally, let me describe the expected relation to representation theory. Let $G_0(x) \subset X$ be a measurable orbit. Consider G_0-homogeneous complex vector bundles $\mathcal{E} \to G_0(x)$ that are holomorphic over every holomorphic arc component of $G_0(x)$. Given $\mathcal{E}$ we form the cohomology groups $H^p(\mathcal{E})$ consisting of the $\mathcal{E}$-valued p-forms ω on $G_0(x)$ such that ω is harmonic and of type $(0, p)$ on every holomorphic arc component and

$$\int_K \int_{k \cdot S_{[x]}} \|\omega\|^2 d\mu_k dk < \infty .$$

Here $d\mu_k$ is the (normalized suitably) invariant Radon measure on the arbitrary holomorphic arc component $k \cdot S_{[x]}$, and dk is Haar measure on a maximal compact subgroup K of G_0. Now the point is to choose $\mathcal{E}$ so that the action of G_0 on $H^p(\mathcal{E})$ is unitary. The evidence is good that measurable open orbits give the discrete series of G_0, and the (necessarily measurable) closed orbit gives the principal series. The process can be viewed as a specialization of Kostant's theory of symplectic manifolds and polarizations ([**4**]; see [**5**] for details) that is set up for semisimple groups. I will go into it in fair detail in [**12**].

The first third (Chapter I) of this paper represents part of some joint work I did with Bertram Kostant, and that work influenced the second third (Chapter II). Also, a result of Masaru Takeuchi [**9**] is included in Chapter III for expository reasons. Thus only about 4/5 of this paper is my own work. For that reason there are chapter notes distributing credit in detail at the ends of the introductions to each of the three chapters.

Chapter I. Decomposition of a complex flag manifold into real group orbits

We establish the basic notions for G_0-orbits on a complex flag manifold $X = G/P$, G_0 a real form of G, in §2; there the root decompositions are set up, the number of G_0-orbits is bounded, open orbits are shown to exist, and dimensions and codimensions of orbits are computed in such a manner that open orbits are easy to manage. In §3 it is shown that there is a unique closed G_0-orbit on X, and various properties are explored for that closed orbit. §4 is the initial assault on open orbits; the structure of open orbits is worked out in detail sufficient to supply an enumeration procedure. Open orbits are studied as homogeneous spaces in §5, where their maximal compact subvarieties are described, where they are proved to be simply connected, and where their holomorphic functions are explicitly described. In §6 we study the existence problem for invariant Radon measures on open orbits and find a complete solution. Finally, in §7, the notion of integrable orbit (which is important for Chapter II and application to group representations) is worked out in detail sufficient for an enumeration.

NOTES FOR CHAPTER I. The results of §2, §3 and §4 were proved in some conversations between Bertram Kostant and this author. Theorem 2.6 is essentially the work of Kostant while Theorem 2.12 is due to this author. The material surrounding Theorem 3.3 is primarily the work of Kostant; Theorem 3.6 and its corollaries are essentially due to this author. Theorem 4.5 is due more to Kostant than to this author, and Theorem 4.9 slightly more to this author than to Kostant.

The results of §5 were worked out by this author. Lemma 5.1 and Theorem 5.4 are easy in the case where $G_0 \cap P_x$ is compact and our proof may be viewed as a reduction to that case. Theorem 5.7 had been suspected by P. A. Griffiths in the case where $G_0 \cap P_x$ is compact and G is simple; he felt that one should be able to join points of $G_0(x)$ by a chain of compact subvarieties under appropriate circumstances; our proof uses a Lie algebra maximality criterion instead.

Theorem 6.3 and its corollaries were worked out jointly in conversations between the author and Bertram Kostant. Then Kostant suggested an examination of the case where G_0 is complex; that case is described at the end of §6; it led to this author working out Theorem 6.7 and its corollary.

The notion of integrable orbit was suggested by Bertram Kostant's work on symplectic homogeneous spaces. Equivalence of (i), (ii) and (iii) in Theorem 7.2 was worked out jointly in conversations between the author and Kostant, and perhaps is due mostly to Kostant. The rest of §7 is due to this author.

The conversations between Bertram Kostant and the author that resulted in material in Chapter I took place between August 1965 and February 1967.

2. **Basic facts on the orbit structure.** Let G be a complex semisimple connected Lie group. $\mathfrak{g}$ is its Lie algebra. Choose a Cartan subalgebra $\mathfrak{h} \subset \mathfrak{g}$ and a system Π of simple roots. If φ is a root we denote its coefficients by

$$\varphi = \sum_{\pi \in \Pi} n\varphi(\pi) \cdot \pi.$$

Suppose $\Phi \subset \Pi$. Then the root system Δ is a disjoint union,

$$\Delta = \Phi^r \cup \Phi^u \cup -\Phi^u = \Phi^* \cup -\Phi^u \tag{2.1a}$$

where we define $\Phi^* = \Phi^r \cup \Phi^u$ and

(2.1b) $\Phi^r = \{\varphi \in \Delta : n_\varphi(\pi) = 0 \quad \text{for all} \quad \pi \in \Pi - \Phi\},$

(2.1c) $\Phi^u = \{\varphi \in \Delta : n_\varphi(\pi) > 0 \quad \text{for some} \quad \pi \in \Pi - \Phi\}.$

That defines subalgebras

(2.2) $\mathcal{P}_\Phi^r = \mathcal{H} + \sum_{\Phi^r} \mathcal{G}_\varphi, \qquad \mathcal{P}_\Phi^u = \sum_{\Phi^u} \mathcal{G}_\varphi, \qquad \mathcal{P}_\Phi = \mathcal{P}_\Phi^r + \mathcal{P}_\Phi^u$

of $\mathcal{G}$. $\mathcal{P}_\Phi^r$ is reductive, $\mathcal{P}_\Phi^u$ is unipotent, $\mathcal{P}_\Phi$ is their semidirect sum; $\mathcal{P}_\Phi$ is the normalizer of $\mathcal{P}_\Phi^u$ in $\mathcal{G}$. $\mathcal{H}_\Phi = \{h \in \mathcal{H} : \Phi(h) = 0\}$ is the center of $\mathcal{P}_\Phi^r$.

2.3. DEFINITION. *$\mathcal{P}_\Phi$ is the* ***parabolic subalgebra*** *of $\mathcal{G}$ defined by $\mathcal{H}$, Π and Φ. If Φ is empty, so $\mathcal{P}_\Phi = \mathcal{H} + \sum_{\varphi>0} \mathcal{G}_\varphi$, then $\mathcal{P}_\Phi$ is the* ***Borel subalgebra*** *of $\mathcal{G}$ defined by $\mathcal{H}$ and Π.*

It is standard that

(i) the Borel subalgebras of $\mathcal{G}$ are just the maximal solvable subalgebras,

(ii) the parabolic subalgebras of $\mathcal{G}$ are just the subalgebras which contain Borel subalgebras of $\mathcal{G}$, and

(iii) given $\mathcal{H}$, Π and a conjugacy class (inner automorphisms) $\{\mathcal{P}\}$ of parabolic subalgebras of $\mathcal{G}$, there is a unique subset $\Phi \subset \Pi$ such that $\mathcal{P}_\Phi \in \{\mathcal{P}\}$.

2.4. DEFINITION. *A* ***Borel subgroup*** *of G is a maximal complex solvable Lie subgroup. A* ***parabolic subgroup*** *of G is a complex Lie subgroup which contains a Borel subgroup.*

It is standard that the following conditions are equivalent for a complex Lie subgroup $L \subset G$:

(i) L is a parabolic subgroup,

(ii) the Lie algebra of L is a parabolic subalgebra of $\mathcal{G}$,

(iii) L is the analytic subgroup of G for a parabolic subalgebra of $\mathcal{G}$, and

(iv) G/L is compact.

The Borel subgroups are the parabolic subgroups whose Lie algebras are Borel subalgebras of $\mathcal{G}$. A parabolic subgroup is connected and equal to its own normalizer in G.

Let $P \subset G$ be a parabolic subgroup with Lie algebra $\mathcal{P}$. Then $\mathcal{G}$ has a Cartan subalgebra $\mathcal{H} \subset \mathcal{P}$, and there is a system Π of simple roots and a subset $\Phi \subset \Pi$, such that $\mathcal{P} = \mathcal{P}_\Phi$. Let P^r and P^u denote the respective analytic subgroups of P for $\mathcal{P}^r = \mathcal{P}_\Phi^r$ and $\mathcal{P}^u = \mathcal{P}_\Phi^u$. Then $P^r \cap P^u = \{1\}$, so $P = P^r \cdot P^u$ semidirect product, and P is the normalizer of P^u in G.

2.5. DEFINITION. *A* ***complex flag manifold*** *is a coset space $X = G/P$, where G is a complex semisimple connected Lie group and P is a parabolic subgroup; X is viewed as a complex manifold and G is viewed as a group of holomorphic transformations of X.*

Let $X = G/P$ be a complex flag manifold. If $x \in X$, say $x = gP$, then $\mathrm{ad}(g)P = gPg^{-1}$ is the isotropy subgroup of G at x. As P is its own normalizer in G we may identify x with $\mathrm{ad}(g)P$, viewing X as the space of parabolic subgroups of G conjugate to P. Whether we do that or not, we often write P_x for the parabolic $\mathrm{ad}(g)P$, $x = gP$, corresponding to x.

Let G_0 be a real form of G. In other words, G_0 is the analytic subgroup for a real form $\mathfrak{g}_0$ of $\mathfrak{g}$. τ denotes complex conjugation of $\mathfrak{g}$ over $\mathfrak{g}_0$, of G over G_0; so G_0 is the identity component of the fixed point set of τ on G.

2.6. THEOREM. *Let $X = G/P$ be a complex flag manifold. Let G_0 be a real form of G and let τ be conjugation of G over G_0.*

1. *If $x \in X$ then the isotropy subalgebra $\mathfrak{g}_0 \cap \mathfrak{P}_x$ is a real form of $\mathfrak{P}_x \cap \tau\mathfrak{P}_x$ and contains a Cartan subalgebra of $\mathfrak{g}_0$. Conversely if $\mathfrak{H}_0$ is a Cartan subalgebra of $\mathfrak{g}_0$ then $\mathfrak{H}_0 \subset \mathfrak{g}_0 \cap \mathfrak{P}_x$ for some $x \in X$.*

2. *Let m be the number of conjugacy classes of Cartan subalgebras of $\mathfrak{g}_0$. Let W_G denote the Weyl group of G. Then G_0 has $\leqq m|W_G|$ orbits on X.*

3. *G_0 has an open orbit on X, and the union of the open G_0-orbits is dense in X.*

PROOF. $\mathfrak{g}_0 \cap \mathfrak{P}_x$ is the isotropy subalgebra of $\mathfrak{g}_0$ at x, and $\mathfrak{g}_0 \cap \mathfrak{P}_x = (\mathfrak{g}_0 \cap \mathfrak{P}_x) \cap \tau(\mathfrak{g}_0 \cap \mathfrak{P}_x) = \mathfrak{g}_0 \cap \mathfrak{P}_x \cap \tau\mathfrak{P}_x$; so $\mathfrak{g}_0 \cap \mathfrak{P}_x$ is a real form of $\mathfrak{P}_x \cap \tau\mathfrak{P}_x$.

Choose a Borel subalgebra $\mathfrak{B} \subset \mathfrak{P}_x$ of $\mathfrak{g}$. Then $\mathfrak{B} \cap \tau\mathfrak{B}$ is τ-stable and contains a Cartan subalgebra of $\mathfrak{g}$, so it contains a τ-stable Cartan subalgebra $\mathfrak{H}$ of $\mathfrak{g}$. Now $\mathfrak{H} \cap \mathfrak{g}_0$ is a Cartan subalgebra of $\mathfrak{g}_0$ contained in $\mathfrak{g}_0 \cap \mathfrak{P}_x$.

Let $\mathfrak{H}_0$ be a Cartan subalgebra of $\mathfrak{g}_0$ and let $\mathfrak{H} = \mathfrak{H}_0^{\mathbb{C}}$, Cartan subalgebra of $\mathfrak{g}$. Choose a system Π of simple roots and a subset $\Phi \subset \Pi$ such that $\mathfrak{P}_\Phi$ is conjugate to $\mathfrak{P}$ in $\mathfrak{g}$. Then $\mathfrak{P}_\Phi = \mathfrak{P}_x$ for some $x \in X$, and $\mathfrak{H}_0 \subset \mathfrak{g}_0 \cap \mathfrak{P}_x$.

This completes the proof of (1).

Let $\{\mathfrak{H}_1, \cdots, \mathfrak{H}_m\}$ represent the conjugacy classes of Cartan subalgebras of $\mathfrak{g}_0$. Given $\mathfrak{H}_i$ we have $|W_G|$ systems Π of simple $\mathfrak{H}_i^{\mathbb{C}}$-roots of $\mathfrak{g}$. Given Π we have a unique subset $\Phi \subset \Pi$ such that $\mathfrak{P}_\Phi = \mathfrak{P}_x$ for some $x \in X$, and that specifies an orbit $G_0(x)$. We have just described $m \cdot |W_G|$ orbits of G_0 on X; they need not be distinct, but we must check that every G_0-orbit is one of them. Let $z \in X$. Then $\mathfrak{g}_0$ has a Cartan subalgebra $\mathfrak{H}_0 \subset \mathfrak{g}_0 \cap \mathfrak{P}_z$. We have $1 \leqq i \leqq m$ and $g \in G$ such that $\mathrm{ad}(g)\mathfrak{H}_0 = \mathfrak{H}_i$; now $\mathfrak{H}_i \subset \mathfrak{g}_0 \cap \mathfrak{P}_{g(z)}$ so we have a Π and a subset $\Phi \subset \Pi$ such that $\mathfrak{P}_{g(z)} = \mathfrak{P}_\Phi$; thus $G_0(z) = G_0(x)$ where $\mathfrak{P}_\Phi = \mathfrak{P}_x$. This proves (2).

Now G_0 has only a finite number of orbits on X. Thus some G_0-orbit has full dimension; hence is open, and the union of the lower dimensional G_0-orbits is a closed lower dimensional subset of X. This proves (3). Q.E.D.

Now the case of compact G_0, which is not new, becomes rather easy:

2.7. COROLLARY. *Let $X = G/P$ be a complex flag manifold. Let G_u be a compact real form of G. Then G_u is transitive on X, so $X = G_u/(G_u \cap P)$, and the isotropy subgroup $G_u \cap P$ is the centralizer of a torus. In particular X is simply connected, the integral cohomology $H^*(X; \mathbf{Z})$ is torsion free, and $H^q(X; \mathbf{Z}) = 0$ for q odd.*

PROOF. Let $G_u(x)$ be an open orbit. It is closed because G_u is compact. Thus $G_u(x) = X$.

Let $\mathcal{H}_u \subset \mathcal{G}_u \cap \mathcal{P}$ be a Cartan subalgebra of $\mathcal{G}_u$, $\mathcal{H} = \mathcal{H}_u^C$, Π a simple root system, and $\Phi \subset \Pi$, such that $\mathcal{P} = \mathcal{P}_\Phi$. $i\mathcal{H}_u$ contains the roots so $\tau\varphi = -\varphi$ for every root φ. Thus $\tau\mathcal{P} \cap \mathcal{P} = \mathcal{P}^r$, centralizer in $\mathcal{G}$ of $\mathcal{H}_\Phi = \{h \in \mathcal{H} : \Phi(h) = 0\}$; now $\mathcal{P}_u^r = \mathcal{G}_u \cap \mathcal{P}$ is a real form of $\mathcal{P}^r$, centralizer in $\mathcal{G}_u$ of $\mathcal{S} = \{h \in \mathcal{H}_u : \Phi(h) = 0\}$. Let S be the toral subgroup of G_u with Lie algebra $\mathcal{S}$. Now the centralizer of S in G_u is connected and contains $G_u \cap P$, and is the identity component of $G_u \cap P$. Thus $G_u \cap P$ is the centralizer of S; in particular X is simply connected and it follows from results of Bott-Samelson [1] that $H^*(X, \mathbf{Z})$ is torsion free, nonzero only in even degrees. Q.E.D.

2.8. COROLLARY. *Let $P_1 \subset P_2$ be parabolic subgroups of G so $X_i = G/P_i$ are complex flag manifolds. Let $\pi : X_1 \to X_2$ be the natural G-equivariant fibration. Then the fibre of π is a complex flag manifold $Y = (P_2^r)'/P_1 \cap (P_2^r)'$, and, relative to any coefficient ring, Poincaré polynomials satisfy*

$$\mathbf{p}(X_1, t) = \mathbf{p}(X_2, t) \cdot \mathbf{p}(Y, t).$$

PROOF. π has fibre $F = P_2/P_1$. Choose a Cartan subalgebra $\mathcal{H} \subset \mathcal{P}_1$ of $\mathcal{G}$, a system Π of simple roots, and subsets $\Phi_1 \subset \Phi_2$ of Π, such that $\mathcal{P}_i = \mathcal{P}_{\Phi_i}$. Then $\mathcal{P}_1^r \subset \mathcal{P}_2^r$ and $\mathcal{P}_1^u \supset \mathcal{P}_2^u$, so P_2^u acts trivially and P_2^r acts transitively on F. As $\mathcal{P}_1$ contains the center of $\mathcal{P}_2^r$, the semisimple part $(P_2^r)'$ acts transitively on F, and its isotropy subgroup $P_1 \cap (P_2^r)'$ is a parabolic subgroup. This realizes F as the complex flag manifold Y. All three of X_1, X_2 and Y now have no integral cohomology in odd degrees and no torsion in even degrees, so the spectral sequence of π is trivial. Q.E.D.

We take a closer look at the G_0-orbits on $X = G/P$. Let $x \in X$, choose a Cartan subalgebra $\mathcal{H}_0 \subset \mathcal{G}_0 \cap \mathcal{P}_x$ of $\mathcal{G}_0$, let $\mathcal{H} = \mathcal{H}_0^C$, and choose a simple root system Π and a subset Φ such that $\mathcal{P}_x = \mathcal{P}_\Phi$. Now we define

$$l = \text{rank}\,\mathcal{G}, \quad \text{i.e.} \quad l = \dim_C \mathcal{H} = \dim_R \mathcal{H}_0, \tag{2.9a}$$

$$r = \text{number of positive roots of } \mathcal{G}, \tag{2.9b}$$

$$r_\Phi = \text{number of positive roots of } \overset{r}{\mathcal{P}}_\Phi. \tag{2.9c}$$

Then $\dim_C G = l + 2r$, $\dim_C \mathcal{P}^r = l + 2r_\Phi$, $\dim_C \mathcal{P}^u = r - r_\Phi$, $\dim_C \mathcal{P} = l + r + r_\Phi$ and $\dim_C X = r - r_\Phi$.

2.10. LEMMA. *$\mathcal{P}_x \cap \tau\mathcal{P}_x$ is the semidirect sum of the reductive algebra $(\mathcal{P}_x \cap \tau\mathcal{P}_x)^r$ given by*

$$\overset{r}{\mathcal{P}}_\Phi \cap \tau\overset{r}{\mathcal{P}}_\Phi = \mathcal{H} + \sum_{\Phi^r \cap \tau\Phi^r} \mathcal{G}_\varphi \tag{2.11a}$$

with the nilpotent ideal $(\mathcal{P}_x \cap \tau\mathcal{P}_x)^u$ given by

$$\begin{aligned}(\overset{r}{\mathcal{P}}_\Phi \cap \tau\overset{u}{\mathcal{P}}_\Phi) + (\overset{u}{\mathcal{P}}_\Phi \cap \tau\overset{r}{\mathcal{P}}_\Phi) + (\overset{u}{\mathcal{P}}_\Phi \cap \tau\overset{u}{\mathcal{P}}_\Phi) \\ = \left\{ \sum_{\Phi^r \cap \tau\Phi^u} + \sum_{\Phi^u \cap \tau\Phi^r} + \sum_{\Phi^u \cap \tau\Phi^u} \right\} (\mathcal{G}_\varphi).\end{aligned} \tag{2.11b}$$

In particular

$$\dim_R(\mathcal{G}_0 \cap \mathcal{P}_x) = \dim_C(\mathcal{P}_x \cap \tau\mathcal{P}_x) = \dim_C \overset{r}{\mathcal{P}}_\Phi + \left| \overset{u}{\Phi} \cap \tau\overset{u}{\Phi} \right|.$$

PROOF. $\mathcal{P}_x \cap \tau\mathcal{P}_x$ is the sum of $\mathcal{H}$ and certain $\mathcal{G}_\varphi$, with $\mathcal{G}_\varphi$ in the reductive part (resp. nilpotent ideal) if and only if $\mathcal{G}_{-\varphi}$ also occurs (resp. does not occur) in $\mathcal{P}_x \cap \tau\mathcal{P}_x$. Thus $(\mathcal{P}_x \cap \tau\mathcal{P}_x)^r$ is the sum of $\mathcal{H}$ and $\{\mathcal{G}_\varphi$: both φ and $-\varphi$ are in $\Phi^* \cap \tau\Phi^*\}$, i.e. the sum of $\mathcal{H}$ and $\{\mathcal{G}_\varphi: \varphi \in \Phi^r \cap \tau\Phi^r\}$, as asserted, and $(\mathcal{P}_x \cap \tau\mathcal{P}_x)^u$ is the sum of all $\mathcal{G}_\varphi$ for $\varphi \in \Phi^* \cap \tau\Phi^*$ but $\varphi \notin \Phi^r \cap \tau\Phi^r$, i.e. $\varphi \in (\Phi^r \cap \tau\Phi^u) \cup (\Phi^u \cap \tau\Phi^r) \cup (\Phi^u \cap \tau\Phi^u)$ as asserted.

Now $\dim_C(\mathcal{P}_x \cap \tau\mathcal{P}_x) = l + |\Phi^r \cap \tau\Phi^r| + |\Phi^r \cap \tau\Phi^u| + |\Phi^u \cap \tau\Phi^r| + |\Phi^u \cap \tau\Phi^u|$. But $\dim_C \mathcal{P}^r_\Phi = l + |\Phi^r| = l + |\Phi^r \cap \tau\Phi^r| + |\Phi^r \cap \tau\Phi^u| + |\Phi^r \cap -\tau\Phi^u|$, and $\Phi^u \cap \tau\Phi^r = -\tau(\Phi^r \cap -\tau\Phi^u)$ implies $|\Phi^r \cap -\tau\Phi^u| = |\Phi^u \cap \tau\Phi^r|$. Thus $\dim_C(\mathcal{P}_x \cap \tau\mathcal{P}_x) - \dim_C \mathcal{P}^r_\Phi = |\Phi^u \cap \tau\Phi^u|$. Q.E.D.

2.12. THEOREM. *Let $X = G/P$ be a complex flag manifold, $x \in X$ and $P_x = P_\Phi$ as above. Then*

(i) $\dim_C X = r - r_\Phi$,

(ii) $|\Phi^u \cap \tau\Phi^u|$ *is the real codimension of the orbit $G_0(x)$ in X, and*

(iii) *$G_0(x)$ is open in X if and only if $\Phi^u \cap \tau\Phi^u$ is empty.*

PROOF. (i) was noted as an immediate consequence of the definitions of r and r_Φ.

For (ii) we calculate

$$\begin{aligned}\operatorname{codim}_R G_0(x) &= \dim_R X - \dim_R G_0(x)\\ &= 2\dim_C X - \{\dim_R G_0 - \dim_R G_0 \cap P_x\}\\ &= (2r - 2r_\Phi) - \{(l+2r) - (l + 2r_\Phi + |\Phi^u \cap \tau\Phi^u|)\}\\ &= |\Phi^u \cap \tau\Phi^u|.\end{aligned}$$

For (ii) we note that $G_0(x)$ is open in X if and only if $\operatorname{codim}_R G_0(x) = 0$. Q.E.D.

Now the orbit $G_0(x)$ has real dimension $2(r-r_\Phi)-|\Phi^u\cap\Phi^u|$, maximized as $2(r-r_\Phi)$ for the open orbits, minimized for a closed orbit.

3. **The closed orbit.** Let $x\in X$. Then either $G_0(x)$ is closed in X or its topological boundary is a nonempty union of lower dimensional orbits. Thus any G_0-orbit of minimal dimension is closed.

Let $G_0(x)$ be a closed orbit. In other words $G_0/(G_0\cap P_x)$ is compact. Thus there is a maximal compact subgroup $K\subset G_0$ and an Iwasawa decomposition $G_0=KAN$, such that AN is contained in the isotropy subgroup $(P_x\cap\tau P_x)_0=G_0\cap P_x$ of G_0 at x. Extend A to a Cartan subgroup $H_0=T\times A$ of G_0, where $T\subset K$ is a maximal torus of the centralizer M of A in K. Then MAN is a minimal parabolic subgroup of G_0 and $(P_x\cap\tau P_x)_0$ contains TAN. We have proved

3.1. LEMMA. *$G_0(x)$ is a closed G_0-orbit in X if and only if there is an Iwasawa decomposition $G_0=KAN$, the corresponding minimal parabolic subgroup MAN of G_0, and a maximal torus T of M, such that $TAN\subset G_0\cap P_x$.*

Now suppose that P is a Borel subgroup of G_0, so $\mathcal{P}_x\cap\tau\mathcal{P}_x$ is solvable. Let $\mathfrak{H}=\mathfrak{H}_0^{\mathbb{C}}$ and let φ be an $\mathfrak{H}$-root. First suppose $\varphi(\mathfrak{A})\neq 0$. If $\mathfrak{G}_\varphi\not\subset\mathfrak{N}^{\mathbb{C}}$ then $\mathfrak{G}_{-\varphi}\subset\mathfrak{N}^{\mathbb{C}}$, so $\mathfrak{G}_\varphi\not\subset\mathcal{P}_x\cap\tau\mathcal{P}_x$ because the latter is solvable. Now suppose $\varphi(\mathfrak{A})=0$ so $\mathfrak{G}_\varphi\subset\mathfrak{M}^{\mathbb{C}}$. If $\mathfrak{G}_\varphi\subset\mathcal{P}_x\cap\tau\mathcal{P}_x$ then $\mathfrak{G}_{-\varphi}\subset\mathcal{P}_x\cap\tau\mathcal{P}_x$ because M is compact, contradicting solvability of $\mathcal{P}_x\cap\tau\mathcal{P}_x$. We have proved

3.2. LEMMA. *Let $X=G/B$ where B is a Borel subgroup. Let $G_0(x)$ be a closed orbit. Then there is an Iwasawa decomposition $G_0=KAN$, the corresponding minimal parabolic subgroup MAN of G_0, and a maximal torus T of M, such that $\mathfrak{G}_0\cap\mathcal{P}_x=\mathfrak{I}+\mathfrak{A}+\mathfrak{N}$.*

Now we can prove uniqueness.

3.3. THEOREM. *Let $X=G/P$ be a complex flag manifold, G_0 a real form of G. Then there is a unique closed G_0-orbit on X.*

PROOF. Let $B\subset P$ be a Borel subgroup of G and let $Y=G/B$ be the corresponding complex flag manifold. We have the natural projection $\pi: Y\to X$ with compact fibres. Let $x\in X$. Then $\pi^{-1}G_0(x)$ is a union of G_0-orbits on Y. If $G_0(x)$ is closed, hence compact, then $\pi^{-1}G_0(x)$ is closed, so $\pi^{-1}G_0(x)$ contains a closed G_0-orbit in Y. If $G_0(x_1)$ and $G_0(x_2)$ are distinct closed G_0-orbits on X, now their respective inverse images contain distinct closed G_0-orbits on Y. Thus we need only prove our uniqueness assertion on Y. In other words the proof is reduced to the case where P is a Borel subgroup of G.

Now P is a Borel subgroup of G. Fix an Iwasawa decomposition $G_0=KAN$, the corresponding minimal parabolic subgroup MAN of G_0, and a maximal torus T of M. If $G_0(x)$ is closed then we have other choices $G_0=K'A'N'$, $M'A'N'$ and $T'\subset M'$, such that $\mathfrak{g}_0\cap\mathfrak{P}_x=\mathfrak{Z}'+\mathfrak{A}'+\mathfrak{N}'$. Any two Iwasawa decompositions of G_0 are conjugate. So we have $g\in G_0$ such that $\mathrm{ad}(g)$ sends K' to K, A' to A and N' to N. Automatically $\mathrm{ad}(g)M'=M$. Any two maximal tori of M are conjugate, and M normalizes both A and N; thus we may also assume $\mathrm{ad}(g)T'=T$. Replacing x by $g(x)$, the orbit $G_0(x)=G_0(gx)$ is not changed, and we may assume $\mathfrak{g}_0\cap\mathfrak{P}_x=\mathfrak{Z}+\mathfrak{A}+\mathfrak{N}$. Now the Borel subalgebra $\mathfrak{P}_x=\mathfrak{L}+\mathfrak{A}^{\mathfrak{c}}+\mathfrak{N}^{\mathfrak{c}}$ where, for some ordering, $\mathfrak{L}$ is the sum of $\mathfrak{Z}^{\mathfrak{c}}$ with the positive $\mathfrak{Z}^{\mathfrak{c}}$-root spaces of $\mathfrak{M}^{\mathfrak{c}}$. Conjugating further by an element $g\in M\subset G_0$ we may assume that ordering to be a given one.

Let $G_0(x_1)$ and $G_0(x_2)$ be closed orbits on X. We have just seen that x_i can be replaced within $G_0(x_i)$ so that $\mathfrak{P}_{x_i}=\mathfrak{L}+\mathfrak{A}^{\mathfrak{c}}+\mathfrak{N}^{\mathfrak{c}}$ in the notation above, i.e. so that $\mathfrak{P}_{x_1}=\mathfrak{P}_{x_2}$, i.e. so that $x_1=x_2$. Thus $G_0(x_1)=G_0(x_2)$. Q.E.D.

An immediate consequence is

3.4. COROLLARY. *Let $X=G/P$ be a complex flag manifold, $x\in X$, and G_0 a real form of G. Then the following conditions are equivalent:*

(i) *$G_0(x)$ is the unique closed G_0-orbit.*

(ii) *$G_0(x)$ is contained in the closure of every G_0-orbit on X.*

(iii) *$G_0(x)$ is contained in every G_0-stable closed subset of X.*

(iv) *Some maximal compact subgroup of G_0 is transitive on $G_0(x)$.*

(v) *Every maximal compact subgroup of G_0 is transitive on $G_0(x)$.*

(vi) *$G_0(x)$ is the lowest-dimensional G_0-orbit on X.*

PROOF. Equivalence of (i), (ii), (iii) and (vi) follows from uniqueness of the closed orbit and the fact that it is on the boundary of every nonclosed orbit. Trivially (v) implies (iv) and (iv) implies (i), so we need only check that (i) implies (v). Choose a Borel subgroup $B\subset P$

of G, let $Y=G/B$ and let $\pi: Y\to X$ be the projection. $G_0(x)$ is closed, so $\pi^{-1}G_0(x)$ is closed, so $\pi^{-1}G_0(x)$ contains a closed orbit $G_0(y)$. In the notation of the lemmas, $\mathfrak{B}_y\cap\mathfrak{G}_0=\mathfrak{Z}+\mathfrak{A}+\mathfrak{N}$ where $G_0=KAN$. Let K' be any maximal compact subgroup of G_0. Then $K'\cap AN=\{1\}$, so

$$\begin{aligned}\dim G_0(y) &\geqq \dim K'(y) = \dim K' - \dim(K'\cap TAN)\\ &\geqq \dim K' - \dim T = \dim K - \dim T\\ &= \dim K(y) = \dim K/T = \dim G_0/TAN\\ &= \dim G_0(y),\end{aligned}$$

proving $K'(y)=G_0(y)$. Now $G_0(x)=\pi G_0(y)=\pi K'(y)=K'(x)$. Q.E.D.

We just made implicit use of the obvious consequence of uniqueness of the closed orbit:

3.5. COROLLARY. *Let $P_1\subset P_2$ be parabolic subgroups of G, $X_i=G/P_i$ the resulting complex flag manifolds, and $\pi: X_1\to X_2$ the natural projection. Let G_0 be a real form of G, $x_i\in X_i$ such that $G_0(x_i)$ is the closed orbit on X_i. Then*

$$\pi G_0(x_1) = G_0(x_2) \quad \textit{and} \quad G_0(x_1) \subset \pi^{-1}G_0(x_2).$$

Now we count dimensions, proving

3.6. THEOREM. *Let $X=G/P$ be a complex flag manifold, $x\in X$, G_0 a real form of G, and $\mathcal{P}_x=\mathcal{P}_\Phi$ relative to a simple root system on a τ-stable Cartan subalgebra $\mathfrak{H}\subset\mathcal{P}_x$ of $\mathfrak{G}$.*

1. $\dim_R G_0(x)\geqq\dim_C X$ $\{=\frac{1}{2}\dim_R X\}$.
2. *The following conditions are equivalent*:

(2a) $\dim_R G_0(x)=\dim_C X$.

(2b) $\tau\Phi^u=\Phi^u$.

(2c) $\tau\mathcal{P}_x=\mathcal{P}_x$, *i.e.* $\tau P_x=P_x$, *i.e. viewing G as a linear algebraic group defined/**R** such that G_0 is the topological identity component of the subgroup $G_{\mathbf{R}}$ of real points, the algebraic subgroup P_x of G is* def/**R**.

(2d) *$G_0(x)$ is the closed orbit; with G* def/**R** *as in* (2c), *some conjugate of P is* def/**R**.

(2e) *X is a complex projective variety* def/**R** *in such a manner that $G_0(x)$ is the set $X_{\mathbf{R}}$ of real points.*

PROOF. $\dim_R G_0(x)=2(r-r_\Phi)-|\Phi^u\cap\tau\Phi^u|\geqq 2(r-r_\Phi)-|\Phi^u|=r-r_\Phi$ $=\dim_C X$, with equality if and only if $\Phi^u=\tau\Phi^u$. This proves (1) and gives equivalence of (2a) and (2b). $\tau\Phi^u=\Phi^u$ if and only if $\tau\mathcal{P}_\Phi^u=\mathcal{P}_\Phi^u$ because $\tau\mathfrak{H}=\mathfrak{H}$, $\tau\mathcal{P}_\Phi^u=\mathcal{P}_\Phi^u$ if and only if $\tau\mathcal{P}_\Phi=\mathcal{P}_\Phi$ because $\mathcal{P}_\Phi$ (resp. $\tau\mathcal{P}_\Phi$) is the normalizer of $\mathcal{P}_\Phi^u$ (resp. $\tau\mathcal{P}_\Phi^u$) in $\mathfrak{G}$; this gives equivalence of (2b) and (2c).

(1), the characterization of the closed orbit as the orbit of lowest dimension, and equivalence of (2a) and (2c), show that (2c) implies (2d). Conversely let $G_0(x)$ be closed and some P_z def/**R**; then $\dim_R G_0(z) = \dim_{\mathbf{C}} X$, so $G_0(z)$ is a lowest dimensional orbit, so $G_0(x) = G_0(z)$ and we have (2a). Thus (2c) and (2d) are equivalent.

(2e) implies (2a), thus (2c). Conversely assume (2c). Then the projective variety X is def/**R** and $G_0(x)$ is a topological component of $X_{\mathbf{R}}$. But $X_{\mathbf{R}}$ is a finite union of topological components $X_{\mathbf{R}}^{(i)}$, G_0 is transitive on each $X_{\mathbf{R}}^{(i)}$, each $X_{\mathbf{R}}^{(i)}$ is closed in X, and there is just one closed G_0-orbit. Now there is just one $X_{\mathbf{R}}^{(i)}$, so $G_0(x) = X_{\mathbf{R}}$ and we have (2e). Thus (2c) and (2e) are equivalent. Q.E.D.

We reformulate condition (2c) of Theorem 3.6 in terms of the action of the Galois group of **C** over **R** on a simple root system.

3.7. Corollary. *Let $X = G/P$ be a complex flag manifold, $x \in X$, G_0 a real form of G, $\mathfrak{H} \subset \mathfrak{P}_x$ a τ-stable Cartan subalgebra of $\mathfrak{g}$, and Π a simple root system such that $\mathfrak{P}_x = \mathfrak{P}_\Phi$ for some $\Phi \subset \Pi$. Define $\nu = w \cdot \tau$ where w is an element of the Weyl group which sends $\tau(\Pi)$ back to Π. Then $\dim_R G_0(x) = \dim_{\mathbf{C}}(X)$ if and only if*

(i) *$\nu\Phi = \Phi$ and*

(ii) *there is a minimal parabolic subgroup $MAN \subset G_0$ such that $\mathfrak{M}^{\mathbf{C}} + \mathfrak{A}^{\mathbf{C}} + \mathfrak{N}^{\mathbf{C}} = \mathfrak{P}_\Psi$ with $\Psi \subset \Phi$.*

Proof. $\{1, \nu\}$ is the action of the Galois group of **C** over **R** on the simple root system Π, and (ii) is the statement that $\mathfrak{P}_x \cap \mathfrak{g}_0$ contains a minimal parabolic subalgebra of $\mathfrak{g}_0$. Now (i) and (ii) just say that $\tau\mathfrak{P}_x = \mathfrak{P}_x$. Q.E.D.

We reformulate condition (2e) of Theorem 3.6 in terms of linear representations. To do that we order the weights by Π and say that a positive weight λ *belongs* to the subset $\Phi \subset \Pi$ if $\Phi = \{\varphi \in \Pi : \langle \lambda, \varphi \rangle = 0\}$. If λ is a positive weight then π_λ is the corresponding irreducible **C**-linear representation of $\mathfrak{g}$, V^λ is the representation space, and v_λ denotes an arbitrary nonzero highest weight vector. We have the usual map $p: V^\lambda \to \mathrm{proj}(V^\lambda)$ onto the complex projective space determined by V^λ, and $\bar{\pi}_\lambda$ denotes the projective representation of G on $\mathrm{proj}(V^\lambda)$ determined by these data. If λ belongs to Φ we note that $\mathfrak{P}_\Phi = \{h \in \mathfrak{g} : \pi_\lambda(h)(v_\lambda)$ is a multiple of $v_\lambda\}$, so $X = G/P_\Phi$ is realized as a subvariety of $\mathrm{proj}(V^\lambda)$ which is an orbit, $X = \bar{\pi}_\lambda(G)(pv_\lambda)$.

3.8. Corollary. *Let $X = G/P$ be a complex flag manifold, $x \in X$, G_0 a real form of G, $\mathfrak{H} \subset \mathfrak{P}_x$ a τ-stable Cartan subalgebra of $\mathfrak{g}$, and Π a simple root system such that $\mathfrak{P}_x = \mathfrak{P}_\Phi$ for some $\Phi \subset \Pi$. Then $\dim_R G_0(x) = \dim_C X$ if and only if there is a positive weight λ which belongs to Φ, such that*

(i) V^λ *has a* $\pi_\lambda(\mathfrak{G}_0)$*-stable real form* $V^\lambda_{\mathbf{R}}$ *and*

(ii) $V^\lambda_{\mathbf{R}}$ *contains a nonzero highest weight vector* v_λ.

In that case, identifying $X=\tilde{\pi}_\lambda(G)(pv_\lambda)$, *the set of real points of* X *is given by*

$$X_{\mathbf{R}} = X \cap p(V^\lambda_{\mathbf{R}}) = \tilde{\pi}_\lambda(G_0)(pv_\lambda).$$

We specialize Theorem 3.6 to hermitian symmetric spaces.

3.9. COROLLARY. *Let* $X=G/P=G_u/K$ *be a hermitian symmetric space of compact type, let* $X_0=G_0/K$ *be the dual bounded symmetric domain, take the Borel embedding* $X_0\subset X$ *given by* $X_0=G_0(x_0)$ *where* $P=P_{x_0}$ *and* $K=G_0\cap P=G_u\cap P$. *Let* $\check{S}$ *be the Bergman-Šilov boundary of* X_0 *in* X. *Then* $\check{S}$ *is the unique closed* G_0*-orbit on* X *and is contained in the closure of every* G_0*-orbit. Let* $x\in\check{S}$ *and let* $c\in G_u$ *be the Cayley transform (Korányi-Wolf* [2]) *carrying* x_0 *to* x. *Then the following conditions are equivalent.*

(i) $\dim_R \check{S}=\dim_{\mathbb{C}} X$.

(ii) X_0 *is a tube domain.*

(iii) *Viewing* G def/$\mathbf{R}$ *with* G_0 *as the identity component of* $G_{\mathbf{R}}$ *we have* ad$(c)P$ *defined over* $\mathbf{R}$.

(iv) X *is a complex projective variety defined over* $\mathbf{R}$ *in such a manner that* $\check{S}=X_{\mathbf{R}}$.

PROOF. It is known [2] that $\check{S}$ is the unique closed G_0-orbit on the topological boundary of X_0 in X. Thus $\check{S}$ is the unique closed G_0-orbit on X.

Equivalence of (i), (iii) and (iv) is given by Theorem 3.6; equivalence of (i) and (ii) is given in [2]. Q.E.D.

In view of Corollary 3.9 one might hope to prove that the closed orbit is always the Bergman-Šilov boundary of an open orbit. But this is far from the case; Theorem 5.7 will show us that most open orbits do not have any nonconstant holomorphic functions.

In order to specialize Theorem 3.6 to the case where P is a Borel subgroup of G we have to be able to recognize the cases in which $\mathfrak{G}$ has a Borel subalgebra $\mathfrak{B}$ stable under τ. That is the case in which $\mathfrak{G}_0$ (resp. G_0) is a "Steinberg normal real form" of $\mathfrak{G}$ (resp. of G).

Recall that $\mathfrak{G}_0$ (resp. G_0) is called the **Cartan normal real form** or **normal real form** of $\mathfrak{G}$ (resp. G) when the following equivalent conditions hold.

(3.10a) G_0 *has a Cartan subgroup* H_0 *that is split*/$\mathbf{R}$, *i.e. that is a real vector group, i.e. such that the* $\mathfrak{H}_0^{\mathbb{C}}$*-roots of* $\mathfrak{G}$ *take real values on* $\mathfrak{H}_0$.

(3.10b) *If* K *is a maximal compact subgroup of* G_0 *then the symmetric space rank of* G_0/K *is equal to the Lie group rank of* G_0.

(3.10c) $\mathcal{G}_0$ *is a direct sum of simple algebras from the list*:

type A_n: Lie algebra of $\mathbf{SL}(n+1, \mathbf{R})$

type B_n: Lie algebra of $\mathbf{SO}^n(2n+1) = \mathbf{SO}(n, n+1)$

type C_n: Lie algebra of $\mathbf{Sp}(n, \mathbf{R})$

type D_n: Lie algebra of $\mathbf{SO}^n(2n) = \mathbf{SO}(n, n)$

type G_2: maximal compactly embedded subalgebra $A_1 \oplus A_1$

type F_4: maximal compactly embedded subalgebra $A_1 \oplus C_3$

type E_6: maximal compactly embedded subalgebra C_4

type E_7: maximal compactly embedded subalgebra A_7

type E_8: maximal compactly embedded subalgebra D_8

The Cartan normal real form of $\mathcal{G}$ is a Steinberg normal real form. For if $H_0^{\mathbf{C}}$ is an $\mathbf{R}$-split Cartan subgroup of G_0 and Δ^+ is a positive $\mathcal{H}_0^{\mathbf{C}}$-root system of $\mathcal{G}$, then $\tau\varphi = \varphi$ for every root, so $\mathcal{B} = \mathcal{H}_0^{\mathbf{C}} + \sum_{\Delta^+} \mathcal{G}_\varphi$ is a τ-stable Borel subalgebra of $\mathcal{G}$.

Let σ be a Cartan involution of $\mathcal{G}_0$ and $\mathcal{G}_0 = \mathcal{K} + \mathfrak{M}$ the Cartan decomposition. Let $\mathcal{H}_0 = \mathcal{H}_T + \mathcal{H}_V$ be a σ-stable Cartan subalgebra of $\mathcal{G}_0$, let $\mathcal{H} = \mathcal{H}_0^{\mathbf{C}}$, choose a system Δ^+ of positive $\mathcal{H}$-roots of $\mathcal{G}$, and let $\mathcal{B} = \mathcal{H} + \sum_{\Delta^+} \mathcal{G}_\varphi$. If $\tau\mathcal{B} = \mathcal{B}$ then we have $h \in i\mathcal{H}_T + \mathcal{H}_V$ such that

$$\{\varphi \in \Delta: \varphi(h) > 0\} = \Delta^+ = \{\varphi \in \Delta: \varphi(\tau h) > 0\}.$$

Then $h = h_T + h_V$, $\tau(h) = -h_T + h_V$, such that

$$\Delta^+ = \{\varphi \in \Delta^+: \varphi(h_V) > 0\}.$$

In other words $\mathcal{H}_V$ contains a regular element of $\mathcal{G}$, and thus $\mathcal{H}_0$ is a maximally split Cartan subalgebra of $\mathcal{G}_0$. Conversely, if $\mathcal{H}_V$ contains a regular element of $\mathcal{G}$, that element defines a τ-stable positive root system Δ^+ for which $\tau\mathcal{B} = \mathcal{B}$.

Now we can say that $\mathcal{G}_0$ (resp. G_0) is a **Steinberg normal real form** of $\mathcal{G}$ (resp. of G) if and only if the following equivalent conditions hold. There $\mathcal{H}_0 = \mathfrak{Z} + \mathfrak{A}$ is a maximally split Cartan subalgebra of $\mathcal{G}_0$.

(3.11a) $\mathcal{G}$ *has a Borel subalgebra* $\mathcal{B}$ *with* $\tau\mathcal{B} = \mathcal{B}$.

(3.11b) $\mathfrak{A}$ *contains a regular element of* $\mathcal{G}$.

(3.11c) *Every* $\mathcal{H}_0^{\mathbf{C}}$-*root has nonzero restriction to* $\mathfrak{A}$.

(3.11d) *Let* K *be a maximal compact subgroup of* G_0 *and let* $\mathcal{S}$ *be the Satake diagram of the symmetric space* G_0/K. *Then* $\mathcal{S}$ *has no black vertices and* $\mathcal{K}$ *contains no simple ideal of* $\mathcal{G}_0$.

(3.11e) $\mathcal{G}_0$ *is a direct sum of simple algebras from the list*:

any type: *complex simple Lie algebra*

any type: *Cartan normal real form*

type A_{2n-1}: *Lie algebra of* $\mathbf{SU}^n(2n) = \mathbf{SU}(n, n)$

type D_n: *Lie algebra of* $\mathbf{SO}^{n-1}(2n) = \mathbf{SO}(n-1, n+1)$

type E_6: *maximal compactly embedded subalgebra* $A_1 \oplus A_5$

Summarizing, we have the following special case of Theorem 3.6.

3.12. COROLLARY. *Let* $X = G/B$, *complex flag manifold where* B *is a Borel subgroup of* G. *Let* $x \in X$ *and* G_0 *a real form of* G, *such that* $G_0(x)$ *is the closed orbit on* X. *Then the following conditions are equivalent*:

(i) $\dim_R G_0(x) = \dim_C X$.

(ii) G *is a Steinberg normal real form of* G.

(iii) $\mathfrak{G}_0$ *satisfies* (3.11e) *above.*

4. **Open orbits: construction, covering and counting.** We work with a complex flag manifold $X = G/P$, a point $x \in X$, a real form G_0 of G, conjugation τ of G over G_0, a τ-stable Cartan subalgebra $\mathfrak{H} \subset \mathfrak{P}_x$ of $\mathfrak{G}$, and a system Π of simple roots such that $\mathfrak{P}_x = \mathfrak{P}_\Phi$ for some $\Phi \subset \Pi$. Theorem 2.12 says that $G_0(x)$ is an open G_0-orbit on X if and only if $\Phi^u \cap \tau\Phi^u$ is empty.

4.1. LEMMA. *Let* $\mathfrak{H}_0$ *be a Cartan subalgebra of* $\mathfrak{G}_0$. *Choose a maximal compact subgroup* K *of* G_0, *and take the Cartan decomposition* $\mathfrak{G}_0 = \mathfrak{K} + \mathfrak{M}$, *such that*

$$(4.2) \qquad \mathfrak{H}_0 = \mathfrak{H}_T + \mathfrak{H}_V, \qquad \mathfrak{H}_T = \mathfrak{H}_0 \cap \mathfrak{K}, \qquad \mathfrak{H}_V = \mathfrak{H}_0 \cap \mathfrak{M},$$

decomposition of $\mathfrak{H}_0$ *into toral and vector parts. Then the following conditions are equivalent*:

(i) $\mathfrak{H}_T$ *is a Cartan subalgebra of* $\mathfrak{K}$.

(ii) $\mathfrak{H}_T$ *contains a regular element of* $\mathfrak{G}$.

(iii) τ *sends some Weyl chamber of* $\mathfrak{H}$ *to its negative.*

(iv) *There is a system* Δ^+ *of positive* $\mathfrak{H}_0^{\mathfrak{C}}$*-roots*, $\Delta^- = -\Delta^+$, *such that* $\tau\Delta^+ = \Delta^-$.

PROOF. (i) implies (ii) because $\mathfrak{K}$ contains a regular element of $\mathfrak{G}$. Let $h_0 \in \mathfrak{H}_T$ be a regular element of $\mathfrak{G}$. Then $\mathfrak{H}_0$ is the centralizer of h_0 in $\mathfrak{G}_0$, so $\mathfrak{H}_T$ is its own centralizer in $\mathfrak{K}$; this shows that (ii) implies (i).

(iii) and (iv) are equivalent; for if $\mathfrak{D} \subset i\mathfrak{H}_T + \mathfrak{H}_V$ is a Weyl chamber and $\Delta^+ = \{\varphi \in \Delta : \varphi > 0 \text{ on } \mathfrak{D}\}$ is the corresponding positive root system, then $\tau(\mathfrak{D}) = -\mathfrak{D}$ if and only if $\tau(\Delta^+) = \Delta^-$.

(i) implies (iv); for if $h_0 \in \mathfrak{H}_T$ is a regular element we have $\tau(ih_0) = -ih_0$ and $\{\varphi \in \Delta : \varphi(ih_0) > 0\}$ is a positive root system. Conversely if $\mathcal{H}_T$ contains no regular element we have a root φ such that $\varphi(\mathcal{H}_T) = 0$. Then $\varphi \in \mathcal{H}_V$, so $\tau(\varphi) = \varphi$, and $\tau\Delta^+ = \Delta^-$ is impossible for every positive root system. Thus (iv) implies (i). Q.E.D.

4.3. DEFINITION. *A Cartan subalgebra $\mathcal{H}_0 \subset \mathcal{G}_0$ is* ***maximally compact*** *if it satisfies the conditions of Lemma* 4.1, *is* ***compact*** *if moreover $\mathcal{H}_0 = \mathcal{H}_T$ in Lemma* 4.1, *i.e. if* rank K = rank G_0.

4.4. LEMMA. *Any two maximally compact Cartan subalgebras of $\mathcal{G}_0$ are conjugate.*

PROOF. Let $\mathcal{H}_0$ and $\mathcal{H}_0'$ be maximally compact Cartan subalgebras of $\mathcal{G}_0$. We have maximal compact subgroups $K, K' \subset G_0$, Cartan decompositions $\mathcal{K} + \mathfrak{M} = \mathcal{G}_0 = \mathcal{K}' + \mathfrak{M}'$, and expressions $\mathcal{H}_0 = \mathcal{H}_T + \mathcal{H}_V$ and $\mathcal{H}_0' = \mathcal{H}_T' + \mathcal{H}_V'$ as in (4.2). By hypothesis $\mathcal{H}_T$ is a Cartan subalgebra of $\mathcal{K}$ and $\mathcal{H}_T'$ is a Cartan subalgebra of $\mathcal{K}'$.

G_0 has an element g such that $\operatorname{ad}(g)\mathcal{K}' = \mathcal{K}$, so we may replace $\mathcal{H}_0'$ by its conjugate $\operatorname{ad}(g)\mathcal{H}_0'$ and assume $\mathcal{K} = \mathcal{K}'$. K has an element k such that $\operatorname{ad}(k)\mathcal{H}_T' = \mathcal{H}_T$, so we may further replace $\mathcal{H}_0'$ by its conjugate $\operatorname{ad}(k)\mathcal{H}_0'$ and assume $\mathcal{H}_T' = \mathcal{H}_T$. Now $\mathcal{H}_0 = \mathcal{H}_T + \mathcal{H}_V$ and $\mathcal{H}_0' = \mathcal{H}_T + \mathcal{H}_V'$. Let $h_0 \in \mathcal{H}_T$ be a regular element of $\mathcal{G}$. Then $\mathcal{H}_0 = \mathcal{L} = \mathcal{H}_0'$ where $\mathcal{L}$ is the centralizer of h_0 in $\mathcal{G}_0$. Q.E.D.

4.5. THEOREM. *Let $X = G/P$ be a complex flag manifold, $x \in X$ and G_0 a real form of G. Then the orbit $G_0(x)$ is open in X if and only if there exist*

(i) *a maximally compact Cartan subalgebra $\mathcal{H}_0 \subset \mathcal{G}_0 \cap \mathcal{P}_x$ of $\mathcal{G}_0$, and*

(ii) *a system* Π *of simple $\mathcal{H}_0^{\mathbb{C}}$-roots such that $\tau\Delta^+ = \Delta^-$ and $\mathcal{P}_x = \mathcal{P}_\Phi$ for some $\Phi \subset \Pi$.*

If $P_1 \subset P_2$ are parabolic subgroups of G, $X_i = G/P_i$ the resulting complex flag manifolds, $x_i \in X_i$ and $\pi: X_1 \to X_2$ the natural projection, then

(1) *if $G_0(x_1)$ is an open orbit on X_1 then $\pi(G_0(x_1)) = G_0(\pi x_1)$ is an open orbit on X_2, and*

(2) *if $G_0(x_2)$ is an open orbit on X_2 then $\pi^{-1}(G_0(x_2))$ contains an open G_0-orbit on X_1.*

PROOF. We first prove the assertions on $\pi: X_1 \to X_2$. If $G_0(x_2)$ is open in X_2 then $\pi^{-1}(G_0(x_2))$ is open in X_1 because π is continuous. As G_0 commutes with π, $\pi^{-1}(G_0(x_2))$ is a (necessarily finite) union of G_0-orbits, so it contains an orbit of full dimension in X_1; that orbit is necessarily open. This proves assertion (2). If $G_0(x_1)$ is open in X_1 then $\pi(G_0(x_1)) = G_0(\pi x_1)$ is open in X_2 because π is an open map. This proves assertion (1).

Suppose for the moment that P is a Borel subgroup of G and let $\mathcal{H} \subset \mathcal{P}_x$ be a τ-stable Cartan subalgebra of $\mathcal{G}$. There is just one simple root system Π such that $\mathcal{P}_x = \mathcal{P}_\Phi$ with $\Phi \subset \Pi$; Φ is empty and Π is characterized by $\Phi^u = \Delta^+$. If $G_0(x)$ is open, so $\Phi^u \cap \tau\Phi^u$ is empty by Theorem 2.12, then we have $\tau\Delta^+ = \Delta^-$, and $\mathfrak{H}_0 = \mathfrak{H} \cap \mathfrak{g}_0$ is a maximally compact Cartan subalgebra of $\mathfrak{g}_0$ by Lemma 4.1. Conversely, if $\mathfrak{H}_0 = \mathfrak{H} \cap \mathfrak{g}_0$ is maximally compact and $\tau\Delta^+ = \Delta^-$, then $\Phi^u \cap \tau\Phi^u$ is empty, so $G_0(x)$ is open in X by Theorem 2.12. The first statement is proved for the case where P is a Borel subgroup of G.

Now in general choose a Borel subgroup $B \subset P$ of G, let $Y = G/B$ be the complex flag manifold, and let $\pi: Y \to X$ be the natural projection. First suppose $G_0(x)$ open in X. Then assertion (2), which was already proved, provides an element $y \in \pi^{-1}(x)$ such that $G_0(y)$ is open in Y. Note $\mathcal{B}_y \subset \mathcal{P}_x$. Choose a τ-stable Cartan subalgebra $\mathfrak{H} \subset \mathcal{B}_y$ of $\mathfrak{g}$ and the simple root system Π such that $\mathcal{B}_y = \mathcal{P}_\Phi$, Φ empty. As was seen, $\mathfrak{H}_o = \mathfrak{H} \cap \mathfrak{g}_0$ is a maximally compact Cartan subalgebra of $\mathfrak{g}_0$, and $\tau\Delta^+ = \Delta^-$, because $G_0(y)$ is open in Y. Now $\mathcal{P}_x = \mathcal{P}_\Psi$ for some $\Psi \subset \Pi$ because $\mathcal{B}_y \subset \mathcal{P}_x$. Thus we have (i) and (ii). Conversely suppose that there is a maximally compact Cartan subalgebra $\mathfrak{H}_0 \subset \mathcal{P}_x$ of $\mathfrak{g}_0$, a system Π of simple $\mathfrak{H}_0^{\mathbb{C}}$-roots such that $\tau\Delta^+ = \Delta^-$, and a subset $\Psi \subset \Pi$ such that $\mathcal{P}_x = \mathcal{P}_\Psi$. $\tau\Delta^+ = \Delta^-$ says that $\Delta^+ \cap \tau\Delta^+$ is empty. $\Psi \subset \Pi$ says that $\Psi^u \subset \Delta^+$, so now $\Psi^u \cap \tau\Psi^u$ is empty. Theorem 2.12 now implies that $G_0(x)$ is open in X. Q.E.D.

Now we set out to construct and count the open G_0-orbits on $X = G/P$. To do this we choose a Borel subgroup $B \subset P$ and let $\pi: Y \to X$ be the projection, $Y = G/B$. First we construct and count the open G_0-orbits on Y, and then we carry the result down to X.

4.6. THEOREM. *Let $Y = G/B$, complex flag manifold where B is a Borel subgroup of G. Let G_0 be a real form of G and choose a maximally compact Cartan subalgebra $\mathfrak{H}_0 = \mathfrak{H}_T + \mathfrak{H}_V$ of $\mathfrak{g}_0$. Let $\{\mathfrak{D}_1, \cdots, \mathfrak{D}_m\}$ be the chambers in $i\mathfrak{H}_T$ cut out*[4] *by the $\mathfrak{H}_0^{\mathbb{C}}$-roots of $\mathfrak{g}$, and let W_K denote the Weyl group of K for the Cartan subalgebra $\mathfrak{H}_T$ of $\mathfrak{K}$.*

1. *Let $1 \leqq \alpha \leqq m$, choose $h_\alpha \in \mathfrak{D}_\alpha$, let Π_α be the system of simple $\mathfrak{H}_0^{\mathbb{C}}$-roots of $\mathfrak{g}$ for the positive root system $\Delta_\alpha^+ = \{\varphi \in \Delta: \varphi(h_\alpha) > 0\}$, let $\mathcal{B}_\alpha = \mathfrak{H}_0^{\mathbb{C}} + \sum_{\Delta_\alpha^+} \mathfrak{g}_\varphi$ be the corresponding Borel subalgebra of $\mathfrak{g}$ and let $y_\alpha \in Y$ be the element such that $\mathcal{B}_{y_\alpha} = \mathcal{B}_\alpha$. Then the open G_0-orbits on Y are just the $G_0(y_\alpha)$, and $G_0(y_\alpha) = G_0(y_\beta)$ if and only if some element of W_K sends $\mathfrak{D}_\alpha$ to $\mathfrak{D}_\beta$.*
2. *W_K acts simply on $\{\mathfrak{D}_1, \cdots, \mathfrak{D}_m\}$.*

---

[4] Given a root φ, $\{h \in i\mathfrak{H}_T : \varphi(h) = 0\}$ is a subspace of codimension $\geqq 1$ in $i\mathfrak{H}_T$. The union of these subspaces, as φ ranges over the root system Δ, is the intersection of $i\mathfrak{H}_T$ with the "diagram" in $i\mathfrak{H}_T + \mathfrak{H}_V$. The $\mathfrak{D}_\alpha$ are the components of the complement of this intersection in $i\mathfrak{H}_T$, i.e. of the set of regular elements of $\mathfrak{g}$ contained in $i\mathfrak{H}_T$.

3. *There are precisely* $m/|W_K|$ *distinct open* G_0*-orbits on* Y.

PROOF. Let $G_0(y)$ be an open orbit on Y. Theorem 4.5(i) gives us a maximally compact Cartan subalgebra $\mathcal{H}_0' \subset \mathcal{B}_y$ of $\mathfrak{g}_0$ and Lemma 4.4 gives us an element $g \in G_0$ such that $\mathrm{ad}(g)\mathcal{H}_0' = \mathcal{H}_0$. Replacing y by $g(y)$, $G_0(y)$ is not changed but we may assume $\mathcal{H}_0 \subset \mathcal{B}_y$. Theorem 4.5(ii) provides a simple root system Π, and the corresponding system Δ^+ of positive roots, such that $\tau\Delta^+ = \Delta^-$ and $\mathcal{B}_y = \mathcal{H}_0^{\mathfrak{c}} + \sum_{\Delta^+} \mathfrak{g}_\varphi$. Let $h \in i\mathcal{H}_T + \mathcal{H}_V$ be a regular element such that $\Delta^+ = \{\varphi \in \Delta: \varphi(h) > 0\}$. Decompose $h = h_T + h_V$, $h_T \in i\mathcal{H}_T$ and $h_V \in \mathcal{H}_V$, so $\varphi(h) = \varphi(h_T) + \varphi(h_V)$ and $(\tau\varphi)(h) = \varphi(\tau h) = -\varphi(h_T) + \varphi(h_V)$ for $\varphi \in \Delta$. Now let $\varphi \in \Delta^+$ and use $\tau\Delta^+ = \Delta^-$; so $\varphi(h_T) + \varphi(h_V) > 0 > -\varphi(h_T) + \varphi(h_V)$, which implies $\varphi(h_T) > 0$. Thus $\Delta^+ = \Delta_\alpha^+$, so $\mathcal{B}_y = \mathcal{B}_\alpha$, so $y = y_\alpha$, so $G_0(y) = G_0(y_\alpha)$, for the index α with $h_T \in \mathfrak{D}_\alpha$. This proves that every open G_0-orbit on Y is one of the $G_0(y_\alpha)$. Conversely, $\tau(h_\alpha) = -h_\alpha$, which implies $\tau\Delta_\alpha^+ = \Delta_\alpha^-$, so $G_0(y_\alpha)$ is an open orbit.

Let $w \in W_K$ send $\mathfrak{D}_\alpha$ to $\mathfrak{D}_\beta$. Represent $w = \mathrm{ad}(k)|\mathcal{H}_0^{\mathfrak{c}}$, $k \in K$. Now $\mathrm{ad}(k)\mathcal{B}_\alpha = \mathcal{B}_\beta$, i.e. $k(y_\alpha) = y_\beta$, so $G_0(y_\beta) = G_0(ky_\alpha) = (G_0k)(y_\alpha) = G_0(y_\alpha)$. Conversely suppose $G_0(y_\alpha) = G_0(y_\beta)$, so $g(y_\alpha) = y_\beta$ for some $g \in G_0$. $\mathcal{H}_0$ is both the centralizer of y_α in $\mathfrak{g}_0$ and the centralizer of y_β in $\mathfrak{g}_0$; now $\mathrm{ad}(g)\mathcal{H}_0 = \mathcal{H}_0$ and $\mathrm{ad}(g)\mathcal{B}_\alpha = \mathcal{B}_\beta$. Thus g represents an element w of the Weyl group W_{G_0} of G_0 relative to $\mathcal{H}_0$, such that $w(\mathfrak{D}_\alpha) = \mathfrak{D}_\beta$. The analytic subgroup for $\mathcal{H}_0$ in G_0 has finite index in its normalizer, for the latter is algebraic and thus has only finitely many components. It follows that $g = kv$ where $k \in K$ represents w and $v \in \exp(\mathcal{H}_V)$. In other words $w \in W_K$. We have proved that $G_0(y_\alpha) = G_0(y_\beta)$ if and only if an element of W_K sends $\mathfrak{D}_\alpha$ to $\mathfrak{D}_\beta$.

Let $w \in W_K$ such that $w(\mathfrak{D}_\alpha) = \mathfrak{D}_\alpha$. As $W_K \subset W_G$, we have $w \in W_G$. $\mathfrak{D}_\alpha = \mathfrak{D} \cap i\mathcal{H}_T$ for some Weyl chamber $\mathfrak{D} \subset i\mathcal{H}_T + \mathcal{H}_V$; as $\mathfrak{D}_\alpha$ is nonempty now $w(\mathfrak{D}) = \mathfrak{D}$. That says $w = 1$. Now W_K acts simply on $\{\mathfrak{D}_1, \cdots, \mathfrak{D}_m\}$.

Now we can count the open G_0-orbits. They are the $G_0(y_\alpha)$, $1 \leqq \alpha \leqq m$, and the action of W_K partitions them into sets of $|W_K|$ elements under the relation of W_K-equivalence. Thus there are $m/|W_K|$ distinct open orbits. Q.E.D.

If rank $K =$ rank G_0, i.e. if $\mathcal{H}_0$ is a compact Cartan subalgebra of $\mathfrak{g}_0$ in the notation of Theorem 4.6, then $\mathcal{H}_0 = \mathcal{H}_T$ and the $\mathfrak{D}_\alpha$ are the Weyl chambers. Then the number m of $\mathfrak{D}_\alpha$ is the order $|W_G|$ of the Weyl group of G, so part 3 of Theorem 4.6 reduces to

4.7. COROLLARY. *Let* $Y = G/B$, *complex flag manifold where* B *is a Borel subgroup of* G. *Let* G_0 *be a real form of* G *whose maximal compact subgroup* K *satisfies* rank $K =$ rank G_0. *Then there are precisely* $|W_G|/|W_K|$ *distinct open* G_0*-orbits on* Y.

On the other hand, if rank $K<$rank G_0, there may be Weyl chambers in $i\mathfrak{H}_T+\mathfrak{H}_V$ which do not meet $i\mathfrak{H}_T$, so the number of $\mathfrak{D}_\alpha$ may be less than $|W_G|$. For example, if $G=\mathbf{SL}(3,\mathbf{C})$ and $G_0=\mathbf{SL}(3,\mathbf{R})$, then $|W_G|=6$, so there are 6 Weyl chambers, but 4 of them are not sent to their negatives by τ; thus $m=2$, $|W_K|=|W_{\mathbf{SO}(3)}|=2$, and so there is just one open G_0-orbit on G/B. This trick for finding m works in general:

4.8. COROLLARY. *Let $Y=G/B$, complex flag manifold where B is a Borel subgroup of G. Let G_0 be a real form of G, let K be a maximal compact subgroup of G_0, let σ denote the involutive automorphism of $\mathfrak{G}_0$ with fixed point set $\mathfrak{K}$, and let $\mathfrak{H}_0=\mathfrak{H}_T+\mathfrak{H}_V$ be a maximally compact Cartan subalgebra of $\mathfrak{G}_0$ such that $\mathfrak{H}_T\subset\mathfrak{K}$. Define*

W_G: *Weyl group of G for the Cartan subalgebra $\mathfrak{H}_0^{\mathfrak{C}}$,*

W_K: *Weyl group of K for the Cartan subalgebra $\mathfrak{H}_T$,*

$W_G^{\mathfrak{H}_0}$: $\{w\in W_G\colon w(\mathfrak{H}_0)=\mathfrak{H}_0\}$, *subgroup of W_G.*

Then $W_G^{\mathfrak{H}_0}$ is simply transitive on the set of Weyl chambers which meet $i\mathfrak{H}_T$ (i.e. which are sent to their negatives by τ), $W_G^{\mathfrak{H}_0}$ is the centralizer of $\sigma|_{i\mathfrak{H}_T+\mathfrak{H}_V}$ in W_G, and G_0 has precisely $|W_G^{\mathfrak{H}_0}|/|W_K|$ distinct open orbits on Y.

PROOF. Let $\mathfrak{D}$ be a Weyl chamber which meets $i\mathfrak{H}_T$, i.e. such that $\tau\mathfrak{D}=-\mathfrak{D}$. Note that $\sigma=-\tau$ on the ambient space $i\mathfrak{H}_T+\mathfrak{H}_V$ of the Weyl chambers. Let $w\in W_G$ and notice $-w(\mathfrak{D})=w(\tau\mathfrak{D})$, so $\tau(w\mathfrak{D})=-w\mathfrak{D}$ if and only if $w\mathfrak{D}=\tau^{-1}w\tau\mathfrak{D}$, i.e. if and only if $w\tau=\tau w$, i.e. if and only if $w\sigma=\sigma w$, i.e. if and only if w preserves $i\mathfrak{H}_T$ and $\mathfrak{H}_V$, i.e. if and only if $w\in W_G^{\mathfrak{H}_0}$. Now our assertions follow from part 3 of Theorem 4.6. Q.E.D.

Now we push Theorem 4.6 down to a general complex flag manifold.

4.9. THEOREM. *Let $X=G/P$ be a complex flag manifold, let $B\subset P$ be a Borel subgroup of G, let $Y=G/B$, and let $\pi\colon Y\to X$ be the natural projection. Let G_0 be a real form of G, let $x\in X$ such that $G_0(x)$ is open in X, and let $\mathfrak{H}_0\subset\mathfrak{G}_0\cap\mathfrak{P}_x$ be a maximally compact Cartan subalgebra of $\mathfrak{G}_0$. Decompose $\mathfrak{P}_x=\mathfrak{P}_x^r+\mathfrak{P}_x^u$ relative to $\mathfrak{H}=\mathfrak{H}_0^{\mathfrak{C}}$ and define $\mathfrak{P}_x^{-u}=\sum_{-\Phi^u}\mathfrak{G}_\varphi$.*

1. *Let $y\in\pi^{-1}(x)$. Then $G_0(y)$ is open in Y if and only if $\mathfrak{B}_y=\mathfrak{B}'+(\mathfrak{P}_x^r\cap\tau\mathfrak{P}_x^{-u})+\mathfrak{P}_x^u$ where $\mathfrak{B}'$ is a Borel subalgebra of $\mathfrak{P}_x^r\cap\tau\mathfrak{P}_x^r=(\mathfrak{P}_x\cap\tau\mathfrak{P}_x)^r$ such that $\mathfrak{B}'\cap\tau\mathfrak{B}'=\mathfrak{H}$.*

2. *The open G_0-orbits in Y which lie over $G_0(x)$ are in one to one correspondence with the open $(P_x^r\cap\tau P_x^r)$-orbits on $(P_x^r\cap\tau P_x^r)/(Borel)$.*

3. *The open G_0-orbits on X are enumerated by the elements of the double coset space $W_K\backslash W_G^{\mathfrak{H}_0}/W_{P_x^r\cap\tau P_x^r}^{\mathfrak{H}_0}$.*

PROOF. Let $y\in\pi^{-1}(x)$ with $G_0(y)$ open in Y. We may assume $\mathcal{H}\subset\mathcal{B}_y$. Then $\mathcal{B}_y=\mathcal{B}''+\mathcal{P}_x^u=\mathcal{H}+\sum_{\Delta^+}\mathfrak{g}_\varphi$ where Δ^+ is a positive root system such that $\Phi^u\subset\Delta^+$ and $\tau\Delta^+=\Delta^-$, and where $\mathcal{B}''$ is a Borel subalgebra of $\mathcal{P}_x^r$. Decompose $\mathcal{P}_x^r$ as the sum of $(\mathcal{P}_x^r\cap\tau\mathcal{P}_x^r)$, $(\mathcal{P}_x^r\cap\tau\mathcal{P}_x^u)$ and $(\mathcal{P}_x^r\cap\tau\mathcal{P}_x^{-u})$. $\Phi^u\subset\Delta^+$ and $\tau\Delta^+=\Delta^-$ imply $\tau\Phi^u\subset\Delta^-$, so $\mathcal{B}''=\mathcal{B}'+(\mathcal{P}_x^r\cap\tau\mathcal{P}_x^{-u})$ where $\mathcal{B}'$ is a Borel subalgebra of $\mathcal{P}_x^r\cap\tau\mathcal{P}_x^r$. Now $\tau\Delta^+=\Delta^-$ implies $\mathcal{B}'\cap\tau\mathcal{B}'=\mathcal{H}$.

Conversely let $y\in\pi^{-1}(x)$ with $\mathcal{B}_y=\mathcal{B}'+(\mathcal{P}_x^r\cap\tau\mathcal{P}_x^{-u})+\mathcal{P}_x^u$ where $\mathcal{B}'$ is a Borel subalgebra of $\mathcal{P}_x^r\cap\tau\mathcal{P}_x^r$ such that $\mathcal{B}'\cap\tau\mathcal{B}'=\mathcal{H}$. Then $\mathcal{B}_y\cap\tau\mathcal{B}_y=\mathcal{H}$, so $G_0(y)$ is open in Y. This completes the proof of (1).

The number of distinct Borel subalgebras $\mathcal{B}'+(\mathcal{P}_x^r\cap\tau\mathcal{P}_x^{-u})+\mathcal{P}_x^u$, $\mathcal{B}'\subset\mathcal{P}_x^r\cap\tau\mathcal{P}_x^r$ Borel subalgebra, $\mathcal{B}'\cap\tau\mathcal{B}'=\mathcal{H}$, is the number of chambers cut out on $i\mathcal{H}_T$ by the $\mathcal{H}$-roots of $\mathcal{P}_x^r\cap\tau\mathcal{P}_x^r=(\mathcal{P}_x\cap\tau\mathcal{P}_x)^r$. Now, as in Theorem 4.6(3), the number of open G_0-orbits on Y over $G_0(x)$ is the number of open $(P_x^r\cap\tau P_x^r)$-orbits on $(P_x^r\cap\tau P_x^r)/(\text{Borel})$.

Finally we count the open G_0-orbits on X. Corollary 4.8 identifies the coset space $W_K\backslash W_G{}^{\mathcal{H}_0}$ with the open G_0-orbits on Y as follows. Choose a Weyl chamber $\mathcal{D}\subset i\mathcal{H}_T+\mathcal{H}_V$ such that $\tau\mathcal{D}=-\mathcal{D}$. Then all other such Weyl chambers are given by the $w(\mathcal{D})$, $w\in W_G^{\mathcal{H}_0}$; and the Borel subalgebras specified by $w_1(\mathcal{D})$ and $w_2(\mathcal{D})$, $w_i\in W_G^{\mathcal{H}_0}$, give the same open G_0-orbit if and only if $w_1=w_Kw_2$ for some $w_K\in W_K$. On the other hand, if $\mathcal{D}$ specifies one of the $\mathcal{B}'+(\mathcal{P}^r\cap\tau\mathcal{P}_x^{-u})+\mathcal{P}_x^u$ of (1), all other such $\mathcal{D}$ are given by the $w(\mathcal{D})$, $w\in W^{\mathcal{H}_0}_{P_x^r\cap\tau P_x^r}$; this follows from (1) and Corollary 4.8. Now (3) follows. Q.E.D.

5. **Open orbits: coset space structure and holomorphic functions.** In order to study open orbits as coset spaces of the real group, and in order to study holomorphic functions on open orbits, we need a remark on compact subvarieties:

5.1. LEMMA. *Let $X=G/P$ be a complex flag manifold, G_0 a real form of G, $x\in X$, such that $G_0(x)$ is open in X, and K a maximal compact subgroup of G_0 such that $\mathfrak{g}_0\cap\mathcal{P}_x$ contains a maximally compact Cartan subalgebra $\mathcal{H}_0=\mathcal{H}_T+\mathcal{H}_V$ of $\mathfrak{g}_0$ with $\mathcal{H}_T\subset\mathcal{K}$. Then $K(x)$ is a compact complex submanifold of $G_0(x)$.*

PROOF. Let $B\subset P$ be a Borel subgroup of G, let $Y=G/B$ and $\pi\colon Y\to X$ the natural projection, and choose $y\in\pi^{-1}(x)$ such that $G_0(y)$ is open in Y and $\mathcal{H}_0=\mathcal{B}_y\cap\tau\mathcal{B}_y$. Now $\pi(K(y))=K(x)$, and both $K(y)$ and $K(x)$ are compact real submanifolds. If $K(y)$ is a complex submanifold of $G_0(y)$, then its every tangent space $K(y)_{k(y)}$ is stable under the almost complex structure operator on Y, so $K(x)_{k(x)}$ is also stable for all $k\in K$, proving that $K(x)$ is a complex submanifold of X. Now the lemma is reduced to the case where P is a Borel subgroup of G.

Let P be a Borel subgroup. Then $\mathcal{P}_x \cap \tau\mathcal{P}_x = \mathcal{H}_0^{\mathbb{C}}$ because $G_0(x)$ is open in X. At x the $(\sqrt{-1})$-eigenspace of the almost complex structure is the holomorphic tangent space $\sum_{\Delta^+} \mathfrak{g}_{-\varphi}$, and the $(-\sqrt{-1})$-eigenspace is the antiholomorphic tangent space $\sum_{\Delta^+} \mathfrak{g}_{\varphi}$, where $\mathcal{P}_x = \mathcal{H}_0^{\mathbb{C}} + \sum_{\Delta^+} \mathfrak{g}_{\varphi}$. Now choose $h_0 \in i\mathcal{H}_T^{\mathbb{C}}$ such that $\Delta^+ = \{\varphi \in \Delta : \varphi(h_0) > 0\}$. Decompose $\mathcal{K} = \mathcal{H}_T + \mathfrak{F}$ where $\mathfrak{F}$ is the orthocomplement under the Killing form of $\mathcal{K}$. Now $\mathfrak{F}^{\mathbb{C}}$ is the sum of the $\mathcal{H}^{\mathbb{C}}{}_T$-root spaces on $\mathcal{K}^{\mathbb{C}}$, $\mathfrak{F}^{\mathbb{C}}$ gives the complexified tangent space to $K(x)$ at x, and $\mathfrak{F}^{\mathbb{C}}$ is the sum of its intersections with the holomorphic and antiholomorphic tangent spaces. It follows (look at h_0) that the real tangent space to $K(x)$ at x, represented by $\mathfrak{F}$, is stable under the almost complex structure. Thus $K(x)$ is a complex submanifold. Q.E.D.

To describe open orbits as coset spaces we need some notation. Let $X = G/P$ be a complex flag manifold, G_0 a real form of G, $x \in X$, and $\mathcal{H} \subset \mathcal{P}_x$ a τ-stable Cartan subalgebra of $\mathfrak{g}$. For a subset Φ of a simple $\mathcal{H}$-root system Π we have $\mathcal{P}_x = \mathcal{P}_\Phi$. Write $\mathcal{P}_x^r$ for $\mathcal{P}_\Phi^r$ and $\mathcal{P}_x^u$ for $\mathcal{P}_\Phi^u$. Recall, from Lemma 2.10, that $\mathcal{P}_x \cap \tau\mathcal{P}_x$ is the semidirect sum $(\mathcal{P}_x \cap \tau\mathcal{P}_x^u)^r + (\mathcal{P}_x \cap \tau\mathcal{P}_x^u)^u$ where $(\mathcal{P}_x \cap \tau\mathcal{P}_x)^r = \mathcal{P}_x^r \cap \tau\mathcal{P}_x^r$ and $(\mathcal{P}_x \cap \tau\mathcal{P}_x)^u = (\mathcal{P}_x^r \cap \tau\mathcal{P}^u) + (\mathcal{P}_x^r \cap \tau\mathcal{P}_x^r) + (\mathcal{P}_x^u \cap \tau\mathcal{P}_x^u)$. We define real forms of $(\mathcal{P}_x \cap \tau\mathcal{P}_x)$, $(\mathcal{P}_x \cap \tau\mathcal{P}_x)^r$ and $(\mathcal{P}_x \cap \tau\mathcal{P}_x)^u$ in the obvious manner, by

(5.2a) $$(\mathcal{P}_x \cap \tau\mathcal{P}_x)_0 = (\mathcal{P}_x \cap \tau\mathcal{P}_x) \cap \mathfrak{g}_0, \quad \text{real form of } \mathcal{P}_x \cap \tau\mathcal{P}_x;$$

(5.2b) $$(\mathcal{P}_x \cap \tau\mathcal{P}_x)_0^r = (\mathcal{P}_x^r \cap \tau\mathcal{P}_x^r) \cap \mathfrak{g}_0, \quad \text{real form of } (\mathcal{P}_x \cap \tau\mathcal{P}_x)^r;$$

(5.2c) $$\begin{aligned}(\mathcal{P}_x \cap \tau\mathcal{P}_x)_0^u &= \{(\mathcal{P}_x^r \cap \tau\mathcal{P}_x^u) + (\mathcal{P}_x^u \cap \tau\mathcal{P}_x^r)\} \cap \mathfrak{g}_0 \\ &\quad + (\mathcal{P}_x^u \cap \tau\mathcal{P}_x^u) \cap \mathfrak{g}_0, \quad \text{real form of } (\mathcal{P}_x \cap \tau\mathcal{P}_x)^u.\end{aligned}$$

Thus $(\mathcal{P}_x \cap \tau\mathcal{P}_x)_0 = (\mathcal{P}_x \cap \tau\mathcal{P}_x)_0^r + (\mathcal{P}_x \cap \tau\mathcal{P}_x)_0^u$ semidirect sum of reductive and nilpotent parts. Rather than proliferate notation, we denote

(5.3a) $(P_x \cap \tau P_x)_0$: analytic subgroup of G_0 for $(\mathcal{P}_x \cap \tau\mathcal{P}_x)_0$,

(5.3b) $(P_x \cap \tau P_x)_0^r$: analytic subgroup of G_0 for $(\mathcal{P}_x \cap \tau\mathcal{P}_x)_0^r$,

(5.3c) $(P_x \cap \tau P_x)_0^u$: analytic subgroup of G_0 for $(\mathcal{P}_x \cap \tau\mathcal{P}_x)_0^u$.

These are all closed connected subgroups of G_0, and $(P_x \cap \tau P_x)_0 = (P_x \cap \tau P_x)_0^r \cdot (P_x \cap \tau P_x)_0^u$ semidirect product of reductive and unipotent parts. By definition, $(P_x \cap \tau P_x)_0$ is the identity component of the isotropy subgroup $G_0 \cap P_x$ of G_0 at x.

5.4. Theorem. *Let $X = G/P$ be a complex flag manifold, G_0 a real form of G, $x \in X$, such that $G_0(x)$ is open in X. Then the open orbit $G_0(x)$ is simply connected and G_0 has connected isotropy subgroup $(P_x \cap \tau P_x)_0$ at x.*

PROOF. We work with a τ-stable Cartan subalgebra $\mathfrak{H}\subset\mathcal{P}_x$ of $\mathfrak{g}$ such that $\mathfrak{H}_0=\mathfrak{H}\cap\mathfrak{g}_0$ is a maximally compact Cartan subalgebra of $\mathfrak{g}_0$. Now $\mathcal{P}_x=\mathcal{P}_\Phi$ as usual. Let $\mathfrak{H}_0=\mathfrak{H}_T+\mathfrak{H}_V$, $\mathfrak{H}_T\subset\mathfrak{K}$, where K is a maximal compact subgroup of G_0. For the moment let L denote the isotropy subgroup of G_0 at x, so $L\cap K$ is the isotropy subgroup of K. Lemma 5.1 gives K-invariant complex structure on $K/(L\cap K)$ via its realization as $K(x)$. As $\operatorname{rank}(L\cap K)=\operatorname{rank} K$ because $\mathfrak{H}_T\subset\mathfrak{L}$, it follows [14] that $L\cap K$ is connected and $K/(L\cap K)$ is simply connected. But $G_0(x)\cong G_0/L$ has a deformation retraction back onto $K/(L\cap K)$; thus $G_0(x)$ is simply connected and it follows that L is connected. Q.E.D.

Let $X=G/P$ be a complex flag manifold and G_0 a real form of G. Then $\mathfrak{g}_0=\mathfrak{g}_o^{(1)}\oplus\cdots\oplus\mathfrak{g}_o^{(t)}$, direct sum of simple ideals, and we have

$$\mathfrak{g}=\mathfrak{g}^{(1)}\oplus\cdots\oplus\mathfrak{g}^{(t)}\quad\text{where}\quad\mathfrak{g}^{(i)}=\mathfrak{g}_0^{(i)\mathbb{C}},$$

$$\mathcal{P}=\mathcal{P}^{(1)}\oplus\cdots\oplus\mathcal{P}^{(t)}\quad\text{where}\quad\mathcal{P}^{(i)}=\mathcal{P}\cap\mathfrak{g}^{(i)},$$

$$X=X_1\times\cdots\times X_t\quad\text{where}\quad X_i=G^{(i)}/P^{(i)}.$$

Let $x\in X$, $x=(x_1,\cdots,x_t)$ with $x_i\in X_i$. Then

$$G_0(x)=G_0^{(1)}(x_1)\times\cdots\times G_0^{(t)}(x_t),\qquad G^{(i)}(x_i)\subset X_i.$$

Of course $G_0(x)$ is open in X if and only if $G_0^{(i)}(x_i)$ is open in X_i for $1\leqq i\leqq t$. Now consider the conditions

(5.5a) $(P_{x_i}^{(i)}\cap\tau P_{x_i}^{(i)})_0$ is compact; hence contained in a maximal compact subgroup $K^{(i)}$ of $G_o^{(i)}$,

(5.5b) $G_o^{(i)}/K^{(i)}$ is a hermitian symmetric coset space,

(5.5c) $G_o^{(i)}/(P_{x_i}^{(i)}\cap\tau P_{x_i}^{(i)})_0\to G_o^{(i)}/K^{(i)}$ is holomorphic for one of the two invariant complex structures on $G_o^{(i)}/K^{(i)}$.

5.6. DEFINITION. *Let $G_0(x)$ be open in X. Then the* ***bounded symmetric domain subordinate to the orbit*** *$G_0(x)$ is the coset space*

$$D(G_0,x)=G_0^{(1)}/L^{(1)}\times\cdots\times G_0^{(t)}/L^{(t)}$$

where

(i) *if the conditions* (5.5) *all hold, then $L^{(i)}=K^{(i)}$ and $G_0^{(i)}/L^{(i)}$ carries the invariant complex structure such that $G_0^{(i)}/(P_{x_i}^{(i)}\cap\tau P_{x_i}^{(i)})_0\to G_0^{(i)}/L_0^{(i)}$ is holomorphic,*

(ii) *if the conditions* (5.5) *do not all hold then $L^{(i)}=G_0^{(i)}$.*

We justify the terminology by observing that

(a) if $G_0^{(i)}/L^{(i)}$ is not a point, then it is an irreducible hermitian symmetric space of noncompact type, so $D(G_0,x)$ is a hermitian symmetric space of noncompact type (and possibly of dimension 0), and

(b) $D(G_0, x)$ is the "largest" hermitian symmetric space of noncompact type such that the natural map $G_0(x)\to D(G_0, x)$ is G_0-equivariant and holomorphic.

5.7. THEOREM. *Let $X=G/P$ be a complex flag manifold, G_0 a real form of G, and $x\in X$ such that $G_0(x)$ is open in X. Let*

$$\psi: G_0(x) \to D(G_0, x)$$

denote the natural projection onto the bounded symmetric domain subordinate to $G_0(x)$. Then the holomorphic functions on $G_0(x)$ are just the functions $\tilde{f}=f\cdot\psi$ where f is a holomorphic function on $D(G_0, x)$.

PROOF. We first consider the case where G_0 is simple. Let $\mathfrak{H}_0=\mathfrak{H}_T+\mathfrak{H}_V\subset\mathfrak{g}_0\cap\mathfrak{P}_x$ be a maximally compact Cartan subalgebra of $\mathfrak{g}_0$; we have $\mathfrak{P}_x^r$ specified by $\mathfrak{H}=\mathfrak{H}_0^C$. Let K be a maximal compact subgroup of G_0 such that $\mathfrak{H}_T\subset\mathfrak{K}$. If $\mathfrak{S}\subset\mathfrak{g}$, $z\in X$ and h is a smooth function on a neighborhood of z, then $\mathfrak{S}(h)_z$ denotes the set of values of derivatives of h at z by vector fields induced from elements of $\mathfrak{S}$.

Suppose that h is a nonconstant holomorphic function on $G_0(x)$. As $K(x)$ is a compact complex submanifold of $G_0(x)$ by Lemma 5.1, h must be constant on $K(x)$, so $\mathfrak{K}(h)_x=0$. The vector fields induced by $(\mathfrak{P}_x\cap\tau\mathfrak{P}_x)_0$ vanish at x, so also $(\mathfrak{P}_x\cap\tau\mathfrak{P}_x)_0(h)_x=0$. If $(P_x\cap\tau P_x)_0\not\subset K$ then $\mathfrak{K}$ and $(\mathfrak{P}_x\cap\tau\mathfrak{P}_x)_0$ generate $\mathfrak{g}_0$, for $\mathfrak{g}_0$ is simple, so $\mathfrak{g}_0(h)_x=0$; then the same argument shows $\mathfrak{g}_0(h)_z=0$ for every $z\in G_0(x)$, so all first derivatives of h vanish, whence h is constant contrary to hypothesis. Thus $(P_x\cap\tau P_x)_0$ is compact and contained in K.

The compactness just proved shows $\mathfrak{H}_0=\mathfrak{H}_T\subset\mathfrak{K}$. Now the root system $\Delta=\Delta_K\cup\Delta_M$ where Δ_K is the root system of $\mathfrak{K}^{\mathbf{C}}$ (compact roots) and Δ_M consists of the remaining roots (noncompact roots). If $\varphi\in\Delta$ then $\tau(\varphi)=-\varphi$ because $\mathfrak{H}_0$ is compact. Thus $\tau\Phi^r=\Phi^r$ and $\tau\Phi^u=-\Phi^u$ where $\mathfrak{P}_x=\mathfrak{P}_\Phi$. In other words $\mathfrak{P}_x\cap\tau\mathfrak{P}_x=\mathfrak{P}_x^r$ and $\mathfrak{K}^{\mathbf{C}}+\mathfrak{P}_x=\mathfrak{K}^{\mathbf{C}}+\sum_{\Delta_M^+}\mathfrak{g}_\varphi$. Define $\mathfrak{Q}=\mathfrak{K}^{\mathbf{C}}+\mathfrak{P}_x$. We have $\mathfrak{K}^{\mathbf{C}}(h)_x=0$ because h is holomorphic and $\mathfrak{K}(h)_x=0$; we have $\mathfrak{P}_x(h)_x=0$ because the vector fields induced by $\mathfrak{P}_x$ vanish at x; now $\mathfrak{Q}(h)_x=0$. If $\mathfrak{Q}$ generates $\mathfrak{g}$, then $\mathfrak{g}(h)_x=0$, so $\mathfrak{g}_0(h)_x=0$ and we contradict nonconstancy of h as above. Thus $\mathfrak{Q}=\mathfrak{K}^{\mathbf{C}}+\sum_{\Delta_M^+}\mathfrak{g}_\varphi$ does not generate $\mathfrak{g}$. As G_0 is simple it follows that G_0/K is hermitian symmetric, $\mathfrak{Q}$ is a parabolic subalgebra of $\mathfrak{g}$ containing $\mathfrak{P}_x$, and (Borel embedding) $gK\to gQ$ embeds G_0/K as an open G_0-orbit on the complex flag mainfold $Z=G/Q$. Let π: $X\to Z$ be the projection, $\pi(gP_x)=gQ$. Then π is holomorphic and its restriction to $G_0(x)$ is the projection $g(x)\to gK$ of $G_0(x)$ to G_0/K. Note that h is the lift of a function on G_0/K because it must be constant on the compact complex submanifolds $gK(x)$, $g\in G_0$, of $G_0(x)$.

Now we consider the general case, using the notation preceding Definition 5.6. Let h be a holomorphic function on $G_0(x)$. Let $1 \leqq i \leqq t$; fix $z_s \in G_0^{(s)}(x_s)$ for $s \neq i$, so $h(z_1, \cdots, z_t) = h_i(z_i)$ is a holomorphic function on $G_0^{(i)}(x_i)$. We have just proved that h_i is constant if the conditions (5.5) do not all hold, and that h_i is the lift of a function on $G_0^{(i)}/K^{(i)}$ if the conditions (5.5) do all hold. This proves that $h = f \cdot \psi$ where $\psi: G_0(x) \rightarrow D(G_0, x)$ is the natural projection and f is a function $D(G_0, x)$. If f is not holomorphic, $\bar{\partial} f \neq 0$, so $\bar{\partial} h = \psi^* \bar{\partial} f \neq 0$ and h is not holomorphic; thus $h = f \cdot \psi$ where f is a holomorphic function on $D(G_0, x)$. Conversely, if f is a holomorphic function on $D(G_0, x)$, then its lift $f \cdot \psi$ is holomorphic because ψ is holomorphic. Q.E.D.

Two immediate consequences are:

5.8. Corollary. *Let $X = G/P$ be a complex flag manifold, G_0 a real form of G, and $x \in X$ such that $G_0(x)$ is open in X.*

1. *The holomorphic functions separate points on $G_0(x)$, if and only if $(P_x \cap \tau P_x)_0$ is a maximal compact subgroup of G_0 and $G_0/(P_x \cap \tau P_x)_0$ is a hermitian symmetric coset space.*

2. *The following conditions are equivalent:*

(2a) *$G_0(x)$ carries a nonconstant holomorphic function.*

(2b) $\dim D(G_0, x) > 0$.

(2c) *Let $\mathfrak{H}_0 \subset \mathfrak{G}_0 \cap \mathfrak{P}_x$ be a maximally compact Cartan subalgebra of $\mathfrak{G}_0$. Then $\mathfrak{G}_0$ has a simple ideal G_0', and $\mathfrak{G}_0'$ has a maximal compactly embedded subalgebra $\mathfrak{K}'$, such that*

(i) *$\mathfrak{H}_0' = \mathfrak{H}_0 \cap \mathfrak{K}'$ is a Cartan subalgebra of $\mathfrak{G}_0'$,*

(ii) *$\mathfrak{G}_0'^{\mathbf{C}} \cap \mathfrak{P}_x = \mathfrak{P}_x'$ is given as $\mathfrak{P}_{\Phi'}'$ where $\Phi' \cup \{\varphi_1, \cdots, \varphi_s\}$ is a system of simple $\mathfrak{H}_0'^{\mathbf{C}}$-roots of $\mathfrak{G}_0'^{\mathbf{C}}$, and*

(iii) *$\mathfrak{K}'^{\mathbf{C}} = \mathfrak{P}_{\Psi'}^{r}$ where $\Psi' = \Phi' \cup \{\varphi_1, \cdots, \varphi_{s-1}\}$ and φ_s has coefficient 1 in the highest root.*

Now we have seen that an extremely special situation is required in order that an open orbit $G_0(x) \subset X$ carry a nonconstant holomorphic function. In particular we cannot expect the closed G_0-orbit to be the Bergman-Šilov boundary of an open orbit except in the usual [2] hermitian symmetric case.

6. Open orbits: invariant measure.

In this section we develop the notion of measurable orbit for the case of an open orbit.

6.1. Definition. *Let $X = G/P$ be a complex flag manifold, $x \in X$, and G_0 a real form of G. Then we say that $G_0(x)$ is a* **measurable open orbit** *if it is open in X and carries a G_0-invariant volume element.*

The elementary tool required is

6.2. LEMMA. *Let $H_0 \subset L \subset G_0$ be connected Lie groups where G_0 is reductive and $\mathfrak{H}_0$ is a Cartan subalgebra of $\mathfrak{G}_0$. Then G_0/L carries an invariant volume element if and only if L is reductive.*

PROOF. $\mathfrak{L}^{\mathbf{c}} = \mathfrak{H} + \sum_{\Lambda} \mathfrak{G}_\varphi$ and $\mathfrak{G} = \mathfrak{H} + \sum_{\Delta} \mathfrak{G}_\varphi$ where $\mathfrak{H} = \mathfrak{H}_0^{\mathbf{c}}$, $\mathfrak{G} = \mathfrak{G}_0^{\mathbf{c}}$ and the root systems $\Lambda \subset \Delta$. The representation χ of L on the tangent space of G_0/L complexifies to the restriction of $\mathrm{ad}_G(L)$ to $\sum_{\Delta-\Lambda} \mathfrak{G}_\varphi$. Now G_0/L has an invariant volume element if and only if $\det \chi = 1$, and $\det \chi = 1$ if and only if $\mathrm{ad}_{\mathfrak{G}}(\mathfrak{L})$ acts on $\sum_{\Delta-\Lambda} \mathfrak{G}_\varphi$ with trace 0. As $\mathrm{ad}_{\mathfrak{G}}(\mathfrak{L})$ acts on $\mathfrak{G}$ with trace 0, because $\mathfrak{G}$ is reductive, now G_0/L carries a G_0-invariant volume element if and only if $\mathrm{ad}_{\mathfrak{L}}(\mathfrak{L})$ has trace 0, i.e. if and only if $\sum_{\Lambda} \varphi = 0$, i.e. if and only if $\varphi \in \Lambda$ implies $-\varphi \in \Lambda$, i.e. if and only if L is reductive. Q.E.D.

Now we have a quick characterization

6.3. THEOREM. *Let $X = G/P$ be a complex flag manifold, $x \in X$, and G_0 a real form of G. Then the following conditions are equivalent, and each implies that $G_0(x)$ is open in X.*

(1a) *$G_0(x)$ is a measurable open orbit.*

(1b) *$G_0(x)$ has a G_0-invariant volume element.*

(1c) *$G_0(x)$ has a G_0-invariant, possibly indefinite, kaehler metric.*

(1d) *The isotropy subgroup $G_0 \cap P_x$ is the centralizer in G_0 of a (compact) toral subgroup $T \subset G_0$.*

(2a) *$\mathfrak{P}_x \cap \tau\mathfrak{P}_x$ is reductive, i.e. $\mathfrak{P}_x \cap \tau\mathfrak{P}_x = \mathfrak{P}_x^r \cap \tau\mathfrak{P}_x^r$.*

(2b) *$\mathfrak{P}_x \cap \tau\mathfrak{P}_x = \mathfrak{P}_x^r$.*

(2c) *$\tau\mathfrak{P}_x^u = \mathfrak{P}_x^{-u}$, i.e. $\mathfrak{G} = \mathfrak{P}_x + \tau\mathfrak{P}_x^u$.*

(2d) *$\mathfrak{P}_x = \mathfrak{P}_\Phi$ with $\tau\Phi^r = \Phi^r$ and $\tau\Phi^u = -\Phi^u$.*

PROOF. Let $\mathfrak{H} \subset \mathfrak{P}_x$ be a τ-stable Cartan subalgebra of $\mathfrak{G}$ and let Π be a simple root system such that $\mathfrak{P}_x = \mathfrak{P}_\Phi$ for a subset $\Phi \subset \Pi$. Lemma 2.10 says $(\mathfrak{P}_x \cap \tau\mathfrak{P}_x)^r = \mathfrak{P}_x^r \cap \tau\mathfrak{P}_x^r$, root system $\Phi^r \cap \tau\Phi^r$, and $(\mathfrak{P}_x \cap \tau\mathfrak{P}_x)^u = (\mathfrak{P}_x^r \cap \tau\mathfrak{P}_x^u) + (\mathfrak{P}_x^u \cap \tau\mathfrak{P}_x^r) + (\mathfrak{P}_x^u \cap \tau\mathfrak{P}_x^u)$, root system $(\Phi^r \cap \tau\Phi^u) \cup (\Phi^u \cap \tau\Phi^r) \cup (\Phi^u \cap \tau\Phi^u)$.

(2d) implies (2b) by summing root spaces and (2b) visibly implies (2a). If $\mathfrak{P}_x \cap \tau\mathfrak{P}_x$ is reductive then $\tau\Phi^u \cap (\Phi^r \cup \Phi^u)$ is empty, so $\tau\Phi^u = -\Phi^u$; thus (2a) implies (2c). If $\tau\Phi^u = -\Phi^u$ then τ preserves $\{\varphi \in \Delta : \varphi \notin \Phi^u \cup -\Phi^u\}$, which is Φ^r; thus (2c) implies (2d). Now (2a), (2b), (2c) and (2d) are equivalent and each implies that $\Phi^u \cap \tau\Phi^u$ is empty, i.e. that $G_0(x)$ is open in X. It follows from Lemma 6.2 that (1a), (1b), (2a), (2b), (2c) and (2d) are equivalent.

Let $G_0(x)$ be a measurable open orbit, i.e. $\tau\Phi^r = \Phi^r$ and $\tau\Phi^u = -\Phi^u$. τ preserves $\mathfrak{P}_x^r$, hence its center $\mathfrak{H}_\Phi = \{h \in \mathfrak{H} : \Phi(h) = 0\}$; thus $\mathfrak{G}_0 \cap \mathfrak{H}_\Phi = \mathfrak{H}_T' + \mathfrak{H}_V'$ sum of toral and vector parts. We define

$$A = \{\varphi \in \Phi^u : \varphi(\mathcal{H}_T') = 0\} \quad \text{and} \quad \alpha = \sum_{\Phi^u} \varphi.$$

$\tau\Phi^u = -\Phi^u$ says $\tau(\alpha) = -\alpha$, so $\alpha \in i\mathcal{H}_T$. If $\varphi \in \Phi^u$ then $\varphi = \varphi_1 + \varphi_2$ with $\varphi_1(\mathcal{H}_\Phi) = 0$ and $\varphi_2 \in i\mathcal{H}_T' + \mathcal{H}_V'$. Projection to the trivial representation space of the Weyl group of $\mathcal{P}_x^r$ on $i\mathcal{H}_T + \mathcal{H}_V$ annihilates φ_1, and that Weyl groups preserves Φ^u, so $\alpha \in i\mathcal{H}_T' + \mathcal{H}_V'$. Now $\alpha \in i\mathcal{H}_T'$. If $\varphi \in \Phi^u$ then $\langle \varphi, \alpha \rangle > 0$. Thus A is empty. Now $\{\varphi \in \Delta : \varphi(\mathcal{H}_T') = 0\}$, which contains Φ^r, contains no element of $\Phi^u \cup -\Phi^u$. In other words $\mathcal{P}_x^r$ is the centralizer of $\mathcal{H}_T'$ in $\mathcal{G}$, so $G_0 \cap P_x = (P_x^r)_0$ is the centralizer in G_0 of the (compact) torus $T = \exp(\mathcal{H}_T')$. We have proved that (1a) implies (1d).

Let $G_0 \cap P_x$ be the centralizer of a torus $T \subset G_0$. Choose $t \in \mathfrak{J}$ such that $\mathcal{G}_0 \cap \mathcal{P}_x$ is the centralizer of t in $\mathcal{G}_0$, and let η denote the left invariant 1-form on G_0 whose value at the identity is given by $\eta(h) = \langle t, h \rangle$. Then $\tilde{\omega} = d\eta$ is a left invariant 2-form on G_0, right invariant by $G_0 \cap P_x$, inducing a nondegenerate G_0-invariant 2-form ω on $G_0(x) = G_0/(G_0 \cap P_x)$. Now the invariant almost complex structure J on $G_0(x)$, together with ω, defines an invariant hermitian metric with kaehler form ω. As $d\tilde{\omega} = d(d\eta) = 0$ is the lift of $d\omega$ to G_0, it follows that $d\omega = 0$ and the metric is kaehler. Thus (1d) implies (1c). And obviously (1c) implies (1b) because a kaehler metric (of any signature) has a volume element. Q.E.D.

6.4. COROLLARY. *Let $X = G/P$ be a complex flag manifold, G_0 a real form of G, and $x \in X$ such that $G_0(x)$ is open in X. If G_0 has its maximal compact subgroups K such that* rank K = rank G_0, *then $G_0(x)$ is a measurable open orbit.*

PROOF. Let $\mathcal{H} \subset \mathcal{P}_x$ be a τ-stable Cartan subalgebra of $\mathcal{G}$, such that $\mathcal{H}_0 = \mathcal{H} \cap \mathcal{G}_0$ is a maximally compact Cartan subalgebra of $\mathcal{G}_0$. If rank K = rank G_0, then $\mathcal{H}_0$ is compact, so $\tau\varphi = -\varphi$ for every $\mathcal{H}$-root φ; then $\tau\Phi^r = \Phi^r$ and $\tau\Phi^u = -\Phi^u$ where $\mathcal{P}_x = \mathcal{P}_\Phi$. Q.E.D.

The condition rank K = rank G_0 is not necessary for an open orbit to be measurable. For example

6.5. COROLLARY. *Let $X = G/B$ be a complex flag manifold where B is a Borel subgroup of G. Let G_0 be a real form of G. Then every open G_0-orbit on X is measurable.*

PROOF. $\mathcal{B} = \mathcal{P}_\Phi$ where Φ is empty. Thus $\tau\Phi^r = \Phi^r$, empty set, and $\tau\Phi^u = \tau\Delta^+ = \Delta^- = -\Phi^u$. Q.E.D.

The main point of Corollaries 6.4 and 6.5 is that measurability does not depend on the choice of open orbit. In order to prove that in general we need the notion of opposed parabolic subgroups.

6.6. DEFINITIONS. *The* **opposition automorphism** *of a reductive complex Lie algebra* $\mathcal{G}$ *relative to a Cartan subalgebra* $\mathcal{H}$ *is the automorphism*

$$\nu: \mathcal{G} \rightarrow \mathcal{G} \quad \text{by} \quad \nu(h) = -h \qquad \text{for } h \subset \mathcal{H} \quad \text{and} \quad \nu(e_\varphi) = e_{-\varphi}$$

where $\{h_\pi; e_\varphi\}$ *is a Weyl basis of* $\mathcal{G}$. *Note that any two opposition automorphisms are conjugate by an inner automorphism. Two subsets* $\mathcal{S}$, $\mathcal{S}' \subset \mathcal{G}$ *are* **opposed** *if* $\mathcal{S}$ *is conjugate to* $\nu(\mathcal{S}')$ *by an inner automorphism, for some (hence any) opposition automorphism of* $\mathcal{G}$.

Let $\mathcal{P}$ be a parabolic subalgebra of a reductive or semisimple complex Lie algebra $\mathcal{G}$. Choose a Cartan subalgebra $\mathcal{H} \subset \mathcal{P}$ of $\mathcal{G}$ and a simple root system Π such that $\mathcal{P} = \mathcal{P}_\Phi$ for a subset $\Phi \subset \Pi$. Let ν be the opposition automorphism for this data and let w be the Weyl group element such that $w(\nu\Pi) = \Pi$. Then of course the parabolic $\nu(\mathcal{P})$ opposed to $\mathcal{P}$ is conjugate to $\mathcal{P}_{w\nu\Phi}$. We can see the action of $w\nu$ on Π via the Dynkin diagram. It preserves each component, for ν preserves each simple ideal of $\mathcal{G}$; if a simple ideal has $-I$ in the Weyl group, i.e. if it is not of type A_l $(l>1)$, D_{2n+1} $(n>0)$ or E_6, then $w = \nu = \nu^{-1}$ on that ideal, so $w\nu$ acts trivially on the corresponding component of the Dynkin diagram; if a simple ideal does not have $-I$ in the Weyl group, then $w\nu$ acts on its Dynkin diagram by:

$A_l, l > 1$:

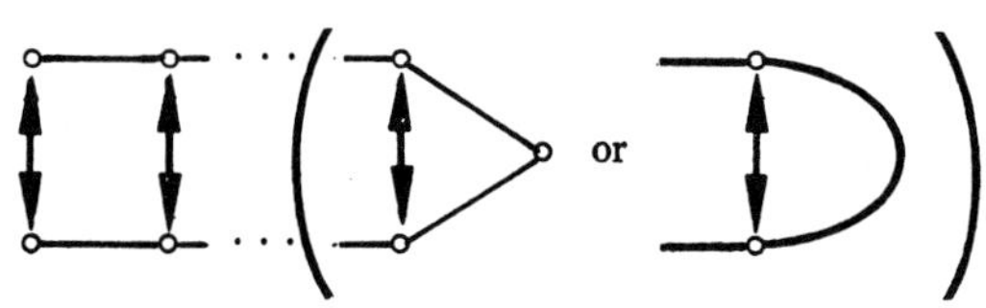

$D_{2n+1}, n > 0$:

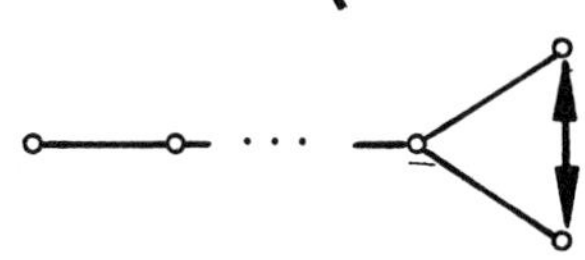

E_6:

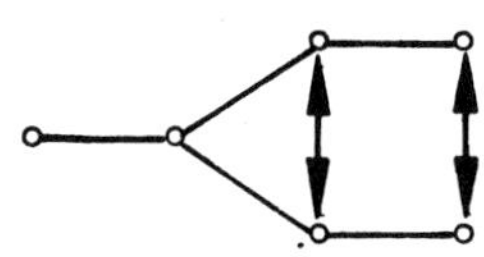

6.7. THEOREM. *Let* $X = G/P$ *be a complex flag manifold and* G_0 *a real form of* G. *Then the following conditions are equivalent*:

(i) $\mathcal{P}$ *and* $\tau\mathcal{P}$ *are opposed parabolic subalgebras of* $\mathcal{G}$.

(ii) *Some open* G_0*-orbit on* X *is measurable.*

(iii) *Every open* G_0*-orbit on* X *is measurable.*

PROOF. (iii) implies (ii) because G_0 has an open orbit.

Assume (ii) and let $G_0(x)$ be a measurable open orbit. $\mathcal{H} \subset \mathcal{P}_x$ is a τ-stable Cartan subalgebra of $\mathfrak{g}$ such that $\mathcal{H}_0 = \mathcal{H} \cap \mathfrak{g}_0$ is a maximally compact Cartan subalgebra of $\mathfrak{g}_0$. Now $\mathcal{P}_x = \mathcal{P}_\Phi$ where Φ is contained in a simple root system Π. We take the opposition automorphism ν for $\mathcal{H}$ and Π. Using Theorem 6.3, $\tau\Phi^u = -\Phi^u = \nu\Phi^u$, so $\tau\mathcal{P}_x$ and $\nu\mathcal{P}_x$ each is the normalizer of $\mathcal{P}_x^{-u}$ in $\mathfrak{g}$, whence $\tau\mathcal{P}_x = \nu\mathcal{P}_x$. Thus $\tau\mathcal{P}$ is conjugate to $\nu\mathcal{P}$, i.e. $\mathcal{P}$ is opposed to $\tau\mathcal{P}$. That shows that (ii) implies (i).

Let $\mathcal{P}$ and $\tau\mathcal{P}$ be opposed, let $G_0(x)$ be an open orbit, and let $\mathcal{H}_0 \subset \mathfrak{g}_0 \cap \mathcal{P}_x$ be a maximally compact Cartan subalgebra of $\mathfrak{g}_0$. Now $\mathcal{P}_x = \mathcal{P}_\Phi$ where Φ is contained in a simple root system Π such that $\tau(\Pi) = -\Pi$. $\tau\mathcal{P}_x$ and $\nu\mathcal{P}_x$ are conjugate parabolic subalgebras of $\mathfrak{g}$ with a common Borel subalgebra $\mathcal{H}_0^{\mathbb{C}} + \sum_{\Delta^+} \mathfrak{g}_{-\varphi}$; thus $\tau\mathcal{P}_x = \nu\mathcal{P}_x$. Now $\mathcal{P}_x \cap \tau\mathcal{P}_x = \mathcal{P}_x \cap \nu\mathcal{P}_x = \mathcal{P}_x^r$, so $G_0(x)$ is measurable by Theorem 6.3. Thus (i) implies (iii). Q.E.D.

We recover Corollary 6.4 from Theorem 6.7 because $\tau\varphi = -\varphi = \nu\varphi$ relative to a compact Cartan subalgebra $\mathcal{H}_0 \subset \mathfrak{g}_0$, and we recover Corollary 6.5 because $\mathcal{B}$, $\tau\mathcal{B}$ and $\nu\mathcal{B}$ are all conjugate. But, in addition, we have

6.8. COROLLARY. *Let $X = G/P$ be a complex flag manifold. Let G_0 be a real form of G such that $\mathfrak{g}_0$ is a direct sum of absolutely simple ideals $\mathfrak{g}_0^{(i)}$ each of which has one or more of the properties*

(i) *$\mathfrak{g}_0^{(i)}$ is not of type A_l $(l > 1)$, D_{2n+1} $(n > 0)$ nor E_6; or*

(ii) *the maximal compactly embedded subalgebras of $\mathfrak{g}_0^{(i)}$ have maximal rank; or*

(iii) *$\mathcal{P} \cap \mathfrak{g}^{(i)}$ is a Borel subalgebra of $\mathfrak{g}^{(i)} = \mathfrak{g}_0^{(i)\mathbb{C}}$.*

Then every open G_0-orbit on X is measurable.

The key proviso of Corollary 6.8 is that the simple ideals of $\mathfrak{g}_0$ are absolutely simple, i.e. that τ preserves every simple ideal of $\mathfrak{g}$. But it is interesting to consider the other extreme.

Let G_0 be a complex semisimple connected Lie group *viewed as a real Lie group*. Then $G = G_0^{\mathbb{C}}$ is $G_0 \times G_0$ and the embedding $G_0 \subset G$ is given by $g \to \delta(g) = (g, g)$. τ acts on G by $\tau(g_1, g_2) = (g_2, g_1)$. $P = P_1 \times P_2$ where P_i are parabolic subgroups of G_0. In particular $\tau(P) = P_2 \times P_1$ where $P = P_1 \times P_2$. The opposition automorphism acts by $\nu(P) = \nu(P_1) \times \nu(P_2)$ where $P = P_1 \times P_2$. *Thus $\mathcal{P}$ and $\tau\mathcal{P}$ are opposed if and only if $\mathcal{P}_1$ and $\mathcal{P}_2$ are opposed.* In other words, *the open G_0-orbits* on G/P *are measurable if and only if $P = P_1 \times P_2$, $\mathcal{P}_1$ and $\mathcal{P}_2$ opposed*, i.e. *if and only if* $P = P_1 \times \nu P_1$. If G_0 has rank r as a complex group, then G has rank $2r$, so G has 2^{2r} conjugacy classes of parabolic subgroups $P = P_1 \times P_2$ and just 2^r of them are of the form $P_1 \times \nu P_1$. Thus "most" parabolic subgroups $P \subset G$ have the property that the open G_0-orbits on G/P are not measurable.

7. **Integrable orbits.** Let $X=G/P$ be a complex flag manifold, $x \in X$ and G_0 a real form of G. We observe $\mathcal{P}_x+\tau\mathcal{P}_x=\mathcal{G} \Leftrightarrow \mathcal{P}_x^{-u} \subset \tau\mathcal{P}$ $\Leftrightarrow \mathcal{P}_x^{-u} \cap \tau\mathcal{P}_x^{-u}=0 \Leftrightarrow \mathcal{P}_x^{u} \cap \tau\mathcal{P}_x^{u}=0 \Leftrightarrow G_0(x)$ is open in X. This leads us to some definitions whose mystery will evaporate in §9.

7.1. DEFINITIONS. *Let $X=G/P$ be a complex flag manifold, $x \in X$ and G_0 a real form of G. If $\mathcal{P}_x+\tau\mathcal{P}_x$ is a subalgebra of $\mathcal{G}$, then we say that*

$\mathcal{P}_x$ is an ***integrable*** *parabolic subalgebra of $\mathcal{G}$,*

P_x is an ***integrable*** *parabolic subgroup of G,*

x is an ***integrable*** *point of X, and*

$G_0(x)$ is an ***integrable*** *G_0-orbit on X.*

These notions are, of course, dependent on choice of the real form G_0.

The basic facts are given by

7.2. THEOREM. *Let $\mathcal{G}$ be a complex semisimple Lie algebra, $\mathcal{P}$ a parabolic subalgebra, and $\mathcal{G}_0$ a real form of $\mathcal{G}$. Denote*

$$\mathcal{U} = \mathcal{P}^u \cap \tau\mathcal{P}^u \quad \text{and} \quad \mathcal{N} \text{ is the normalizer of } \mathcal{U} \text{ in } \mathcal{G}.$$

Then the following conditions are equivalent:

(i) *$\mathcal{P}$ is integrable, i.e. $\mathcal{P}+\tau\mathcal{P}$ is a subalgebra of $\mathcal{G}$.*

(ii) *$\mathcal{P}+\tau\mathcal{P}=\mathcal{N}$.*

(iii) *$\mathcal{P}+\tau\mathcal{P}$ is an algebra and $\mathcal{U}$ is its nilpotent radical.*

(iv) *$\mathcal{U}$ is the nilpotent radical of $\mathcal{N}$.*

(v) *$\mathcal{P}$ normalizes $\mathcal{U}$, i.e. $\mathcal{P} \subset \mathcal{N}$.*

PROOF. Relative to a τ-stable Cartan subalgebra $\mathcal{H} \subset \mathcal{P}$ of $\mathcal{G}$ and a subset Φ of a simple root system such that $\mathcal{P}=\mathcal{P}_\Phi$, we have $\mathcal{P}+\tau\mathcal{P}$ $=\mathcal{H}+\sum_\Gamma \mathcal{G}_\varphi$ where $\Gamma=\Phi^r \cup \tau\Phi^r \cup \Phi^u \cup \tau\Phi^u$. If $\varphi \in \Gamma$ and $-\varphi \notin \Gamma$ it follows that $\varphi \in \Phi^u \cup \tau\Phi^u$, $\varphi \notin \Phi^u \cap \tau\Phi^r$ and $\varphi \notin \Phi^r \cap \tau\Phi^u$; then $\varphi \in \Phi^u$ $\cap \tau\Phi^u$ or $\varphi \in \Phi^u \cap -\tau\Phi^u$ or $\varphi \in -\Phi^u \cap \tau\Phi^u$; if $\varphi>0$ and $\tau\varphi>0$ we conclude that $\varphi \in \Phi^u \cap \tau\Phi^u$.

Let $\mathcal{P}+\tau\mathcal{P}$ be an algebra. The discussion above shows that it has nilpotent radical $\mathcal{U}$. Thus (i) implies (iii). A parabolic subalgebra is the normalizer of its nilpotent radical; thus (iii) implies (ii). And visibly (ii) implies (i). Now (i), (ii) and (iii) are equivalent.

If $\mathcal{P}+\tau\mathcal{P}=\mathcal{N}$ then $\mathcal{P} \subset \mathcal{N}$; thus (ii) implies (v). If $\mathcal{P} \subset \mathcal{N}$ then $\mathcal{N}$ is parabolic; $\tau\mathcal{U}=\mathcal{U}$ implies $\tau\mathcal{N}=\mathcal{N}$ so $\mathcal{P}+\tau\mathcal{P} \subset \mathcal{N}$; now the discussion of the first paragraph of the proof shows that $\mathcal{U}$ contains the nilpotent radical of $\mathcal{N}$; thus (v) implies (iv). Now we need only prove that (iv) implies (ii).

Let $\mathcal{U}$ be the nilpotent radical of $\mathcal{N}$. If $\mathcal{N}$ is a parabolic subalgebra of $\mathcal{G}$, then $\mathcal{N}=\mathcal{P}_\Psi$ with $\Phi \subset \Psi$ such that $\Psi^u=\Phi^u \cap \tau\Phi^u$. As Δ^+ is the disjoint union of $\Phi^r \cap \Delta^+$, $\Phi^u \cap (\tau\Phi^r \cup -\tau\Phi^u)$ and $\Phi^u \cap \tau\Phi^u$, and also is the disjoint union of $\Psi^r \cap \Delta^+$ and $\Psi^u = \Phi^u \cap \tau\Phi^u$, it follows that $\Psi^r \cap \Delta^+ = (\Phi^r \cap \Delta^+) \cup (\Phi^u \cap \tau\Phi^r) \cup (\Phi^u \cap -\tau\Phi^u)$ so $\mathfrak{N} = \mathcal{P}^r + \mathcal{P}^u + \tau\mathcal{P}^r + \tau\mathcal{P}^u = \mathcal{P} + \tau\mathcal{P}$. Thus (iv) implies (ii) provided that (iv) forces $\mathfrak{N}$ to be parabolic. In other words, the theorem is reduced to

7.3. LEMMA. *Let $\mathcal{G}$ be a complex semisimple Lie algebra, let $\mathfrak{U}$ be a nilpotent*[5] *subalgebra, and let $\mathfrak{N}$ be the normalizer of $\mathfrak{U}$ in $\mathcal{G}$. Suppose that $\mathfrak{N}$ contains a Cartan subalgebra $\mathfrak{K}$ of $\mathcal{G}$ and that $\mathfrak{U}$ is the nilpotent*[5] *radical of $\mathfrak{N}$. Then $\mathfrak{N}$ is a parabolic subalgebra of $\mathcal{G}$.*

PROOF. $\mathfrak{N} = \mathcal{L} + \mathfrak{U}$ semidirect where $\mathfrak{K} \subset \mathcal{L}$ and $\mathcal{L}$ is reductive. $\mathfrak{U} = \sum_A \mathcal{G}_\varphi$ for some set A of $\mathfrak{K}$-roots of $\mathcal{G}$. $\mathcal{L} = \mathcal{L}' \oplus \mathcal{S}$ where $\mathcal{S} \subset \mathfrak{K}$ is the center of $\mathcal{L}$ and $\mathcal{L}' = [\mathcal{L}, \mathcal{L}]$ is the semisimple part. $\mathfrak{K} + \mathfrak{U}$ is contained in a Borel subalgebra of $\mathcal{G}$, so we order the roots such that $\varphi > 0$ for every $\varphi \in A$.

Let $\mathcal{L}$ be semisimple, let W_L be its Weyl group and define $\alpha = \sum_A \varphi$. Then α is a W_L-invariant positive sum of positive roots. The action of W_L on $\mathfrak{K}$ has no trivial subrepresentation. Thus $\alpha = 0$ and A is empty. It follows that $\mathfrak{U} = 0$, so $\mathfrak{N} = \mathcal{G}$, which is a parabolic subalgebra of $\mathcal{G}$.

Now we may assume $\mathcal{L}$ not semisimple, so its center $\mathcal{S} \neq 0$. Define $B = \{\varphi \in A : \varphi(\mathcal{S}) = 0\}$ and $\beta = \sum_B \varphi$. As above it follows that $\beta = 0$ so B is empty. Let $\mathcal{Z}$ be the centralizer of $\mathcal{S}$ in $\mathcal{G}$ and order the weights of $\mathcal{S}$ so that positive weight spaces are sums of positive root spaces. Then the corresponding parabolic subalgebra $\mathcal{Q}$ of $\mathcal{G}$ has $\mathcal{Z} = \mathcal{Q}^r$, and $\mathfrak{U} \subset \mathcal{Q}^u$ because $\mathcal{Z} \cap \mathfrak{U} = 0$. Let $\mathcal{W} = \mathfrak{N} \cap \mathcal{Q}^u$; so $\mathfrak{U} \subset \mathcal{W} \subset \mathcal{Q}^u$. If $\mathfrak{U} \neq \mathcal{Q}^u$ then $\mathfrak{U} \neq \mathcal{W}$ because a proper nilpotent subalgebra of a nilpotent Lie algebra cannot be its own normalizer. But $\mathfrak{N} = \mathcal{L} + \mathfrak{U}$ with $\mathcal{L} \subset \mathcal{Q}^r$ and $\mathfrak{U}$ is the nilpotent radical of $\mathfrak{N}$; thus $\mathfrak{U} = \mathfrak{N} \cap \mathcal{Q}^u = \mathcal{W}$; now $\mathfrak{U} = \mathcal{Q}^u$ so $\mathfrak{N} = \mathcal{Q}$. Q.E.D.

Now we go on to construct all integrable G_0-orbits on $X = G/P$.

Let $\mathcal{G}$ be a complex semisimple Lie algebra, $\mathfrak{K}$ a Cartan subalgebra and Π a simple root system. We know that each conjugacy class of parabolic subalgebras of $\mathcal{G}$ is represented by just one algebra $\mathcal{P}_\Phi$, $\Phi \subset \Pi$; in particular, if $\mathcal{P}$ and $\mathcal{Q}$ are parabolic subalgebras of $\mathcal{G}$, then the conditions

(i) $\mathcal{Q}$ contains a conjugate of $\mathcal{P}$,
(ii) $\mathcal{P}$ is contained in a conjugate of $\mathcal{Q}$, and
(iii) $\mathcal{P}$ is conjugate to $\mathcal{P}_\Phi$, $\mathcal{Q}$ to $\mathcal{P}_\Psi$, with $\Phi \subset \Psi \subset \Pi$,

are equivalent.

---

[5] This means that $\mathfrak{U}$ is represented by nilpotent matrices in the adjoint representation of $\mathcal{G}$.

7.4. Definition. *We write* $\mathcal{P} \prec \mathcal{Q}$ *to denote the equivalent conditions* (i), (ii) *and* (iii) *above.*

Now let $\mathfrak{g}_0$ be a real form of $\mathfrak{g}$ and assume $\mathfrak{H}$ stable under the complex conjugation τ of $\mathfrak{g}$ over $\mathfrak{g}_0$. Let $w \in W_G$ be the Weyl group element which carries the simple root system $\tau(\Pi)$ back to Π. We have a $(w\tau)$-stable subset $\Psi_0 \subset \Pi$ determined by: $(\mathfrak{M}+\mathfrak{A}+\mathfrak{N})^{\mathfrak{c}}$ is conjugate to $\mathcal{P}_{\Psi_0}$ where $\mathfrak{M}+\mathfrak{A}+\mathfrak{N}$ is the minimal parabolic subalgebra of $\mathfrak{g}_0$. Now the conjugacy classes of parabolic subalgebras, such that the class contains an algebra defined/R (i.e. τ-stable), are represented by the $\mathcal{P}_\Psi$ such that

$$(7.5) \qquad \text{(a) } (w\tau)\Psi = \Psi \quad \text{and} \quad \text{(b) } \Psi_0 \subset \Psi \subset \Pi.$$

In order to apply the criterion (7.5), one uses the fact [15, p. 909] that, if $\mathfrak{g}$ is simple, then $w\tau$ acts trivially on Π except in the cases:

$\mathfrak{g}_0$	action of $w\tau$ on Π
$\mathfrak{SU}^k(n)$	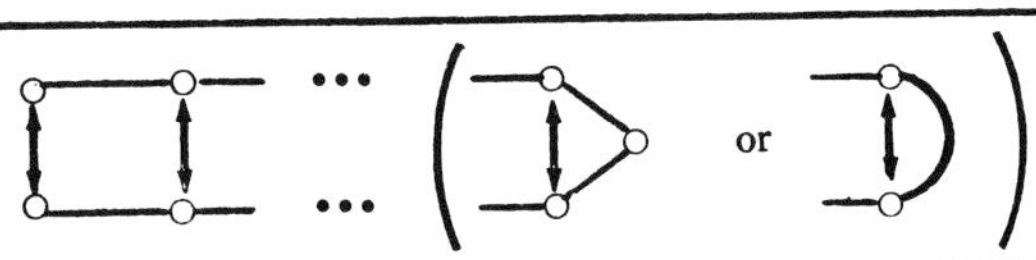
$\mathfrak{SO}^*(4n+2)$ $\mathfrak{SO}^{2k}(4n+2)$ $\mathfrak{SO}^{2k+1}(4n)$	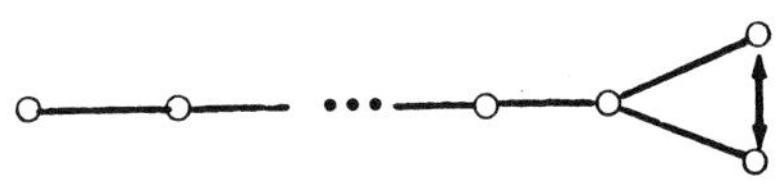
$\mathfrak{E}_{6,F_4}$; $\mathfrak{E}_{6,C_4}$	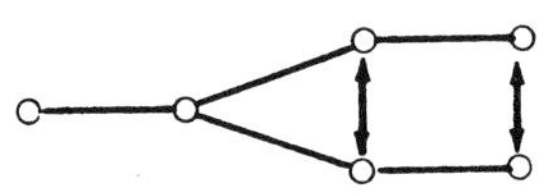

Now suppose that we have a fixed parabolic subalgebra $\mathcal{P} \subset \mathfrak{g}$. We look for the τ-stable parabolic subalgebras $\mathcal{Q} \subset \mathfrak{g}$ such that $\mathcal{P} \prec \mathcal{Q}$. Given such an algebra $\mathcal{Q}$ we have a conjugate $\mathcal{P}_x$ of $\mathcal{P}$ such that $\mathcal{P}_x + \tau\mathcal{P}_x \subset \mathcal{Q}$. Suppose that $\mathcal{P}$ is conjugate to $\mathcal{P}_\Phi$ and $\mathcal{Q}$ is conjugate to $\mathcal{P}_\Psi$; then $\Phi \subset \Psi$ and Ψ satisfies (7.5). In other words

$$(7.6) \quad \text{(a) } (w\tau)\Psi = \Psi \quad \text{and} \quad \text{(b) } \{\Psi_0 \cup \Phi \cup (w\tau)\Phi\} \subset \Psi \subset \Pi.$$

Conversely, if Ψ satisfies (7.6), then $\mathcal{P}_\Psi$ has a τ-stable conjugate $\mathcal{Q}$ because Ψ satisfies (7.5), and $\mathcal{P} \prec \mathcal{Q}$ because $\Phi \subset \Psi$. This proves the first statement of

7.7. THEOREM. *Let $X = G/P$ be a complex flag manifold, G_0 a real form of G, $\mathcal{H}$ a τ-stable Cartan subalgebra of $\mathcal{G}$, Π a simple $\mathcal{H}$-root system, and $\Phi \subset \Pi$ such that $\mathcal{P}$ is conjugate to $\mathcal{P}_\Phi$. Let $w \in W_G$ such that $(w\tau)\Pi = \Pi$ and define*

$$\Psi_X = \Psi_0 \cup \Phi \cup (w\tau)\Phi. \tag{7.8}$$

1. *The $\mathcal{G}$-conjugacy classes of parabolic subalgebras $\mathcal{Q} \subset \mathcal{G}$ such that*

$$\mathcal{P} \prec \mathcal{Q} \quad \text{and} \quad \tau(\mathcal{Q}) = \mathcal{Q}$$

are the classes represented by algebras $\mathcal{P}_\Psi$ such that

$$(w\tau)\Psi = \Psi \quad \text{and} \quad \Psi_X \subset \Psi \subset \Pi.$$

2. *If $\mathcal{Q} \subset \mathcal{G}$ is a τ-stable parabolic subalgebra then the following conditions are equivalent:*

(2a) $\mathcal{P} \prec \mathcal{Q}$.

(2b) $\mathcal{Q} = \mathcal{P}_x + \tau\mathcal{P}_x$ *for some (necessarily integrable)* $x \in X$.

(2c) $\mathcal{Q}$ *is conjugate to* $\mathcal{P}_\Psi$ *with* $\Psi_X \subset \Psi \subset \Pi$.

PROOF. Statement (1) was proved in the discussion preceding the theorem.

If $\tau(\mathcal{Q}) = \mathcal{Q}$ and $\mathcal{P} \prec \mathcal{Q}$ then (1) implies that $\mathcal{Q}$ is conjugate to $\mathcal{P}_\Psi$ with $\Psi_X \subset \Psi \subset \Pi$. If $\tau(\mathcal{Q}) = \mathcal{Q}$ and $\mathcal{Q}$ is conjugate to $\mathcal{P}_\Psi$ with $\Psi_X \subset \Psi \subset \Pi$ then $\Phi \subset \Psi$ so $\mathcal{P} \prec \mathcal{Q}$. Now (2a) and (2c) are equivalent. As (2b) trivially implies (2a), we now need only check that (2a) implies (2b).

Let $\tau(\mathcal{Q}) = \mathcal{Q}$ and $\mathcal{P} \prec \mathcal{Q}$. If we have a Borel subalgebra $\mathcal{B} \subset \mathcal{Q}$ of $\mathcal{G}$ such that $\mathcal{B} + \tau\mathcal{B} = \mathcal{Q}$, then we choose a conjugate $\mathcal{P}_x$ of $\mathcal{P}$ such that $\mathcal{B} \subset \mathcal{P}_x \subset \mathcal{Q}$; that gives $\mathcal{Q} = \mathcal{B} + \tau\mathcal{B} \subset \mathcal{P}_x + \tau\mathcal{P}_x \subset \mathcal{Q} + \tau\mathcal{Q} = \mathcal{Q}$, and we conclude $\mathcal{P}_x + \tau\mathcal{P}_x = \mathcal{Q}$. Now the proof of Theorem 7.7 comes down to the proof of

7.9 LEMMA. *Let $\mathcal{G}$ be a complex semisimple Lie algebra, $\mathcal{G}_0$ a real form of $\mathcal{G}$, and $\mathcal{Q}$ a τ-stable parabolic subalgebra of $\mathcal{G}$. Then there is a Borel subalgebra $\mathcal{B} \subset \mathcal{Q}$ of $\mathcal{G}$ with $\mathcal{B} + \tau\mathcal{B} = \mathcal{Q}$.*

PROOF. $\mathcal{Q} = \mathcal{Q}^r + \mathcal{Q}^u$ with each summand stable under τ. Now $\mathcal{Q}_0^r = \mathcal{G}_0 \cap \mathcal{Q}^r$ is a real form of $\mathcal{Q}^r$ and is a reductive subalgebra of maximal rank in $\mathcal{G}_0$. Let $\mathcal{H}_0 = \mathcal{H}_T + \mathcal{H}_V$ be a maximally compact Cartan subalgebra of $\mathcal{Q}_0^r$. Apply Theorem 4.5 to the action of Q_0^r on $Q^r/(\text{Borel subgroup})$; it shows that $\mathcal{Q}^r$ has a Borel subalgebra $\mathcal{B}_1$ such that $\mathcal{B}_1 + \tau\mathcal{B}_1 = \mathcal{Q}^r$.

Now $\mathcal{B}=\mathcal{B}_1+\mathcal{Q}^u$ is a Borel subalgebra of $\mathcal{G}$, and $\mathcal{B}+\tau\mathcal{B}=\mathcal{Q}^r+\mathcal{Q}^u=\mathcal{Q}$.

Q.E.D.

We extend the proof that (2a) implies (2b) in Theorem 7.7 to an existence and enumeration theorem for integrable orbits.

7.10. THEOREM. *Let $X=G/P$ be a complex flag manifold, G_0 a real form of G, and $\mathcal{H}$ a τ-stable Cartan subalgebra of $\mathcal{G}$. Let Π be a simple $\mathcal{H}$-root system, $w\in W_G$ such that $(w\tau)\Pi=\Pi$, $\Phi\subset\Pi$ such that $\mathcal{P}$ is conjugate to $\mathcal{P}_\Phi$, and $\Psi_X=\Phi\cup(w\tau)\Phi\cup\Psi_0$, $(w\tau)$-stable subset of Π. For every set Ψ such that $\Psi_X\subset\Psi\subset\Pi$ and $(w\tau)\Psi=\Psi$, choose a τ-stable parabolic subalgebra $\mathcal{Q}_\Psi\subset\mathcal{G}$ which is conjugate to $\mathcal{P}_\Psi$, and choose a maximally compact Cartan subalgebra $\mathcal{H}_{\Psi,0}$ of the real form $\mathcal{G}_0\cap\mathcal{Q}_\Psi^r$ of $\mathcal{Q}_\Psi^r$.*

1. *If $\mathcal{B}_1$ is a Borel subalgebra of $\mathcal{Q}_\Psi^r$ which contains $\mathcal{H}_\Psi=\mathcal{H}_{\Psi,0}^{\mathbf{C}}$ and such that $\mathcal{B}_1+\tau\mathcal{B}_1=\mathcal{Q}_\Psi^r$, and if $\mathcal{B}$ denotes the Borel subalgebra $\mathcal{B}_1+\mathcal{Q}_\Psi^u$ of $\mathcal{G}$, and if $\mathcal{P}_x$ is the conjugate of $\mathcal{P}$ such that $\mathcal{B}\subset\mathcal{P}_x\subset\mathcal{Q}_\Psi$, then*
   (1a) *x is an integrable point of X,*
   (1b) *$G_0(x)$ is an integrable G_0-orbit on X, and*
   (1c) *$\mathcal{P}_x+\tau\mathcal{P}_x=\mathcal{Q}_\Psi$.*

2. *If $z\in X$ is an integrable point, and if the parabolic subalgebra $\mathcal{P}_z+\tau\mathcal{P}_z$ is conjugate to $\mathcal{P}_\Psi$, then in* (1) *we can choose the Borel subalgebra $\mathcal{B}_1$ of $\mathcal{Q}_\Psi^r$ in such a manner that $z\in G_0(x)$.*

3. *For each set Ψ, $\Psi_X\subset\Psi\subset\Pi$ and $(w\tau)\Psi=\Psi$, choose one point $x_\Psi\in X$ as constructed in* (1). *Let $Q_{\Psi,0}^r$ denote the identity component of $G_0\cap Q_\Psi^r$. Then $Q_\Psi(x_\Psi)=Q_\Psi^r(x_\Psi)=Q_\Psi^r/(Q_\Psi^r\cap P_{x_\Psi})$ is a complex flag manifold X_Ψ, and the open $Q_{\Psi,0}^r$-orbits on X_Ψ are in one to one correspondence with the integrable orbits $G_0(x)$ such that $\mathcal{P}_x+\tau\mathcal{P}_x$ is conjugate to $\mathcal{P}_\Psi$.*

4. *Let K_Ψ be a maximal compact subgroup of $Q_{\Psi,0}^r$ such that $\mathcal{K}_\Psi$ contains the compact part of $\mathcal{H}_{\Psi,0}$ and is orthogonal to the noncompact part. Then the integrable orbits $G_0(x)\subset X$ such that $\mathcal{P}_x+\tau\mathcal{P}_x$ is conjugate to $\mathcal{P}_\Psi$ are enumerated by the double coset space*

$$W_{K_\Psi}\backslash W^{\mathcal{H}_{\Psi,0}}_{Q_\Psi^r}/W^{\mathcal{H}_{\Psi,0}}_{P^r_{x_\Psi}\cap\tau P^r_{x_\Psi}}$$

where superscript $\mathcal{H}_{\Psi,0}$ means the subgroup of the Weyl group which preserves $\mathcal{H}_{\Psi,0}$ in $\mathcal{H}_\Psi$.

PROOF. Statement (1) is the construction employed in proving that (2a) implies (2b) in Theorem 7.7.

Let $\mathcal{P}_z+\tau\mathcal{P}_z$ be conjugate to $\mathcal{P}_\Psi$. Then $\mathcal{Q}_\Psi$ and $\mathcal{P}_z+\tau\mathcal{P}_z$ are parabolic subalgebras of $\mathcal{G}$ which are conjugate by an element of G and are defined/$\mathbf{R}$; so they are conjugate by an element of G_0. Replacing z by $g(z)$ for some $g\in G_0$ we may assume $\mathcal{P}_z+\tau\mathcal{P}_z=\mathcal{Q}_\Psi$. Further replacing z

by $g(z)$ for some $g \in G_0$ we may also assume $\mathcal{H}_{\Psi,0} \subset \mathcal{P}_z$. Let $\mathcal{B}_2$ be a Borel subalgebra of $\mathcal{P}_z^r \cap \tau\mathcal{P}_z^r$ containing $\mathcal{H}_\Psi$ such that $\mathcal{B}_2 + \tau\mathcal{B}_2 = \mathcal{P}_z^r \cap \tau\mathcal{P}_z^r$; such $\mathcal{B}_2$ exist because $\mathcal{H}_{\Psi,0}$ is maximally compact in $\mathcal{G}_0 \cap \mathcal{Q}_\Psi^r$, hence in $\mathcal{G}_0 \cap (\mathcal{P}_z^r \cap \tau\mathcal{P}_z^r)$. Now $\mathcal{B} = \mathcal{B}_2 + (\mathcal{P}_z^r \cap \tau\mathcal{P}_z^{-u}) + \mathcal{P}_z^u$ is a Borel subalgebra of $\mathcal{G}$ such that $\mathcal{B} \subset \mathcal{P}_z \subset \mathcal{Q}_\Psi$ and $\mathcal{B} + \tau\mathcal{B} \supset \mathcal{P}_z$. As $\mathcal{P}_z + \tau\mathcal{P}_z = \mathcal{Q}_\Psi$ now $\mathcal{B} + \tau\mathcal{B} = \mathcal{Q}_\Psi$. Decompose $\mathcal{B} = \mathcal{B}_1 + \mathcal{Q}_\Psi^u$ where $\mathcal{B}_1$ is a Borel subalgebra of $\mathcal{Q}_\Psi^r$ containing $\mathcal{H}_\Psi$; now $\mathcal{B}_1 + \tau\mathcal{B}_1 = \mathcal{Q}_\Psi^r$. Thus, having replaced z by $g(z)$ for some $g \in G_0$, we end up with z as one of the points $x \in X$ constructed in (1). In other words, we started with $z \in G_0(x)$. Statement (2) is proved.

The choice available in the construction of statement (1) is the choice of $\mathcal{B}_1$. That choice is free, without changing $G_0(x)$, $\mathcal{Q}_\Psi$ or $\mathcal{H}_{\Psi,0}$, up to conjugation by an element of $Q_{\Psi,0}$ which normalizes $\mathcal{H}_{\Psi,0}$, i.e. up to conjugation by an element of $G_0 \cap Q_\Psi^r$, i.e. up to conjugation by an element of $Q_{\Psi,0}^r$. This proves (3), and (4) follows from Theorem 4.9. Q.E.D.

Theorems 7.7 and 7.10 give a complete description of the integrable G_0-orbits on X in terms of the isotropy subalgebras of $\mathcal{G}$ at integrable points. But they do not "locate" the integrable orbits on X.

The complex flag manifold $X = G/P$ always has some integrable G_0-orbits, for at the beginning of this §7 we observed that $\mathcal{P}_x + \tau\mathcal{P}_x = \mathcal{G}$ if and only if $G_0(x)$ is open in X. But it can happen that the open orbits are the only integrable G_0-orbits:

7.11. PROPOSITION. *Let $X = G/P$ be a complex flag manifold where P is a maximal parabolic subgroup*[6] *of G. Let $x \in X$ and let G_0 be a real form of G. Then the orbit $G_0(x)$ is integrable if and only if*

(a) *$G_0(x)$ is open in X, or*

(b) *$G_0(x)$ is the closed orbit and* $\dim_R G_0(x) = \dim_C X$.

PROOF. Let $G_0(x)$ be integrable. If $\tau\mathcal{P}_x = \mathcal{P}_x$ we are in case (b). If $\tau\mathcal{P}_x \neq \mathcal{P}_x$ then $\mathcal{P}_x + \tau\mathcal{P}_x = \mathcal{G}$ by maximality of P, so we are in case (a). Q.E.D.

7.12. COROLLARY. *Let $X_0 = G_0/K$ be a hermitian symmetric space of noncompact type, $G = G_0^{\mathbb{C}}$, and $X = G_u/K = G/P$ the compact dual of X. Decompose X_0 as a product of irreducible factors $X_0^{(i)} = G_0^{(i)}/K^{(i)}$, G as a product of simple groups $G^{(i)} = G_0^{(i)C}$, and X as a product of irreducible factors $X^{(i)} = G_u^{(i)}/K^{(i)} = G^{(i)}/P^{(i)}$, so that the G_0-orbits on X are just the products of $G_0^{(i)}$-orbits on the $X^{(i)}$. Let $x = (x_1, \cdots, x_t) \in X^{(1)} \times \cdots$*

[6] This means that $P \neq G$ and there is no parabolic subgroup $Q \subset G$ with $P \subsetneq Q \subsetneq G$. In other words, if $\mathcal{P} = \mathcal{P}_\Phi$ where Φ is in a simple root system Π, then $|\Pi| = |\Phi| + 1$.

$\times X^{(t)}=X$. Then the orbit $G_0(x)\subset X$ is integrable if, and only if, for each i,

either $G_0^{(i)}(x_i)$ is open in $X^{(i)}$

or $X_0^{(i)}$ is of tube type and $G_0^{(i)}(x_i)$ is its Bergman-Šilov boundary in $X^{(i)}$.

PROOF. We may assume X_0 irreducible. Then $t=1$ and P is a maximal parabolic subgroup of G, and the two cases of Corollary 7.12 correspond to the respective cases (a) and (b) of Proposition 7.11. Q.E.D.

CHAPTER II. DECOMPOSITION OF A REAL GROUP ORBIT INTO COMPLEX ANALYTIC PIECES

Let $X=G/P$ be a complex flag manifold and let G_0 be a real form of G. §8 and §9 are an analysis of the "amount" of complex structure inherited from X by an arbitrary orbit $G_0(x)\subset X$.

In §8 the orbit $G_0(x)$ is decomposed into "holomorphic arc components" $S_{[z]}$. They are equivalence classes for the G_0-invariant relation: $y\in S_{[z]}$ if and only if it can be joined to z by a chain of holomorphic (in X) arcs in $G_0(x)$. The holomorphic arc components are real analytic submanifolds of $G_0(x)$. In fact,

$$N_{[z],0} = \text{identity component of } \{g\in G_0\colon gS_{[z]} = S_{[z]}\}$$

is a closed analytic subgroup of G_0 that is transitive on $S_{[z]}$. We approach analysis of the holomorphic arc components by studying the algebra $\mathfrak{N}_{[z]}=\mathfrak{N}_{[z],0}^{\mathbf{C}}$.

Theorem 8.5, Proposition 8.7 and Theorem 8.9 determine subspaces $\mathfrak{Q}_{[z]}\subset\mathfrak{M}_{[z]}\subset\{\mathfrak{N}_{[z]}\cap(\mathfrak{P}_z+\tau\mathfrak{P}_z)\}$ stable under τ, such that $\mathfrak{Q}_{[z]}$ is a parabolic subalgebra of $\mathfrak{g}$ (so $N_{[z],0}$ is the identity component of a parabolic subgroup of G_0), $S_{[z]}$ is a complex submanifold of X if and only if $\mathfrak{M}_{[z]}=\mathfrak{N}_{[z]}$, and $\mathfrak{M}_{[z]}=\mathfrak{N}_{[z]}$ if and only if $\mathfrak{M}_{[z]}$ is an algebra. Thus one can decide whether the holomorphic arc components of $G_0(x)$ are complex submanifolds of X, and in that case their structure is determined in principle. If they are not complex then their submanifolds $Q_{[z],0}(z)\subset S_{[z]}$, where $Q_{[z],0}$ is the analytic group for $\mathfrak{Q}_{[z]}\cap\mathfrak{g}_0$, are acceptable substitutes which are complex submanifolds of X; we call them the "algebraic arc components" of $G_0(x)$.

If $P_1\subset P_2$ are parabolic subgroups of G, $X_i=G/P_i$ and $\pi\colon X_1\to X_2$ the projection, $x_i\in X_i$ with $\pi(x_1)=x_2$, then Proposition 8.14 relates the holomorphic and algebraic arc component structure of the $G_0(x_i)$.

§8 ends with a description of the space of holomorphic arc components of $G_0(x)$.

In §9 we study certain global conditions on the holomorphic arc components of $G_0(x)$. They are

(i) partially complex: $S_{[x]}$ is a complex manifold,

(ii) flag type: $N_{[x]}(x)$ is a complex flag manifold,

(iii) measurable: $S_{[x]}$ has an $N_{[x],0}$-invariant measure,

(iv) polarized: we can arrange $\mathcal{H}$ so that $\tau\mathcal{P}_x^r = \mathcal{P}_x^r$,

(v) integrable: $\mathcal{P}_x + \tau\mathcal{P}_x$ is an algebra.

The important ones are "partially complex" and "measurable," and the others are useful in their study. Theorem 9.2 gives the basic analysis of measurable orbits, describing $\mathfrak{N}_{[x]}$ and proving them partially complex and of flag type. Theorem 9.7 performs a similar task for orbits of flag type. Then Theorem 9.9 shows that, if $G_0(x)$ is polarized, then the conditions "measurable," "integrable," and "partially complex and of flag type" coincide. Various corollaries complete the general part of §9.

The specific part of §9 starts with Theorem 9.12, which says that the closed G_0-orbit on X is automatically measurable. That result is related [**12**] to the existence of principal series representations of G_0. Proposition 9.18 and Corollary 9.21 extend the method of Theorem 9.12.

§9 ends with a description of the case where X is a hermitian symmetric space and G_0 is an *arbitrary* real form of G. Proposition 9.24 gives measurability criteria for the open G_0-orbits, and one criterion is that they be indefinite-kaehler symmetric spaces. The surprising corollary is that if the maximal compact subgroups of G_0 are of rank equal to that of G_0, then the open G_0-orbits on X are indefinite-kaehler symmetric spaces.

NOTES FOR CHAPTER II. From August 1965 to May 1968, this author had a number of conversations with Bertram Kostant in which we tried to find the structure of the holomorphic arc components. We succeeded in the integrable case and saw that the normalizer of $\mathcal{P}_x^u \cap \tau\mathcal{P}_x^u$ should play a role. Statement 1 of Theorem 8.5 is due to Kostant, and the general case as worked out in §8 was undoubtedly guided by the background of discussions with him.

Lemma 9.22 and Corollary 9.25 are due to Kostant.

Except for the basic contribution of Kostant described above, the material of §8 and §9 was worked out by this author.

8. **Holomorphic arc components.** Given a complex flag manifold $X = G/P$ and a real form G_0 of G, the open G_0-orbits on X are complex submanifolds. If $G_0(x) \subset X$ is the Borel embedding of an Hermitian symmetric space of noncompact type, then [**15**] one also knows that

every G_0-orbit on the boundary of $G_0(x)$ has a natural decomposition into complex submanifolds of X. In this section we give an analogous decomposition of an arbitrary G_0-orbit into "holomorphic arc components," including an explicit criterion and characterization of the cases where those components are complex submanifolds of X.

Let D be a subset of a complex analytic space V. By **holomorphic arc in** D we mean a holomorphic map

$$f: \{z \in \mathbf{C}: |z| < 1\} \to V$$

of the unit disc into V such that D contains the image of f. By **chain of holomorphic arcs in** D we mean a finite sequence $\{f_1, \cdots, f_k\}$ of holomorphic arcs in D such that the image of f_i meets the image of f_{i+1} for $1 \leqq i < k$.

8.1. DEFINITION. *Let D be a subset of a complex analytic space V. Then the* **holomoprphic arc components** *of D in V are the equivalence classes of elements of D under the relation: $v_1 \sim v_2$ if there is a chain $\{f_1, \cdots, f_k\}$ of holomorphic arcs in D such that v_1 is in the image of f_1 and v_2 is in the image of f_k.*

We draw some simple consequences from this definition.

8.2. LEMMA. *Let V be a complex analytic space, let A be a group of holomorphic diffeomorphisms of V, and let D be an A-stable subset of V. Let S be a holomorphic arc component of D in V and define $N_A(S) = \{a \in A : a(S) = S\}$.*

1. *If $a \in A$ and $a(S)$ meets S, then $a(S) = S$ i.e. $a \in N_A(S)$.*
2. *If D is an A-orbit then $N_A(S)$ is transitive on S.*
3. *If A is a Lie transformation group on V, and if D is an A-orbit, then $N_A(S)$ is a Lie subgroup of A, and S and D are embedded real submanifolds of V.*

PROOF. For (1) we have $s_1, s_2 \in S$ with $a(s_1) = s_2$. If $s \in S$ we have a chain $\{f_1, \cdots, f_k\}$ of holomorphic arcs in D from s_1 to s; now $\{a \cdot f_1, \cdots, a \cdot f_k\}$ is a chain of holomorphic arcs in D from $a(s_1) = s_2$ to $a(s)$; thus $a(s) \in S$; that proves $a(S) \subset S$. Similarly $a^{-1}(S) \subset S$. Thus $a(S) = S$.

For (2) let $s_1, s_2 \in S$. As $s_i \in D$ we have $a \in A$ with $a(s_1) = s_2$. Now $a \in N_A(S)$ by (1). Thus $N_A(S)$ has an element that sends s_1 to s_2.

For (3) we first note that S is arcwise connected. Lifting arcs via $N_A(S) \to S$ we see that the identity component $N_A(S)^0$ is arcwise connected and is transitive on S; thus $N_A(S)^0$ is an analytic subgroup of A, so $N_A(S)$ is a Lie subgroup, and S is an embedded real submanifold of the embedded submanifold D. Q.E.D.

8.3. NOTATION. Let $X=G/P$ be a complex flag manifold, G_0 a real form of G, and $x\in X$. Denote

(8.3a) $S_{[x]}$: holomorphic arc component of $G_0(x)$ in X that contains x.

(8.3b) $N_{[x],0}$: identity component of $\{g\in G_0: gS_{[x]}=S_{[x]}\}$.

(8.3c) $\mathfrak{N}_{[x],0}$ is the Lie algebra of $N_{[x],0}$, and $\mathfrak{N}_{[x]}=\mathfrak{N}^{\mathbb{C}}_{[x],0}$.

(8.3d) $N_{[x]}$ is the complex analytic subgroup of G for $\mathfrak{N}_{[x]}$.

Lemma 8.2 says that $S_{[x]}$ is an embedded real submanifold of X, $N_{[x],0}$ is a Lie subgroup of G_0, and $N_{[x],0}$ is transitive on $S_{[x]}$. Note $\tau\mathfrak{N}_{[x]}=\mathfrak{N}_{[x]}$ by construction.

8.4. LEMMA. *Let $X=G/P$ be a complex flag manifold, G_0 a real form of G, $x\in X$, $\mathfrak{H}\subset\mathfrak{P}_x$ a τ-stable Cartan subalgebra of $\mathfrak{G}$, and $\mathfrak{L}$ a complex subalgebra of $\mathfrak{G}$ such that*

$$\tau\mathfrak{L}=\mathfrak{L}, \qquad [\mathfrak{H},\mathfrak{L}]\subset\mathfrak{L} \quad \textit{and} \quad \mathfrak{L}\subset\mathfrak{P}_x+\tau\mathfrak{P}_x.$$

Let $\mathfrak{L}_0=\mathfrak{L}\cap\mathfrak{G}_0$, real form of $\mathfrak{L}$; let L and L_0 denote the analytic subgroups of G and G_0 for $\mathfrak{L}$ and $\mathfrak{L}_0$.

1. *$L_0(x)$ and $L(x)$ are complex submanifolds of X with $L_0(x)$ open in $L(x)$.*

2. *$L_0(x)\subset S_{[x]}$.*

3. *$\mathfrak{L}_0\subset\mathfrak{N}_{[x],0}$ and $L_0\subset N_{[x],0}$; $\mathfrak{L}\subset\mathfrak{N}_{[x]}$ and $L\subset N_{[x]}$.*

PROOF. Choice of $\mathfrak{H}$ specifies $\mathfrak{P}_x^r$ and $\mathfrak{P}_x^{\pm u}$. As $\mathfrak{L}$ is complex and normalized by $\mathfrak{H}$, it is $\mathfrak{L}\cap\mathfrak{H}$ plus certain root spaces; now $\mathfrak{L}\subset(\mathfrak{P}_x+\tau\mathfrak{P}_x)$ shows

$$\mathfrak{L}=\mathfrak{L}^0+\mathfrak{L}^++\mathfrak{L}^-, \qquad \mathfrak{L}^0=\mathfrak{L}\cap(\mathfrak{P}_x\cap\tau\mathfrak{P}_x), \qquad \mathfrak{L}^+=\mathfrak{L}\cap\tau\mathfrak{P}_x^{-u}\subset\mathfrak{P}_x$$

and (using $\tau\mathfrak{L}=\mathfrak{L}$)

$$\mathfrak{L}^-=\mathfrak{L}\cap\mathfrak{P}_x^{-u}\subset\tau\mathfrak{P}_x \qquad \text{with } \mathfrak{L}^-=\tau\mathfrak{L}^+.$$

$L(x)$ is a complex submanifold of X because L is a complex Lie subgroup of G. Now

$$\begin{aligned}\dim_R L_0(x) &= \dim_R\mathfrak{L}_0-\dim_R\mathfrak{L}_0\cap\mathfrak{P}_x\\ &= \dim_R\mathfrak{L}_0-\dim_R\mathfrak{L}_0\cap\mathfrak{P}_x\cap\tau\mathfrak{P}_x\\ &= \dim_{\mathbb{C}}\mathfrak{L}-\dim_{\mathbb{C}}\mathfrak{L}^0\\ &= \dim_{\mathbb{C}}\mathfrak{L}^++\dim_{\mathbb{C}}\mathfrak{L}^-=2\dim_{\mathbb{C}}\mathfrak{L}^-\\ &= 2\{\dim_{\mathbb{C}}\mathfrak{L}-\dim_{\mathbb{C}}\mathfrak{L}\cap\mathfrak{P}_x\}\\ &= 2\dim_{\mathbb{C}}L(x)=\dim_R L(x)\end{aligned}$$

shows that $L_0(x)$ is open in $L(x)$. That proves (1).

Covering by complex local coordinate neighborhoods one sees that $L_0(x)$ has just one holomorphic arc component. Thus $L_0(x)\subset S_{[x]}$, proving (2), and (3) follows immediately. Q.E.D.

We use Lemma 8.4 to prove $\mathfrak{N}_{[x]}$ parabolic by constructing a parabolic subalgebra $\mathfrak{Q}_{[x]}\subset\mathfrak{N}_{[x]}$ of $\mathfrak{G}$. That is the first step in the determination of $S_{[x]}$ in algebraic terms.

8.5. THEOREM. *Let $X=G/P$ be a complex flag manifold, G_0 a real form of G, $x\in X$, $\mathfrak{H}\subset\mathfrak{P}_x$ a τ-stable Cartan subalgebra of $\mathfrak{G}$, and Φ a subset of a simple $\mathfrak{H}$-root system such that $\mathfrak{P}_x=\mathfrak{P}_\Phi$.*

Define a τ-stable linear form δ_Φ on $\mathfrak{H}$ by

$$(8.6a)\qquad \delta_\Phi = \sum_{\Phi^u\cap\tau\Phi^u} \varphi.$$

Define τ-stable sets of $\mathfrak{H}$-roots of $\mathfrak{G}$ by

$$(8.6b)\qquad \Gamma^r = \{\varphi\in\Delta:\langle\varphi,\delta_\Phi\rangle = 0\},\qquad \Gamma^u = \{\varphi\in\Delta:\langle\varphi,\delta_\Phi\rangle > 0\},\qquad \Gamma^* = \Gamma^r\cup\Gamma^u.$$

Define a τ-stable parabolic subalgebra $\mathfrak{Q}_{[x]}\subset\mathfrak{G}$ by

$$(8.6c)\qquad \mathfrak{Q}^r_{[x]} = \mathfrak{H}+\sum_{\Gamma^r}\mathfrak{G}_\varphi,\qquad \mathfrak{Q}^u_{[x]} = \sum_{\Gamma^u}\mathfrak{G}_\varphi,\qquad \mathfrak{Q}_{[x]} = \mathfrak{Q}^r_{[x]}+\mathfrak{Q}^u_{[x]}.$$

1. *The normalizer $\mathfrak{N}_\mathfrak{G}(\mathfrak{U}_x)$ of $\mathfrak{U}_x=\mathfrak{P}^u_x\cap\tau\mathfrak{P}^u_x$ satisfies $\mathfrak{N}_\mathfrak{G}(\mathfrak{U}_x)^r\subset\mathfrak{Q}^r_{[x]}$ and $\mathfrak{N}_\mathfrak{G}(\mathfrak{U}_x)^u\subset\mathfrak{Q}^u_{[x]}$.*

2. $\mathfrak{U}_x\subset\mathfrak{N}_\mathfrak{G}(\mathfrak{U}_x)\subset\mathfrak{Q}_{[x]}\subset\{\mathfrak{N}_{[x]}\cap(\mathfrak{P}_x+\tau\mathfrak{P}_x)\}$.

3. *$\mathfrak{N}_{[x]}$ is a parabolic subalgebra of $\mathfrak{G}$ and $N_{[x],0}$ is the identity component of the parabolic subgroup $N_{[x]}\cap G_0$ of G_0.*

PROOF. Note $\mathfrak{H}+\mathfrak{U}_x\subset\mathfrak{N}_\mathfrak{G}(\mathfrak{U}_x)$, so $\mathfrak{N}_\mathfrak{G}(\mathfrak{U}_x)$ is the sum of $\mathfrak{H}$ with certain root spaces. Let φ be a root with $\mathfrak{G}_\varphi\subset\mathfrak{N}_\mathfrak{G}(\mathfrak{U}_x)$, let $\mathfrak{G}[\varphi]$ denote the three dimensional simple algebra $h_\varphi\mathbf{C}+\mathfrak{G}_\varphi+\mathfrak{G}_{-\varphi}$, and let $\mathfrak{Z}$ denote the centralizer of $\mathfrak{G}[\varphi]$ in $\mathfrak{G}$. Then

$$\mathfrak{Z} = \{h\in\mathfrak{H}:\varphi(h)=0\}+\sum_{\psi\perp\!\!\!\perp\varphi}\mathfrak{G}_\psi$$

where $\psi\perp\!\!\!\perp\varphi$ denotes strong orthogonality, i.e. denotes that neither of $\varphi\pm\psi$ is a root. And further

$$\mathfrak{G} = \mathfrak{Z}+\sum\mathfrak{M}_i,\qquad \mathfrak{M}_1 = \mathfrak{G}[\varphi],$$

where the $\mathfrak{M}_i$ are the nontrivial irreducible $\mathfrak{G}[\varphi]$-modules under $\mathrm{ad}_\mathfrak{G}$. Thus for $i\neq 1$ we have $\mathfrak{M}_i$ of the form

$$\mathfrak{M}_i = \sum_{j=-r}^{s} \mathfrak{g}_{\psi+j\varphi}, \qquad \psi \in \Delta, \qquad \frac{2\langle\psi, \varphi\rangle}{\langle\varphi, \varphi\rangle} = r - s,$$

for the maximal string $S_\psi = \{\psi + j\varphi\}_{-r \leq j \leq s}$ of roots.

Suppose that S_ψ meets $\Phi^u \cap \tau\Phi^u$, say $\psi + j_0\varphi \in \Phi^u \cap \tau\Phi^u$. Then $\mathfrak{g}_\varphi \subset \mathfrak{N}\mathfrak{g}(\mathfrak{U}_x)$ implies $\psi + j\varphi \in \Phi^u \cap \tau\Phi^u$ for $j_0 \leqq j \leqq s$, so we have an integer j_ψ, $-r \leqq j_\psi \leqq s$, such that $S_\psi \cap (\Phi^u \cap \tau\Phi^u) = \{\psi + j\varphi : j_\psi \leqq j \leqq s\}$. Note $S_\psi \subset \Phi^u \cap \tau\Phi^u$ if and only if $j_\psi = -r$. But $j' > j''$ implies $\langle\varphi, \psi + j'\varphi\rangle > \langle\varphi, \psi + j''\varphi\rangle$ because $\langle\varphi, \varphi\rangle > 0$; and $\mathrm{ad}(h_\varphi)$ has trace 0 $= \langle\varphi, \sum_{-r}^{s} (\psi + j\varphi)\rangle$ on $\mathfrak{M}_i$ because $\mathfrak{g}[\varphi]$ is semisimple; thus

either $S_\psi \subset \Phi^u \cap \tau\Phi^u$ and $\langle\varphi, \sum_{S_\psi \cap \Phi^u \cap \tau\Phi^u} \gamma\rangle = 0$

or $S_\psi \not\subset \Phi^u \cap \tau\Phi^u$ and $\langle\varphi, \sum_{S_\psi \cap \Phi^u \cap \tau\Phi^u} \gamma\rangle > 0$.

On the other hand, $\mathfrak{g}_{-\varphi} \subset \mathfrak{N}\mathfrak{g}(\mathfrak{U}_x)$ if and only if $\mathfrak{g}_{-\varphi}$ normalizes each $\mathfrak{U}_x \cap \mathfrak{M}_i$, and, as above, that is equivalent to $S_\psi \subset \Phi^u \cap \tau\Phi^u$ whenever S_ψ meets $\Phi^u \cap \tau\Phi^u$. Thus for $\varphi \notin \Phi^u \cap \tau\Phi^u$, we have $\langle\varphi, \delta_\Phi\rangle = 0$ if $\mathfrak{g}_\varphi \subset \mathfrak{N}\mathfrak{g}(\mathfrak{U}_x)^r$ and $\langle\varphi, \delta_\Phi\rangle > 0$ if $\mathfrak{g}_\varphi \subset \mathfrak{N}\mathfrak{g}(\mathfrak{U}_x)^u$. And for φ in $\Phi^u \cap \tau\Phi^u$ we have $\mathfrak{g}_\varphi \subset \mathfrak{N}\mathfrak{g}(\mathfrak{U}_x)^u$ with $\langle\varphi, \delta_\Phi\rangle > 0$. That proves $\mathfrak{N}\mathfrak{g}(\mathfrak{U}_x)^u \subset \mathfrak{Q}_{[x]}^u$ and $\mathfrak{N}\mathfrak{g}(\mathfrak{U}_x)^r \subset \mathfrak{Q}_{[x]}^r$.

We have just proved $\mathfrak{U}_x \subset \mathfrak{N}\mathfrak{g}(\mathfrak{U}_x) \subset \mathfrak{Q}_{[x]}$. And $\mathfrak{U}_x \subset \mathfrak{N}\mathfrak{g}(\mathfrak{U}_x)^u \subset \mathfrak{Q}_{[x]}^u$ proves $\mathfrak{Q}_{[x]} \subset \mathcal{P}_x + \tau\mathcal{P}_x$. As $\mathfrak{H} \subset \mathfrak{Q}_{[x]} = \tau\mathfrak{Q}_{[x]}$, Lemma 8.4 says $\mathfrak{Q}_{[x]} \subset \mathfrak{N}_{[x]}$. That proves (2), and (3) follows because $\mathfrak{Q}_{[x]}$ is parabolic in $\mathfrak{g}$. Q.E.D.

The next step in the algebraic determination of S consists of expanding $\mathfrak{Q}_{[x]}$ out toward $\mathfrak{N}_{[x]}$ as follows.

8.7. Proposition. *Let $X = G/P$ be a complex flag manifold, G_0 a real form of G, $x \in X$, $\mathfrak{H} \subset \mathcal{P}_x$ a τ-stable Cartan subalgebra of $\mathfrak{g}$, and Φ a subset of a simple $\mathfrak{H}$-root system such that $\mathcal{P}_x = \mathcal{P}_\Phi$. Retain the notation* (8.6). *Define a τ-stable subset $\Gamma^0 \subset (\Phi^* \cup \tau\Phi^*) \backslash \Gamma^*$ by*

(8.8a) $\Gamma^0 = \{\varphi \in \Delta : -\varphi \notin \Phi^u \cap \tau\Phi^u, \langle\varphi, \delta_\Phi\rangle < 0, \varphi + \tau\varphi \notin \Delta\}$.

Define a τ-stable subspace $\mathfrak{M}_{[x]}$ of $\mathfrak{g}_0$ by

(8.8b) $$\mathfrak{M}_{[x]} = \mathfrak{Q}_{[x]} + \sum_{\Gamma^0} \mathfrak{g}_\varphi.$$

1. $\mathfrak{Q}_{[x]} \subset \mathfrak{M}_{[x]} \subset \mathfrak{N}_{[x]} \cap (\mathcal{P}_x + \tau\mathcal{P}_x)$.
2. *If $\varphi \in \Gamma^0$, then $-\varphi \neq \tau\varphi \neq \varphi$, and either*

(2a) *$\varphi \perp \tau\varphi$, $\tau\varphi - \varphi \notin \Delta$, and $h_\varphi\mathbf{C} + h_{\tau\varphi}\mathbf{C} + \mathfrak{g}_\varphi + \mathfrak{g}_{-\varphi} + \mathfrak{g}_{\tau\varphi} + \mathfrak{g}_{-\tau\varphi}$ is a τ-stable subalgebra of type $A_1 \oplus A_1$ of $\mathfrak{g}$ contained in $\mathfrak{N}_{[x]} \cap (\mathcal{P}_x + \tau\mathcal{P}_x)$; or*

(2b) *$\langle\varphi, \tau\varphi\rangle = \frac{1}{2}\langle\varphi, \varphi\rangle$, $\tau\varphi - \varphi \in \Delta$, and $h_\varphi\mathbf{C} + h_{\tau\varphi}\mathbf{C} + \mathfrak{g}_\varphi + \mathfrak{g}_{-\varphi} + \mathfrak{g}_{\tau\varphi} + \mathfrak{g}_{-\tau\varphi} + \mathfrak{g}_{\tau\varphi-\varphi} + \mathfrak{g}_{\varphi-\tau\varphi}$ is a τ-stable subalgebra of type A_2 in $\mathfrak{g}$ contained in $\mathfrak{N}_{[x]} \cap (\mathcal{P}_x + \tau\mathcal{P}_x)$.*

3. *$\mathfrak{N}_{[x]} \subset \mathcal{P}_x + \tau\mathcal{P}_x$ if and only if $\mathfrak{N}_{[x]} = \mathfrak{M}_{[x]}$.*

PROOF. Decompose the real form $\mathcal{H}_0 = \mathcal{H} \cap \mathcal{G}_0$ of $\mathcal{H}$ as $\mathcal{H}_0 = \mathcal{H}_T + \mathcal{H}_V$ such that $i\mathcal{H}_T + \mathcal{H}_V$ is the real span of the roots. If $\varphi \in \Delta$ then $\varphi = \varphi_T + \varphi_V$ where $\varphi_T(\mathcal{H}_V) = 0 = \varphi_V(\mathcal{H}_T)$, and then $\tau\varphi = -\varphi_T + \varphi_V$. Now suppose $\langle \varphi, \delta_\Phi \rangle < 0$ with $-\varphi \notin \Phi^u \cap \tau\Phi^u$. If $\tau\varphi = \varphi$ then $\mathcal{G}_\varphi \subset \mathcal{P}_x + \tau\mathcal{P}_x$ implies $\mathcal{G}_\varphi \subset \mathcal{P}_x \cap \tau\mathcal{P}_x \subset \mathfrak{M}_{\mathcal{G}}(\mathcal{U}_x)$, so $\langle \varphi, \delta_\Phi \rangle \geqq 0$, contradicting $\langle \varphi, \delta_\Phi \rangle < 0$. If $\tau\varphi = -\varphi$ then $\langle \tau\varphi, \delta_\Phi \rangle = \langle \tau\varphi, \tau\delta_\Phi \rangle = \langle \varphi, \delta_\Phi \rangle = -\langle \tau\varphi, \delta_\Phi \rangle$, contradicting $\langle \varphi, \delta_\Phi \rangle < 0$. Thus $-\varphi \neq \tau\varphi \neq \varphi$, i.e. $\varphi_T \neq 0 \neq \varphi_V$. In particular $|\langle \varphi, \tau\varphi \rangle| < |\langle \varphi, \varphi \rangle|$ so the integer $2\langle \varphi, \tau\varphi \rangle / \langle \varphi, \varphi \rangle = 2\langle \varphi, \tau\varphi \rangle / \langle \tau\varphi, \tau\varphi \rangle$ is -1, 0 or $+1$.

Suppose further that $\varphi + \tau\varphi$ is not a root, i.e. that $\varphi \in \Gamma^0$. Then the maximal chain $\{\varphi + j\tau\varphi\}_{-r \leq j \leq s}$ of roots has $s = 0$, so $2\langle \varphi, \tau\varphi \rangle / \langle \tau\varphi, \tau\varphi \rangle = r \geqq 0$, proving that $2\langle \varphi, \tau\varphi \rangle / \langle \varphi, \varphi \rangle$ is 0 (in which case $\varphi \perp \tau\varphi$ and $\tau\varphi - \varphi \notin \Delta$) or is 1 (in which case $\langle \varphi, \tau\varphi \rangle = \frac{1}{2}\langle \varphi, \varphi \rangle$, $\tau\varphi - \varphi \in \Delta$, and $\tau\varphi - 2\varphi \notin \Delta$). Let $\mathcal{L}$ denote $h_\varphi \mathbf{C} + h_{\tau\varphi}\mathbf{C} + \mathcal{G}_\varphi + \mathcal{G}_{-\varphi} + \mathcal{G}_{\tau\varphi} + \mathcal{G}_{-\tau\varphi}$ in the first case, $h_\varphi \mathbf{C} + h_{\tau\varphi}\mathbf{C} + \mathcal{G}_\varphi + \mathcal{G}_{-\varphi} + \mathcal{G}_{\tau\varphi} + \mathcal{G}_{-\tau\varphi} + \mathcal{G}_{\varphi-\tau\varphi} + \mathcal{G}_{\tau\varphi-\varphi}$ in the second case. Then $\mathcal{L}$ is a τ-stable subalgebra of type $A_1 \oplus A_1$ or A_2 in $\mathcal{G}$, $[\mathcal{H}, \mathcal{L}] \subset \mathcal{L}$, and $\mathcal{L} \subset \mathcal{P}_x + \tau\mathcal{P}_x$; the latter is obvious in the first case and follows from $\langle \tau\varphi - \varphi, \delta_\Phi \rangle = 0 = \langle \varphi - \tau\varphi, \delta_\Phi \rangle$ in the second case. Lemma 8.4 says $\mathcal{L} \subset \mathfrak{N}_{[x]}$. That completes the proof of (2), and (1) is an immediate consequence.

If $\mathfrak{N}_{[x]} = \mathfrak{M}_{[x]}$, then $\mathfrak{M}_{[x]} \subset \mathcal{P}_x + \tau\mathcal{P}_x$ implies $\mathfrak{N}_{[x]} \subset \mathcal{P}_x + \tau\mathcal{P}_x$. Conversely suppose $\mathfrak{N}_{[x]} \subset \mathcal{P}_x + \tau\mathcal{P}_x$. Then $\mathfrak{M}_{[x]} \subset \mathfrak{N}_{[x]}$ implies that $\mathfrak{N}_{[x]} = \mathfrak{M}_{[x]} + \sum_A \mathcal{G}_\varphi$ where A is some τ-stable set of roots such that

$$\text{if } \alpha \in A \quad \text{then } -\alpha \notin \Phi^u \cap \tau\Phi^u, \qquad \langle \alpha, \delta_\Phi \rangle < 0 \quad \text{and} \quad \alpha + \tau\alpha \in \Delta.$$

If A is not empty, choose $\alpha \in A$ and observe that $\mathcal{G}_{\alpha+\tau\alpha} \subset [\mathfrak{N}_{[x]}, \mathfrak{N}_{[x]}] \subset \mathfrak{N}_{[x]} \subset \mathcal{P}_x + \tau\mathcal{P}_x$, so $\tau(\alpha + \tau\alpha) = \alpha + \tau\alpha$ implies $\mathcal{G}_{\alpha+\tau\alpha} \subset \mathcal{P}_x \cap \tau\mathcal{P}_x$; that implies $\langle \alpha + \tau\alpha, \delta_\Phi \rangle \geqq 0$, contradicting

$$\langle \alpha + \tau\alpha, \delta_\Phi \rangle = \langle \alpha, \delta_\Phi \rangle + \langle \tau\alpha, \tau\delta_\Phi \rangle = 2\langle \alpha, \delta_\Phi \rangle < 0.$$

Thus A is empty and $\mathfrak{N}_{[x]} = \mathfrak{M}_{[x]}$. Q.E.D.

We apply Theorem 8.5 and Proposition 8.7 to the question of whether holomorphic arc components of $G_0(x)$ are complex submanifolds of X. Note that, in Theorem 8.9, (5) is the algebraic criterion for (1); it then specifies $\mathfrak{N}_{[x]}$ by (4), thus specifying $S_{[x]} = N_{[x],0}(x) \cong N_{[x],0}/(P_x \cap \tau P_x)_0$.

8.9. THEOREM. *Let $X = G/P$ be a complex flag manifold, G_0 a real form of G, $x \in X$, $\mathcal{H} \subset \mathcal{P}_x$ a τ-stable Cartan subalgebra of $\mathcal{G}$, and Φ a subset of a simple $\mathcal{H}$-root system such that $\mathcal{P}_x = \mathcal{P}_\Phi$. Retain the notation* (8.6) *and* (8.8). *Then the following conditions are equivalent*:

1. *Every holomorphic arc component of $G_0(x)$ in X is a complex submanifold of X.*

2. $S_{[x]}$ *is a complex submanifold of* X.
3. $\mathfrak{N}_{[x]} \subset \mathcal{P}_x + \tau\mathcal{P}_x$.
4. $\mathfrak{N}_{[x]} = \mathfrak{M}_{[x]}$, *given by* (8.6) *and* (8.8).
5. $\mathfrak{M}_{[x]}$ *is an algebra.*

PROOF. (1) implies its special case (2). Conversely, if $g \in G_0$, then $gS_{[x]} = S_{[gx]}$, so (2) implies (1).

Lemma 8.4 says that (3) implies (2). For the converse we define a G_0-invariant distribution $T = \{T_y\}_{y \in G_0(x)}$ on $G_0(x)$ as follows. If $v \in \mathfrak{g}$ then v^* denotes the corresponding real tangent vector field on X, and its value at $y \in X$ is denoted v_y^*. Now define, for every $y \in G_0(x)$,

(8.10a) $T_y = \{v_y^*: v \in (\mathcal{P}_y + \tau\mathcal{P}_y) \cap \mathfrak{g}_0\}$, real subspace of $G_0(x)_y$.

We will prove that every holomorphic arc in $G_0(x)$ is tangent to T, i.e. that

(8.10b) if $f: D \to X$ is a holomorphic arc in $G_0(x)$ then $f_* D_z \subset T_{f(z)}$ for every $z \in D$.

Assume (8.10b). As (2) says that each tangent space to $S_{[x]}$ is filled out by the tangent spaces of holomorphic arcs in $G_0(x)$, it follows that $\{v_y^*: v \in \mathfrak{N}_{[y],0}\} \subset T_y$ for every $y \in G_0(x)$; it follows that $\mathfrak{N}_{[x],0} \subset (\mathcal{P}_x + \tau\mathcal{P}_x) \cap \mathfrak{g}_0$, so $\mathfrak{N}_{[x]} \subset \mathcal{P}_x + \tau\mathcal{P}_x$. Thus (2) implies (3) modulo the truth of (8.10b).

We prove (8.10b). Let $z \in D$, $\beta \in D_z$. Let J_D and J_X be the almost complex structures on D and X, so $f_*(J_D\beta) = J_X f_*\beta$ because f is holomorphic. $f(D) \subset G_0(x)$ implies that both $f_*\beta$ and $f_*J_D\beta$ are in $G_0(x)_{f(z)}$. Thus we have $v \in \mathfrak{g}_0$ with $f_*\beta = v_{f(z)}^*$, and $f_*J_D\beta = J_X f_*\beta = J_X v_{f(z)}^* = (iv)_{f(z)}^*$. That says $iv \in \mathfrak{g}_0 + \mathcal{P}_{f(z)}$, say $iv = u + w$ where $u \in \mathfrak{g}_0$ and $w \in \mathcal{P}_{f(z)}$. Now $\tau v = v$ implies

$$2iv = iv - \tau(iv) = (u + w) - (\tau u + \tau w) = w - \tau w.$$

Thus $2iv \in \mathcal{P}_{f(z)} + \tau\mathcal{P}_{f(z)}$, so $v \in \mathfrak{g}_0 \cap (\mathcal{P}_{f(z)} + \tau\mathcal{P}_{f(z)})$, proving $f_*\beta = v_{f(z)}^* \in T_{f(z)}$. (8.10b) is proved.

Proposition 8.7 contains the equivalence of (3) and (4). Summarizing now, we have proved (8.10b) and have shown that (1), (2), (3) and (4) are equivalent. (4) implies (5) because $\mathfrak{N}_{[x]}$ is an algebra. Now we only need to prove that (5) implies (4).

Suppose that $\mathfrak{M}_{[x]}$ is an algebra. Let $\mathfrak{M}_{[x],0}$ denote its real form $\mathfrak{M}_{[x]} \cap \mathfrak{g}_0$; let $M_{[x]}$ and $M_{[x],0}$ denote the analytic subgroups of G and G_0 for $\mathfrak{M}_{[x]}$ and $\mathfrak{M}_{[x],0}$. Define $\mathfrak{M}_{[gx]} = \mathrm{ad}(g)\mathfrak{M}_{[x]}$, then $\mathfrak{M}_{[gx],0}$, $M_{[gx]}$ and $M_{[gx],0}$ as above, for every $g \in G_0$; those algebras and groups are well defined because $M_{[x],0} \supset (P_x \cap \tau P_x)_0 = G_0 \cap P_x$. Now we have a distribution $M = \{M_y\}$ on $G_0(x)$ given by

$$M_y = \{v_y^*: v \in \mathfrak{M}_{[y],0}\}.$$

The distribution is integrable because M_y is the tangent space to $M_{[y],0}(y)$ at y. We will prove that every holomorphic arc in $G_0(x)$ is tangent to M. As M is integrable, and as its integral manifolds $M_{[y],0}(y)$ are complex submanifolds of X by Lemma 8.4 and $M_{[gx],0}(gx)=gM_{[x],0}(x)$ for $g\in G_0$, we will conclude that the holomorphic arc components of $G_0(x)$ are complex submanifolds of X. Then (5) implies (4) and the theorem is proved.

Now we show that every holomorphic arc in $G_0(x)$ is tangent to M. As $M_{g(y)}=g_*M_y$ for $y\in G_0(x)$ and $g\in G_0$, it suffices to consider holomorphic arcs $f\colon D\to X$ in $G_0(x)$ with $f(0)=x$, and to prove $f_*D_0\subset M_x$. If $f_*D_0=0$ this is automatic. Now suppose $f_*D_0\neq 0$. The implicit function theorem says that f maps a neighborhood of 0 diffeomorphically onto its image, so we cut D down and assume that $f\colon D\to f(D)$ is a diffeomorphism. Let η be a real vector field on D, let J_D and J_X be the almost complex structures on D and X, and let $\xi=f_*\eta$. Then (8.10b) ξ and $J_X\xi=f_*J_D\eta$ are tangent to $T|_{f(D)}$, and so is $[\xi, J_X\xi]$ $=f_*[\eta, J_D\eta]$.

Given $v\in\mathfrak{g}_0$, the vector field v^* induced on $G_0(x)$ is induced from the 1-parameter group $\{\exp tv\}\subset G_0$ acting on $G_0(x)$; it is the projection of a right invariant vector field on G_0. Let σ be a smooth local section of $G_0\to G_0/(P_x\cap\tau P_x)_0=G_0(x)$ in a neighborhood W of x. If $v\in\mathfrak{g}_0$, and if $\tilde{v}$ denotes the corresponding left invariant vector field on G_0, then $\tilde{v}|_{\sigma(W)}$ projects to a vector field $v^\dagger$ on W. Observe

$$T_y=\{v_y^*\colon v\in\mathfrak{g}_0\cap(\mathcal{P}_y+\tau\mathcal{P}_y)\}\text{ is }\{v_y^\dagger\colon v\in\mathfrak{g}_0\cap(\mathcal{P}_x+\tau\mathcal{P}_x)\},$$

$$M_y=\{v_y^*\colon v\in\mathfrak{M}_{[y],0}\}\text{ is }\{v_y^\dagger\colon v\in\mathfrak{M}_{[x],0}\},$$

$$N_y=\{v_y^*\colon v\in\mathfrak{N}_{[y],0}\}\text{ is }\{v_y^\dagger\colon v\in\mathfrak{N}_{[x],0}\},$$

from G_0-invariance of the corresponding distributions. Now cut W down if necessary, so that the tangent vector field ξ to $f(D)\subset S_{[x]}$ $=N_{[x],0}(x)$ has representation

$$\xi_y=\zeta(y)_y^\dagger\quad\text{where}\quad\zeta\colon f(D)\cap W\to\mathfrak{N}_{[x],0}\text{ is smooth.}$$

As ξ is tangent to T, ζ takes values in $(\mathcal{P}_x+\tau\mathcal{P}_x)\cap\mathfrak{N}_{[x],0}$.

Let A' denote the set of all $\mathfrak{H}$-roots φ such that $\mathfrak{g}_\varphi\subset(\mathcal{P}_x+\tau\mathcal{P}_x)$ $\cap\mathfrak{N}_{[x]}$ but $\mathfrak{g}_\varphi\not\subset\mathfrak{M}_{[x]}$. Then $(\mathcal{P}_x+\tau\mathcal{P}_x)\cap\mathfrak{N}_{[x]}=\mathfrak{M}_{[x]}+\mathfrak{A}$ direct sum where $\mathfrak{A}=\sum_{A'}\mathfrak{g}_\varphi$, and ζ takes values in $\mathfrak{M}_{[x],0}+\mathfrak{A}_0$ where $\mathfrak{A}_0=\mathfrak{A}\cap\mathfrak{g}_0$. Decompose $\zeta=\mu+\alpha$ where μ takes values in $\mathfrak{M}_{[x],0}$ and α takes values in $\mathfrak{A}_0$. That decomposes $\xi=m+a$ and $J_X\xi=J_Xm+J_Xa$ where

$$m_y=\mu(y)_y^\dagger\quad\text{and}\quad a_y=\alpha(y)_y^\dagger.$$

Now

$$\begin{aligned}[a, J_X a]_x &= [\xi - m, J_X\xi - J_X m]_x \\ &= [\xi, J_X\xi]_x - [m, J_X\xi]_x - [\xi, J_X m]_x + [m, J_X m]_x.\end{aligned}$$

We have already seen $[\xi, J_X\xi]_x \in T_x$. We have $[m, J_X\xi]_x \in T_x$ and $[\xi, J_X m]_x \in T_x$ because m and $J_X m$ are tangent to M and $M_{[x],0}$ preserves the real span of the tangent spaces to holomorphic arcs. And finally $[m, J_X m]_x \in M_x \subset T_x$ because $\mathfrak{M}_{[x],0}$ is an algebra. Thus $[a, J_X a]_x \in T_x$.

$A' = A \cup \tau A$ disjoint where $A = A' \cap \tau\Phi^* = A' \cap -\Phi^u$. If $\varphi \in A$ choose $0 \neq e_\varphi \in \mathfrak{g}_\varphi$. Define $p_\varphi = e_\varphi + \tau e_\varphi \in \mathfrak{g}_0$ and $q_\varphi = ie_\varphi + \tau(ie_\varphi) = ie_\varphi - i\tau e_\varphi \in \mathfrak{g}_0$. As $\mathcal{P}_x^{-u}$ is the holomorphic tangent space to X at x, $J_X p_\varphi^\dagger = q_\varphi^\dagger$ and $J_X q_\varphi^\dagger = -p_\varphi^\dagger$, for $\varphi \in A$, at x. Invariance then gives $J_X p_\varphi^\dagger = q_\varphi^\dagger$ and $J_X q_\varphi^\dagger = -p_\varphi^\dagger$ on $G_0(x)$ for $\varphi \in A$. Now express $\alpha(y) = \sum_A (s_\varphi(y) p_\varphi + t_\varphi(y) q_\varphi)$, so that $a = \sum_A (s_\varphi p_\varphi^\dagger + t_\varphi q_\varphi^\dagger)$ and $J_X a = \sum_A (-t_\varphi p_\varphi^\dagger + s_\varphi q_\varphi^\dagger)$. Then we compute $[a, J_X a] = \sum_{\varphi,\psi\in A} [s_\varphi^\dagger p_\varphi + t_\varphi q_\varphi^\dagger, -t_\psi p_\psi^\dagger + s_\psi q_\psi^\dagger]$ and

$$\begin{aligned}&[s_\varphi p_\varphi^\dagger + t_\varphi q_\varphi^\dagger, -t_\psi p_\psi^\dagger + s_\psi q_\psi^\dagger] \\ &\qquad = - s_\varphi \cdot p_\varphi^\dagger(t_\psi) \cdot p_\psi^\dagger + t_\psi \cdot p_\psi^\dagger(s_\varphi) \cdot p_\varphi^\dagger - s_\varphi t_\psi [p_\varphi, p_\psi]^\dagger \\ &\qquad\quad + s_\varphi \cdot p_\varphi^\dagger(s_\psi) \cdot q_\psi^\dagger - s_\psi \cdot q_\psi^\dagger(s_\varphi) \cdot p_\varphi^\dagger + s_\varphi s_\psi [p_\varphi, q_\psi]^\dagger \\ &\qquad\quad - t_\varphi \cdot q_\varphi^\dagger(t_\psi) \cdot p_\psi^\dagger + t_\psi \cdot p_\psi^\dagger(t_\varphi) \cdot q_\varphi^\dagger - t_\varphi t_\psi [q_\varphi, p_\psi]^\dagger \\ &\qquad\quad + t_\varphi \cdot q_\varphi^\dagger(s_\psi) \cdot q_\psi^\dagger - s_\psi \cdot q_\psi^\dagger(t_\varphi) \cdot q_\varphi^\dagger + t_\varphi s_\psi [q_\varphi, q_\psi]^\dagger.\end{aligned}$$

As $\lambda \in A$ implies $p_{\lambda x}^\dagger, q_{\lambda x}^\dagger \in T_x$, and as $[a, J_X a]_x \in T_x$, it follows that

$$\begin{aligned}0 &\equiv [a, J_X a]_x \bmod. T_x \\ &\equiv \sum_{\varphi,\psi\in A} \{ -s_\varphi(x) t_\psi(x) [p_\varphi, p_\psi]_x^\dagger + s_\varphi(x) s_\psi(x) [p_\varphi, q_\psi]_x^\dagger \\ &\qquad - t_\varphi(x) t_\psi(x) [q_\varphi, p_\psi]_x^\dagger + t_\varphi(x) s_\psi(x) [q_\varphi, q_\psi]_x^\dagger \} \bmod. T_x \\ &= \sum_{\varphi,\psi\in A} \{ -s_\varphi(x) t_\psi(x) ([e_\varphi, e_\psi] + \tau[e_\varphi, e_\psi] + [e_\varphi, \tau e_\psi] + [\tau e_\varphi, e_\psi])_x^\dagger \\ &\qquad + s_\varphi(x) s_\psi(x) (i[e_\varphi, e_\psi] - i\tau[e_\varphi, e_\psi] - i[e_\varphi, \tau e_\psi] + i[\tau e_\varphi, e_\psi])_x^\dagger \\ &\qquad - t_\varphi(x) t_\psi(x) (i[e_\varphi, e_\psi] - i\tau[e_\varphi, e_\psi] + i[e_\varphi, \tau e_\psi] - i[\tau e_\varphi, e_\psi])_x^\dagger \\ &\qquad + t_\varphi(x) s_\psi(x) (-[e_\varphi, e_\psi] - \tau[e_\varphi, e_\psi] + [e_\varphi, \tau e_\psi] + [\tau e_\varphi, e_\psi])_x^\dagger \} \\ &= -2 \sum_{\varphi,\psi\in A} s_\varphi(x) t_\psi(x) ([e_\varphi, \tau e_\psi] + [\tau e_\varphi, e_\psi])_x^\dagger \\ &\qquad + \sum_{\varphi,\psi\in A} \{ s_\varphi(x) s_\psi(x) + t_\varphi(x) t_\psi(x) \} \{ i([\tau e_\varphi, e_\psi] + [\tau e_\psi, e_\varphi])_x^\dagger \}.\end{aligned}$$

Let $\beta \in A$. Then $\langle \beta, \delta_\Phi \rangle < 0$, $-\beta \notin \Phi^u \cap \tau\Phi^u$ and (because $\mathfrak{g}_\beta \not\subset \mathfrak{M}_{[x]}$) $\beta + \tau\beta$ is a root. If $\mathfrak{g}_{\beta+\tau\beta} \subset \mathcal{P}_x + \tau\mathcal{P}_x$, then $\tau\mathfrak{g}_{\beta+\tau\beta} = \mathfrak{g}_{\beta+\tau\beta}$ would imply $\mathfrak{g}_{\beta+\tau\beta} \subset \mathcal{P}_x \cap \tau\mathcal{P}_x$; that would say $\langle \beta+\tau\beta, \delta_\Phi \rangle \geqq 0$, contradicting $\langle \beta+\tau\beta, \delta_\Phi \rangle = 2\langle \beta, \delta_\Phi \rangle < 0$. Thus $\mathfrak{g}_{\beta+\tau\beta} \subset \mathcal{P}_x^{-u} \cap \tau\mathcal{P}_x^{-u}$. Define $E_{\beta+\tau\beta} = \{v_x^\dagger : v_x \in \mathfrak{g}_0 \cap \mathfrak{g}_{\beta+\tau\beta}\}$, 2-dimensional real subspace of the tangent space to $G_0(x)$ at x. Then the projection of $[a, Jxa]_x$ to $E_{\beta+\tau\beta}$ is

$$\begin{aligned} 0 = {} & \{s_\beta(x)^2 + t_\beta(x)^2\}\{2i[\tau e_\beta, e_\beta]\}_x^\dagger \\ & + \sum_{\substack{\varphi \neq \psi \text{ in } A \\ \varphi+\tau\psi = \beta+\tau\beta}} (\{is_\varphi(x)s_\psi(x) + it_\varphi(x)t_\psi(x) - 2s_\varphi(x)t_\psi(x)\}[\tau e_\varphi, e_\psi])_x^\dagger \\ & + \sum_{\substack{\varphi \neq \psi \text{ in } A \\ \varphi+\tau\psi = \beta+\tau\beta}} (\{is_\varphi(x)s_\psi(x) + it_\varphi(x)t_\psi(x) + 2s_\varphi(x)t_\psi(x)\}[\tau e_\psi, e_\varphi])_x^\dagger . \end{aligned}$$

Suppose that the holomorphic arc f does not satisfy $f_* D_0 \subset M_x$. Then we can choose η such that $\xi_x \notin M_x$, i.e. such that $\alpha(x) \neq 0$, so there is a root $\beta \in A$ for which $s_\beta(x)^2 + t_\beta(x)^2 \neq 0$. Recall that $\mathfrak{A}$ is disjoint from the parabolic subalgebra $\mathfrak{Q}_{[x]}$, in fact that $\mathfrak{A} \subset \mathfrak{Q}_{[x]}^{-u}$. So there is an ordering of the $\mathfrak{H}$-roots of $\mathfrak{g}$ such that A consists of negative roots. Now choose β to be the highest root in A such that $s_\beta(x)^2 + t_\beta(x)^2 \neq 0$, i.e. such that the real numbers $s_\beta(x)$ and $t_\beta(x)$ are not both zero. Now suppose $\varphi, \psi \in A$ with $\varphi \neq \psi$ and $\varphi + \tau\psi = \beta + \tau\beta$. Then either $s_\varphi(x) = 0 = t_\varphi(x)$ or $s_\psi(x) = 0 = t_\psi(x)$; for otherwise $\varphi < \beta$ and $\psi < \beta$, whence $\varphi + \tau\psi < \beta + \tau\beta$. Thus

$$s_\varphi(x)s_\psi(x) = t_\varphi(x)t_\psi(x) = s_\varphi(x)t_\psi(x) = 0,$$

and our calculation above reduces to

$$\{s_\beta(x)^2 + t_\beta(x)^2\}\{2i[\tau e_\beta, e_\beta]\}_x^\dagger = 0.$$

As $\{i[\tau e_\beta, e_\beta]\}_x^\dagger \neq 0$ from $\tau\mathfrak{g}_{\beta+\tau\beta} = \mathfrak{g}_{\beta+\tau\beta}$, it follows that $s_\beta(x)^2 + t_\beta(x)^2 = 0$, which contradicts our choice of β. We conclude that $f_* D_0 \subset M_x$. That completes the proof of the theorem. Q.E.D.

8.11. COROLLARY. *Let $X = G/P$ be a complex flag manifold, G_0 a real form of G, $x \in X$, $\mathfrak{H} \subset \mathcal{P}_x$ a τ-stable Cartan subalgebra of $\mathfrak{g}$, and Φ a subset of a simple $\mathfrak{H}$-root system such that $\mathcal{P}_x = \mathcal{P}_\Phi$. Then the following conditions are equivalent*:

(i) $\mathfrak{N}_{[x]} = \mathfrak{Q}_{[x]}$, *parabolic subalgebra defined by* (8.6).

(ii) $\mathfrak{Q}_{[x]} = \mathfrak{M}_{[x]}$, *parabolic subalgebra defined by* (8.8).

(iii) *If φ is an $\mathfrak{H}$-root such that $\langle \varphi, \delta_\Phi \rangle < 0$ and $-\varphi \notin \Phi^u \cap \tau\Phi^u$ then $\varphi + \tau\varphi$ is a root.*

If those conditions hold, the holomorphic arc components of $G_0(x)$ are complex submanifolds of X.

PROOF. (iii) implies (ii) by (8.8). Given (ii), $\mathfrak{M}_{[x]}$ is an algebra, so $\mathfrak{N}_{[x]}=\mathfrak{M}_{[x]}$ by Theorem 8.9, whence (ii) says $\mathfrak{N}_{[x]}=\mathfrak{Q}_{[x]}$. Given (i), $\mathfrak{Q}_{[x]}\subset\mathfrak{M}_{[x]}\subset\mathfrak{N}_{[x]}$ implies (ii), and (iii) follows by (8.8). Q.E.D.

8.12. EXAMPLE. Let $G=\mathbf{SL}(2m, \mathbf{C})$ and either $G_0=\mathbf{SL}(m, \mathbf{Q})$ with maximal compact subgroup $\mathbf{Sp}(m)$, or $G_0=\mathbf{SL}(2m, \mathbf{R})$ with maximal compact subgroup $\mathbf{SO}(2m)$. Choose a τ-stable Cartan subalgebra $\mathfrak{H}\subset\mathfrak{g}$ and a τ-stable simple root system Ψ. Then τ acts on Ψ by

τ:

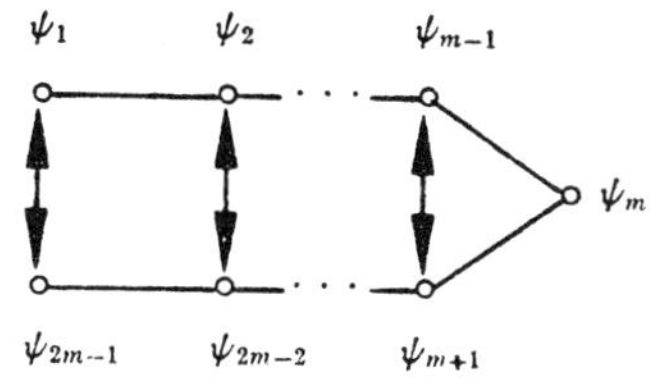

$(\tau\psi_i = \psi_{2m-i})$.

Let $\mathcal{P}=\mathcal{P}_x=\mathcal{P}_\Phi$ where $\Phi=\{\psi_1, \cdots, \psi_{2m-2}\}$. Then

$$X = G/P \text{ is complex projective } (2m-1)\text{-space,}$$

and we note

$$\Phi^u = \{\psi_i+\psi_{i+1}+\cdots+\psi_{2m-1}: 1\leqq i\leqq 2m-1\},$$
$$\tau\Phi^u = \{\psi_1+\psi_2+\cdots+\psi_j: 1\leqq j\leqq 2m-1\},$$

so

$$\Phi^u\cap\tau\Phi^u = \{\mu\} \quad \text{where} \quad \mu=\psi_1+\cdots+\psi_{2m-1} \text{ maximal root.}$$

Thus

$$\delta_\Phi = \mu \quad \text{and} \quad \mathfrak{Q}_{[x]} = \mathcal{P}_\Gamma, \qquad \Gamma = \{\psi_2, \psi_3, \cdots, \psi_{2m-2}\}.$$

Now let B consist of all roots φ such that $\mathfrak{g}_\varphi\subset\mathcal{P}_x+\tau\mathcal{P}_x$ and $\mathfrak{g}_\varphi\not\subset\mathfrak{Q}_{[x]}$. Then

$$B = \{-(\psi_1+\cdots+\psi_i), -(\psi_j+\cdots+\psi_{2m-1}):$$
$$1\leqq i<2m-1, 1<j\leqq 2m-1\}.$$

If $\varphi\in B$, now $\varphi+\tau\varphi$ is of the form

$$-[(\psi_1+\cdots+\psi_i)+(\psi_{2m-i}+\cdots+\psi_{2m-1})],$$

which cannot be a root because we cannot have $2m-i=i+1$. Thus

$$\text{if } \varphi\in B \quad \text{then } \varphi+\tau\varphi \text{ is not a root.}$$

In other words, $B=\Gamma^0$, so now

$$\mathfrak{M}_{[x]} = \mathcal{P}_x+\tau\mathcal{P}_x = \mathfrak{H}+\sum_{\varphi\neq-\mu}\mathfrak{g}_\varphi, \quad \text{not an algebra.}$$

Thus the holomorphic arc components of $G_0(x)$ are not complex submanifolds of X. In fact, as $\mathfrak{Q}_{[x]}$ is maximal among the τ-stable subalgebras of $\mathfrak{G}$, one expects that the $Q_{[y],0}(y)$, $y \in G_0(x)$, would be maximal among the connected complex submanifolds of X that are contained in $G_0(x)$.

The example above is based on the situation that τ preserves a simple root system Ψ but acts nontrivially on Ψ. The latter occurs if and only if τ induces an outer automorphism on the compact real form of G, i.e. if and only if the maximal compact subgroups K of G_0 have rank $K <$ rank G_0. Later we will see, for the case where X is a compact hermitian symmetric space and the inclusion $G_0(x_0) \subset X$ of some open orbit is the Borel embedding of the noncompact dual of X, that every holomorphic arc component of every G_0-orbit on X is a hermitian symmetric submanifold (the signature of metric can be indefinite there). In particular, if X is complex projective space as in the example, and if G_0 is not one of the real forms of G considered above, then it will follow that every holomorphic arc component of every G_0-orbit is a complex submanifold of X.

8.13. LEMMA/DEFINITION. *Let $X = G/P$ be a complex flag manifold, G_0 a real form of G, $x \in X$, $\mathfrak{H} \subset \mathcal{P}_x$ a τ-stable Cartan subalgebra of $\mathfrak{G}$, and Φ a subset of a simple $\mathfrak{H}$-root system such that $\mathcal{P}_x = \mathcal{P}_\Phi$. Let $\mathfrak{Q}_{[x]}$ be the τ-stable parabolic subalgebra of $\mathfrak{G}$ defined by* (8.6), *$\mathfrak{Q}_{[x],0}$ the real form $\mathfrak{Q}_{[x]} \cap \mathfrak{G}_0$ of $\mathfrak{Q}_{[x]}$, and $Q_{[x],0}$ the analytic subgroup of G_0 for $\mathfrak{Q}_{[x],0}$. Then*

(i) *$Q_{[x],0}(x) \subset S_{[x]}$ is a complex submanifold of X,*

(ii) *$\mathfrak{Q}_{[x]}$, $\mathfrak{Q}_{[x],0}$ and $Q_{[x],0}(x)$ are independent of choices of $\mathfrak{H}$ and Φ,*

(iii) *if $y \in Q_{[x],0}(x)$ then $Q_{[y],0}(y) = Q_{[x],0}(x)$, and*

(iv) *if $g \in G_0$ then $Q_{[gx],0}(gx) = gQ_{[x],0}(x)$.*

We will refer to the $Q_{[y],0}(y)$, $y \in G_0(x)$, as the **algebraic arc components** *of $G_0(x)$.*

REMARK. Lemma 8.13 says that the algebraic arc components of $G_0(x)$ enjoy the same formal properties as the holomorphic arc components. Example 8.12 shows that they are the best substitute for the holomorphic arc components in case the latter are not complex submanifolds of X. Corollary 8.11 gives the criterion for the algebraic arc components of $G_0(x)$ to coincide with the holomorphic arc components.

PROOF. Assertion (i) is immediate from Lemma 8.4.

Suppose that $Q_{[x],0}(x)$ does not depend on the choices of $\mathfrak{H}$ and Φ. If $g \in G_0$, then we define $Q_{[gx],0}(gx)$ from the choices $\mathrm{ad}(g)\mathfrak{H} \subset \mathrm{ad}(g)\mathcal{P}_x = \mathcal{P}_{gx}$ and $\mathrm{ad}(g^{-1})^*\Phi$; then $\mathfrak{Q}_{[gx]} = \mathrm{ad}(g)\mathfrak{Q}_{[x]}$, so $\mathfrak{Q}_{[gx],0} = \mathrm{ad}(g)\mathfrak{Q}_{[x],0}$, implying the statement $Q_{[gx],0}(gx) = g \cdot Q_{[x],0}(x)$ of (iv). If $g \in Q_{[x],0}$ we obtain $Q_{[gx],0} = Q_{[x],0}$, so $Q_{[gx],0}(gx) = Q_{[x],0}(gx) = Q_{[x],0}(x)$, proving (iii). Thus we are reduced to proving (ii).

Given $\mathfrak{H}$, $\delta_\Phi = \sum_{\Phi^u \cap \tau\Phi^u} \varphi = \sum_{-\Phi^u \cap -\tau\Phi^u}(-\psi)$ is the sum of all $\mathfrak{H}$-roots $-\psi$ such that $\mathfrak{g}_\psi \not\subset \mathcal{P}_x + \tau\mathcal{P}_x$; thus δ_Φ, $\mathfrak{Q}_{[x]}$, $\mathfrak{Q}_{[x],0}$ and $Q_{[x],0}(x)$ are independent of choice of the set Φ for which $\mathcal{P}_x = \mathcal{P}_\Phi$, and they depend only on the choice of $\mathfrak{H}$.

Let $\mathfrak{H}'$ and $\mathfrak{H}''$ be two choices of $\mathfrak{H}$. They are τ-stable Cartan subalgebras of respective reductive parts $(\mathcal{P}_x \cap \tau\mathcal{P}_x)'$ and $(\mathcal{P}_x \cap \tau\mathcal{P}_x)''$ of $\mathcal{P}_x \cap \tau\mathcal{P}_x$. Let $Q'_{[x]}$ and $Q''_{[x]}$ be the analytic subgroups of G for the corresponding algebras $\mathfrak{Q}'_{[x]}$ and $\mathfrak{Q}''_{[x]}$; let L', L'' and L denote the respective analytic subgroups of G for $(\mathcal{P}_x \cap \tau\mathcal{P}_x)'$, $(\mathcal{P}_x \cap \tau\mathcal{P}_x)''$ and $(\mathcal{P}_x \cap \tau\mathcal{P}_x)$. Then $(L' \cup L'') \subset L \subset (Q'_{[x]} \cap Q''_{[x]})$. As any two reductive parts of L are conjugate, we have $g_1 \in L$ such that $\mathrm{ad}(g_1)L'' = L'$; as any two Cartan subgroups of L' are conjugate, we have $g_2 \in L'$ such that $\mathrm{ad}(g_2g_1)\mathfrak{H}'' = \mathfrak{H}'$. Now $g = g_2g_1$ is an element of $Q''_{[x]}$ such that $\mathrm{ad}(g)Q''_{[x]} = Q'_{[x]}$. Thus $\mathfrak{Q}''_{[x]} = \mathfrak{Q}'_{[x]}$. It follows that $\mathfrak{Q}''_{[x],0} = \mathfrak{Q}'_{[x],0}$ and $Q''_{[x],0}(x) = Q'_{[x],0}(x)$. Q.E.D.

We note the relations of holomorphic and algebraic arc components with projections of flag manifolds.

8.14. Proposition. *Let $P_1 \subset P_2$ be parabolic subgroups of G, $X_i = G/P_i$ the resulting complex flag manifolds, and $\pi: X_1 \to X_2$ the projection $gP_1 \to gP_2$. Let $x_2 \in X_2$ and let G_0 be a real form of G.*

1. *There exist points $x_1 \in \pi^{-1}(x_2)$ such that $G_0(x_1)$ is an open subset of $\pi^{-1}G_0(x_2)$.*

2. *If $x_1 \in \pi^{-1}(x_2)$, then the following conditions are equivalent:*

(2a) *$G_0(x_1)$ is open in $\pi^{-1}G_0(x_2)$.*

(2b) $(\mathcal{P}_1)_{x_1} + \tau(\mathcal{P}_1)_{x_1} = (\mathcal{P}_2)_{x_2} + \tau(\mathcal{P}_2)_{x_2}$.

(2c) $\mathrm{codimension}_R(G_0(x_1) \subset X_1) = \mathrm{codimension}_R(G_0(x_2) \subset X_2)$.

(2d) *Let $\mathfrak{H} \subset (\mathcal{P}_1)_{x_1}$ be a τ-stable Cartan subalgebra of $\mathfrak{g}$, $\Phi_1 \subset \Phi_2$ subsets of a simple $\mathfrak{H}$-root system such that $(\mathcal{P}_i)_{x_i} = \mathcal{P}_{\Phi_i}$; then $\Phi_1^u \cap \tau\Phi_1^u = \Phi_2^u \cap \tau\Phi_2^u$.*

3. *Let $x_1 \in \pi^{-1}(x_2)$ with $G_0(x_1)$ open in $\pi^{-1}G_0(x_2)$. Choose $\mathfrak{H}$ and $\Phi_1 \subset \Phi_2$ as in* (2d). *Then*

(3a) *the linear forms $\delta_{\Phi_1} = \delta_{\Phi_2}$, the parabolic subalgebras $\mathfrak{Q}_{[x_1]} = \mathfrak{Q}_{[x_2]}$, and the algebraic arc components of the $G_0(x_i)$ are related by*

$$Q_{[x_2],0}(x_2) = \pi Q_{[x_1],0}(x_1) \quad \textit{and} \quad Q_{[x_1],0}(x_1) = G_0(x_1) \cap \pi^{-1}Q_{[x_2],0}(x_2);$$

(3b) *the spaces $\mathfrak{M}_{[x_1]} = \mathfrak{M}_{[x_2]}$, and in particular one is an algebra if and only if the other is an algebra;*

(3c) *the holomorphic arc components of $G_0(x_1)$ are complex submanifolds of X_1 if and only if the holomorphic arc components of $G_0(x_2)$ are complex submanifolds of X_2;*

(3d) *holomorphic arc components $S_{[x_i]} \subset G_0(x_i) \subset X_i$ are related by*

$$S_{[x_2]} = \pi S_{[x_1]} \quad \textit{and} \quad S_{[x_1]} = G_0(x_1) \cap \pi^{-1} S_{[x_2]};$$

(3e) *the algebras* $\mathfrak{N}_{[x_1]} = \mathfrak{N}_{[x_2]}$ *and the groups* $N_{[x_1],0} = N_{[x_2],0}$.

PROOF. The projection $\pi: X_1 \to X_2$ is G_0-invariant, so $\pi^{-1}G_0(x_2)$ is a union of G_0-orbits. As there are only finitely many G_0-orbits on X_1, now $\pi^{-1}G_0(x_2)$ is a finite union of G_0-orbits, so at least one of them has full dimension in $\pi^{-1}G_0(x_2)$. That proves (1).

If $G_0(x_1)$ is open in $\pi^{-1}G_0(x_2)$, then

$$\begin{aligned}
\text{codimension}_R(G_0(x_1) \subset X_1) &= \dim_R X_1 - \dim_R G_0(x_1) \\
&= \dim_R X_1 - \dim_R \pi^{-1} G_0(x_2) \\
&= [\dim_R X_2 + \{\dim_R X_1 - \dim_R X_2\}] \\
&\quad - [\dim_R G_0(x_2) + \{\dim_R X_1 - \dim_R X_2\}] \\
&= \dim_R X_2 - \dim_R G_0(x_2) \\
&= \text{codimension}_R(G_0(x_2) \subset X_2).
\end{aligned}$$

Conversely, if those codimensions are equal, the argument shows that $\dim_R G_0(x_1) = \dim_R \pi^{-1}G_0(x_2)$, so $G_0(x_1)$ is open in $\pi^{-1}G_0(x_2)$. Thus (2a) and (2c) are equivalent. Theorem 2.12 (ii) says that (2c) is equivalent to $|\Phi_1^u \cap \tau\Phi_1^u| = |\Phi_2^u \cap \tau\Phi_2^u|$. As $\Phi_2^u \subset \Phi_1^u$, the latter is equivalent to $\Phi_1^u \cap \tau\Phi_1^u = \Phi_2^u \cap \tau\Phi_2^u$. Thus (2c) and (2d) are equivalent. We reformulate (2d) as $-\Phi_1^u \cap -\tau\Phi_1^u = -\Phi_2^u \cap -\tau\Phi_2^u$, i.e. as $(\mathcal{P}_1)_{x_1}^{-u} \cap \tau(\mathcal{P}_1)_{x_1}^{-u} = (\mathcal{P}_2)_{x_2}^{-u} \cap \tau(\mathcal{P}_2)_{x_2}^{-u}$, i.e. as $(\mathcal{P}_1)_{x_1} + \tau(\mathcal{P}_1)_{x_1} = (\mathcal{P}_2)_{x_2} + \tau(\mathcal{P}_2)_{x_2}$, which is (2b). That completes the proof of (2).

Let $x_1 \in \pi^{-1}(x_2)$ with $G_0(x_1)$ open in $\pi^{-1}G_0(x_2)$. Then we have (2b), (2c) and (2d). From (2d), $\delta_{\Phi_1} = \delta_{\Phi_2}$, so $\mathfrak{Q}_{[x_1]} = \mathfrak{Q}_{[x_2]}$, $Q_{[x_1],0} = Q_{[x_2],0}$ and $Q_{[x_2],0}(x_2) = \pi Q_{[x_1],0}(x_1)$; and $\mathcal{P}_{x_1} \cap \mathfrak{g}_0 \subset \mathfrak{Q}_{[x_1],0}$ implies further that $Q_{[x_1],0}(x_1) = G_0(x_1) \cap \pi^{-1} Q_{[x_2],0}(x_2)$, completing the proof of (3a). $\mathfrak{Q}_{[x_1]} = \mathfrak{Q}_{[x_2]}$ together with (2b) implies $\mathfrak{M}_{[x_1]} = \mathfrak{M}_{[x_2]}$, proving (3b). Theorem 8.9 and (3b) imply (3c). If the holomorphic arc components are complex manifolds then (3d) follows from (3b) and Theorem 8.9. But in general we must look at holomorphic arcs.

If $f: D \to X_1$ is a holomorphic arc in $G_0(x_1)$ then $(\pi \cdot f): D \to X_2$ is a holomorphic arc in $\pi G_0(x_1) = G_0(x_2)$. Thus $\pi S_{[x_1]} \subset S_{[x_2]}$. Let $S = G_0(x_1) \cap \pi^{-1} S_{[x_2]}$, $p = \pi|_S$, and ξ the fibration $p: S \to S_{[x_2]}$. Then ξ is locally trivial and its fibres carry complex manifold structures as open subsets of π-fibres. If $y \in S$ this implies $S_{[y]} = p^{-1} p S_{[y]}$. Thus $S_{[x_2]}$ is a disjoint union of sets $pS_{[y]} = \pi S_{[y]}$, $y \in S$, and S is disjoint union of the corresponding $p^{-1}\pi S_{[y]}$. As the images of holomorphic arcs in $G_0(x_2)$ fill out a neighborhood in $S_{[x_2]}$ of every point $p(y) \in S_{[x_2]}$, and as those holomorphic arcs lift to S, the sets $p^{-1}\pi S_{[y]}$ are open in S, so

$S=p^{-1}\pi S_{[x_1]}$ by connectedness. That proves (3d). It follows that $\mathfrak{N}_{[x_1],0}\equiv\mathfrak{N}_{[x_2],0}$ modulo the isotropy subalgebras of $\mathfrak{g}_0$ at x_1 and x_2. As those isotropy subalgebras $(\mathcal{P}_i)_{x_i}\cap\mathfrak{g}_0=(\mathcal{P}_i)_{x_i}\cap\tau(\mathcal{P}_i)_{x_i}\cap\mathfrak{g}_0\subset\mathfrak{Q}_{[x_i]}\cap\mathfrak{g}_0\subset\mathfrak{N}_{[x_i],0}$, now $\mathfrak{N}_{[x_1],0}=\mathfrak{N}_{[x_2],0}$; (3e) follows. Q.E.D.

We describe the "space of holomorphic arc components" of a G_0-orbit on X.

8.15. THEOREM. *Let $X=G/P$ be a complex flag manifold, $x\in X$, G_0 a real form of G, and K a maximal compact subgroup of G_0. Define projections*

$$\text{(8.16a)}\quad \sigma\colon G_0(x)\to G_0/(G_0\cap N_{[x]})\qquad \text{by } \sigma(gx)=g(G_0\cap N_{[x]}),$$

$$\text{(8.16b)}\quad \kappa\colon G_0(x)\to G_0/(G_0\cap Q_{[x]})\qquad \text{by } \kappa(gx)=g(G_0\cap Q_{[x]}),$$

and

$$\text{(8.16c)}\quad \nu\colon G_0/(G_0\cap Q_{[x]})\to G_0/(G_0\cap N_{[x]})\qquad \text{by } g(G_0\cap Q_{[x]})\to g(G_0\cap N_{[x]}).$$

1. *σ is well defined and* (8.16a) *is a G_0-equivariant differentiable fibre bundle with*

(1a) *structure group $G_0\cap N_{[x]}$, the parabolic subgroup of G_0 with identity component $N_{[x],0}$,*

(1b) *fibre over $\sigma(gx)$ equal to the holomorphic arc component $S_{[gx]}=gS_{[x]}$ of $G_0(x)$ through $g(x)$, and*

(1c) *compact base $G_0/(G_0\cap N_{[x]})=K/(K\cap N_{[x]})$, real flag manifold that parameterizes the holomorphic arc components of $G_0(x)$.*

2. *κ is well defined and* (8.16b) *is a G_0-equivariant differentiable fibre bundle with*

(2a) *structure group $G_0\cap Q_{[x]}$, the parabolic subgroup of G_0 with identity component $Q_{[x],0}$,*

(2b) *fibre over $\kappa(gx)$ equal to the algebraic arc component $Q_{[gx],0}(gx)=gQ_{[x],0}(x)$ of $G_0(x)$ through $g(x)$, and*

(2c) *compact base $G_0/(G_0\cap Q_{[x]})=K/(K\cap Q_{[x]})$, real flag manifold that parameterizes the algebraic arc components of $G_0(x)$.*

3. *$\nu=\sigma\cdot\kappa^{-1}$, well-defined projection of real flag manifolds, and* (8.16c) *has fibre over $\nu(gx)$ parameterizing the algebraic arc components of $G_0(x)$ that are contained in the holomorphic arc component $S_{[gx]}$.*

PROOF. $\mathcal{P}_x\cap\tau\mathcal{P}_x\subset\mathfrak{Q}_{[x]}\subset\mathfrak{N}_{[x]}$ implies $G_0\cap P_x\subset G_0\cap Q_{[x]}\subset G_0\cap N_{[x]}$, so σ, κ and $\nu=\sigma\cdot\kappa^{-1}$ are well defined, and (8.16a, b, c) are G_0-equivariant differentiable fibre bundles with structure groups as described in (1a) and (2a).

Let F be the complex flag manifold $G/N_{[x]}$ or $G/Q_{[x]}$. View G as a linear algebraic group def/R such that G_0 is the topological identity component of the set $G_{\mathbf{R}}$ of real points. Then F is def/R and Theorem 3.6(2) says that the G_0-orbit of the identity coset on F is $F_{\mathbf{R}}$, and that it is the unique closed G_0-orbit on F. That proves that the base spaces of (8.16a) and (8.16b) are real flag manifolds and that K acts transitively on them. Now it only remains to prove (1b) and (2b); for then (1c), (2c) and (3c) will follow immediately.

Choose a τ-stable Cartan subalgebra $\mathfrak{H}\subset\mathfrak{P}_x$ of $\mathfrak{G}$ and decompose $\mathfrak{H}_0=\mathfrak{H}\cap\mathfrak{G}_0$ as $\mathfrak{H}_T+\mathfrak{H}_V$, toroidal and vector parts relative to a Cartan involution θ of $\mathfrak{G}_0$. The choice of $\mathfrak{H}$ determines reductive parts $\mathfrak{P}^r_x$, $\mathfrak{Q}^r_{[x]}$ and $\mathfrak{N}^r_{[x]}$; let $\mathfrak{Z}_P\subset\mathfrak{H}$, $\mathfrak{Z}_Q\subset\mathfrak{H}$ and $\mathfrak{Z}_N\subset\mathfrak{H}$ denote their respective centers, and let $\mathfrak{Z}\subset\mathfrak{H}$ denote the center of $\mathfrak{P}^r_x\cap\tau\mathfrak{P}^r_x$; note $\tau\mathfrak{Z}=\mathfrak{Z}$, $\tau\mathfrak{Z}_Q=\mathfrak{Z}_Q$, $\tau\mathfrak{Z}_N=\mathfrak{Z}_N$ and $\mathfrak{Z}_N\subset\mathfrak{Z}_Q\subset\mathfrak{Z}$.

$\mathfrak{N}^r_{[x]}$ is the centralizer of $\mathfrak{Z}_N$ in $\mathfrak{G}$, $\mathfrak{Q}^r_{[x]}$ is the centralizer of $\mathfrak{Z}_Q$ in $\mathfrak{G}$, and $\mathfrak{P}^r_x$ is the centralizer of $\mathfrak{Z}_P$ in $\mathfrak{G}$, because $\mathfrak{N}_{[x]}$, $\mathfrak{Q}_{[x]}$ and $\mathfrak{P}_x$ are parabolic subalgebras of $\mathfrak{G}$. Now $\mathfrak{P}^r_x\cap\tau\mathfrak{P}^r_x$ is the centralizer of $\mathfrak{Z}_P+\tau\mathfrak{Z}_P$ in $\mathfrak{G}$, and it follows that $\mathfrak{P}^r_x\cap\tau\mathfrak{P}^r_x$ is the centralizer of $\mathfrak{Z}$ in $\mathfrak{G}$. Decompose real forms:

$$\mathfrak{Z}_{N,0} = \mathfrak{Z}_N\cap\mathfrak{G}_0 \quad\text{and}\quad \mathfrak{Z}_{N,0} = \mathfrak{Z}_{N,T}+\mathfrak{Z}_{N,V},$$

$$\mathfrak{Z}_{Q,0} = \mathfrak{Z}_Q\cap\mathfrak{G}_0 \quad\text{and}\quad \mathfrak{Z}_{Q,0} = \mathfrak{Z}_{Q,T}+\mathfrak{Z}_{Q,V},$$

$$\mathfrak{Z}_0 = \mathfrak{Z}\cap\mathfrak{G}_0 \quad\text{and}\quad \mathfrak{Z}_0 = \mathfrak{Z}_T+\mathfrak{Z}_V$$

into toroidal and vector parts by intersecting with $\mathfrak{H}_T$ and $\mathfrak{H}_V$. If $\mathfrak{S}\subset\mathfrak{G}$ then $Z(\mathfrak{S})$ denotes the centralizer of $\mathfrak{S}$ in G and $Z_0(\mathfrak{S})$ denotes the centralizer $G_0\cap Z(\mathfrak{S})$ of $\mathfrak{S}$ in G_0. Now we have $N^r_{[x]}=Z(\mathfrak{Z}_{N,T})\cap Z(\mathfrak{Z}_{N,V})$, $Q^r_{[x]}=Z(\mathfrak{Z}_{Q,T})\cap Z(\mathfrak{Z}_{Q,V})$ and $P^r_x\cap\tau P^r_x=Z(\mathfrak{Z}_T)\cap Z(\mathfrak{Z}_V)$; for $Z(\mathfrak{S})=Z(\mathfrak{S}\cap\mathfrak{G}_0)$ whenever $\mathfrak{S}$ is a τ-stable subspace of $\mathfrak{G}$. Thus

$$G_0\cap N^r_{[x]} = Z_0(\mathfrak{Z}_{N,T})\cap Z_0(\mathfrak{Z}_{N,V}),$$

$$G_0\cap Q^r_{[x]} = Z_0(\mathfrak{Z}_{Q,T})\cap Z_0(\mathfrak{Z}_{Q,V}),$$

and

$$G_0\cap P^r_x\cap\tau P^r_x = Z_0(\mathfrak{Z}_T)\cap Z_0(\mathfrak{Z}_V).$$

It follows that

$$G_0\cap\exp(i\mathfrak{Z}_{N,V}) \text{ meets every component of } G_0\cap N^r_{[x]},$$

$$G_0\cap\exp(i\mathfrak{Z}_{Q,V}) \text{ meets every component of } G_0\cap Q^r_{[x]},$$

and

$$G_0\cap\exp(i\mathfrak{Z}_V) \text{ meets every component of } G_0\cap P^r_x\cap\tau P^r_x.$$

Using $(\mathcal{P}_x \cap \tau\mathcal{P}_x)^r = \mathcal{P}_x^r \cap \tau\mathcal{P}_x^r$, we also have

$$G_0 \cap \exp(i\mathfrak{Z}_V) \text{ meets every component of } G_0 \cap (P_x \cap \tau P_x)^r.$$

Connectivity of real unipotent groups says that $G_0 \cap N_{[x]}^r$ meets every component of $G_0 \cap N_{[x]}$, $G_0 \cap Q_{[x]}^r$ meets every component of $G_0 \cap Q_{[x]}$, and $G_0 \cap (P_x \cap \tau P_x)^r$ meets every component of $G_0 \cap (P_x \cap \tau P_x) = G_0 \cap P_x$. Thus

$$G_0 \cap \exp(i\mathfrak{Z}_{N,V}) \text{ meets every component of } G_0 \cap N_{[x]},$$

$$G_0 \cap \exp(i\mathfrak{Z}_{Q,V}) \text{ meets every component of } G_0 \cap Q_{[x]},$$

and

$$G_0 \cap \exp(i\mathfrak{Z}_V) \text{ meets every component of } G_0 \cap P_x.$$

$\mathfrak{Z}_N \subset \mathfrak{Z}_Q \subset \mathfrak{Z}$ implies $i\mathfrak{Z}_{N,V} \subset i\mathfrak{Z}_{Q,V} \subset i\mathfrak{Z}_V$, so $G_0 \cap \exp(i\mathfrak{Z}_{N,V}) \subset G_0 \cap \exp(i\mathfrak{Z}_{Q,V}) \subset G_0 \cap \exp(i\mathfrak{Z}_V)$. Thus $G_0 \cap P_x$ meets every component of $G_0 \cap Q_{[x]}$ and also meets every component of $G_0 \cap N_{[x]}$. This proves that $(G_0 \cap Q_{[x]})(x)$ and $(G_0 \cap N_{[x]})(x)$ are connected, hence equal to $Q_{[x],0}(x)$ and $N_{[x],0}(x) = S_{[x]}$, respectively. That proves (1b) and (2b), completing the proof of the theorem. Q.E.D.

8.17. DEFINITIONS. *In the notation of Theorem* 8.15, *the base space* $G_0/(G_0 \cap N_{[x]})$ *of the fibre bundle* (8.16a) *will be called the* **space of holomorphic arc components** *of* $G_0(x)$, *and the base* $G_0/(G_0 \cap Q_{[x]})$ *of the bundle* (8.16b) *will be called the* **space of algebraic arc components** *of* $G_0(x)$. *This terminology is based on results* (1c) *and* (2c) *of Theorem* 8.15.

8.18. COROLLARY. *Let* $P_1 \subset P_2$ *be parabolic subgroups of* G, $X_i = G/P_i$ *the resulting complex flag manifolds,* $\pi: X_1 \to X_2$ *the projection* $gP_1 \mapsto gP_2$, G_0 *a real form of* G, $x_2 \in X_2$, *and* $x_1 \in \pi^{-1}(x_2)$ *such that* $G_0(x_1)$ *is open in* $\pi^{-1}G_0(x_2)$.

1. π *induces a diffeomorphism* $S_{[x]} \mapsto \pi S_{[x]} = S_{[\pi x]}$ *from the space of holomorphic arc components of* $G_0(x_1)$ *onto the space of holomorphic arc components of* $G_0(x_2)$.

2. π *induces a diffeomorphism* $Q_{[x],0}(x) \mapsto \pi Q_{[x],0}(x) = Q_{[\pi x],0}(\pi x)$ *from the space of algebraic arc components of* $G_0(x_1)$ *onto the space of algebraic arc components of* $G_0(x_2)$.

PROOF. This is immediate from Theorem 8.15 and the results $Q_{[x_1]} = Q_{[x_2]}$ and $N_{[x_1]} = N_{[x_2]}$ of Proposition 8.14(3). Q.E.D.

8.19. COROLLARY. *Let* $X = G/P$ *be a complex flag manifold,* $x \in X$, *and* G_0 *a real form of* G. *Then the following conditions are equivalent*:

(i) $G_0(x)$ *is the unique closed* G_0*-orbit on* X.

(ii) $G_0(x)$ *is compact.*

(iii) *The holomorphic arc components of* $G_0(x)$ *are compact.*
(iv) *The algebraic arc components of* $G_0(x)$ *are compact.*

PROOF. We have noted equivalence of (i) and (ii) from compactness of X. Now consider the fibrations (8.16)

$$\sigma: G_0(x) \to G_0/(G_0 \cap N_{[x]}),$$

$$\kappa: G_0(x) \to G_0/(G_0 \cap Q_{[x]})$$

whose base spaces, both compact, are the respective spaces of holomorphic and algebraic arc components of $G_0(x)$. As the base is compact, the total space is compact if and only if the fibres are compact. That given equivalence of (ii) with (iii) and with (iv). Q.E.D.

8.20. COROLLARY. *Let* $X=G/P$ *be a complex flag manifold,* G_0 *a real form of* G, *and* $x\in X$. *If the holomorphic arc components of* $G_0(x)$ *are points, then* $G_0(x)$ *is the closed* G_0*-orbit on* X. *Semiconversely, if* $G_0(x)$ *is the closed* G_0*-orbit on* X, *and if* $\dim_R G_0(x)=\dim_C X$, *then the holomorphic arc components of* $G_0(x)$ *are points.*

PROOF. The first assertion follows from Corollary 8.19 because points are compact. Now suppose $\dim_R G_0(x)=\dim_C X$. Then $\mathcal{P}_x=\tau\mathcal{P}_x$ by Theorem 3.6, so $\mathcal{P}_x\cap\tau\mathcal{P}_x=\mathcal{P}_x+\tau\mathcal{P}_x$, and the construction (8.8) yields $\mathfrak{M}_{[x]}$ as the algebra $\mathcal{P}_x$. Theorem 8.9 then says $\mathfrak{N}_{[x]}=\mathcal{P}_x$, so $S_{[x]}=N_{[x],0}(x)=(G_0\cap P_x)(x)=x$. Q.E.D.

9. Global conditions for the components of an orbit.

We will study certain conditions related to complexity and measurability for the holomorphic arc components of a G_0-orbit. Those conditions are defined as follows.

9.1. DEFINITIONS. *Let* $X=G/P$ *be a complex flag manifold,* G_0 *a real form of* G, $x\in X$, $S_{[x]}$ *the holomorphic arc component of* $G_0(x)$ *through* x, $N_{[x],0}$ *the identity component of the normalizer of* $S_{[x]}$ *in* G_0, *and* $N_{[x]}=N^C_{[x],0}$.

1. *The orbit* $G_0(x)$ *is called* ***partially complex*** *if its holomorphic arc components are complex submanifolds of* X, *i.e. if* $S_{[x]}$ *is a complex submanifold of* X.

2. *The orbit* $G_0(x)$ *is said to be of* ***flag type*** *if the* $N_{[x']}(x')$, $x'\in G_0(x)$, *are complex flag manifolds, i.e. if* $N_{[x]}(x)$ *is a complex flag manifold, i.e. if* $N_{[x]}$ *has a* τ*-stable reductive part* $N^r_{[x]}$ *that is transitive on* $N_{[x]}(x)$ *and that has* $N^r_{[x]}\cap P_x$ *as a parabolic subgroup.*

3. *The orbit* $G_0(x)$ *is called* ***measurable*** *if its holomorphic arc components carry invariant positive Radon measures, i.e. if* $S_{[x]}$ *carries an* $N_{[x],0}$*-invariant positive Radon measure.*

4. *The orbit $G_0(x)$ is said to be* **polarized** *if the $\mathcal{P}_{x'}$, $x' \in G_0(x)$, have τ-stable reductive parts, i.e. if $\tau\mathcal{P}_x^r = \mathcal{P}_x^r$ for some choice of reductive part $\mathcal{P}_x^r \subset \mathcal{P}_x$.*

5. *The orbit $G_0(x)$ is called* **integrable** *if the distribution T of* (8.10) *is integrable, i.e. if $\mathcal{P}_x + \tau\mathcal{P}_x$ is a subalgebra of $\mathfrak{g}$.*

Theorem 8.9 gives several conditions for $G_0(x)$ to be partially complex. The analogous condition for algebraic arc components is automatic (Lemma 8.13), so it will not be given a name.

The notion of flag type is a useful hereditary condition. The analogous condition for algebraic arc components is that $Q_{[x]}(x)$ be a complex flag manifold. In that case we will say that $G_0(x)$ is of **flag type relative to its algebraic arc components.**

The notion of measurable orbit is the primary object of this paper. In §6 we obtained the basic facts for measurable open orbits, and the definition here specializes to that of §6. The analogous condition for algebraic arc components is that $Q_{[x],0}(x)$ carry a $Q_{[x],0}$-invariant positive Radon measure. In that case we will say that $G_0(x)$ is **measurable relative to its algebraic arc components.**

The notion of polarized orbit will be useful in studying the three concepts above. Let $\mathcal{H} \subset \mathcal{P}_x$ be any τ-stable Cartan subalgebra of $\mathfrak{g}$. That determines a choice of $\mathcal{P}_x^r$, and Lemma 2.10 says $(\mathcal{P}_x \cap \tau\mathcal{P}_x)^r = \mathcal{P}_x^r \cap \tau\mathcal{P}_x^r$. Thus $G_0(x)$ is polarized if and only if $\dim(\mathcal{P}_x \cap \tau\mathcal{P}_x)^r = \dim \mathcal{P}_x^r$. The latter can be expressed as $\dim\{(\mathcal{P}_x \cap \tau\mathcal{P}_x)/(\mathcal{P}_x \cap \tau\mathcal{P}_x)^u\} = \dim\{\mathcal{P}_x/\mathcal{P}_x^u\}$, which is independent of choice of $\mathcal{H}$. Thus $G_0(x)$ is polarized if and only if $\mathcal{P}_x^r = \tau\mathcal{P}_x^r$ relative to every τ-stable Cartan subalgebra $\mathcal{H} \subset \mathcal{P}_x$ of $\mathfrak{g}$.

The notion of integrable orbit is recalled from §7, where it was studied in some detail.

9.2. Theorem. *Let $X = G/P$ be a complex flag manifold, G_0 a real form of G, $x \in X$, $\mathcal{H} \subset \mathcal{P}_x$ a τ-stable Cartan subalgebra of $\mathfrak{g}$, and Φ a subset of a simple $\mathcal{H}$-root system such that $\mathcal{P}_x = \mathcal{P}_\Phi$. Define*

$$(9.3)\quad \mathcal{V}_x^+ = \sum_{\Phi^u \cap -\tau\Phi^u} \mathfrak{g}_\varphi, \qquad \mathcal{V}_x^- = \tau\mathcal{V}_x^+ = \sum_{-\Phi^u \cap \tau\Phi^u} \mathfrak{g}_\varphi, \qquad \mathcal{V}_x = \mathcal{V}_x^+ + \mathcal{V}_x^-.$$

Then $G_0(x)$ is measurable if and only if

$$(9.4)\quad \begin{gathered} \mathfrak{N}_{[x]}^u = (\mathcal{P}_x \cap \tau\mathcal{P}_x)^u, \qquad \mathfrak{N}_{[x]}^r = (\mathcal{P}_x \cap \tau\mathcal{P}_x)^r + \mathcal{V}_x, \\ \mathfrak{N}_{[x]} = (\mathcal{P}_x \cap \tau\mathcal{P}_x) + \mathcal{V}_x. \end{gathered}$$

If $G_0(x)$ is measurable, then

(i) *$G_0(x)$ is partially complex;*

(ii) *$G_0(x)$ is of flag type;*

(iii) *$G_0(x)$ is polarized if and only if it is integrable; and*

(iv) *the invariant Radon measure on $S_{[x]}$ is the volume element of an $N_{[x],0}$-invariant kaehler metric (nondegenerate, but not necessarily positive definite).*

PROOF. Let A be the set of $\mathcal{H}$-roots such that $\mathfrak{N}_{[x]}=(\mathcal{P}_x\cap\tau\mathcal{P}_x)+\sum_A \mathcal{G}_\varphi$ and define $\alpha=\sum_A\varphi$. If $G_0(x)$ is measurable, then $S_{[x]}$ has a positive $N_{[x],0}$-invariant Radon measure μ. Then the coset space $N_{[x],0}/(N_{[x],0}\cap P_x)=N_{[x],0}/(P_x\cap\tau P_x)_0$ representing $S_{[x]}$ has μ as invariant measure, so $(P_x\cap P_x)_0$ acts with determinant ± 1 on its tangent space, i.e. $(\mathcal{P}_x\cap\tau\mathcal{P}_x)_0$ acts with trace 0 on $\mathfrak{N}_{[x],0}/(\mathcal{P}_x\cap\tau\mathcal{P}_x)_0$, i.e. $\mathcal{P}_x\cap\tau\mathcal{P}_x$ acts with trace 0 on $\mathfrak{N}_{[x]}/(\mathcal{P}_x\cap\tau\mathcal{P}_x)$, i.e. $(\mathcal{P}_x\cap\tau\mathcal{P}_x)^r$ acts with trace 0 on $\sum_A\mathcal{G}_\varphi$, i.e. $\mathcal{H}$ acts with trace 0 on $\sum_A\mathcal{G}_\varphi$, i.e. $\alpha=0$. The argument is reversible. Thus $G_0(x)$ is measurable if and only if $\alpha=0$.

Suppose $G_0(x)$ measurable. If $\varphi\in A$ with $\mathcal{G}_\varphi\subset\mathfrak{N}^u_{[x]}$, then α has positive inner product with the sum of the roots of $\mathfrak{N}^u_{[x]}$, contradicting $\alpha=0$. Thus $\sum_A\mathcal{G}_\varphi\subset\mathfrak{N}^r_{[x]}$ and it follows that $\mathfrak{N}^u_{[x]}\subset(\mathcal{P}_x\cap\tau\mathcal{P}_x)$. Now $A=-A$ by Lemma 6.2, for $\mathfrak{N}^r_{[x]}\cap(\mathcal{P}_x\cap\tau\mathcal{P}_x)$ is reductive. As A is disjoint from $\Phi^*\cap\tau\Phi^*$, i.e. as

$$A\subset(\Phi^r\cap-\tau\Phi^u)\cup(-\Phi^u\cap\tau\Phi^r)\cup(\Phi^u\cap-\tau\Phi^u)\cup(-\Phi^u\cap\tau\Phi^u),$$

this says $\sum_A\mathcal{G}_\varphi\subset\mathcal{V}_x$. As $\mathfrak{N}_{[x]}$ is parabolic now $\sum_A\mathcal{G}_\varphi=\mathcal{V}_x$. Thus $\mathfrak{N}_{[x]}=(\mathcal{P}_x\cap\tau\mathcal{P}_x)+\mathcal{V}_x$. That proves (9.4).

Conversely assume (9.4). Then $N^u_{[x],0}$ acts trivially on $S_{[x]}$ and $N^r_{[x],0}$ acts transitively with reductive isotropy subgroup $(P^r_x\cap\tau P^r_x)_0$, so Lemma 6.2 says that $G_0(x)$ is measurable.

Suppose $G_0(x)$ measurable. Then (9.4) implies $\mathfrak{N}_{[x]}\subset(\mathcal{P}_x+\tau\mathcal{P}_x)$, so Theorem 8.9 says that $G_0(x)$ is partially complex; that proves (i). $\mathfrak{N}^u_{[x]}\subset\mathcal{P}_x$ by (9.4), so $N^u_{[x]}$ acts trivially on $N_{[x]}(x)$, whence $N^r_{[x]}$ is transitive on $N_{[x]}(x)$; the isotropy subalgebra of $\mathfrak{N}^r_{[x]}$ at x is $\mathfrak{N}^r_{[x]}\cap\mathcal{P}_x=(\mathcal{P}_x\cap\tau\mathcal{P}_x)^r+\mathcal{V}^+_x$, parabolic subalgebra; thus $N_{[x]}(x)$ is a complex flag manifold; that proves (ii). If $G_0(x)$ is polarized then $\tau\Phi^r=\Phi^r$, so $(\mathcal{P}_x\cap\tau\mathcal{P}_x)=\mathcal{P}^r_x+\tau\mathcal{P}^r_x+(\mathcal{P}^u_x\cap\tau\mathcal{P}^u_x)$ so $(\mathcal{P}_x\cap\tau\mathcal{P}_x)+\mathcal{V}_x=(\mathcal{P}_x+\tau\mathcal{P}_x)$, and (9.4) says that $\mathfrak{N}_{[x]}=(\mathcal{P}_x+\tau\mathcal{P}_x)$; thus $\mathcal{P}_x+\tau\mathcal{P}_x$ is an algebra and $G_0(x)$ is integrable. If $G_0(x)$ is integrable then $\mathfrak{N}_{[x]}=(\mathcal{P}_x+\tau\mathcal{P}_x)$ by Theorem 8.9, for $\mathcal{Q}_{[x]}=(\mathcal{P}_x\cap\tau\mathcal{P}_x)$; then (9.4) says that $\Phi^r\cap-\tau\Phi^u$ and $-\Phi^u\cap\tau\Phi^r$ are empty, so $\tau\Phi^r=\Phi^r$ and $G_0(x)$ is polarized. That proves (iii).

Suppose $G_0(x)$ measurable. By (ii), $Y=N_{[x]}(x)$ is a complex flag manifold $N^r_{[x]}/(N^r_{[x]}\cap P_x)$ on which $N^u_{[x]}$ acts trivially. Lemma 8.4 says that $S_{[x]}=N^r_{[x],0}(x)$ is an open $N^r_{[x],0}$-orbit on Y. Now $S_{[x]}$ is a measurable open $N^r_{[x],0}$-orbit on the complex flag manifold Y, so

Theorem 6.3 says that its measure is the volume element of a (possibly indefinite) $N^r_{[x],0}$-invariant kaehler metric ds^2. As $N^u_{[x],0}$ acts trivially on $S_{[x]}$ it must preserve ds^2. Thus ds^2 is $N_{[x],0}$-invariant. That proves (iv), completing the proof of the theorem. Q.E.D.

9.5. COROLLARY. *Let $X=G/P$ be a complex flag manifold, G_0 a real form of G, $x\in X$, and $\mathcal{V}_x$ as defined in* (9.3). *Then $G_0(x)$ is measurable relative to its algebraic arc components if and only if*

$$(9.6)\qquad \begin{aligned} \mathcal{Q}^u_{[x]} &= (\mathcal{P}_x \cap \tau\mathcal{P}_x)^u, \qquad \mathcal{Q}^r_{[x]} = (\mathcal{P}_x \cap \tau\mathcal{P}_x)^r + \mathcal{V}_x, \\ \mathcal{Q}_{[x]} &= (\mathcal{P}_x \cap \tau\mathcal{P}_x) + \mathcal{V}_x. \end{aligned}$$

In that case,

(i) *$G_0(x)$ is of flag type relative to its algebraic arc components;*

(ii) *$G_0(x)$ is polarized if and only if it is integrable;*

(iii) *the invariant Radon measure on $Q_{[x],0}(x)$ is the volume element of a $Q_{[x],0}$-invariant, possibly indefinite, kaehler metric.*

PROOF. $\mathcal{Q}_{[x]}$ can be substituted for $\mathfrak{N}_{[x]}$ throughout the proof of Theorem 9.2. Q.E.D.

To decide whether a given $G_0(x)$ is measurable, compute δ_Φ and $\mathcal{Q}_{[x]}$ by (8.6), then $\mathfrak{M}_{[x]}$ by (8.8), and finally $\mathcal{W}_{[x]}=(\mathcal{P}_x\cap\tau\mathcal{P}_x)+\mathcal{V}_x$ by (9.3). Then $G_0(x)$ is measurable if and only if (i) $\mathfrak{M}_{[x]}=\mathcal{W}_{[x]}$ and (ii) $\mathcal{W}_{[x]}$ is an algebra. It is easy to check (i), and it follows from (9.3) that (ii) is equivalent to $[\mathcal{V}_x^+, \mathcal{V}_x^-]\subset\mathcal{W}_{[x]}$. The latter inclusion can be unpleasant to check, but it is much more easily decidable than is the inclusion $[\mathfrak{M}_{[x]}, \mathfrak{M}_{[x]}]\subset\mathfrak{M}_{[x]}$ that says $\mathfrak{M}_{[x]}$ is an algebra. So measurability is easier to check than is partial complexity.

9.7. THEOREM. *Let $X=G/P$ be a complex flag manifold, G_0 a real form of G, $x\in X$, $\mathcal{H}\subset\mathcal{P}_x$ a τ-stable Cartan subalgebra of $\mathcal{G}$, Φ a subset of a simple $\mathcal{H}$-root system such that $\mathcal{P}_x=\mathcal{P}_\Phi$, and $\mathcal{V}_x$ as in* (9.3).

1. *$G_0(x)$ is of flag type if and only if $\mathcal{V}_x\subset\mathfrak{N}_{[x]}$.*

2. *If $G_0(x)$ is of flag type, then $G_0(x)$ is partially complex if and only if $S_{[x]}=N^r_{[x],0}(x)$ is an open $N^r_{[x],0}$-orbit on the complex flag manifold $N^r_{[x]}(x)$.*

3. *If $G_0(x)$ is partially complex and of flag type, then* $\dim_C S_{[x]} \geqq |\Phi^u\cap -\tau\Phi^u|$, *with equality if and only if $G_0(x)$ is measurable.*

PROOF. Suppose $G_0(x)$ of flag type, so $N^r_{[x]}$ is transitive on $Y=N_{[x]}(x)$ and $N^r_{[x]}\cap P_x$ is parabolic in $N^r_{[x]}$. As $N^u_{[x]}$ is solvable it has a fixed point $y\in Y$. If $g\in N_{[x]}$ then $N^u_{[x]}(gy)=gN^u_{[x]}(y)=gy$; thus $N^u_{[x]}$ acts trivially on Y; now $\mathfrak{N}^u_{[x]}\subset\mathcal{P}_x$, so $\tau\mathfrak{N}^u_{[x]}=\mathfrak{N}^u_{[x]}$ gives us $\mathfrak{N}^u_{[x]}\subset(\mathcal{P}_x\cap\tau\mathcal{P}_x)$. If $\varphi\in\Phi^u\cap-\tau\Phi^u$, it follows that neither φ nor $-\varphi$ is a root of $\mathfrak{N}^u_{[x]}$, so both are roots of $\mathfrak{N}^r_{[x]}$; thus $\mathcal{V}_x\subset\mathfrak{N}_{[x]}$.

Conversely suppose $\mathcal{V}_x \subset \mathfrak{N}_{[x]}$. Then $\mathfrak{N}_{[x]}^r$ contains $(\mathcal{P}_x^r + \tau\mathcal{P}_x^r) + \mathcal{V}_x$, and $\mathfrak{N}_{[x]}$ contains $(\mathcal{P}_x \cap \tau\mathcal{P}_x)^u$; it follows that $\mathfrak{N}_{[x]}^u \subset (\mathcal{P}_x \cap \tau\mathcal{P}_x)^u$, so $N_{[x]}^r$ is transitive on $N_{[x]}(x)$. The isotropy subalgebra $\mathfrak{N}_{[x]}^r \cap \mathcal{P}_x$ contains $(\mathcal{P}_x \cap \tau\mathcal{P}_x) + \mathcal{V}_x^+$, so it is parabolic in $\mathfrak{N}_{[x]}^r$. Thus $G_0(x)$ is of flag type.

Let $G_0(x)$ be of flag type, $Y = N_{[x]}(x) = N_{[x]}^r(x)$. Then $S_{[x]} = N_{[x],0}(x) = N_{[x],0}^r(x)$. If $G_0(x)$ is partially complex, then $\mathfrak{N}_x \subset (\mathcal{P}_x + \tau\mathcal{P}_x)$ by Theorem 8.9, so $S_{[x]}$ is open in Y by Lemma 8.4. If $S_{[x]}$ is open in Y, then it is a complex submanifold of X, so $G_0(x)$ is partially complex.

Let $G_0(x)$ be partially complex and of flag type. Then $\mathfrak{N}_{[x]} = (\mathcal{P}_x \cap \tau\mathcal{P}_x) + \mathcal{V}_x + \mathcal{W}_x$ where $\mathcal{W}_x$ is a τ-stable sum of root spaces $\mathfrak{g}_\varphi$, $\varphi \in (\Phi^r \cap -\tau\Phi^u) \cup (-\Phi^u \cap \tau\Phi^r)$. Now $\dim_C S_{[x]} = \dim_C N_{[x]}(x) = \dim_C \mathcal{V}_x^- + \dim_C(\mathcal{W}_x \cap \tau\mathcal{P}_x) = |\Phi^u \cap -\tau\Phi^u| + \dim_C(\mathcal{W}_x \cap \tau\mathcal{P}_x) \geqq |\Phi^u \cap -\tau\Phi^u|$, and equality means $\mathcal{W}_x = 0$, i.e. $\mathfrak{N}_{[x]} = (\mathcal{P}_x \cap \tau\mathcal{P}_x) + \mathcal{V}_x$, i.e. $G_0(x)$ measurable. Q.E.D.

9.8. Corollary. *Let $X = G/P$ be a complex flag manifold, G_0 a real form of G, $x \in X$, $\mathfrak{H} \subset \mathcal{P}_x$ a τ-stable Cartan subalgebra of $\mathfrak{g}$, Φ a subset of a simple $\mathfrak{H}$-root system such that $\mathcal{P}_x = \mathcal{P}_\Phi$, and $\mathcal{V}_x$ as in* (9.3).

1. *$G_0(x)$ is of flag type relative to its algebraic arc components if and only if $\mathcal{V}_x \subset \mathfrak{Q}_{[x]}$.*

2. *If $G_0(x)$ is of flag type relative to its algebraic arc components, then*

(2a) *$G_0(x)$ is of flag type,*

(2b) *the algebraic arc component of $G_0(x)$ through x is an open $Q_{[x],0}^r$-orbit on the complex flag manifold $Q_{[x]}^r(x)$,*

(2c) *the algebraic arc components of $G_0(x)$ have complex dimension $\geqq |\Phi^u \cap -\tau\Phi^u|$, with equality if and only if $G_0(x)$ is measurable relative to its algebraic arc components.*

Proof. $\mathfrak{Q}_{[x]}$ can be substituted for $\mathfrak{N}_{[x]}$ throughout the proof of Theorem 9.7. Q.E.D.

Recall Example 8.12. There $G = \mathbf{SL}(2m, \mathbf{C})$, G_0 is $\mathbf{SL}(m, \mathbf{Q})$ or $\mathbf{SL}(2m, \mathbf{R})$, the simple root system $\Psi = \{\psi_1, \cdots, \psi_{2m-1}\}$, $\tau\psi_i = \psi_{2m-i}$, and $\mathcal{P}_x = \mathcal{P}_\Phi$ where $\Phi = \{\psi_1, \cdots, \psi_{2m-2}\}$. So X is complex projective $(2m-1)$-space and we calculated

$\mathfrak{Q}_{[x]} = \mathcal{P}_\Gamma$ where $\Gamma = \{\psi_2, \cdots, \psi_{2m-2}\}$ and

$\mathfrak{M}_{[x]} = \mathcal{P}_x + \tau\mathcal{P}_x = \sum_{\varphi \neq -\mu} \mathfrak{g}_\varphi$ where $\mu = \psi_1 + \psi_2 + \cdots + \psi_{2m-1}$.

Thus $\mathfrak{M}_{[x]}$ generates $\mathfrak{g}$. As $\mathfrak{M}_{[x]} \subset \mathfrak{N}_{[x]}$ by Proposition 8.7, it follows that $\mathfrak{N}_{[x]} = \mathfrak{g}$. Thus $G_0(x)$ is of flag type, but it is not partially complex.

9.9. Theorem. *Let $X = G/P$ be a complex flag manifold, G_0 a real form of G, and $x \in X$ such that the orbit $G_0(x)$ is polarized. Then the following conditions are equivalent:*

(i) *$G_0(x)$ is integrable.*

(ii) *$G_0(x)$ is measurable.*

(iii) *$G_0(x)$ is measurable relative to its algebraic arc components.*

(iv) *$G_0(x)$ is partially complex and of flag type.*

(v) *$G_0(x)$ is partially complex and is of flag type relative to its algebraic arc components.*

If the above conditions hold, then $\mathfrak{N}_{[x]} = (\mathcal{P}_x \cap \tau\mathcal{P}_x) + \mathcal{V}_x = \mathcal{Q}_{[x]}$, so the algebraic arc components of $G_0(x)$ coincide with the holomorphic arc components.

PROOF. If $G_0(x)$ is measurable, Theorem 9.2 says that it is integrable. If $G_0(x)$ is measurable relative to its algebraic arc components, Corollary 9.5 says that it is integrable.

Suppose $G_0(x)$ integrable, so (8.6) and Theorem 8.9 show that $\mathfrak{N}_{[x]} = (\mathcal{P}_x + \tau\mathcal{P}_x) = \mathcal{Q}_{[x]}$. As $G_0(x)$ is polarized, $(\mathcal{P}_x + \tau\mathcal{P}_x) = (\mathcal{P}_x \cap \tau\mathcal{P}_x) + \mathcal{V}_x$. Thus $\mathfrak{N}_{[x]} = (\mathcal{P}_x \cap \tau\mathcal{P}_x) + \mathcal{V}_x = \mathcal{Q}_{[x]}$, and Theorem 9.2 says that $G_0(x)$ both is measurable, and is measurable relative to its algebraic arc components.

If $G_0(x)$ is measurable, Theorem 9.2 says it is partially complex and of flag type. If $G_0(x)$ is partially complex and of flag type, Theorems 8.9 and 9.7 say that $(\mathcal{P}_x \cap \tau\mathcal{P}_x) + \mathcal{V}_x \subset \mathfrak{N}_{[x]} \subset (\mathcal{P}_x \cap \tau\mathcal{P}_x)$, while $(\mathcal{P}_x \cap \tau\mathcal{P}_x) + \mathcal{V}_x = (\mathcal{P}_x + \tau\mathcal{P}_x)$ because $G_0(x)$ is polarized; thus $\mathfrak{N}_{[x]} = (\mathcal{P}_x \cap \tau\mathcal{P}_x) + \mathcal{V}_x$ and $G_0(x)$ is measurable by Theorem 9.2.

If $G_0(x)$ is measurable relative to its algebraic arc components, then it is measurable because it is polarized, so it is partially complex by Theorem 9.2; and $\mathcal{V}_x \subset \mathcal{O}_{[x]}$ by Corollary 9.5, so $G_0(x)$ is of flag type relative to its algebraic arc components by Corollary 9.8. If $G_0(x)$ is partially complex and is of flag type relative to its algebraic arc components, Theorem 8.9 and Corollary 9.8 say that $(\mathcal{P}_x \cap \tau\mathcal{P}_x) + \mathcal{V}_x \subset \mathcal{Q}_{[x]} \subset \mathfrak{N}_{[x]} \subset (\mathcal{P}_x + \tau\mathcal{P}_x)$, while $(\mathcal{P}_x \cap \tau\mathcal{P}_x) + \mathcal{V}_x = (\mathcal{P}_x + \tau\mathcal{P}_x)$ because $G_0(x)$ is polarized; thus $\mathcal{Q}_{[x]} = (\mathcal{P}_x \cap \tau\mathcal{P}_x) + \mathcal{V}_x$ and Corollary 9.5 says that $G_0(x)$ is measurable relative to its algebraic arc components.

Q.E.D.

9.10. COROLLARY. *Let B be a Borel subgroup of G, $Y = G/B$ a complex flag manifold, G_0 a real form of G, and $y \in Y$. Then the following conditions are equivalent, and they imply that the algebraic arc components of $G_0(y)$ coincide with the holomorphic arc components:*

(i) *$G_0(y)$ is integrable.*

(ii) *$G_0(y)$ is measurable.*

(iii) *$G_0(y)$ is measurable relative to its algebraic arc components.*

(iv) *$G_0(y)$ is partially complex and of flag type.*

(v) *$G_0(y)$ is partially complex and is of flag type relative to its algebraic arc components.*

PROOF. Let $\mathfrak{H} \subset \mathfrak{B}_y$ be a τ-stable Cartan subalgebra of $\mathfrak{g}$. Then $\mathfrak{B}_y^r = \mathfrak{H} = \tau\mathfrak{H} = \tau\mathfrak{B}_y^r$, so $G_0(y)$ is polarized and Theorem 9.9 applies. Q.E.D.

9.11. COROLLARY. *Let $X = G/P$ be a complex flag manifold, G_0 a real form of G, and $x \in X$ such that the orbit $G_0(x)$ is integrable.*

1. *$\mathfrak{N}_{[x]} = (\mathfrak{P}_x + \tau\mathfrak{P}_x) = \mathfrak{Q}_{[x]}$, so the algebraic arc components of $G_0(x)$ coincide with the holomorphic arc components.*
2. *$G_0(x)$ is partially complex and of flag type.*
3. *$G_0(x)$ is measurable if and only if it is polarized.*

PROOF. Assertion (1) is immediate from (8.6) and Theorem 8.9. Now Theorem 8.9 proves $G_0(x)$ partially complex and Theorem 9.7 says that it is of flag type. If $G_0(x)$ is measurable, Theorem 9.2 says it is polarized; if $G_0(x)$ is polarized, Theorem 9.9 says it is measurable. Q.E.D.

If $G_0(x)$ is an open orbit, then $S_{[x]} = G_0(x)$, so it is rather trivial that $G_0(x)$ is partially complex and of flag type. Criteria for measurability of $G_0(x)$ are given by Theorems 6.3 and 6.7, the latter being that $\mathfrak{P}$ and $\tau\mathfrak{P}$ are opposed. We now see that closed orbits are somewhat better behaved.

9.12. THEOREM. *Let $X = G/P$ be a complex flag manifold, G_0 a real form of G, and $x \in X$ such that $G_0(x)$ is the closed G_0-orbit on X. Then $G_0(x)$ is measurable, hence partially complex and of flag type.*

PROOF. Let $B \subset P$ be a Borel subgroup of G, $Y = G/B$ the resulting complex flag manifold, and $\pi: Y \to X$ the projection. Choose $y \in \pi^{-1}(x)$ such that $G_0(y)$ is the closed G_0-orbit on Y.

Let $\mathfrak{H}_0$ be the Lie algebra of a maximally split Cartan subgroup of G_0, $\mathfrak{H}_0 = \mathfrak{T} + \mathfrak{A}$ the decomposition into toroidal and vector parts, and $\mathfrak{H} = \mathfrak{H}_0^C$. The $\mathfrak{H}$-root system $\Delta = \Delta_0 \cup \Delta_r$ where Δ_0 consists of those roots that vanish on $\mathfrak{A}$ and Δ_r consists of those that do not. Ordering the roots, we have

$$\text{Iwasawa decomposition } \mathfrak{g}_0 = \mathfrak{K} + \mathfrak{A} + \mathfrak{N}, \qquad \mathfrak{N}^C = \sum_{\Delta_r^+} \mathfrak{g}_\varphi$$

$$\text{minimal parabolic } \mathfrak{L} = \mathfrak{M} + \mathfrak{A} + \mathfrak{N}, \qquad \mathfrak{M}^C = \mathfrak{T}^C + \sum_{\Delta_0} \mathfrak{g}_\varphi.$$

Now define a Borel subalgebra $\mathfrak{B}' \subset \mathfrak{g}$ by

$$\mathfrak{B}' = (\mathfrak{T} + \mathfrak{A} + \mathfrak{N})^C + \sum_{\Delta_r^+} \mathfrak{g}_\varphi.$$

Then

$$\tau\mathcal{B}' = (\mathfrak{J} + \mathfrak{A} + \mathfrak{N})^C + \sum_{\Delta_0^-} \mathfrak{g}_\varphi.$$

Thus

(i) $\mathcal{B}' \cap \tau\mathcal{B}' = (\mathfrak{J}+\mathfrak{A}+\mathfrak{N})^C$ and

(ii) $\mathcal{B}' + \tau\mathcal{B}' = \mathfrak{L}^C$.

Let $y' \in Y$ be the point such that $\mathcal{B}_{y'} = \mathcal{B}'$. Then (ii) says that $G_0(y')$ is integrable, hence measurable by Corollary 9.10, hence partially complex and of flag type by Theorem 9.2 or Corollary 9.10. On the other hand, (i) shows that the compact group K with Lie algebra $\mathfrak{K}$ is transitive on $G_0(y')$. Thus $G_0(y')$ is the closed G_0-orbit on Y, i.e. $G_0(y') = G_0(y)$. That proves our assertions for the case where P is a Borel subgroup of G.

Conjugate $\mathcal{B}'$ by an element of G_0 that sends y' to y. That done, we may assume that $\mathcal{B}_y$ is given by the description of $\mathcal{B}'$ above. Now $\mathcal{P}_x = \mathcal{B}_y + \mathfrak{D}_0 + \mathfrak{D}_r$ where, for certain subsets $D_0 \subset \Delta_0^-$ and $D_r \subset \Delta_r^-$,

$$\mathfrak{D}_0 = \sum_{D_0} \mathfrak{g}_\varphi \quad \text{and} \quad \mathfrak{D}_r = \sum_{D_r} \mathfrak{g}_\varphi.$$

Thus

$$\mathcal{P}_x = \mathcal{B}_y + \mathfrak{D}_0 + \mathfrak{D}_r \quad \text{and} \quad \overline{\mathcal{B}}_y^{-u} = \overline{\mathcal{P}}_x^{-u} + \mathfrak{D}_0 + \mathfrak{D}_r,$$

and, defining

$$\mathfrak{D}_0^- = \sum_{-D_0} \mathfrak{g}_\varphi, \qquad \mathfrak{D}_r^- = \sum_{-D_r} \mathfrak{g}_\varphi, \qquad \mathfrak{N}^- = \Big\{ \sum_{\Delta_r^-} \mathfrak{g}_\varphi \Big\} \cap \mathfrak{g}_0,$$

we have

$$\tau\mathfrak{D}_0 = \mathfrak{D}_0^- \subset \mathcal{B}_y \quad \text{and} \quad \mathfrak{D}_r + \tau\mathfrak{D}_r \subset (\mathfrak{N}^-)^C.$$

In particular,

$$\mathfrak{D}_r \cap \tau\mathcal{B}_y^u = \tau\mathfrak{D}_r \cap \mathcal{B}_y^u = \mathfrak{D}_r^- \cap \tau\overline{\mathcal{B}}_y^{-u} = \tau\mathfrak{D}_r^- \cap \overline{\mathcal{B}}_y^{-u} = 0.$$

Retaining the notation (9.3), we now have

$$\begin{aligned}
\mathcal{V}_y &= (\mathcal{B}_y^u \cap \tau\overline{\mathcal{B}}_y^{-u}) + (\overline{\mathcal{B}}_y^{-u} \cap \tau\mathcal{B}_y^u) \\
&= \{(\mathcal{P}_x^u + \mathfrak{D}_0^-) \cap (\tau\overline{\mathcal{P}}_x^{-u} + \tau\mathfrak{D}_0)\} + \{(\overline{\mathcal{P}}_x^{-u} + \mathfrak{D}_0) \cap (\tau\mathcal{P}_x^u + \tau\mathfrak{D}_0^-)\} \\
&= (\mathcal{P}_x^u \cap \tau\overline{\mathcal{P}}_x^{-u}) + (\overline{\mathcal{P}}_x^{-u} \cap \tau\mathcal{P}_x^u) + \mathfrak{D}_0 + \mathfrak{D}_0^- = \mathcal{V}_x + \mathfrak{D}_0 + \mathfrak{D}_0^-.
\end{aligned}$$

We also compute

$$\begin{aligned}
(\mathcal{P}_x \cap \tau\mathcal{P}_x) &= (\mathcal{B}_y + \mathfrak{D}_0 + \mathfrak{D}_r) \cap (\tau\mathcal{B}_y + \tau\mathfrak{D}_0 + \tau\mathfrak{D}_r) \\
&= (\mathcal{B}_y \cap \tau\mathcal{B}_y) + (\mathfrak{D}_0 + \mathfrak{D}_0^-) + (\mathfrak{D}_r \cap \tau\mathfrak{D}_r).
\end{aligned}$$

As $G_0(y)$ is measurable, $\mathfrak{N}_{[y]}=(\mathfrak{B}_y\cap\tau\mathfrak{B}_y)+\mathfrak{V}_y$ by Theorem 9.2. Now define

(9.13a) $\mathfrak{W}_{[x]}=(\mathfrak{P}_x\cap\tau\mathfrak{P}_x)+\mathfrak{V}_x$.

Our calculations above say that

$$\begin{aligned}\mathfrak{W}_{[x]} &= (\mathfrak{B}_y\cap\tau\mathfrak{B}_y)+(\mathfrak{D}_0+\bar{\mathfrak{D}}_0)+(\mathfrak{D}_r\cap\tau\mathfrak{D}_r)+\mathfrak{V}_x\\ &= (\mathfrak{B}_y\cap\tau\mathfrak{B}_y)+\mathfrak{V}_y+(\mathfrak{D}_r\cap\tau\mathfrak{D}_r).\end{aligned}$$

In other words,

(9.13b) $\mathfrak{W}_{[x]}=\mathfrak{N}_{[y]}+(\mathfrak{D}_r\cap\tau\mathfrak{D}_r)$.

As both $\mathfrak{N}_{[y]}$ and $(\mathfrak{P}_x\cap\tau\mathfrak{P}_x)$ are subalgebras of $\mathfrak{W}_{[x]}$, and $(\mathfrak{P}_x\cap\tau\mathfrak{P}_x)+\mathfrak{V}_x=\mathfrak{W}_{[x]}=\mathfrak{N}_{[y]}+(\mathfrak{D}_r\cap\tau\mathfrak{D}_r)$ by (9.13 a, b), the condition for $\mathfrak{W}_{[x]}$ to be an algebra is

$$[\mathfrak{V}_x,\ \mathfrak{D}_r\cap\tau\mathfrak{D}_r]\subset\mathfrak{W}_{[x]}.$$

Observing $\mathfrak{D}_r\subset\mathfrak{P}_x^r$, we calculate

$$\begin{aligned}[\mathfrak{V}_x^+,\ \mathfrak{D}_r\cap\tau\mathfrak{D}_r] &\subset [\mathfrak{P}_x^u\cap\tau\mathfrak{P}_x^{-u},\ \mathfrak{P}_x^r\cap\tau\mathfrak{P}_x^r]\\ &\subset\{[\mathfrak{P}_x^u,\mathfrak{P}_x^r]\cap[\tau\mathfrak{P}_x^{-u},\tau\mathfrak{P}_x^r]\}\subset(\mathfrak{P}_x^u\cap\tau\mathfrak{P}_x^{-u})\subset\mathfrak{V}_x^+,\end{aligned}$$

and similarly $[\mathfrak{V}_x^-,\ \mathfrak{D}_r\cap\tau\mathfrak{D}_r]\subset\mathfrak{V}_x^-$. Thus $[\mathfrak{V}_x,\ \mathfrak{D}_r\cap\tau\mathfrak{D}_r]\subset\mathfrak{V}_x\subset\mathfrak{W}_{[x]}$. That proves $\mathfrak{W}_{[x]}$ to be an algebra. As it contains the parabolic subalgebra $\mathfrak{N}_{[y]}$ of $\mathfrak{G}$, and as $\tau\mathfrak{W}_{[x]}=\mathfrak{W}_{[x]}$, now

(9.13c) $\mathfrak{W}_{[x]}$ is a τ-stable parabolic subalgebra of $\mathfrak{G}$.

Comparing (9.4) and (9.13a), Theorem 9.2 says that $G_0(x)$ is measurable if and only if $\mathfrak{W}_{[x]}=\mathfrak{N}_{[x]}$. As $\mathfrak{W}_{[x]}\subset\mathfrak{M}_{[x]}$ by Lemma 8.4 and Proposition 8.7, Theorem 8.9 says we need only check that $\mathfrak{W}_{[x]}=\mathfrak{M}_{[x]}$. Define A to be the set of roots such that $\mathfrak{M}_{[x]}=\mathfrak{W}_{[x]}+\sum_A\mathfrak{G}_\varphi$ (direct); now the assertion of our theorem is reduced to the statement that A is empty.

Express $\mathfrak{W}_{[x]}^u=\sum_M\mathfrak{G}_\varphi$ and define $\mu=\sum_M\varphi$. Then (9.13c) says that a root space $\mathfrak{G}_\varphi\subset\mathfrak{W}_{[x]}^r$ if and only if $\langle\varphi,\mu\rangle=0$, $\mathfrak{G}_\varphi\subset\mathfrak{W}_{[x]}^u$ if and only if $\langle\varphi,\mu\rangle>0$, and $\mu=\tau\mu$. In particular, $\langle\varphi,\mu\rangle<0$ for every $\varphi\in A$, and $\langle\varphi,\mu\rangle\geqq 0$ whenever $\mathfrak{G}_\varphi\subset(\mathfrak{P}_x\cap\tau\mathfrak{P}_x)$.

Let $\varphi\in A$. If $\varphi+\tau\varphi$ is a root, then $\mathfrak{G}_{\varphi+\tau\varphi}\subset(\mathfrak{P}_x+\tau\mathfrak{P}_x)$ implies $\mathfrak{G}_{\varphi+\tau\varphi}\subset(\mathfrak{P}_x\cap\tau\mathfrak{P}_x)$ because $\tau(\varphi+\tau\varphi)=\varphi+\tau\varphi$; then $0\leqq\langle\varphi+\tau\varphi,\mu\rangle=\langle\varphi,\mu\rangle+\langle\tau\varphi,\mu\rangle=\langle\varphi,\mu\rangle+\langle\tau\varphi,\tau\mu\rangle=2\langle\varphi,\mu\rangle<0$, which is absurd; thus $\mathfrak{G}_{\varphi+\tau\varphi}\subset\mathfrak{P}_x^{-u}\cap\tau\mathfrak{P}_x^{-u}$, so φ is not an $\mathfrak{H}$-root of $\mathfrak{M}_{[x]}$; that contradicts $\varphi\in A$.

Let $\varphi\in A$. Now $\varphi+\tau\varphi$ is not a root. As $(\mathfrak{B}_y+\tau\mathfrak{B}_y)=\mathfrak{N}_{[y]}\subset\mathfrak{W}_{[x]}$ because $G_0(y)$ is integrable, we have $\mathfrak{G}_\varphi\not\subset(\mathfrak{B}_y+\tau\mathfrak{B}_y)$. However $\mathfrak{G}_\varphi\subset(\mathfrak{P}_x+\tau\mathfrak{P}_x)$. Replace φ by $\tau\varphi$ if necessary so $\mathfrak{G}_\varphi\subset\mathfrak{P}_x=\mathfrak{B}_y+\mathfrak{D}_0+\mathfrak{D}_r$. As $\mathfrak{D}_0\subset\tau\mathfrak{B}_y$ now $\mathfrak{G}_\varphi\subset\mathfrak{D}_r$. As $\mathfrak{D}_r\cap\tau\mathfrak{D}_r\subset\mathfrak{W}_{[x]}$ now $\mathfrak{G}_\varphi\not\subset\tau\mathfrak{P}_x$. Thus

$\mathfrak{g}_\varphi \subset \mathcal{P}_x^r \cap \tau\mathcal{P}_x^{-u}$. Also, $\mathfrak{g}_\varphi \subset \mathfrak{D}_r$ shows $\langle \varphi, \tau\varphi\rangle \neq 0$, so Proposition 8.7(2) gives us the fact that

$$(9.14a) \quad \mathfrak{Z} = h_\varphi \mathbf{C} + h_{\tau\varphi}\mathbf{C} + \mathfrak{g}_\varphi + \mathfrak{g}_{\tau\varphi} + \mathfrak{g}_{-\varphi} + \mathfrak{g}_{-\tau\varphi} + \mathfrak{g}_{\tau\varphi-\varphi} + \mathfrak{g}_{\varphi-\tau\varphi}$$

is a τ-stable subalgebra of type A_2 in $\mathfrak{g}$ that is contained in $\mathfrak{M}_{[x]}$ and has Cartan subalgebra $\mathfrak{H}' = h_\varphi\mathbf{C} + h_{\tau\varphi}\mathbf{C}$. Define real forms $\mathfrak{Z}_0 = \mathfrak{Z} \cap \mathfrak{g}_0$ and $\mathfrak{H}_0' = \mathfrak{H}' \cap \mathfrak{g}_0 = \mathfrak{H} \cap \mathfrak{Z}_0$. Then $\mathfrak{H}_0' = \mathfrak{T}' + \mathfrak{A}'$ where

$$(9.14b) \quad \mathfrak{T}' = \mathfrak{T} \cap \mathfrak{H}' = i(h_\varphi - h_{\tau\varphi})\mathbf{R} \text{ and } \mathfrak{A}' = \mathfrak{A} \cap \mathfrak{H}' = (h_\varphi + h_{\tau\varphi})\mathbf{R}.$$

Note that $\mathfrak{T}'$ contains a regular element $i(h_\varphi - h_{\tau\varphi})$ of $\mathfrak{Z}$. Now Lemma 4.1 says that $\mathfrak{H}_0'$ is a maximally compact Cartan subalgebra of $\mathfrak{Z}_0$. As $\mathfrak{Z}$ is of type A_2, i.e. $\mathfrak{Z} = \mathfrak{S}\mathfrak{L}(3, \mathbf{C})$, its real forms are $\mathfrak{S}\mathfrak{L}(3, \mathbf{R})$ and the Lie algebras of the groups $\mathbf{SU}(3)$ and $\mathbf{SU}^1(3)$. In the latter two cases, a maximally compact Cartan subalgebra of the real form is the Lie algebra of a (compact) torus. Thus $\mathfrak{Z}_0 \subset \mathfrak{Z}$ is $\mathfrak{S}\mathfrak{L}(3, \mathbf{R}) \subset \mathfrak{S}\mathfrak{L}(3, \mathbf{C})$. That implies that $\mathfrak{Z}_0$ has a split Cartan subalgebra $\mathfrak{A}''$. Now consider the algebra $\mathfrak{H}_0 + \mathfrak{Z}_0$. It is reductive, and $\mathfrak{H}_0$ is its maximally split Cartan subalgebra. Now there exists an element g in the analytic subgroup of G_0 for $\mathfrak{H}_0 + \mathfrak{Z}_0$, such that $\operatorname{ad}(g)\mathfrak{A}'' \subset \mathfrak{A}$. Replacing x by $g^{-1}(x)$ we obtain (9.14a) with $\mathfrak{H}_0' = \mathfrak{A}''$ split, contradicting (9.14b). That contradicts $\varphi \in A$.

We have proved A empty, so $\mathfrak{M}_{[x]} = \mathfrak{W}_{[x]}$ algebra, whence $\mathfrak{N}_{[x]} = \mathfrak{W}_{[x]}$ by Theorem 8.9, so $G_0(x)$ is measurable by (9.13a) and Theorem 9.2. Q.E.D.

9.15. Corollary. *Let $P_1 \subset P_2$ be parabolic subgroups of G, $X_i = G/P_i$ the resulting complex flag manifolds, and $\pi: X_1 \to X_2$ the projection $gP_1 \to gP_2$. Let G_0 be a real form of G. Choose $x_i \in X_i$ such that $\pi(x_1) = x_2$ and $G_0(x_i)$ is the closed G_0-orbit on X_i. Then holomorphic arc components satisfy $\pi S_{[x_1]} = S_{[x_2]}$.*

Proof. Let $B \subset P_1$ be a Borel subgroup of G, $Y = G/B$, and $\nu_i: Y \to X_i$ the projections. Then $\nu_2 = \pi \cdot \nu_1$. Let $y \in \nu_1^{-1}(x_1)$ such that $G_0(y)$ is the closed G_0-orbit on Y. Then $\nu_i(y) = x_i$. Recall (9.13b) from the proof of Theorem 9.12; it says $\mathfrak{W}_{[x_i]} - \mathfrak{N}_{[y]} + $(a certain subspace of $(\mathcal{P}_i)_{x_i} \cap \tau(\mathcal{P}_i)_{x_i}$). The subsequent portion of the proof of Theorem 9.12 consisted of showing $\mathfrak{W}_{[x_i]} = \mathfrak{N}_{[x_i]}$. Now $\mathfrak{N}_{[x_i]} = \mathfrak{N}_{[y]} + $(subspace of $(\mathcal{P}_i)_{x_i} \cap \tau(\mathcal{P}_i)_{x_i}$). That proves $\nu_i S_{[y]} = S_{[x_i]}$. As $\pi = \nu_2 \cdot \nu_1^{-1}$, it follows that $\pi S_{[x_1]} = S_{[x_2]}$. Q.E.D.

The (rather minor) portion of Theorem 9.12 which says that closed orbits are partially complex, allows us to strengthen Corollary 8.19.

9.16. Corollary. *Let $X = G/P$ be a complex flag manifold, $x \in X$, and G_0 a real form of G. Then the following conditions are equivalent.*

(i) *$G_0(x)$ is the closed G_0-orbit on X.*

(ii) *The holomorphic arc components of $G_0(x)$ are compact.*

(iii) *The holomorphic arc components of $G_0(x)$ are complex flag manifolds.*

(iv) *$\mathfrak{N}^u_{[x]} \subset (\mathcal{P}_x \cap \tau\mathcal{P}_x)$, and $\mathfrak{N}^r_{[x]} = \mathfrak{L} \oplus \mathfrak{J}$ direct sum of τ-stable ideals, where $\mathfrak{J} \subset (\mathcal{P}_x \cap \tau\mathcal{P}_x)$ and $\mathfrak{L} \cap \mathfrak{g}_0$ is a compactly embedded subalgebra of $\mathfrak{g}_0$.*

PROOF. Corollary 8.19 gives equivalence of (i) and (ii). As complex flag manifolds are compact, (iii) implies (ii). Conversely (ii) implies (iii) because $G_0(x)$ is of flag type and $S_{[x]}$ is an open $N^r_{[x],0}$-orbit on the complex flag $N^r_{[x]}(x)$.

Given (iv), $\mathfrak{J} + \mathfrak{N}^u_{[x]}$ is an ideal of $\mathfrak{N}_{[x]}$ contained in the isotropy subalgebra at x, so the subgroup $J_0 \cdot N^u_{[x],0}$ of G_0 for the subalgebra $(\mathfrak{J} + \mathfrak{N}^u_{[x]}) \cap \mathfrak{g}_0$ must act trivially on $S_{[x]}$. Thus the subgroup L_0 for $\mathfrak{L} \cap \mathfrak{g}_0$ is transitive on $S_{[x]}$. But $\mathfrak{L} \cap \mathfrak{g}_0$ is compactly embedded, and G_0 has finite center, so L_0 is compact. Thus $S_{[x]}$ is compact. Now (iv) implies (ii).

Given (iii), we have $\mathfrak{N}^u_{[x]} \subset (\mathcal{P}_x \cap \tau\mathcal{P}_x)$ and $N^r_{[x],0}$ transitive on $S_{[x]}$. Let J_0 be the kernel of its action, so the Lie algebra $\mathfrak{J}_0$ is an ideal of $\mathfrak{N}^r_{[x],0}$ contained in $(\mathcal{P}_x \cap \tau\mathcal{P}_x)_0$. Let $\mathfrak{L}_0$ be a τ-stable complementary ideal, L_0 the corresponding subgroup of $N^r_{[x],0}$. Then L_0 is transitive on the flag manifold $S_{[x]}$, acting with finite kernel, so L_0 is compact. Now (iv) follows with $\mathfrak{L} = \mathfrak{L}_0^C$ and $\mathfrak{J} = \mathfrak{J}_0^C$. Q.E.D.

We can now complete the results of Corollary 8.20.

9.17. COROLLARY. *Let $X = G/P$ be a complex flag manifold, G_0 a real form of G, and $x \in X$. Then the holomorphic arc components of $G_0(x)$ are points if and only if* (a) *$G_0(x)$ is the closed orbit and* (b) *$\tau\mathcal{P}^u_x \subset \mathcal{P}_x$.*

PROOF. If the holomorphic arc components of $G_0(x)$ are points, they are compact, so $G_0(x)$ is the closed orbit by Corollary 8.19 or Corollary 9.16. Now let $G_0(x)$ be the closed orbit, so $\mathfrak{N}_{[x]} = (\mathcal{P}_x \cap \tau\mathcal{P}_x) + \mathcal{V}_x$ by Theorem 9.2; thus $S_{[x]}$ is a point if and only if $\mathcal{V}_x = 0$. As $\mathcal{V}_x = (\mathcal{P}^u_x \cap \tau\mathcal{P}^{-u}_x) + \tau(\mathcal{P}^u_x \cap \tau\mathcal{P}^{-u}_x)$, it is 0 if and only if $\tau\mathcal{P}^u_x \cap \mathcal{P}^{-u}_x = 0$. Thus $S_{[x]}$ is a point if and only if $\tau\mathcal{P}^u_x \subset \mathcal{P}_x$. Q.E.D.

It turns out that the proof of Theorem 9.12 does not apply exclusively to the case where $G_0(x)$ is closed.

9.18. PROPOSITION. *Let $X = G/P$ be a complex flag manifold, G_0 a real form of G, and $x \in X$. Suppose that $\mathfrak{g}_0 \cap \mathcal{P}_x$ contains a maximally split Cartan subalgebra $\mathfrak{H}_0$ of $\mathfrak{g}_0$. Then the following conditions are equivalent:*

(i) *$G_0(x)$ is measurable.*

(ii) *$\mathfrak{N}_{[x]} = (\mathcal{P}_x \cap \tau\mathcal{P}_x) + \mathcal{V}_x$.*

(iii) *$(\mathcal{P}_x \cap \tau\mathcal{P}_x) + \mathcal{V}_x$ is an algebra.*

PROOF. Theorem 9.2 is the equivalence of (i) and (ii), and of course (ii) implies (iii). Thus we need only prove that (iii) implies (ii).

Assume (iii) and define $\mathcal{W}_{[x]}=(\mathcal{P}_x\cap\tau\mathcal{P}_x)+\mathcal{V}_x$. The latter agrees with (9.13a). Now $\mathcal{W}_{[x]}$ is an algebra. As $\tau\mathcal{W}_{[x]}=\mathcal{W}_{[x]}$ and $(\mathcal{P}_x\cap\tau\mathcal{P}_x)\subset\mathcal{W}_{[x]}\subset(\mathcal{P}_x+\tau\mathcal{P}_x)$, Lemma 8.4 and Proposition 8.7 say $\mathcal{W}_{[x]}\subset\mathfrak{M}_{[x]}$. Here we work with the Cartan subalgebra $\mathcal{H}=\mathcal{H}_0^C$ of $\mathcal{G}$. Define A to be the set of $\mathcal{H}$-roots such that $\mathfrak{M}_{[x]}$ is the direct sum of $\mathcal{W}_{[x]}$ and $\sum_A \mathcal{G}_\varphi$. As in Theorem 9.12, we will prove that A is empty.

Let φ be a root with $\mathcal{G}_\varphi\not\subset\mathcal{W}_{[x]}$. Then φ must be in $\Phi^r\cap-\tau\Phi^u$, $-\Phi^u\cap\tau\Phi^r$ or $-\Phi^u\cap-\tau\Phi^u$ where $\mathcal{P}_x=\mathcal{P}_\Phi$. Thus $\mathcal{G}_{-\varphi}\subset\mathcal{W}_{[x]}$. That proves $\mathcal{W}_{[x]}$ parabolic in $\mathcal{G}$. Define μ to be the sum of all roots φ with $\mathcal{G}_\varphi\subset\mathcal{W}^u_{[x]}$. Then $\mu=\tau\mu$ and, for any root φ,

$$\begin{aligned}
&\text{if } \langle\varphi,\mu\rangle>0 &&\text{then } \mathcal{G}_\varphi\subset\mathcal{W}^u_{[x]},\\
&\text{if } \langle\varphi,\mu\rangle=0 &&\text{then } \mathcal{G}_\varphi\subset\mathcal{W}^r_{[x]},\\
&\text{if } \langle\varphi,\mu\rangle<0 &&\text{then } \mathcal{G}_\varphi\not\subset\mathcal{W}_{[x]}.
\end{aligned}$$

In particular, $\langle\varphi,\mu\rangle<0$ for every $\varphi\in A$, and $\langle\varphi,\mu\rangle\geqq 0$ whenever $\mathcal{G}_\varphi\subset(\mathcal{P}_x\cap\tau\mathcal{P}_x)$.

Let $\varphi\in A$. If $\varphi+\tau\varphi$ is a root, then $\mathcal{G}_{\varphi+\tau\varphi}\subset(\mathcal{P}_x+\tau\mathcal{P}_x)$ would imply $\mathcal{G}_{\varphi+\tau\varphi}\subset(\mathcal{P}_x\cap\tau\mathcal{P}_x)$, so $0\leqq\langle\varphi+\tau\varphi,\mu\rangle=2\langle\varphi,\mu\rangle<0$, which is absurd; and $\mathcal{G}_{\varphi+\tau\varphi}\not\subset(\mathcal{P}_x+\tau\mathcal{P}_x)$ would contradict $\mathcal{G}_\varphi\subset\mathfrak{M}_{[x]}$; thus $\varphi+\tau\varphi$ is not a root. As in Proposition 8.7, it follows that there are just two possibilities, given as follows.

Case 1. $\varphi\neq\tau\varphi\neq-\varphi$, $\langle\varphi,\tau\varphi\rangle=0$, $\tau\varphi-\varphi$ is not a root, and

$$\mathcal{Z}=h_\varphi\mathbf{C}+h_{\tau\varphi}\mathbf{C}+\mathcal{G}_\varphi+\mathcal{G}_{-\varphi}+\mathcal{G}_{\tau\varphi}+\mathcal{G}_{-\tau\varphi}\tag{9.19a}$$

is a τ-stable semisimple subalgebra of type $A_1\oplus A_1$ in $\mathcal{W}_{[x]}$.

Case 2. $\varphi\neq\tau\varphi\neq-\varphi$, $\langle\varphi,\tau\varphi\rangle=\frac{1}{2}\langle\varphi,\varphi\rangle$, $\tau\varphi-\varphi$ is a root, and

$$\mathcal{Z}=h_\varphi\mathbf{C}+h_{\tau\varphi}\mathbf{C}+\mathcal{G}_\varphi+\mathcal{G}_{-\varphi}+\mathcal{G}_{\tau\varphi}+\mathcal{G}_{-\tau\varphi}+\mathcal{G}_{\tau\varphi-\varphi}+\mathcal{G}_{\varphi-\tau\varphi}\tag{9.19b}$$

is a τ-stable semisimple subalgebra of type A_2 in $\mathcal{W}_{[x]}$.

In both cases, $\mathcal{H}'=h_\varphi\mathbf{C}+h_{\tau\varphi}\mathbf{C}$ is a Cartan subalgebra of $\mathcal{Z}$, and we define real forms $\mathcal{Z}_0=\mathcal{Z}\cap\mathcal{G}_0$ and $\mathcal{H}_0'=\mathcal{H}\cap\mathcal{Z}_0=\mathcal{H}'\cap\mathcal{G}_0$. Then, as in (9.14b), $\mathcal{H}_0'=\mathfrak{I}'+\mathcal{A}'$ where

$$\mathfrak{I}'=i(h_\varphi-h_{\tau\varphi})\mathbf{R}\quad\text{and}\quad\mathcal{A}'=(h_\varphi+h_{\tau\varphi})\mathbf{R}.\tag{9.19c}$$

Checking the two cases separately one sees that $i(h_\varphi-h_{\tau\varphi})\in\mathfrak{I}'$ is a regular element of $\mathcal{Z}$, so Lemma 4.1 assures us that $\mathcal{H}_0'$ is a maximally compact Cartan subalgebra of $\mathcal{Z}_0$. We now use the hypothesis that $\mathcal{H}_0$ is maximally split in $\mathcal{G}_0$; it allows us to use the argument following (9.14b), and this eliminates Case 2.

Now we are in Case 1 with $\mathcal{Z}$ given by (9.19a). As $\varphi \in A$ we have $\mathcal{G}_\varphi \subset (\mathcal{P}_x + \tau\mathcal{P}_x)$. Replace φ by $\tau\varphi$ if necessary for the normalization $\mathcal{G}_\varphi \subset \mathcal{P}_x$. As $\langle \varphi, \mu \rangle < 0$ implies $\mathcal{G}_\varphi \not\subset (\mathcal{P}_x \cap \tau\mathcal{P}_x)$ we have $\mathcal{G}_\varphi \subset (\mathcal{P}_x \cap \tau\mathcal{P}_x^{-u})$. Now $\mathcal{G}_\varphi \subset \mathcal{W}_{[x]}$ gives us $\mathcal{G}_\varphi \subset (\mathcal{P}_x^r \cap \tau\mathcal{P}_x^{-u})$. It follows that

$$\begin{aligned} \mathcal{Z} \cap \mathcal{P}_x &= \mathcal{H}' + \mathcal{G}_\varphi + \mathcal{G}_{-\varphi} + \mathcal{G}_{-\tau\varphi} \quad \text{and} \\ \mathcal{Z} \cap (\mathcal{P}_x \cap \tau\,\mathcal{P}_x) &= \mathcal{H}' + \mathcal{G}_{-\varphi} + \mathcal{G}_{-\tau\varphi}. \end{aligned} \tag{9.19d}$$

The first equation of (9.19d) says that $\mathcal{Z} \cap \mathcal{P}_x$ is parabolic in $\mathcal{Z}$, i.e. that the analytic subgroup Z of G for $\mathcal{Z}$ has the property that $Z(x)$ is a complex flag manifold. Let Z_0 denote the analytic subgroup of G_0 for $\mathcal{Z}_0$. Now (9.19d) gives us $\dim_R Z(x) = 2\dim_C Z(x) = 2\{\dim_C \mathcal{Z} - \dim_C(\mathcal{Z} \cap \mathcal{P}_x)\} = 2\{6-5\} = 2 = 6-4 = \dim_C \mathcal{Z} - \dim_C(\mathcal{Z} \cap \mathcal{P}_x \cap \tau\mathcal{P}_x) = \dim_R \mathcal{Z}_0 - \dim_R(\mathcal{Z}_0 \cap \mathcal{P}_x) = \dim_R Z_0(x)$. Thus

(9.19e) $Z_0(x)$ is open in the complex flag manifold $Z(x)$.

On the other hand, (9.19a) and (9.19c) show that $\mathcal{H}_0'$ is a maximally split Cartan subalgebra of $\mathcal{Z}_0$. The roots positive on $-(h_\varphi + h_{\tau\varphi}) \in \mathcal{A}'$ are $-\varphi$ and $-\tau\varphi$. Thus we have an Iwasawa decomposition $\mathcal{Z}_0 = \mathcal{K}' + \mathcal{A}' + \mathcal{N}'$, $\mathcal{N}'^C = \mathcal{G}_{-\varphi} + \mathcal{G}_{-\tau\varphi}$, with $(\mathcal{J}' + \mathcal{A}' + \mathcal{N}') \subset \mathcal{Z} \cap (\mathcal{P}_x \cap \tau\mathcal{P}_x)$, so Lemma 3.1 says that $Z_0(x)$ is the closed Z_0-orbit on $Z(x)$. Combining this with (9.19e), we see that

$$Z_0 \text{ is the compact real form of } Z,$$

so the maximally compact Cartan subalgebras of $\mathcal{Z}_0$ are compact. That contradicts (9.19c). We conclude that $\varphi \in A$ cannot exist, i.e. that A is empty.

Now A is empty, so $\mathcal{M}_{[x]} = \mathcal{W}_{[x]}$ algebra; thus $\mathcal{M}_{[x]} = \mathcal{N}_{[x]}$ by Theorem 8.9, so $\mathcal{W}_{[x]} = \mathcal{N}_{[x]}$. Thus (iii) implies (ii), completing the proof of the proposition. Q.E.D.

Let K be a maximal compact subgroup of G_0. Then every Cartan subalgebra of $\mathcal{G}_0$ is maximally split, if and only if rank $\mathcal{G}_0$ = rank $\mathcal{K}$ + rank G_0/K, the latter being the symmetric space rank. The only such cases for which the symmetric space G_0/K is irreducible with G_0 acting almost-effectively are

(9.20a) G_0 of type E_6 with K of type F_4,

(9.20b) G_0 of type A_{2n-1} with K of type C_n,

(9.20c) G_0 of type D_{n+1} with K of type B_n, and

(9.20d) G_0 complex simple but viewed as a real group.

9.21. Corollary. *Let $X = G/P$ be a complex flag manifold. Let G_0 be a real form of G that is locally isomorphic to product of compact simple* *groups and groups given by* (9.20) *above. Let* $x \in X$. *Then the following conditions are equivalent*:

(i) $G_0(x)$ *is measurable.*

(ii) $\mathfrak{N}_{[x]} = (\mathcal{P}_x \cap \tau\mathcal{P}_x) + \mathcal{V}_x$.

(iii) $(\mathcal{P}_x \cap \tau\mathcal{P}_x) + \mathcal{V}_x$ *is an algebra.*

PROOF. Let $\mathfrak{H} \subset \mathcal{P}_x$ be a Cartan subalgebra of $\mathcal{G}$ stable under τ. Then the Cartan subalgebra $\mathfrak{H}_0 = \mathfrak{H} \cap \mathcal{G}_0$ of $\mathcal{G}_0$ is a maximally split Cartan subalgebra of $\mathcal{G}_0$, by hypothesis on G_0. Now Proposition 9.18 applies. Q.E.D.

We end §9 with a short study of the case where the complex flag manifold $X = G/P$ is a hermitian symmetric space G_u/P_u^r of compact type relative to the compact real form G_u of G, and where G_0' is any real form of G. Here we use the notation G_0' for purposes of reference from §10, where G_0 will denote a real form of G such that G_0/K is the noncompact hermitian symmetric space dual to $X = G_u/K$.

9.22. LEMMA. *Let* $X = G/P$ *be a complex flag manifold. Then the following conditions are equivalent*:

(i) X *is a hermitian symmetric space* G_u/K.

(ii) $\mathcal{P}^u$ *is abelian, i.e.* $[\mathcal{P}^u, \mathcal{P}^u] = 0$.

(iii) $[\mathcal{P}^u, \mathcal{P}^{-u}] \subset \mathcal{P}^r$.

PROOF. Let $X = G_u/K$ hermitian symmetric. Then $\mathfrak{K}$ has a central element z that defines the almost complex structure on X by

$$\mathrm{ad}(z) = \sqrt{-1} \text{ on } \mathcal{P}^{-u}, \qquad \mathrm{ad}(z) = -\sqrt{-1} \text{ on } \mathcal{P}^u,$$
$$\mathrm{ad}(z) = 0 \text{ on } \mathfrak{K}^C = \mathcal{P}^r.$$

If $v, w \in \mathcal{P}^u$ now $\mathrm{ad}(z)[v, w] = [z, [v, w]] = [[z, v], w] + [v, [z, w]] = [\mathrm{ad}(z)v, w] + [v, ad(z)w] = -2\sqrt{-1}[v, w]$, while $-2\sqrt{-1}$ is not an eigenvalue of $\mathrm{ad}(z)$; thus $[v, w] = 0$; that shows $\mathcal{P}^u$ abelian. If $v \in \mathcal{P}^u$ and $w \in \mathcal{P}^{-u}$ then $\mathrm{ad}(z)[v, w] = [\mathrm{ad}(z)v, w] + [v, \mathrm{ad}(z)w] = 0$, so $[v, w] \in \mathcal{P}^r$; that shows $[\mathcal{P}^u, \mathcal{P}^{-u}] \subset \mathcal{P}^r$. Now (i) implies (ii) and (iii).

Let $\mathfrak{H} = \mathfrak{J}^C$ be a Cartan subalgebra of $\mathcal{G}$ where $\mathfrak{J} \subset \mathcal{G}_u \cap \mathcal{P}$. Choose a subset Φ of a simple $\mathfrak{H}$-root system Π such that $\mathcal{P} = \mathcal{P}_\Phi$. Decompose $\mathcal{G}$ as direct sum of simple ideals $\mathcal{L}_i$, so $\mathfrak{H} = \sum \mathfrak{H}_i$ where $\mathfrak{H}_i = \mathfrak{H} \cap \mathcal{L}_i$ is a Cartan subalgebra of $\mathcal{L}_i$, and $\Pi = \bigcup \Pi_i$ where Π_i is a simple $\mathfrak{H}_i$-root system for $\mathcal{L}_i$. Let μ_i be the maximal root of $\mathcal{L}_i$, so $\mu_i = \sum_{\Pi_i} n_\psi \cdot \psi$, n_ψ integers $\geqq 1$. If $\Pi_i \backslash \Phi$ contains a root ψ_0 with $n_{\psi_0} > 1$, we have positive roots

$$\alpha = \sum_{\Pi_i} a_\psi \cdot \psi, \qquad \beta = \sum_{\Pi_i} b_\psi \cdot \psi, \qquad \gamma = \alpha + \beta$$

with $a_{\psi_0}>0$ and $b_{\psi_0}>0$. If $\Pi_i \backslash \Phi$ contains two roots ψ_1 and ψ_2, then we have positive roots α, β, γ as above with $a_{\psi_1}>0$ and $b_{\psi_2}>0$. In either case $\alpha, \beta, \gamma \in \Phi^u$, so

$\mathfrak{g}_\gamma=[\mathfrak{g}_\alpha, \mathfrak{g}_\beta] \subset [\mathcal{P}^u, \mathcal{P}^u]$ shows that (ii) fails, and

$\mathfrak{g}_\alpha=[\mathfrak{g}_\gamma, \mathfrak{g}_{-\beta}] \subset [\mathcal{P}^u, \mathcal{P}^{-u}] \cap \mathcal{P}^u$ shows that (iii) fails.

Thus (ii) and (iii) each implies that, for every i, either $\Pi_i \subset \Phi$ or $\Pi_i \backslash \Phi=\{\psi_0\}$ with $n_{\psi_0}=1$. It is immediate [10, Lemma 3] that $G_u/G_u \cap P=G_u/P_u^r$ is a realization of X as a hermitian symmetric space of compact type. Q.E.D.

9.23. REMARK. Let $X=G/P$ be a complex flag manifold that is hermitian symmetric. Let G_0' be *any* real form of G, τ' conjugation of $\mathfrak{g}$ over $\mathfrak{g}_0'$. If $x \in X$ then Lemma 9.22 shows that $\mathcal{W}_{[x]}=(\mathcal{P}_x \cap \tau'\mathcal{P}_x)+(\mathcal{P}_x^u \cap \tau'\mathcal{P}_x^{-u})+(\mathcal{P}_x^{-u} \cap \tau'\mathcal{P}_x^u)$ is a subalgebra of $\mathfrak{g}$. As

$$(\mathcal{P}_x \cap \tau'\mathcal{P}_x) \subset \mathcal{W}_{[x]} = \tau'\mathcal{W}_{[x]} \subset (\mathcal{P}_x + \tau'\mathcal{P}_x),$$

Lemma 8.4 says that $\mathcal{W}_{[x]} \subset \mathfrak{N}_{[x]}$, so Theorem 9.7 says that $G_0'(x)$ is of flag type. However, Example 8.12 shows that one cannot expect $G_0'(x)$ to be partially complex without further restriction on X, x, or G_0'. In that context we prove

9.24. PROPOSITION. *Let $X=G/P$ be a complex flag manifold that is a hermitian symmetric space. Let G_0' be any real form of G. Decompose*

$$\mathfrak{g} = \mathfrak{g}^1 \oplus \cdots \oplus \mathfrak{g}^r, \qquad \mathcal{P} = \mathcal{P}^1 \oplus \cdots \oplus \mathcal{P}^r, \qquad X = X^1 \times \cdots \times X^r$$

where the $\mathfrak{g}^i$ are the minimal τ'-stable ideals of $\mathfrak{g}$, $\mathcal{P}^i=\mathcal{P}\cap\mathfrak{g}^i$, and $X^i=G^i/P^i$ quotient of the analytic subgroups of G for $\mathfrak{g}^i$ and $\mathcal{P}^i$; so the $\mathfrak{g}_0'^i=\mathfrak{g}_0' \cap \mathfrak{g}^i$ are the simple ideals of $\mathfrak{g}_0'$. Then the following conditions are equivalent:

(1) *Some open G_0'-orbit on X is a symmetric coset space of G_0', with invariant nondegenerate kaehler metric that may be of indefinite signature.*

(1′) *Every open G_0'-orbit on X is a symmetric coset space of G_0', with invariant nondegenerate kaehler metric that may be of indefinite signature.*

(2) *Some open G_0'-orbit on X is measurable.*

(2′) *Every open G_0'-orbit on X is measurable.*

(3) *For each index i, $0 \leqq i \leqq r$, if $\mathfrak{g}^i$ is simple then either*

(3a) *the maximal compactly embedded subalgebras $\mathfrak{K}^i \subset \mathfrak{g}_0'^i$ satisfy* rank $\mathfrak{K}^i=$rank $\mathfrak{g}_0'^i$, *or*

(3b) *X^i is the grassmannian* $\mathbf{U}(2n)/\mathbf{U}(n)\times\mathbf{U}(n)$ *of n-planes through the origin in $\mathbf{C}^{2n}$ with $\mathfrak{g}^i=\mathfrak{SL}(2n, \mathbf{C})$ and either $\mathfrak{g}_0'^i=\mathfrak{SL}(2n, \mathbf{R})$ or $\mathfrak{g}_0'^i=\mathfrak{SL}(n, \mathbf{Q})$, or*

(3c) *X^i is the complex quadric* $\mathbf{SO}(2n+2)/\mathbf{SO}(2n)\times\mathbf{SO}(2)$ *with $\mathfrak{g}^i=\mathfrak{SO}(2n+2, \mathbf{C})$ and $\mathfrak{g}_0'^i=\mathfrak{SO}^{2t+1}(2n+2)$ where $0 \leqq t \leqq [n/2]$.*

(4) *For each index* i, $\mathcal{G}^i$ *simple, either*

(4a) rank $\mathcal{K}^i = \text{rank } \mathcal{G}_0'^i$, or

(4b) *the noncompact dual of the irreducible compact hermitian symmetric space* X^i *is of tube type, but not isomorphic to an* $\mathbf{SO}^*(2l)/\mathbf{U}(l)$ *for* $l > 4$.

PROOF. Let $x \in X$ such that $G_0'(x)$ is open in X. If $G_0'(x)$ is symmetric, say with symmetry σ at x, then the isotropy subalgebra $\mathcal{G}_0' \cap \mathcal{P}_x = \{v \in \mathcal{G}_0' : \sigma(v) = v\}$ is reductive in $\mathcal{G}_0'$, so Lemma 6.2 says that $G_0'(x)$ is measurable. If $G_0'(x)$ is measurable, Theorem 6.3 says

(i) there is an invariant, possibly indefinite, kaehler metric on $G_0'(x)$ and

(ii) $\tau' \mathcal{P}_x^r = \mathcal{P}_x^r$ and $\tau' \mathcal{P}_x^u = \mathcal{P}_x^{-u}$.

Choose a compact real form G_u of G such that $\mathcal{G}_0'$ is invariant under the automorphism σ of $\mathcal{G}$ induced from conjugation of $\mathcal{G}_u$ by the symmetry at x; σ is $+1$ on $\mathcal{G}_u \cap \mathcal{P}_x = (\mathcal{P}_x^r)_u$ and -1 on a complement; thus σ is $+1$ on $\mathcal{P}_x^r$ and -1 on $\mathcal{P}_x^u + \mathcal{P}_x^{-u}$; now $\mathcal{G}_0' \cap \mathcal{P}_x = (\text{real form of } \mathcal{P}_x^r) = \{v \in \mathcal{G}_0' : \sigma(v) = v\}$, so $G_0'(x)$ is symmetric. We have proved equivalence of (1) and (2) for $G_0'(x)$. As (2) and (2′) are equivalent by Theorem 6.7, we now have equivalence of (1), (1′), (2) and (2′).

In proving (2), (3) and (4) equivalent, it suffices to consider the case where $\mathcal{G}_0'$ is simple. For if $G_0(x)$ is open in X, and $x = (x^1, \cdots, x^r)$ with $x^i \in X^i$, then Theorem 6.7 shows that $G_0'(x)$ is measurable if and only if each $G_0'^i(x^i)$ is measurable. So we now assume that $\mathcal{G}_0'$ is simple, and we fix $x \in X$ with $G_0'(x)$ open in X.

If $\mathcal{G}$ is not simple, i.e. if $\mathcal{G}_0'$ is not absolutely simple, then $\mathcal{G}_0'$ is the real structure of a complex simple Lie algebra $\mathcal{L}$, $\mathcal{G} = \mathcal{L} \oplus \mathcal{L}$ with $\mathcal{G}_0'$ embedded diagonally, and τ' acts on $\mathcal{G}$ by $(u, v) \to (v, u)$. Lemma 9.22 shows that $(\mathcal{P}_x \cap \tau' \mathcal{P}_x) + (\mathcal{P}_x^u \cap \tau' \mathcal{P}_x^{-u}) + (\mathcal{P}_x^{-u} \cap \tau' \mathcal{P}_x^u)$ is an algebra; that is condition (iii) of Corollary 9.21; thus $G_0'(x)$ is measurable.

Now let $\mathcal{G}$ be simple. If the maximal compactly embedded subalgebras $\mathcal{K} \subset \mathcal{G}_0'$ satisfy rank $\mathcal{K} = \text{rank } \mathcal{G}_0'$, then Corollary 6.4 says that $G_0'(x)$ is measurable.

Finally suppose $\mathcal{G}$ simple with rank $\mathcal{K} < \text{rank } \mathcal{G}_0'$. If $G_0'(x)$ is measurable, we have the symmetry σ of $X = G_u/P_u^r$ at x, such that $\sigma \mathcal{G}_0' = \mathcal{G}_0'$ and $\mathcal{G}_0' \cap \mathcal{P}_x = \{v \in \mathcal{G}_0' : \sigma(v) = v\}$, real form of $\mathcal{P}_x^r$. Let α be a Cartan involution of $\mathcal{G}_0'$ that commutes with σ, $\mathcal{G}_0' = \mathcal{K} + \mathfrak{M}_0$ the Cartan decomposition under α. Then $\mathcal{G}_u = \mathcal{K} + \sqrt{-1}\mathfrak{M}_0$, and rank $\mathcal{K} < \text{rank } \mathcal{G}_u$, so α carries over to an involutive outer automorphism of $\mathcal{G}_u$ that preserves $(\mathcal{P}_x^r)_u$, thus acts on $X = G_u/(P_x^r)_u$, and preserves the invariant complex structure on X. We apply [**11**, Theorem 13.3] to α and $X = G_u/(P_x^r)_u$, concluding that either $X = \mathbf{SU}(2n)/\mathbf{S}(\mathbf{U}(n) \times \mathbf{U}(n))$ with α interchanging the two $\mathbf{U}(n)$ factors, or $X = \mathbf{SO}(2n+2)/\mathbf{SO}(2n)$

$\times \mathbf{SO}(2)$ with α acting on $\mathbf{SO}(2n)\times\mathbf{SO}(2)\subset\mathbf{O}(2n)\times\mathbf{SO}(2)$ as conjugation by some (Q, P) with $\det Q=-1$. The other cases of [**11**, Theorem 13.3(2)] are eliminated because they are not symmetric. Thus measurability of $G_0'(x)$ implies alternative (3b) or (3c).

Conversely, suppose that we have alternative (3b) or (3c). Then rank $\mathfrak{K}<$rank $\mathfrak{G}_0'$ shows that τ' induces an outer automorphism of a compact real form $\mathfrak{G}_u$ of $\mathfrak{G}$. Let $\mathfrak{J}$ denote a τ'-invariant Cartan subalgebra of $\mathfrak{G}_u$. Choose a τ'-invariant Weyl chamber for the Cartan subalgebra $\mathfrak{H}=\mathfrak{J}^C$ of $\mathfrak{G}$. Then τ' acts on the corresponding simple root system Π by

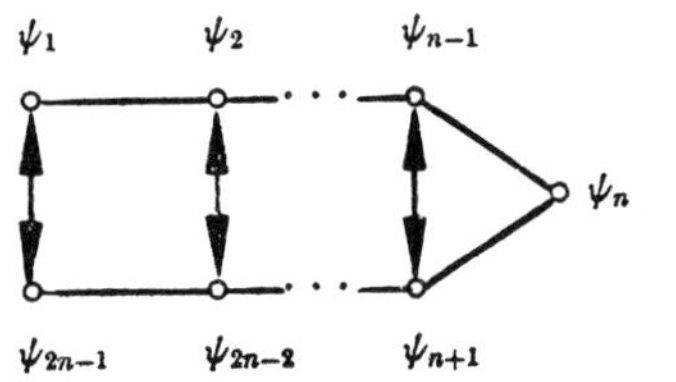

in the case of alternative (3b),

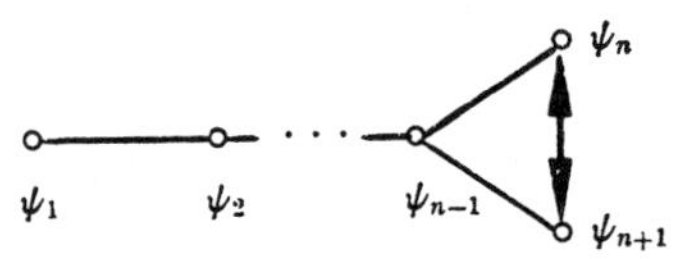

in the case of alternative (3c).

On the other hand, $\mathcal{P}$ is conjugate to $\mathcal{P}_\Phi$ where, in the notation just above,

$$\Phi = \{\psi_1, \cdots, \psi_{n-1}; \psi_{n+1}, \cdots, \psi_{2n-1}\} \text{ in alternative (3b),}$$

$$\Phi = \{\psi_2, \psi_3, \cdots, \psi_{n-1}, \psi_n, \psi_{n+1}\} \text{ in alternative (3c).}$$

Thus $\tau'\Phi=\Phi$, so $\tau'\mathcal{P}_\Phi=\mathcal{P}_\Phi$. But $\mathcal{P}_\Phi$ is opposed to itself. Thus $\mathcal{P}$ and $\tau'\mathcal{P}$ are opposed. Theorem 6.7 now says that $G_0'(x)$ is measurable.

We have proved that (2) and (3) are equivalent.

Suppose $\mathfrak{G}$ simple with rank $\mathfrak{K}<$rank $\mathfrak{G}_0'$. Then $\mathfrak{G}$ has an involutive outer automorphism, so it is of type A_l, D_l or E_6. Let X_0 denote the noncompact dual of $X=G_u/P_u^\tau$. If $\mathfrak{G}$ is of type A_l, then X is the grassmannian of k-planes through the origin in $\mathbf{C}^{l+1}$, $2k\leqq l+1$, and X_0 is of tube type if and only if $2k=l+1$. If $\mathfrak{G}$ is of type D_l, then either $X=\mathbf{SO}(2l)/\mathbf{U}(l)$ where X_0 is of tube type if and only if l is even, or $X=\mathbf{SO}(2l)/\mathbf{SO}(2l-2)\times\mathbf{SO}(2)$ complex quadric with X_0 of tube type. If $\mathfrak{G}$ is of type E_6 then X_0 is not of tube type. That proves equivalence of (3) and (4), completing the proof of the proposition. Q.E.D.

9.25. Corollary. *Let $X=G/P$ be a complex flag manifold that is a* *hermitian symmetric space. Let G_0' be any real form of G such that the maximal compact subgroups $K \subset G_0'$ satisfy* rank K = rank G_0'. *Then every open G_0'-orbit on X is a symmetric coset space of G_0' with invariant, nondegenerate, possibly indefinite, kaehler metric.*

For that is a special case of alternative (3a) of Proposition 9.24.

Chapter III. Hermitian symmetric spaces

Let $X=G/P$ be a complex flag manifold that is a hermitian symmetric space of compact type. In other words $X=G_u/K$, where G_u is the compact real form of G and $K=P\cap G_u$ (necessarily a compact real form of P^r) is a symmetric subgroup of G_u. Let G_0 denote the real form of G such that $X_0=G_0/K$ is the hermitian symmetric space of noncompact type dual to X. §10 is a complete analysis of the G_0-orbit structure of X and the holomorphic arc component structure of an arbitrary G_0-orbit. §11 carries that analysis over to a decomposition theory for an arbitrary orbit $G_0(x)$ that extends the Siegel domain theory for the convex open orbit X_0. A summary of the Siegel domain theory for X_0 and its reformulation as a theory of holomorphic fibrations of X_0 over its boundary components, is collected in the Appendix to §11 for use in our decomposition of $G_0(x)$.

In §10 we pick out certain points $x_{\Gamma,\Sigma}\in X$ by means of the Wolf-Korányi partial Cayley transform theory [15], and certain complex subgroups $G_{\Psi\setminus\Gamma}\subset G$ and their G_0-real forms $G_{\Psi\setminus\Gamma,0}$, such that the G_0-orbits on X are just the $G_0(x_{\Gamma,\Sigma})$ and the holomorphic arc components of $G_0(x_{\Gamma,\Sigma})$ are just the $k\cdot G_{\Psi\setminus\Gamma,0}(x_{\Gamma,\Sigma})$ where $k\in K$. More precisely let Ψ be a maximal set of strongly orthogonal noncompact positive roots of $\mathfrak{G}$ relative to a Cartan subalgebra $\mathfrak{J}^C$ with $\mathfrak{J}\subset\mathfrak{K}$. Any two choices of $(\mathfrak{J}, \Psi)$ are $\mathrm{ad}_G(K)$-conjugate, $|\Psi|$ = rank X_0, and $X_0=K\cdot G_0[\Psi](x_0)$ where x_0 is the base point and $G_0[\Psi](x_0) = \prod_\Psi G_0[\psi](x_0)$ polydisc in X_0. If $\psi\in\Psi$ then c_ψ is the Cayley transform on the Riemann sphere $G[\psi](x_0)$, rotation by $\pi/2$ carrying the south pole x_0 up to the equator. If $B\subset\Psi$ then $c_B=\prod_B c_\psi$. If $\Gamma, \Sigma\subset\Psi$ then $x_{\Gamma,\Sigma}=c_\Gamma c_\Sigma^2 x_0$ and $G_{\Psi\setminus\Gamma}$ is the derived group of the centralizer of $\mathfrak{J}^C+\sum_{\psi\perp\Gamma}\mathfrak{G}_\psi$ in G.

Theorem 10.6 says that the G_0-orbits on X are the $G_0(x_{\Gamma,\Sigma})$, Γ and Σ disjoint in Ψ, and gives criteria for $G_0(x_{\Gamma',\Sigma'})=G_0(x_{\Gamma,\Sigma})$ and $G_0(x_{\Gamma',\Sigma'})\subset\overline{G_0(x_{\Gamma,\Sigma})}$. If $\{r_1, \cdots, r_p\}$ are the ranks of the irreducible factors of X_0, it follows that there are precisely $\frac{1}{2}\sum_{1\leq i\leq p}(r_i+1)(r_i+2)$ G_0-orbits on X. The open G_0-orbits are the $G_0(x_{\phi,\Sigma})$; they are $\sum_{1\leq j\leq p}(r_j+1)=|\Psi|+p$ in number, are indefinite-hermitian symmetric spaces of G_0, and are indefinite metric versions of X_0. Theorem

10.9 says that $G_{\Psi\setminus\Gamma}(x_{\Gamma,\Sigma}) = G_{\Psi\setminus\Gamma}(x_{\Gamma,\phi})$ is a compact totally geodesic hermitian symmetric subspace of rank $|\Psi\setminus\Gamma|$ in X, that the $G_{\Psi\setminus\Gamma,0}(x_{\Gamma,\Sigma})$, $\Gamma\cap\Sigma=\phi$, are the open $G_{\Psi\setminus\Gamma,0}$-orbits on it, that the holomorphic arc components of $G_0(x_{\Gamma,\Sigma})$ are the $k\cdot G_{\Psi\setminus\Gamma,0}(x_{\Gamma,\Sigma})$ with $k\in K$, and that the complex Lie algebra normalizer $\mathfrak{N}_{[x_{\Gamma,\Sigma}]} = \mathfrak{N}_{\Gamma,\Sigma}$ of $G_{\Psi\setminus\Gamma,0}(x_{\Gamma,\Sigma})$ depends only on Γ. The method is rather delicate; Theorem 8.9 is used to show $\mathfrak{N}_{\Gamma,\Sigma}$ independent of Σ, and results from Wolf-Korányi [**15**] are used to determine $\mathfrak{N}_{\Gamma,\phi}$. In particular every holomorphic arc component of every G_0-orbit is indefinite-hermitian symmetric, so every G_0-orbit is measurable.

In §11 we work out a decomposition:

$$\begin{array}{ccccc} & & G_0(x_{\Gamma,\Sigma}) & & \\ & \swarrow^{\gamma} & \downarrow^{\beta} & \searrow^{\alpha} & \\ W\cdot G^{B'}_{E',0}(x_{\Gamma,\Sigma}) & \longrightarrow & K(x_{\Gamma,\Sigma}) & \longrightarrow & G_0/N_{\Gamma,\Sigma,0} \end{array}$$

There α is the K-equivariant fibration whose fibres are the holomorphic arc components of $G_0(x_{\Gamma,\Sigma})$. β is a K-equivariant fibration whose restriction to a holomorphic arc component is a holomorphic fibration over the maximal compact subvariety of that component; the fibres of β are hermitian symmetric spaces $kc_\Gamma c^2_\Sigma\{G_{\Psi\setminus(\Gamma\cup\Sigma),0}(x_0)\times G^{\Sigma}_{\Sigma,0}(x_0)\}$ of noncompact type whose second factor is the tube part of $G_{\Sigma,0}(x_0)$. And γ is an a.e. analytic measurable bundle whose restriction to $c_\Gamma c^2_\Sigma\{G_{\Psi\setminus(\Gamma\cup\Sigma),0}(x_0)\times G^{\Sigma}_{\Sigma,0}(x_0)\}$ is a holomorphic fibration over a bounded symmetric domain, any of the Siegel domain realizations of type III for a boundary component of $G_{\Psi\setminus(\Gamma\cup\Sigma),0}(x_0)\times G^{\Sigma}_{\Sigma,0}(x_0)$. The Siegel domain theory used in γ is in the Appendix to §11. The fibration β plays two roles. It gets rid of compact subvarieties of holomorphic arc components so that some function theory is possible despite the injunction of Theorem 5.7, and it allows construction of the "generalized Siegel domain" γ in a manner whose discontinuities do not interfere with integration.

NOTES FOR CHAPTER III. The results of Chapter III depend in an essential manner, both on the results of Chapter II, and on the earlier work of Wolf-Korányi [**15**]. The latter contains the theorems of §10 for those G_0-orbits that are in the topological closure of X_0.

Enumeration of the G_0-orbits on X (Theorem 10.6) was first worked out by Takeuchi [**9**]. Using the partial Cayley transforms of Wolf-Korányi [**15**] it is not difficult to see that every G_0-orbit must be one of the $G_0(x_{\Gamma,\Sigma})$, so the problem is reduced to that of finding a criterion for $G_0(x_{\Gamma',\Sigma'}) = G_0(x_{\Gamma,\Sigma})$. There, a little serious thought reduces the question to the case $\Gamma'=\phi=\Gamma$ of open orbits. Takeuchi [**9**] settles the latter by picking out the maximal compact subvarieties of open orbits and counting them with the aid of some Morse theory results on cell decompositions from his thesis [8]. We find it more convenient to apply Theorem 4.9 and count some double cosets.

The fibration $\beta: G_0(x_{\Gamma,\Sigma}) \to K(x_{\Gamma,\Sigma})$ was worked out by Takeuchi [9] for the case $\Gamma = \phi$ of open orbits, where holomorphic arc components are not needed; part 3 of Theorem 11.8 answers (in general) the triviality question on β that Takeuchi raised in his special case. A key to our result Theorem 11.8 on β is Lemma 11.6, which was proved by Takeuchi [9], for the case $\Gamma = \phi$ of open orbits, in a different way.

Except for the overlap with Wolf-Korányi [15] and Takeuchi [9] described above, the material of §10 and §11 was worked out by this author in the fall and winter of 1968.

10. Hermitian symmetric spaces: orbit structure and holomorphic arc components. Let $X = G_u/\mathcal{K}$ be a hermitian symmetric space of compact type. Then X is a flag manifold G/P where G_u is the compact real form of G and $K = P \cap G_u$ is the compact real form of P^r. Now let $X_0 = G_0/K$ be the dual hermitian symmetric space of noncompact type; if $\mathfrak{G}_u = \mathcal{K} + \mathfrak{M}_u$ Cartan decomposition, then $\mathfrak{G}_0 = \mathcal{K} + \mathfrak{M}_0$, $\mathfrak{M}_0 = \sqrt{-1}\mathfrak{M}_u$, real form of $\mathfrak{G}$ such that $K = P \cap G_0$, maximal compact subgroup of G_0. The "Borel embedding" of X_0 in X is the embedding $b: X_0 \to X$ of X_0 as an open G_0-orbit on X, given by $b(gK) = gP$ for $g \in G_0$. Let x_0 denote the identity coset in $X = G/P$, so $b(X_0) = G_0(x_0)$. We map $\mathcal{P}^{-u}$ onto a dense open subset of X by $\xi(z) = \exp(z) \cdot x_0$. Then the "Harish-Chandra embedding" of X_0 as a bounded domain D in a complex euclidean space $\mathcal{P}^{-u}$ is given by $D = \kappa(X_0)$ where $\kappa(gK) = \xi^{-1}b(gK)$, and $\xi: D \cong G_0(x_0)$ relates the two embeddings. One views the inclusions $\{\xi(D) = G_0(x_0)\} \subset \xi(\mathcal{P}^{-u}) \subset X$ as a generalization of the classical (unit disc) $\subset$ (complex line) $\subset$ (Riemann sphere).

Bott-Korányi (see [2, Theorem 3.6]) used convexity properties of D in $\mathcal{P}^{-u}$ to identify the Bergman-Šilov boundary of $G_0(x_0)$ in X as the unique closed G_0-orbit in the topological boundary $\partial G_0(x_0)$ of $G_0(x_0)$ in X. Korányi-Wolf [2] studied the structure of the Bergman-Šilov boundary and its role in the realization of X_0 as a Siegel Domain of type II. Then Wolf-Korányi [15] made full use of the convexity of D in $\mathcal{P}^{-u}$ to obtain the G_0-orbit structure of $\partial G_0(x_0)$, the holomorphic arc components of $\partial G_0(x_0)$ (there called the "boundary components" of $G_0(x_0)$ in X), and the realizations of X_0 as Siegel Domains of type III. There, it was shown that every holomorphic arc component of $\partial G_0(x_0)$ lay in a boundary orbit; so the holomorphic arc components of $\partial G_0(x)$ are the same as the holomorphic arc components of orbits $G_0(x)$, $x \in \partial G_0(x_0)$. Having seen Korányi-Wolf [2] but not Wolf-

Korányi [15], Maebashi [6] enumerated the G_0-orbits on X for the case where X_0 is the Siegel upper half plane $\mathbf{Sp}(n, \mathbf{R})/\mathbf{U}(n)$. Finally, Takeuchi [3] extended part of Wolf-Korányi [15], enumerating the G_0-orbits on X in general by means of a recursive procedure based on a counting of the open G_0-orbits; the latter consisted of describing a set consisting of just one maximal compact subvariety from each open G_0-orbit as the critical set of a certain function on X. Nonconvexity of the general open G_0-orbit on X prevented Takeuchi from extending the boundary component theory. That difficulty is avoided by our notion of holomorphic arc component developed in §§8 and 9, and this allows us to give a complete treatment of the G_0-orbit structure of X.

We establish notation for hermitian symmetric spaces.

X_0: hermitian symmetric space of noncompact type.
G_0: largest connected group of isometries of X_0.
x_0: fixed "base point" in X_0.
K: isotropy subgroup of G_0 at x_0, maximal compact in G_0.
σ: Cartan involution $\mathrm{ad}(s)$, where s is the symmetry of X_0 at x_0.

Let $\mathcal{K}\subset\mathcal{G}_0$ denote the Lie algebras of $K\subset G_0$. Then

$\mathcal{G}_0=\mathcal{K}+\mathfrak{M}_0$: Cartan decomposition into (± 1)-eigenspaces of σ.
$\mathcal{G}=\mathcal{G}_0^C=\mathcal{K}^C+\mathfrak{M}$ complexification where $\mathfrak{M}=\mathfrak{M}_0^C$.
$\mathcal{G}_u=\mathcal{K}+\mathfrak{M}_u$, $\mathfrak{M}_u=i\mathfrak{M}_0$, compact real form of $\mathcal{G}$.
G: adjoint group of $\mathcal{G}$, so G_0 is the analytic subgroup for $\mathcal{G}_0$.
G_u: analytic subgroup of G for $\mathcal{G}_u$, compact real form of G.
X: G_u/K, hermitian symmetric space, compact dual of X_0.
$\mathfrak{J}$: Cartan subalgebra of $\mathcal{G}_0$ (or $\mathcal{G}_u$) contained in $\mathcal{K}$.
Δ: $\mathfrak{J}^C$-root system of $\mathcal{G}$.
Δ_K: compact roots, i.e. $\mathfrak{J}^C$-root system of $\mathcal{K}^C$.
Δ_M: noncompact roots, i.e. roots φ such that $\mathcal{G}_\varphi\subset\mathfrak{M}$.
z: central element of $\mathcal{K}$ such that $J=\mathrm{ad}(z)|_{\mathfrak{M}}$ is the almost complex structure of X_0 and X.
$\mathfrak{M}^\pm$: $(\mp\sqrt{-1})$-eigenspaces of $J=\mathrm{ad}(z)|_{\mathfrak{M}}$.
$\mathcal{P}$: $\mathcal{K}^C+\mathfrak{M}^+$, parabolic subalgebra of $\mathcal{G}$ defined by iz.
P: parabolic subgroup of G with Lie algebra $\mathcal{P}$.

We observe $\mathcal{P}^r=\mathcal{K}^C$, $\mathcal{P}^u=\mathfrak{M}^+$ and $\mathcal{P}^{-u}=\mathfrak{M}^-$. Theorem 5.4 says $G_0\cap P=K=G_u\cap P$. Thus G_u is transitive on the complex flag manifold G/P with isotropy subgroup K. That identifies $X=G_u/K$ with the complex flag manifold G/P, providing the Borel embedding b: $X_0\to X$ by $b(gK)=gP$ for $g\in G_0$. Now we identify X_0 with $b(X_0)$, thus identifying the base point x_0 with the identity coset of X, thus identifying X_0 with the open orbit $G_0(x_0)$.

Choose a simple root system Π such that $\mathcal{P}=\mathcal{P}_\Phi$ for some subset $\Phi\subset\Pi$. Then Φ consists of the compact simple roots and $\Pi\backslash\Phi$ the noncompact simple roots. There is one noncompact simple root for each simple ideal of $\mathcal{G}$. Note

$$\Phi^r = \Delta_K, \qquad \Phi^u = \Delta_M^+, \qquad -\Phi^u = \Delta_M^-.$$

Two roots φ and ψ are *strongly orthogonal* if neither of $\varphi\pm\psi$ is a root; in that case $\langle\varphi, \psi\rangle=0$. Now we define

Ψ: maximal set of strongly orthogonal positive noncompact roots by $\Psi=\{\Psi_1, \cdots, \Psi_r\}$ where Ψ_{j+1} is the lowest element of Δ_M^+ strongly orthogonal to each of $\{\psi_1, \cdots, \psi_j\}$. Thus, for each simple ideal of $\mathcal{G}$, the noncompact simple root is the first element of Ψ that is a root of the ideal. The number $r=|\Psi|$ is the common symmetric space rank of X_0 and X. Define a real basis of $\mathfrak{M}_0$ consisting of all

$$x_{\varphi,0} = e_\varphi + e_{-\varphi} \quad\text{and}\quad y_{\varphi,0} = -i(e_\varphi - e_{-\varphi}), \varphi \quad \in \Delta_M^+,$$

where $e_{\pm\varphi}\in\mathcal{G}_{\pm\varphi}$ are normalized by $[e_\varphi, e_{-\varphi}]=h_\varphi\in i\mathfrak{Z}$. Then $\mathfrak{M}_u$ has a real basis consisting of all

$$x_\varphi = i(e_\varphi + e_{-\varphi}) \quad\text{and}\quad y_\varphi = e_\varphi - e_{-\varphi}, \qquad \varphi\in\Delta_M^+.$$

That gives us

$$\mathfrak{A}_0: \sum_\Psi \mathbf{R}x_{\psi,0}, \qquad \text{maximal abelian subspace of } \mathfrak{M}_0.$$

$$\mathfrak{A}_u: i\mathfrak{A}_0 = \sum_\Psi \mathbf{R}x_\psi, \quad \text{maximal abelian subspace of } \mathfrak{M}_u.$$

For every noncompact positive root φ we compute

$$Jx_{\varphi,0} = [z, x_{\varphi,0}] = y_{\varphi,0}, \quad Jy_{\varphi,0} = [z, y_{\varphi,0}] = -x_{\varphi,0}, \quad [x_{\varphi,0}, y_{\varphi,0}] = 2ih_\varphi,$$
$$Jx_\varphi = [z, x_\varphi] = y_\varphi, \qquad Jy_\varphi = [z, y_\varphi] = -x_\varphi, \qquad [x_\varphi, y_\varphi] = -2ih_\varphi.$$

In particular, using strong orthogonality of the roots in Ψ,

$$[\mathfrak{A}_0, J\mathfrak{A}_0] - [\mathfrak{A}_u, J\mathfrak{A}_u] = \sum_\Psi (ih_\psi)\mathbf{R}.$$

That gives

(10.1a) $\mathfrak{Z}=\mathfrak{Z}^+ + \mathfrak{Z}^-$ orthogonal direct sum, where

(10.b) $\mathfrak{Z}^-=[\mathfrak{A}_0, J\mathfrak{A}_0]$ and $\mathfrak{Z}^+$ is the centralizer of $\mathfrak{A}_0$ in $\mathfrak{Z}$.

For every subset $\Gamma\subset\Psi$ we define

(10.2a) $\mathfrak{Z}_\Gamma^- = \sum_{\Psi\backslash\Gamma}(ih_\psi)\mathbf{R}$, so $\mathfrak{Z}_\phi^-=\mathfrak{Z}^-$ and $\mathfrak{Z}_\Psi^-=0$;

(10.2b) $\mathfrak{A}_\Gamma = \sum_\Gamma(x_{\gamma,0})\mathbf{R}$, so $\mathfrak{A}_\phi=0$ and $\mathfrak{A}_\Psi=\mathfrak{A}_0$;

(10.2c) $\mathcal{H}_\Gamma=\mathfrak{Z}^+ + \mathfrak{Z}_\Gamma^- + \mathfrak{A}_\Gamma$, so $\mathcal{H}_\phi=\mathfrak{Z}$ and $\mathcal{H}_\Psi=\mathfrak{Z}^+ + \mathfrak{A}_0$.

Note that (10.2c) is an orthogonal direct sum. The $\mathfrak{H}_\Gamma$ are the Cartan subalgebras of $\mathfrak{g}_0$ with which we shall work. They are related by the partial Cayley transforms [15] defined by

(10.3a) $c_\psi = \exp((\pi/4)y_\psi) \in G_u$ for every $\psi \in \Psi$, and

(10.3b) $c_\Gamma = \prod_\Gamma c_\gamma \in G_u$ for every $\Gamma \subset \Psi$.

Now we have

(10.4a) $\mathrm{ad}(c_\psi): y_\psi \to y_\psi,\ x_\psi \to 2ih_\psi,\ 2ih_\psi \to -x_\psi$; so

(10.4b) $\mathrm{ad}(c_{\Gamma' \setminus \Gamma})\mathfrak{H}_\Gamma^C = \mathfrak{H}_{\Gamma'}^C$ for $\Gamma \subset \Gamma' \subset \Psi$, and

(10.4c) $\mathrm{ad}(c_\Sigma)^2 \mathfrak{H}_\Gamma = \mathfrak{H}_\Gamma$ for $\Gamma, \Sigma \subset \Psi$.

We use the partial Cayley transforms to define points

(10.5) $x_{\Gamma,\Sigma} = c_\Gamma c_\Sigma^2 x_0 \in X$ whenever $\Gamma, \Sigma \subset \Psi$.

Now we can state the decomposition of X into G_0-orbits.

10.6 THEOREM. *Decompose* $\mathfrak{g} = \mathfrak{g}^1 \oplus \cdots \oplus \mathfrak{g}^p$ *direct sum of simple ideals, so* $\Delta = \Delta^1 \cup \cdots \cup \Delta^p$ *where* Δ^k *is the root system of* $\mathfrak{g}^k$.

1. *The* G_0*-orbits on* X *are just the* $G_0(x_{\Gamma,\Sigma})$, Γ *and* Σ *disjoint.*

2. *Orbits* $G_0(x_{\Gamma,\Sigma})$ *and* $G_0(x_{\Gamma',\Sigma'})$ *are equal if and only if, for* $1 \leqq k \leqq p$, $|(\Sigma' \setminus \Gamma') \cap \Delta^k| = |(\Sigma \setminus \Gamma) \cap \Delta^k|$ *and* $|\Gamma' \cap \Delta^k| = |\Gamma \cap \Delta^k|$. *In particular, the number of* G_0*-orbits on* X *is* $\frac{1}{2}\sum_{1 \leqq k \leqq p}(r_k+1)(r_k+2)$ *where* $r_k = |\Psi \cap \Delta^k|$.

3. $G_0(x_{\Gamma',\Sigma'})$ *is in the closure of* $G_0(x_{\Gamma,\Sigma})$ *if and only if, for* $1 \leqq k \leqq p$, $|(\Sigma' \setminus \Gamma') \cap \Delta^k| \leqq |(\Sigma \setminus \Gamma) \cap \Delta^k|$ *and* $|(\Sigma \cup \Gamma) \cap \Delta^k| \leqq |(\Sigma' \cup \Gamma') \cap \Delta^k|$.

4. $G_0(x_{\Gamma,\Sigma})$ *is open in* X *if and only if* $\Gamma = \phi$ *empty set. In particular there are precisely* $\sum_{1 \leqq k \leqq p}(r_k+1) = |\Psi| + p$ *open* G_0*-orbits on* X.

5. $G_0(x_{\Psi,\phi})$ *is the closed* G_0*-orbit on* X, *Bergman-Šilov boundary of* X_0 *in* X, *contained in the closure of every* G_0*-orbit.*

The corresponding result for Cartan subalgebras is

10.7. THEOREM. *Let* $\mathfrak{g}^k$ *and* Δ^k *be as in Theorem* 10.6.

1. $\mathfrak{H}_\Gamma$ *is maximally compact among the Cartan subalgebras of* $\mathfrak{g}_0$ *contained in* $\mathfrak{P}_{x_{\Gamma,\Sigma}}$, $\Gamma \cap \Sigma = \phi$, *and every Cartan subalgebra of* $\mathfrak{g}_0$ *is conjugate to some* $\mathfrak{H}_\Gamma$.

2. $\mathfrak{H}_\Gamma$ *and* $\mathfrak{H}_{\Gamma'}$ *are conjugate in* $\mathfrak{g}_0$, *if and only if, for* $1 \leqq k \leqq p$, $|\Gamma \cap \Delta^k| = |\Gamma' \cap \Delta^k|$.

3. $\mathfrak{H}_\Gamma$ *centralizes a* $\mathfrak{g}_0$*-conjugate of* $\mathfrak{J}_{\Gamma'}^-$, *if and only if, for* $1 \leqq k \leqq p$, $|\Gamma \cap \Delta^k| \leqq |\Gamma' \cap \Delta^k|$.

4. $\mathfrak{H}_\Gamma$ *is maximally compact in* $\mathfrak{g}_0$ *if and only if* $\Gamma = \phi$.

5. $\mathfrak{H}_\Gamma$ *is maximally split in* $\mathfrak{g}_0$ *if and only if* $\Gamma = \Psi$.

We will prove Theorems 10.6 and 10.7 as we decompose a G_0-orbit into holomorphic arc components. To express that decomposition, note that

$$\mathfrak{J}^C + \sum_{\varphi \perp \Gamma} \mathfrak{g}_\varphi \text{ is the centralizer of } \mathfrak{J}_{\Psi \setminus \Gamma}^- \text{ in } \mathfrak{g}.$$

Now define

(10.8a) $\mathfrak{g}_{\Psi\backslash\Gamma}$ is the derived algebra of $\mathfrak{Z}^C+\sum_{\varphi\perp\Gamma}\mathfrak{g}_\varphi$,

(10.8b) $\mathfrak{g}_{\Psi\backslash\Gamma,0}$ is the real form $\mathfrak{g}_0\cap\mathfrak{g}_{\Psi\backslash\Gamma}$ of $\mathfrak{g}_{\Psi\backslash\Gamma}$, and

(10.8c) $G_{\Psi\backslash\Gamma}$ and $G_{\Psi\backslash\Gamma,0}$ are the analytic subgroups for $\mathfrak{g}_{\Psi\backslash\Gamma}$ and $\mathfrak{g}_{\Psi\backslash\Gamma,0}$.

10.9. THEOREM. *Let Γ and Σ be arbitrary subsets of Ψ.*

1. $G_{\Psi\backslash\Gamma}(x_{\Gamma,\Sigma})=c_\Gamma c_\Sigma^2 G_{\Psi\backslash\Gamma}(x_0)$, *compact totally geodesic hermitian symmetric subspace of X; its symmetric space rank is $|\Psi\backslash\Gamma|$; relative to the Cartan subalgebra $\mathfrak{Z}\cap\mathfrak{g}_{\Psi\backslash\Gamma,0}$, $\Psi\backslash\Gamma$ is a maximal set of strongly orthogonal noncompact roots of $\mathfrak{g}_{\Psi\backslash\Gamma,0}$.*

2. $G_{\Psi\backslash\Gamma,0}(x_{\Gamma,\Sigma})$ *is an open $G_{\Psi\backslash\Gamma,0}$-orbit on the symmetric complex flag manifold $G_{\Psi\backslash\Gamma}(x_{\Gamma,\Sigma})$. It is an indefinite-hermitian symmetric space of $G_{\Psi\backslash\Gamma,0}$; if $\Gamma\cap\Sigma=\phi$, then its invariant kaehler metric is positive definite if and only if Σ is a (possibly empty) union of sets $(\Psi\backslash\Gamma)\cap\Delta^k$ for which $G_0^k(x_0)$ is of tube type. Its isotropy subalgebra $\mathfrak{g}_{\Psi\backslash\Gamma,0}\cap\mathcal{P}_{x_{\Gamma,\Sigma}}$ at $x_{\Gamma,\Sigma}$ has complexification* $\mathrm{ad}(c_\Sigma^2)\{\mathfrak{g}_{\Psi\backslash\Gamma}\cap\mathcal{P}_{x_{\Gamma,\phi}}\}$.

3. *The holomorphic arc components of $G_0(x_{\Gamma,\Sigma})$ are just the $g\cdot G_{\Psi\backslash\Gamma,0}(x_{\Gamma,\Sigma})$, $g\in K$. If*

(10.10a) $N_{\Gamma,\Sigma,0}=\{g\in G_0\colon g\cdot G_{\Psi\backslash\Gamma,0}(x_{\Gamma,\Sigma})=G_{\Psi\backslash\Gamma,0}(x_{\Gamma,\Sigma})\}$,

(10.10b) $Z_{\Gamma,\Sigma,0}=\{g\in N_{\Gamma,\Sigma,0}\colon g$ *acts trivially on* $G_{\Psi\backslash\Gamma,0}(x_{\Gamma,\Sigma})\}$,

then

(10.10c) $N_{\Gamma,\Sigma,0}=G_{\Psi\backslash\Gamma,0}\cdot Z_{\Gamma,\Sigma,0}$,

so $G_{\Psi\backslash\Gamma,0}$ and $N_{\Gamma,\Sigma,0}$ have the same action on the holomorphic arc component $G_{\Psi\backslash\Gamma,0}(x_{\Gamma,\Sigma})$ of $G_0(x_{\Gamma,\Sigma})$.

4. *The parabolic subalgebra $\mathfrak{N}_{[x_{\Gamma,\Sigma}]}$ of $\mathfrak{g}$, complexification of the Lie algebra of $N_{\Gamma,\Sigma,0}$, is just the parabolic subalgebra that is the sum of the nonnegative eigenspaces of* $\mathrm{ad}(\{\sum_{\Gamma\backslash\Sigma}x_{\gamma,0}\}-\{\sum_{\Gamma\cap\Sigma}x_{\gamma,0}\})$; *it is independent of Σ if we require that Γ and Σ be disjoint.*

10.11. COROLLARY. *Every G_0-orbit on X is measurable.*

PROOF OF THEOREM 10.6. If φ is a noncompact root then $\mathfrak{g}[\varphi]$ denotes the simple algebra $h_\varphi\mathbf{C}+\mathfrak{g}_\varphi+\mathfrak{g}_{-\varphi}$, and $\mathfrak{g}_0[\varphi]$ denotes its real form $\mathfrak{g}_0\cap\mathfrak{g}[\varphi]$; $G[\varphi]$ and $G_0[\varphi]$ are the analytic subgroups for $\mathfrak{g}[\varphi]$ and $\mathfrak{g}_0[\varphi]$; and $S[\varphi]$ is the Riemann sphere $G[\varphi](x_0)$. The $G_0[\varphi]$-orbits on $S[\varphi]$ are:

$$\text{lower hemisphere } G_0[\varphi]\left(\left\{\exp\frac{\pi}{4}y_\varphi\right\}^n x_0\right),\qquad n\equiv 0 \bmod 4,$$

$$\text{equator } G_0[\varphi]\left(\left\{\exp\frac{\pi}{4}y_\varphi\right\}^n x_0\right),\qquad n\equiv 1 \bmod 2,$$

$$\text{upper hemisphere } G_0[\varphi]\left(\left\{\exp\frac{\pi}{4}y_\varphi\right\}^n x_0\right),\qquad n\equiv 2 \bmod 4.$$

Define $\mathfrak{g}[\Psi]=\sum_\Psi \mathfrak{g}[\psi]$, so the real form $\mathfrak{g}_0[\Psi]=\mathfrak{g}_0\cap\mathfrak{g}[\Psi]$ is $\sum_\Psi \mathfrak{g}_0[\psi]$; then the analytic subgroups for $\mathfrak{g}[\Phi]$ and $\mathfrak{g}_0[\Psi]$ are $G[\Psi] = \prod_\Psi G[\psi]$ and $G_0[\Psi]=\prod_\Psi G_0[\psi]$, and $S[\Psi]=G[\Psi](x_0)$ is the product $\prod_\Psi S[\Psi]$ of $|\Psi|$ Riemann spheres. Now the $G_0[\Psi]$-orbits on $S[\Psi]$ are the products $\prod_\Psi D_\psi$ where D_ψ is a $G_0[\psi]$-orbit on $S[\psi]$. Given such an orbit, define

$$\Gamma = \{\psi \in \Psi\colon D_\psi \text{ is the equator of } S[\psi]\}, \quad \text{and}$$

$$\Sigma = \{\psi \in \Psi\colon D_\psi \text{ is the upper hemisphere of } S[\psi]\}.$$

Then the orbit is $G_0[\Psi](c_\Gamma c_\Sigma^2 x_0)=G_0[\Psi](x_{\Gamma,\Sigma})$, and Γ and Σ are disjoint.

The equator of $S[\psi]$ has descriptions $G_0[\psi](c_\psi x_0)=G_0[\psi](c_\psi^3 x_0)$. If Γ and Σ are arbitrary subsets of Ψ, it follows that $x_{\Gamma,\Sigma}=c_{\Gamma\setminus\Sigma}\cdot c_{\Gamma\cap\Sigma}^3 \cdot c_{\Sigma\setminus\Gamma}^2 x_0 \in G_0[\Psi](x_{\Gamma,\Sigma\setminus\Gamma})$. Thus

(10.12a) $G_0[\Psi](x_{\Gamma,\Sigma})=G_0[\Psi](x_{\Gamma,\Sigma\setminus\Gamma})$.

If we also have subsets Γ' and Σ' of Ψ, it follows that

(10.12b) $G_0[\Psi](x_{\Gamma,\Sigma})=G_0[\Psi](x_{\Gamma',\Sigma'})$ if and only if $(\Sigma\setminus\Gamma)=(\Sigma'\setminus\Gamma')$ and $\Gamma=\Gamma'$.

It also follows from (10.12a) that $G_0[\Psi](x_{\Gamma',\Sigma'})$ is in the closure of $G_0[\Psi](x_{\Gamma,\Sigma})$ if and only if

(i) $\Gamma\subset\Gamma'$,

(ii) $(\Sigma\setminus\Gamma)\subset(\Sigma'\cup\Gamma')$, and

(iii) $\{\Psi\setminus(\Sigma\cup\Gamma)\}\subset\{[\Psi\setminus(\Sigma'\cup\Gamma')]\cup\Gamma'\}$.

Those three conditions can be phrased

(a) $(\Sigma'\setminus\Gamma')\subset(\Sigma\setminus\Gamma)$ and (b) $(\Sigma\cup\Gamma)\subset(\Sigma'\cup\Gamma')$.

Thus

(10.12c) $G_0[\Psi](x_{\Gamma',\Sigma'})$ is in the closure of $G_0[\Psi](x_{\Gamma,\Sigma})$ if and only if $(\Sigma'\setminus\Gamma')\subset(\Sigma\setminus\Gamma)$ and $(\Sigma\cup\Gamma)\subset(\Sigma'\cup\Gamma')$.

Finally we observe that

(10.12d) $G_0[\Psi](x_{\Gamma,\Sigma})$ is open in $S[\Psi]$ if and only if $\Gamma=\phi$,

(10.12e) $G_0[\Psi](x_{\Gamma,\Sigma})$ is closed in $S[\Psi]$ if and only if $\Gamma=\Psi$.

To carry results from $G_0[\Psi]$-orbits on $S[\Psi]$ to G_0-orbits on X, we define $A=\exp J\mathfrak{a}_u\subset G_u$. As $J\mathfrak{a}_u$ is a maximal abelian subspace of $\mathfrak{M}_u$ we have $G_u=KAK$. Thus

$$X = G_u(x_0) = KAK(x_0) = KA(x_0) = K\cdot S[\Psi].$$

If $x\in X$ we thus have $x=k(s)$, $s\in S[\Psi]$, so $G_0(x)=G_0(s)=G_0\cdot G_0[\Psi](s)$. We have disjoint sets Γ and Σ in Ψ with $G_0[\Psi](s)=G_0[\Psi](x_{\Gamma,\Sigma})$. It follows that $G_0(x)=G_0(x_{\Gamma,\Sigma})$. That proves part 1 of Theorem 10.6.

Theorem 3.3 says that X has just one closed G_0-orbit. In [2, Theorem 3.6], $G_0(x_{\Psi,\phi})$ is shown to be the closed G_0-orbit on the boundary of X_0 and is identified as the Bergman-Šilov boundary of X_0 in X. That proves part 5 of Theorem 10.6.

Let $\Sigma \subset \Psi$. We have $\mathfrak{J} \subset \mathfrak{G}_0 \cap \mathcal{P}_{x_0}$, and (10.4c) with $\Gamma = \phi$ says $\mathfrak{J} = \mathrm{ad}(c_\Sigma^2)\mathfrak{J} \subset \mathrm{ad}(c_\Sigma^2)\mathcal{P}_{x_0} = \mathcal{P}_{c^2_\Sigma x_0} = \mathcal{P}_{x_{\phi,\Sigma}}$. Thus $\mathfrak{G}_0 \cap \mathcal{P}_{x_{\phi,\Sigma}}$ contains a compact Cartan subalgebra $\mathfrak{J}$ of $\mathfrak{G}_0$, so Theorem 4.5 says that $G_0(x_{\phi,\Sigma})$ is open in X. Conversely, if $G_0(x_{\Gamma,\Sigma})$ is open in X, then $G_0(x_{\Gamma,\Sigma}) \cap S[\Psi]$ is an open $G_0[\Psi]$-invariant subset of $S[\Psi]$ that contains $x_{\Gamma,\Sigma}$. As it is open and $G_0[\Psi]$-invariant, it is a union of sets $G_0[\Psi](x_{\phi,B})$, $B \subset \Psi$. Thus we have $B \subset \Psi$ with $x_{\Gamma,\Sigma} \in G_0[\Psi](x_{\phi,B})$. It follows that the open G_0-orbits on X are just the $G_0(x_{\phi,\Sigma})$ with $\Sigma \subset \Psi$.

Let Σ and Σ' be two subsets of Ψ. If $|\Sigma \cap \Delta^k| = |\Sigma' \cap \Delta^k|$ for $1 \leqq k \leqq p$, then [7, Theorem 2] there is an element $g \in K$ normalizing $\mathfrak{J}$ such that $\mathrm{ad}(g)\Sigma = \Sigma'$, so $G_0(x_{\phi,\Sigma'}) = G_0(c_{\Sigma'}^2 x_0) = G_0(g c_\Sigma^2 g^{-1})x_0 = G_0(c_\Sigma^2 x_0) = G_0(x_{\phi,\Sigma})$. Conversely suppose $G_0(x_{\phi,\Sigma}) = G_0(x_{\phi,\Sigma'})$. Then, as both orbits are open, Theorem 4.9 says that c_Σ^2 and $c_{\Sigma'}^2$ represent the same double coset in a certain space $W_K \backslash W_G^{\mathfrak{K}_0} / W^{\mathfrak{K}_0}_{P^r \cap \tau P^r}$. There $\mathfrak{K}_0 = \mathfrak{J}$, so $W_G^{\mathfrak{K}_0} = W_G$; and $P^r = \tau P^r = K^C$, so $W^{\mathfrak{K}_0}_{P^r \cap \tau P^r} = W_K$. Thus $c_{\Sigma'}^2 \in W_K \cdot c_\Sigma^2 \cdot W_K$, so we have $w_i \in W_K$ with $\mathrm{ad}(c_{\Sigma'}^2) = w_1 \cdot \mathrm{ad}(c_\Sigma^2) \cdot w_2$. If z denotes the central element of $\mathfrak{K}$ such that $J = \mathrm{ad}(z)|\mathfrak{M}$, and if $\Sigma'' \subset \Psi$, then $\mathrm{ad}(c_{\Sigma''}^2)z - z = \sum_{\psi \in \Sigma''}(ih_\psi)$. Thus

$$
\begin{aligned}
-\sum_{\psi \in \Sigma'} h_\psi &= i\{\mathrm{ad}(c_{\Sigma'}^2)z - z\} = i\{w_1 \cdot \mathrm{ad}(c_\Sigma^2) \cdot w_2 z - z\} \\
&= i\{w_1 \cdot \mathrm{ad}(c_\Sigma^2)z - z\} = w_1 \cdot i\{\mathrm{ad}(c_\Sigma^2)z - z\} \\
&= w_1\Bigl(-\sum_{\psi \in \Sigma} h_\psi\Bigr).
\end{aligned}
$$

As $\frac{1}{2}\sum_{\psi \in \Sigma''} x_{\psi,0} = -(i/2)\sum_{\psi \in \Sigma''} x_\psi = \mathrm{ad}(c_{\Sigma''})(-\sum_{\psi \in \Sigma''} h_\psi)$, it follows that $\frac{1}{2}\sum_{\psi \in \Sigma'} x_{\psi,0}$ and $\frac{1}{2}\sum_{\psi \in \Sigma} x_{\psi,0}$ are conjugate elements of $\mathfrak{G}$. Thus the corresponding parabolic subalgebras of $\mathfrak{G}$ are conjugate. But [15, Theorem 6.5] those parabolic subalgebras of $\mathfrak{G}$ are the complex normalizers of the boundary components of types $\Psi \backslash \Sigma$ and $\Psi \backslash \Sigma'$ of X_0 in X, so [15, Corollary 6.9] $G_0(x_{\Sigma,\phi}) = G_0(x_{\Sigma',\phi})$, and it follows [15, Lemma 5.1] that $|\Sigma \cap \Delta^k| = |\Sigma' \cap \Delta^k|$ for $1 \leqq k \leqq p$. In conclusion, open orbits $G_0(x_{\phi,\Sigma})$ and $G_0(x_{\phi,\Sigma'})$ are equal if and only if, for $1 \leqq k \leqq p$, $|\Sigma \cap \Delta^k| = |\Sigma' \cap \Delta^k|$. In particular, if $r_k = |\Psi \cap \Delta^k|$ then there are precisely $\sum_{1 \leqq k \leqq p}(r_k + 1) = |\Psi| + p$ open G_0-orbits on X. That completes the proof of part 4 of Theorem 10.6.

Consider an open orbit $G_0(x_{\phi,\Sigma})$. The intersection with $S[\Psi]$ is $G_0[\Psi]$-invariant and open in $S[\Psi]$, hence a union of sets $G_0[\Psi](x_{\phi,B})$. The sets B there are just the subsets of Ψ such that $G_0(x_{\phi,\Sigma}) = G_0(x_{\phi,B})$.

Thus

(10.13a) $G_0(x_{\phi,\Sigma})\cap S[\Psi]=\bigcup_{\mathcal{B}} G_0[\Psi](x_{\phi,B})$

where

(10.13b) $\mathcal{B}=\{B\subset\Psi:|B\cap\Delta^k|=|\Sigma\cap\Delta^k| \quad \text{for } 1\leqq k\leqq p\}$.

Now consider an orbit $G_0(x_{\Gamma',\Sigma'})$. It is in the closure of $G_0(x_{\phi,\Sigma})$ if and only if $x_{\Gamma',\Sigma'}$ is in that closure. So (10.12c) says that

(10.13c) $G_0(x_{\Gamma',\Sigma'})$ is in the closure of $G_0(x_{\phi,\Sigma})$ if and only if, for some $B\in\mathcal{B}$, $(\Sigma'\backslash\Gamma')\subset B\subset(\Sigma'\cup\Gamma')$.

In view of (10.13b), we reformulate (10.13c) as

(10.13d) $G_0(x_{\Gamma',\Sigma'})$ is in the closure of $G_0(x_{\phi,\Sigma})$ if and only if, for $1\leqq k\leqq p$, $|(\Sigma'\backslash\Gamma')\cap\Delta^k|\leqq|\Sigma\cap\Delta^k|\leqq|(\Sigma'\cup\Gamma')\cap\Delta^k|$.

Now consider arbitrary orbits $G_0(x_{\Gamma,\Sigma})$ and $G_0(x_{\Gamma',\Sigma'})$. According to (10.13d), $G_0(x_{\Gamma,\Sigma})$ is in the closure of an open orbit $G_0(x_{\phi,\Sigma^*})$ if and only if

$$|(\Sigma\backslash\Gamma)\cap\Delta^k| \leqq |\Sigma^*\cap\Delta^k| \leqq |(\Sigma\cup\Gamma)\cap\Delta^k| \qquad \text{for } 1\leqq k\leqq p.$$

In particular, $G_0(x_{\Gamma,\Sigma})$ is in the closures of the open orbits

$$G_0(x_{\phi,\Sigma\backslash\Gamma}) \quad \text{and} \quad G_0(x_{\phi,\Sigma\cup\Gamma}).$$

If $G_0(x_{\Gamma',\Sigma'})$ is in the closure of $G_0(x_{\Gamma,\Sigma})$, then it is in the closures of $G_0(x_{\phi,\Sigma\backslash\Gamma})$ and $G_0(x_{\phi,\Sigma\cup\Gamma})$, and (10.13d) says that $|(\Sigma'\backslash\Gamma')\cap\Delta^k|\leqq|(\Sigma\backslash\Gamma)\cap\Delta^k|$ and $|(\Sigma\cup\Gamma)\cap\Delta^k|\leqq|(\Sigma'\cup\Gamma')\cap\Delta^k|$ for $1\leqq k\leqq p$. Conversely, if we have those numerical inequalities, then [7, Theorem 2] there is an element $g\in K$ normalizing $\mathfrak{Z}$ such that $\mathrm{ad}(g)$ sends $\Sigma'\backslash\Gamma'$ into $\Sigma\backslash\Gamma$ and sends $\Sigma'\cup\Gamma'$ to a set containing $\Sigma\cup\Gamma$; as $G_0(x_{\Gamma',\Sigma'})=G_0(x_{\mathrm{ad}(g)\Gamma',\mathrm{ad}(g)\Sigma'})$ we thus may assume $(\Sigma'\backslash\Gamma')\subset(\Sigma\backslash\Gamma)$ and $(\Sigma\cup\Gamma)\subset(\Sigma'\cup\Gamma')$, so (10.12c) says that $G_0(x_{\Gamma',\Sigma'})$ is in the closure of $G_0(x_{\Gamma,\Sigma})$. That completes the proof of part 3 of Theorem 10.6.

Orbits $G_0(x_{\Gamma,\Sigma})$ and $G_0(x_{\Gamma',\Sigma'})$ are equal if and only if each is in the closure of the other. According to part 3 just proved, that can be formulated

$$|(\Sigma'\backslash\Gamma')\cap\Delta^k| = |(\Sigma\backslash\Gamma)\cap\Delta^k| \quad \text{and}$$
$$|(\Sigma'\cup\Gamma')\cap\Delta^k| = |(\Sigma\cup\Gamma)\cap\Delta^k|,$$

which is equivalent to the criterion

$$|(\Sigma'\backslash\Gamma')\cap\Delta^k| = |(\Sigma\backslash\Gamma)\cap\Delta^k| \quad \text{and} \quad |\Gamma'\cap\Delta^k| = |\Gamma\cap\Delta^k|$$

asserted in part 2. In particular, if we enumerate $\Psi\cap\Delta^k=\{\psi_{k,1},\cdots,\psi_{k,r_k}\}$, and the choices of $\Gamma\cap\Delta^k$ and $\Sigma\cap\Delta^k$ leading to distinct G_0-orbits are

$$\Gamma\cap\Delta^k = \{\psi_{k,1},\cdots,\psi_{k,j}\}, \qquad 0\leqq j\leqq r_k,$$
$$\Sigma\cap\Delta^k = \{\psi_{k,j+1},\cdots,\psi_{k,l}\}, \qquad j\leqq l\leqq r_k.$$

The number of such choices is

$$\sum_{j=0}^{r_k}(r_k+1-j)=(r_k+1)^2-\sum_{j=0}^{r_k} j=(r_k+1)^2-\tfrac{1}{2}r_k(r_k+1)$$
$$=\tfrac{1}{2}(r_k+1)(r_k+2).$$

Thus there are precisely $\frac{1}{2}\sum_{k=0}^{p}(r_k+1)(r_k+2)$ G_0-orbits on X. That proves part 2 of Theorem 10.6, completing the proof of the theorem. Q.E.D.

PROOF OF THEOREM 10.7. Decompositions (10.1) and (10.2) show that $\mathfrak{H}_\Gamma$ is an abelian subalgebra of $\mathfrak{H}_0$ with the same dimension as $\mathfrak{T}$. If σ is the Cartan involution of $\mathfrak{g}_0$ with fixed point set $\mathfrak{K}$, then $\sigma(\mathfrak{H}_\Gamma)=\mathfrak{H}_\Gamma$ by (10.2), so $\mathfrak{H}_\Gamma$ is a reductive subalgebra of $\mathfrak{g}_0$. It follows that $\mathfrak{H}_\Gamma$ is a Cartan subalgebra of $\mathfrak{g}_0$. Note $\mathfrak{H}_\Gamma^C=\mathrm{ad}(c_\Gamma)\mathfrak{T}^C=\mathrm{ad}(c_\Gamma)\mathrm{ad}(c_\Sigma^2)\mathfrak{T}^C\subset\mathrm{ad}(c_\Gamma c_\Sigma^2)\mathcal{P}_{x_0}=\mathcal{P}_{c_\Gamma c_\Sigma^2 x_0}=\mathcal{P}_{x_{\Gamma,\Sigma}}$; thus $\mathfrak{H}_\Gamma$ is a Cartan subalgebra of $\mathfrak{g}_0$ contained in $\mathcal{P}_{x_{\Gamma,\Sigma}}$. It is standard [3] that every Cartan subalgebra of $\mathfrak{g}_0$ is conjugate to one of the $\mathfrak{H}_\Gamma$.

Let $\mathfrak{C}$ be maximally compact among the Cartan subalgebras of $\mathfrak{g}_0$ contained in $\mathcal{P}_{x_{\Gamma,\Sigma}}$. Conjugating by an element of $G_0\cap P_{x_{\Gamma,\Sigma}}$, we may assume

$$\mathfrak{C}=(\mathfrak{C}\cap\mathfrak{K})+(\mathfrak{C}\cap\mathfrak{M}_0),\quad(\mathfrak{T}^{+}+\mathfrak{T}_\Gamma^{-})\subset(\mathfrak{C}\cap\mathfrak{K}),\quad(\mathfrak{C}\cap\mathfrak{M}_0)\subset\mathfrak{A}_\Gamma.$$

If the latter two inclusions are not equalities, then we have $0\neq v\in\mathfrak{C}\cap\mathfrak{T}_{\Gamma\backslash\Psi}^{-}$. Thus $0\neq v\in\mathcal{P}_{x_{\Gamma,\Sigma}}\cap\mathfrak{T}_{\Psi\backslash\Gamma}$. But $\mathrm{ad}(c_\Gamma^{-1})\mathrm{ad}(c_\Sigma^{-2})$ sends $\mathcal{P}_{x_{\Gamma,\Sigma}}$ to $\mathcal{P}$ and sends $\mathfrak{T}_{\Psi\backslash\Gamma}^{-}$ to $i\mathfrak{A}_\Gamma$; thus $0\neq\mathrm{ad}(c_\Gamma^{-1})\mathrm{ad}(c_\Sigma^{-2})v\in\mathcal{P}\cap i\mathfrak{A}_\Gamma$, contradicting $\mathcal{P}\cap i\mathfrak{A}_\Gamma=0$. Thus

$$(\mathfrak{C}\cap\mathfrak{K})=(\mathfrak{T}^{+}+\mathfrak{T}_\Gamma^{-})\quad\text{and}\quad(\mathfrak{C}\cap\mathfrak{M}_0)=\mathfrak{A}_\Gamma,$$

proving $\mathfrak{C}=\mathfrak{H}_\Gamma$. Now the first part of Theorem 10.7 is proved.

If $\mathfrak{H}_\Gamma$ and $\mathfrak{H}_{\Gamma'}$ are conjugate in $\mathfrak{g}_0$, then for $1\leqq k\leqq p$ their split parts $\mathfrak{A}_\Gamma\cap\mathfrak{g}^k$ and $\mathfrak{A}_{\Gamma'}\cap\mathfrak{g}^k$ along the simple ideal $\mathfrak{g}_0\cap\mathfrak{g}^k$ of $\mathfrak{g}_0$ must have the same dimension; those dimensions are $|\Gamma\cap\Delta^k|$ and $|\Gamma'\cap\Delta^k|$, which thus must be equal. Conversely, if $|\Gamma\cap\Delta^k|=|\Gamma'\cap\Delta^k|$ for $1\leqq k\leqq p$, then [7, Theorem 2], we have $g\in K$ normalizing $\mathfrak{T}$ with $\mathrm{ad}(g)\mathfrak{H}_{\Gamma'}=\mathfrak{H}_\Gamma$. That proves part 2 of Theorem 10.7. Parts 4 and 5 are immediate, for $\mathfrak{H}_\Gamma$ is maximally compact if and only if it is conjugate to $\mathfrak{T}=\mathfrak{H}_\phi$, maximally split if and only if it is conjugate to $\mathfrak{T}^{+}+\mathfrak{A}_0=\mathfrak{H}_\Psi$.

If $\mathfrak{H}_\Gamma$ centralizes a G_0-conjugate $\mathrm{ad}(g)\mathfrak{T}_{\Gamma'}^{-}$, then $\mathrm{ad}(g)\mathfrak{T}_\Gamma^{-}\subset\mathfrak{H}_\Gamma$. As $\mathfrak{H}_\Gamma$ is commutative, now $\mathrm{ad}(g)\mathfrak{T}_{\Gamma'}^{-}\subset\mathfrak{K}\cap\mathfrak{H}_\Gamma=\mathfrak{T}^{+}+\mathfrak{T}_\Gamma^{-}\subset\mathfrak{T}$, so we can assume that $g\in K$ and $\mathrm{ad}(g)\mathfrak{T}=\mathfrak{T}$; thus $\mathfrak{A}_\Gamma+J\mathfrak{A}_\Gamma+[\mathfrak{A}_\Gamma, J\mathfrak{A}_\Gamma]=\mathfrak{g}_0[\Gamma]$ centralizes $\mathrm{ad}(g)\mathfrak{T}_\Gamma^{-}$, so $\mathfrak{g}_0[\Gamma]\subset\mathrm{ad}(g)(\mathfrak{g}_{\Gamma',0})$. The former has Γ, and the latter has $\mathrm{ad}(g)\Gamma'$, as maximal set of strongly orthogonal noncompact roots. Thus, for $1 \leqq k \leqq p$, $|\Gamma \cap \Delta^k| \leqq |\mathrm{ad}(g)\Gamma' \cap \Delta^k| = |\Gamma' \cap \Delta^k|$. Conversely, if $|\Gamma \cap \Delta^k| \leqq |\Gamma' \cap \Delta^k|$ for $1 \leqq k \leqq p$, then we conjugate Γ' by an element of W so that $\Gamma \subset \Gamma'$, and then $\mathfrak{H}_\Gamma$ centralizes $\mathfrak{Z}_{\overline{\Gamma'}}$. That proves part 3, completing the proof of Theorem 10.7. Q.E.D.

PROOF OF THEOREM 10.9. Our assertions all are known for G_0-orbits on the boundary of $X_0 = G_0(x_0)$ in X: in Wolf-Korányi [**15**] it is shown that $G_{\Psi\setminus\Gamma}(x_0)$ is a compact totally geodesic hermitian symmetric subspace of X (Lemma 4.2); that $G_{\Psi\setminus\Gamma,0}(x_0) \subset G_{\Psi\setminus\Gamma}(x_0)$ is the Borel embedding of the noncompact dual (Lemma 4.2); that $\Psi\setminus\Gamma$ is a maximal set of strongly orthogonal roots of $\mathfrak{g}_{\Psi\setminus\Gamma,0}$ relative to $\mathfrak{Z} \cap \mathfrak{g}_{\Psi\setminus\Gamma,0}$ (Lemma 4.2); that the holomorphic arc components of the topological boundary ∂X_0 of X_0 in X are just the sets $g \cdot c_\Gamma G_{\Psi\setminus\Gamma,0}(x_0)$ with $g \in K$ and $\phi \neq \Gamma \subset \Psi$ (Theorem 4.8); and that the set $B_{\Psi\setminus\Gamma}$ of all elements of G_0 that preserve $c_\Gamma G_{\Psi\setminus\Gamma,0}(x_0)$ is the parabolic subgroup of G_0 whose Lie algebra $\mathfrak{B}_{\Psi\setminus\Gamma}$ is the sum of the nonnegative eigenspaces of $\mathrm{ad}(\sum_\Gamma x_{\gamma,0})$ on $\mathfrak{g}_0$ and satisfies $\mathfrak{B}_{\Psi\setminus\Gamma} = \mathfrak{g}_{\Psi\setminus\Gamma,0} + \mathfrak{Z}_{\Psi\setminus\Gamma}$ where $\mathfrak{Z}_{\Psi\setminus\Gamma}$ is the ideal that is the Lie algebra of the kernel of the action of $B_{\Psi\setminus\Gamma}$ on $c_\Gamma G_{\Psi\setminus\Gamma,0}(x_0)$ (Theorems 6.5 and 6.8). Now we make the key observation that every holomorphic arc component of ∂X_0 is contained in some G_0-orbit on ∂X_0, and thus is a holomorphic arc component of that G_0-orbit. Thus the case $\Sigma = \phi$ of Theorem 10.9, i.e. the case of G_0-orbits in the closure of X_0, is contained in the results of Wolf-Korányi [**15**].

Now we go to an arbitrary orbit $G_0(x_{\Gamma,\Sigma})$. To minimize subscripts on subscripts, we adopt the notation

$$\mathcal{P}_{\Gamma,\Sigma} = \mathcal{P}_{x_{\Gamma,\Sigma}} \quad \text{and} \quad \mathfrak{N}_{\Gamma,\Sigma} = \mathfrak{N}_{[x_{\Gamma,\Sigma}]}.$$

Recall the Cartan subalgebra $\mathfrak{H}_\Gamma \subset \mathcal{P}_{\Gamma,\Sigma}$ of $\mathfrak{g}_0$. All roots in this paragraph will be $\mathfrak{H}_\Gamma^C$-roots of $\mathfrak{g}$. Now

$$\Phi_{\Gamma,\Sigma} \text{ denotes the root system of } \mathcal{P}_{\Gamma,\Sigma},$$
$$\Lambda_{\Gamma,\Sigma} \text{ denotes the root system of } \mathfrak{N}_{\Gamma,\Sigma}.$$

From (10.3) and (10.4) we have

$$\mathrm{ad}(c_\Sigma^2)\mathfrak{H}_\Gamma = \mathfrak{H}_\Gamma \quad \text{and} \quad \tau \cdot \mathrm{ad}(c_\Sigma^2)|_{\mathfrak{H}_\Gamma} = \mathrm{ad}(c_\Sigma^2) \cdot \tau|_{\mathfrak{H}_\Gamma}.$$

This and $\mathrm{ad}(c_\Sigma^2)\mathcal{P}_{\Gamma,\phi} = \mathcal{P}_{\Gamma,\Sigma}$ imply

$$\text{(10.14a)} \qquad \mathrm{ad}(c_\Sigma^2)\Phi_{\Gamma,\phi} = \Phi_{\Gamma,\Sigma} \quad \text{and} \quad \mathrm{ad}(c_\Sigma^2)\tau\Phi_{\Gamma,\phi} = \tau\Phi_{\Gamma,\Sigma};$$

in particular

$$\text{(10.14b)} \qquad \mathrm{ad}(c_\Sigma^2)\{\Phi_{\Gamma,\phi} \cap \tau\Phi_{\Gamma,\phi}\} = \Phi_{\Gamma,\Sigma} \cap \tau\Phi_{\Gamma,\Sigma},$$

(10.14c) $$\mathrm{ad}(c_\Sigma^2)\{\Phi_{\Gamma,\phi} \cup \tau\Phi_{\Gamma,\phi}\} = \Phi_{\Gamma,\Sigma} \cup \tau\Phi_{\Gamma,\Sigma}.$$

Recall Theorem 8.5, let $\delta_{\Gamma,\Sigma}$ denote the linear form (8.6a) on $\mathfrak{H}_\Gamma^C$, and let $\mathfrak{Q}_{\Gamma,\Sigma}$ denote the parabolic subalgebra (8.6c) for $x_{\Gamma,\Sigma}$. Then (10.14a) gives us

$$\mathrm{ad}(c_\Sigma^2)\delta_{\Gamma,\phi} = \mathrm{ad}(c_\Sigma^2) \sum_{\overset{u}{\Phi}_{\Gamma,\phi}\cap\tau\overset{u}{\Phi}_{\Gamma,\phi}} \varphi = \sum_{\overset{u}{\Phi}_{\Gamma,\Sigma}\cap\tau\overset{u}{\Phi}_{\Gamma,\Sigma}} \varphi = \delta_{\Gamma,\Sigma}$$

so $\mathrm{ad}(c_\Sigma^2)\mathfrak{Q}_{\Gamma,\phi} = \mathfrak{Q}_{\Gamma,\Sigma}$. Recall Proposition 8.7 and let $\mathfrak{M}_{\Gamma,\Sigma}$ denote the linear space (8.8b) for $x_{\Gamma,\Sigma}$. In view of (10.14c) it follows that $\mathrm{ad}(c_\Sigma^2)\mathfrak{M}_{\Gamma,\phi} = \mathfrak{M}_{\Gamma,\Sigma}$. As $G_0(x_{\Gamma,\phi})$ is partially complex, Theorem 8.9 says $\mathfrak{N}_{\Gamma,\phi} = \mathfrak{M}_{\Gamma,\phi}$, so $\mathfrak{M}_{\Gamma,\phi}$ is an algebra, and thus $\mathfrak{M}_{\Gamma,\Sigma} = \mathrm{ad}(c_\Sigma^2)\mathfrak{M}_{\Gamma,\phi}$ is an algebra; now $\mathfrak{N}_{\Gamma,\Sigma} = \mathfrak{M}_{\Gamma,\Sigma}$ by Theorem 8.9. Thus

(10.15a) $$\mathfrak{N}_{\Gamma,\Sigma} = \mathrm{ad}(c_\Sigma^2)\mathfrak{N}_{\Gamma,\phi} \quad \text{and} \quad N_{\Gamma,\Sigma} = \mathrm{ad}(c_\Sigma^2)N_{\Gamma,\phi}$$

where the latter are analytic subgroups of G. Now let $Z_{\Gamma,\Sigma}$ be the kernel of the action of $N_{\Gamma,\Sigma}$ on $N_{\Gamma,\Sigma}(x_{\Gamma,\Sigma})$, $\mathfrak{Z}_{\Gamma,\Sigma}$ its Lie algebra. Then (10.15a) gives us

(10.15b) $$\mathfrak{Z}_{\Gamma,\Sigma} = \mathrm{ad}(c_\Sigma^2)\mathfrak{Z}_{\Gamma,\phi} \quad \text{and} \quad Z_{\Gamma,\Sigma} = \mathrm{ad}(c_\Sigma^2)Z_{\Gamma,\phi}.$$

As (10.4a) says $\mathrm{ad}(c_\Sigma^2)\mathfrak{Z}_{\Psi\backslash\Gamma}^- = \mathfrak{Z}_{\Psi\backslash\Gamma}^-$, (10.8a) yields

(10.15c) $$\mathrm{ad}(c_\Sigma^2)\mathfrak{G}_{\Psi\backslash\Gamma} = \mathfrak{G}_{\Psi\backslash\Gamma}.$$

We already know $\mathfrak{N}_{\Gamma,\phi} = \mathfrak{G}_{\Psi\backslash\Gamma} + \mathfrak{Z}_{\Gamma,\phi}$; thus (10.15a, b, c) combine for

(10.15d) $$\mathfrak{N}_{\Gamma,\Sigma} = \mathfrak{G}_{\Psi\backslash\Gamma} + \mathfrak{Z}_{\Gamma,\Sigma} \quad \text{and} \quad N_{\Gamma,\Sigma} = G_{\Psi\backslash\Gamma}\cdot Z_{\Gamma,\Sigma}.$$

Now (10.10c) follows. That proves part 3 of Theorem 10.9. As $\mathfrak{N}_{\Gamma,\phi}$ is the sum of the nonnegative eigenspaces of $\mathrm{ad}\{\sum_\Gamma x_{\gamma,0}\}$ on $\mathfrak{G}$, (10.15a) says that $\mathfrak{N}_{\Gamma,\Sigma}$ is the sum of the nonnegative eigenspaces of

$$\mathrm{ad}\left\{\mathrm{ad}(c_\Sigma^2)\sum_\Gamma x_{\gamma,0}\right\} = \mathrm{ad}\left(\left\{\sum_{\Gamma\backslash\Sigma} x_{\gamma,0}\right\} - \left\{\sum_{\Gamma\cap\Sigma} x_{\gamma,0}\right\}\right).$$

That proves part 4 of Theorem 10.9.

Note that c_Γ centralizes $G_{\Psi\backslash\Gamma}$. Using (10.15c), we have

$$G_{\Psi\backslash\Gamma}(x_{\Gamma,\Sigma}) = c_\Sigma^2 G_{\Psi\backslash\Gamma}(x_{\Gamma,\phi}) = c_\Gamma c_\Sigma^2 G_{\Psi\backslash\Gamma}(x_0).$$

We already know that $G_{\Psi\backslash\Gamma}(x_{\Gamma,\phi})$ is a compact totally geodesic hermitian symmetric subspace of X, rank $|\Psi\backslash\Gamma|$ with $\Psi\backslash\Gamma$ as maximal set of strongly orthogonal noncompact roots of $\mathfrak{G}_{\Psi\backslash\Gamma,0}$ relative to $\mathfrak{Z}\cap\mathfrak{G}_{\Psi\backslash\Gamma,0}$. Now the same statements follow for $G_{\Psi\backslash\Gamma}(x_{\Gamma,\Sigma})$. That proves part 1 of Theorem 10.9.

$G_{\Psi\backslash\Gamma,0}(x_{\Gamma,\Sigma})$ is open in $G_{\Psi\backslash\Gamma}(x_{\Gamma,\Sigma})$ by (10.15d) and the fact that $G_0(x_{\Gamma,\Sigma})$ is partially complex. $G_{\Psi\backslash\Gamma,0}(x_{\Gamma,\Sigma})$ is symmetric with invariant kaehler metric by Corollary 9.25. If $\Gamma\cap\Sigma=\phi$ and Σ is a (possibly empty) union of sets $(\Psi\backslash\Gamma)\cap\Delta^k$ for which $G_0^k(x_0)$ is of tube type, then $G_{\Psi\backslash\Gamma,0}$ has the same (thus compact) isotropy subgroups at $x_{\Gamma,\phi}$ and $x_{\Gamma,\Sigma}$, so the metric is positive definite. If $\Gamma\cap\Psi=\phi$ and the metric is positive definite, then $G_{\Psi\backslash\Gamma,0}$ has the same isotropy subgroups at $x_{\Gamma,\phi}$ and $x_{\Gamma,\Sigma}$, so [15, Lemma 5.6] the (-1)-eigenspace of $\mathrm{ad}(c_\Sigma^4)$ on $\mathfrak{g}_{\Psi\backslash\Gamma}$ is zero, and the last statement of [15, Theorem 5.7] says that Σ is a union of sets $(\Psi\backslash\Gamma)\cap\Delta^k$ for which $G_{\Psi\backslash\Gamma,0}^k(x_0)$ is of tube type. However, [15, Corollary 4.11] says that the latter is equivalent to $G_0^k(x_0)$ being of tube type. That proves part 2, completing the proof of Theorem 10.9. Q.E.D.

11. Hermitian symmetric spaces: compact subvarieties and Siegel domain realizations. Retain the notation and terminology of §10. We will apply our explicit analysis of the G_0-orbit structure of X to obtain certain "generalized Siegel domain" realizations of an arbitrary orbit $G_0(x_{\Gamma,\Sigma})\subset X$. Those realizations are composed of several decompositions of $G_0(x_{\Gamma,\Sigma})$ as follows.

1. α: $G_0(x_{\Gamma,\Sigma})\to G_0/N_{\Gamma,\Sigma,0}$ is the K-equivariant fibration whose fibres are the holomorphic arc components $k\cdot G_{\Psi\backslash\Gamma,0}(x_{\Gamma,\Sigma})=kc_\Gamma G_{\Psi\backslash\Gamma,0}(x_{\phi,\Sigma})$, $k\in K$, of $G_0(x_{\Gamma,\Sigma})$, and whose base is the space $G_0/N_{\Gamma,\Sigma,0}=K/(K\cap N_{\Gamma,\Sigma,0})$ of holomorphic arc components of $G_0(x_{\Gamma,\Sigma})$.

2. β: $G_0(x_{\Gamma,\Sigma})\to K(x_{\Gamma,\Sigma})$ is a K-equivariant fibration whose restriction β_k to a holomorphic arc component $k\cdot G_{\Psi\backslash\Gamma,0}(x_{\Gamma,\Sigma})$ is a holomorphic fibration of that arc component over its maximal compact subvariety $k\cdot(K\cap G_{\Psi\backslash\Gamma,0})(x_{\Gamma,\Sigma})$. Assume Γ and Σ disjoint. Then the fibre of β (and also of β_k) over $kk'(x_{\Gamma,\Sigma})$, $k'\in K\cap G_{\Psi\backslash\Gamma,0}$, is a certain hermitian symmetric space

$$\begin{aligned} kk'\cdot \overset{\Sigma}{G}_{\Psi\backslash\Gamma,0}(x_{\Gamma,\Sigma}) &= kk'c_\Gamma \overset{2}{c}_\Sigma\cdot \overset{\Sigma}{G}_{\Psi\backslash\Gamma,0}(x_0)\\ &= kk'c_\Gamma \overset{2}{c}_\Sigma\{G_{\Psi\backslash(\Gamma\cup\Sigma),0}(x_0)\times \overset{\Sigma}{G}_{\Sigma,0}(x_0)\}\end{aligned}$$

of noncompact type and rank $|\Psi\backslash\Gamma|$ where $G_{\Sigma,0}^{\Sigma}(x_0)$ denotes the tube domain part (c.f. [2]) of the domain $G_{\Sigma,0}(x_0)$. α factors through β, i.e. $\alpha\cdot\beta^{-1}$: $K(x_{\Gamma,\Sigma})\to G_0/N_{\Gamma,\Sigma,0}$ is a well-defined K-equivariant fibration.

3. $\gamma^{A,B}$: $G_0(x_{\Gamma,\Sigma})\to W\cdot G_{E',0}^{B'}(x_{\Gamma,\Sigma})$ where W is a certain subset of K and we have disjoint decompositions

$$\Sigma = A\cup A',\quad \Psi\backslash(\Gamma\cup\Sigma) = B\cup B',\quad E = A\cup B,\quad E' = A'\cup B'.$$

Given $k \in W$, the restriction $\gamma_k^{A,B}$ of $\gamma^{A,B}$ to $(\gamma^{A,B})^{-1}(k \cdot G_{E',0}^{B'}(x_{\Gamma,\Sigma}))$ is a holomorphic fibration $\gamma_k^{A,B}: k \cdot G_{\Psi\setminus\Gamma,0}^{\Sigma}(x_{\Gamma,\Sigma}) \to k \cdot G_{E',0}^{B'}(x_{\Gamma,\Sigma})$ equivalent to a certain projection that is implicit (but made explicit in the appendix to this §11) in the intrinsic realization (c.f. [15] or the appendix to this §11) of $G_{\Psi\setminus\Gamma,0}^{\Sigma}(x_0)$ as a Siegel domain of type III over its boundary component $G_{E',0}^{B'}(x_{E,\Phi})$. $\gamma^{A,B}$ is not necessarily continuous, but it is real analytic almost everywhere, and β factors through $\gamma^{A,B}$.

The fibration $\alpha: G_0(x_{\Gamma,\Sigma}) \to G_0/N_{\Gamma,\Sigma,0}$ is the subject of Theorem 8.15 with details supplied by Theorem 10.9. The fibration $\beta: G_0(x_{\Gamma,\Sigma}) \to K(x_{\Gamma,\Sigma})$ is described in Theorem 11.8 below, while $\gamma^{A,B}$ is described by Proposition 11.20 and Theorem 11.23. Proposition 11.20 is a transcription of Proposition 11.3A (A for appendix) which extends the Wolf-Korányi description [15] of the realizations of hermitian symmetric spaces as Siegel domains of type III.

Let Γ and Σ be disjoint subsets of Ψ. We decompose the real form $\mathfrak{g}_{\Psi\setminus\Gamma,0} = \mathfrak{g}_0 \cap \mathfrak{g}_{\Psi\setminus\Gamma}$ of $\mathfrak{g}_{\Psi\setminus\Gamma}$ by

$$\text{(11.1a)} \quad \mathfrak{g}_{\Psi\setminus\Gamma,0} = \mathcal{K}_{\Psi\setminus\Gamma} + \mathfrak{M}_{\Psi\setminus\Gamma,0}, \quad \mathcal{K}_{\Psi\setminus\Gamma} = \mathcal{K} \cap \mathfrak{g}_{\Psi\setminus\Gamma} \quad \text{and} \quad \mathfrak{M}_{\Psi\setminus\Gamma,0} = \mathfrak{M}_0 \cap \mathfrak{g}_{\Psi\setminus\Gamma}.$$

If $\psi \in \Psi$ then $\operatorname{ad}(c_\psi)$ has order 4 on the odd dimensional irreducible representation spaces of $\operatorname{ad}_{\mathfrak{g}}(\mathfrak{g}[\psi])$, order 8 on the even dimensional ones; so $\operatorname{ad}(c_\psi^4)$ has square 1. Now

$$\operatorname{ad}(c_\psi^4) = \operatorname{ad}(c_\psi^4)^{-1} = \tau^{-1} \cdot \operatorname{ad}(c_\psi^4) \cdot \tau$$

shows that $\operatorname{ad}(c_\psi^4)$ preserves $\mathfrak{g}_0$, and

$$\operatorname{ad}(c_\psi^4) = \operatorname{ad}(c_\psi^4)^{-1} = \sigma^{-1} \cdot \operatorname{ad}(c_\psi^4) \cdot \sigma$$

shows further that $\operatorname{ad}(c_\psi^4)$ preserves $\mathcal{K}$ and $\mathfrak{M}_0$. Define

$$\text{(11.1b)} \quad \mathfrak{g}_{\Psi\setminus\Gamma}^{\Sigma} = \{v \in \mathfrak{g}_{\Psi\setminus\Gamma} : \operatorname{ad}(c_\Sigma^4)v = v\} \quad \text{and} \quad \mathfrak{g}_{\Psi\setminus\Gamma,0}^{\Sigma} = \mathfrak{g}_0 \cap \mathfrak{g}_{\Psi\setminus\Gamma}^{\Sigma}.$$

Now $\mathfrak{g}_{\Psi\setminus\Gamma,0}^{\Sigma}$ is a real form of $\mathfrak{g}_{\Psi\setminus\Gamma}^{\Sigma}$, and in fact

$$\text{(11.1c)} \quad \begin{aligned} &\mathfrak{g}_{\Psi\setminus\Gamma,0}^{\Sigma} = \mathcal{K}_{\Psi\setminus\Gamma}^{\Sigma} + \mathfrak{M}_{\Psi\setminus\Gamma,0}^{\Sigma}, \quad \mathcal{K}_{\Psi\setminus\Gamma}^{\Sigma} = \mathcal{K} \cap \mathfrak{g}_{\Psi\setminus\Gamma}^{\Sigma} \quad \text{and} \\ &\mathfrak{M}_{\Psi\setminus\Gamma,0}^{\Sigma} = \mathfrak{M}_0 \cap \mathfrak{g}_{\Psi\setminus\Gamma}^{\Sigma}. \end{aligned}$$

Finally, we define

(11.1d) $\mathfrak{Q}_{\Psi\setminus\Gamma}^{\Sigma}$ is the (-1)-eigenspace of $\operatorname{ad}(c_\Sigma^4)$ on $\mathcal{K}_{\Psi\setminus\Gamma}$,

(11.1e) $\mathfrak{R}_{\Psi\setminus\Gamma,0}^{\Sigma}$ is the (-1)-eigenspace of $\operatorname{ad}(c_\Sigma^4)$ on $\mathfrak{M}_{\Psi\setminus\Gamma,0}$.

That gives us symmetric decompositions under $\mathrm{ad}(c_\Sigma^4)$:

(11.1f) $\mathcal{K}_{\Psi\backslash\Gamma}=\overset{\Sigma}{\mathcal{K}}_{\Psi\backslash\Gamma}+\overset{\Sigma}{\mathcal{Q}}_{\Psi\backslash\Gamma}$ and $\mathcal{G}_{\Psi\backslash\Gamma,0}=\overset{\Sigma}{\mathcal{G}}_{\Psi\backslash\Gamma,0}+(\overset{\Sigma}{\mathcal{Q}}_{\Psi\backslash\Gamma}+\overset{\Sigma}{\mathcal{R}}_{\Psi\backslash\Gamma,0})$.

Roman letters denote corresponding analytic groups.

We collect some information in a sequence of lemmas.

11.2. LEMMA. *$K_{\Psi\backslash\Gamma}=K\cap G_{\Psi\backslash\Gamma,0}$, maximal compact subgroup of $G_{\Psi\backslash\Gamma,0}$.*

11.3. LEMMA. *$K^\Sigma_{\Psi\backslash\Gamma}=K\cap G^\Sigma_{\Psi\backslash\Gamma,0}$, maximal compact subgroup of $G^\Sigma_{\Psi\backslash\Gamma,0}$. Both $K^\Sigma_{\Psi\backslash\Gamma}$ and $G^\Sigma_{\Psi\backslash\Gamma,0}$ are normalized by $c_\Gamma c_\Sigma^2$, and $K^\Sigma_{\Psi\backslash\Gamma}$ is the isotropy subgroup of $G^\Sigma_{\Gamma\backslash\Psi,0}$ both at x_0 and at $x_{\Gamma,\Sigma}$. In particular*

$$\overset{\Sigma}{G}_{\Psi\backslash\Gamma,0}(x_{\Gamma,\Sigma}) = c_\Gamma \overset{2}{c}_\Sigma \overset{\Sigma}{G}_{\Psi\backslash\Gamma,0}(x_0) \cong \overset{\Sigma}{G}_{\Psi\backslash\Gamma,0}/\overset{\Sigma}{K}_{\Psi\backslash\Gamma},$$

and that orbit is a totally geodesic hermitian symmetric subspace of noncompact type and rank $|\Psi\backslash\Gamma|$ in $c_\Gamma c_\Sigma^2 X_0$.

11.4. LEMMA. *$K^\Sigma_{\Psi\backslash\Gamma}$ is the isotropy subgroup of $K_{\Psi\backslash\Gamma}$ at $x_{\Gamma,\Sigma}$, and the orbit $K_{\Psi\backslash\Gamma}(x_{\Gamma,\Sigma})\cong K_{\Psi\backslash\Gamma}/K^\Sigma_{\Psi\backslash\Gamma}$ is a totally geodesic hermitian symmetric subspace of compact type in X.*

11.5. LEMMA. *The isotropy subgroup of $G_{\Psi\backslash\Gamma,0}$ at $x_{\Gamma,\Sigma}$ is the analytic subgroup with Lie algebra $\mathcal{K}^\Sigma_{\Psi\backslash\Gamma}+\mathcal{R}^\Sigma_{\Psi\backslash\Gamma,0}=\mathrm{ad}(c_\Gamma c_\Sigma^2)\mathcal{K}^C_{\Psi\backslash\Gamma}\cap\mathcal{G}_{\Psi\backslash\Gamma,0}$.*

11.6. LEMMA. *Define $f: K_{\Psi\backslash\Gamma}\times\mathfrak{M}^\Sigma_{\Psi\backslash\Gamma,0}\times\mathcal{R}^\Sigma_{\Psi\backslash\Gamma,0}\to G_{\Psi\backslash\Gamma,0}$ by $f(k, v_1, v_2)=k\cdot\exp v_1\cdot\exp v_2$. Then f is a diffeomorphism.*

PROOF OF LEMMAS. $G_{\Psi\backslash\Gamma,0}(x_0)\cong G_{\Psi\backslash\Gamma,0}/(K\cap G_{\Psi\backslash\Gamma,0})$ is a hermitian symmetric subspace of noncompact type and rank $|\Psi\backslash\Gamma|$ in X_0. Thus $K\cap G_{\Psi\backslash\Gamma,0}$ is connected and maximal compact in $G_{\Psi\backslash\Gamma,0}$. As $K\cap G_{\Psi\backslash\Gamma,0}$ has Lie algebra $\mathcal{K}_{\Psi\backslash\Gamma}$, now $K\cap G_{\Psi\backslash\Gamma,0}=K_{\Psi\backslash\Gamma}$ and Lemma 11.2 is proved.

Note $\mathcal{A}_{\Psi\backslash\Gamma,0}\subset\mathcal{G}^\Sigma_{\Psi\backslash\Gamma,0}\subset\mathcal{G}_{\Psi\backslash\Gamma}$ and $\mathrm{ad}(z)\mathfrak{M}^\Sigma_{\Psi\backslash\Gamma,0}=\mathfrak{M}^\Sigma_{\Psi\backslash\Gamma,0}$. It follows that $G^\Sigma_{\Psi\backslash\Gamma,0}(x_0)$ is a totally geodesic hermitian symmetric submanifold of noncompact type and rank $|\Psi\backslash\Gamma|$ in X_0. As above, now $K^\Sigma_{\Psi\backslash\Gamma}=K\cap G^\Sigma_{\Psi\backslash\Gamma,0}$ isotropy subgroup at x_0 and maximal compact subgroup of $G^\Sigma_{\Psi\backslash\Gamma,0}$.

As c_Γ centralizes $G_{\Psi\backslash\Gamma}$ it centralizes the subgroups $K^\Sigma_{\Psi\backslash\Gamma}$ and $G^\Sigma_{\Psi\backslash\Gamma,0}$. As $\mathrm{ad}(c_\Sigma^2)$ preserves $\mathcal{G}_{\Psi\backslash\Gamma}$ and commutes with $\mathrm{ad}(c_\Sigma^4)$, it preserves $\mathcal{G}^\Sigma_{\Psi\backslash\Gamma}$. Now $\mathrm{ad}(c_\Sigma^2)$ preserves $\mathcal{G}^\Sigma_{\Psi\backslash\Gamma,0}$ because

$$\mathrm{ad}(c_\Sigma^2) = \mathrm{ad}(c_\Sigma^2)^{-1} = \tau^{-1}\cdot\mathrm{ad}(c_\Sigma^2)\cdot\tau \quad \text{on } \overset{\Sigma}{\mathcal{G}}_{\Psi\backslash\Gamma}.$$

Furthermore

$$\mathrm{ad}(c_\Sigma^2) = \mathrm{ad}(c_\Sigma^2)^{-1} = \sigma^{-1}\cdot\mathrm{ad}(c_\Sigma^2)\cdot\sigma \quad \text{on } \overset{\Sigma}{\mathcal{G}}_{\Psi\backslash\Gamma,0},$$

so $\mathrm{ad}(c_\Sigma^2)$ preserves both $\mathcal{K}_\Psi^\Sigma\backslash_\Gamma$ and $\mathfrak{M}_\Psi^\Sigma\backslash_\Gamma$. Now $G_\Psi^\Sigma\backslash_{\Gamma,0}$ and $K_\Psi^\Sigma\backslash_\Gamma$ are centralized by $\mathrm{ad}(c_\Gamma)$ and normalized by $\mathrm{ad}(c_\Sigma^2)$, so they are normalized by $\mathrm{ad}(c_\Gamma c_\Sigma^2)$. As $K_\Psi^\Sigma\backslash_\Gamma$ is the isotropy subgroup of $G_\Psi^\Sigma\backslash_{\Gamma,0}$ at x_0, it thus is the isotropy subgroup at $c_\Gamma c_\Sigma^2 x_0 = x_{\Gamma,\Sigma}$. Lemma 11.3 is proved.

The symmetry of X at $x_{\Gamma,\Sigma}$ is $\mathrm{ad}(c_\Gamma c_\Sigma^2)\cdot\sigma\cdot\mathrm{ad}(c_\Gamma c_\Sigma^2)^{-1} = \mathrm{ad}(c_\Gamma c_\Sigma^2)^2\cdot\sigma = \mathrm{ad}(c_\Gamma^2)\cdot\mathrm{ad}(c_\Sigma^4)\cdot\sigma$. It acts on $G_\Psi\backslash_{\Gamma,0}$ as $\mathrm{ad}(c_\Sigma^4)\cdot\sigma$. Thus it normalizes $K_\Psi\backslash_\Gamma$, so $K_\Psi\backslash_\Gamma(x_{\Gamma,\Sigma})$ is a compact totally geodesic symmetric subspace of X. It is a complex submanifold of X by Lemma 5.1, so it is hermitian symmetric. The isotropy subgroup of $K_\Psi\backslash_\Gamma$ at $x_{\Gamma,\Sigma}$ contains $K_\Psi^\Sigma\backslash_\Gamma$, has Lie algebra $\mathcal{K}_\Psi^\Sigma\backslash_\Gamma$ that contains the center of $\mathcal{K}_\Psi\backslash_\Gamma$; thus $K_\Psi^\Sigma\backslash_\Gamma(x_0)$ is of compact type and necessarily simply connected; now $K_\Psi^\Sigma\backslash_\Gamma$ is the isotropy subgroup of $K_\Psi\backslash_\Gamma$ at $x_{\Gamma,\Sigma}$. Lemma 11.4 is proved.

The isotropy subalgebra of $\mathcal{G}_\Psi\backslash_{\Gamma,0}$ at $c_\Gamma x_0$ is $\mathcal{K}_\Psi\backslash_\Gamma$, so the isotropy subalgebra at $c_\Sigma^2(c_\Gamma x_0) = x_{\Gamma,\Sigma}$ is $\mathrm{ad}(c_\Sigma^2)\mathcal{K}_\Psi^C\backslash_\Gamma \cap \mathcal{G}_\Psi\backslash_{\Gamma,0}$. Now $\mathrm{ad}(c_\Sigma^2)$ preserves $\mathcal{K}_\Psi^\Sigma\backslash_\Gamma$, as seen above, and sends $\mathcal{Q}_\Psi^\Sigma\backslash_\Gamma$ to $i\mathcal{R}_\Psi^\Sigma\backslash_\Gamma$ by [**15**, Lemma 5.6]. Thus $\mathcal{G}_\Psi\backslash_{\Gamma,0}$ has isotropy subalgebra $\mathcal{K}_\Psi^\Sigma\backslash_\Gamma + \mathcal{R}_\Psi^\Sigma\backslash_{\Gamma,0}$ at $x_{\Gamma,\Sigma}$. The isotropy subgroup of $G_\Psi\backslash_{\Gamma,0}$ at $x_{\Gamma,\Sigma}$ is connected by Theorem 5.4, so it must be the analytic subgroup for $\mathcal{K}_\Psi^\Sigma\backslash_\Gamma + \mathcal{R}_\Psi^\Sigma\backslash_{\Gamma,0}$. Lemma 11.5 is proved.

Define φ: $\mathfrak{M}_\Psi\backslash_{\Gamma,0} \to G_\Psi\backslash_{\Gamma,0}(x_0)$ by $\varphi(v) = \exp(-v_2)\cdot\exp(-v_1)\cdot x_0$ where $v = v_1 + v_2$, $v_1 \in \mathfrak{M}_\Psi^\Sigma\backslash_{\Gamma,0}$ and $v_2 \in \mathcal{R}_\Psi^\Sigma\backslash_{\Gamma,0}$. We view φ as a differentiable map from the euclidean space $\mathfrak{M}_\Psi\backslash_{\Gamma,0}$ to the symmetric space $G_\Psi\backslash_{\Gamma,0}(x_0)$ of noncompact type. We will prove $\|\varphi_*\xi\| \geqq \|\xi\|$ for every tangent vector ξ to $\mathfrak{M}_\Psi\backslash_{\Gamma,0}$. Then [**13**, Theorem 1] φ is surjective and covering, hence a diffeomorphism because $G_\Psi\backslash_{\Gamma,0}(x_0)$ is simply connected, and it follows that from $f(k, v_1, v_2)^{-1}x_0 = \varphi(v_1+v_2)$ that f defines a diffeomorphism of $K \times \mathfrak{M}_\Psi^\Sigma\backslash_{\Gamma,0} \times \mathcal{R}_\Psi^\Sigma\backslash_{\Gamma,0}$ onto $G_\Psi\backslash_{\Gamma,0}$.

Let $v_1 \in \mathfrak{M}_\Psi^\Sigma\backslash_{\Gamma,0}$ be fixed, $v_2 \in \mathcal{R}_\Psi^\Sigma\backslash_{\Gamma,0}$ variable. Then $\varphi(v_1+v_2) = \exp(-v_2)\ \exp(-v_1)x_0 = \exp(-v_1)\cdot\exp(v_1)\cdot\exp(-v_2)\cdot\exp(-v_1)x_0 = \exp(-v_1)\cdot\{\mathrm{ad}(\exp v_1)\cdot\exp(-v_2)\}x_0 = \exp(-v_1)\cdot\{\exp(-\mathrm{ad}(\exp v_1)\cdot v_2)\}x_0$. Now $\exp(-v_1)$ is an isometry on $G_\Psi\backslash_{\Gamma,0}(x_0)$, and

$$\mathrm{ad}(\exp v_1)\cdot v_2 = \exp(\mathrm{ad}\, v_1)\cdot v_2 = \sum_{n\geqq 0} \frac{1}{n!}\,\mathrm{ad}(v_1)^n\cdot v_2.$$

Note $\mathrm{ad}(v_1)^n v_2 \in \mathcal{R}_\Psi^\Sigma\backslash_{\Gamma,0}$ for n even, $\in \mathcal{Q}_\Psi^\Sigma\backslash_\Gamma$ for n odd; so

$$\mathrm{ad}(\exp v_1)\cdot v_2 = \sum_{n\geqq 0} \frac{1}{(2n)!}\,\mathrm{ad}(v_1)^{2n}\cdot v_2 + \sum_{n\geqq 0} \frac{1}{(2n+1)!}\,\mathrm{ad}(v_1)^{2n+1}\cdot v_2$$

with the first sum in $\mathcal{R}_\Psi^\Sigma\backslash_{\Gamma,0}$ and the second in $\mathcal{Q}_\Psi^\Sigma\backslash_\Gamma$. As $\mathrm{ad}(v_1)$ is symmetric, so $\mathrm{ad}(v_1)^2$ is positive semidefinite, now

$$\|(\operatorname{ad}(\exp v_1)\cdot v_2)_{\mathfrak{M}_{\Psi\setminus\Gamma,0}}\| \geqq \|v_2\|.$$

Thus φ does not decrease lengths of tangent vectors on $v_1+\mathfrak{R}^{\Sigma}_{\Psi\setminus\Gamma,0}$.

Let $v_2\in\mathfrak{R}^{\Sigma}_{\Psi\setminus\Gamma,0}$ be fixed, $v_1\in\mathfrak{M}^{\Sigma}_{\Psi\setminus\Gamma,0}$ variable. Then $\varphi(v_1+v_2)=\exp(-v_2)\exp(-v_1)x_0$ is the composition of -1 on $\mathfrak{M}^{\Sigma}_{\Psi\setminus\Gamma,0}$, exponential map in the symmetric space $G^{\Sigma}_{\Psi\setminus\Gamma,0}(x_0)$ of noncompact type, and left translation by $\exp(-v_2)$. As $G^{\Sigma}_{\Psi\setminus\Gamma,0}(x_0)$ has curvature $\leqq 0$, now φ does not decrease lengths of tangent vectors to $\mathfrak{M}^{\Sigma}_{\Psi\setminus\Gamma,0}+v_2$.

Fix $v_1\in\mathfrak{M}^{\Sigma}_{\Psi\setminus\Gamma,0}$ and let w_1 be variable there; fix $v_2\in\mathfrak{R}^{\Sigma}_{\Psi\setminus\Gamma,0}$ and let w_2 be variable there. Now

$$\varphi(v_1+\mathfrak{R}^{\Sigma}_{\Psi\setminus\Gamma,0}) = \{\exp(-v_2-w_2)\cdot\exp(-v_1)\cdot x_0\},$$
$$\varphi(\mathfrak{M}^{\Sigma}_{\Psi\setminus\Gamma,0}+v_2) = \{\exp(-v_2)\cdot\exp(-v_1-w_1)\cdot x_0\}.$$

Those manifolds intersect at $\varphi(v_1+v_2)=\exp(-v_2)\cdot\exp(-v_1)\cdot x_0$, and we want to show the intersection there to be orthogonal. For that, translate by the isometry $\exp(v_1)$ sending $\exp(-v_2)\exp(-v_1)x_0$ to $\exp(-\operatorname{ad}(\exp v_1)\cdot v_2)x_0\in\varphi(\mathfrak{R}^{\Sigma}_{\Psi\setminus\Gamma,0})$, and note first that $\exp(v_1)\varphi(\mathfrak{R}^{\Sigma}_{\Psi\setminus\Gamma,0})$ consists of all

$$\exp(-\operatorname{ad}(\exp v_1)\cdot v_2-\operatorname{ad}(\exp v_1)w_2)\cdot x_0$$

and second that $\exp(v_1)\varphi(\mathfrak{M}^{\Sigma}_{\Psi\setminus\Gamma,0})$ consists of all

$$\exp(-\operatorname{ad}(\exp v_1)\cdot v_2)\exp(-w_1)\cdot x_0.$$

Thus the question of orthogonality is reduced to the case $v_1=0$, where it follows from first

$$\exp(v_2)\varphi(\mathfrak{R}^{\Sigma}_{\Psi\setminus\Gamma,0}) = \{\exp(-w_2)x_0\} = \varphi(\mathfrak{R}^{\Sigma}_{\Psi\setminus\Gamma,0})$$

because $\exp(v_2)\exp(-v_2-w_2)\equiv\exp(-w_2) \bmod K^{\Sigma}_{\Psi\setminus\Gamma}$, and second from

$$\exp(v_2)\varphi(\mathfrak{M}^{\Sigma}_{\Psi\setminus\Gamma,0}+v_2) = \{\exp(-w_1)x_0\} = \varphi(\mathfrak{M}^{\Sigma}_{\Psi\setminus\Gamma,0}).$$

Now let ξ be tangent to $\mathfrak{M}_{\Psi\setminus\Gamma,0}$ at v. Decompose $\xi=\xi_1+\xi_2$ along $\mathfrak{M}_{\Psi\setminus\Gamma,0}=\mathfrak{M}^{\Sigma}_{\Psi\setminus\Gamma,0}+\mathfrak{R}^{\Sigma}_{\Psi\setminus\Gamma,0}$. We have proved

$$\|\varphi_*\xi_1\|\geqq\|\xi_1\|,\qquad \|\varphi_*\xi_2\|\geqq\|\xi_2\|,\qquad \varphi_*\xi_1\perp\varphi_*\xi_2.$$

Thus

$$\|\varphi_*\xi\|^2=\|\varphi_*\xi_1+\varphi_*\xi_2\|^2=\|\varphi_*\xi_1\|^2+\|\varphi_*\xi_2\|^2\geqq\|\xi_1\|^2+\|\xi_2\|^2=\|\xi\|^2.$$

That completes the proof of Lemma 11.6. Q.E.D.

We need notation for the factorization of Lemma 11.6. The notation will be, for every $g\in G_{\Psi\setminus\Gamma,0}$, that

(11.7) $g = g_K \cdot g_M \cdot g_R$ where $g_K \in K_{\Psi \backslash \Gamma}, g_M \in \exp \mathfrak{M}^{\Sigma}_{\Psi \backslash \Gamma, 0}, g_R \in \exp \mathfrak{R}^{\Sigma}_{\Psi \backslash \Gamma, 0}$.

11.8. THEOREM. *Let* Γ *and* Σ *be disjoint subsets of* Ψ. *Given* $k \in K$, *define*

$$\beta_k : k \cdot G_{\Psi \backslash \Gamma, 0}(x_{\Gamma, \Sigma}) \to k \cdot K_{\Psi \backslash \Gamma}(x_{\Gamma, \Sigma}) \quad by\ \beta_k(kgx_{\Gamma, \Sigma}) = kg_K(x_{\Gamma, \Sigma}).$$

Now define

$$\beta : G_0(x_{\Gamma, \Sigma}) \to K(x_{\Gamma, \Sigma}) \quad by\ \beta |_{k \cdot G_{\Psi \backslash \Gamma, 0}(x_{\Gamma, \Sigma})} = \beta_k.$$

1. $\beta_k : k \cdot G_{\Psi \backslash \Gamma, 0}(x_{\Gamma, \Sigma}) \to k \cdot K_{\Psi \backslash \Gamma}(x_{\Gamma, \Sigma})$ *is a well-defined holomorphic bre bundle with*

(1a) *structure group: the connected reductive complex Lie group* $K^{\Sigma, C}_{\Psi \backslash \Gamma}$;

(1b) *total space: the homomorphic arc component* $k \cdot G_{\Psi \backslash \Gamma, 0}(x_{\Gamma, \Sigma})$; *of* $G_0(x_{\Sigma, \Gamma})$;

(1c) *base space: the maximal compact subvariety* $k \cdot K_{\Psi \backslash \Gamma}(x_{\Gamma, \Sigma})$ *which is a compact totally geodesic hermitian symmetric submanifold of* X;

(1d) *fibre over* $kk'(x_{\Gamma, \Sigma})$, $k' \in K_{\Psi \backslash \Gamma}$: *the totally geodesic hermitian symmetric submanifold*

$$\begin{aligned} kk' \cdot G^{\Sigma}_{\Psi \backslash \Gamma, 0}(x_{\Gamma, \Sigma}) &= kk' c_\Gamma c^2_\Sigma \cdot G^{\Sigma}_{\Psi \backslash \Gamma, 0}(x_0) \\ &= kk' c_\Gamma c^2_\Sigma \{ G_{\Psi \backslash (\Gamma \cup \Sigma), 0}(x_0) \times G^{\Sigma}_{\Sigma, 0}(x_0) \} \end{aligned}$$

of noncompact type and rank $|\Psi \backslash \Gamma|$ *in* $kk' c_\Gamma c^2_\Sigma \cdot G_0(x_0)$.

2. *Let* ν *be the holomorphic normal bundle to* $k \cdot K_{\Psi \backslash \Gamma}(x_{\Gamma, \Sigma})$ *in* $k \cdot G_{\Psi \backslash \Gamma, 0}(x_{\Gamma, \Sigma})$. *Then* ν *is the homogeneous holomorphic vector bundle over the hermitian symmetric coset space* $K_{\Psi \backslash \Gamma} / K^{\Sigma}_{\Psi \backslash \Gamma} \cong k \cdot K^{\Sigma}_{\Psi \backslash \Gamma}(x_{\Gamma, \Sigma})$ *defined by the representation of* $K^{\Sigma}_{\Psi \backslash \Gamma}$ *on the holomorphic tangent space of* $G^{\Sigma}_{\Psi \backslash \Gamma, 0} / K^{\Sigma}_{\Psi \backslash \Gamma} \cong G^{\Sigma}_{\Psi \backslash \Gamma, 0}(x_{\Gamma, \Sigma})$, *and the bundle* β_k *is holomorphically fibre-equivalent to a relatively compact tubular neighborhood of the zero-section of* ν.

3. *Let* $\det \nu$ *denote the determinant bundle of the holomorphic normal bundle* ν. *If* $\dim_C k \cdot K_{\Psi \backslash \Gamma}(x_{\Gamma, \Sigma}) > 0$, *then* $\det \nu$ *is a positive holomorphic line bundle (in the sense of K. Kodaira), and the bundles* β_k, ν *and* $\det \nu$ *are all topologically nontrivial. Furthermore, the following conditions are equivalent*:

(3a) β_k *is holomorphically trivial*;

(3b) β_k *is topologically trivial*;

(3c) β_k *has base reduced to a point*;

(3d) $G_{\Psi \backslash \Gamma, 0}(x_{\Gamma, \Sigma})$ *is hermitian symmetric*;

(3e) $\mathrm{ad}(c_\Sigma)$ *has order* 4 *on* $G_{\Psi\setminus\Gamma}$; *and*

(3f) *If* Δ^m *are the root systems of the simple ideals of* $\mathcal{G}$, *then for each* m, *either* $\Sigma\cap\Delta^m=\phi$, *or* $(\Psi\cap\Delta^m)\subset(\Gamma\cup\Sigma)$ *with* $G_{\Psi\cap\Delta^m,0}(x_0)$ *of tube type.*

4. β: $G_0(x_{\Gamma,\Sigma})\to K(x_{\Gamma,\Sigma})$ *is a well-defined real analytic fibre bundle whose fibres and structure group coincide with those of the restrictions* β_k *to the holomorphic arc components* $k\cdot G_{\Psi\setminus\Gamma,0}(x_{\Gamma,\Sigma})$ *of* $G_0(x_{\Gamma,\Sigma})$. *The bundle* β *is real analytically equivalent to the homogeneous complex vector bundle over* $K(x_{\Gamma,\Sigma})$ *with fibre equal to the holomorphic tangent space of* $G^{\Sigma}_{\Psi\setminus\Gamma,0}(x_{\Gamma,\Sigma})$. *The following conditions are equivalent*:

(4a) β *is real analytically trivial*;

(4b) β *is topologically trivial*;

(4c) *for each irreducible factor of* X, *the corresponding factor of* β *has base or fibre reduced to a point*; *and*

(4d) *if* Δ^m *are the root systems for the simple ideals of* $\mathcal{G}$, *then for each* m, *either*

(4di) $\Gamma\cap\Delta^m=\phi=\Sigma\cap\Delta^m$, *or*

(4dii) $\Gamma\cap\Delta^m=\Psi\cap\Delta^m$ *and* $\Sigma\cap\Delta^m=\phi$, *or*

(4diii) $\Gamma\cap\Delta^m=\phi$ *and* $\Sigma\cap\Delta^m=\Psi\cap\Delta^m$ *with the bounded domain* $G_{\Psi\cap\Delta^m,0}(x_0)$ *of tube type.*

PROOF. Fix $k\in K$. Then β_k is well defined by the factorization (11.7) resulting from Lemma 11.6. The total space $k\cdot G_{\Psi\setminus\Gamma,0}(x_{\Gamma,\Sigma})$ is a holomorphic arc component of $G_0(x_{\Gamma,\Sigma})$ by Theorem 10.9, and the base space $kK_{\Psi\setminus\Gamma}(x_{\Gamma,\Sigma})\cong K_{\Psi\setminus\Gamma}/K^{\Sigma}_{\Psi\setminus\Gamma}$ is a totally geodesic hermitian symmetric submanifold of X by Lemma 11.4. That realizes β_k as a real analytic fibre bundle with structure group $K^{\Sigma}_{\Psi\setminus\Gamma}$; now we may view $K^{\Sigma,C}_{\Psi\setminus\Gamma}$ as the structure group.

Let $k'\in K_{\Psi\setminus\Gamma}$. The β_k-fibre over $kk'(x_{\Gamma,\Sigma})$ consists of all $kg(x_{\Gamma,\Sigma})$ such that $kk'(x_{\Gamma,\Sigma})=kg_K(x_{\Gamma,\Sigma})$. The latter condition is $k'(x_{\Gamma,\Sigma})=g_K(x_{\Gamma,\Sigma})$, which is $g_K\in k'K^{\Sigma}_{\Psi\setminus\Gamma}$ by Lemma 11.4. Thus the fibre consists of all $kk'k_1g_Mg_Rx_{\Gamma,\Sigma}$ with $g_M\in\exp\mathfrak{M}^{\Sigma}_{\Psi\setminus\Gamma,0}$ and $g_R\in\exp\mathfrak{R}^{\Sigma}_{\Psi\setminus\Gamma,0}$ Lemma 11.5 says $g_Rx_{\Gamma,\Sigma}=x_{\Gamma,\Sigma}$, and we have $k_1g_Mx_{\Gamma,\Sigma}=\mathrm{ad}(k_1)g_Mx_{\Gamma,\Sigma}$ which is in $\exp(\mathfrak{M}^{\Sigma}_{\Psi\setminus\Gamma,0})(x_{\Gamma,\Sigma})$. Thus the fibre consists of all $kk'g_Mx_{\Gamma,\Sigma}$, and now Lemma 11.3 says that it is the totally geodesic hermitian symmetric subspace $kk'G^{\Sigma}_{\Psi\setminus\Gamma,0}(x_{\Gamma,\Sigma})=kk'c_\Gamma c^2_\Sigma G^{\Sigma}_{\Psi\setminus\Gamma,0}(x_0)$ of noncompact type and rank $|\Psi\setminus\Gamma|$ in $kk'c_\Gamma c^2_\Sigma X_0$.

In the notation of [15], it is shown [15, §5.4] that $\mathcal{G}^{\Psi\setminus\Sigma}=\{[\mathcal{P}_1^{\Psi\setminus\Sigma},\mathcal{P}_1^{\Psi\setminus\Sigma}]+\mathcal{P}_1^{\Psi\setminus\Sigma}\}\oplus\mathcal{L}_2^{\Psi\setminus\Sigma}$ where the second ideal is in $\mathcal{K}$, it is shown [15, Lemma 4.4] that $\mathcal{P}_1^{\Psi\setminus\Sigma}=\mathcal{P}_{\Psi\setminus\Sigma}+\mathcal{P}_{\Sigma,1}$, and it is shown [15, §6.6] that $[\mathcal{G}_{\Psi\setminus\Sigma},\mathcal{G}_{\Sigma,1}]=0$. The

$$\mathcal{G}^{\Psi\setminus\Sigma} \text{ of [15] is our } \mathrm{ad}\left(\exp\frac{\pi}{4}z\right)\{\mathcal{G}_u \cap \mathcal{G}^{\Sigma}\},$$

$$\mathcal{P}_1{}^{\Psi\setminus\Sigma} \text{ of [15] is our } \mathrm{ad}\left(\exp\frac{\pi}{4}z\right)\{\mathfrak{M}_u \cap \mathcal{G}^{\Sigma}\},$$

$$\mathcal{P}_{\Psi\setminus\Sigma} \text{ of [15] is our } \mathrm{ad}\left(\exp\frac{\pi}{4}z\right)\{\mathfrak{M}_u \cap \mathcal{G}_{\Psi\setminus\Sigma}\},$$

$$\mathcal{P}_{\Sigma,1} \text{ of [15] is our } \mathrm{ad}\left(\exp\frac{\pi}{4}z\right)\{\mathfrak{M}_u \cap \mathcal{G}_{\Sigma}^{\Sigma}\},$$

$$\mathcal{G}_{\Psi\setminus\Sigma} \text{ of [15] is our } \mathrm{ad}\left(\exp\frac{\pi}{4}z\right)\{\mathcal{G}_u \cap \mathcal{G}_{\Psi\setminus\Sigma}\},$$

$$\mathcal{G}_{\Sigma,1} \text{ of [15] is defined as } [\mathcal{P}_{\Sigma,1}, \mathcal{P}_{\Sigma,1}] + \mathcal{P}_{\Sigma,1}.$$

Using that dictionary we conclude that

$$\mathcal{G}_{\Psi}^{\Sigma} = \mathcal{G}_{\Psi\setminus\Sigma} \oplus \mathcal{G}_{\Sigma}^{\Sigma} \oplus \mathcal{L}_{\Sigma}, \qquad \mathcal{L}_{\Sigma} \subset \mathcal{K}^{C}.$$

Taking intersection with $\mathcal{G}_{\Psi\setminus\Gamma}$, now

$$\mathcal{G}_{\Psi\setminus\Gamma}^{\Sigma} = \mathcal{G}_{\Psi\setminus(\Gamma\cup\Sigma)} \oplus \mathcal{G}_{\Sigma}^{\Sigma} \oplus \mathcal{L}_{\Gamma,\Sigma}, \qquad \mathcal{L}_{\Gamma,\Sigma} \subset \mathcal{K}^{C}.$$

All three ideals are τ-stable. Thus

$$\mathcal{G}_{\Psi\setminus\Gamma,0}^{\Sigma} = \mathcal{G}_{\Psi\setminus(\Gamma\cup\Sigma),0} \oplus \mathcal{G}_{\Sigma,0}^{\Sigma} \oplus \mathcal{L}_{\Gamma,\Sigma,0}, \qquad \mathcal{L}_{\Gamma,\Sigma,0} \subset \mathcal{K}.$$

It follows that

$$G_{\Psi\setminus\Gamma,0}^{\Sigma}(x_0) = G_{\Psi\setminus(\Gamma\cup\Sigma),0}(x_0) \times G_{\Sigma,0}^{\Sigma}(x_0).$$

We have proved every part of (1) except for the fact that the bundle β_k is holomorphic. However we have seen that its total space, base and fibres are complex submanifolds of X, and that its structure group is a complex Lie group acting holomorphically on the typical fibre, so we need only demonstrate the existence of holomorphic transition functions. Thus, to complete the proof of (1) it suffices to complete the proof of (2).

$G_{\Psi\setminus\Gamma,0}(x_0)$ has holomorphic tangent space $\mathfrak{M}_{\Psi\setminus\Gamma}^{-} = \mathfrak{M}^{-} \cap \mathfrak{M}_{\Psi\setminus\Gamma,0}^{C}$ at x_0, and $\mathfrak{M}_{\Psi\setminus\Gamma}^{\Sigma-} = \mathfrak{M}^{-} \cap \mathfrak{M}_{\Psi\setminus\Gamma,0}^{\Sigma C}$ is the subspace that is the holomorphic tangent space to $G_{\Psi\setminus\Gamma,0}^{\Sigma}(x_0)$. Define $\mathfrak{R}_{\Psi\setminus\Gamma}^{\Sigma-} = \mathfrak{M}^{-} \cap \mathfrak{R}_{\Psi\setminus\Gamma,0}^{\Sigma C}$, so $\mathfrak{M}_{\Psi\setminus\Gamma}^{-} = \mathfrak{M}_{\Psi\setminus\Gamma}^{\Sigma-} + \mathfrak{R}_{\Psi\setminus\Gamma}^{\Sigma-}$. Recalling that $\mathrm{ad}(c_{\Sigma}^{2})$ interchanges $\mathfrak{R}_{\Psi\setminus\Gamma,0}^{\Sigma C}$ with $\mathfrak{Q}_{\Psi\setminus\Gamma}^{\Sigma C}$ and noting that $\mathrm{ad}(c_{\Sigma}^{4})\mathfrak{R}_{\Psi\setminus\Gamma}^{\Sigma-} = \mathfrak{R}_{\Psi\setminus\Gamma}^{\Sigma-}$, we define $\mathfrak{Q}_{\Psi\setminus\Gamma}^{\Sigma-} = \mathrm{ad}(c_{\Sigma}^{2})\mathfrak{R}_{\Psi\setminus\Gamma}^{\Sigma-}$; then $\mathrm{ad}(c_{\Sigma}^{2})\mathfrak{M}_{\Psi\setminus\Gamma}^{-} = \mathrm{ad}(c_{\Sigma}^{2})\mathfrak{M}_{\Psi\setminus\Gamma}^{\Sigma-} + \mathfrak{Q}_{\Psi\setminus\Gamma}^{\Sigma-}$. Now Lemmas 11.3, 11.4 and 11.5 say that, at $x_{\Gamma,\Sigma}$,

$$G_{\Psi\backslash\Gamma,0}(x_{\Gamma,\Sigma}) \text{ has holomorphic tangent space } \operatorname{ad}(c_\Sigma^2)\mathfrak{M}_{\Psi\backslash\Gamma}^{\Sigma-} + \mathfrak{Q}_{\Psi\backslash\Gamma}^{\Sigma-},$$
$$G^{\Sigma}_{\Psi\backslash\Gamma,0}(x_{\Gamma,\Sigma}) \text{ has holomorphic tangent space } \operatorname{ad}(c_\Sigma^2)\mathfrak{M}_{\Psi\backslash\Gamma}^{\Sigma-},$$
$$K_{\Psi\backslash\Gamma}(x_{\Gamma,\Sigma}) \text{ has holomorphic tangent space } \mathfrak{Q}_{\Psi\backslash\Gamma}^{\Sigma-}.$$

Now the holomorphic normal space of $K_{\Psi\backslash\Gamma}(x_{\Gamma,\Sigma})$ in $G_{\Psi\backslash\Gamma,0}(x_{\Gamma,\Sigma})$ at $x_{\Gamma,\Sigma}$ is $\operatorname{ad}(c_\Sigma^2)\mathfrak{M}_{\Psi\backslash\Gamma}^{\Sigma-}$, and $K_{\Psi\backslash\Gamma}^{\Sigma C}$ preserves it and acts there by its representation on the holomorphic tangent space to $G^{\Sigma}_{\Psi\backslash\Gamma,0}(x_{\Gamma,\Sigma})$. That identifies the holomorphic normal bundle ν of $k\cdot K_{\Psi\backslash\Gamma}(x_{\Gamma,\Sigma})$ in $k\cdot G_{\Psi\backslash\Gamma}(x_{\Gamma,\Sigma})$ with the homogeneous vector bundle over $K_{\Psi\backslash\Gamma}/K^{\Sigma}_{\Psi\backslash\Gamma}$ with fibre $\operatorname{ad}(c_\Sigma^2)\mathfrak{M}_{\Psi\backslash\Gamma}^{\Sigma-}$ that has $K^{\Sigma}_{\Psi\backslash\Gamma}$-module structure from ad_G. The Harish-Chandra realization of $G^{\Sigma}_{\Psi\backslash\Gamma,0}(x_{\Gamma,\Sigma})$ as a bounded domain in $\operatorname{ad}(c_\Sigma^2)\mathfrak{M}_{\Psi\backslash\Gamma}^{\Sigma-}$ identifies β_k with a relatively compact open sub-bundle of ν. Now (2) is proved, and that completes the proof of (1).

Let κ be the canonical (holomorphic) line bundle over $k\cdot K_{\Psi\backslash\Gamma}(x_{\Gamma,\Sigma})$. It is the holomorphic line bundle whose sections are the $(d, 0)$-forms, $d=\dim_C k\cdot K_{\Psi\backslash\Gamma}(x_{\Gamma,\Sigma})$. Thus its dual κ^* has fibre $\Lambda^d\mathfrak{Q}_{\Psi\backslash\Gamma}^{\Sigma-}$ where of course $d=\dim_C \mathfrak{Q}_{\Psi\backslash\Gamma}^{\Sigma-}$. Let δ denote the representation of $K^{\Sigma}_{\Psi\backslash\Gamma}$ on the fibre of κ, so its dual δ^* is the representation on $\Lambda^d\mathfrak{Q}_{\Psi\backslash\Gamma}^{\Sigma-}$. Define $z_{\Psi\backslash\Gamma}=(i/2)\sum_{\Psi\backslash\Gamma} h_\psi \in \mathfrak{K}^{\Sigma}_{\Psi\backslash\Gamma}$. Then [15, Lemma 6.2] $\operatorname{ad}(z_{\Psi\backslash\Gamma})$ acts on $\mathfrak{R}_{\Psi\backslash\Gamma}^{\Sigma-}$ as scalar multiplication by $i/2$. As $\operatorname{ad}(c_\Sigma^2)$ preserves $\mathfrak{K}^{\Sigma}_{\Psi\backslash\Gamma}$ and interchanges $\mathfrak{R}_{\Psi\backslash\Gamma}^{\Sigma-}$ with $\mathfrak{Q}_{\Psi\backslash\Gamma}^{\Sigma-}$, now $\operatorname{ad}(c_\Sigma^2)(2z_{\Psi\backslash\Gamma})$ is an element of $\mathfrak{K}^{\Sigma}_{\Psi\backslash\Gamma}$ that acts on $\mathfrak{Q}_{\Psi\backslash\Gamma}^{\Sigma-}$ under $\operatorname{ad}_{\mathfrak{G}}$ as scalar multiplication by i. Thus

$$z^{\Sigma}_{\Psi\backslash\Gamma} = \operatorname{ad}(c_\Sigma^2)(2z_{\Psi\backslash\Gamma}) \text{ is central in } \mathfrak{K}^{\Sigma}_{\Psi\backslash\Gamma}$$

and

$$\operatorname{ad}(z^{\Sigma}_{\Psi\backslash\Gamma})\Big|_{\mathfrak{Q}^{\Sigma C}_{\Psi\backslash\Gamma}} \text{ is the almost complex structure of } k\cdot K_{\Psi\backslash\Gamma}(x_{\Gamma,\Sigma}).$$

Let $\mathcal{E}\subset\mathfrak{K}^{\Sigma}_{\Psi\backslash\Gamma}$ denote the Lie algebra of the kernel of the action of $K_{\Psi\backslash\Gamma}$ on $K_{\Psi\backslash\Gamma}(x_{\Gamma,\Sigma})\cong K_{\Psi\backslash\Gamma}/K^{\Sigma}_{\Psi\backslash\Gamma}$. Now δ^* can be described by: $\delta^*(z_{\Psi\backslash\Gamma})=\mathrm{id}$, $\delta^*(\mathcal{E})=0$, and δ^* annihilates the orthocomplement of the image of $z^{\Sigma}_{\Psi\backslash\Gamma}$ in $\mathfrak{K}^{\Sigma}_{\Psi\backslash\Gamma}/\mathcal{E}$. Thus we describe κ as the homogeneous line bundle on $K_{\Psi\backslash\Gamma}/K^{\Sigma}_{\Psi\backslash\Gamma}$ induced by the character δ on $K^{\Sigma}_{\Psi\backslash\Gamma}$ such that $\delta(z^{\Sigma}_{\Psi\backslash\Gamma})=-\mathrm{id}$, $\delta(\mathcal{E})=0$, and δ annihilates the orthocomplement of the image of $z^{\Sigma}_{\Psi\backslash\Gamma}$ in $\mathfrak{K}^{\Sigma}_{\Psi\backslash\Gamma}/\mathcal{E}$.

Let ν over $k\cdot K_{\Psi\backslash\Gamma}(x_{\Gamma,\Sigma})$ denote the holomorphic normal bundle in $k\cdot G_{\Psi\backslash\Gamma}(x_{\Gamma,\Sigma})$. We will describe the determinant bundle $\det\nu$ over $k\cdot K_{\Psi\backslash\Gamma}(x_{\Gamma,\Sigma})$ so as to be able to compare it with the canonical bundle κ. $\det\nu$ has fibre $\Lambda^a \operatorname{ad}(c_\Sigma^2)\mathfrak{M}_{\Psi\backslash\Gamma}^{\Sigma-}$ where $a=\dim_C \mathfrak{M}_{\Psi\backslash\Gamma}^{\Sigma-}$. Let α denote the representation of $K^{\Sigma}_{\Psi\backslash\Gamma}$ on that fibre. We know [15, Lemma 4.4] that $\operatorname{ad}(z_{\Psi\backslash\Gamma})$ is scalar multiplication by i on $\mathfrak{M}_{\Psi\backslash\Gamma}^{\Sigma-}$; thus $\operatorname{ad}(z^{\Sigma}_{\Psi\backslash\Gamma})$ is scalar multiplication by $2i$ on $\operatorname{ad}(c_\Sigma^2)\mathfrak{M}_{\Psi\backslash\Gamma}^{\Sigma-}$; we conclude that $\alpha(z^{\Sigma}_{\Psi\backslash\Gamma})$

$=2ia$. Let $\mathcal{E}'\subset\mathcal{K}^{\Sigma}_{\Psi\backslash\Gamma}$ denote the Lie algebra of the kernel of the action of $G^{\Sigma}_{\Psi\backslash\Gamma,0}$ on $G^{\Sigma}_{\Psi\backslash\Gamma,0}(x_{\Gamma,\Sigma})\cong G^{\Sigma}_{\Psi\backslash\Gamma,0}/K^{\Sigma}_{\Psi\backslash\Gamma}$. Then $\mathcal{E}'$ is orthogonal to the subalgebra $\mathcal{G}_0[\Psi\backslash\Gamma]$ of $\mathcal{G}^{\Sigma}_{\Psi\backslash\Gamma,0}$ that contains $z^{\Sigma}_{\Psi\backslash\Gamma}$. Thus we can specify: $\alpha(z^{\Sigma}_{\Psi\backslash\Gamma})=2ia$ and α annihilates the orthocomplement of $z^{\Sigma}_{\Psi\backslash\Gamma}$ in $\mathcal{K}^{\Sigma}_{\Psi\backslash\Gamma}$. Comparing representations we now conclude that $(\det \nu)^d=\kappa^{-2a}$ as holomorphic line bundles.

Recall $d=\dim_C k\cdot K_{\Psi\backslash\Gamma}(x_{\Gamma,\Sigma})$ and $a=\dim_C G^{\Sigma}_{\Psi\backslash\Gamma,0}(x_{\Gamma,\Sigma})$. If $a=0$, then the noncompact type symmetric space $G^{\Sigma}_{\Psi\backslash\Gamma,0}(x_{\Gamma,\Sigma})$ is reduced to a point, so its rank $|\Psi\backslash\Gamma|=0$; then $\Gamma=\Psi$ so $G_{\Psi\backslash\Gamma,0}(x_{\Gamma,\Sigma})=x_{\Gamma,\Sigma}=K_{\Psi\backslash\Gamma}(x_{\Gamma,\Sigma})$, whence $d=0$. In other words, if $d>0$, then $a>0$.

Suppose $d>0$. Then the canonical bundle κ of $K_{\Psi\backslash\Gamma}/K^{\Sigma}_{\Psi\backslash\Gamma}$ is negative in the sense of K. Kodaira; as $a>0$ it follows that κ^{-2a} is positive; thus $\det\nu$ is a positive line bundle. Let $[c(\det\nu)]\in H^{1,1}(K_{\Psi\backslash\Gamma}/K^{\Sigma}_{\Psi\backslash\Gamma};\mathbf{Z})$ denote the Chern class of $\det \nu$. Now $[c(\det \nu)]$ is represented by a $K_{\Psi\backslash\Gamma}$-invariant positive definite (1, 1)-form $c(\det \nu)$. As it is positive definite, $c(\det \nu)\neq 0$; as it is $K_{\Psi\backslash\Gamma}$-invariant and $K_{\Psi\backslash\Gamma}/K^{\Sigma}_{\Psi\backslash\Gamma}$ is symmetric, now $[c(\det \nu)]\neq 0$; thus $\det \nu$ is not topologically trivial. It follows that ν is not topologically trivial. As the bundles ν and β_k are topologically equivalent, now β_k is not topologically trivial.

If $d=0$, then $K_{\Psi\backslash\Gamma}=K^{\Sigma}_{\Psi\backslash\Gamma}$, so $\mathcal{Q}^{\Sigma}_{\Psi\backslash\Gamma}=0$; then $\mathcal{R}^{\Sigma}_{\Psi\backslash\Gamma,0}=\operatorname{ad}(c^2_{\Sigma})\cdot(i\mathcal{Q}^{\Sigma}_{\Psi\backslash\Gamma})=0$, so $\mathcal{G}_{\Psi\backslash\Gamma,0}=\mathcal{G}^{\Sigma}_{\Psi\backslash\Gamma,0}$. That says that $\operatorname{ad}(c_{\Sigma})$ has order 4 on $\mathcal{G}_{\Psi\backslash\Gamma}$. Let $\mathcal{G}=\sum\mathcal{G}^m$ decomposition into simple ideals, $\Delta=\bigcup\Delta^m$ corresponding decomposition of the root system of $\mathcal{G}$. Then [**15**, Corollary 4.10] the $\mathcal{G}^m_{\Psi\backslash\Gamma}=\mathcal{G}_{(\Psi\backslash\Gamma)\cap\Delta^m}$ are the simple ideals of $\mathcal{G}_{\Psi\backslash\Gamma}$ (possibly with some of them equal to zero); further [**15**, Corollary 4.11] if $G_{(\Psi\backslash\Gamma)\cap\Delta^m,0}(x_0)$ is of tube type and positive dimension, then $G_{\Psi\cap\Delta^m,0}(x_0)$ is of tube type; finally [**15**, Theorem 5.7] $\operatorname{ad}(c_{\Sigma})$ has order 4 on $\mathcal{G}_{\Psi\backslash\Gamma}$ if and only if, for each simple ideal $\mathcal{G}_{(\Psi\backslash\Gamma)\cap\Delta^m}$ of $\mathcal{G}_{\Psi\backslash\Gamma}$, either

(i) $\Sigma\cap(\Psi\backslash\Gamma)\cap\Delta^m$ is empty or

(ii) $G_{(\Psi\backslash\Gamma)\cap\Delta^m,0}(x_0)$ is of tube type and $(\Psi\backslash\Gamma)\cap\Delta^m\subset\Sigma$.

We conclude, using disjointness of Γ and Σ, that $\operatorname{ad}(c_{\Sigma})$ has order 4 on $\mathcal{G}_{\Psi\backslash\Gamma}$ if and only if, for each m, either

(i) $\Sigma\cap\Delta^m=\phi$ or

(ii) $G_{\Psi\cap\Delta^m,0}(x_0)$ is of tube type and $\Psi\cap\Delta^m\subset(\Gamma\cup\Sigma)$.

That completes the proof of (3).

We will prove $\beta:G_0(x_{\Gamma,\Sigma})\to K(x_{\Gamma,\Sigma})$ well-defined. In other words, given k, $k'\in K$ with $k\cdot G_{\Psi\backslash\Gamma,0}(x_{\Gamma,\Sigma})=k'\cdot G_{\Psi\backslash\Gamma,0}(x_{\Gamma,\Sigma})$, we will prove $\beta_k=\beta_{k'}$. Now suppose

$$x\in k\cdot G_{\Psi\backslash\Gamma,0}(x_{\Gamma,\Sigma}) = k'\cdot G_{\Psi\backslash\Gamma,0}(x_{\Gamma,\Sigma}); \qquad k,k'\in K.$$

Express

$$x = kg(x_{\Gamma,\Sigma}) = k'g'(x_{\Gamma,\Sigma}); \qquad g,g'\in G_{\Psi\backslash\Gamma,0}.$$

$k^{-1}k' \in K$ preserves the holomorphic arc component $G_{\Psi\backslash\Gamma,0}(x_{\Gamma,\Sigma})$ of $G_0(x_{\Gamma,\Sigma})$; that holomorphic arc component being a symmetric space of noncompact type, now $k^{-1}k' = k_1k_2$ where $k_1 \in K_{\Psi\backslash\Gamma}$ and k_2 acts trivially on $G_{\Psi\backslash\Gamma,0}(x_{\Gamma,\Sigma})$. Thus we may ignore k_2 and assume

$$k' = kk_1 \quad \text{with } k_1 \in K_{\Psi\backslash\Gamma}.$$

As in (11.7) we decompose

$$g = g_K g_M g_R \quad \text{and} \quad g' = g'_K g'_M g'_R.$$

Using $g_R x_{\Gamma,\Sigma} = x_{\Gamma,\Sigma} = g'_R x_{\Gamma,\Sigma}$, now we have

$$x = k g_K g_M x_{\Gamma,\Sigma} = k k_1 g'_K g'_M x_{\Gamma,\Sigma}.$$

Thus $g_K g_M h = k_1 g'_K g'_M$ where $h \in G_{\Psi\backslash\Gamma,0}$ fixes $x_{\Gamma,\Sigma}$. Lemma 11.5 says $h = h_K h_R$, $h_K \in K^{\Sigma}_{\Psi\backslash\Gamma}$ and $h_R \in \exp \mathfrak{R}^{\Sigma}_{\Psi\backslash\Gamma,0}$. Now

$$(k_1 g'_K) g'_M = g_K g_M h_K h_R = (g_K h_K)(h_K^{-1} g_M h_K) h_R.$$

Uniqueness of (11.7) thus says

$$k_1 g'_K = g_K h_K, \qquad g'_M = h_K^{-1} g_M h_K, \qquad 1 = h_R.$$

Thus

$$\beta_k(x) = k g_K(x_{\Gamma,\Sigma}) = k g_K h_K(x_{\Gamma,\Sigma}) = k k_1 g'_K(x_{\Gamma,\Sigma}) = k' g'_K(x_{\Gamma,\Sigma}) = \beta_{k'}(x),$$

and β is proved to be well defined.

Now $\beta: G_0(x_{\Gamma,\Sigma}) \to K(x_{\Gamma,\Sigma})$ is a well-defined real analytic map. If $x \in G_0(x_{\Gamma,\Sigma})$, then x and $\beta(x)$ are in the same holomorphic arc component of $G_0(x_{\Gamma,\Sigma})$. It follows that β is a real analytic bundle whose fibres coincide with those of the restrictions β_k. The structure group of β is the isotropy subgroup of K at $x_{\Gamma,\Sigma}$. That isotropy group preserves the holomorphic arc component $G_{\Psi\backslash\Gamma,0}(x_{\Gamma,\Sigma})$, so its every element has factorization $k = k_1k_2$ with $k_1 \in G_{\Psi\backslash\Gamma,0} \cap K = K_{\Psi\backslash\Gamma}$ and k_2 acting trivially on the fibre $G_{\Psi\backslash\Gamma,0}(x_{\Gamma,\Sigma})$ of β. Note $k_1 \in K^{\Sigma}_{\Psi\backslash\Gamma}$ because $k_1(x_{\Gamma,\Sigma}) = x_{\Gamma,\Sigma}$. Thus we can reduce the structure group of β to the structure group $K^{\Sigma}_{\Psi\backslash\Gamma}$ of its restrictions β_k.

Consider the special case $\beta: G_0(x_{\Gamma,\phi}) \to K(x_{\Gamma,\phi})$ where G_0 is simple and $\phi \neq \Gamma \neq \Psi$. The fibre of β over $k(x_{\Gamma,\phi})$ is the holomorphic arc component $k \cdot G_{\Psi\backslash\Gamma,0}(x_{\Gamma,\phi})$ of $G_0(x_{\Gamma,\phi})$. The space $K(x_{\phi,\Gamma})$, maximal compact subvariety of the open orbit $G_0(x_{\phi,\Gamma})$, has been seen to be a totally geodesic hermitian symmetric subspace of compact type in X. We will construct a positive holomorphic line bundle λ over $K(x_{\phi,\Gamma})$, and a fibration $f: K(x_{\Gamma,\phi}) \to K(x_{\phi,\Gamma})$, such that

$$f^*\lambda \text{ is the determinant bundle of } \beta\colon G_0(x_{\Gamma,\phi}) \to K(x_{\Gamma,\phi}).$$

For that construction we use the notation

$$L^\Gamma\text{: isotropy subgroup of } K \text{ at } x_{\Gamma,\phi};$$
$$K^\Gamma\text{: isotropy subgroup of } K \text{ at } x_{\phi,\Gamma}.$$

Lemma 11.4 says that $K^\Gamma = K_\Psi^\Gamma$, centralizer of c_Γ^4 in K; the decomposition $\mathcal{G}_{\Psi,\phi}^\Gamma = \mathcal{L}_{\Gamma,\phi,0} \oplus \mathcal{G}_{\Psi\backslash\Gamma,0} + \mathcal{G}_{\Gamma,0}^\Gamma$ informs us first that $L^\Gamma = L_{\Gamma,\phi,0}\cdot K_{\Psi\backslash\Gamma}$ local direct product where the first factor acts trivially on $G_{\Psi\backslash\Gamma,0}(x_{\Gamma,\phi})$, and second that $K^\Gamma = L^\Gamma\cdot K_\Gamma^\Gamma$ local direct product. Now we define

$$f\colon K(x_{\Gamma,\phi}) = K/L^\Gamma \to K/K^\Gamma = K(x_{\phi,\Gamma}) \quad \text{by } f(kL^\Gamma) = kK^\Gamma.$$

View f as identification under the right action of K_Γ^Γ on K/L^Γ. That action commutes with the action of L^Γ on the fibres of β. Thus, in the category of topological vector bundles,

$$\beta = f^*\gamma \quad \text{for some vector bundle } \gamma \text{ over } K(x_{\phi,\Gamma}).$$

As K^Γ acts on the fibres $k\cdot G_{\Psi\backslash\Gamma,0}(x_{\Gamma,\phi})$ of γ via the action of $K_{\Psi\backslash\Gamma}^C$ on $G_{\Psi\backslash\Gamma,0}(x_{\Gamma,\phi})$, which is holomorphic,

$$\gamma \text{ over } K(x_{\phi,\Gamma}) \text{ is the } K\text{-homogeneous holomorphic}$$
$$\text{vector bundle with fibre } \mathfrak{M}_{\Psi\backslash\Gamma}^- \text{ over } x_{\phi,\Gamma}.$$

Now we compare $\det\gamma$ with the canonical bundle κ over $K(x_{\phi,\Gamma})$. $\operatorname{ad}(c_\Gamma^2)z$ acts on the fibre $\Lambda^a\mathfrak{M}_{\Psi\backslash\Gamma}^-$, $a = \dim_C \mathfrak{M}_{\Psi\backslash\Gamma}^-$, of $\det\gamma$, as multiplication by ia. It acts on the fibre $(\Lambda^d\mathcal{Q}_\Psi^{\Gamma-})^* = (\Lambda^d \operatorname{ad}(c_\Gamma^2)\mathcal{R}_\Psi^{\Gamma-})^*$, $d = \dim_C \mathcal{Q}_\Psi^{\Gamma-} = \dim_C K(x_{\phi,\Gamma})$, of κ, as multiplication by $-\mathrm{id}$. Thus as before,

$$(\det\gamma)^d = \kappa^{-a}.$$

Our hypothesis that G_0 be simple and $\phi \neq \Gamma \neq \Psi$ ensures $d \neq 0 \neq a$. As κ is negative, now $(\det\gamma)^d$ is positive, so $\det\gamma$ is positive. Define $\lambda = \det\gamma$, and we have

$$f^*\lambda = f^* \det\gamma = \det f^*\gamma = \det(\beta\colon G_{\Psi\backslash\Gamma,0}(x_{\Gamma,\phi}) \to K(x_{\Gamma,\phi})).$$

That completes the construction.

Suppose G_0 simple. We will prove that $\beta\colon G_0(x_{\Gamma,\Sigma}) \to K(x_{\Gamma,\Sigma})$ is topologically trivial if and only if

(i) $\Gamma = \phi = \Sigma$, so β is fibration of $G_0(x_0) = X_0$ over the single point x_0, or

(ii) $\Gamma = \Psi$ and $\Sigma = \phi$, so $\beta\colon G_0(x_{\Psi,\phi}) \to K(x_{\Psi,\phi})$ is the fibration of the Bergman-Šilov boundary of X_0 in X over itself with fibre reduced to a point; or

(iii) $\Gamma=\phi$, $\Sigma=\Psi$, and $X_0=G_0(x_0)$ is of tube type, so β is fibration of $G_0(x_{\phi,\Psi})=c_\Psi^2 X_0$ over a single point $x_{\phi,\Psi}=c_\Psi^2 x_0$.

First note that β is topologically trivial in the three cases described just above. Conversely, now assume β topologically trivial. Then the restrictions β_k to the holomorphic arc components $k\cdot G_{\Psi\setminus\Gamma,0}(x_{\Gamma,\Sigma})$ of $G_0(x_{\Gamma,\Sigma})$ are topologically trivial, so either $\Sigma=\phi$, or $\Psi=\Gamma\cup\Sigma$ with X_0 of tube type. First suppose $\Sigma=\phi$, so $\beta: G_0(x_{\Gamma,\phi})\to K(x_{\Gamma,\phi})$, and recall the map $f: K(x_{\Gamma,\phi})\to K(x_{\phi,\Gamma})$ of the paragraph before last. If $\phi\neq\Gamma\neq\Psi$, we saw that the determinant bundle of β, viewing β topologically as a complex vector bundle, is of the form $f^*\lambda$ where λ over $K(x_{\phi,\Gamma})$ is a positive holomorphic line bundle. As its Chern class is nonzero, λ is not topologically trivial, so $f^*\lambda=\det\beta$ is not topologically trivial, contradicting topological triviality of β. Thus either $\Gamma=\phi$ and we are in case (i) above, or $\Gamma=\Psi$ and we are in case (ii) above. Now suppose $\Sigma\neq\phi$, so $\Psi=\Gamma\cup\Sigma$ with X_0 of tube type. As X_0 is of tube type, i.e. as $c_\Psi^4=1$, we have K and G_0 stable under $\mathrm{ad}(c_\Psi^2)$. Now, using the consequence $c_\Psi=c_\Gamma c_\Sigma$ of $\Psi=\Gamma\cup\Sigma$ and $\Gamma\cap\Sigma=\phi$, we have

$$c_\Psi^2: G_0(x_{\Gamma,\Sigma})\to G_0(c_\Psi^2 x_{\Gamma,\Sigma}) = G_0(c_\Gamma^3 c_\Sigma^4 x_0) = G_0(x_{\Gamma,\phi}),$$

and that the map has the properties

$$c_\Psi^2 K(x_{\Gamma,\Sigma}) = K(x_{\Gamma,\phi}) \quad\text{and}\quad c_\Psi^2 k\cdot G_{\Psi\setminus\Gamma,0}(x_{\Gamma,\Sigma}) = k'\cdot G_{\Psi\setminus\Gamma,0}(x_{\Gamma,\phi}).$$

Thus c_Ψ^2 defines a fibre bundle isomorphism between the fibrations β given by

$$G_0(x_{\Gamma,\Sigma})\to K(x_{\Gamma,\Sigma}) \quad\text{and}\quad G_0(x_{\Gamma,\phi})\to K(x_{\Gamma,\phi}).$$

The first being topologically trivial by hypothesis, the second must be trivial, and we have just seen that to imply either $\Gamma=\phi$ or $\Gamma=\Psi$. But $\Psi=\Gamma\cup\Sigma$, $\Gamma\cap\Sigma=\phi$ and $\Sigma\neq\phi$; thus $\Gamma=\phi$, $\Sigma=\Psi$ and we are in case (iii) above.

Let $\mathfrak{g}^m$ denote the simple ideals of $\mathfrak{g}$, Δ^m the root system of $\mathfrak{g}^m$, $\mathfrak{g}_0^m=\mathfrak{g}_0\cap\mathfrak{g}^m$, G^m and G_0^m the analytic subgroups for $\mathfrak{g}^m$ and $\mathfrak{g}_0^m$, and $X^m=G^m(x_0)$ and $X_0^m=G_0^m(x_0)$. Then $G_0(x_{\Gamma,\Sigma})$ is the product of the $G_0^m(x_{\Gamma\cap\Delta^m,\Sigma\cap\Delta^m})$, and the base and fibres of $\beta: G_0(x_{\Gamma,\Sigma})\to K(x_{\Gamma,\Sigma})$ decompose similarly. Thus β is topologically trivial if and only if each of the "factors"

$$G_0^n(x_{\Gamma\cap\Delta^m,\Sigma\cap\Delta^m})\to(K\cap G_0^m)(x_{\Gamma\cap\Delta^m,\Sigma\cap\Delta^m})$$

is topologically trivial. According to the paragraph before the last one, β is topologically trivial if and only if, for each m,

(i) $\Psi\cap\Delta^m=\phi=\Sigma\cap\Delta^m$, or

(ii) $\Gamma\cap\Delta^m=\Psi\cap\Delta^m$ and $\Sigma\cap\Delta^m=\phi$, or

(iii) $\Gamma\cap\Delta^m=\phi$ and $\Sigma\cap\Delta^m=\Psi\cap\Delta^m$ with X_0^m of tube type.

That completes the proof of (4), completing the proof of Theorem 11.8. Q.E.D.

11.9. COROLLARY. *Let $\Gamma\subset\Psi$. Then the fibration β: $G_0(x_{\Gamma,\phi})\rightarrow K(x_{\Gamma,\phi})$ of Theorem 11.8 is given by $\beta(k\cdot G_{\Psi\backslash\Gamma,0}(x_{\Gamma,\phi}))=k(x_{\Gamma,\phi})$ for $k\in K$, and it is real analytically equivalent to a complex homogeneous vector bundle μ over $K(x_{\Gamma,\phi})$. There is a fibration $f: K(x_{\Gamma,\phi})\rightarrow K(x_{\phi,\Gamma})$ defined by $f(kx_{\Gamma,\phi})=kx_{\phi,\Gamma}$, whose fibres are compact totally geodesic submanifolds $k\cdot K_\Psi^\Gamma(x_{\Gamma,\phi})$ of X, and whose base is the totally geodesic hermitian symmetric subspace $K(x_{\phi,\Gamma})$ of X; $\mu=f^*\gamma$ for a certain holomorphic homogeneous vector bundle γ over $K(x_{\phi,\Gamma})$. If $\phi\neq\Gamma\cap\Delta^m\neq\Psi\cap\Delta^m$ where Δ^m is the root system of some simple ideal of $\mathfrak{g}$, then* $\dim_C K(x_{\phi,\Gamma})>0$ *and* $\det\gamma$ *is a positive homogeneous holomorphic line bundle over $K(x_{\phi,\Gamma})$; in particular, in that case* $\det\gamma$, γ, $\det\mu$, μ *and β all are topologically nontrivial.*

11.10. COROLLARY. *Let $\Gamma\subset\Psi$ and $\Sigma=\Psi\backslash\Gamma$ with X_0 of tube type. Then the fibration $\beta: G_0(x_{\Gamma,\Sigma})\rightarrow K(x_{\Gamma,\Sigma})$ of Theorem 11.8 is given by mapping $c_\Psi^2 kG_{\Psi\backslash\Gamma,0}(x_{\Gamma,\phi})=\mathrm{ad}(c_\Psi^2)k\cdot G_{\Psi\backslash\Gamma,0}(x_{\Gamma,\Sigma})$ to $c_\Psi^2 kx_{\Gamma,\phi}=\mathrm{ad}(c_\Psi^2)k\cdot x_{\Gamma,\Sigma}$, and it is real analytically equivalent to a complex homogeneous vector bundle μ' over $K(x_{\Gamma,\Sigma})$. There is a fibration $f': K(x_{\Gamma,\Sigma})\rightarrow K(x_{\phi,\Sigma})$ defined by $f'(kx_{\Gamma,\Sigma})=kx_{\phi,\Sigma}$, whose fibres are compact totally geodesic submanifolds $k\cdot K_\Psi^\Gamma(x_{\Gamma,\Sigma})=k\cdot K_\Psi^\Sigma(x_{\Gamma,\Sigma})$ of X, and whose base is the totally geodesic hermitian symmetric subspace $K(x_{\phi,\Sigma})$ of X; $\mu'=f'^*\gamma'$ where γ' is a certain holomorphic homogeneous vector bundle over $K(x_{\phi,\Sigma})$. If $\phi\neq\Sigma\cap\Delta^m\neq\Psi\cap\Delta^m$ where Δ^m is the root system of some simple ideal of $\mathfrak{g}$, then* $\dim_C K(x_{\phi,\Sigma})>0$ *and* $\det\gamma'$ *is a positive homogeneous holomorphic line bundle over $K(x_{\phi,\Sigma})$; in particular, in that case* $\det\gamma'$, γ', $\det\mu'$, μ' *and β are all topologically nontrivial.*

Corollaries 10.9 and 10.10 are related by: c_Ψ^2 is a bundle isomorphism of $G_0(x_{\Gamma,\Sigma})\rightarrow K(x_{\Gamma,\Sigma})$ onto $G_0(x_{\Gamma,\phi})\rightarrow K(x_{\Gamma,\phi})$ and maps $K(x_{\phi,\Sigma})$ onto $K(x_{\phi,\Gamma})$, inducing the fibrations f' from f, μ' from μ and γ' from γ.

PROOF OF COROLLARIES. Corollary 11.9 is contained in the proof of part 4 of Theorem 11.8 for the case where $\mathfrak{g}$ is simple, and the general case follows. The second paragraph of the statement of Corollary 11.10 is also contained in that proof, and the first paragraph then follows from Corollary 11.9. Q.E.D.

The special case of an open orbit is worth separate examination.

11.11. COROLLARY. *Every open G_0-orbit on X has a holomorphic fibration*

$$\beta: G_0(x_{\phi,\Sigma}) \to K(x_{\phi,\Sigma}) \quad by \quad \beta(gx_{\phi,\Sigma}) = g_K x_{\phi,\Sigma}$$

where

$$g = g_K g_M g_R, \qquad g_K \in K, \qquad g_M \in \exp \overset{\Sigma}{\mathfrak{M}}_{\Psi,0}, \qquad g_R \in \exp \overset{\Sigma}{\mathfrak{R}}_{\Psi,0}.$$

1. *The bundle β has structure group K_Ψ^Σ, has total space $G_0(x_{\phi,\Sigma})$, has base space $K(x_{\phi,\Sigma})$ which is a compact totally geodesic hermitian symmetric submanifold of X, and has fibre over $k(x_{\phi,\Sigma})$ which is the totally geodesic hermitian symmetric submanifold*

$$k\cdot \overset{\Sigma}{G}_{\Psi,0}(x_{\phi,\Sigma}) = kc_\Sigma^2 \overset{\Sigma}{G}_{\Psi,0}(x_0) = kc_\Sigma^2\{G_{\Psi\setminus\Sigma,0}(x_0) \times \overset{\Sigma}{G}_{\Sigma,0}(x_0)\}$$

of noncompact type and rank $|\Psi|$ in $kc_\Sigma^2 G_0(x_0)$.

2. *The holomorphic normal ν bundle of $K(x_{\phi,\Sigma})$ in $G_0(x_{\phi,\Sigma})$ is the homogeneous vector bundle over $K(x_{\phi,\Sigma}) \cong K/K_\Psi^\Sigma$ associated to the representation of K_Ψ^Σ on the holomorphic tangent space of $G_{\Psi,0}^\Sigma/K_\Psi^\Sigma$. The bundle β is holomorphically equivalent to a relatively compact open subbundle of that holomorphic normal bundle ν. If* $\dim_C K(x_{\phi,\Sigma}) > 0$, *i.e. if the root system Δ^m of some simple ideal of $\mathfrak{g}$ satisfies $\Sigma \cap \Delta^m \neq \phi$, and satisfies $\Sigma \cap \Delta^m \neq \Psi \cap \Delta^m$ in case $G_{\Psi\cap\Delta m,0}(x_0)$ is of tube type, then* det ν *is a positive homogeneous holomorphic line bundle over $K(x_{\phi,\Sigma})$, and the bundles* det ν, ν, *and β all are topologically nontrivial.*

PROOF. This is the case $\Gamma = \phi$ of Theorem 11.8. Q.E.D.

The case of the "opposite" open orbit is of further interest because it gives new characterizations of tube domains.

11.12. COROLLARY. *The "opposite" open orbit $G_0(x_{\phi,\Psi}) \subset X$ has a holomorphic fibration*

$$\beta: G_0(x_{\phi,\Psi}) \to K(x_{\phi,\Psi}) \quad by \quad \beta(gx_{\phi,\Psi}) = g_K x_{\phi,\Psi}$$

where

$$g = g_K g_M g_R, \qquad g_K \in K, \qquad g_M \in \exp \overset{\Psi}{\mathfrak{M}}_{\Psi,0}, \qquad g_R \in \exp \overset{\Psi}{\mathfrak{R}}_{\Psi,0}.$$

1. *The bundle β has structure group $K_\Psi^{\Psi C}$, total space $G_0(x_{\phi,\Psi})$, base space $K(x_{\phi,\Psi})$ which is a maximal compact subvariety of $G_0(x_{\phi,\Psi})$ and is a compact totally geodesic hermitian symmetric submanifold of X, and has fibre over $k(x_{\phi,\Psi})$ which is the "tube part" (c.f.* [2])

$$k\cdot \overset{\Psi}{G}_{\Psi,0}(x_{\phi,\Psi}) = kc_\Psi^2 \cdot \overset{\Psi}{G}_{\Psi,0}(x_0)$$

of $kc_\Psi^2 \cdot G_0(x_0)$.

2. *The bundle* β *is a relatively compact open subbundle of the holomorphic normal bundle* ν *of* $K(x_{\phi,\Psi})$ *in* $G_0(x_{\phi,\Psi})$. ν *is the homogeneous holomorphic vector bundle over* $K(x_{\phi,\Psi})=K/K_\Psi^\Psi$ *associated to the representation of* K_Ψ^Ψ *on the holomorphic tangent space of the tube part* $G_{\Psi,0}^\Psi(x_0)$ *of* $G_0(x_0)$.

3. *The following conditions are equivalent*:

(3a) β *is holomorphically trivial.*

(3b) ν *is holomorphically trivial.*

(3c) $\det \nu$ *is holomorphically trivial.*

(3d) β *is topologically trivial.*

(3e) ν *is topologically trivial.*

(3f) $\det \nu$ *is topologically trivial.*

(3g) $K(x_{\phi,\Psi})$ *is the single point* $x_{\phi,\Psi}$.

(3h) c_Ψ *has order* 4 *on* X.

(3i) *The bounded symmetric domain* $X_0=G_0(x_0)$ *is of tube type, i.e. it has a holomorphic realization as a tube domain over a homogeneous self dual cone.*

(3j) *The Bergman-Šilov boundary* $G_0(x_{\Psi,\phi})$ *of* X_0 *in* X *is a totally geodesic submanifold of* X.

(3k) $\dim_{\mathbf{R}} G_0(x_{\Psi,\phi})=\dim_C X$.

(3l) *Viewing* G_0 *as the identity component of the group* $G_{\mathbf{R}}$ *of real points in the linear algebraic group* G, *we have* $P_{x_{\Psi,\phi}}$ *defined*/$\mathbf{R}$, *i.e.* X *is defined*/$\mathbf{R}$.

PROOF. Statements 1 and 2, and equivalence of (3a) through (3h), are the case $\Sigma=\Psi$ of Corollary 11.11. Equivalence of (3h), (3i), (3j) and (3k) is contained in [2, Theorems 4.9 and 6.8], and Theorem 3.6 says that (3k) is equivalent to (3l). Q.E.D.

We recapitulate the decomposition of an arbitrary G_0-orbit $G_0(x_{\Gamma,\Sigma})\subset X$. Here Γ and Σ are disjoint subsets of Ψ. First we have the fibration of Theorem 11.8,

$$\beta: G_0(x_{\Gamma,\Sigma}) \to K(x_{\Gamma,\Sigma}),$$

that maps an arbitrary holomorphic arc component $k\cdot G_{\Psi\setminus\Gamma,0}(x_{\Gamma,\Sigma})$ holomorphically onto its maximal compact subvariety $k\cdot K_{\Psi\setminus\Gamma}(x_{\Gamma,\Sigma})$; the fibre over $kk'(x_{\Gamma,\Sigma})$, $k\in K$ and $k'\in K_{\Psi\setminus\Gamma}$, is

$$kk'\cdot G_{\Psi\setminus\Gamma,0}^{\Sigma}(x_{\Gamma,\Sigma}) = kk'c_\Gamma c_\Sigma^2\{G_{\Psi\setminus(\Gamma\cup\Sigma),0}(x_0)\times G_{\Sigma,0}^{\Sigma}(x_0)\}.$$

We then follow β by $\alpha\cdot\beta^{-1}$ where

$$\alpha: G_0(x_{\Gamma,\Sigma}) \to G_0/N_{\Gamma,\Sigma,0}$$

is the fibration of the orbit over the space of its holomorphic arc components. This amounts to collapsing each of the maximal compact subvarieties $k\cdot K_{\Psi\backslash\Gamma}(x_{\Gamma,\Sigma})\subset k\cdot G_{\Psi\backslash\Gamma,0}(x_{\Gamma,\Sigma})$ to a point. That collapsing is, of course, already done when, whenever Δ^m is the root system of a simple ideal of $\mathfrak{g}$, either $\Sigma\cap\Delta^m=\phi$, or $(\Gamma\cup\Sigma)\cap\Delta^m=\Psi\cap\Delta^m$ with $G_{\Psi\cap\Delta^m,0}(x_0)$ of tube type. Now we come to the final stage of the decomposition, which is realization of the β-fibre

$$G^{\Sigma}_{\Psi\backslash\Gamma,0}(x_{\Gamma,\Sigma}) = G_{\Psi\backslash(\Gamma\cup\Sigma),0}(x_{\Gamma,\Sigma}) \times G^{\Sigma}_{\Sigma,0}(x_{\Gamma,\Sigma})$$

as a Siegel domain of type III over any of its boundary components. That Siegel domain realization is the content of Wolf-Korányi [**15**, §7]; the appendix to this §11 transcribes it into our notation, and explicitly works out the holomorphic fibration over the boundary component (Proposition 11 A.3) that is implicit in the Siegel domain formulation, for the hermitian symmetric space $X_0=G_0(x_0)$. We now recast the results of that appendix to this §11 to cover our case of the hermitian symmetric space

$$\begin{aligned} G^{\Sigma}_{\Psi\backslash\Gamma,0}(x_{\Gamma,\Sigma}) &= G_{\Psi\backslash(\Gamma\cup\Sigma),0}(x_{\Gamma,\Sigma}) \times G^{\Sigma}_{\Sigma,0}(x_{\Gamma,\Sigma}) \\ &= c_\Gamma c_\Sigma^2\{G_{\Psi\backslash(\Gamma\cup\Sigma),0}(x_0) \times G^{\Sigma}_{\Sigma,0}(x_0)\} = c_\Gamma c_\Sigma^2 G^{\Sigma}_{\Psi\backslash\Gamma,0}(x_0). \end{aligned}$$

$\Psi\backslash\Gamma$ is the maximal set of strongly orthogonal noncompact positive roots of $\mathfrak{g}^{\Sigma}_{\Psi\backslash\Gamma,0}$. Let E be any subset. Then we have disjoint unions

$$\Psi\backslash(\Gamma\cup\Sigma) = A\cup A', \quad \Sigma = B\cup B', \quad E = A\cup B, \quad E' = A'\cup B'.$$

Now observe

$$\begin{aligned} \mathfrak{M}^{\Sigma-}_{\Psi\backslash\Gamma}\cap\mathfrak{M}^{E-}_{E} &= \mathfrak{M}^{A-}_{A} + \mathfrak{M}^{B-}_{B}, \\ \mathfrak{M}^{E-}_{\Psi\backslash\Gamma}\cap\mathfrak{R}^{E-}_{\Psi} &= \mathfrak{R}^{A-}_{\Psi\backslash(\Gamma\cup\Sigma)} + (\mathfrak{R}^{B-}_{\Sigma}\cap\mathfrak{M}^{\Sigma-}_{\Sigma}), \quad \text{and} \\ \mathfrak{M}^{\Sigma-}_{\Psi\backslash\Gamma}\cap\mathfrak{M}^{-}_{\Psi\backslash E} &= \mathfrak{M}^{-}_{A'} + \mathfrak{M}^{B'-}_{B'}. \end{aligned}$$

Thus we define

(11.13a) $\quad V_1' = \mathfrak{M}^{A-}_{A}, \quad V_2' = \mathfrak{R}^{A-}_{\Psi\backslash(\Gamma\cup\Sigma)}$ and $V_3' = \mathfrak{M}^{-}_{A'}$;

(11.13b) $\quad V_1'' = \mathfrak{M}^{B-}_{B}, \quad V_2'' = (\mathfrak{R}^{B-}_{\Sigma}\cap\mathfrak{M}^{\Sigma-}_{\Sigma})$ and $V_3'' = \mathfrak{M}^{B'-}_{B'}$; and

(11.13c) $\quad V_1 = V_1'\oplus V_1'', \quad V_2 = V_2'\oplus V_2''$ and $V_3 = V_3'\oplus V_3''$.

Now every $v\in\mathfrak{M}^{\Sigma-}_{\Psi\backslash\Gamma}$ has unique decomposition

(11.13d) $\quad v = v_1' + v_1'' + v_2' + v_2'' + v_3' + v_3'', \qquad v_i'\in V_i'$ and $v_j''\in V_j''$

that we use without further explanation.

If $P\subset Q\subset\Psi$ we have the bounded domain $D_Q^P=\xi^{-1}G_{Q,0}^P(x_0)$ in $\mathfrak{M}_Q^{P-}$. Now the bounded domains in V_3', V_3'' and V_3 that serve our Siegel domains, are

$$(11.14a)\quad D_{A'}=\xi^{-1}G_{A',0}(x_0)\subset V_3',\qquad D_{B'}^{B'}=\xi^{-1}G_{B',0}^{B'}(x_0)\subset V_3'',\quad\text{and}$$

$$(11.14b)\quad D_{A'}\times D_{B'}^{B'}=D_{E'}^{B'}=\xi^{-1}G_{E',0}^{B'}(x_0)\subset V_3.$$

Note that they are the respective intersections of V_3', V_3'' and V_3 with the bounded domain $D_{\Psi\setminus E,0}(x_0)\subset\mathfrak{M}_{\Psi\setminus E}^-$.

The real forms of V_1', V_1'' and V_1 are

$$(11.15a)\quad \begin{aligned} U_1' &= \mathrm{ad}(c_A)\{\mathfrak{g}_0\cap\mathrm{ad}(c_A^{-1})V_1'\},\\ U_1'' &= \mathrm{ad}(c_B)\{\mathfrak{g}_0\cap\mathrm{ad}(c_B^{-1})V_1''\}\quad\text{and}\end{aligned}$$

$$(11.15b)\quad U_1'\oplus U_1''=U_1=\mathrm{ad}(c_E)\{\mathfrak{g}_0\cap\mathrm{ad}(c_E^{-1})V_1\}.$$

They contain the self dual cones

$$(11.16a)\quad \mathfrak{C}_A=\mathrm{ad}(K_A^{A*})\cdot\xi^{-1}(x_{A,\phi})\subset U_1'\quad\text{where}\quad \xi^{-1}(x_{A,\phi})=\textstyle\sum_{Ae-\psi},$$

$$(11.16b)\quad \mathfrak{C}_B=\mathrm{ad}(K_B^{B*})\cdot\xi^{-1}(x_{B,\phi})\subset U_1''\quad\text{where}\quad \xi^{-1}(x_{B,\phi})=\textstyle\sum_{Be-\psi},$$

$$(11.16c)\quad \begin{aligned}&\mathfrak{C}_A\times\mathfrak{C}_B=\mathfrak{C}_E=\mathrm{ad}(K_E^{E*})\cdot\xi^{-1}(x_{E,\phi})\subset U_1\quad\text{where}\\ &\qquad\xi^{-1}(x_{E,\phi})=\textstyle\sum_{Ee-\psi}.\end{aligned}$$

If $w\in D_{E'}^{B'}$ we have the conjugate-linear map $\mu(w)$: $V_2\to V_2$ of operator norm <1, given by

$$(11.17a)\quad \begin{aligned}\mu(w)v &= \mathrm{ad}(w)\cdot\mathrm{ad}(c_E^2)\cdot\tau_u(v)=\mathrm{ad}(w')\cdot\mathrm{ad}(c_A^2)\cdot\tau_u(v')\\ &\quad+\mathrm{ad}(w'')\cdot\mathrm{ad}(c_B^2)\cdot\tau_u(v'').\end{aligned}$$

It specifies the "bilinear plus hermitian" map F_w: $V_2\times V_2\to V_1$ by

$$(11.17b)\quad F_w(u,v)=-(i/2)[u,\mathrm{ad}(c_E^2)\cdot\tau_u\cdot(I-\mu(w))^{-1}v].$$

The splitting in (11.17a) says that $\mu(w)$: $V_2\to V_2$ is the direct sum of operators

$$\mu(w'):\ V_2'\to V_2'\quad\text{and}\quad \mu(w''):\ V_2''\to V_2''.$$

Thus $(I-\mu(w))^{-1}$: $V_2\to V_2$ splits as the direct sum of

$$(I-\mu(w'))^{-1}:\ V_2'\to V_2'\quad\text{and}\quad (I-\mu(w''))^{-1}:\ V_2''\to V_2''.$$

It follows that F_w: $V_2\times V_2\to V_1$ is the direct sum of maps

$$F_{w'}:\ V_2'\times V_2'\to V_1'\quad\text{and}\quad F_{w''}:\ V_2''\times V_2''\to V_1''$$

in the sense that, for $w \in D_{E'}^{B'}$ and $u, v \in V_2$,

$$F_w(u, v) = F_{w'}(u', v') + F_{w''}(u'', v''). \tag{11.17c}$$

Carrying (11A.1) back to our situation, now

$$\mathfrak{S}_{\Psi \setminus (\Gamma \cup \Sigma), A} = \{v' \in \mathfrak{M}^{-}_{\Psi \setminus (\Gamma \cup \Sigma)} : v_3' \in D_{A'} \text{ and } \operatorname{Im} v_1' - F_{v_3'}(v_2', v_2') \in \mathfrak{C}_A\}, \tag{11.18a}$$

$$\overset{\Sigma}{\mathfrak{S}}_{\Sigma, B} = \{v'' \in \mathfrak{M}^{\Sigma -}_{\Sigma} : v_3'' \in D^{B'}_{B'} \text{ and } \operatorname{Im} v_1'' = F_{v_3''}(v_2'', v_2'') \in \mathfrak{C}_B\}, \tag{11.18b}$$

$$\overset{\Sigma}{\mathfrak{S}}_{\Psi \setminus \Gamma, E} = \{v \in \mathfrak{M}^{\Sigma -}_{\Psi \setminus \Gamma} : v_3 \in D^{B'}_{E'} \text{ and } \operatorname{Im} v_1 - F_{v_3}(v_2, v_2) \in \mathfrak{C}_E\} \tag{11.18c}$$

are Siegel domains of type III that satisfy

$$\overset{\Sigma}{\mathfrak{S}}_{\Psi \setminus \Gamma, E} = \mathfrak{S}_{\Psi \setminus (\Gamma \cup \Sigma), A} \times \overset{B}{\mathfrak{S}}_{\Sigma, B} \subset \mathfrak{M}^{-}_{\Psi \setminus (\Gamma \cup \Sigma)} \oplus \mathfrak{M}^{\Sigma -}_{\Sigma} = \mathfrak{M}^{\Sigma -}_{\Psi \setminus \Gamma}. \tag{11.18d}$$

Furthermore, (11A.2) in our context says that

$$c_A G_{\Psi \setminus (\Gamma \cup \Sigma), 0}(x_0) \subset \xi(\mathfrak{M}^{-}_{\Psi \setminus (\Gamma \cup \Sigma)}) \text{ and } \xi^{-1} : c_A G_{\Psi \setminus (\Gamma \cup \Sigma, 0}(x_0) \cong \mathfrak{S}_{\Psi \setminus (\Gamma \cup \Sigma), A}, \tag{11.19a}$$

$$c_B \overset{\Sigma}{G}_{\Sigma, 0}(x_0) \subset \xi(\mathfrak{M}^{\Sigma -}_{\Sigma}) \text{ and } \xi^{-1} : c_B \overset{\Sigma}{G}_{\Sigma, 0}(x_0) \cong \overset{\Sigma}{\mathfrak{S}}_{\Sigma, B}, \tag{11.19b}$$

$$c_E \overset{\Sigma}{G}_{\Psi \setminus \Gamma, 0}(x_0) \subset \xi(\mathfrak{M}^{\Sigma -}_{\Psi \setminus \Gamma}) \text{ and } \xi^{-1} : c_E \overset{\Sigma}{G}_{\Psi \setminus \Gamma, 0}(x_0) \cong \overset{\Sigma}{\mathfrak{S}}_{\Psi \setminus \Gamma, E}, \tag{11.19c}$$

such that the analytic equivalence of (11.19c) is the direct product of those of (11.19a) and (11.19b). Finally, now, the coarse structure portion of Proposition 11A.3 becomes:

11.20. PROPOSITION. *Let Γ and Σ be disjoint subsets of Ψ, let $E \subset \Psi \setminus \Gamma$, and define disjoint union decompositions by*

$$E = A \cup B, \quad E' = A' \cup B', \quad \Psi \setminus (\Gamma \cup \Sigma) = A \cup A' \text{ and } \Sigma = B \cup B'.$$

Retain the notation (11.13) *through* (11.18). *Then we have holomorphic fibre bundles*

$$\pi' : \mathfrak{S}_{\Psi \setminus (\Gamma \cup \Sigma), A} \to D_{A'}, \qquad \pi'' : \overset{\Sigma}{\mathfrak{S}}_{\Sigma, B} \to D^{B'}_{B'} \text{ and } \pi : \overset{\Sigma}{\mathfrak{S}}_{\Psi \setminus \Gamma, E} \to D^{B'}_{E'}$$

given by

$$\pi'(v') = v_3', \qquad \pi''(v'') = v_3'' \text{ and } \pi(v) = v_3 = v_3' + v_3''.$$

1. *There are connected Lie groups $H_{\pi'}$, $H_{\pi''}$ and H_π that act transitively on the respective total spaces of π', π'' and π by holomorphic bundle*

automorphisms. They have semidirect product structure

$$H_{\pi'} = G_{A',0}\cdot Z_{\pi'}, \qquad H_{\pi''} = G^{B'}_{B',0}\cdot Z_{\pi''} \quad \text{and} \quad H_\pi = G^{B'}_{E',0}\cdot Z_\pi$$

where the first factor is equivariant for its natural transitive action on the base, and where the normal subgroups composing the second factor are the respective identity components of

$$G_{\Psi\backslash(\Gamma\cup\Sigma)}\cap \operatorname{ad}(c_A)Z_{A,\phi,0}, \quad G^\Sigma_\Sigma\cap \operatorname{ad}(c_B)Z_{B,\phi,0} \quad \text{and} \quad G^\Sigma_{\Psi\backslash\Gamma}\cap \operatorname{ad}(c_E)Z_{E,\phi,0}.$$

The latter are trivial on the base and transitive on every fibre. Finally, H_π *acts on* $\mathfrak{S}^\Sigma_{\Psi\backslash\Gamma,E}=\mathfrak{S}_{\Psi\backslash(\Gamma\cup\Sigma),A}\times\mathfrak{S}^\Sigma_{\Sigma,B}$ *as* $H_{\pi'}\times H_{\pi''}$.

2. *The fibres of* π', π'' *and* π *are holomorphically equivalent to hermitian symmetric spaces of noncompact type with respective ranks* $|A|$, $|B|$ *and* $|E|=|A|+|B|$ *and with respective tube parts* $G^A_{A,0}(x_0)$, $G^B_{B,0}(x_0)$ *and* $G^E_{E,0}(x_0)=G^A_{A,0}(x_0)\times G^B_{B,0}(x_0)$. *The realizations of the fibres over* 0 *as Siegel domains of type* II *are*

$$\pi'^{-1}(0) = \{v_1'+v_2'\in \mathfrak{M}^{A-}_A + \mathfrak{R}^{A-}_{\Psi\backslash(\Gamma\cup\Sigma)}\colon \operatorname{Im} v_1' - F_0(v_2',v_2')\in \mathfrak{C}_A\},$$

$$\pi''^{-1}(0) = v\{_1''+v_2''\in \mathfrak{M}^{B-}_B + (\mathfrak{R}^{B-}_\Sigma\cap \mathfrak{M}^{\Sigma-}_\Sigma)\colon$$

$$\operatorname{Im} v_1'' - F_0(v_2'',v_2'')\in \mathfrak{C}_B\},$$

$$\pi^{-1}(0) = \pi'^{-1}(0)\times\pi''^{-1}(0) = \{v\in V_1\oplus V_2\colon \operatorname{Im} v_1 - F_0(v_2,v_2)\in\mathfrak{C}_E\}.$$

3. *The bundles* π', π'' *and* π *are real-analytically trivial. Let* $\{\Delta_k'\}$ *be the root systems of the simple ideals of* $\mathfrak{G}_{\Psi\backslash(\Gamma\cup\Sigma)}$, $\{\Delta_l''\}$ *the root systems of the simple ideals of* $\mathfrak{G}^\Sigma_\Sigma$. *Then*

(3a) π' *is holomorphically trivial if and only if A is a union of sets of the form* $[\Psi\backslash(\Gamma\cup\Sigma)]\cap\Delta_k'$,

(3b) π'' *is holomorphically trivial if and only if B is a union of sets of the form* $\Sigma\cap\Delta_l''$, *and*

(3c) π *is holomorphically trivial if and only if both* π' *and* π'' *are holomorphically trivial.*

We come to the final step of the decomposition of $G_0(x_{\Gamma,\Sigma})$. Recall $\beta\colon G_0(x_{\Gamma,\Sigma})\to K(x_{\Gamma,\Sigma})$ with $\beta^{-1}k(x_{\Gamma,\Sigma})=k\cdot G^\Sigma_{\Psi\backslash\Gamma,0}(x_{\Gamma,\Sigma})$. Proposition 11.20 fibres

$$\beta^{-1}(x_{\Gamma,\Sigma}) = G^\Sigma_{\Psi\backslash\Gamma,0}(x_{\Gamma,\Sigma}) \to G^{B'}_{E',0}(x_{\Gamma,\Sigma}) \tag{11.21a}$$

by

$$c_\Gamma c^2_\Sigma c_E^{-1}\xi(v) \to c_\Gamma c^2_\Sigma\xi(\pi v) \qquad \text{for every } v\in\mathfrak{S}^\Sigma_{\Psi\backslash\Gamma,E}. \tag{11.21b}$$

As seen from the case where the unit disc is fibered over one of its (one point) boundary components, $G^{B'}_{E',0}(x_{\Gamma,\Sigma})$ is generally not $K^{\Sigma}_{\Psi\backslash\Gamma}$-stable, so we cannot just extend (11.21) to $G_0(x_{\Gamma,\Sigma})$ by K-equivariance. That problem is avoided by use of a measurable, a.e. real analytic, "section" of the principal $K^{\Sigma}_{\Psi\backslash\Gamma}$-bundle $K\to K/K^{\Sigma}_{\Psi\backslash\Gamma}$.

Let $\mathcal{V}^{\Sigma}_{\Psi\backslash\Gamma}$ denote the orthocomplement of $\mathcal{K}^{\Sigma}_{\Psi\backslash\Gamma}$ in $\mathcal{K}$; define $V^{\Sigma}_{\Psi\backslash\Gamma}=\exp(\mathcal{V}^{\Sigma}_{\Psi\backslash\Gamma})\subset K$. Then every $k\in K$ has expression $k=vk'$, $v\in V^{\Sigma}_{\Psi\backslash\Gamma}$ and $k'\in K^{\Sigma}_{\Psi\backslash\Gamma}$, but that factorization need not be unique. To approach uniqueness, let $K/K^{\Sigma}_{\Psi\backslash\Gamma}$ carry a riemannian metric induced by a bi-invariant metric on K, so the exponential map

$$\exp_{\mathcal{O}}\colon \mathcal{V}^{\Sigma}_{\Psi\backslash\Gamma}\to K/K^{\Sigma}_{\Psi\backslash\Gamma} \qquad \text{by } \exp_{\mathcal{O}}(v) = \exp(v)\cdot\mathcal{O}$$

where $\mathcal{O}$ is the base point. Now let $\mathcal{W}^{\Sigma}_{\Psi\backslash\Gamma}$ consist of all $v\in\mathcal{V}^{\Sigma}_{\Psi\backslash\Gamma}$ such that $\exp_{\mathcal{O}}(tv)$, $0\leqq t\leqq 1$, is the unique minimizing geodesic from $\mathcal{O}$ to $\exp_{\mathcal{O}}(v)$. Define $W^{\Sigma}_{\Psi\backslash\Gamma}=\exp(\mathcal{W}^{\Sigma}_{\Psi\backslash\Gamma})\subset V^{\Sigma}_{\Psi\backslash\Gamma}$. Then

(11.22a) $W^{\Sigma}_{\Psi\backslash\Gamma}(\mathcal{O})$ is a dense open subset of $K/K^{\Sigma}_{\Psi\backslash\Gamma}$, and

(11.22b) $\exp_{\mathcal{O}}\colon \mathcal{W}^{\Sigma}_{\Psi\backslash\Gamma}\to W^{\Sigma}_{\Psi\backslash\Gamma}(\mathcal{O})$ is a diffeomorphism.

That gives us a "section" $s\colon K/K^{\Sigma}_{\Psi\backslash\Gamma}\to K$ by

(11.22c) $s(\exp_{\mathcal{O}}(w))=\exp(w)\in W^{\Sigma}_{\Psi\backslash\Gamma}$ for $w\in\mathcal{W}^{\Sigma}_{\Psi\backslash\Gamma}$,

(11.22d) $s(k\cdot\mathcal{O})\in kK^{\Sigma}_{\Psi\backslash\Gamma}$ arbitrary for $k\cdot\mathcal{O}\notin W^{\Sigma}_{\Psi\backslash\Gamma}(\mathcal{O})$.

Although s may be discontinuous, it is real analytic on the dense open subset $W^{\Sigma}_{\Psi\backslash\Gamma}(\mathcal{O})\subset K/K^{\Sigma}_{\Psi\backslash\Gamma}$ whose complement has measure zero.

11.23. THEOREM. *Let Γ and Σ be disjoint subsets of Ψ, let $E\subset\Psi\backslash\Gamma$, and consider disjoint unions*

$$E = A\cup B,\quad E' = A'\cup B',\quad \Psi\backslash(\Gamma\cup\Sigma) = A\cup A',\quad \Sigma = B\cup B'.$$

If $k\in K$, define

$$\gamma_k^{A,B}\colon k\cdot G^{\Sigma}_{\Psi\backslash\Gamma,0}(x_{\Gamma,\Sigma})\to k\cdot G^{B'}_{E',0}(x_{\Gamma,\Sigma})$$

by

$$\gamma_k^{A,B}(kc_\Gamma c_\Sigma^2 c_E^{-1}\xi v) = kc_\Gamma c_\Sigma^2\xi(\pi v) \qquad \textit{for } v\in\mathfrak{S}^{\Sigma}_{\Psi\backslash\Gamma,E}.$$

Retain the notation (11.22) *and define*

$$\gamma^{A,B}\colon G_0(x_{\Gamma,\Sigma})\to s(K/K^{\Sigma}_{\Psi\backslash\Gamma})\cdot G^{B'}_{E',0}(x_{\Gamma,\Sigma})$$

by

$$\gamma^{A,B}\Big|_{kG^{\Sigma}_{\Psi\backslash\Gamma,0}(x_{\Gamma,\Sigma})} = \gamma^{A,B}_{s(k\cdot\mathfrak{O})}.$$

1. *Given* $k\in K$, $\gamma_k^{A,B}$ *is a holomorphic fibre bundle that is equivalent to the bundle* $\pi\colon \mathfrak{S}^{\Sigma}_{\Psi\backslash\Gamma,E}\to D^{B'}_{E'}$ *of Proposition* 11.20 *under the maps*:

$$\begin{array}{ccc} k\cdot G^{\Sigma}_{\Psi\backslash\Gamma,0}(x_{\Gamma,\Sigma}) & \xrightarrow[\xi^{-1}c_E c_\Sigma^{-2} c_\Gamma^{-1} k^{-1}]{} & \mathfrak{S}^{\Sigma}_{\Psi\backslash\Gamma,E} \\ \Big\downarrow \gamma_k^{A,B} & & \Big\downarrow \pi \\ k\cdot G^{B'}_{E',0}(x_{\Gamma,\Sigma}) & \xrightarrow{\xi^{-1} c_\Sigma^{-2} c_\Gamma^{-1} k^{-1}} & D^{B'}_{E'} \end{array}$$

2. $\gamma^{A,B}$ *is a well-defined, but not necessarily continuous, "fibre bundle" whose fibres coincide with those of its restrictions* $\gamma^{A,B}_{s(k\cdot\mathfrak{O})}$, *and which is real analytic over* $W^{\Sigma}_{\Psi\backslash\Gamma}\cdot G^{B'}_{E',0}(x_{\Gamma,\Sigma})$.

3. *The fibration* $\beta\colon G_0(x_{\Gamma,\Sigma})\to K(x_{\Gamma,\Sigma})$ *of Theorem* 11.8 *comes from* $\gamma^{A,B}$ *by*

$$\beta = \{\beta\cdot(\gamma^{A,B})^{-1}\}\cdot\gamma^{A,B} \quad \textit{where} \quad \{\beta\cdot(\gamma^{A,B})^{-1}\}\cdot kG^{B'}_{E'}(x_{\Gamma,\Sigma}) = k(x_{\Gamma,\Sigma}).$$

PROOF. Statement (1) is immediate from Proposition 11.20. $\gamma_k^{A,B}$ sends the β-fibre $k\cdot G^{\Sigma}_{\Psi\backslash\Gamma,0}(x_{\Gamma,\Sigma}) = s(k\cdot\mathfrak{O})\cdot G^{\Sigma}_{\Psi\backslash\Gamma,0}(x_{\Gamma,\Sigma})$ over $k(x_{\Gamma,\Sigma}) = s(k\cdot\mathfrak{O})(x_{\Gamma,\Sigma})$ into its hermitian symmetric subspace $k\cdot G^{B'}_{E',0}(x_{\Gamma,\Sigma})$. As β is well defined, (3) follows.

We check that $\gamma^{A,B}$ is well defined. Let $x\in G_0(x_{\Gamma,\Sigma})$. Suppose $\beta(x)=k(x_{\Gamma,\Sigma})$, $k\in K$. Then $x\in\beta^{-1}k(x)=s(k\cdot\mathfrak{O})\cdot G^{\Sigma}_{\Psi\backslash\Gamma,0}(x_{\Gamma,\Sigma})$, so $x=s(k\cdot\mathfrak{O})g(x_{\Gamma,\Sigma})$ for some $g\in G^{\Sigma}_{\Psi\backslash\Gamma,0}$. Now, by definition, $\gamma^{A,B}(x) = \gamma^{A,B}_{s(k\cdot\mathfrak{O})}(x) = s(k\cdot\mathfrak{O})c_\Gamma c_\Sigma\xi(\pi v)$ where $v\in\mathfrak{S}^{\Sigma}_{\Psi\backslash\Gamma,E}$ such that $g(x_{\Gamma,\Sigma}) = c_\Gamma c_\Sigma^2 c_E^{-1}\xi(v)$. That procedure is unique, first because k is specified modulo $K_{\Psi\backslash\Gamma}$ because β is well defined, and then because k is specified modulo $K^{\Sigma}_{\Psi\backslash\Gamma}$ by Lemma 11.4.

$(\gamma^{A,B})^{-1}(s(k\cdot\mathfrak{O})(x_{\Gamma,\Sigma})) = (\gamma^{A,B}_{s(k\cdot\mathfrak{O})})^{-1}(s(k\cdot\mathfrak{O})(x_{\Gamma,\Sigma}))$ by construction and well definition of $\beta\cdot(\gamma^{A,B})^{-1}$.

Over $W^{\Sigma}_{\Psi\backslash\Gamma}\cdot G^{B'}_{E',0}(x_{\Gamma,\Sigma})$, $\gamma^{A,B}$ is given by

$$\gamma^{A,B}(kc_\Gamma c_\Sigma^2 c_E^{-1}\xi(v)) = kc_\Gamma c_\Sigma^2\xi(\pi v), \qquad k\in W^{\Sigma}_{\Psi\backslash\Gamma} \text{ and } v\in\mathfrak{S}^{\Sigma}_{\Psi\backslash\Gamma,E}.$$

That is real analytic. Q.E.D.

We now have the triple fibration

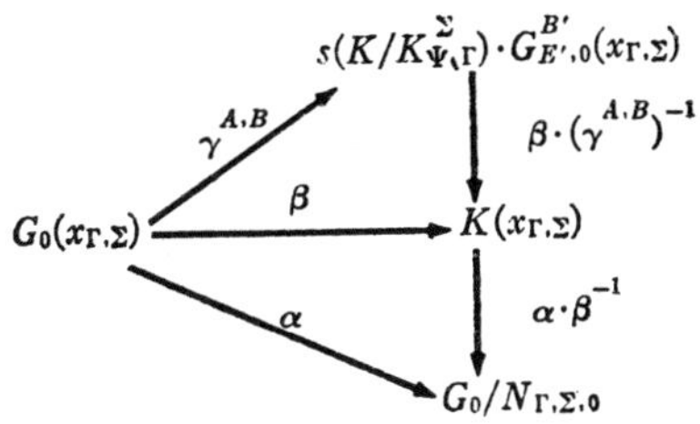

of an arbitrary orbit $G_0(x_{\Gamma,\Sigma})\subset X$ as described at the beginning of this §11.

Appendix to §11—Siegel domains. We first recall the concepts of cone and Siegel domain.

Let U be a real vector space with a positive definite inner product $\langle\ ,\ \rangle$. Then a *self dual cone* in U means a subset $\mathfrak{C}\subset U$ such that

$$\mathfrak{C} = \{u\in U\colon \text{if } 0\neq v\in\bar{\mathfrak{C}} \text{ then } \langle u, v\rangle > 0\}.$$

It is automatic that a self dual cone is a cone (nonempty, and $u\in\mathfrak{C}$ and $r>0$ imply $ru\in\mathfrak{C}$), is open, is convex, and does not contain a straight line. "Domain of positivity" is another term for self dual cone.

Let $\mathfrak{C}\subset U$ be a convex open cone that does not contain a straight line. Define $V=U+iU$ complexification; if $v\in V$ than $v=(\operatorname{Re} v)+i(\operatorname{Im} v)$ with $\operatorname{Re} v$ and $\operatorname{Im} v$ in U. The **tube domain** over $\mathfrak{C}$ is defined to be the set $\{v\in V\colon \operatorname{Im} v\in\mathfrak{C}\}$. The Poincaré and Siegel upper half planes are examples. By **Siegel domain of type I** we mean a tube domain over a convex open cone that does not contain a straight line. A hermitian symmetric space is of **tube type** if it is holomorphically diffeomorphic to a Siegel domain of type I; in that case the cone is self dual. A standard classification result says that a hermitian symmetric space is of tube type if and only if it is a product, each of whose factors is one of

$$\begin{array}{lll} \mathbf{SU}(m,m)/\mathbf{S}(\mathbf{U}(m)\times\mathbf{U}(m)), & m\geqq 1; & \text{or}\\ \mathbf{SO}^*(4m)/\mathbf{U}(2m), & m\geqq 2; & \text{or}\\ \mathbf{Sp}(m;\mathbf{R})/\mathbf{U}(m), & m\geqq 1; & \text{or}\\ \mathbf{SO}(2,m)/\mathbf{SO}(2)\times\mathbf{SO}(m), & m\geqq 1; & \text{or}\\ \mathbf{E}_{7,\,\text{index}-25}/\mathbf{E}_6\cdot\mathbf{SO}(2). & & \end{array}$$

Let V_1 and V_2 be complex vector spaces, U_1 a real form of V_1, $\mathfrak{C}\subset U_1$ a cone as above, and $F\colon V_2\times V_2\to V_1$ an hermitian map that is positive definite in the sense

$$0 \neq v_2 \in V_2 \qquad \text{implies } 0 \neq F(v_2, v_2) \in \overline{\mathfrak{C}}.$$

That data defines the domain

$$\{(v_1, v_2) \in V_1 \oplus V_2: \operatorname{Im} v_1 - F(v_2, v_2) \in \mathfrak{C}\}.$$

Those domains are the **Siegel domains of type II.** Siegel domains of type I are the special case $V_2=0$. Every hermitian symmetric space of noncompact type has a standard realization (Korányi-Wolf [2]) as a Siegel domain of type II, and that Siegel domain is of type I if and only if the hermitian symmetric space is of tube type; the cone is self dual.

Let V_1, V_2 and V_3 be complex vector spaces, U_1 a real form of V_1, $\mathfrak{C} \subset U_1$ a cone as above, and $D \subset V_3$ a bounded domain. Suppose further that, for every $w \in D$, we have

$$\overset{h}{F_w}: V_2 \times V_2 \to V_1 \text{ hermitian and}$$
$$\overset{b}{F_w}: V_2 \times V_2 \to V_1 \text{ bilinear}$$

depending smoothly on $w \in D$, such that

$$\overset{b}{F_0} = 0 \quad \text{and} \quad \overset{h}{F_0} \text{ is positive definite relative to } \mathfrak{C}.$$

Then we define

$$F_w: V_2 \times V_2 \to V_1 \qquad \text{by } F_w = \overset{h}{F_w} + \overset{b}{F_w}.$$

Assume that for every $w \in D$, F_w is nondegenerate in the sense that each of $F_w(v, V_2)=0$ and $F_w(V_2, v)=0$ implies $v=0$. Then the domain

$$\{(v_1, v_2, v_3) \in V_1 \oplus V_2 \oplus V_3: v_3 \in D \text{ and } \operatorname{Im} v_1 - \operatorname{Re} F_{v_3}(v_2, v_2) \in \mathfrak{C}\}$$

is called a **Siegel domain of type III** with **base** D. Siegel domains of type II are the special case $V_3=0$, i.e. the case where the base is reduced to a point.

In [15, §7] Wolf and Korányi describe $X_0=G_0(x_0)$ as a Siegel domain of type III in $\mathfrak{M}^-$ whose base is the bounded domain in $\mathfrak{M}^-_{\Psi\setminus E}$ equivalent to an arbitrary boundary component $G_{\Psi\setminus E,0}(x_{E,\phi})$ of X_0. That description involves the holomorphic injection

$$\xi: \mathfrak{M}^- \to X \qquad \text{by } \xi(v) = \exp(v)\cdot x_0$$

of $\mathfrak{M}^-$ as a dense open subset of X, and is given in our notation as follows.

First note $\mathfrak{M}^- = \mathfrak{M}^{E-}_{\Psi} + \mathfrak{R}^{E-}_{\Psi} = \mathfrak{M}^-_{\Psi\setminus E} + \mathfrak{M}^{E-}_{E} + \mathfrak{R}^{E-}_{\Psi}$ direct. Now if $v \in \mathfrak{M}^-$, we will decompose

$$v = v_1 + v_2 + v_3 \qquad \text{with } v_1 \in \mathfrak{M}^{E-}_E,\ v_2 \in \mathfrak{R}^{E-}_{\Psi},\ v_3 \in \mathfrak{M}^-_{\Psi\setminus E}.$$

Our bounded domain in $\mathfrak{M}^-_{\Psi\setminus E}$ is

$$D_{\Psi\setminus E} = \xi^{-1} G_{\Psi\setminus E,0}(x_0).$$

The real form of $\mathfrak{M}_E^{E-}$ is $\mathrm{ad}(c_E)\{\mathcal{G}_0 \cap \mathrm{ad}(c_E^{-1})\mathfrak{M}_E^{E-}\}$, and the positive definite inner product on that real form is induced by the positive definite hermitian inner product $(u, v) = -\langle u, \sigma\tau v\rangle$ on $\mathcal{G}$. The self dual cone $\mathfrak{C}_E \subset \mathrm{ad}(c_E)\{\mathcal{G}_0 \cap \mathrm{ad}(c_E^{-1})\mathfrak{M}_E^{E-}\}$ requires some notation. If $F \subset \Psi$ we have $K_F = K \cap G_{F,0}$ and the fixed point set K_F^E of $\mathrm{ad}(c_E^4)$ on K_F; now $\mathrm{ad}(c_E^2)$ restricts to an involutive automorphism of $\mathcal{K}_F^E$, and that defines an algebra

$$\mathcal{K}_F^{E*} = \{v \in \mathcal{K}_F^E : \mathrm{ad}(c_E^2)v = v\} + i\{v \in \mathcal{K}_F^E : \mathrm{ad}(c_E^2)v = -v\}.$$

Its analytic group is denoted K_F^{E*}. Now the cone is given, in our notation, by

$$\mathfrak{C}_E = \mathrm{ad}(K_E^{E*}) \cdot \xi^{-1}(x_{E,\phi}) \quad \text{and} \quad \xi^{-1}(x_{E,\phi}) = \textstyle\sum_E e_{-\psi}.$$

If $w \in \mathfrak{M}^-_{\Psi\setminus E}$ then $\mu(w): \mathcal{R}_\Psi^{E-} \to \mathcal{R}_\Psi^{E-}$ is defined by $\mu(w)v = \mathrm{ad}(w) \cdot \mathrm{ad}(c_E^2) \cdot \sigma\tau(v)$; it has operator norm <1 if and only if $w \in D_{\Psi\setminus E}$. Thus if $w \in D_{\Psi\setminus E}$, $I + \mu(w)$ is invertible and we have $F_w: \mathcal{R}_\Psi^{E-} \times \mathcal{R}_\Psi^{E-} \to \mathfrak{M}_E^{E-}$ by

$$F_w(u, v) = -(i/2)[u, \mathrm{ad}(c_E^2) \cdot \sigma\tau \cdot (I + \mu(w))^{-1} v].$$

The main result of [15, §7] says first that

(11A.1) $\mathfrak{S}_E = \{v \in \mathfrak{M}^- : v_3 \in D_{\Psi\setminus E}$ and $\mathrm{Im}\, v_1 - \mathrm{Re}\, F_{v_3}(v_2, v_2) \in \mathfrak{C}_E\}$ is a Siegel domain of type III with base $D_{\Psi\setminus E}$, and second that

(11A.2) $\qquad c_E X_0 \subset \xi(\mathfrak{M}^-)$ and $\xi^{-1}: c_E X_0 \cong \mathfrak{S}_E$.

In proving the results stated above, Wolf and Korányi use certain symmetry properties

$$\mathrm{ad}(k) F_w(u, v) = F_{\mathrm{ad}(k)w}(\mathrm{ad}(k)u, \mathrm{ad}(k)v) \qquad \text{for } k \in K_\Psi^{E*},$$

$$F_w(u, (I + \mu(w))v) = F_0(u, v) \quad \text{and} \quad F_0(u, \mu(w)v) = F_0(v, \mu(w)u)$$

of the F_w and rely on the ξ^{-1}-equivariant actions of $\mathrm{ad}(c_E)Z_{E,\phi,0}$ and $\mathrm{ad}(c_E)N_{E,\phi,0}$ (cf.(10.10)) from the natural action on $c_E X_0$ to the (defined by equivariance) action on $\mathfrak{M}^-$. For if N^E denotes the common unipotent radical of those groups,

$$N^E = \{\mathrm{ad}(c_E) N_{E,\phi,0}\}^u = \{\mathrm{ad}(c_E) Z_{E,\phi,0}\}^u,$$

then [**15**, Theorem 6.8] N^E is 2-step nilpotent, its Lie algebra $\mathfrak{N}^E$ has complexification

$$\mathfrak{N}^{EC} = \mathrm{ad}(c_E)\mathfrak{N}^u_{E,\phi} = \mathrm{ad}(c_E)\mathfrak{Z}^u_{E,\phi} = (\mathfrak{Q}_\Psi^{E+} + \mathfrak{R}_\Psi^{E-}) + \mathfrak{M}_E^{E-},$$

and $\mathrm{ad}(c_E)Z_{E,\phi,0}$ and $\mathrm{ad}(c_E)N_{E,\phi,0}$ have respective identity components

$$\{L_{E,\phi,0}\cdot K_E^{E*}\}\cdot N^E \quad \text{and} \quad \{G_{\Psi\setminus E,0}\cdot L_{E,\phi,0}\cdot K_E^{E*}\}\cdot N^E;$$

here we recall $\mathfrak{G}^E_{\Psi,0} = \mathfrak{G}_{\Psi\setminus E,0} \oplus \mathfrak{L}_{E,\phi,0} \oplus \mathfrak{G}^E_{E,0}$, $\mathfrak{L}_{E,\phi,0} \subset \mathfrak{K}$, so that

$$K_\Psi^{E*} = K_{\Psi\setminus E}\cdot L_{E,\phi,0}\cdot K_E^{E*}.$$

The action of K_Ψ^{E*} on $\mathfrak{M}^-$ is induced by ad_G; it preserves each of $\mathfrak{M}_E^{E-}$, $\mathfrak{R}_\Psi^{E-}$ and $\mathfrak{M}^-_{\Psi\setminus E}$, the real form specified of $\mathfrak{M}_E^{E-}$ and the cone $\mathfrak{C}_E$ there, the bilinear maps F_w and the domain $D_{\Psi\setminus E}$, and thus also the Siegel domain $\mathfrak{S}_E$. The action of N^E on $\mathfrak{M}^-$ is based on the unique representation of every $g \in N^E$ in the form $\exp(u+(I-\mathrm{ad}(c_E^2)\sigma\tau)v)$ with $u \in \mathfrak{M}_E^{E-}$ and $v \in \mathfrak{R}_\Psi^{E-}$; if

$$f: \mathfrak{R}_\Psi^{E-} \times \bar{\mathfrak{M}}_{\Psi\setminus E} \to \mathfrak{R}_\Psi^{E-} \quad \text{by} \quad (v, w) \to f_v(w) = v + \mu(w)v,$$

then N^E acts on $\mathfrak{M}^-$ by

$$g(e) = e + u + f_v(e_3) + 2iF_{e_3}(e_2, f_v(e_3)) + iF_{e_3}(f_v(e_3), f_v(e_3));$$

that action preserves the Siegel domain $\mathfrak{S}_E$. The action of $G_{\Psi\setminus E,0}$ on $\mathfrak{M}^-$ or $\mathfrak{S}_E$ is not made explicit in [**15**], but the action of $K_{\Psi\setminus E} \subset K_\Psi^{E*}$ is described above, and $G_{\Psi\setminus E,0}$ acts on $D_{\Psi\setminus E}$ in the natural fashion.

11A.3. PROPOSITION. *Let $\mathfrak{S}_E$ be the Siegel domain* (11A.1) *with base $D_{\Psi\setminus E}$ that is holomorphically equivalent (by $c_E^{-1}\cdot\xi$) to X_0. Define*

$$\pi: \mathfrak{S}_E \to D_{\Psi\setminus E} \quad by \quad \pi(v) = v_3.$$

1. *$\pi: \mathfrak{S}_E \to D_{\Psi\setminus E}$ is a holomorphic fibre bundle.*
2. *The identity component $\{G_{\Psi\setminus E,0}\cdot L_{E,\phi,0}\cdot K_E^{E*}\}\cdot N^E$ of $\mathrm{ad}(c_E)N_{E,\phi,0}$ acts transitively on $\mathfrak{S}_E$ by holomorphic bundle automorphisms, with $G_{\Psi\setminus E,0}$ transitive on the base and the identity component $\{L_{E,\phi,0}\cdot K_E^{E*}\}\cdot N^E$ of $\mathrm{ad}(c_E)Z_{E,\phi,0}$ trivial on the base and transitive on every fibre.*
3. *The fibre $\pi^{-1}(0)$ is the Siegel domain $\{v \in \mathfrak{M}_E^{E-} + \mathfrak{R}_\Psi^{E-}: \mathrm{Im}\, v_1 - F_0(v_2, v_2) \in \mathfrak{C}_E\}$ of type* II *homogeneous under $\{L_{E,\phi,0}\cdot K_E^{E*}\}\cdot N^E$, and it is holomorphically equivalent to a hermitian symmetric space of noncompact type that has rank $|E|$ and tube part $G^E_{E,0}(x_0)$.*

4. *The bundle* $\pi:\mathfrak{S}_E\to D_{\Psi\setminus E}$ *is real-analytically trivial, and it is holomorphically trivial if and only if* E *is a union of sets* $\Psi\cap\Delta^k$ *where the* Δ^k *are the root systems of the simple ideals* $\mathcal{G}^k$ *of* $\mathcal{G}$.

PROOF. Let $e=e_1+e_2+e_3\in\mathfrak{S}_E$ and represent $g\in N^E$ in the form $\exp(u+(I-\operatorname{ad}(c_E^2)\sigma\tau)v)$ with $u\in\mathfrak{M}_E^{E-}$ and $v\in\mathfrak{R}_\Psi^{E-}$. Let $e'=g(e)$; then

$$e' = e + u + f_v(e_3) + 2iF_{e_3}(e_2, f_v(e_3)) + iF_{e_3}(f_v(e_3), f_v(e_3)),$$

so $e_3'=e_3$, $e_2'=e_2+f_v(e_3)$ and

$$e_1' = e_1 + u + 2iF_{e_3}(e_2, f_v(e_3)) + iF_{e_3}(f_v(e_3), f_v(e_3)).$$

Thus the action of N^E on $\mathfrak{S}_E$ preserves every π-fibre. If $k\in L_{E,\phi,0}\cdot K_E^{E*}$, then $k(e)=\operatorname{ad}(k)e$, so $k(e)_3=\operatorname{ad}(k)e_3=e_3$. Thus the action of $L_{E,\phi,0}\cdot K_E^{E*}$, on $\mathfrak{S}_E$ preserves every π-fibre. Define

$$v = -(I+\mu(e_3))^{-1}e_2 \quad\text{and}\quad g = \exp((I-\operatorname{ad}(c_E^2)\sigma\tau)v)\in N^E.$$

Then $g(e)=e_1'+0+e_3$. Now define

$$u = -\mathfrak{Re}\, e_1' \quad\text{and}\quad g' = \exp(u)\in N^E.$$

Then $g'g(e)=e_1''+0+e_3$ with ie_1'' real. In fact $g'g(e)\in\mathfrak{S}_E$ gives us $\operatorname{Im} e_1''-\operatorname{Re} F_{e_3}(0,0)=\operatorname{Im} e_1''\in\mathfrak{C}_E$. Thus there exists $k\in K_E^{E*}$ such that $\operatorname{ad}(k)\cdot\operatorname{Im} e_1''=\xi^{-1}(x_{E,\phi})$. Now $kg'g\in K_E^{E*}\cdot N^E$ has the property that

$$(kg'g)(e) = i\xi^{-1}(x_{E,\phi}) + e_3.$$

We have proved that the π-fibres are just the orbits of $\{L_{E,\phi,0}\cdot K_E^{E*}\}\cdot N^E$ on $\mathfrak{S}_E$.

Let $g\in G_{\Psi\setminus E,0}$. Then g normalizes $\{L_{E,\phi,0}\cdot K_E^{E*}\}\cdot N^E$ in $\{G_{\Psi\setminus E,0}\cdot L_{E,\phi,0}\cdot K_E^{E*}\}\cdot N^E$, hence permutes its orbits on $\mathfrak{S}_E$. Now g is a π-fibre map of $\mathfrak{S}_E$. Thus (2) is proved subject to (1) being demonstrated.

Let Q denote the identity component of $\operatorname{ad}(c_E)N_{E,\phi,0}$, Y its isotropy subgroup at $i\xi^{-1}(x_{E,\phi})\in\mathfrak{S}_E$, and Z the subgroup preserving $\pi^{-1}(0)$. Then $Y\subset Z=\{K_{\Psi\setminus E}\cdot L_{E,\phi,0}\cdot K_E^{E*}\}\cdot N^E=K_\Psi^{E*}\cdot N^E$ and π is projection $Q/Y\to Q/Z$ by $qY\to qZ$. Thus π is a differentiable fibre bundle with structure group Z. But π visibly is a holomorphic map, and the fibre over $g(0)$, $g\in G_{\Psi\setminus E,0}$, is the complex submanifold

$$g\cdot\pi^{-1}(0) = g\cdot\{v\in\mathfrak{M}_E^{E-}+\mathfrak{R}_\Psi^{E-}: \operatorname{Im} v_1 - F_0(v_2,v_2)\in\mathfrak{C}_E\}$$

of $\mathfrak{S}_E$. As Z acts by holomorphic transformations of $\pi^{-1}(0)$, it follows that $\pi:\mathfrak{S}_E\to D_{\Psi\setminus E}$ is a holomorphic fibre bundle. That completes the proof of (1) and (2).

Every π-fibre is of the form $g\cdot\pi^{-1}(0)$, $g\in G_{\Psi\setminus E,0}$, and $G_{\Psi\setminus E,0}$ normalizes the transitive group $\{L_{E,\phi,0}\cdot K_E^{E*}\}\cdot N^E$ of transformations of the fibre; thus an arbitrary π-fibre has the same properties as $\pi^{-1}(0)$ as complex homogeneous space of $\{L_{E,\phi,0}\cdot K_E^{E*}\}\cdot N_E^E$. Now we observe

$$\pi^{-1}(0) = \{v \in \mathfrak{M}_E^{E-} + \mathfrak{R}_\Psi^{E-} : \operatorname{Im} v_1 - F_0(v_2, v_2) \in \mathfrak{C}_E\}$$

where $F_0:\mathfrak{R}_\Psi^{E-}\times\mathfrak{R}_\Psi^{E-}\to\mathfrak{M}_E^{E-}$ is the positive definite (relative to $\mathfrak{C}_E$) hermitian map given by

$$F_0(u, v) = -(i/2)[u, \operatorname{ad}(c_E^2)\sigma\tau v].$$

Thus $\pi^{-1}(0)$, hence every fibre, is a homogeneous Siegel domain of type II. Comparison with [2, §6] suggests that it should be symmetric, and we prove that symmetry as follows.

Observe that $\{v\in\pi^{-1}(0): v_2=0\}$ is the tube domain over the self dual cone $\mathfrak{C}_E$; in fact that is the realization of $G_{E,0}^E(x_0)$ as a Siegel domain of type I. Thus there is an analytic automorphism δ of order 2 on that tube domain with $i\xi^{-1}(x_{E,\phi})$ as the only fixed point. Given $e\in\pi^{-1}(0)$, define

$$v = -e_2 \quad\text{and}\quad g_e = \exp((I - \operatorname{ad}(c_E^2)\sigma\tau)v) \in N^E.$$

Then $e'=g_e(e)$ satisfies $e_3'=0$, $e_2'=e_2+f_{-e_2}(0)$ and

$$e_1' = e_1 + 2iF_0(e_2, f_{-e_2}(0)) + iF_0(f_{-e_2}(0), f_{-e_2}(0)).$$

As $f_{-e_2}(0)=-e_2+\mu(0)(-e_2)=-e_2$, now $e_1'=e_1+2iF_0(e_2, -e_2)+iF_0(-e_2, -e_2)=e_1-iF_0(e_2, e_2)$, $e_2'=0$ and $e_3'=0$. Thus

$$g_e(e) = e_1 - iF_0(e_2, e_2) \in \{v \in \pi^{-1}(0): v_2 = 0\}.$$

Now define $e^*=g_e\delta g_e(e)$. It satisfies $e_1^*=\delta g_e(e)+2iF_0(0, f_{-e_2}(0))+iF_0(f_{-e_2}(0), f_{-e_2}(0))=\delta g_e(e)+iF_0(e_2, e_2)$ and $e_2^*=f_{-e_2}(0)=-e_2$. Thus

$$e^* = g_e\delta g_e(e) = \delta(e_1 - iF_0(e_2, e_2)) + iF_0(e_2, e_2) - e_2.$$

Now

$$\begin{aligned}(e^*)^* &= \delta(\delta(e_1 - iF_0(e_2, e_2)) + iF_0(e_2, e_2) - iF_0(e_2, e_2)) + iF_0(e_2, e_2) + e_2\\ &= \delta^2(e_1 - iF_0(e_2, e_2)) + iF_0(e_2, e_2) + e_2\\ &= e_1 - iF_0(e_2, e_2) + iF_0(e_2, e_2) + e_2 = e_1 + e_2 = e.\end{aligned}$$

Thus $e\to e^*$ is an analytic automorphism of $\pi^{-1}(0)$ with square 1. If $e=e^*$ then

$$e_1 = \delta(e_1 - iF_0(e_2, e_2)) + iF_0(e_2, e_2) \quad\text{and}\quad e_2 = -e_2.$$

Thus $e_2=0$ and $e_1=\delta(e_1)$. The latter says $e_1=i\xi^{-1}(x_{E,\phi})$. Thus $e=e^*$ implies $e=i\xi^{-1}(x_{E,\phi})$. Now $e\rightarrow e^*$ is an analytic automorphism of order 2 on $\pi^{-1}(0)$ with $i\xi^{-1}(x_{E,\phi})$ as unique fixed point. As $\pi^{-1}(0)$ is a homogeneous complex manifold holomorphically equivalent to a bounded domain, it follows that $\pi^{-1}(0)$ is a hermitian symmetric space of noncompact type. The tube part of that symmetric space is $\{v\in\pi^{-1}(0): v_2=0\}$, tube domain over $\mathfrak{C}_E$, holomorphically equivalent to $G^E_{E,0}(x_0)$; in particular $\pi^{-1}(0)$ has rank $|E|$. Statement (3) is proved.

$G_{\Psi\backslash E,0}$ has Iwasawa decomposition $K_{\Psi\backslash E}\cdot A_{\Psi\backslash E}\cdot U_{\Psi\backslash E}$ and the solvable group $B=A_{\Psi\backslash E}\cdot U_{\Psi\backslash E}$ is simply transitive on $D_{\Psi\backslash E}$. Now define

$$\lambda: \pi^{-1}(0)\times D_{\Psi\backslash E}\rightarrow \mathfrak{S}_E \quad \text{by} \quad \lambda(t, b(0)) = b(t).$$

Then λ is a real analytic trivialization of $\pi:\mathfrak{S}_E\rightarrow D_{\Psi\backslash E}$.

Let $\mathfrak{g}^k$ denote the simple ideals of $\mathfrak{g}$, Δ^k the root system of $\mathfrak{g}^k$. If E is a union of sets $\Psi\cap\Delta^k$, then $G_0(x_0)=G_{\Psi\backslash E,0}(x_0)\times G_{E,0}(x_0)$ and π projects to the first factor. Conversely suppose π holomorphically trivial. Let D'_E denote the bounded symmetric domain in $\mathfrak{M}^{E-}_E+\mathfrak{R}^{E-}_\Psi$ holomorphically equivalent to $\pi^{-1}(0)$. Then there is a holomorphic equivalence of $G_0(x_0)\cong\mathfrak{S}_E$ with $D_{\Psi\backslash E}\times D'_E$. As G_0 is the largest connected group of holomorphic automorphisms of $G_0(x_0)$, as $G_{\Psi\backslash E,0}$ is the largest connected group for $D_{\Psi\backslash E}$, and as the Bergman metric on $D_{\Psi\backslash E}\times D'_E$ is the direct sum of the Bergman metrics of $D_{\Psi\backslash E}$ and of D'_E, it follows that $G_0=G_{\Psi\backslash E,0}\times G_{E,0}$ where $G_{E,0}$ is the largest connected group of holomorphic automorphisms of D'_E. Thus $\mathfrak{g}_{\Psi\backslash E}$ is a sum of simple ideals of $\mathfrak{g}$, so $\Psi\backslash E$ is a union of sets $\Psi\cap\Delta^k$, and we conclude that E is a union of sets $\Psi\cap\Delta^k$. Statement (4) is proved, and that completes the proof of Proposition 11A.3. Q.E.D.

References

1. R. Bott and H. Samelson, *The cohomology ring of G/T*, Proc. Nat. Acad. Sci. U.S.A. **41** (1955), 490–493.

2. A. Korányi and J. A. Wolf, *Realization of hermitian symmetric spaces as generalized half-planes*, Ann. of Math. (2) **81** (1965), 265–288.

3. B. Kostant, *On the conjugacy of real Cartan subalgebras*, Proc. Nat. Acad. Sci. U.S.A. **41** (1955), 967–970.

4. ———, *Orbits, symplectic structures and representation theory*, Proc. U.S. Japan Seminar in Differential Geometry (Kyoto, 1965), Nippon Hyoronsho, Tokyo, 1966.

5. ———, *Lectures on symplectic manifolds and unitary representations*, Course given at M.I.T. in Fall, 1965; class notes by N. Iwahori.

6. T. Maebashi, *A fixpoint theorem on* Sp(*n*)/U(*n*), Math. J. Okayama Univ. **12** (1966), 91–105.

7. C. C. Moore, *Compactifications of symmetric spaces*. II: *The Cartan domains*, Amer. J. Math. **86** (1964), 358–378.

8. M. Takeuchi, *Cell decompositions and Morse inequalities on certain symmetric spaces*, J. Fac. Sci. Univ. Tokyo **12** (1965), 81–192.

9. ———, *On orbits in a compact hermitian symmetric space*, Amer. J. Math. **90** (1968), 657–680.

10. J. A. Wolf, *On the classification of hermitian symmetric spaces*, J. Math. Mech. **13** (1964), 489–496.

11. ———, *The geometry and structure of isotropy irreducible homogeneous spaces*, Acta Math. **120** (1968), 59–148.

12. ———, *The action of a real semisimple group on a complex flag manifold.* II: *Unitary representations on partially holomorphic cohomology spaces* (in preparation).

13. J. A. Wolf and P. A. Griffiths, *Complete maps and differentiable coverings*, Michigan Math. J. **10** (1963), 253–255.

14. J. A. Wolf and A. Gray, *Homogeneous spaces defined by Lie group automorphisms*, J. Differential Geometry **2** (1968), 77–159.

15. J. A. Wolf and A. Korányi, *Generalized Cayley transformations of bounded symmetric domains*, Amer. J. Math. **87** (1965), 899–939.

UNIVERSITY OF CALIFORNIA, BERKELEY, CALIFORNIA 94720

HARMONIC ANALYSIS ON SEMISIMPLE LIE GROUPS

BY HARISH-CHANDRA

1. **Introduction.** Let G be a locally compact group which we assume to be separable and unimodular. Let dx denote the Haar measure on G. If π is a unitary representation of G on a Hilbert space $\mathfrak{H}$ and $f \in L_1(G)$, we write

$$\pi(f) = \int_G f(x)\pi(x)dx.$$

Then $\pi(f)$ is a bounded operator on $\mathfrak{H}$ and

$$\pi(f * g) = \pi(f)\pi(g) \quad (f, g \in L_1(G)),$$

where $f * g$ denotes the convolution of f and g.

Let A be a bounded linear operator on $\mathfrak{H}$. We say that A is of the trace class if the series

$$\sum_i |(\psi_i, A\psi_i)|$$

converges for every orthonormal base $\{\psi_i\}_{i \in J}$ of $\mathfrak{H}$. Moreover if this is so, we define

$$\operatorname{tr} A = \sum_i (\psi_i, A\psi_i).$$

Then tr A is actually independent of the choice of this base.

Let V_π denote the set of all $f \in L_1(G)$ such that $\pi(f)$ is of the trace class. Then V_π is a linear subspace of $L_1(G)$. Put

$$\Theta_\pi(f) = \operatorname{tr} \pi(f) \qquad (f \in V_\pi).$$

Then Θ_π is a linear function on V_π which we may call the character of π. Of course this concept would be useful only when the space V_π is fairly large.

Let $\mathcal{E}(G)$ denote the set of all equivalence classes of irreducible unitary representations of G. It is easy to see that for any representation π, V_π and Θ_π depend only on the class ω of π. Hence we may de-

A Colloquium Lecture delivered before the 74th Summer meeting of the Society at Eugene, Oregon, on August 26, 1969; received by the editors October 29, 1969.

AMS Subject Classifications. Primary 2265, 2260.

Key Words and Phrases. Characters, invariant eigendistributions, parabolic subgroups, Schwartz space, cusp forms, discrete series, Fourier transform of a Schwartz function, Plancherel formula, p-adic groups.

note them by V_ω and Θ_ω respectively. Put

$$V_0 = \bigcap_{\omega\in\mathcal{E}(G)} V_\omega.$$

For any function $f\in L_1(G)$ and $x\in G$, define the function f^x by $f^x(y)=f(x^{-1}yx)$. It is clear that $f^x\in V_0$ whenever $f\in V_0$ and therefore the group G operates on V_0. Now, in order to describe the main problem of harmonic analysis, we have to fix a certain topological vector space V satisfying the following conditions:[1]

(1) $V\subset V_0\cap C(G)$, where $C(G)$ is the space of all complex-valued continuous functions on G.

(2) V is stable under G and every $x\in G$ defines a continuous endomorphism of V.

(3) For any $\omega\in\mathcal{E}(G)$, the restriction of Θ_ω on V is a continuous linear function on V, which determines ω uniquely.

For any $f\in V$, define $\hat{f}(\omega)=\Theta_\omega(f)$ $(\omega\in\mathcal{E}(G))$. Then $\hat{f}$ is a complex-valued function on $\mathcal{E}(G)$, which may be regarded as the Fourier transform of f. Let us agree to call a continuous linear function on V, a distribution. For any distribution T and $x\in G$, define the distribution T^x by $T^x(f)=T(f^{x^{-1}})$ $(f\in V)$. We say that T is invariant if $T^x=T$ for all $x\in G$.

The central question in harmonic analysis may now be formulated as follows. *Given an invariant distribution T on G, how to express it as a "linear combination" of the characters Θ_ω $(\omega\in\mathcal{E}(G))$?* In fact, to every such T, we would like to associate a "distribution" $\hat{T}$ on $\mathcal{E}(G)$ in such a way that $T(f)=\hat{T}(\hat{f})$ for all $f\in V$. (Here by a "distribution" on $\mathcal{E}(G)$, we mean some sort of linear functional on a suitable space of functions on $\mathcal{E}(G)$.) If δ is the Dirac measure, so that $\delta(f)=f(1)$ $(f\in V)$, the determination of $\hat{\delta}$ is just the problem of the explicit Plancherel formula for G.

For an arbitrary G, our understanding of $\mathcal{E}(G)$ is still very rudimentary (see [**5**]) and therefore no serious attempt to attack the above problem can yet be contemplated. However the situation for real semisimple (or reductive) Lie groups seems rather encouraging [**4**(**a**), (**d**)] and there is reason to hope that some substantial results on reductive $\mathfrak{p}$-adic groups may be attainable within a few years. Therefore I shall limit myself to these two cases in these lectures. The other extreme case, when G is a nilpotent or solvable Lie group, has been studied extensively by Dixmier, Kirillov, Moore, and Pukánszky (see [**6**], [**7**]).

[1] We do not claim here that such a space V necessarily exists. Of course the crucial condition is (3).

Our entire approach to harmonic analysis on reductive groups is based on the philosophy of cusp forms (see [2] and [4(c), (d)]) and the success of this method depends on a good knowledge of the discrete series (see §3) for such groups. In the case of real groups the characters of the discrete series are all known (see Theorem 8) and therefore the theory works reasonably well. But for $\mathfrak{p}$-adic groups our understanding of the supercuspidal representations (§14) is still very poor. Nevertheless the resemblance between the two cases is so striking that one cannot give up the hope that the same philosophy would eventually work in both cases.

2. **Characters in the real case.** Suppose G is a real, connected, semisimple Lie group with finite center. Fix a maximal compact subgroup K of G. If $\omega \in \mathcal{E}(G)$ and $\mathfrak{d} \in \mathcal{E}(K)$, we define $[\omega:\mathfrak{d}]$ as follows. Fix a representation $\pi \in \omega$ and let π_K denote the restriction of π on K. Then $[\omega:\mathfrak{d}]$ is the multiplicity of $\mathfrak{d}$ in π_K. Since K is compact, every irreducible representation of K is finite-dimensional. We denote by $d(\mathfrak{d})$ the degree of a representation in the class $\mathfrak{d}$.

THEOREM 1. *There exists an integer $N \geqq 1$ such that*

$$[\omega:\mathfrak{d}] \leqq Nd(\mathfrak{d})$$

for all $\omega \in \mathcal{E}(G)$ and $\mathfrak{d} \in \mathcal{E}(K)$.

Let $C_c^\infty(G)$ denote the space of all complex-valued C^∞ functions on G with compact support, taken with its usual topology [9]. The following result is an easy consequence of Theorem 1.

THEOREM 2. *Let π be an irreducible unitary representation of G. Then for any $f \in C_c^\infty(G)$, $\pi(f)$ is of the trace class. Put*

$$\Theta_\pi(f) = \operatorname{tr} \pi(f) \qquad (f \in C_c^\infty(G)).$$

Then Θ_π is a distribution on G in the sense of Schwartz [9].

As we have seen in §1, Θ_π depends only on the class ω of π. Hence we may denote it by Θ_ω.

THEOREM 3. *Let ω_1, ω_2 be two elements in $\mathcal{E}(G)$. Then $\Theta_{\omega_1} = \Theta_{\omega_2}$ if and only if $\omega_1 = \omega_2$.*

This shows that if we take $V = C_c^\infty(G)$, then all the conditions of §1 are fulfilled.

Let u be a differential operator on G. Then its adjoint u^* is the differential operator given by the relation

$$\int_G u^*f \cdot g\,dx = \int_G f \cdot ug\,dx \qquad (f, g \in C_c^\infty(G)).$$

If T is a distribution on G, the mapping

$$f \to T(u^*f) \qquad (f \in C_c^\infty(G)),$$

is also a distribution which we denote by uT. Let F be a locally summable function on G. Put

$$T_F(f) = \int Ffdx \qquad (f \in C_c^\infty(G)).$$

Then T_F is a distribution on G. We say that $T=F$ if $T=T_F$.

Let $\mathfrak{Z}$ be the algebra of all differential operators on G which commute with both left and right translations of G. Then $\mathfrak{Z}$ is abelian. A distribution T is called $\mathfrak{Z}$-finite if the space of all distributions of the form zT ($z \in \mathfrak{Z}$) has finite dimension. Moreover T is said to be an eigendistribution of $\mathfrak{Z}$, if this space has dimension $\leqq 1$. Finally we say that T is K-finite, if the left and right translates of T by elements of K, span a finite-dimensional space.

Fix an indeterminate t and let $l=\text{rank } G$. For any $x \in G$, let $D(x)$ denote the coefficient of t^l in the polynomial $\det(t+1-\text{Ad}(x))$. Then D is an analytic function on G which is not identically zero. A point $x \in G$ is called regular if $D(x) \neq 0$. Let G' denote the set of all regular points of G. Then G' is an open dense subset of G whose complement is of measure zero.

THEOREM 4. *Let Θ be an invariant and $\mathfrak{Z}$-finite distribution on G. Then there exists an analytic function F on G' such that F is locally summable on G and $\Theta = F$.*

Fix $\omega \in \mathcal{E}(G)$ and let Θ_ω denote, as before, the character of ω. It is easy to see that Θ_ω is an invariant eigendistribution of $\mathfrak{Z}$. Therefore it follows from Theorem 4 that Θ_ω is a function.

3. **The discrete series.** We return to the assumptions of §1. Let π be an irreducible unitary representation of G on a Hilbert space $\mathfrak{H}$. We say that π is of type L_p ($p \geqq 1$), if there exist nonzero elements $\phi, \psi \in \mathfrak{H}$ such that

$$\int_G |(\phi, \pi(x)\psi)|^p dx < \infty.$$

In case π is of type L_2 (or, as we sometimes say, square-integrable), there exists a positive number $d(\pi)$ such that[2]

[2] conj c ($c \in C$) denotes the complex conjugate of c.

$$\int_G (\phi_1, \pi(x)\psi_1)\cdot\mathrm{conj}(\phi_2, \pi(x)\psi_2)dx = d(\pi)^{-1}(\phi_1, \phi_2)\cdot\mathrm{conj}(\psi_1, \psi_2)$$

for all $\phi_i, \psi_i \in \mathfrak{H}$ $(i=1, 2)$ (see [**3**] and [**5**, p. 640]). These are called the Schur orthogonality relations for π and $d(\pi)$ the formal degree of π. Let $\mathcal{E}_p(G)$ denote the set of all classes $\omega\in\mathcal{E}(G)$ which contain representations of type L_p. If $\omega\in\mathcal{E}_2(G)$, we put $d(\omega)=d(\pi)$ for $\pi\in\omega$. The set $\mathcal{E}_2(G)$ is called the discrete series for G.

Now suppose K is an *open and compact* subgroup of G. Normalize dx in such a way that the total measure of K is 1. For any $\omega\in\mathcal{E}(G)$ and $\mathfrak{d}\in\mathcal{E}(K)$, let $[\omega:\mathfrak{d}]$ denote, as before (see §2), the multiplicity of $\mathfrak{d}$ in ω and $d(\mathfrak{d})$ the degree of $\mathfrak{d}$.

THEOREM 5. *Fix* $\mathfrak{d}\in\mathcal{E}(K)$. *Then*

$$d(\mathfrak{d}) \geqq \sum_{\omega\in\mathcal{E}_2(G)} d(\omega)[\omega:\mathfrak{d}].$$

This is a simple consequence of the Schur orthogonality relations.

COROLLARY. *Fix* $\omega\in\mathcal{E}_2(G)$. *Then*

$$[\omega:\mathfrak{d}] \leqq d(\omega)^{-1}d(\mathfrak{d})$$

for all $\mathfrak{d}\in\mathcal{E}(K)$.

4. Characters of the discrete series (real case). Define G as in §2.

THEOREM 6. $\mathcal{E}_2(G)\neq\emptyset$ *if and only if* G *has a compact Cartan subgroup.*

Let us now assume that G has a compact Cartan subgroup B. Then any two such Cartan subgroups are conjugate in G. Let $\mathfrak{g}$ be the Lie algebra of G and $\mathfrak{g}_c$ its complexification. Let G_c be the simply connected complex-analytic group corresponding to $\mathfrak{g}_c$. For simplicity we assume that G is the real analytic subgroup of G_c corresponding to $\mathfrak{g}$.

Let $\mathfrak{b}$ be the Lie algebra of B and $\mathfrak{b}^*$ the dual of $\mathfrak{b}$. Fix some order on the real vector space $(-1)^{1/2}\mathfrak{b}^*$. Let $W=W(\mathfrak{g}/\mathfrak{b})$ denote the Weyl group of[3] $(\mathfrak{g}_c, \mathfrak{b}_c)$ and B^* the character group of B. Then W operates on $\mathfrak{b}$ and B and therefore by duality, it also operates on $\mathfrak{b}^*$ and B^*. Let $\langle b^*, b\rangle$ $(b^*\in B^*, b\in B)$ denote the value of the character b^* at b. Then we have the relations

[3] For any finite-dimensional vector space V over $\mathbf{R}$, we denote by V^* its dual and by V_c its complexification. Moreover $V_c^*=(V^*)_c$.

$$(\exp H)^s = \exp(sH), \qquad \langle b^{*s}, b\rangle = \langle b^*, b^{s^{-1}}\rangle$$

for $H \in \mathfrak{b}$ and $s \in W$.

Put

$$\varpi = \prod_{\alpha \in P} H_\alpha$$

where P is the set of all positive roots of $(\mathfrak{g}, \mathfrak{b})$ and H_α the element in $\mathfrak{b}_c$, which is dual to α under the Killing form so that

$$\alpha(H) = \operatorname{tr}(\operatorname{ad} H \cdot \operatorname{ad} H_\alpha) \qquad (H \in \mathfrak{b}).$$

Then ϖ is a polynomial function on $\mathfrak{b}_c^*$ and $\varpi^s = \epsilon(s)\varpi$ $(s \in W)$ where $\epsilon(s) = 1$ or -1. Moreover there exists an analytic function Δ on B such that

$$\Delta(\exp H) = \prod_{\alpha \in P} (e^{\alpha(H)/2} - e^{-\alpha(H)/2}) \qquad (H \in \mathfrak{b}).$$

For any $b^* \in B^*$, let $\log b^*$ denote the element $\lambda \in (-1)^{1/2}\mathfrak{b}^*$ such that

$$\langle b^*, \exp H\rangle = e^{\lambda(H)} \qquad (H \in \mathfrak{b}),$$

and put

$$\varpi(b^*) = \varpi(\log b^*) \qquad (b^* \in B^*).$$

We say that b^* is regular if $\varpi(b^*) \neq 0$.

Let $\tilde{B}$ be the normalizer of B in G. Then $W(G/B) = \tilde{B}/B$ may be regarded as a subgroup of W.

THEOREM 7. *Fix $b^* \in B^*$. Then there exists an invariant eigendistribution Θ of $\mathfrak{Z}$ on G such that*[4]

(1) $\sup_{x \in G'} |D(x)|^{1/2} |\Theta(x)| < \infty$,

(2) $\Theta = \Delta^{-1} \sum_{s \in W(G/B)} \epsilon(s) b^{*s}$ *pointwise on* $B \cap G'$.

Moreover Θ *is unique if* $\varpi(b^*) \neq 0$.

Let $B^{*\prime}$ be the set of all regular elements in B^* and for any $b^* \in B^{*\prime}$, let Θ_{b^*} denote the corresponding distribution of Theorem 7. Put $\epsilon(b^*) = \operatorname{sign} \varpi(b^*)$ and $q = \frac{1}{2}\dim G/K$.

THEOREM 8. *Given $b^* \in B^{*\prime}$, there exists a unique class $\omega(b^*) \in \mathcal{E}_2(G)$ such that*

$$(-1)^q \epsilon(b^*) \Theta_{b^*}$$

is the character of $\omega(b^*)$. The mapping $b^* \to \omega(b^*)$ from $B^{*\prime}$ to $\mathcal{E}_2(G)$ is

[4] Here we have to keep in mind Theorem 4.

surjective. Moreover $\omega(b_1^*)=\omega(b_2^*)$ *if and only if* b_1^*, b_2^* *are conjugate under* $W(G/B)$.

Let σ be an automorphism of G such that $\sigma B=B$. Then σ also operates on B^* by the relation

$$\langle \sigma b^*, b\rangle = \langle b^*, \sigma^{-1}b\rangle \qquad (b^* \in B^*,\ b \in B).$$

Define $\epsilon(\sigma)=\pm 1$ by the rule $\varpi^\sigma=\epsilon(\sigma)\varpi$. Also if T is a distribution on G, we define the distribution T^σ by

$$T^\sigma(f) = T(f') \qquad (f \in C_c^\infty(G)),$$

where $f'(x)=f(\sigma x)$ $(x\in G)$.

LEMMA 1. *Let* σ *be an automorphism of* G *such that* $\sigma B=B$. *Then*

$$(\Theta_{b^*})^\sigma = \epsilon(\sigma)\Theta_{\sigma b^*} \qquad (b^* \in B^{*\prime}).$$

Actually it is possible to define Θ_{b^*} for all $b^*\in B^*$ in such a way that both Theorem 7 and Lemma 1 hold for all σ.

5. **Parabolic subgroups.** Let G be as in §2. By a parabolic subgroup (psgp) P of G, we mean a closed subgroup of G with the following two properties.

(1) If $\mathfrak{p}$ is the Lie algebra of P, then $\mathfrak{p}_c$ contains a Borel subalgebra (i.e. a maximal solvable subalgebra) of $\mathfrak{g}_c$.

(2) P is the normalizer of $\mathfrak{p}$ in G.

By the radical N of P, we mean the maximal normal subgroup of P such that $\mathrm{Ad}(n)$ is unipotent for every $n\in N$. An abelian subgroup A of G is said to split if, for every $a\in A$, $\mathrm{Ad}(a)$ can be diagonalized over $\boldsymbol{R}$. Fix K as in §2 and let θ denote the Cartan involution of G corresponding to K. Put $M_1=P\cap\theta(P)$ and let A be a maximal connected split abelian subgroup lying in the center of M_1. Then A is unique and M_1 is the centralizer of A in G. Let $X(M_1)$ denote the group of all continuous homomorphisms of M_1 into the multiplicative group $\boldsymbol{R}^\times$ of real numbers. Put

$$M = \bigcap_{\chi\in X(M_1)} \ker|\chi|,$$

where $|\chi|(m)=|\chi(m)|$ $(m\in M_1)$. Then $M_1=MA$, $P=MAN$ and the corresponding mapping of $M\times A\times N$ into P is a diffeomorphism. We call this the Langlands decomposition of P and A the split component of P. Let $\mathfrak{a}$ be the Lie algebra of A. Then the exponential mapping defines a bijection of $\mathfrak{a}$ on A and we denote its inverse by log. By the parabolic rank of P we mean the dimension of A and denote it by prk P.

The group M is, in general, neither connected nor semisimple. However it is always reductive. We say that the psgp P is cuspidal if M has a compact Cartan subgroup.

Let $\mathfrak{h}$ be a Cartan subalgebra of $\mathfrak{g}$. We denote by $\mathfrak{h}_R$ the set of all points in $\mathfrak{h}$ where every root of $(\mathfrak{g}, \mathfrak{h})$ takes a real value.

LEMMA 2. *A psgp $P=MAN$ is cuspidal if and only if there exists a θ-stable Cartan subalgebra $\mathfrak{h}$ of $\mathfrak{g}$ such that $\mathfrak{a}=\mathfrak{h}_R$.*

Let $P=MAN$ be a *prgp* of G. Then $G=KP$. Hence every element $x\in G$ can be written in the form $x=kman$ $(k\in K, m\in M, a\in A, n\in N)$. The element a is uniquely determined and we put $H_P(x)=\log a$. Let $\mathfrak{a}^*$ be the dual of $\mathfrak{a}$ and $\mathfrak{n}$ the Lie algebra of N. Then we define $\rho_P\in\mathfrak{a}^*$ by

$$\rho_P(H) = \tfrac{1}{2}\,\mathrm{tr}(\mathrm{ad}\, H)_{\mathfrak{n}} \qquad (H\in\mathfrak{a}),$$

where $(\mathrm{ad}\, H)_{\mathfrak{n}}$ denotes the restriction of ad H on $\mathfrak{n}$. By a root of (P, A) (or P) we mean an element $\alpha\in\mathfrak{a}^*$ with the following property. There should exist an element $X\neq 0$ in $\mathfrak{n}$ such that $[H, X]=\alpha(H)X$ for all $H\in\mathfrak{a}$. Let $\mathfrak{a}^+$ be the set of all $H\in\mathfrak{a}$ such that $\alpha(H)\geqq 0$ for every root α of (P, A). We put $A^+=\exp\mathfrak{a}^+$.

Two psgps P_1, P_2 are said to be associated if their split components A_1, A_2 are conjugate under G (or equivalently under K).

A psgp P is called minimal if it is minimal among all psgps of G. The following three conditions are mutually equivalent.

(1) $P=MAN$ is a minimal psgp.

(2) $M\subset K$.

(3) prk $P=\mathrm{rank}\; G/K$.

Any two minimal psgps are conjugate under K.

6. **The functions Ξ and σ.** Fix a minimal psgp $P=MAN$ of G. Then $G=KAN$ and this is an Iwasawa decomposition of G. Put

$$\Xi(x) = \int_K e^{-\rho(H(xk))}dk \qquad (x\in G)$$

where $H(x)=H_P(x)$, $\rho=\rho_P$ and the Haar measure dk on K is so normalized that the total measure of K is 1.

Define a norm on $\mathfrak{g}$ by setting

$$\|X\|^2 = -\,\mathrm{tr}(\mathrm{ad}\, X\; \mathrm{ad}\, \theta(X)) \qquad (X\in\mathfrak{g}).$$

Then since $G=KAK$, there exists a unique function σ on G such that

(1) $\sigma(k_1xk_2)=\sigma(x)$ $\quad(k_1, k_2\in K, x\in G)$,

(2) $\sigma(\exp H)=\|H\|$ $\quad(H\in\mathfrak{a})$.

The functions Ξ and σ are both independent of the choice of P and they have the following remarkable properties.

(1) $\int_K \Xi(xky)dk = \Xi(x)\Xi(y) \qquad (x, y \in G)$.

(2) There exist numbers $c, d \geqq 0$ such that

$$1 \leqq \Xi(a)e^{\rho(\log a)} \leqq c(1 + \sigma(a))^d \qquad (a \in A^+).$$

(3) We can choose $r_0 \geqq 0$ such that

$$\int_G \Xi(x)^2(1 + \sigma(x))^{-r_0}dx < \infty.$$

(4) There exists a number $r \geqq 0$ such that[5]

$$\int_G | D(x) |^{-1/2}\Xi(x)(1 + \sigma(x))^{-r}dx < \infty.$$

Let $\mathfrak{q}$ be any linear subspace of $\mathfrak{g}$. Then we denote by $d\mathfrak{q}$ the Euclidean measure on $\mathfrak{q}$ corresponding to the above Euclidean norm on $\mathfrak{g}$. It is convenient to normalize the Haar measure dx on G as follows. Let da and dn denote the Haar measures on A and N which corespond to $d\mathfrak{a}$ and $d\mathfrak{n}$ respectively under the exponential mapping. Then if $x = kan$ ($k \in K$, $a \in A$, $n \in N$),

$$dx = e^{2\rho(\log a)}dkdadn$$

where dk is the normalized Haar measure of K. This normalization of dx is independent of the choice of the minimal psgp P.

7. **The Schwartz space and the cusp forms.** Let $\mathfrak{D}$ be the algebra of all differential operators on G. Let $\mathfrak{D}_l$ and $\mathfrak{D}_r$ denote the subalgebras consisting of those $D \in \mathfrak{D}$ which commute with the left and right translations of G respectively. Let $\mathfrak{D}_0$ be the subalgebra of $\mathfrak{D}$ generated by $\mathfrak{D}_l \cup \mathfrak{D}_r$.

For $f \in C^\infty(G)$, $D \in \mathfrak{D}_0$ and $r \geqq 0$, put

$$\nu_{D,r}(f) = \sup_G |Df| (1 + \sigma)^r \Xi^{-1}.$$

Let $\mathcal{C}(G)$ denote the space of all $f \in C^\infty(G)$ such that $\nu_{D,r}(f) < \infty$ for all $D \in \mathfrak{D}_0$ and $r \geqq 0$. We topologize $\mathcal{C}(G)$ by means of the set of seminorms $\nu_{D,r}$ ($D \in \mathfrak{D}_0$, $r \geqq 0$). Then $\mathcal{C}(G)$ is a Hausdorff, locally convex and complete space and we call it the Schwartz space of G. We list here some of the properties of this space.

(1) The inclusion mapping of $C_c^\infty(G)$ into $\mathcal{C}(G)$ is continuous and the image is dense in $\mathcal{C}(G)$.

[5] Here D has the same meaning as in §2.

(2) $\mathcal{C}(G) \subset L_2(G)$.

(3) $\mathcal{C}(G)$ is closed under convolution and the mapping $(f, g) \to f * g$ $(f, g \in \mathcal{C}(G))$ is continuous.

For $f \in \mathcal{C}(G)$ and $P = MAN$ a psgp of G, define

$$f^P(x) = \int_N f(xn)dn \qquad (x \in G)$$

where dn is the Haar measure on N. (This integral always exists.) We say that f is a cusp form if $f^P = 0$ for every psgp $P \neq G$. Let ${}^\circ\mathcal{C}(G)$ denote the space of all cusp forms on G. Then ${}^\circ\mathcal{C}(G)$ is a closed subspace of $\mathcal{C}(G)$.

THEOREM 9. *Every K-finite and $\mathfrak{Z}$-finite function in $L_2(G)$ is a cusp form. Conversely K-finite eigenfunctions of $\mathfrak{Z}$ in $\mathcal{C}(G)$ span a dense subspace of ${}^\circ\mathcal{C}(G)$.*

Combining this with Theorem 6 we get the following result.

COROLLARY. ${}^\circ\mathcal{C}(G) \neq \{0\}$ *if and only if* rank G = rank K.

Let τ be a unitary double representation of K on a finite-dimensional Hilbert space V. We denote by $\mathcal{C}(G, \tau)$ the space of all $f \in \mathcal{C}(G) \otimes V$ such that $f(k_1xk_2) = \tau(k_1)f(x)\tau(k_2)$ $(k_1, k_2 \in K, x \in G)$. Put

$${}^\circ\mathcal{C}(G, \tau) = \mathcal{C}(G, \tau) \cap ({}^\circ\mathcal{C}(G) \otimes V).$$

THEOREM 10. dim ${}^\circ\mathcal{C}(G, \tau) < \infty$.

Fix a psgp $P = MAN$ of G. Then M is a reductive group with finitely many connected components. Moreover $K_M = K \cap M$ is a maximal compact subgroup of M. Let τ_M denote the restriction of τ on K_M. Then it is not difficult to define the spaces $\mathcal{C}(M)$, ${}^\circ\mathcal{C}(M)$ and ${}^\circ\mathcal{C}(M, \tau_M)$ in the same way as above.

Let $f \in \mathcal{C}(G)$. Then we write $f^P \sim 0$ if

$$\int_M \operatorname{conj} \phi(m) \cdot f^P(xm)dm = 0$$

for all $x \in G$ and $\phi \in {}^\circ\mathcal{C}(M)$. (Here dm is the Haar measure on M and the integral always exists.)

THEOREM 11. *Let f be an element in $\mathcal{C}(G)$ such that $f^P \sim 0$ for all cuspidal subgroups P of G. Then $f = 0$.*

Let $\mathfrak{h}_i$ $(1 \leq i \leq r)$ be a complete set of θ-stable Cartan subalgebras of $\mathfrak{g}$, no two of which are conjugate under G. Put $\mathfrak{a}_i = \mathfrak{h}_{i,R}$ and $A_i = \exp \mathfrak{a}_i$. Let $\mathcal{C}_i(G)$ denote the set of all $f \in \mathcal{C}(G)$ with the following property.

If $P=MAN$ is any cuspidal subgroup of G, then $f^P\sim 0$ unless A is conjugate to A_i under K. Then $\mathcal{C}_i(G)$ is a closed subspace of $\mathcal{C}(G)$.

THEOREM 12. *$\mathcal{C}(G)=\sum_{1\leq i\leq r}\mathcal{C}_i(G)$ where the sum is direct and smooth.*

Let E_i denote the projection of $\mathcal{C}(G)$ on $\mathcal{C}_i(G)$ corresponding to the above direct sum. Then smoothness means that E_i are continuous endomorphisms of $\mathcal{C}(G)$. Let $\mathfrak{H}_i$ denote the closure of $\mathcal{C}_i(G)$ in $\mathfrak{H}=L_2(G)$. Then $\mathfrak{H}$ is the orthogonal sum of $\mathfrak{H}_i$ $(1\leq i\leq r)$ and E_if $(f\in\mathcal{C}(G))$ is, actually, the orthogonal projection of f in $\mathfrak{H}_i$.

A distribution T on G is called tempered if it extends to a continuous linear function on $\mathcal{C}(G)$.

LEMMA 3. *Let Θ be an invariant and $\mathfrak{Z}$-finite distribution on G. Then Θ is tempered if and only if we can choose c, $r\geq 0$ such that*[4,5]

$$|D(x)|^{1/2}|\Theta(x)| \leqq c(1+\sigma(x))^r$$

for all $x\in G'$.

COROLLARY. *Suppose Θ is tempered in the above lemma. Then*

$$\Theta(f) = \int_G f\cdot\Theta dx$$

for $f\in\mathcal{C}(G)$.

This follows immediately from (4) of §6. We write

$$(\Theta,f) = \int_G \operatorname{conj}\Theta\cdot f dx \qquad (f\in\mathcal{C}(G)).$$

Let f be a continuous function on G. We say that f satisfies the weak inequality if there exist numbers c, $r\geq 0$ such that

$$|f(x)| \leqq c\Xi(x)(1+\sigma(x))^r$$

for all $x\in G$.

LEMMA 4. *Let T be a tempered distribution on G which is both K-finite and $\mathfrak{Z}$-finite. Then T satisfies the weak inequality.*

Here we have to observe that a K-finite and $\mathfrak{Z}$-finite distribution is necessarily an analytic function.

8. **The projection of $\mathcal{C}(G)$ on ${}^\circ\mathcal{C}(G)$.** We now return to the assumptions and notation of §4. Put $B'=B\cap G'$ and

$$F_f(b) = \Delta(b)\int_G f(xbx^{-1})dx \qquad (b\in B')$$

for $f\in \mathcal{C}(G)$. This integral is always convergent and F_f is a C^∞ function on B'.

THEOREM 13. *Let $f\in {}^\circ\mathcal{C}(G)$. Then F_f extends to a C^∞ function on B and*

$$F_f(b) = \sum_{b^*\in B^{*\prime}} (\Theta_{b^*}, f)\langle b^*, b\rangle \qquad (b\in B).$$

Now

$$\varpi = \prod_{\alpha>0} H_\alpha$$

may be regarded as a differential operator on B. Then, for any $f\in \mathcal{C}(G)$, ϖF_f extends to a continuous function B.

LEMMA 5. *There exists a number $c_G>0$ such that*

$$\varpi F_f(1) = (-1)^q c_G f(1)$$

for all $f\in\mathcal{C}(G)$.

It is possible to determine c_G explicitly as follows. Let $\mathfrak{k}$ be the Lie algebra of K. We assume, as we may, that $\mathfrak{k}\supset\mathfrak{b}$. Let P be the set of all positive roots of $(\mathfrak{g}, \mathfrak{b})$ and P_k the subset of those $\alpha\in P$ which can be regarded as roots of $(\mathfrak{k}, \mathfrak{b})$. Put

$$\varpi_k = \prod_{\alpha\in P_k} H_\alpha, \qquad \rho_k = \frac{1}{2}\sum_{\alpha\in P_k}\alpha$$

and normalize the measure dx as in §6.

LEMMA 6. *With the above normalization of the Haar measure of G, we have*[6]

$$c_G = [W(G/B)](2\pi)^q 2^{n/2}\varpi_k(\rho_k)$$

where

$$n = \dim(G/K) - \operatorname{rank}(G/K).$$

THEOREM 14. *For any $f\in\mathcal{C}(G)$ and $b^*\in B^*$, define*

$$f_{b^*}(x) = (-1)^q c_G^{-1}\varpi(b^*)(\Theta_{b^*}, r(x)f) \qquad (x\in G)$$

where r denotes the right-regular representation of G on $\mathcal{C}(G)$. Then

[6] $[S]$ denotes the number of elements in a finite set S.

$f_{b^*} \in {}^\circ\mathcal{C}(G)$ and the series

$$\sum_{b^* \in B^*} f_{b^*}$$

converges absolutely [4(a), p. 8] *in $\mathcal{C}(G)$. Let ${}^\circ f$ denote its sum. Then $f \to {}^\circ f$ is a continuous projection of $\mathcal{C}(G)$ onto ${}^\circ\mathcal{C}(G)$.*

Let ${}^\circ\mathfrak{H}$ denote the closure of ${}^\circ\mathcal{C}(G)$ in $\mathfrak{H} = L_2(G)$ and l the left-regular representation of G on $\mathfrak{H}$.

THEOREM 15. *${}^\circ\mathfrak{H}$ is the closure of an orthogonal sum of closed subspaces which are invariant and irreducible under l. Conversely every irreducible subspace of $\mathfrak{H}$ is contained in ${}^\circ\mathfrak{H}$. Let ${}^\circ E$ denote the orthogonal projection of $\mathfrak{H}$ on ${}^\circ\mathfrak{H}$. Then ${}^\circ f = {}^\circ Ef$ for $f \in \mathcal{C}(G)$.*

9. **The space $\mathfrak{A}(G, \tau)$.** Fix G and τ as in §2 and §7. We denote by $\mathfrak{A}(G, \tau)$ the space of all functions $f \in C^\infty(G) \otimes V$ satisfying the following three conditions:

(1) f is τ-spherical i.e. $f(k_1 x k_2) = \tau(k_1) f(x) \tau(k_2)$ $(k_1, k_2 \in K,\ x \in G)$.

(2) f is $\mathfrak{Z}$-finite.

(3) $|f|$ satisfies the weak inequality.

For a psgp $P = MAN$ of G, put

$$d_P(m) = |\det(Ad(m))_{\mathfrak{n}}|^{1/2} \qquad (m \in MA).$$

Let a be a variable element in A. We say that $a \to_P \infty$ if $\sigma(a) \to \infty$ and there exists a number $\epsilon > 0$ such that $\alpha(\log a) \geqq \epsilon\sigma(a)$ for every root α of (P, A). Since MA is a reductive group with finitely many connected components and $K_M = K \cap M$ is a maximal compact subgroup of MA, the space $\mathfrak{A}(MA, \tau_M)$ can be defined without difficulty.

THEOREM 16. *For any $f \in \mathfrak{A}(G, \tau)$, there exists a unique element $f_P \in \mathfrak{A}(MA, \tau_M)$ such that*

$$\lim_{\substack{a \to \infty \\ P}} \{d_P(ma) f(ma) - f_P(ma)\} = 0$$

for $m \in MA$ and $a \in A$.

Given $f \in \mathfrak{A}(G, \tau)$, we write $f_P \sim 0$ if

$$\int_M (\phi(m), f_P(ma))_V \, dm = 0$$

for all $\phi \in {}^\circ\mathcal{C}(M, \tau_M)$ and $a \in A$. Here the scalar product is in V and the integral is always convergent.

LEMMA 7. *Let f be an element in $\mathfrak{A}(G, \tau)$ such that $f_P \sim 0$ for every* *cuspidal subgroup P of G. Then $f=0$.*

Here the analogy with Theorem 11 is obvious.

For any integer q, let $\mathfrak{A}_q(G, \tau)$ denote the space of all $f \in \mathfrak{A}(G, \tau)$ with the following property. If P is a psgp of G, then $f_P \sim 0$ unless prk $P=q$.

THEOREM 17. $\mathfrak{A}_0(G, \tau) = {}^\circ\mathcal{C}(G, \tau)$. *Fix $f \neq 0$ in $\mathfrak{A}(G, \tau)$ and choose a* psgp $P=MAN$ *of G with the following two properties*:

(1) $f_P \not\sim 0$.

(2) *P is minimal with respect to condition* (1).

Then for any fixed $a \in A$, the function $m \to f_P(ma)$ $(m \in M)$ lies in ${}^\circ\mathcal{C}(M, \tau_M)$.

10. **The Eisenstein integral.** Fix a cuspidal subgroup $P=MAN$ of G. Given $\psi \in {}^\circ\mathcal{C}(M, \tau_M)$, we extend it to a function on G as follows:

$$\psi(kman) = \tau(k)\psi(m) \qquad (k \in K,\ m \in M,\ a \in A,\ n \in N).$$

For $\nu \in \mathfrak{a}_c^*$ and $\psi \in {}^\circ\mathcal{C}(M, \tau_M)$, define

$$E(P:\psi:\nu:x) = \int_K \psi(xk)\tau(k^{-1}) \exp\{((-1)^{1/2}\nu - \rho_P)(H_P(xk))\} dk$$

for $x \in G$. We call this the Eisenstein integral. For fixed P and ψ, it is an analytic function of (ν, x) on $\mathfrak{a}_c^* \times G$ which is holomorphic in ν.

LEMMA 8. *$E(P:\psi:\nu) \in \mathfrak{A}(G, \tau)$ for $\nu \in \mathfrak{a}^*$. If P' is another psgp of G, then*

$$E_{P'}(P:\psi:\nu) \sim 0 \qquad (\nu \in \mathfrak{a}^*)$$

unless P' is associated to P.

Let $N(A)$ and $Z(A)$ respectively denote the normalizer and centralizer of A in G. Then $Z(A)=MA$ and

$$\mathfrak{w} = \mathfrak{w}(A) = N(A)/Z(A)$$

is a finite group which operates on $\mathfrak{a}$ and $\mathfrak{a}^*$ in the usual way. Let $P'=MAN'$ be a psgp with the same split component A. Then if $\nu \in \mathfrak{a}^*$ is not too special (i.e. does not lie on a finite set of hyperplanes passing through the origin), there exist unique endomorphisms $c_{P'|P}(s:\nu)$ $(s \in \mathfrak{w})$ of the finite-dimensional space $L = {}^\circ\mathcal{C}(M, \tau_M)$ (Theorem 10) such that

$$E_{P'}(P:\psi:\nu:ma) = \sum_{s \in \mathfrak{w}} (c_{P'|P}(s:\nu)\psi)(m) \exp((-1)^{1/2}s\nu(\log a))$$

for all $m\in M$, $a\in A$ and $\psi\in L$. Regarded as functions of ν, $c_{P'|P}(s:\nu)$ are meromorphic on $\mathfrak{a}^*+(-1)^{1/2}U$ where U is an open neighborhood of zero in $\mathfrak{a}^*$. However it seems likely that they are actually meromorphic in the whole complex space $\mathfrak{a}_c^*$. Here the analogy with Eisenstein series is quite clear (see [**4(b)**, Chapter V] and [**4(d)**]).

11. **The Fourier transform of a Schwartz function.** From now on we shall assume for simplicity that G satisfies the conditions of §4. Let $\mathfrak{a}$ be a θ-stable Cartan subalgebra of $\mathfrak{g}$ and A the corresponding Cartan subgroup of G. Put $A_I=A\cap K$ and $A_R=\exp \mathfrak{a}_R$. Let A^*, A_I^* and A_R^* denote the character groups of A, A_I and A_R respectively. Then we can identify A_R^* with $\mathfrak{a}_R^*$ and therefore A^* with $A_I^*\times\mathfrak{a}_R^*$.

Let $\mathcal{P}(\mathfrak{a}_R)$ denote the set of all psgps P of G with split component A_R. Then $P=MA_RN$ and A_I is a compact Cartan subgroup of M. Let $\mathfrak{Z}_M$ denote the algebra of all differential operators on M which commute with left and right translations of M. Then for every $a^*\in A_I^*$, we have (see §4) a tempered and invariant eigendistribution θ_{a^*} of $\mathfrak{Z}_M$ on M.

Fix $a^*\in A^*$. Then $a^*=(a_0^*, \nu)$ where $a_0^*\in A_I^*$ and $\nu\in\mathfrak{a}_R^*$. We now define a tempered distribution $\Theta_{a^*}=\Theta_{a_0^*,\nu}$ on G as follows:

$$\Theta_{a^*}(f) = \theta_{a_0^*}(g_{f,\nu}) \qquad (f\in \mathcal{C}(G)).$$

Here

$$g_{f,\nu}(m) = \int_{A_R\times N} \bar{f}(man)\exp((-1)^{1/2}\nu+\rho)(\log a)\,da\,dn, \qquad (m\in M)$$

$\rho=\rho_P$ and

$$\bar{f}(x) = \int_K f(kxk^{-1})\,dk \qquad (x\in G).$$

The Haar measures da and dn are normalized as in §6 by $da=d\mathfrak{a}_R$ and $dn=d\mathfrak{n}$. One can show that $g_{f,\nu}\in\mathcal{C}(M)$ and Θ_{a^*} is a tempered and invariant eigendistribution of $\mathfrak{Z}$ on G. Moreover the definition of Θ_{a^*} is independent of the choice of P in $\mathcal{P}(\mathfrak{a}_R)$.

Let $\tilde{A}$ be the normalizer of A in G. Put $W(G/A)=\tilde{A}/A$. Then $W(G/A)$ is a finite group which operates in the usual way on $\mathfrak{a}$, A and A^*. Let Q be the set of all positive roots of $(\mathfrak{g}, \mathfrak{a})$ (under some order) and Q_I the subset of those $\alpha\in Q$ which vanish identically on $\mathfrak{a}_R$. Put

$$\varpi = \prod_{\alpha\in Q} H_\alpha, \qquad {}_I\varpi = \prod_{\alpha\in Q_I} H_\alpha,$$

where H_α denotes, as usual, the element in $\mathfrak{a}_c$ dual to α under the Killing form. Let $\mathfrak{a}_I$ be the Lie algebra of A_I and λ the linear function on $\mathfrak{a}_I$ given by

$$\langle a_0^*, \exp H\rangle = e^{\lambda(H)} \qquad (H \in \mathfrak{a}_I).$$

We define $\log a_0^* = \lambda$, $\log a^* = \lambda + (-1)^{1/2}\nu$ and $\varpi_I(a^*) = \varpi_I(a_0^*) = \varpi_I(\lambda)$, $\varpi(a^*) = \varpi(a_0^*:\nu) = \varpi(\lambda + (-1)^{1/2}\nu)$. Since ϖ_I^2 is invariant under $W(G/A)$, $\varpi_I^s = \epsilon_I(s)\varpi_I$ $(s \in W(G/A))$ where $\epsilon_I(s) = \pm 1$. Let $A_I^{*\prime}$ denote the set of all $a_0^* \in A_I^*$ such that $\varpi_I(a_0^*) \neq 0$.

LEMMA 9. *$\Theta_{sa^*} = \epsilon_I(s)\Theta_{a^*}$ for $s \in W(G/A)$ and $a^* \in A^*$.*

This follows from Lemma 1 (§4).

Since A_I^* is a discrete group, $A^* = A_I^* \times \mathfrak{a}_R^*$ is a Lie group. By a polynomial function p on A^*, we mean a function of the form

$$p(a^*) = q(\log a^*) \qquad (a^* \in A^*),$$

where q is a polynomial function on $\mathfrak{a}_c^*$. Let $\mathfrak{D}(\mathfrak{a}_R^*)$ denote the algebra of all translation-invariant differential operators on $\mathfrak{a}_R^*$ and $\mathfrak{D}(A^*)$ the algebra of differential operators on A^* generated by $\mathfrak{D}(\mathfrak{a}_R^*)$ and the polynomial functions on A^*. Let $\mathcal{C}(A^*)$ be the space of all functions $f \in C^\infty(A^*)$ such that

$$\mu_D(f) = \sup_{A^*} |Df| < \infty$$

for all $D \in \mathfrak{D}(A^*)$. We topologize $\mathcal{C}(A^*)$ by means of the seminorms μ_D $(D \in \mathfrak{D}(A^*))$. Then $C_c^\infty(A^*)$ is dense in $\mathcal{C}(A^*)$. By a tempered distribution on A^*, we mean a continuous linear function on $\mathcal{C}(A^*)$.

For any $f \in \mathcal{C}(G)$, put (see §7)

$$\hat{f}_A(a^*) = (\Theta_{a^*}, f) \qquad (a^* \in A^*).$$

Then $\hat{f}_A \in \mathcal{C}(A^*)$ and $f \to \hat{f}_A$ is a continuous mapping of $\mathcal{C}(G)$ into $\mathcal{C}(A^*)$.

LEMMA 10. *Let $f \in \mathcal{C}(G)$. Then*

$$\hat{f}_A(sa^*) = \epsilon_I(s)\hat{f}_A(a^*) \qquad (s \in W(G/A),\ a^* \in A^*).$$

This is an immediate consequence of Lemma 9.

Let $A^*(S)$ denote the set of all points $a^* \in A^*$ where $\varpi(a^*) = 0$.

Now fix a complete set $\mathfrak{a}_i$ $(1 \leq i \leq r)$ of θ-stable Cartan subalgebras of $\mathfrak{g}$, no two of which are conjugate under G. Let A_i denote the corresponding Cartan subgroup. Put $\hat{f}_i = \hat{f}_{A_i}$ $(f \in \mathcal{C}(G))$ and define $\epsilon_{i,I}(s)$ $(s \in W(G/A_i))$ as above.

THEOREM 18. *Suppose we are given, for each* i $(1 \leqq i \leqq r)$, *a tempered distribution* T_i *on* A_i^* *such that:*[7]

(1) $T_i^s = \epsilon_{i,I}(s) T_i$ $(s \in W(G/A_i))$,

(2) $\sum_{1 \leq i \leq r} T_i(\hat{f}_i) = 0$ *for all* $f \in \mathcal{C}(G)$.

Then[8] Supp $T_i \subset A_i^*(S)$ $(1 \leqq i \leqq r)$.

We think of $(\hat{f}_i)_{1 \leq i \leq r}$ as the Fourier transform of $f \in \mathcal{C}(G)$.

12. **The Plancherel formula for** G. Define $\mathfrak{b}$, B, $W = W(\mathfrak{g}/\mathfrak{b})$ and ϖ as in §4. (We assume that $B \subset K$.) For any $f \in \mathcal{C}(G)$, put

$$ {}^*F_f(b) = [W]^{-1} \sum_{s \in W} \varpi F_f(b^s) \qquad (b \in B). $$

Then (see §8) ${}^*F_f$ is a continuous function on B. We now use the notation of Theorem 18. Let $d_i a^*$ denote the Haar measure on A_i^*.

THEOREM 19. *There exist unique continuous functions* C_i *on* $A_i^* \times B$ $(1 \leqq i \leqq r)$ *with the following properties:*

(1) $C_i(a^*: b) = 0$ *if* $\varpi_{i,I}(a^*) = 0$ $(a^* \in A_i^*, b \in B)$.

(2) $C_i(sa^*: b) = \epsilon_{i,I}(s) C_i(a^*: b)$ $(s \in W(G/A_i))$.

(3) *We can choose numbers* c, $p \geqq 0$ *such that*[9]

$$ \left| C_i(a^*:b) \right| \leqq c(1 + \|\log a^*\|)^p $$

for all $a^* \in A_i^*$ *and* $b \in B$.

(4) $$ {}^*F_f(b) = \sum_{1 \leq i \leq r} \int_{A_i^*} C_i(a^*: b) \hat{f}_i(a^*) d_i a^* $$

for $f \in \mathcal{C}(G)$ *and* $b \in B$.

The uniqueness is an immediate consequence of Theorem 18. Combining this with Lemma 5, we get the following result.

COROLLARY. $(-1)^q c_G f(1) = \sum_{1 \leq i \leq r} \int_{A_i^*} C_i(a^*: 1) \hat{f}_i(a^*) d_i a^*$ *for* $f \in \mathcal{C}(G)$.

This is substantially the Plancherel formula for G.

The proof of Theorem 19 is based on a long induction on the rank of G. It makes essential use of the differential equations connecting the functions F_f and F_{zf} $(z \in \mathfrak{Z}, f \in \mathcal{C}(G))$. The basic idea of utilizing these differential equations to obtain a theorem of the above type,

---

[7] $W(G/A_i)$ operates on the space of distributions on A_i^* in the obvious way.

[8] Supp T_i denotes the support of T_i.

[9] The Euclidean norm on $\mathfrak{g}$ (see §6) defines the structure of a Hilbert space on $\mathfrak{g}_c$ and hence also on $\mathfrak{a}_c$ and $\mathfrak{a}_c^*$ for $\mathfrak{a} = \mathfrak{a}_i$.

goes back to R. P. Langlands who had worked out the case of SL(2, $\boldsymbol{R}$) by this method in 1962.

Fix i $(1 \leqq i \leqq r)$, $a_0^* \in A_{i,I}^{*\prime}$ and put $\mathfrak{F}_i = \mathfrak{a}_{i,R}^*$. Then the function

$$C_i(a_0^*:\nu:b) \qquad (\nu \in \mathfrak{F}_i,\ b \in B)$$

is of class C^∞ on $\mathfrak{F}_i \times B'$ and it is analytic in ν. In fact it is a rather simple function which, for a fixed $b \in B'$, is meromorphic on $\mathfrak{F}_{i,c}$.

13. **The relation between the Plancherel measure and the asymptotic behaviour of certain Eisenstein integrals.** Fix i $(1 \leqq i \leqq r)$ and drop it from the notation. Now put

$$q(a_0^*:\nu) = C_i(a_0^*:\nu:1) \qquad (a_0 \in A_I^{*\prime},\ \nu \in \mathfrak{F}).$$

We wish to obtain a formula for $q(a_0^*:\nu)$.

For any $\nu \in \mathfrak{F}_c$, put $\nu = \nu_R + (-1)^{1/2}\nu_I$ $(\nu_R, \nu_I \in \mathfrak{F})$. Fix $P = MA_RN$ in $\mathcal{P}(\mathfrak{a}_R)$ and for any $\delta > 0$ let $\mathfrak{F}_c(\delta)$ denote the set of all $\nu \in \mathfrak{F}_c$ such that $\|\nu_I\| < \delta$. Then if δ is sufficiently small, we have defined (see §10) the meromorphic function $c_{P|P}(1:\nu)$ on $\mathfrak{F}_c(\delta)$ with values in the space of endomorphisms of ${}^\circ\mathcal{C}(M, \tau_M)$. Let $\alpha_1, \cdots, \alpha_l$ be all the distinct simple roots of (P, A_R). Define $H_j \in \mathfrak{a}_R$ by $\alpha_i(H_j) = \delta_{ij}$ $(1 \leqq i, j \leqq l)$. Let $\mathfrak{F}_c(P)$ denote the set of all $\nu \in \mathfrak{F}_c$ such that $\nu_I(H_j) < 0$ $(1 \leqq j \leqq l)$. Then $c_{P|P}(1:\nu)$ is actually holomorphic on $\mathfrak{F}_c(P)$. Put $\bar{N} = \theta(N)$. Given $x \in G$, we can write it uniquely in the form $x = kman$ $(k \in K, m \in M, a \in A, n \in N)$ where $\theta(m) = m^{-1}$. Define $\kappa(x) = k$ and $\mu(x) = m$.

LEMMA 11. *Put $\rho = \rho_P$ and $H(x) = H_P(x)$ $(x \in G)$. Then*

$$(c_{P|P}(1:\nu)\psi)(m)$$
$$= \int_{\bar{N}} \psi(m\mu(\bar{n})^{-1})\tau(\kappa(\bar{n}))^{-1} \exp[-((-1)^{1/2}\nu + \rho)(H(\bar{n}))]d_0\bar{n}$$

for $\psi \in {}^\circ\mathcal{C}(M, \tau_M)$, $\nu \in \mathfrak{F}_c(P)$ and $m \in M$. Here the integral is convergent and the Haar measure $d_0\bar{n}$ on $\bar{N}$ is so normalized that

$$\int_{\bar{N}} e^{-2\rho(H(\bar{n}))} d_0\bar{n} = 1.$$

Let V_0 be the subspace of all $v_0 \in V$ such that $\tau(k)v_0 = v_0\tau(k)$ for all $k \in K$. Fix $a_0^* \in A_I^{*\prime}$, $v_0 \in V_0$ and define $\psi(a_0^*: v_0) = \psi \in {}^\circ\mathcal{C}(M, \tau_M)$ by

$$\psi(m) = \int_{K_M} \theta_{a_0^*}(mk^{-1})\tau(k)v_0 d_0 k \qquad (m \in M),$$

where d_0k is the normalized Haar measure on K_M and $\theta_{a_0}^*$ has the same meaning as in §11. Let $\mathfrak{F}'(a_0^*)$ denote the set of all $\nu \in \mathfrak{F}$ such that $\varpi(a_0^*: \nu) \neq 0$. Then $c_{P|P}(1: \nu)\psi$ is holomorphic in ν for $\nu \in \mathfrak{F}'(a_0^*)$. Put

$$\|\phi\|_M^2 = \int_M |\phi(m)|^2 dm \qquad (\phi \in {}^\circ\mathcal{C}(M, \tau_M)),$$

$q_G = \frac{1}{2} \dim G/K$ and $q_M = \frac{1}{2} \dim M/K_M$.

THEOREM 20. *There exists a constant* $c_0(P) > 0$ *(independent of* a_0^* *and* τ*) such that*

$$c_0(P) q(a_0^*:\nu) \|c_{P|P}(1:\nu)\psi\|_M^2 = (-1)^{q_G + q_M} \varpi_I(a_0^*) \|\psi\|_M^2$$

for $\nu \in \mathfrak{F}'(a_0^*)$ *and* $\psi = \psi(a_0^*: v_0)$ $(v_0 \in V_0)$.

Actually it is possible to determine $c_0(P)$ explicitly. Moreover it is not difficult to show that for any given $a_0^* \in A_I^{*\prime}$, we can choose τ in such a way that $\psi(a_0^*: v_0) \neq 0$ for some $v_0 \in V_0$.

14. The $\mathfrak{p}$-adic case. First we recall some standard facts about algebraic groups [1].

Let Ω be a field. By an Ω-group, we mean a (linear) algebraic group defined over Ω. Let $\boldsymbol{G}$ be a connected and reductive Ω-group. By a parabolic subgroup $\boldsymbol{P}$ of $\boldsymbol{G}$, we mean an algebraic subgroup which contains a Borel subgroup of $\boldsymbol{G}$. We say that $\boldsymbol{P}$ is Ω-parabolic if it is parabolic and defined over Ω. Let $\boldsymbol{N}$ denote the (unipotent) radical of $\boldsymbol{P}$. Then $\boldsymbol{N}$ is an Ω-subgroup and $\boldsymbol{P}$ is the normalizer of $\boldsymbol{N}$ in $\boldsymbol{G}$.

Now assume that Ω is a $\mathfrak{p}$-adic field (i.e. a locally compact field with a discrete valuation). Let G denote the set of all Ω-rational points of $\boldsymbol{G}$. Then $G \subset \mathrm{GL}(n, \Omega)$ for a suitable $n \geq 1$. $\mathrm{GL}(n, \Omega)$, being an open subset of a vector space over Ω of dimension n^2, is a locally compact group. Since G is closed in $\mathrm{GL}(n, \Omega)$, it is also locally compact. Moreover it is easy to verify that it is separable and unimodular.

By a parabolic subgroup P of G, we mean a subgroup of the form $P = G \cap \boldsymbol{P}$, where $\boldsymbol{P}$ is an Ω-parabolic subgroup of $\boldsymbol{G}$. Then $\boldsymbol{P}$ is uniquely determined by P. Put $N = G \cap \boldsymbol{N}$ where $\boldsymbol{N}$ is the radical of $\boldsymbol{P}$. Let $C_c(G)$ denote the space of all continuous complex-valued functions on G with compact support. For any $f \in C_c(G)$, define

$$f^P(x) = \int_N f(xn)dn \qquad (x \in G),$$

where dn is the Haar measure of the unimodular group N. f is said to be a cusp form if $f^P=0$ for all parabolic subgroups $P \neq G$. Let ${}^\circ C_c(G)$ denote the space of all cusp forms in $C_c(G)$.

Let π be an irreducible unitary representation of G and $\mathfrak{H}_\pi$ the representation space of π. We say that π is supercuspidal if there exist nonzero elements ϕ, ψ in $\mathfrak{H}_\pi$ such that the function $x \rightarrow (\phi, \pi(x)\psi)$ $(x \in G)$ lies in ${}^\circ C_c(G)$. Let ${}^\circ\mathcal{E}(G)$ denote the set of all classes $\omega \in \mathcal{E}(G)$, which contain supercuspidal representations. It is obvious that ${}^\circ\mathcal{E}(G) \subset \mathcal{E}_2(G)$.

Let G_0 be an open subset of G and $C_c^\infty(G_0)$ the space of all locally constant complex-valued functions on G_0 with compact support. By a distribution T on G_0, we mean a linear mapping of $C_c^\infty(G_0)$ into $\mathbf{C}$. Let F be a function on G_0 which is locally summable with respect to the Haar measure dx of G. Then we say that $T=F$ if

$$T(f) = \int_G F \cdot f dx$$

for all $f \in C_c^\infty(G_0)$.

THEOREM 21. *Let π be a square-integrable representation of G. Then for any $f \in C_c^\infty(G)$, the operator*

$$\pi(f) = \int_G f(x)\pi(x)dx$$

is of the trace class.

Since every neighborhood of 1 in G, contains an open compact subgroup, this is an immediate consequence of the corollary of Theorem 5.

Define

$$\Theta_\pi(f) = \operatorname{tr} \pi(f) \qquad (f \in C_c^\infty(G)).$$

Then Θ_π is a distribution on G which depends only on the class ω of π. Hence we may denote it by Θ_ω. Then Θ_ω is invariant and the mapping $\omega \rightarrow \Theta_\omega$ $(\omega \in \mathcal{E}_2(G))$ is injective.

THEOREM 22. *Let $\mathfrak{H}_\pi$ denote the representation space of a square-integrable representation π of G. Fix $\phi, \psi \in \mathfrak{H}_\pi$. Then*

$$(\phi, \psi)\Theta_\pi(f) = d(\pi)\int_G (\phi, \pi(x)\pi(f)\pi(x^{-1})\psi)dx$$

$$= d(\pi)\int_G dx \int_G f(y)(\phi, \pi(xyx^{-1})\psi)dy$$

for $f \in C_c^\infty(G)$.

This is an easy consequence of the Schur orthogonality relations.

The function D and the regular set G' are defined in exactly the same way as in §2. Then D is a regular function on $\mathbf{G}$ and G' is an open and dense subset of G whose complement is of measure zero.

THEOREM 23. *Fix $\omega \in {}^\circ\mathcal{E}(G)$. Then there exists a locally constant function F on G' such that $\Theta_\omega = F$ on G'.*

It seems very likely that F is actually locally summable on G and $\Theta_\omega = F$. In any case, the analogy with the results of §2 is obvious.

For any $\alpha \in \Omega$, we define its absolute value $|\alpha|$ in the usual way so that

$$d(\alpha t) = |\alpha| \, dt$$

for any additive Haar measure dt on Ω.

LEMMA 12. *If Ω has characteristic zero, the function $|D|^{-1/2}$ is locally summable on G.*

Put $\Phi = |D|^{1/2} F$ in the notation of Theorem 23. It seems reasonable to expect that Φ remains bounded on G'.

By a Cartan subgroup A of G, we mean a subgroup of the form $A = G \cap \mathbf{A}$ where $\mathbf{A}$ is a maximal Ω-torus in $\mathbf{G}$. Let Φ_A denote the restriction of Φ on $A' = A \cap G'$.

LEMMA 13. *There exists a compact subset C of A such that $\Phi_A = 0$ outside C. Moreover if A is compact and Ω has characteristic zero, then Φ_A remains bounded on A'.*

When $\mathbf{G} = \mathrm{SL}(2)$ and the characteristic of the residue field of Ω is not 2, the characters of G have been explicitly computed by Sally and Shalika [8] and their results agree with the above statements.

15. Two problems. Let me conclude these lectures by discussing two problems (or conjectures) in the $\mathfrak{p}$-adic case.

Fix an open compact subgroup K_0 of G and let ${}^\circ C_c(G//K_0)$ denote the space of all functions in ${}^\circ C_c(G)$ which are constant on double cosets of K_0. Then one would like to show that, in case prk $G = 0$,

$$\dim {}^\circ C_c(G//K_0) < \infty.$$

Here the analogy with Theorem 10 is obvious.

In view of Theorem 5, it would be enough to prove that

$$\inf_{\omega \in {}^\circ\mathcal{E}(G)} d(\omega) > 0.$$

Actually I am inclined to believe that the numbers $d(\omega)$ ($\omega \in \mathcal{E}_2(G)$) are all integers under a suitable normalization of the Haar measure of G.

K being any open compact subgroup of G, an affirmative answer to the above question would imply that

$$\sup_{\omega\in\mathcal{E}(G)} [\omega:\mathfrak{d}] < \infty$$

for $\mathfrak{d}\in\mathcal{E}(K)$, provided Ω has characteristic zero.

In order to state the second problem, we have to recall a recent theorem of Bruhat and Tits. Let $(P_0, \mathbf{A}_0)$ be a mincuspidal pair in G and $P_0=M_0N_0$ the corresponding Levi decomposition [4(c), §2]. Put $A_0=G\cap\mathbf{A}_0$ and for any root α of $(P_0, \mathbf{A}_0)$, let ξ_α denote the corresponding character of $\mathbf{A}_0$. Let A_0^+ be the set of all points $a\in A_0$ where $|\xi_\alpha(a)|\geqq 1$ for every root α of $(P_0, \mathbf{A}_0)$. Define $\mathfrak{w}(\mathbf{A}_0)$ as in [4(c), §2].

THEOREM (BRUHAT-TITS). *We can choose an open and compact subgroup K of G with the following properties*:

(1) $G=KP_0$.

(2) $G=KA_0^+CK$, *where C is a finite subset of* M_0.

(3) *Every element of $\mathfrak{w}(\mathbf{A}_0)$ has a representative in K.*

(4) *If $(P, \mathbf{A})\succ(P_0, \mathbf{A}_0)$ is a cuspidal pair and $P=MN$ the corresponding Levi decomposition, then $P\cap K=(M\cap K)(N\cap K)$.*

(5) *Put $K_M=K\cap M$ and $^*P_0=M\cap P_0$ in the notation of* (4). *Then if we replace $(G, P_0, \mathbf{A}_0, K)$ by $(M, {}^*P_0, \mathbf{A}_0, K_M)$, the above four conditions are again fulfilled.*

We keep to the notation of the above theorem. Let τ be a unitary representation of K on a finite-dimensional Hilbert space V and let $C_c(G, \tau)$ denote the space of all functions f from G to $\mathrm{End}(V)$ with compact support such that

$$f(k_1xk_2) = \tau(k_1)f(x)\tau(k_2) \qquad (k_1, k_2 \in K,\ x \in G).$$

Then $C_c(G, \tau)$ is an algebra under convolution. On the other hand, let V_P be the subspace of all $v\in V$ such that $\tau(n)v=v$ for all $n\in N\cap K$. Let E_P denote the orthogonal projection of V on V_P. Put

$$\tau_M(m) = \tau(m)E_P \qquad (m \in K_M).$$

Then τ_M may be regarded as a representation of K_M on V_P and so we can consider the algebra $C_c(M, \tau_M)$.

Let d_rp denote a right-invariant Haar measure on P and define the function δ_P on P by the relation

$$d_r(qp) = \delta_P(q)d_rp \qquad (q \in P).$$

For any $f \in C_c(G, \tau)$, put

$$f^{(P)}(m) = \delta_P(m)^{1/2} \int_N f(mn) E_P dn \qquad (m \in M),$$

where dn is a Haar measure on N. Then $f^{(P)} \in C_c(M, \tau_M)$ and the mapping $\mu_P: f \to f^{(P)}$ is actually a homomorphism of $C_c(G, \tau)$ into $C_c(M, \tau_M)$. We would now like to assert that *$C_c(M, \tau_M)$ is a finite right-module over $\mu_P(C_c(G, \tau))$.* If this is true, much of the theory for real groups [**4**(a), Part II] can be imitated in the $\mathfrak{p}$-adic case.

References

1. A. Borel and J. Tits, *Groupes rèductifs*, Inst. Hautes Études Sci. Publ. Math No. 27 (1965), 55–150. MR **34** #7527.

2. I. M. Gelfand, *Automorphic functions and the theory of representations*, Proc. Internat. Congress Math. (1962) Inst. Mittag-Leffler, Djursholm, 1963, pp. 74–85. MR **28** #1.

3. R. Godement, *Sur les relations d'orthogonalité de V. Bargmann.* I: *Démonstration générale*, C. R. Acad. Sci. Paris **225** (1947), 657–659. MR **9**, 134.

4. Harish-Chandra, (a) *Discrete series for semisimple Lie groups.* II, Acta Math. **116** (1966), 1–111. MR **36** #2745. (b) *Automorphic forms on semisimple Lie groups*, Lecture notes in Math., no. 62, Springer-Verlag, Berlin and New York, 1968. MR **38** #1216. (c) *Eisenstein series over finite fields*, Proc. Stone Jubilee Conf., Chicago, 1968. (d) *Some applications of the Schwartz space of a semisimple Lie group*, Lecture Series, Modern Anal. and Appl., George Washington Univ., 1968.

5. G. W. Mackey, *Infinite-dimensional group representations*, Bull. Amer. Math. Soc. **69** (1963), 628–686. MR **27** #3745.

6. C. C. Moore, *Decomposition of unitary representations defined by discrete subgroups of nilpotent groups*, Ann. of Math. (2) **82** (1965), 146–182. MR **31** #5928.

7. L. Pukánszky, *Leçons sur les représentations des groupes*, Monographies de la Société Mathématique de France, no. 2, Dunod, Paris, 1967. MR **36** #311.

8. P. J. Sally, Jr. and J. A. Shalika, *Characters of the discrete series of representations of* SL(2) *over a local field*, Proc. Nat. Acad. Sci. U.S.A. **61** (1968), 1231–1237.

9. L. Schwartz, *Théorie des distributions.* Tome I, Actualités Sci. Indust., no. 1091, Hermann, Paris, 1950. MR **12**, 31.

Institute for Advanced Study, Princeton, New Jersey 08540

ON THE SELBERG TRACE FORMULA IN THE CASE OF COMPACT QUOTIENT

BY NOLAN R. WALLACH

1. **Introduction.** Let G be a connected unimodular Lie group. Let Γ be a discrete subgroup of G so that $\Gamma\backslash G$ is compact. We fix a Haar measure, dg, on G. Then dg induces a G-invariant measure on $\Gamma\backslash G$. We can then form a unitary representation $(\pi_\Gamma, L^2(\Gamma\backslash G))$ where $(\pi_\Gamma(g)f)(x)=f(xg)$ for $f\in L^2(\Gamma\backslash G)$, $x\in\Gamma\backslash G$, $g\in G$. If $\phi\in C_c^\infty(G)$ (the space of all C^∞ compactly supported complex valued functions on G) we can form

$$(\pi_\Gamma(\phi)f)(x)=\int_G \phi(g)f(xg)\,dg.$$

It is a standard fact (see §2) that $\pi_\Gamma(\phi)$ is of trace class. In particular, $\pi_\Gamma(\phi)$ is completely continuous for $\phi\in C_c^\infty(G)$. This implies that $L^2(\Gamma\backslash G)$ decomposes into an orthogonal direct sum of irreducible invariant subspaces, $\{H_j\}_{j=1}^\infty$ and for each i there are only a finite number of k so that H_j is equivalent with H_k as a representation of G (cf. Gelfand, Graev, Pyateckiĭ-Shapiro [**9**]). Let $\hat{G}$ denote the set of equivalence classes of irreducible representations of G. Then we have observed that

$$\pi_\Gamma=\sum_{\omega\in\hat{G}} N_\Gamma(\omega)\omega$$

where $N_\Gamma(\omega)$ is a nonnegative integer. If $\omega\in\hat{G}$ we say that ω is of trace class if for each $(\pi, H)\in\omega$, $\phi\in C_c^\infty(G)$, $\pi(\phi)=\int_G\phi(g)\pi(g)\,dg$ is a trace class operator on H. If $\omega\in\hat{G}$ is of trace class, then set $\Theta_\omega(\phi)=\operatorname{tr}\pi(\phi)$ for $(\pi, H)\in\omega$. The above observations imply that if $\omega\in\hat{G}$ and $N_\Gamma(\omega)\neq 0$, then ω is of trace class. We therefore see that if $\phi\in C_c^\infty(G)$, then

$$\operatorname{tr}\pi_\Gamma(\phi)=\sum_{\omega\in\hat{G}} N_\Gamma(\omega)\Theta_\omega(\phi).$$

The numbers $N_\Gamma(\omega)$ have been the subject of a great deal of investigation in the last few years. In this article we will give a short survey of various techniques that have been used to study these integers. We will concentrate our attention on semisimple Lie groups, G. We will also, for most of the article, look at the easiest groups Γ. These groups have no elements of finite order other than the identity. Without this assumption many (interesting)

AMS(MOS) subject classifications (1970). Primary 10D20, 22E40, 43A15.

This address was given at the annual meeting of the American Mathematical Society in Washington, D.C. on January 24, 1975; received by the editors May 8, 1975.

technicalities occur. We apologize to the reader for our avoidance of these technicalities.

It should also be pointed out that much of the important research on $L^2(\Gamma\backslash G)$ has been on the case where Γ is discrete and $\Gamma\backslash G$ has finite volume relative to the measure on $\Gamma\backslash G$ induced by dg (see Harish-Chandra [**12**], Duflo-Lebesse [**4**], Arthur [**7**]). The most noteworthy example is $G = SL(2, \mathbf{R})$, $\Gamma = SL(2, \mathbf{Z})$ ($\mathbf{R}$ the reals, $\mathbf{Z}$ the integers).

A serious reader of this article will be irritated with a noteworthy omission in this article. We will never give an example of a discrete group Γ so that $\Gamma\backslash G$ is compact. The best we can say is that there are many of them. (See Mostow [**29**], Raghunathan [**30**].)

The author was introduced to the subject matter of this article by Professor Paul Sally. Many of the ideas in §9 are an outgrowth of joint research with Sally. We also thank Professor Rioshi Hotta for having taught the author Matsushima's work on the Betti numbers of locally symmetric spaces. Finally, we thank Professor Kenneth Johnson for patiently teaching us his work on the Paley-Wiener problem for semisimple Lie groups.

2. **The trace formula.** In the Introduction we asserted that if $\phi \in C_c^\infty(G)$, then $\pi_\Gamma(\phi)$ is of trace class. We also computed a formula for $\operatorname{tr} \pi_\Gamma(\phi)$ in terms of the $N_\Gamma(\omega)$. We now compute another such formula first observed by Selberg. Let $f \in C^\infty(\Gamma\backslash G)$.

$$\begin{aligned}\pi_\Gamma(\phi)f(\Gamma x) &= \int_G f(\Gamma xg)\phi(g)\,dg = \int_G f(\Gamma g)\phi(x^{-1}g)\,dg \\ &= \int_{\Gamma\backslash G} \sum_{\gamma\in\Gamma} f(\Gamma\gamma g)\phi(x^{-1}\gamma g)\,d\dot{g} \\ &= \int_{\Gamma\backslash G} f(\Gamma g)\Big(\sum_{\gamma\in\Gamma}\phi(x^{-1}\gamma g)\Big)\,d\dot{g}.\end{aligned}$$

Thus if we set $K_{\phi,\Gamma}(x, g) = \sum_{\gamma\in\Gamma}\phi(x^{-1}\gamma g)$, then we note

$$K_{\phi,\Gamma}(\tau x, g) = K_{\phi,\Gamma}(x, \tau g) = K_{\phi,\Gamma}(x, g)$$

for $\tau \in \Gamma$. Thus $K_{\phi,\Gamma}: \Gamma\backslash G \times \Gamma\backslash G \to C$ is a C^∞ function. We have

$$(\pi_\Gamma(\phi)f)(x) = \int_{\Gamma\backslash G} K_{\Gamma,\phi}(x, y)f(y)\,dy.$$

Standard theory now implies the trace class assertion and

$$\operatorname{tr} \pi_\Gamma(\phi) = \int_{\Gamma\backslash G} K_{\Gamma,\phi}(\dot{x}, x)\,dx = \int_{\Gamma\backslash G}\Big(\sum_{\gamma\in\Gamma}\phi(x^{-1}\gamma x)\Big)\,d\dot{x}.$$

Arguing now as in Gelfand, Graev, Pyateckiĭ-Shapiro [**9**, p. 30] we can put this in the following form:

THEOREM 2.1 (THE TRACE FORMULA). *If $\phi \in C_c^\infty(G)$, then*

$$\operatorname{tr} \pi_\Gamma(\phi) = \sum_{[\gamma]\in[\Gamma]} \operatorname{vol}(\Gamma_\gamma\backslash G_\gamma)\int_{G_\gamma\backslash G}\phi(g^{-1}\gamma g)\,d\dot{g}.$$

Here $[\Gamma]$ is the set of Γ-equivalence classes in Γ and $[\gamma]$ is the Γ-equivalence class of γ. G_γ is the centralizer of a representative of $[\gamma]=\{\tau\gamma\tau^{-1} \mid \tau\in\Gamma\}$. $\Gamma_\gamma=\Gamma\cap G_\gamma$. Here $\operatorname{vol}(\Gamma_\gamma\backslash G_\gamma)$ is the total measure of $\Gamma_\gamma\backslash G_\gamma$ and the measures are normalized by

$$\int_G \psi(g)\,dg=\int_{G_\gamma\backslash G}\int_{G_\gamma}\psi(xg)\,dx\,d\dot g$$

and

$$\int_{G_\gamma}\eta(g)\,dg=\int_{\Gamma_\gamma\backslash G_\gamma}\sum_{\tau\in\Gamma_\gamma}\eta(\tau g)\,d\dot g.$$

3. **How to use the trace formula.** In this section we will assume that G is a semisimple Lie group with finite center. We take $K\subset G$ a maximal compact subgroup. Then $X=G/K$ is the most general symmetric space of noncompact type.

For simplicity we assume that Γ has no elements of finite order. This assumption implies (cf. Mostow [**30**]) that:

(a) If $\gamma\in\Gamma$ then $\operatorname{Ad}(\gamma)$ is a semisimple automorphism of the Lie algebra of G.

(b) Γ acts freely on X.

Let $\hat G_d$ be the set of all equivalence classes of irreducible unitary representations equivalent with a subrepresentation of $L^2(G)$. Harish-Chandra [**13**] has shown that $\hat G_d\neq\varnothing$ if and only if there is a maximal torus, T, of K which is maximal abelian in G. In [**13**], Harish-Chandra also proves

THEOREM 3.1. *Let $\omega\in\hat G_d$, $(\pi, H)\in\omega$. Let v, $w\in H$ be K-finite (that is $\pi(K)v$ and $\pi(K)w$ are contained in finite dimensional subspaces of H) and set $\psi(g)=\langle\pi(g)v, w\rangle$. Then if γ is an element of G so that* $\operatorname{Ad}(\gamma)$ *is semisimple, $\int_{G_\gamma\backslash G}\psi(g^{-1}\gamma g)\,d\dot g$ converges and is zero unless $\gamma\in\{gTg^{-1}\mid g\in G\}$.*

Let $\hat G_d'$ be the set of $\omega\in\hat G_d$ such that there are v, $w\in H$ $((\pi, H)\in\omega)$ so that $\psi(g)$ (as above) is absolutely integrable. In Borel [**2**] it is proved that $\pi_\Gamma(\psi)$ is trace class and the trace formula applies.

Now if ψ corresponds as above to $\omega\in G_d'$, then $\Theta_\eta(\psi)=0$ if $\eta\neq\omega$. Thus $\operatorname{tr}\pi_\Gamma(\psi)=N_\Gamma(\omega)\Theta_\omega(\psi)$. On the other hand,

$$\operatorname{tr}\pi_\Gamma(\psi)=\sum_{\gamma\in[\Gamma]}\operatorname{vol}(\Gamma_\gamma\backslash G_\gamma)\int_{G_\gamma\backslash G}\psi(g^{-1}\gamma g)\,d\dot g.$$

But $\gamma\in\Gamma$ is conjugate to an element of K if and only if $\gamma=I$. Thus we have

$$N_\Gamma(\omega)\Theta_\omega(\psi)=\operatorname{vol}(\Gamma\backslash G)\psi(I).$$

Now the Schur orthogonality relations (see Harish-Chandra [**10**]) imply $\psi(I)=d(\omega)\Theta_\omega(\psi)$. Here $d(\omega)$ is defined by

$$\int_G|\langle\pi(g)v, w\rangle|^2\,dg=\langle v, v\rangle\langle w, w\rangle d(\omega)^{-1}$$

for v, $w\in H$, $(\pi, H)\in\omega$, $d(\omega)$ is called the formal degree of ω. Clearly, if we take $v=w$ and $\langle v, v\rangle=1$ we have

THEOREM 3.2 (LANGLANDS [**26**]). *Suppose that Γ is as above. If $\omega \in \hat{G}'_d$, then $N_\Gamma(\omega)=d(\omega)\mathrm{vol}(\Gamma\backslash G)$. (Here, of course, we take the same normalization of Haar measure to define $d(\omega)$ and $d\dot{g}$ on $\Gamma\backslash G$. Then the product $d(\omega)\mathrm{vol}(\Gamma\backslash G)$ is independent of normalization.)*

If $\omega \in \hat{G}_d$, but not in $\hat{G}'_d$, then this argument breaks down. The formula of 3.2 is so nice that one might hope that it is true for $\hat{G}_d$. Unfortunately, it is not. If we take $G=PSL(2, \mathbf{R})$ (the group of holomorphic automorphisms of the upper half plane) and $\Gamma\subset G$ as above, then $\Gamma\backslash G/K$ ($K=SO(2)$) is the most general Riemann surface of genus $g\geqq 2$. The Gauss-Bonnet theorem implies that we can normalize dg so that $\mathrm{vol}(\Gamma\backslash G)=-\chi(\Gamma\backslash G/K)=2g-2$.

$\hat{G}_d$ is naturally parametrized as $\{\omega_n | n\in\mathbf{Z}, n\neq 0\}$ and relative to this parametrization $\hat{G}_d=\{\omega_1, \omega_{-1}\}\cup \hat{G}'_d$, $d(\omega_n)=|n|/2$.

We therefore see that 3.2 implies

(i) $$N_\Gamma(\omega_n)=|n|\,(g-1)=d(\omega_n)\mathrm{vol}(\Gamma\backslash G) \quad \text{if } |n|\geqq 2.$$

On the other hand, Langlands has shown (we will see this also in §4) that

(ii) $$N_\Gamma(\omega_{\pm 1})=g=d(\omega_{\pm 1})\mathrm{vol}(\Gamma\backslash G)+1.$$

This certainly shows that the formula of Theorem 3.2 breaks down if $\omega \notin \hat{G}'_d$. How do we interpret the "defect", $|d(\omega)\mathrm{vol}(\Gamma\backslash G)-N_\Gamma(\omega)|=1$ in formula (ii)? We will see that the interpretation of 1 is $N_\Gamma(1)$, the multiplicity of the trivial representation.

The numbers $N_\Gamma(\omega)$, $\omega\in\hat{G}_d$ have been studied, using cohomological methods, by W. Schmid [**33**] and Hotta and Parthasarathy [**17**]. In the latter paper it is shown that the formula of Theorem 3.2 is true for a large class of elements of $\hat{G}_d$ not in $\hat{G}'_d$.

4. **Relations with the topology of** $\Gamma\backslash G/K$. In the last section we noted that if $G=PSL(2, \mathbf{R})$ and Γ satisfies the conditions of the last section, then

(i) $$N_\Gamma(\omega_1)=N_\Gamma(\omega_{-1})=g$$

where g is the genus of $\Gamma\backslash G/K$.

Matsushima [**27**] has a generalization of this formula which we will now describe.

Let (τ, V) be the complexification of the isotropy representation of K on $T(G\backslash K)_0$ (0 the coset $I\cdot K$). Let J_p be the set of equivalence classes of irreducible representations of K that appear as a subrepresentation of $\Lambda^p V$. If $\omega\in\hat{G}$ and $\gamma\in\hat{K}$, let $[\omega|_K:\gamma]$ denote the multiplicity of γ as a subrepresentation of any representative of ω. Finally, let Ω be the Casimir operator of G (Ω is defined as follows: let $x_1, \cdots, x_n$ be a basis of $\mathfrak{G}$, the Lie algebra of G, let $x^1, \cdots, x^n$ be defined by $\mathrm{tr}\,\mathrm{ad}\,x_i\,\mathrm{ad}\,x^j=\delta_{ij}$; then $\Omega=\sum x_i x^i$). Let $\hat{G}_0=\{\omega\in\hat{G}|\pi(\Omega)=0$ if $(\pi, H)\in\omega\}$.

THEOREM 4.1 (MATSUSHIMA [**27**]). *Let $b_p(\Gamma\backslash G/K)=\dim H^p(\Gamma\backslash G/K, C)$. Then*

$$b_p(\Gamma\backslash G/K)=\sum_{\omega\in\hat{G}_0} N_\Gamma(\omega)\Big(\sum_{\gamma\in J_p}[\omega|_K:\gamma][\Lambda^p V:\gamma]\Big).$$

This formula suggests that one should find all $\omega \in \hat{G}_0$ so that $[\omega|_K : \gamma] \neq 0$ for some $\gamma \in J_p$. Once such ω are found, then the next task it suggests is to find $N_\Gamma(\omega)$. We note that this formula implies that if for any $\omega \in \hat{G}_0$, $\gamma \in J_p$, $[\omega|_K : \gamma] = 0$, then $b_p(\Gamma \backslash G/K) = 0$. Let us show how one can use this observation to prove that Betti numbers vanish.

Suppose that G/K has a G-invariant complex structure. Then V (above) splits into $V^+ \oplus V^-$ (corresponding to the holomorphic and antiholomorphic tangent spaces). $\Gamma \backslash G/K$ is then a projective algebraic variety (cf. Morrow [**29**]). We define $J_{p,q} \subset \hat{K}$ to be the set of equivalence classes of irreducible representations of K appearing in $\Lambda^p V^+ \otimes \Lambda^q V^-$. Then one has the analogous formula:

$$\text{(ii)} \qquad b_{p,q}(\Gamma \backslash G/K) = \sum_{\omega \in \hat{G}_0} N_\Gamma(\omega) \Big(\sum_{\gamma \in J_{p,q}} [\omega|_K : \gamma][\Lambda^p V^+ \otimes \Lambda^q V^- : \gamma] \Big).$$

THEOREM 4.2 (HOTTA AND WALLACH [**18**]). *Suppose that G is simple and that G/K has a G-invariant complex structure. Let $l = \operatorname{rank}(G/K) =$ split $\operatorname{rank}(G)$ (we will define this term in the next section). If $0 < p < l$, then $\{\omega \in \hat{G}_0 | [\omega|_K : \gamma] \neq 0$ for $\gamma \in J_{0,p}\} = \varnothing$.*

Using the above observations we have

COROLLARY 4.3. *$b_{0,p}(\Gamma \backslash G/K) = 0$ for $0 < p < \operatorname{rank}(G/K)$. Since*

$$b_1(\Gamma \backslash G/K) = b_{0,1}(\Gamma \backslash G/K) + b_{1,0}(\Gamma \backslash G/K) = 2b_{0,1}(\Gamma \backslash G/K),$$

we have

COROLLARY 4.4 (MATSUSHIMA [**26**]). *If G/K has a G-invariant complex structure, G is simple, and $\operatorname{rank}(G/K) > 1$, then $b_1(\Gamma \backslash G/K) = 0$.*

Actually this theorem has been generalized by Kazdan [**21**] as follows:

THEOREM 4.5. *If G is a simple Lie group with split rank larger than 1 and if $\Gamma \backslash G$ is a discrete subgroup of G so that $\operatorname{vol}(\Gamma \backslash G) < \infty$, then $\Gamma/[\Gamma, \Gamma]$ is finite.*

This is a generalization of Corollary 4.4 since $\operatorname{rank}(\Gamma/[\Gamma, \Gamma]) = b_1(\Gamma \backslash G/K)$ if Γ is as in §3.

The proof of Theorem 4.2 also has implications for the rank 1 case. If G is simple and G/K has a G-invariant complex structure, then G is locally isomorphic with $SU(n, 1)$. $SU(n, 1)$ is the subgroup of $SL(n+1, C)$ leaving the Hermitian form $\sum_{i=1}^n |z_i|^2 - |z_{n+1}|^2$ invariant.

G/K is the unit ball in C^n under the action $g \cdot z = (\langle z, c \rangle + d)^{-1}(Az + b)$ where

$$g = \begin{bmatrix} A & b \\ c^* & d \end{bmatrix}$$

with A, $n \times n$; b; $n \times 1$; c, $n \times 1$; d, 1×1. Here C^n is looked upon as column vectors and c^* is the conjugate transpose of c. $SU(1, 1)$ is the twofold covering group of $PSL(2, \mathbf{R})$.

One proves in this case

PROPOSITION 4.6. *There are elements $\omega_{p,0}$ and $\omega_{0,p}$ in $\hat{G}_0$ for*

$p=0,1,2,\cdots,n$ such that $\{\omega\in\hat{G}_0|[\omega|_K:\gamma]\neq 0$ for some $\gamma\in J_{p,q}\}=\{\omega_{p,q}\}$ for p or $q=0$. Furthermore, $J_{0,p}$ and $J_{p,0}$ are singletons and $[\omega_{0,p}|_K:\gamma]=1$ for $\gamma\in J_{0,p}$.

In the specific case $n=1$ we find $\omega_{0,1}=\omega_1$ and $\omega_{1,0}=\omega_{-1}$. Thus (ii) implies $N_\Gamma(\omega_{\pm 1})=g$. This gives another proof of this formula of Langlands.

5. **An example.** In this section we study the analogue of $\omega_{0,1}$ or $\omega_{1,0}$ for the Lorentz groups. Let $G=SO(n,1)$ be the group of all $g\in SL(n+1,\mathbf{R})$ that leave the form $\sum_{i=1}^n x_i^2-x_{n+1}^2$ invariant. Then if $K=G\cap SO(n+1)$, G/K can be identified with the unit ball in $\mathbf{R}^n$, $B^n_{\mathbf{R}}$. This can be done via the action $g\cdot x=(\langle x,c\rangle+d)^{-1}(Ax+b)$ where

$$g=\begin{bmatrix} A & b\\ {}^tc & d\end{bmatrix}$$

with A, $n\times n$; c, b, $n\times 1$; d, 1×1. As before, $\mathbf{R}^n$ will be looked at as $n\times 1$ column vectors. It is easy to see that G also acts on S^{n-1}, the unit sphere. For $f\in C^\infty(S^{n-1})$ let $(\pi_1(g)f)(x)=f(g^{-1}\cdot x)$. If $f\in C^\infty(S^{n-1})$, then $f=\sum_{p=0}^\infty f_p$ with f_p a spherical harmonic of degree p. Let $(\ ,\ )$ be the L^2 inner product on $C^\infty(S^{n-1})$ relative to the normalized standard volume element on S^{n-1}. Define for f, $g\in C^\infty(S^{n-1})$,

$$\langle f,g\rangle=\sum_{p=1}^\infty\binom{n+p-2}{n-1}(f_p,g_p).$$

Then $\langle\ ,\ \rangle$ defines a pre-Hilbert space structure on $C^\infty(S^{n-1})/C\cdot 1$, 1 the constant function with value 1 on S^{n-1}. It can be shown that $\langle\pi_1(x)f,g\rangle=\langle f,\pi_1(x)^{-1}g\rangle$ for f, $g\in C^\infty(S^{n-1})$ and that if H_1 is the Hilbert space completion of $(C^\infty(S^{n-1})/C\cdot 1,\langle\ ,\ \rangle)$, then (π_1,H_1) is a unitary representation of G. For details see Johnson and Wallach [**20**].

PROPOSITION 5.1 (HOTTA AND WALLACH [**18**]). *$b_1(\Gamma\backslash G_0/K_0)=N_\Gamma(\pi_1)$. G_0 is the identity component of G, and $K_0=G_0\cap K$.*

A unitary representation (π,H) of G, a semisimple Lie group, is called tempered if Θ_π defines a tempered distribution on G (that is, it extends to a continuous linear functional on the Schwartz space of G; see G. Warner [**38**]).

LEMMA 5.2. *π_1 is a tempered representation of $(SO(n,1))_0$ if and only if $n=2$ or 3.*

See Johnson and Wallach [**20**] for a proof of this result.

If G is a semisimple linear group, then a discrete subgroup, Γ, of G is said to be arithmetic (see Borel [**3**]) if there is an injective finite dimensional real representation (ρ,V) of G and a basis of V so that $\rho^{-1}(GL(N,Q))$ is dense in G, and if $\Gamma_1=\rho^{-1}(GL(N,\mathbf{Z}))$ ($N=\dim V$), then $\Gamma\cap\Gamma_1$ is of finite index in Γ and Γ_1.

THEOREM 5.3 (VINBERG [**34**]). *If $n=3,4$, or 5 there is $\Gamma\backslash G_0$ ($G=SO(n,1)$) so that $\Gamma\backslash G_0$ is compact, Γ is arithmetic and $\Gamma/[\Gamma,\Gamma]$ is infinite.*

Using Theorem 5.3, we see that Proposition 5.1 implies that $N_p(\pi_1)\neq 0$

for the Γ of Theorem 5.3. Lemma 5.2 says that π_1 is not tempered for $n \geq 4$. This gives the first example of a *specific* nontempered representation, ω, with $N_p(\omega) \neq 0$. (Recently, Millson has shown that such Γ exist for all n.)

6. **The Selberg trace formula for (rank 1).** Let G be a connected semisimple Lie group with finite center. Let K be as above and let A, N be such that $G=KAN$ is an Iwasawa decomposition of G (see, e.g., Helgason [**15**] or Wallach [**35**]). Then A is a maximal vector subgroup of G and N is a maximal unipotent subgroup of G.

DEFINITION 6.1. G is said to be split rank 1 if $\dim A = 1$.

We assume that G is split rank 1. Let M be the centralizer of A in K. We set $P=MAN$. Then P is a *parabolic* subgroup of G. Since G has split rank 1, *all* parabolic subgroups not equal to G are gotten in this manner. (For the general definition see Warner [**38**, Chapter 1].)

Let $\mathfrak{a}$ be the Lie algebra of A. Then $\mathfrak{a} = \mathbf{R}H$. Let $\mathfrak{n}$ be the Lie algebra of N. We can normalize H so that $\operatorname{ad} H|_{\mathfrak{n}}$ has eigenvalues 1 and possibly 2 (cf. Wallach [**36**]). Let $A^+ = \{\exp tH \mid t>0\}$.

Let p be the dimension of the eigenspace with eigenvalue 1 for $\operatorname{ad} H|_{\mathfrak{n}}$, q the dimension of the eigenspace with eigenvalue 2 for $\operatorname{ad} H|_{\mathfrak{n}}$. Then $p>0$ and $0 \leqq q < p$. Set $\rho = \frac{1}{2}(p+2q)$. If $\nu \in \mathbf{R}$ and $\xi \in \hat{M}$, let H^ξ denote the space of functions

$$f: K \to H_\xi \qquad ((\xi, H_\xi) \in \xi)$$

$$f(km) = \xi(m)^{-1} f(k) \quad \text{and} \quad \int_K \|f(k)\|^2 \, dk = \|f\|^2 < \infty.$$

If $f \in H^\xi$ let $f_\nu(k \exp tHn) = e^{-(i\nu+\rho)t} f(k)$, $k \in K$, $t \in \mathbf{R}$, $n \in N$. Set $(\pi_{\xi,\nu}(g)f)(k) = f_\nu(g^{-1}k)$. Then $(\pi_{\xi,\nu}, H^\xi)$ is a unitary representation of G.

If $f \in C_0^\infty(G)$ we define

$$F_f(m \exp tH) = e^{t\rho} \int_{K \times N} f(kmank^{-1}) \, dn \, dk$$

(here dn is normalized so that

$$dg = e^{2t\rho} \, dk \, dt \, dn, \qquad g = k \exp tHn).$$

THEOREM 6.2. *Let $\Theta_{\pi_{\xi,\nu}} = \Theta_{\xi,\nu}$ be the character of $\pi_{\xi,\nu}$. Then*

$$\Theta_{\xi,\nu}(f) = \int_M \int_{-\infty}^{\infty} F_f(m \exp tH) \operatorname{tr} \xi(m) e^{i\nu t} \, dt \, dm.$$

Applying the Fourier inversion theorem and the Peter-Weyl theorem we have

LEMMA 6.3.

$$F_f(m \exp tH) = \frac{1}{2\pi} \sum_{\xi \in \hat{M}} \int_{-\infty}^{\infty} \Theta_{\xi,\nu}(f) e^{-i\nu t} \overline{\operatorname{tr} \xi(m)} \, d\nu.$$

Now define for $m \in M$, $t \in \mathbf{R}$,

$$D(m \exp tH) = e^{-t\rho} \left|\det\left((\operatorname{Ad}(ma_t)^{-1} - I)|_{\mathfrak{n}}\right)\right|.$$

Clearly $D(m \exp tH) \neq 0$ if $t \neq 0$. A basic relation between F_f and the trace formula is

LEMMA 6.4 (CF. WALLACH [**35**, 7.7.10]). *If $t \neq 0$, then*

$$F_f(ma) = D(ma) \int_{G/A} f(gmag^{-1})\, d\dot{g}.$$

(Here $d\dot{g}$ on G/A is defined by

$$\int \phi(g)\, dg = \int_{G/A} \int_{-\infty}^{\infty} \phi(g \exp tH)\, dt\, d\dot{g}.\Big)$$

LEMMA 6.5. *If $g \in G$ and $\mathrm{Ad}(g)$ is semisimple, then g is either conjugate to an element of K or an element of MA^+.*

This follows from the fact that G has at most two conjugacy classes of Cartan subgroups (see Warner [**38**] for details).

Suppose now that $\Gamma \subset G$ satisfies the conditions of §3. If $\gamma \in \Gamma$, $\gamma \neq I$, then there is $x \in G$ so that $x\gamma x^{-1} = m_\gamma \exp t_\gamma$, $t_\gamma > 0$, $t_\gamma \in \mathbf{R}$, $m_\gamma \in M$.

LEMMA 6.6. *t_γ depends only on γ (not on the choice of x or Iwasawa decomposition). Also m_γ is determined up to conjugacy in M.*

PROOF. Since γ is conjugate to an element of MA^+ we see that the eigenvalues of $\mathrm{ad}\,\gamma$ are of the form δ, $e^{t_\gamma}\lambda$, $e^{-t_\gamma}\mu$, $e^{2t_\gamma}\eta$ or $e^{-2t_\gamma}\psi$ with $1 = |\lambda| = |\mu| = |\eta| = |\psi| = |\delta|$. Thus e^{t_γ} is uniquely described as $(\max\{|\lambda| \mid \lambda \text{ an eigenvalue of } \mathrm{Ad}\,\gamma\})^{1/2}$ if $\mathrm{ad}\,H$ has eigenvalue 2 or $\max\{|\lambda| \mid \lambda$ an eigenvalue of $\mathrm{Ad}(\gamma)\}$ otherwise. The second assertion is equally easy and we leave it to the reader.

Lemma 6.6 says that if $\gamma \in \Gamma$, $\xi \in \hat{M}$ and $\gamma \neq I$, then $D(\gamma) = D(m_\gamma \exp t_\gamma H)$, t_γ and $\mathrm{tr}\,\xi(m_\gamma)$ are well defined (independent of $x \in G$ so that $x\gamma x^{-1} \in MA$).

If $h \in MA^+$ and $G_h = \{g \in G \mid ghg^{-1} = h\}$, then

$$\int_{G/A} f(ghg^{-1})\, dg = \int_{G_h \backslash G} f(g^{-1}hg)\, dg \operatorname{vol}(G_h/A)$$

since G_h/A is compact. Let $u(\gamma) = \operatorname{vol}(G_{m_\gamma} \exp t_\gamma H/A)$. Combining Theorem 6.2, Lemmas 6.3, 6.5, 6.6 and the preceding observations, we have

THEOREM 6.7 (THE SELBERG TRACE FORMULA). *If $f \in C_c^\infty(G)$, then*

$$\sum_{\omega \in \hat{G}} N_\Gamma(\omega)\Theta_\omega(f) = \operatorname{vol}(\Gamma\backslash G)f(I) + \frac{1}{2\pi} \sum_{[\gamma]\in[\Gamma]-[I]} \operatorname{vol}(\Gamma_\gamma\backslash G_\gamma) D(\gamma)^{-1} \cdot u(\gamma)$$

$$\cdot \sum_{\xi \in \hat{M}} \overline{\mathrm{tr}\,\xi(m_\gamma)} \int_{-\infty}^{\infty} \Theta_{\xi,\nu}(f) \cdot e^{-it_\gamma \nu}\, d\nu.$$

Theorem 6.7 says that the following problem is quite important to the computation of the $N_\Gamma(\omega)$.

Problem 6.8 (The Paley-Wiener problem). Describe the functions $\nu \to \Theta_{\xi,\nu}(f)$, $\xi \in \hat{M}$, $\nu \in \mathbf{R}$ for $f \in C_c^\infty(G)$.

This problem has been solved by K. Johnson [**19**] up to a fairly touchy technical problem. There is, however, one case where the answer is exactly what one wishes. (For another, see §8.)

THEOREM 6.9 (S. HELGASON [**16**], R. GANGOLLI [**7**]). *If $\psi \in C_c^\infty(\mathbf{R})$ define*

$$\tilde{\phi}(\nu) = \int_{-\infty}^{\infty} \phi(t) e^{i\nu t}\, dt.$$

A necessary and sufficient condition that a function $\psi : \mathbf{R} \to \mathbf{R}$ be of the form $\Theta_{1,\nu}(f) = \psi(\nu)$ for $f \in I_c^\infty(G) = \{f \in C_c^\infty(G) | f(k_1 g k_2) = f(g)$ for all $k_1, k_2 \in K\}$ is that $\psi = \tilde{\phi}$ for $\phi \in C_c^\infty(\mathbf{R})$, and $\phi(-t) = \phi(t)$ for all $t \in \mathbf{R}$.

If $\psi \in C_c^\infty(\mathbf{R})$ let $f_\psi \in I_c^\infty(G)$ be such that $\Theta_{1,\nu}(f_\psi) = \tilde{\psi}(\nu)$. The Plancherel theorem for spherical functions says

THEOREM 6.10 (HARISH-CHANDRA [**11**], [**13**]). *$f_\phi(I) = \int_{-\infty}^{\infty} \tilde{\phi}(\nu) \cdot \mu_1(\nu)\, d\nu$ with $\mu_1 \in C^\infty(\mathbf{R})$.*

In order to derive a generalization of Selberg's original result for $PSL(2, \mathbf{R})$, we need one more theorem.

THEOREM 6.11 (KOSTANT [**24**]). *To each $0 \leqq d_{p,q} \leqq \nu < c_{p,q} < \infty$ there exists an irreducible unitary representation of G, $\pi_{i\nu}$, so that if $f \in I_c^\infty(G)$, $f = f_\phi$ for $\phi \in C_c^\infty(\mathbf{R})$, ϕ even, then*

$$\Theta_{\pi_{i\nu}}(f) = \tilde{\phi}(i\nu).$$

If $\omega \in \hat{G}$ and $\Theta_\omega(f) \neq 0$ for some $f \in I_c^\infty(G)$, then $\omega = \pi_{1,\nu}$ for some $\nu \in \mathbf{R}$, $\omega = 1$ the trivial representation, or $\omega = \pi_{i\nu}$ for some $0 \leqq d_{p,q} \leqq \nu < c_{p,q}$. (Notice that $d_{p,q}$ and $c_{p,q}$ depend only on p, q.)

Combining all of the above results we have

COROLLARY 6.12. *Let $\phi \in C_c^\infty(\mathbf{R})$ be an even function. Then*

$$\int_{-\infty}^{\infty} \phi(t)\, dt + \sum_{\nu \in \mathbf{R}} N_\Gamma(\pi_{1,\nu}) \tilde{\phi}(\nu) + \sum_{d_{p,q} \leqq \nu < c_{p,q}} N_\Gamma(\pi_{i\nu}) \tilde{\phi}(i\nu)$$
$$= \mathrm{vol}(\Gamma \backslash G) \int_{-\infty}^{\infty} \tilde{\phi}(\nu) \mu_1(\nu)\, d\nu + \sum_{[\gamma] \in [\Gamma] - [I]} \mathrm{vol}(\Gamma_\gamma \backslash G_\gamma) D(\gamma)^{-1} u(\gamma) \phi(t_\gamma).$$

Although we will not give any applications of this result in this article (see Gelfand, et al. [**9**] for a discussion of this formula in the case $G = PSL(2, \mathbf{R})$), we show how to use the results leading to this formula to prove that if $G = PSL(2, \mathbf{R})$, then

$$N_\Gamma(\omega_1) = d(\omega_1) \mathrm{vol}(\Gamma \backslash G) + 1.$$

In §9 we will show how to use this technique for $SU(2, 1)$ and certain elements of $G_d - G_d'$. The following technique is due to R. P. Langlands. Paul Sally taught the author this technique.

In this case $K = SO(2)/\pm I$, $M = \{I\}$. For $\vartheta \in \mathbf{R}$ let $k(\vartheta)$ be the rotation of $\mathbf{R}^2$ through the angle ϑ. If $n \in \mathbf{Z}$ let $\xi_n(\vartheta) = e^{in\vartheta}$. Then $\omega_1|_K = \sum_{n \geqq 1} \xi_{2n}$.

Choose $f \in C_c^\infty(G)$ so that $f(k(\vartheta_1) g k(\vartheta_2)) = e^{-i(\vartheta_1 + \vartheta_2)} f(g)$ and $\Theta_{\omega_1}(f) = 1$. This is clearly possible. Now

$$F_f(\exp tH) = e^{t\rho} \int_{K \times N} f(k \exp tHnk^{-1})\, dk\, dn \quad \text{for } m \in M,\ t \in \mathbf{R}.$$

Thus $F_f(\exp tH)$ is an element of $C_c^\infty(\mathbf{R})$. Theorem 6.9 says that there is $h \in I_c^\infty(G)$ so that $F_h = F_f$. Thus $F_{f-h}=0$.

According to the classification of irreducible unitary representatives of $PSL(2, \mathbf{R})$ (see P. Sally [**32**] or Gelfand, et. al. [**91**]), if $\omega \in \hat{G}$ and $\Theta_\omega(f-h) \neq 0$, then $\omega = \omega_1$ or 1. Thus we find using the arguments above that

$$\int_G (f-h)(x)\,dx + N_\Gamma(\omega_1)\Theta_{\omega_1}(f-h) = \operatorname{vol}(\Gamma\backslash G)(f-h)(I).$$

Now $\Theta_{\omega_1}(h)=0$ ($[\omega|_K : 1]=0$), $\Theta_{\omega_1}(f)=1$. Furthermore, it can be shown that

$$\int_{-\infty}^{\infty} F_f(\exp tH) e^{ipt}\,dt = \Theta_{\omega_1}(f) + \Theta_{\omega_{-1}}(f) + \int_G f(g)\,dg$$

for $f \in C_c^\infty(G)$. Hence,

$$0 = \Theta_{\omega_1}(f-h) + \int_G (f-h)(g)\,dg.$$

We therefore see that $\int_G (f-h)(g)\,dg = -1$. Finally,

$$(f-h)(I) = \sum_{\omega \in \hat{G}_d} d(\omega)\Theta_\omega(f-h)$$

by the Plancherel theorem for $PSL(2, \mathbf{R})$ ($F_{f-h}=0$). We have already observed that $\Theta_\omega(f-g)=0$ if $\omega \neq \omega_1$, $\omega \in \hat{G}_d$ and $\Theta_{\omega_1}(f-g)=1$. Hence, we have

$$-1 + N_\Gamma(\omega_1) = d(\omega_1)\operatorname{vol}(\Gamma\backslash G).$$

This is the asserted formula.

7. $\hat{G}$ **for** $G=SU(2,1)$. In this section we give a list of the elements of $\hat{G}$ for $G=SU(2,1)$. We first describe the nonunitary principal series for $SU(2,1)=G$. Let G act on $S^3=\{z \in C^2 \mid |z|=1\}$ as follows:

$$g \cdot z = (\langle z, c\rangle + d)^{-1}(Az+b), \quad g = \begin{bmatrix} A & b \\ c^* & d \end{bmatrix}$$

(See §4.) Set $a(g,z)=\bar{d}-\langle z, b\rangle$ for $z \in S^3$, $g \in G$. If $k_1, k_2 \in C$ and $k_1-k_2 \in \mathbf{Z}$ (the integers), define

$$(\pi_{k_1,k_2}(g)f)(z) = a(g,z)^{k_1}\,\overline{a(g,z)}^{k_2} f(g^{-1}\cdot z),$$

for $f \in C^\infty(S^3)$, $g \in G$.

Then $\pi_{k_1,k_2}(g)$ extends to a bounded operator on $L^2(S^3)=\mathscr{H}$ and $(\pi_{k_1,k_2}, \mathscr{H})$ defines a continuous representation of G for all $(k_1, k_2) \in C$ such that $k_1-k_2 \in \mathbf{Z}$.

LEMMA 7.1. *π_{k_1,k_2} is reducible if and only if $(k_1,k_2) \in \mathbf{Z}^2$ and $(k_1,k_2) \neq (-1,-1)$.*

We also note

LEMMA 7.2. *$(\pi_{k_1,k_2}, \mathscr{H})$ is a unitary representation (relative to the L^2-norm on $\mathscr{H}$) if and only if $-k_1-k_2=2+i\nu$, $\nu \in \mathbf{R}$.*

The representations of Lemma 7.2 are just a reparametrization of the $\pi_{\xi,\nu}$ of §6.

Before going on with the analysis of the π_{k_1,k_2}, we should explain the notation. Let

$$B=\begin{bmatrix} 2^{-1/2} & 0 & 2^{-1/2} \\ 0 & 1 & 0 \\ 2^{-1/2} & 0 & -2^{-1/2} \end{bmatrix}.$$

Then $B^2=I$. $B(MA)B$ is a real form of the group of diagonal matrices in $SL(3, C)$. Let $\mathfrak{h}$ be the space of all trace zero, diagonal, 3×3, complex matrices. Then if $\mathfrak{G}_C=sl(3, C)=\{X|X, 3\times 3, \operatorname{tr} X=0\}$, $\mathfrak{h}$ is a Cartan subalgebra of $\mathfrak{G}_C$. We say that $\Lambda\in\mathfrak{h}^*$ is G-integral if $h\to\Lambda(BhB)$ is the differential of a quasi-character, ξ_Λ, of MA. Let H be the element

$$H=\begin{bmatrix} 0 & 0 & 1 \\ 0 & 0 & 0 \\ 1 & 0 & 0 \end{bmatrix}.$$

Then $\mathfrak{a}=\mathbf{R}H$. Let Δ be the root system of $\mathfrak{G}_C$ relative to $\mathfrak{h}$. Let $\Delta^+=\{\alpha\in\Delta|\alpha(BhB)>0\}$. Let α_1, α_2 be the simple roots in Δ^+. Let Λ_1, Λ_2 be the basic highest weights for this order. That is

$$2\langle\Lambda_i, \alpha_j\rangle/\langle\alpha_j, \alpha_j\rangle=\delta_{ij}, \qquad 1\leqq i, j\leqq 2.$$

Here $\langle\ ,\ \rangle$ is the dual bilinear form on $\mathfrak{h}^*$ corresponding to the Killing form on $\mathfrak{G}_C$. Then $\Lambda\in\mathfrak{h}^*$ is G-integral if and only if $\Lambda=k_1\Lambda_1+k_2\Lambda_2$, $k_i\in C$, $i=1, 2$, $k_1-k_2\in\mathbf{Z}$. For Λ, G-integral, let X^Λ be the space of all $f\in C^\infty(G)$ such that

(1) $f(gma)=\xi_\Lambda(ma)f(g)$,
(2) $(R_Zf)(g)=0$ for $Z\in B_{\mathfrak{n}^+}B$.

Here $\mathfrak{N}^+$ is the Lie algebra of upper triangular matrices with zeros on the diagonal. If $X\in\mathfrak{G}$,

$$(R_Xf)(g)=\frac{d}{dt}f(g\exp tX)|_{t=0}$$

if $X\in\mathfrak{G}_c$,

$$R_{X_1+iX_2}f=R_{X_1}f+iR_{X_2}f, \qquad X=X_1+iX_2,\ X_1,\ X_2\in\mathfrak{G}.$$

If $f\in X^\Lambda$, $g\subset G$, dcfinc $(T_\Lambda(g)f)(X)=f(g^{-1}X)$. Let

$$K_1=\left\{\left[\begin{array}{c|c} u & 0 \\ \hline 0 & 1\end{array}\right]\ \middle|\ u\in SU(2)\right\}.$$

If $u\in SU(2)$,

$$u=\begin{bmatrix} z_1 & -\bar{z}_2 \\ z_2 & \bar{z}_1\end{bmatrix}=u(z_1, z_2) \quad \text{for } (z_1, z_2)\in S^3.$$

It is easily seen that if $f\in X^\Lambda$ then $f|_{K_1}$ determines f. If $f\in X^\Lambda$ define $\tilde{f}(z)=f(u(z))$, $z\in S^3$. Then if $\Lambda=k_1\Lambda_1+k_2\Lambda_2$, $(T_\Lambda(g)f)u(z)=(\pi_{k_1,k_2}(g)\tilde{f})(z)$ for $g\in G$, $z\in S^3$. Let $\rho=\frac{1}{2}\sum_{\alpha\in\Delta^+}\alpha$. We note that $\rho(BHB)=2$. We will now denote π_{k_1,k_2} by π_Λ, $\Lambda=k_1\Lambda_1+k_2\Lambda_2$. We say Λ is *integral* if $\Lambda=k_1\Lambda_1+k_2\Lambda_2$, $k_1, k_2\in\mathbf{Z}$. We can rephrase Lemma 7.1 to say

LEMMA 7.1′. *π_Λ is reducible if and only if Λ is integral and $\Lambda \neq -\rho$.*

Now let $H^{p,q}$ be the space of all polynomials f on $\mathbf{C}^2$ which are homogeneous of degree p in z_1, z_2, q in $\bar{z}_1, \bar{z}_2$ (that is, $f(\lambda z) = \lambda^p \bar{\lambda}^q f(z)$) and such that

$$\Delta f = \left(\frac{\partial^2}{\partial z_1 \, \partial \bar{z}_1} + \frac{\partial^2}{\partial z_2 \, \partial \bar{z}_2}\right) f = 0.$$

Set $\mathcal{H}^{p,q} = H^{p,q}|_{S^3}$. Then $\mathcal{H} = \sum_{p,q \geqq 0} \mathcal{H}^{p,q}$ a unitary direct sum. Furthermore, $(\pi_\Lambda|_K, \mathcal{H}^{p,q})$ is irreducible.

For $p, q \geqq 0$ set

$$a_{p,q}(\Lambda) = \prod_{j=1}^{p} \frac{\langle \Lambda + (j+1)\rho, \alpha_2 \rangle}{\langle -\Lambda + (j-1)\rho, \alpha_1 \rangle} \cdot \prod_{j=1}^{q} \frac{\langle \Lambda + (j+1)\rho, \alpha_1 \rangle}{\langle -\Lambda + (j-1)\rho, \alpha_2 \rangle}.$$

Here, as usual, $\prod_{j=u}^{v} a_j = 1$ if $u > v$. (Note that if π_Λ is irreducible then $a_{p,q}(\Lambda)$ is defined.)

For $f, g \in C^\infty(S^3)$ let

$$f = \sum f_{p,q}, \quad g = \sum g_{p,q}, \qquad f_{p,q}, g_{p,q} \in \mathcal{H}^{p,q},$$
$$\langle f, g \rangle_\Lambda = \sum_{p,q} a_{p,q}(\Lambda) \langle f_{p,q}, g_{p,q} \rangle.$$

LEMMA 7.3. *Suppose that π_Λ is irreducible. Then there exists a pre-Hilbert space structure on $C^\infty(S^3)$ so that $(\pi_\Lambda, C^\infty(S^3))$ completes to a unitary representation if and only if one of the following holds:*

(1) $\quad -\Lambda(H) = 2 + i\nu, \qquad \nu \in \mathbf{R};$

(2) $\quad \Lambda(H_{\alpha_1}) = \Lambda(H_{\alpha_2})$ *and* $\quad |2\langle \Lambda + \rho, \alpha_1 + \alpha_2 \rangle / \langle \alpha_1 + \alpha_2, \alpha_1 + \alpha_2 \rangle| < 2;$

(3) $\quad \Lambda(H_{\alpha_1}) - \Lambda(H_{\alpha_2})$ *is odd and* $|2\langle \Lambda + \rho, \alpha_1 + \alpha_2 \rangle / \langle \alpha_1 + \alpha_2, \alpha_1 + \alpha_2 \rangle| < 1.$

In the latter two cases the pre-Hilbert space structure is defined by $\langle \ , \ \rangle_\Lambda$.

We are now left with an analysis of the reducible π_Λ. In this case it is easier to use the T_Λ realization. For $f \in C^\infty(G)$ let $R_1 f = R_{Z_1} f$, $R_2 f = R_{Z_2} f$ where

$$Z_1 = B \begin{bmatrix} 0 & 0 & 0 \\ 1 & 0 & 0 \\ 0 & 0 & 0 \end{bmatrix} B, \qquad Z_2 = B \begin{bmatrix} 0 & 0 & 0 \\ 0 & 0 & 0 \\ 0 & 1 & 0 \end{bmatrix} B.$$

LEMMA 7.4. *Set $2\langle \Lambda + \rho, \alpha_i \rangle / \langle \alpha_i, \alpha_i \rangle = m_i$, $i = 1, 2$. (Note that $m_i = k_i + 1$, $i = 1, 2$ if $\Lambda = k_1 \Lambda_1 + k_2 \Lambda_2$.) If $m_i \in \mathbf{Z}$, $m_i > 0$, then $R_i^{m_i} X^\Lambda \subset X^{S_i(\Lambda+\rho)-\rho}$. ($S_i \mu = \mu - (2\langle \mu, \alpha_i \rangle / \langle \alpha_i, \alpha_i \rangle)\alpha_i$.) Furthermore,*

$$R_i^{m_i} \circ T_\Lambda(g) = T_{S_i(\Lambda+\rho)-\rho}(g) \circ R_i^{m_i}.$$

Using the map $f \to \tilde{f}$ of X^Λ onto $C^\infty(S^3)$, we see that if $m_i > 0$, $m_i \in \mathbf{Z}$ we can define $R_i^{m_i} f$ for $f \in C^\infty(S^3)$ by $(R_i^{m_i} f)^\sim = R_i^{m_i} \tilde{f}$. Then

$$R_i^{m_i} \circ \pi_\Lambda(g) = \pi_{S_i(\Lambda+\rho)-\rho}(g) \circ R_i^{m_i}.$$

LEMMA 7.5. *For $j \geqq 0$, $j \in \mathbf{Z}$, let*

$V_j^+ = \{f \in C^\infty(S^3) \mid f_{p,q} = 0 \text{ if } q > j\}$,

$V_j^- = \{f \in C^\infty(S^3) \mid f_{p,q} = 0 \text{ if } p > j\}$,

$W_j^+ = \{f \in C^\infty(S^3) \mid f_{p,q} = 0 \text{ if } q \leqq j\}$,

$W_j^- = \{f \in C^\infty(S^3) \mid f_{p,q} = 0 \text{ if } p \leqq j\}$.

Suppose that $\Lambda \in \mathfrak{h}^*$, $\Lambda = k_1\Lambda_1 + k_2\Lambda_2$.

(1) *If $k_1 \geqq 0$, $k_1 \in \mathbf{Z}$, then* $\ker R_1^{k_1+1} = V_{k_1}^-$, $R_1^{k_1+1}C^\infty(S^3) = W_{k_1}^+$.

(2) *If $k_2 \geqq 0$, $k_2 \in \mathbf{Z}$, then* $\ker R_2^{k_2+1} = V_{k_2}^+$, $R_2^{k_2+1}C^\infty(S^3) = W_{k_2}^-$.

Let $\mathfrak{Z}$ be the center of the universal enveloping algebra of $\mathfrak{G}_C$. If $\Lambda \in \mathfrak{h}^*$, Λ G-integral, it is not hard to see that $\pi_\Lambda(z)f = \chi_\Lambda(z)f$, $z \in \mathfrak{Z}$, $\chi_\Lambda : \mathfrak{Z} \to C$ an algebra homomorphism. Let W_C be the Weyl group of Δ. Then W_C is generated by S_1, S_2. The general theory of the universal enveloping algebra implies that $\chi_\Lambda = \chi_{S(\Lambda+\rho)-\rho}$ for $S \in W_C$. We are almost ready to list the remaining elements of $\hat{G}$. We first need to look at certain systems of positive roots. Let

$$\Delta_1^+ = S_1S_2\Delta^+, \qquad \Delta_2^+ = S_2S_1\Delta^+,$$

and let

$$S_1S_2S_1\Delta^+ = S_2S_1S_2\Delta^+ = -\Delta^+ = \Delta_3^+.$$

Let $\mathscr{F}$ denote the G-integral elements of $\mathfrak{h}^*$.

I. THE HOLOMORPHIC DISCRETE SERIES. For $\Lambda \in \mathscr{F}$ and $\langle \Lambda+\rho, \alpha\rangle > 0$ for $\alpha \in \Delta_1^+$ let $(D_\Lambda^+, V_+^\Lambda)$ be defined as follows: $\Lambda = k_1\Lambda_1 + k_2\Lambda_2$ with $k_1 < 0$, $k_2 \geqq 0$. Let V_+^Λ be the Hilbert space completion of $V_{k_2}^+$ relative to $\langle f, g\rangle_\Lambda = \sum a_{p,q}(\Lambda)\langle f_{p,q}, g_{p,q}\rangle$. $D_\Lambda^+(g) = \pi_\Lambda(g)|_{V_+^\Lambda}$.

II. THE ANTIHOLOMORPHIC DISCRETE SERIES. For $\Lambda \in \mathscr{F}$ and $\langle \Lambda+\rho, \alpha\rangle > 0$ for $\alpha \in \Delta_2^+$ let $(D_\Lambda^-, V_-^\Lambda)$ be defined as follows: $\Lambda = k_1\Lambda_1 + k_2\Lambda_2$, $k_1 \geq 0$, $k_2 < 0$. V_-^Λ is the Hilbert space completion of $V_{k_1}^-$ relative to $\langle f, g\rangle_\Lambda = \sum a_{p,q}(\Lambda)\langle f_{p,q}\, g_{p,q}\rangle$. $D_\Lambda^-(g) = \pi_\Lambda(g)|_{V_-^\Lambda}$.

III. THE NONHOLOMORPHIC DISCRETE SERIES. For $\Lambda \in \mathscr{F}$ and $\langle \Lambda+\rho, \alpha\rangle > 0$ for $\alpha \in \Delta_3^+$ let (D_Λ, W^Λ) be defined as follows: set $\mathscr{W}^\Lambda = W_{-k_1-2}^+ \cap W_{-k_2-2}^-$. Define

$$b_{p,q}(\Lambda) = \prod_{j=-k_2}^{p} \frac{\langle \Lambda + (j+1)\rho, \alpha_2\rangle}{\langle -\Lambda + (j-1)\rho, \alpha_1\rangle} \prod_{j=-k_1}^{q} \frac{\langle \Lambda + (j+1)\rho, \alpha_1\rangle}{\langle -\Lambda + (j-1)\rho, \alpha_2\rangle},$$

$p \geqq -k_2 - 1$, $q \geqq -k_1 - 1$. Let W^Λ be the Hilbert space completion of $\mathscr{W}^\Lambda$ relative to $(f, g)_\Lambda = \sum b_{p,q}(\Lambda)\langle f_{p,q}, g_{p,q}\rangle$. $D_\Lambda(g) = \pi_\Lambda(g)|_{W^\Lambda}$.

This gives a complete parametrization of $\hat{G}_d$. It is useful to give an alternate parametrization. We first observe that if $\Lambda \in \mathscr{F}^+$, $\Lambda = k_1\Lambda_1 + k_2\Lambda_2$, then $\ker R_1^{k_1+1} \cap \ker R_2^{k_2+1}$ is an invariant finite dimensional subspace of $\mathscr{H}$. Set $V^\Lambda = \ker R_1^{k_1+1} \cap \ker R_2^{k_2+1}$. Then (π_Λ, V^Λ) is the finite dimensional irreducible representation of G with lowest weight $-\Lambda$.

LEMMA 7.6. *If $\Lambda \in \mathscr{F}^+ = \{\Lambda \in \mathscr{F} | \langle \Lambda, \alpha\rangle \geqq 0$ for $\alpha \in \Delta^+\}$, let $G_{d,\Lambda} = \{\omega \in \hat{G}_d |$ if*

$(\pi_1 H) \in \omega$, $\pi(z)=\chi_\Lambda(z)I$ *for* $z \in \mathfrak{Z}\}$. *Then*

$$\hat{G}_{d,\Lambda} = \{D^+_{s_1s_2(\Lambda+\rho)-\rho}, D^-_{s_2s_1(\Lambda+\rho)-\rho}, D_{s_1s_2s_1(\Lambda+\rho)-\rho}\}.$$

We can also describe $\hat{D}'_d$.

LEMMA 7.7. $\hat{G}'_d \cap \hat{G}_{d,\Lambda} = \hat{G}_{d,\Lambda}$ *if* $\Lambda = k_1\Lambda_1 + k_2\Lambda_2$, $k_1, k_2 \geqq 2$. *If* $k_1 \geqq 2$, $0 \leqq k_2 < 2$, *then* $\hat{G}'_d \cap \hat{G}_{d,\Lambda} = \{D^-_{s_2s_1(\Lambda-\rho)-\rho}\}$. *If* $0 \leqq k_1 < 2$ *and* $k_2 \geqq 2$, *then* $\hat{G}'_d \cap \hat{G}_{d,\Lambda} = \{D^+_{s_1s_2(\Lambda+\rho)-\rho}\}$. *If* $0 \leqq k_1, k_2 < 2$, *then* $\hat{G}'_d \cap \hat{G}_{d,\Lambda} = \varnothing$.

The next class of elements of $\hat{G}$ we describe are the irreducible constituents of reducible (unitary) principal series representations. From Lemmas 7.1 and 7.2 we see that the reducible unitary principal series consists of the $(\pi_\Lambda, \mathscr{H})$ with Λ integral, $\Lambda \neq -\rho$ and $\Lambda = k_1\Lambda_1 + k_2\Lambda_2$, $k_1 + k_2 = -2$. We note that k_1 and k_2 cannot both be negative since Λ is integral, $\Lambda \neq -\rho = -\Lambda_1 - \Lambda_2$ and $k_1 + k_2 = -2$. There are thus two cases:

(i) $\Lambda = k_1\Lambda_1 + k_2\Lambda_2$, $k_1 \geqq 0$, $k_1 + k_2 = -2$. In this case if H^-_Λ is the L^2-completion of $V^-_{k_1}$, H^-_Λ is an invariant subspace, and if H^+_Λ is the Hilbert space completion of $W^+_{k_1}$, then H^+_Λ is also invariant. $\mathscr{H} = H^-_\Lambda \oplus H^+_\Lambda$, $\pi_\Lambda = \pi^+_\Lambda \oplus \pi^-_\Lambda$, $\pi^\pm_\Lambda(g) = \pi_\Lambda(g)|_{H^\pm_\Lambda}$.

(ii) $\Lambda = k_1\Lambda_1 + k_2\Lambda_2$, $k_2 \geqq 0$, $k_1 + k_2 = -2$, $k_i \in \mathbf{Z}$. This time take H^+_Λ to be the L^2-completion of $V^+_{k_2}$ and H^-_Λ the L^2-completion of $W^-_{k_2}$. Then $\mathscr{H} = H^+_\Lambda \oplus H^-_\Lambda$ and $\pi_\Lambda = \pi^+_\Lambda \oplus \pi^-_\Lambda$.

At this point we have described all of the tempered representations of G.

LEMMA 7.8. *The tempered representations of G consist of*

(1) $\hat{G}_d$;

(2) *the irreducible principal series*: $(\pi_\Lambda, \mathscr{H})$ *with*

$$\operatorname{Re} 2\langle \Lambda, \alpha_1 + \alpha_2 \rangle / \langle \alpha_1 + \alpha_2, \alpha_1 + \alpha_2 \rangle = -2$$

and $\Lambda = -\rho$ *or* Λ *not integral;*

(3) *the irreducible constituents of reducible principal series*: Λ *integral*, $\Lambda \neq -\rho$, $2\langle \Lambda, \alpha_1 + \alpha_2 \rangle / \langle \alpha_1 + \alpha_2, \alpha_1 + \alpha_2 \rangle = -2$ *and* π^+_Λ, π^-_Λ.

If G were $SL(2, \mathbf{R})$ then the analogous list of Lemmas 7.8 and 7.3 with the addition of the trivial representation would completely describe $\hat{G}$. For $SU(2, 1)$ there are more such "trivial representations". They are analogous to the representations of $SO(n, 1)$ of §5. We now describe these extra representations.

(T^+) For $k \in \mathbf{Z}$, $k \geqq -1$, we define on V^+_0 the Hermitian form

$$\langle f, g \rangle_k = \sum_{p=0}^{\infty} \binom{p+1}{k+2} \langle f_{p,0}, g_{p,0} \rangle;$$

then since $\binom{p+1}{k+2} = 0$ if $p < k+1$, we see that $\langle\ ,\ \rangle_k$ induces a pre-Hilbert space structure on $V^+_0 / V^-_k \cap V^+_0$ ($V^-_{-1} = 0$). Let Z^+_k denote the Hilbert space completion of $V^+_0 / V^-_k \cap V^+_0$. Let $T^+_k(g)$ be the operator induced by $\pi_{k\Lambda_1}(g)|_{V^+_0}$. Then (T^+_k, Z^+_k) defines a unitary representation of G.

(T^-) For $k \in \mathbf{Z}$, $k \geqq -1$, define on V^-_0 the form

$$(f, g)_k = \sum_{p=0}^{\infty} \binom{p+1}{k+2} \langle f_{0,p}, g_{0,p} \rangle.$$

Take Z_k^- to be the Hilbert space completion of $V_0^-/V_0^-\cap V_k^+$ ($V_{-1}^+=0$) relative to $(\ ,\ )_k$. Let T_k^- be defined as in (T^+) (with Λ_1 replaced by Λ_2).

Then (T_k^-, Z_k^-) is a unitary representation of G.

The representations listed combined with the trivial representation completely describe $\hat{G}$.

In §9 we will use this description of $\hat{G}$ in the trace formula. To do this we need a list of the K-types in each element of $\hat{G}$. We will also need to know where the various elements of $\hat{G}$ appear in the composition series of the π_Λ (ignoring the unitary structures). To describe the K-types we first note that

$$K=\left\{\left[\begin{array}{c|c} u & 0 \\ \hline 0 & (\det u)^{-1}\end{array}\right] \middle|\, u\in U(2)\right\}.$$

Thus $K=S^1\cdot K_1$. That is, if $u\in K$, then

$$u=\left[\begin{array}{c|c} e^{i\vartheta}u_1 & 0 \\ \hline 0 & e^{-2i\vartheta}\end{array}\right], \qquad u_1\in SU(2),\ \vartheta\in \mathbf{R}.$$

The representations of $K_1=SU(2)$ are parametrized by their dimensions. Let τ_p denote the irreducible representation of K_1 of dimension $p+1$. Let τ_p^l be defined by the following:

(1) $$\tau_p^l|_{K_1}=\tau_p,$$

(2) $$\tau_p^l\left(\left[\begin{array}{c|c} e_I^{i\vartheta} & 0 \\ \hline 0 & e^{-2i\vartheta}\end{array}\right]\right)=e^{-il\vartheta}I.$$

Of course the only τ_p^l that are representations of $U(2)=K$ are the ones such that $l\equiv p \bmod 2$.

LEMMA 7.9 *Let $\Lambda\in\mathfrak{h}^*$ be G-integral. Set $l=k_2-k_1$, $\Lambda=k_1\Lambda_1+k_2\Lambda_2$ (then $l\in\mathbf{Z}$). Then*

$$(\pi_\Lambda|_K, \mathcal{H}^{p,q})\equiv\tau_{p+q}^{2l+3(p-q)}$$

($\equiv$ indicates K-equivalence).

This lemma allows one to determine completely the K-types in each $\omega\in\hat{G}$. If π is a representation of G (not necessarily unitary) then we say that $\pi\subset\pi_\Lambda$ if π is infinitesimally equivalent with a subquotient of π_Λ. For our purposes this can be taken to mean that if H is the representation space for π, then there is a dense subspace $H_0\subset H$ and $\mathcal{H}\supset V_1\supset V_2$, V_i closed invariant spaces of $\mathcal{H}$ and $A: H_0\to V_1/V_2$ an injective intertwining operator of π with the induced action of π_Λ on V_1/V_2 so that $A(H_0)$ is dense in V_1/V_2.

Set $S_0=S_1S_2S_1=S_2S_1S_2$: Then if π_Λ is irreducible and if $\pi_\Lambda\subset\pi_\mu$ for some μ, then $\mu=S_0(\Lambda+\rho)-\rho$. This can be proved in this case by comparing representations of K_1.

We may therefore confine our attention to $\hat{G}_d$, the π_Λ^+, π_Λ^- and the $T_k^\pm$, $k\geqq-1$.

LEMMA 7.10. *If $\Lambda=k_1\Lambda_1+k_2\Lambda_2$, $k_i\geqq0$, $k_i\in\mathbf{Z}$, then*

(1) *$D_{S_0(\Lambda+\rho)-\rho}\subset\pi_\mu$ if and only if $\mu=S(\Lambda+\rho)-\rho$ for some $s\in W_C$.*

(2) $D^+_{S_1S_2(\Lambda+\rho)-\rho} \subset \pi_\mu$ *if and only if* $\mu = S_1(\Lambda+\rho)-\rho$, *or* $\mu = S_1S_2(\Lambda+\rho)-\rho$.
(3) $D^-_{S_2S_1(\Lambda+\rho)-\rho} \subset \pi_\mu$ *if and only if* $\mu = S_2(\Lambda+\rho)-\rho$ *or* $\mu = S_2S_1(\Lambda+\rho)-\rho$.

LEMMA 7.11 *Let* $\Lambda = k_1\Lambda_1 + k_2\Lambda_2$, $k_1, k_2 \in \mathbf{Z}$, $k_1+k_2=-2$ *and* $\Lambda \neq -\rho$.

(1) *Suppose* $k_1 \geqq 0$. *Then* $\pi^-_\Lambda \subset \pi_\mu$ *if and only if* $\mu = \Lambda$. $\pi^+_\Lambda \subset \pi_\mu$ *if and only if* $\mu = S(\Lambda+\rho)-\rho$ *for some* $S \in W_C$ (*note that* $S_0(\Lambda+\rho)-\rho = \Lambda$).

(2) *Suppose* $k_2 \geqq 0$. *Then* $\pi^+_\Lambda \subset \pi_\mu$ *if and only if* $\mu = \Lambda$. $\pi^-_\Lambda \subset \pi_\mu$ *if and only if* $\mu = S(\Lambda+\rho)-\rho$ *for some* $S \in W_C$ ($S_0(\Lambda+\rho)-\rho=\Lambda$).

We are left with the representations T^+_k *and* T^-_k.

LEMMA 7.12. (1) *If* $k \geq 0$ *then* $T^+_k \subset \pi_\Lambda$ *if and only if* $\Lambda = S_1(k\Lambda_1+\rho)-\rho$, $S_0(k\Lambda_1+\rho)-\rho$, $S_1S_2(k\Lambda_2+\rho)-\rho$ *or* $k\Lambda_1$.

(2) *If* $k \geqq 0$ *then* $T^-_k \subset \pi_\Lambda$ *if and only if* $\Lambda = S_2(k\Lambda_2+\rho)-\rho$, $S_0(k\Lambda_2+\rho)-\rho$, $S_2S_1(k\Lambda_2+\rho)-\rho$ *or* $k\Lambda_2$.

(3) $T^+_{-1} \subset \pi_\Lambda$ *if and only if* $\Lambda = S(-\Lambda_1+\rho)-\rho$ *for some* $S \in W_C$.

(4) $T^-_{-1} \subset \pi_\Lambda$ *if and only if* $\Lambda = S(-\Lambda_2+\rho)-\rho$ *for some* $S \in W_C$.

8. **The image of F_f.** In order to apply the results of §7 to the trace formula, we will need some more information on F_f. We assume that G is a simple Lie group of split rank 1 and has finite center. We retain the notation of §6. The main result of this section is

THEOREM 8.1. *Let* τ *be a one dimensional representation of* K. *If* $\phi \in C_c^\infty(MA)$ *and* $\phi(ma) = \tau(m)\phi(a)$, $\phi(a) = \phi(a^{-1})$ *for* $a \in A$, *then there exists* $f \in C_c^\infty(G)$ *with* $f(k_1gk_2) = \tau(k_1)f(g)\tau(k_2)$ *so that* $F_f = \phi$.

This result has been proved in the case $G = SU(2,1)$ by Gupta (Thesis, University of Washington). His proof is substantially the same as the one outlined below.

We sketch a proof of this result. The idea of the proof is a modification of Helgason's technique for proving Theorem 6.9. We first note that if τ is the trivial representation, then Theorem 8.1 is a special case of Theorem 6.9.

We therefore assume that τ is nontrivial (the technique we describe works for the trivial representation and is actually easier in that case). Since τ is nontrivial, K cannot be semisimple. Since G is simple and split rank 1, this implies G is locally isomorphic with $SU(n,1)$. Let $\xi = \tau|_M$. For $\nu \in \mathbf{R}$ let $(\pi_{\xi,\nu}, H^\xi)$ be as in the beginning of §6.3. We note that for $\nu \in C$, $(\pi_{\xi,\nu}, H^\xi)$ makes perfectly good sense; it is just not necessarily a unitary representation. Let $C_c^\infty(\tau : G : \tau)$ be the space of all C_c^∞ functions f such that $f(k_1gk_2) = \tau(k_1)f(g)\tau(k_2)$. Let $E_\tau : H^\xi \to H^\xi_\tau$ be the projection onto the (one dimensional) subrepresentation of $(\pi_{\xi,\nu}|_K, H^\xi)$ equivalent to τ.

Let $E_\tau(\nu : g) = E_\tau \pi_{\xi,\nu}(g) E_\tau$ for $v \in C$. Then the Plancherel formula for G says that there is a positive function, μ_τ, on $\mathbf{R}$ so that (see Harish-Chandra [**14**], G. Warner [**37**, Epilogue])

$$f(I) = \sum_{\omega \in \hat{G}_d} \Theta_\omega(R_g f) + \int_{-\infty}^{\infty} \alpha(\nu) E_\tau(v : g)\mu_\tau(\nu)\, d\nu.$$

Here $\alpha(\nu)$ is, up to a constant independent of ν, τ, given as

$$\alpha(\nu)=\int_{-\infty}^{\infty} F_f(a_t)e^{it\nu}\,dt.$$

Since τ is one dimensional it is easily checked that $\alpha(\nu)=\alpha(-\nu)$. Since $F_f|_A$ has compact support it follows that α extends to an entire function on C and there is $T>0$ so that if $\varepsilon>0$, $k>0$ are given, there is $C_{k,\varepsilon}$ such that

(i) $|\alpha(\nu)|\leqq C_{k,\varepsilon}(1+|\nu|)^{-k}\exp((T+\varepsilon)|\text{Im}\,\nu|)$,

(ii) $\alpha(\nu)=\alpha(-\nu)$.

Let $\mathcal{P}(T)$ be the space of all entire functions on C satisfying (i), (ii). To prove Theorem 8.1 we need only show that if $\alpha\in\mathcal{P}(T)$ and if

$$\text{(iii)}\qquad f_\alpha(g)=\int_{-\infty}^{\infty}\alpha(\nu)E_\tau(\nu:g)\mu_\tau(\nu)\,d\nu,$$

then there exist $\omega_1,\cdots,\omega_k\in\hat{G}_d$ and $\phi_1,\cdots,\phi_k$ matrix entries of $\omega_1,\cdots,\omega_k$ so that $f_\alpha-\sum\phi_i\in C_c^\infty(\tau:G:\tau)$. Here we use the fact that for such ϕ_j, $F_{\phi_j}=0$ (Harish-Chandra [**13**]).

To do this we study in more detail the functions $E_\tau(\nu:g)$. In Wallach [**35**] it is shown that if $\Delta(t)=(\sin ht)^{2n-2}\sin h2t$, and if $\Phi_\tau(\nu:t)=\Delta(t)^{1/2}E_\tau(\nu:a_t)$ for $t>0$, then there is $Q_\tau\in C^\infty((0,\infty))$ so that

$$\text{(1)}\qquad Q_\tau(t)=\sum_{k=1}^{\infty}a_k e^{-kt}\quad\text{for } t\geqq 1,$$

$$\text{(2)}\qquad -\frac{d^2}{dt^2}\Phi_\tau(v:t)+Q_\tau(t)\Phi_\tau(v:t)=\nu^2\Phi_\tau(v:t).$$

In Wallach [**35,** Appendix], the following result was proved (see also Dunford and Schwartz [**5**]).

LEMMA 8.2. *There exists an* $\varepsilon>0$ *and a continuous function* $\sigma:\bar{H}_\varepsilon\times(0,\infty)\to C$ $(H_\varepsilon=\{z\in C|\text{Im}\,z>-\varepsilon\})$ *so that*

(1) $t\to\sigma(\nu:t)$ *is* C^∞ *for* $\nu\in\bar{H}_\varepsilon$, $t>0$.

(2) $|\sigma(\nu:t)|\leqq 2$ *for* $t\geqq a_0$ *for* $\nu\in\bar{H}_0$, a_0 *independent of* ν.

$$\text{(3)}\qquad -\frac{d^2}{dt^2}(e^{i\nu t}\sigma(\nu:t))+Q_\tau(t)e^{i\nu t}\sigma(\nu:t)=\nu^2e^{i\nu t}\sigma(\nu:t),\qquad \nu\in\bar{H}_\varepsilon.$$

(4) $\lim_{t\to\infty,t>0}\sigma(\nu:t)=1$ *for* $\nu\in\bar{H}_\varepsilon$.

(5) $\nu\to\sigma(\nu:t)$ *is holomorphic for* $\nu\in H_\varepsilon$.

This result was used in Wallach [**35**] to show that if $\nu\in\mathbf{R}$, $v\neq 0$

$$\Phi_\tau(\nu:a_t)=e^{i\nu t}C_\tau(\nu)\sigma(\nu:t)+e^{-i\nu t}C_\tau(-\nu)\sigma(-\nu:t)$$

for $t>0$. It was also shown that if $\text{Im}\,\nu<0$, then

$$C_\tau(\nu)=\int_{\bar{N}}\exp(-(\rho+i\nu)(H(\bar{n})))\tau(K(\bar{n}))^{-1}\,d\bar{n}.$$

(Here $\bar{N}=\vartheta(N)$, $\vartheta(g)={}^t\bar{g}^{-1}$ for $g\in SU(n,1)$ for an appropriate choice of N. Also if $g\in G$, $g=K(g)\exp(H(g)H)n(g)$ with $K(g)\in K$, $n(g)\in N$.) Now K is locally isomorphic to $U(n)$. Let τ_k denote the kth power of the determinant function (for coverings of $SU(n, 1)$, k can be a rational number). Then τ_k describes the typical one dimensional representation of K.

LEMMA 8.3. *There is a constant, $c(n)=C$, depending only on n so that*

$$C_{\tau_k}(\nu)=C\frac{\Gamma(i\nu/2)\Gamma(i\nu/2+\frac{1}{2})}{\Gamma((n+i\nu+k)/2)\Gamma((n+i\nu-k)/2)}$$

where Γ is the classical gamma function (cf. Whittaker and Watson [**38**]).

LEMMA 8.4 (HARISH-CHANDRA [**14**], G. WARNER [**37**]). $\mu_\tau(\nu)=(C_\tau(-\nu)C_\tau(\nu))^{-1}$ *for $\nu\in\mathbf{R}$.*

Lemma 8.4 is a very special case of the Maas-Selberg relations of Harish-Chandra. A proof of this result can also be found in Knapp-Stein [22].

The above results say that

$$\Delta(t)^{1/2}E_\tau(\nu:a_t)\mu_\tau(\nu)=e^{i\nu t}C_\tau(-\nu)^{-1}\sigma(\nu:t)+e^{-i\nu t}C_\tau(\nu)^{-1}\sigma(-\nu:t)$$

for $\nu\in\mathbf{R}$. Thus,

$$\Delta(t)^{1/2}f_\alpha(a_t)=\int_{-\infty}^{\infty}e^{i\nu t}C_\tau(-\nu)^{-1}\alpha(\nu)\sigma(\nu:t)\,d\nu$$

$$+\int_{-\infty}^{\infty}e^{-i\nu t}C_\tau(\nu)^{-1}\alpha(\nu)\sigma(-\nu:t)\,d\nu;$$

since $\alpha(-\nu)=\alpha(\nu)$ we see

$$\Delta(t)^{1/2}f_\alpha(a_t)=2\int_{-\infty}^{\infty}e^{-i\nu t}C_\tau(\nu)^{-1}\alpha(\nu)\sigma(-\nu:t)\,d\nu.$$

Using classical results on the Γ-function we know that if $|\nu|>R$, $\operatorname{Im}\nu\leqq 0$, then

$$|C_\tau(\nu)^{-1}|\leqq C(1+|\nu|)^N$$

for some $C>0$, $N>0$. Integrating $e^{-i\nu t}C_\tau(\nu)^{-1}\alpha(\nu)\sigma(-\nu:t)$ about the contour Γ_R:

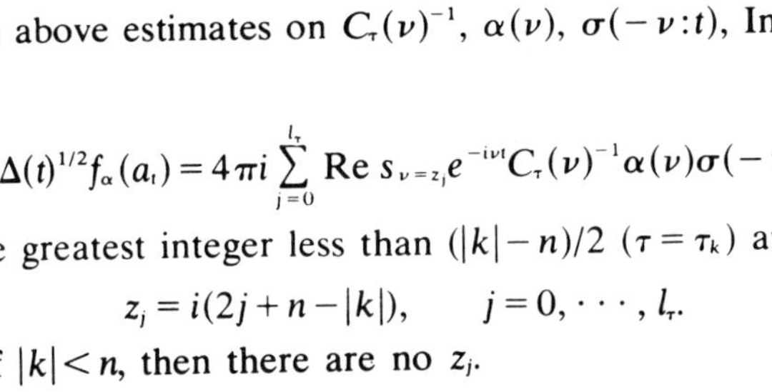

and using the above estimates on $C_\tau(\nu)^{-1}$, $\alpha(\nu)$, $\sigma(-\nu:t)$, $\operatorname{Im}\nu\leqq 0$ we find that if $t>T$,

$$\Delta(t)^{1/2}f_\alpha(a_t)=4\pi i\sum_{j=0}^{l_\tau}\operatorname{Res}_{\nu=z_j}e^{-i\nu t}C_\tau(\nu)^{-1}\alpha(\nu)\sigma(-\nu:t).$$

Here l_τ is the greatest integer less than $(|k|-n)/2$ $(\tau=\tau_k)$ and

$$z_j=i(2j+n-|k|),\qquad j=0,\cdots,l_\tau.$$

Notice that if $|k|<n$, then there are no z_j.

To complete the proof we need

LEMMA 8.5. *If* $\nu=z_j$, $j=0, 1, \cdots, l_\tau$, *then*

$$\Delta(t)^{1/2}E_\tau(\nu : a_t)=e^{i\nu t}C_\tau(-\nu)\sigma(-\nu : t).$$

This lemma is proved by a careful analysis of where the holomorphic discrete series appears in the $\pi_{\xi,\nu}$, $\nu\in C$. Using the fact that $f_\alpha\in L^2(G)$, we therefore see that $g\to E_\tau(\nu : g)$ is a matrix entry of an element of $\hat{G}_d$ for $\nu=z_j$, $j=0, 1, \cdots, l_\tau$. Thus, if

$$\phi_j(g)=4\pi i \operatorname*{res}_{\nu=z_j} C_\tau(\nu)^{-1}\alpha(\nu)E_\tau(-\nu : t),$$

we see that $f_\alpha-\sum \phi_j\in C_c^\infty(\tau : G : \tau)$. This completes the proof of Theorem 8.1.

9. **Applications to $\Gamma\backslash SU(2, 1)$.** We return to the notation of §7. We will study the case $G=SU(2, 1)$ in this section.

PROPOSITION 9.1. *Let $\omega\in\hat{G}$ be such that if $\gamma\in\hat{K}$ and γ is one dimensional, then $[\omega|_K : \gamma]=0$. Let $\tau\in\hat{K}$ be such that $[\omega|_K : \tau]\neq 0$. Set for $\theta\in\mathbf{R}$,*

$$m(\theta)=\begin{bmatrix} e^{i\theta} & 0 & 0 \\ 0 & e^{-2i\theta} & 0 \\ 0 & 0 & e^{i\theta}\end{bmatrix}, \qquad M=\{m(\theta)\mid\theta\in\mathbf{R}\}.$$

Suppose that $\tau(m(\theta))$ diagonalizes with diagonal entries $e^{ik_j\theta}$, $j=1, \cdots, r$. Then there exists $f\in C_c^\infty(G)$ so that

(1) $\Theta_\omega(f)=1$;

(2) $F_f=0$;

(3) *if $\eta\in\hat{G}$ has a different infinitesimal character than ω, then $\Theta_\eta(f)=0$.*

(4) *if $\eta\in\hat{G}$ and $[\eta|_K : \tau]=[\eta|_K : \tau_0^{2k_j}]=0$, $j=1, \cdots, r$, then $\Theta_\eta(f)=0$ (here we use the parametrization of representations of $U(2)$ given in §7).*

PROOF. Let τ^* be the complex conjugate representation to τ. Let (τ, V) be an irreducible unitary representation in the class of τ. Choose

$$h : G\to \operatorname{End}(V) \quad \text{so that} \quad h(k_1gk_2)=\tau^*(k_1)h(g)\tau^*(k_2)$$

for k_1, $k_2\in K$, $g\in G$, h of class C_c^∞ and $v\in V$ so that if $\phi(g)=\langle h(g)v, v\rangle$ ($\langle\ ,\ \rangle$ the K-invariant inner product on V), then $\Theta_\omega(\phi)=1$.

We compute $F_\phi(m(\theta)a)$.

$$\begin{aligned} F_\phi(ma) &= e^{\rho(\log a)}\int_{K\times N}\phi(kmank^{-1})\,dk\,dn \\ &= e^{\rho(\log a)}\int_N\left(\left\langle\int_K h(kmank^{-1})\,dk\cdot v, v\right\rangle\right)dn \\ &= e^{\rho(\log a)}\int_K (\langle\tau^*(k)h(man)\tau^*(k)^{-1}\,dk\cdot v, v\rangle)\,dn. \end{aligned}$$

The Schur orthogonality relations for K say that if $T\in\operatorname{End}(V)$, then

$$\int_K \tau^*(k)T\tau^*(k)^{-1}\,dk=\frac{1}{d_\tau}\operatorname{tr} T,$$

where d_τ is the dimension of τ. Hence,

$$F_\Phi(ma) = \frac{1}{d_\tau} e^{\rho(\log a)}\langle v, v\rangle \cdot \operatorname{tr}\left(\int_N h(man)\, dn\right).$$

Now if $m \in M$, then $n \to mnm^{-1}$ is a diffeomorphism of N preserving dn. Hence,

$$\tau^*(m)\int_N h(an)\, dn = \int_N h(an)\, dn\, \tau^*(m)$$

for $m \in M$. Thus, if $V = \sum_{j=1}^r V_j$ with $\tau(m(\theta))|_{V_j} = e^{ik_j\theta} I$, and if $P_j : V \to V_j$ is the orthogonal projection, then

$$\int_N h(an)\, dn = \sum_{j=1}^r P_j \int_N h(an)\, dn\, P_j.$$

Set

$$\psi_j(m(\theta)a) = \frac{1}{d_\tau} e^{\rho(\log a)}\langle v, v\rangle e^{-ik_j} \cdot \operatorname{tr} P_j \int_N h(an)\, dn\, P_j.$$

Then $F_\Phi(ma) = \sum_{j=1}^r \psi_j(ma)$, $m \in M$, $a \in A$. Now

$$F_\Phi(ma) = F_\Phi(ma^{-1}), \qquad \psi_j(m(\theta)a) = e^{-ik_j\theta}\psi_j(a),$$

and

$$\psi_j(a) = \frac{1}{2\pi}\int_0^{2\pi} e^{ik_j\theta} F_\Phi(m(\theta)a)\, d\theta.$$

Hence, $\psi_j(a) = \psi_j(a^{-1})$. Furthermore, it is easily seen that $\tau^{-2k_j}(m(\theta)) = e^{-ik_j\theta}$. Theorem 8.1 now implies that for each $j = 1, \cdots, r$ there exists $\psi_j \in C_c^\infty(\tau_0^{-2k_j} : G : \tau_0^{-2k_j})$ so that $F_{\Phi_j} = \psi_j$. Set $u = \phi - \sum_{j=1}^r \psi_j$. Then u satisfies (1), (2), (4) of the proposition. From the results of §7 it is easily seen that the set $S = \{\eta \in \hat{G} | \Theta_\eta(u) \neq 0\}$ is finite. Let $\chi_1, \cdots, \chi_m$ be the distinct infinitesimal characters of the elements of S. Let $S_i = \{\eta \in S | \eta$ has infinitesimal character $\chi_i\}$. We may (and do) assume that $\omega \in \chi_1$. Let $\mathfrak{Z}$ be the center of the complexified universal enveloping algebra, U, of G. Then if $\chi \to {}'\chi$ is the antiautomorphism of U defined by $'1 = 1$, $'x = -x$ for $x \in \mathfrak{G}$, the Lie algebra of G, $'(xy) = {}'y\,{}'x$. Then if $z \in \mathfrak{Z}$, $'z \in \mathfrak{Z}$, and if $\eta \in S_i$, $\Theta_\eta(z \cdot u) = \chi_i({}'z)\Theta_\eta(u)$. For each i, let $z_i \in \mathfrak{Z}$, $i = 2, \cdots, m$, be such that $\chi_1(z_i) \neq \chi_i(z_i)$ (this is possible since $\chi_1, \cdots, \chi_m$ are distinct). Let

$$z = \frac{(z_2 - \chi_2(z_2))(z_3 - \chi_3(z_3)) \cdots (z_m - \chi_m(z_m))}{(\chi_1(z_2) - \chi_2(z_2)) \cdots (\chi_1(z_m) - \chi_m(z_m))}.$$

Then $\chi_1(z) = 1$, $\chi_i(z) = 0$, $i \geqq 2$. $f = {}'z \cdot u$ satisfies (1), (2), (3), and (4).

We now give some implications of Proposition 9.1. First of all let $\omega = T_0^+$. Then the K-types of T_0^+ are τ_p^{3p}, $p \geqq 1$. Hence, T_0^+ satisfies the hypothesis of Proposition 9.1. Take $\tau = \tau_1^3$. Then $\tau_1^3(m(\theta))$ diagonalizes with entries 1 and $e^{3i\theta}$. Thus $k_1 = 0$, $k_2 = 3$. The representations with the same infinitesimal character as T_0^+ are 1, T_0^+, T_0^-, $D^+_{S_1S_2\rho-\rho}$, $D^-_{S_2S_1\rho-\rho}$, $D_{-2\rho}$. Let f be as in

Proposition 9.1. Now

$$[D_{-2\rho}|_K : 1] = [D_{-2\rho}|_K : \tau_0^6] = [D_{-2\rho}|_K : \tau_1^3] = 0,$$
$$[D^-_{s_2 s_1 \rho-\rho}|_K : 1] = [D^-_{s_2 s_1 \rho-\rho}|_K : \tau_0^6] = [D^-_{s_2 s_1 \rho-\rho}|_K : \tau_1^3] = 0,$$
$$[T_0^-|_K : 1] = [T_0^-|_K : \tau_0^6] = [T_0^-|_K : \tau_1^3] = 0.$$

Hence, Proposition 9.1 combined with Theorem 6.7 implies that if $\Gamma \subset G$ is a discrete subgroup without elements of finite order other than I so that $\Gamma\backslash G$ is compact, then

$$N_\Gamma(1)\Theta_1(f) + N_\Gamma(T_0^+)\Theta_{T_0^+}(f) + N_\Gamma(D^+_{s_1 s_2 \rho-\rho})\Theta_{s_1 s_2 \rho-\rho}(f) = f(I)\mathrm{vol}(\Gamma\backslash G).$$

Since $F_f = 0$, $f(I) = \sum_{\omega\in \hat{G}_d} d(\omega)\Theta_\omega(f)$. Thus,

$$f(I) = d(D^+_{s_1 s_2 \rho-\rho})\Theta_{D^+_{s_1 s_2 \rho-\rho}}(f).$$

Now $\Theta_{T_0^+}(f) = 1$, π_0 contains 1 and T_0^+ and $\Theta_{\pi_0}(f) = 0$ since $F_f = 0$. Hence, $\Theta_1(f) = -1$. $\pi_{s_1 s_2 \rho-\rho}$ contains T_0^+, $D^+_{s_1 s_2 \rho-\rho}$ but not 1. Hence, $\Theta_{D^+_{s_1 s_2 \rho-\rho}}(f) = -1$. Clearly $N_\Gamma(1) = 1$. Hence, we have

(1) $$1 - N_\Gamma(T_0^+) + N_\Gamma(D^+_{s_1 s_2 \rho-\rho}) = d(D^+_{s_1 s_2 \rho-\rho})\mathrm{vol}(\Gamma\backslash G).$$

If we do the same argument starting with $\omega = D_{-2\rho}$, then we find

(2) $$N_\Gamma(D_{-2\rho}) - N_\Gamma(D^+_{s_1 s_2 \rho-\rho}) - N_\Gamma(D^-_{s_2 s_1 \rho-\rho}) - 1 = -d(D_{-2\rho})\mathrm{vol}(\Gamma\backslash G).$$

Here we note that if $\omega_1, \omega_2 \in \hat{G}_d$ and ω_1, ω_2 have the same infinitesimal character, then $d(\omega_1) = d(\omega_2)$ (see Harish-Chandra [**13**]). Using Matsushima's formula (§4, (ii)) and the list of §7, we find

LEMMA 9.2.

$$b_{0,1}(\Gamma\backslash G/K) = N_\Gamma(T_0^+), \qquad b_{1,0}(\Gamma\backslash G/K) = N_\Gamma(T_0^-),$$
$$b_{0,2}(\Gamma\backslash G/K) = N_\Gamma(D^+_{s_1 s_2 \rho-\rho}), \qquad b_{2,0}(\Gamma\backslash G/K) = N_\Gamma(D^-_{s_2 s_1 \rho-\rho}),$$
$$b_{1,1}(\Gamma\backslash G/K) = 1 + N_\Gamma(D_{-2\rho}).$$

Now the Gauss-Bonnet-Chern theorem (cf. Kobayashi-Nomizu [**23,** VII, §5]) implies that there is a normalization of dg so that $\mathrm{vol}(\Gamma\backslash G) = \chi(\Gamma\backslash G/K)$ (the Euler number of $\Gamma\backslash G/K$). This normalization is the Euler-Poincaré measure of Serre [**33**]. This normalization of dg is independent of Γ. Now

$$\chi(\Gamma\backslash G/K) = 1 - b_1 + b_2 - b_3 + b_4 \quad (b_i = b_i(\Gamma\backslash G/K)) = 2 - 2b_1 + b_2$$

by Poincaré duality. Now $\Gamma\backslash G/K$ is Kähler. Hence, $b_{p,q} = b_{q,p}$. Hence, $b_1 = 2b_{0,1}$, $b_2 = b_{1,1} + 2b_{0,2}$. Thus,

(3) $$2 - 4b_{0,1} + 2b_{0,2} + b_{1,1} = \mathrm{vol}(\Gamma\backslash G).$$

This says (in light of Lemma 9.2)

(4) $$3 - 4N_\Gamma(T_0^+) + 2N_\Gamma(D^+_{s_1 s_2 \rho-\rho}) + N_\Gamma(D_{-2\rho}) = \mathrm{vol}(\Gamma\backslash G).$$

Now using Lemma 9.2 on (1) and (2) and the fact that

$d(D^+_{s_1s_2\rho-\rho})=d(D_{-2\rho})$, we see

$$(1') \qquad 1-b_{0,1}+b_{0,2}=d(D_{-2\rho})\mathrm{vol}(\Gamma\backslash G),$$

$$(2') \qquad b_{1,1}-1-b_{0,2}-b_{2,0}=-d(D_{-2\rho})\mathrm{vol}(\Gamma\backslash G).$$

Combining (1′), (2′) and (3), we find $d(D_{-2\rho})=\frac{1}{3}$.

Computing Chern classes from the G-invariant metric on G/K, we find that $C_1^2+C_2=4E$ where C_1, C_2 are the first and second Chern classes and E is the Euler class (actually this computation can be done on $P^2(C)$).

We have therefore proved

THEOREM 9.3 (MAX NOETHER).

$$1-b_{0,1}(\Gamma\backslash G/K)+b_{0,2}(\Gamma\backslash G/K)=(C_1^2+C_2)[\Gamma\backslash G/K]/12.$$

Of course, this theorem is well known. If $\omega\in\hat{G}_d$ and if ω has infinitesimal character χ_Λ, $\Lambda\in\mathscr{F}^+$ (see Lemma 7.6), then $d(\omega)=C\,\Pi_{\alpha\in\Delta^+}(\langle\Lambda+\rho,\alpha\rangle/\langle\rho,\alpha\rangle)$ with C depending only on dg (see Harish-Chandra [**13**]). Since $d(D_{-2\rho})=\frac{1}{3}$ relative to the Euler-Poincaré normalization and $D_{-2\rho}$ has infinitesimal character χ_0, we see that $C=\frac{1}{3}$. We have also proved

LEMMA 9.4. *Let dg be given the Euler-Poincaré normalization. If $\omega\in\hat{G}_d$ and ω has infinitesimal character χ_Λ, $\Lambda\in\mathscr{F}^+$, then* $d(\omega)=\frac{1}{3}\Pi_{\alpha\in\Delta^+}(\langle\Lambda+\rho,\alpha\rangle/\langle\alpha,\alpha\rangle)$.

We now use the same technique on T_k^+, $k\geqq1$, and find:

(5) If $k\geqq1$, then

$$N_\Gamma(D^+_{s_1s_2(k\Lambda_1+\rho)-\rho})-N_\Gamma(T_k^+)=\tfrac{1}{6}(k+1)(k+2)\mathrm{vol}(\Gamma\backslash G).$$

Also if we start with the "lowest K-type" of $D_{s_1s_2s_1(k\Lambda_1+\rho)-\rho}$, we find

$$(6)\qquad \begin{aligned} &N_\Gamma(D^+_{s_1s_2(k\Lambda_1+\rho)-\rho})+N_\Gamma(D^-_{s_2s_1(k\Lambda_1+\rho)-\rho})\\ &\qquad -N_\Gamma(D_{s_1s_2s_1(k\Lambda_1+\rho)-\rho})=\tfrac{1}{6}(k+1)(k+2)\mathrm{vol}(\Gamma\backslash G).\end{aligned}$$

Now if $k\geqq2$, $D^-_{s_2s_1(k\Lambda_1+\rho)-\rho}$ is in G'_d (see Lemma 7.7). Hence, Theorem 3.2 implies:

(7) If $k\geqq2$ then

$$N_\Gamma(D^+_{s_1s_2(k\Lambda_1+\rho)-\rho})=N_\Gamma(D_{s_1s_2s_1(k\Lambda_1+\rho)-\rho}).$$

We now give another application of this technique. Let $\omega=\pi^-_{-2\Lambda_2}$. Then $\pi_{-2\Lambda_2}=\pi^+_{-2\Lambda_2}\oplus\pi^-_{-2\Lambda_2}$. The elements of $\hat{G}$ with the same infinitesimal character as $\pi^-_{-2\Lambda_2}$ are $\pi^-_{-2\Lambda_2}$, $\pi^+_{-2\Lambda_2}$, T^+_{-1}. Arguing as above we find

LEMMA 9.5. $N_\Gamma(\pi^-_{-2\Lambda_2})-N_\Gamma(\pi^+_{-2\Lambda_2})=N_\Gamma(T^+_{-1})$.

We note that (5) says for $k=1$,

$$N_\Gamma(D^+_{s_1s_2(\Lambda_1+\rho)-\rho})=N_\Gamma(T_1^+)+d(D^+_{s_1s_2(\Lambda_1+\rho)-\rho})\mathrm{vol}(\Gamma\backslash G).$$

This is the reason why the series of representations $T_k^{\pm}$ was labeled with a T. It stands for "trash". The philosophy is that once the trash is "disposed of" (i.e., $N_\Gamma(T)=0$) then the formula for $N_\Gamma(\omega)$ should be $d(\omega)\mathrm{vol}(\Gamma\backslash G)$ for $\omega\in\hat{G}_d$. This "philosophy" is borne out by the following result.

THEOREM 9.6.

$$\begin{aligned} N_\Gamma(D^+_{s_1s_2(k_1\Lambda_1+k_2\Lambda_2+\rho)-\rho}) &= N_\Gamma(D^-_{s_2s_1(k_1\Lambda_1+k_2\Lambda_2+\rho)-\rho}) \\ &= N_\Gamma(D_{s_1s_2s_1(k_1\Lambda_1+k_2\Lambda_2+\rho)-\rho}) \\ &= \frac{1}{3}\prod_{\alpha>0}\frac{\langle k_1\Lambda_1+k_2\Lambda_2+\rho,\alpha\rangle}{\langle\rho,\alpha\rangle}\mathrm{vol}(\Gamma\backslash G) \end{aligned}$$

if $k_1\geqq 1$, *and* $k_2\geqq 1$.

The proof uses the same technique but starts with the "lowest k-type" of a nonunitary representation. Similar results were derived for more general groups in Hotta-Parthasarathy [**17**], using geometric techniques.

10. **Asymptotic formulas.** We conclude this article by discussing another technique for studying the $N_\Gamma(\omega)$. In this technique one studies the distribution of the $N_\Gamma(\omega)$ as ω varies over $\hat{G}$. Let G be a connected semisimple Lie group with finite center. Let $\Gamma\subset G$ be a discrete subgroup so that $\Gamma\backslash G$ is compact. Let K be a maximal compact subgroup of G. Let $Z(G)$ be the center of G, $Z(\Gamma)=Z(G)\cap\Gamma$. Let $\hat{K}_\Gamma$ be the set of all $\tau\in\hat{K}$ so that $\tau|_{Z(\Gamma)}$ is the identity. For $\omega\in\hat{G}$, let $\lambda_\omega I$ be the value of the Casimir element on any representative of ω.

THEOREM 10.1 (GELFAND [**8**], GANGOLLI [**6**], WALLACH [**36**]). *There is a constant C_G depending only on G so that if $\tau\in\hat{K}_\Gamma$ and if $[Z(\Gamma)]$ is the number of elements in $Z(\Gamma)$ and $d=\dim G/K$, then*

$$\sum_{\omega\in\hat{G}} N_\Gamma(\omega)[\omega|_K:\tau]e^{t\lambda_\omega} = C_G d_\tau[Z(\Gamma)](4\pi t)^{-d/2}\mathrm{vol}(\Gamma\backslash G)+o(t^{-d/2})$$

as $t\to 0$, $t>0$.

The value of the constant C_G is just the volume of K relative to the Haar measure that corresponds to the bi-invariant metric on K gotten by restricting the negative of the Killing form of G to the Lie algebra of K.

This formula was conjectured by Gelfand [**8**] and a proof of it for $G=SL(2,\mathbf{R})$ was sketched in Gelfand, Graev, Pyateckiĭ-Shapiro [**9**]. It was proved by Gangolli for the case $\tau=1$ and G complex semisimple in [**6**]. Gangolli also conjectured the general form of this theorem in [**6**].

The idea of the proof is to observe that if Γ has no noncentral elements of finite order, then $\Gamma\backslash G/K=X$ is a manifold. $\Gamma\backslash G\to\Gamma\backslash G/K$ is a principal bundle with structure group K. If $\tau\in\hat{K}_\Gamma$ then we can form the associated unitary vector bundle $\Gamma\backslash G\times_\tau V_\tau=\mathbf{V}_\tau$. There is a natural connection on $\mathbf{V}_\tau$ and the connection Laplacian is just the Casimir operator, plus a scalar depending only on τ. In this case the result is then a fairly easy generalization of the heat equation method (see McKean-Singer [**28**]) for studying the distribution of eigenvalues for a Laplacian. The general case is proved by developing an analogous theory for $G\backslash M$ where M is a compact manifold and G is a finite group acting on M. For details see Wallach [**36**].

This idea of using the heat equation method to study the $N_\Gamma(\omega)$ is due to Gangolli.

References

1. J. Arthur, *The trace formula for groups of Q-rank* 1, Ann. of Math. (to appear).

2. A. Borel, *Introduction to automorphic forms*, Proc. Sympos. Pure Math., vol. 9, Amer. Math. Soc., Providence, R.I., 1966, pp. 199–210. MR **34** #7465.

3. ———, *Introduction aux groupes arithmétiques*, Publ. Inst. Math. Univ. Strasbourg, XV, Actualités Sci. Indust., no. 1341, Hermann, Paris, 1969. Mr **39** #5577.

4. M. Duflo and J-P. Labesse, *Sur la formule des traces de Selberg*, Ann. Sci. École Norm. Sup. (4) **4** (1971), 193–284.

5. N. Dunford and J. Schwartz, *Linear operators*. Vol. 3, Interscience, New York, 1972.

6. R. Gangolli, *Asymptotic behaviour of spectra of compact quotients of certain symmetric spaces*, Acta Math. **121** (1968), 151–192. MR **39** #360.

7. ———, *On the Plancherel formula and the Paley-Wiener theorem for spherical functions on semisimple Lie groups*, Ann. of Math. (2) **93** (1971), 150–165. MR **44** #6912.

8. I. M. Gel'fand, *Automorphic functions and the theory of representations*, Proc. Internat. Congr. Mathematicians (Stockholm, 1962), Inst. Mittag-Leffler, Djursholm, 1963, pp. 74–85. MR **31** #273.

9. M. I. Gel'fand, M. I. Graev, and I. I. Pjateckiĭ-Šhapiro, *Generalized functions*. Vol. 6: *Theory of representations and automorphic functions*, "Nauka", Moscow, 1966; English transl., Saunders, Philadelphia, Pa., 1969. MR **36** #3725; **38** #2093.

10. Harish-Chandra, *Representations of semi-simple Lie groups*. VI. *Integrable and square-integrable representations*, Amer. J. Math. **78** (1956), 564–628. MR **18,** 490.

11. ———, *Spherical functions on a semi-simple Lie group*. I, II, Amer. J. Math. **80** (1958), 241–310, 553–613. MR **20** #925; **21** #92.

12. ———, *Automorphic forms on semi-simple Lie groups*, Lecture Notes in Math., no. 62, Springer-Verlag, New York and Berlin, 1968. MR **38** #1216.

13. ———, *Discrete series for semi-simple Lie Groups*. II. *Explicit determination of the characters*, Acta Math. **116** (1966), 1–111. MR **36** #2745.

14. ———, *On the theory of the Eisenstein integral*, Conf. on Harmonic Analysis (College Park, Md., 1971), Lecture Notes in Math., vol. 266, Springer-Verlag, Berlin and New York, 1972.

15. S. Helgason, *Differential geometry and symmetric spaces*, Pure and Appl. Math., vol. 12, Academic Press, New York and London, 1962. MR **26** #2986.

16. ———, *An analogue of the Paley-Wiener theorem for the Fourier transform on certain symmetric spaces*, Math. Ann. **165** (1966), 297–308. MR **36** #6545.

17. R. Hotta and R. Parthasarathy, *Multiplicity formulae for discrete series*, Invent. Math. **26** (1974), 133–178.

18. R. Hotta and N. R. Wallach, *On Matsushima formula for the Betti numbers of a locally symmetric space*, Osaka J. Math. (to appear).

19. K. Johnson, *Paley-Wiener theorems for rank one groups* (to appear).

20. H. Johnson and N. R. Wallach, *Intertwining operators and composition series for the spherical principal series*. I, Trans. Amer. Math. Soc. (to appear).

21. D. A. Kazdan, *Connection of the dual space of a group and the structure of its closed subgroups*, Funkcional. Anal. i Priložen. **1** (1967), no. 1, 71–74-Functional Anal. Appl. **1** (1967), 63–65. MR **35** #288.

22. A. W. Knapp and E. M. Stein, *Intertwining operators for semi-simple Lie groups*, Ann. of Math. (2) **93** (1971), 489–578.

23. S. Kobayashi and K. Nomizu, *Foundations of differential geometry*. Vol. II, Interscience

Tracts in Pure and Appl. Math., no. 15, Interscience, New York and London, 1969. MR **38** #6501.

24. B. Kostant, *On the existence and irreducibility of certain series of representations*, Bull. Amer. Math. Soc. **75** (1969), 627–642. MR **39** #7031.

25. R. P. Langlands, *Dimension of spaces of automorphic forms*, Proc. Sympos. Pure Math., vol. 9, Amer. Math. Soc., Providence, R. I., 1966, pp. 253–257. MR **35** #3010.

26. Y. Matsushima, *On the first Betti number of compact quotient spaces of higher-dimensional symmetric spaces*, Ann. of Math. (2) **75** (1962), 312–330. MR **28** #1629.

27. ———, *A formula for the Betti numbers of compact locally symmetric Riemannian manifolds*, J. Differential Geometry **1** (1967), 99–109. MR **36** #5958.

28. H. P. McKean, Jr. and I. M. Singer, *Curvature and the eigenvalues of the Laplacian*, J. Differential Geometry, **1** (1967), no. 1, 43–69. MR **36** #828.

29. G. D. Mostow, *Strong rigidity for locally symmetric spaces*, Ann. of Math. Studies, no. 78, Princeton Univ. Press, Princeton, N.J., 1973.

30. M. Raghunathan, *Discrete subgroups of Lie groups*, Ergebnisse der Mathematik und ihrer Grenzgebiete, vol. 68, Springer-Verlag, Berlin, 1972.

31. P. J. Sally, Jr., *Analytic continuation of the irreducible unitary representations of the universal covering group of* $SL(2, \mathbf{R})$, Mem. Amer. Math. Soc. No. 69 (1967). MR **38** #3380.

32. W. Schmid, *On a conjecture of Langlands*, Ann. of Math. (2) **93** (1971), 1–42.

33. J.-P. Serre, *Cohomologie des groupes discrets*, Prospects in Mathematics, Princeton Univ. Press, Princeton, N.J., 1971.

34. È. B. Vinberg, *Some examples of crystallographic groups on Lobačevskiĭ spaces*, Mat. Sb. **78 (120)** (1969), 633–639 = Math. USSR Sb **7** (1969), 617–622. MR **39** #7497.

35. N. R. Wallach, *Harmonic analysis on homogeneous spaces*, Dekker, New York, 1973.

36. ———, *On an asymptotic formula of Gelfand and Gangolli for the spectrum of* $\Gamma\backslash G$, J. Differential Geometry (to appear).

37. G. Warner, *Harmonic analysis on semi-simple Lie groups*. II, Die Grundlehren der math. Wissenschaften, vol. 189, Springer-Verlag, Berlin, 1972.

38. E. T. Whittaker and G. N. Watson, *A course of modern analysis*, 4th ed., Cambridge Univ. Press, Cambridge, 1927.

39. J. A. Morrow and K. Kodaira, *Complex manifolds*, Holt, Rinehart and Winston, New York, 1971.

DEPARTMENT OF MATHEMATICS, RUTGERS UNIVERSITY, NEW BRUNSWICK, NEW JERSEY 08903

BULLETIN OF THE
AMERICAN MATHEMATICAL SOCIETY
Volume 83, Number 5, September 1977

INVARIANT DIFFERENTIAL EQUATIONS ON HOMOGENEOUS MANIFOLDS

BY SIGURDUR HELGASON[1]

1. Historical origins of Lie group theory. Nowadays when Lie groups enter in a profound way into so many areas of mathematics, their historical origin is of considerable general interest. The connection between Lie groups and differential equations is not very pronounced in the modern theory of Lie groups, so in this introduction we attempt to describe some of the foundational work of S. Lie, W. Killing and É. Cartan at the time when the interplay with differential equations was significant. In fact, the actual construction of the exceptional simple Lie groups seems to have been accomplished first by means of differential equations.

Although motion groups in $\mathbf{R}^3$ had occurred in the work of C. Jordan prior to 1870, Lie group theory as a general structure theory for the transformation groups themselves originated around 1873 with Lie's efforts about that time to use group theoretic methods on differential equations as suggested by Galois' theory for algebraic equations. It seems that a lecture by Sylow in 1863 (when Lie was 20) on Galois theory[2] (Lie and Engel [**9**, vol. 3, p. XXII]) and his collaboration with F. Klein, 1870, on curves and transformations (Klein and Lie [**6**], Engel [**3b**, p. 35]) were particularly instrumental in suggesting to him the following:

PROBLEM (LIE [**8a**]). *Given a system of differential equations how can knowledge about its invariance group be utilized towards its integration?*

Since the solutions of a differential equation are functions, not just numbers as for an algebraic equation, one can take two different viewpoints for an analogy with Galois theory.

Analytic viewpoint (*Lie* (1871–1874)). For a system of differential equations, consider the group of diffeomorphisms of the underlying space leaving the system stable (i.e., permuting the solutions).

Algebraic viewpoint (*Picard* (1883), *Vessiot* (1891)). For a given differential equation consider the group of automorphisms of the field generated by the solutions, fixing the elements of the coefficient field.

To indicate the flavor of the resulting theories I just recall a couple of the best known results. In ordinary Galois theory one has the fundamental result that an algebraic equation is solvable by radicals if and only if the Galois group is solvable. In the Picard-Vessiot theory one introduces similarly the

An expanded version of an address given at the annual meeting of the American Mathematical Society in Washington, D. C. on January 23, 1975; received by the editors October 20, 1976.

AMS (MOS) subject classifications (1970). Primary 22E30, 35C15, 43A85, 58G99; Secondary 22-03, 22-02, 43A90.

[1] Partially supported by NSF Grant MCS 75-21415.

[2] See "Sources" at the end of this section.

Galois group of a linear homogeneous ordinary differential equation, the so-called *differential Galois group* (since the automorphisms are assumed to commute with differentiation). The solvability of this group is then necessary and sufficient for the equation to be solvable by quadratures, $f(x) \to \int f(x)\,dx$, and exponentiation, $g(x) \to e^{g(x)}$.

To indicate the rudiments of Lie's theory, consider a differential equation

$$dy/dx = Y(x,y)/X(x,y) \tag{1}$$

in the plane. It is called *stable* under a 1-parameter group ϕ_t ($t \in \mathbf{R}$) of diffeomorphisms if each ϕ_t permutes the integral curves (all concepts are here local).

EXAMPLE.

$$\frac{dy}{dx} = \frac{y + x(x^2+y^2)}{x - y(x^2+y^2)}. \tag{2}$$

The equation can be written

$$\left(\frac{dy}{dx} - \frac{y}{x}\right) \Big/ \left(1 + \frac{y}{x}\frac{dy}{dx}\right) = x^2 + y^2,$$

and since the left-hand side is the tangent of the angle between the integral curve and the radius vector, it is clear that the integral curves intersect each circle $x^2 + y^2 = r^2$ under a fixed angle. The group of rotations around the origin therefore permutes the integral curves, i.e., leaves the equation stable.

For a 1-parameter group ϕ_t of transformations in the plane with ϕ_0 the identity let T denote the induced vector field,

$$T_p = \left\{\frac{d(\phi_t \cdot p)}{dt}\right\}_{t=0} = \xi(p)\frac{\partial}{\partial x} + \eta(p)\frac{\partial}{\partial y}, \qquad p \in \mathbf{R}^2. \tag{3}$$

THEOREM 1.1 (LIE [**8a**]). *Equation* (1) *is stable under* ϕ_t *if and only if the vector field* $Z = X\partial/\partial x + Y\partial/\partial y$ *satisfies*

$$[T, Z] = \lambda Z \tag{4}$$

where λ *is a function. In this case* $(X\eta - Y\xi)^{-1}$ *is an integrating factor for the equation* $X\,dy - Y\,dx = 0$.

Thus, knowing a stability group for a differential equation provides a way to solve it. For the example above, equation (2) is stable under the group

$$\phi_t\colon (x,y) \to (x\cos t - y\sin t,\ x\sin t + y\cos t)$$

for which

$$T = -y\frac{\partial}{\partial x} + x\frac{\partial}{\partial y},$$

and the theorem gives the solution $y = x\tan(\frac{1}{2}(x^2+y^2) + C)$, C a constant.

Generalizing the 1-parameter group (ϕ_t) above, let

$$x_i' = f_i(x_1, \ldots, x_n; t_1, \ldots, t_r) \tag{5}$$

be an r-parameter local transformation group of n-space, the origin $(t_1, \ldots, t_r) = (0, \ldots, 0)$ representing the identity transformation and rank $(\partial f_i/\partial t_k) = r$. In analogy with (3) define the vector fields ("infinitesimal transformations")

$$T_k = \sum_{i=1}^{n} \left(\frac{\partial f_i}{\partial t_k} \right)_{t=0} \frac{\partial}{\partial x_i} \qquad (1 \leqslant k \leqslant r)$$

on $\mathbf{R}^n$. Passing to the second derivatives of the f_i, Lie proved that the group law for (5) implies that

$$[T_k, T_l] = \sum_{p=1}^{r} C_{kl}^p T_p, \tag{6}$$

where the C_{kl}^p are *constants* satisfying

$$C_{kl}^p = -C_{lk}^p, \qquad \sum_{q=1}^{r} (C_{kq}^p C_{lm}^q + C_{mq}^p C_{kl}^q + C_{lq}^p C_{mk}^q) = 0. \tag{7}$$

This device, which through Lie's so-called "three fundamental theorems" relates the study of the local transformation group (5) to the algebraic study of the Lie algebra (6), (7) (the third theorem amounts to the statement that every system of constants C_{kl}^p satisfying (7) arises in this way), forms the foundation of Lie's theory of transformation groups. His first proof of (6) (cf. [**8c**, p. 52]) was incomplete (cf. loc. cit., p. 617) but was completed in his paper [**8d**, p. 462]; in the modern theory of Lie groups relation (6) amounts to the fact that when a Lie group G acts on a manifold M there is induced a homomorphism of the corresponding Lie algebra $\mathfrak{g}$ into the Lie algebra of vector fields on M.

In one generalization of Theorem 1.1 from $\mathbf{R}^2$ to $\mathbf{R}^n$ (cf. Lie [**8b**]) one considers a differential equation

$$\sum_{i=1}^{n} X_i \frac{\partial f}{\partial x_i} = 0, \tag{8}$$

where $X_i \in C^\infty(\mathbf{R}^n)$. Assuming the equation stable under a *solvable* $(n-1)$-parameter transformation group, its solutions can be found by quadratures. (Here the term "solvable" has replaced the older term "integrable" and the term quadrature is used for integration $f(x) \to \int f(x)\,dx$ in analogy with taking square roots in the analogous result for an algebraic equation.)

Such results, and their generalization to systems, suggested the problem of classifying all local transformation groups of $\mathbf{R}^n$. Lie solved this for $n = 1$ where the local groups are

$$x \to x + a, \qquad a \in \mathbf{R} \text{ const}, \tag{9}$$

$$x \to ax + b, \qquad a \in \mathbf{R} - (0), b \in \mathbf{R}, \tag{10}$$

$$x \to (ax + b)/(cx + d), \qquad a, b, c, d \in \mathbf{R}, ad - bc = 1. \tag{11}$$

For $n = 2$ the possibilities are already quite numerous (cf. [**8f**, vol. V, p. 768]) so the assumption of *primitivity* (no invariant decomposition of the space into lower-dimensional submanifolds) was introduced. Then there are for $n = 2$ just 3 possibilities, namely the 8-parameter analog of (11) (the projective group), the 6-parameter analog of (10) (the affine group), and its 5-parameter subgroup of area preserving transformations (analog of (9), the special affine group).[3] Lie [**8e**] and, his student in Leipzig, Page [**11**] settled the cases $n = 3$,

[3] In [**10**] Mostow determines all global transitive Lie transformation groups of surfaces.

4 respectively; later Kowalewski and Beutner worked out the cases $n = 5$ and $n = 6$, respectively. The direct attack on the problem was not continued further because the complexities had become rather formidable, and since the problem had taken a new direction through the work of Killing and Cartan.

W. Killing, who wrote his dissertation with Weierstrass in 1872, began in 1877–1878 geometric investigations which, without knowledge of Lie's work, led him to concepts close to Lie's infinitesimal transformations and to relations equivalent to (6). (Cf. Lie and Engel [**9**, vol. 3, p. 768].) Even before getting acquainted with Lie's work, Killing had set himself the problem of finding all possible "Zusammensetzungen" of r-parameter groups. In other words, he wanted to find all possible ways, up to isomorphism, in which an r-dimensional vector space can be turned into a Lie algebra.

Thus Lie's classification problem (which, as indicated, arose from his study of differential equations) consisted of Killing's algebraic problem together with the problem of classifying the various representations of a given group as a transformation group. This viewpoint was decisive for the theory of Lie groups, but separated it gradually from differential equations.

On 18 October 1887 Killing wrote to F. Engel, who was then Lie's assistant in Leipzig, that he had succeeded in finding a complete classification of the simple Lie algebra $\mathfrak{g}$ over **C**. In this work [**5**] Killing introduced many of the fundamental concepts for the theory of simple Lie algebras, in particular, the following:

(a) The *rank* l of $\mathfrak{g}$.

(b) For the linear transformation ad X: $Y \to [X, Y]$ the *characteristic equation*

$$(12) \qquad \det(\omega - \operatorname{ad} X) \equiv \omega^r - \psi_2(X)\omega^{r-2} + \cdots \pm \psi_{r-l}(X)\omega^l = 0.$$

The coefficient $2\psi_2(X)$ which equals $\operatorname{Trace}(\operatorname{ad} X)^2$ is now called the *Killing form*. Equation (12) had also been used extensively by Lie.

(c) The *roots* of $\mathfrak{g}$ which, by definition, are the functions $\omega(X)$ on $\mathfrak{g}$ satisfying (12).

(d) A basis $\omega_1, \ldots, \omega_l$ of roots of which all roots are integral linear combinations (with all coefficients of the same sign) and the associated matrix (a_{ij}) where

$$(13) \qquad -a_{ij} = \text{the largest integer } q \text{ such that } \omega_j + q\omega_i \text{ is a root.}$$

This matrix is now called the *Cartan matrix*.

In this remarkable work, Killing finds all possibilities for the matrix (a_{ij}) and writes down the corresponding roots $\omega(X)$ (cf. [**5**, II, §15]). Thus he arrives at the statement that apart from the classical simple Lie algebras

$$A_l\ (l \geqslant 1), \quad B_l\ (l \geqslant 2), \quad C_l\ (l \geqslant 3), \quad D_l\ (l \geqslant 3)$$

(known from Lie's work), there are only six more, of ranks and dimension, respectively,

$$l = 2, 4, 4, 6, 7, 8,$$
$$r = 14, 52, 52, 78, 133, 248.$$

These exceptional Lie algebras are denoted G_2, E_4, F_4, E_6, E_7, E_8, respectively. Killing denoted G_2 by (IIC); he observed that $A_3 = D_3$, but did not notice that $E_4 = F_4$, although, as Cartan remarked, this is immediate from his root tables in [5, II, pp. 30–31]. Killing's work [5] was immediately recognized as a breakthrough. However, it was also criticized, particularly by Lie [9, vol. 3, pp. 768–771], for serious gaps and inaccuracies. É. Cartan set himself the problem of giving genuine proofs of the results stated by Killing. Apart from pointing out many errors in detail, Cartan found the following gaps in Killing's papers particularly significant: (a) In [5, II] Killing makes the explicit assumption that the nonzero roots of (12) are simple. While this is correct, his justification in [5, III] of this assumption was incorrect. (b) A proof that at most one $\mathfrak{g}$ can correspond to a given matrix (a_{ij}) was lacking. Cartan showed, case-by-case, that this is so; an *a priori* proof was given much later by van der Waerden [13].

The actual existence of the exceptional Lie algebras is another major weakness of Killing's work. He indicates in [5, II, §18] how the structural constants C^i_{jk} can be determined from his root tables. Then the Jacobi identity (7) has to be verified; Killing does this for G_2, but for the others his indications [5, II, p. 48] seem unconvincing.

In his thesis [1b], É. Cartan gave a complete proof of the classification results stated by Killing; in outline his method follows Killing's program. He determined the matrices (a_{ij}), the roots $\omega(X)$ and a basis for each of the exceptional Lie algebras with respect to which the structural constants have a simple and symmetric form [1b, §§18–20] whereby the Jacobi identity (7) is (presumably) simple to verify.[4] But he was also interested in realizing the exceptional Lie groups by transformations, like e.g. the classical algebra C_l is the Lie algebra of the linear group leaving invariant the Pfaffian form

$$x_1dy_1 - y_1dx_1 + \cdots + x_ldy_l - y_ldx_l.$$

Killing had been led to expect that G_2 could be realized as a transformation group in $\mathbf{R}^5$, but not in a lower-dimensional space. Engel and Cartan showed that it can be realized as the stability group of the system

$$\begin{aligned} dx_3 + x_1dx_2 - x_2dx_1 &= 0, \\ dx_4 + x_3dx_1 - x_1dx_3 &= 0, \\ dx_5 + x_2dx_3 - x_3dx_2 &= 0, \end{aligned}$$

in $\mathbf{R}^5$ (Engel [3a], Cartan [1b, p. 281], Lie and Engel [9, vol. 3, p. 764]).

Cartan represented F_4 similarly by the Pfaffian system in $\mathbf{R}^{15}$ given by

$$(14) \qquad dz = \sum_1^4 y_i dx_i, \qquad dx_{ij} = x_i dx_j - x_j dx_i + y_h dy_k - y_k dy_h,$$

where $z, x_i, y_j, x_{ij} = -x_{ji}$ $(i \neq j,\ i, j = 1, 2, 3, 4)$ are coordinates in $\mathbf{R}^{15}$ and in (14) i, j, h, k is an even permutation [1a, p. 418]. Similar results for E_6 in $\mathbf{R}^{16}$, E_7 in $\mathbf{R}^{27}$ and E_8 in $\mathbf{R}^{29}$ as contact transformations are indicated in [1a]. Unfortunately, detailed proofs of these remarkable representations of the exceptional groups do not seem to be available.

[4] In [15], Witt gives an explicit geometric construction of the 5 exceptional Lie algebras and proves a priori that to each Cartan matrix (a_{ij}) corresponds a simple Lie algebra (provided this is so for $l \leqslant 4$). Chevalley [2] indicates a general algebraic proof (without this proviso); see also Harish-Chandra [4].

SOURCES

1a. É. Cartan, *Über die einfachen Transformationsgruppen*, Leipzig Ber., 1893, pp. 395–420; reprint, *Oeuvres complètes*, Vol. I, no. 1, Gauthier-Villars, Paris, 1952, pp. 107–132.

1b. ______, *Sur la structure des groupes de transformations finis et continus*, Thèse, Paris Nony, 1894; reprint, *Oeuvres complètes*, Vol. I, no. 1, *Groupes de Lie*, Gauthier-Villars, Paris, 1952, pp. 137–287. MR **14**, 343.

1c. ______, *Oeuvres complètes*, Gauthier-Villars, Paris, 1952. MR **14**, 343.

2. C. Chevalley, *Sur la classification des algèbres de Lie simples et de leurs représentations*, C. R. Acad. Sci. Paris **227** (1948), 1136–1138. MR **10**, 280.

3a. F. Engel, *Sur un groupe simple à quatorze paramètres*, C. R. Acad. Sci. Paris **116** (1893), 786–788.

3b. ______, *Sophus Lie* (obituary), Jber. Deutsch. Math.-Verein **8** (1900), 30–46.

3c. ______, *Wilhelm Killing* (obituary), Jber. Deutsch. Math.-Verein **39** (1930), 140–154.

4. Harish-Chandra, *On some applications of the universal enveloping algebra of a semisimple Lie algebra*, Trans. Amer. Math. Soc. **70** (1951), 28–96. MR **13**, 428.

5. W. Killing, *Die Zusammensetzung der stetigen endlichen Transformationsgruppen*. I, II, III, Math. Ann. **31** (1888), 252–290; ibid. **33** (1888), 1–48; ibid. **34** (1889), 57–122.

6. F. Klein and S. Lie, *Uber diejenigen ebenen Kurven, welche durch ein geschlossenen System von einfach unendlich vielen vertauschbaren linearen Transformationen in sich übergehen*, Math. Ann. **4** (1871), 50–84.

7. G. Kowalewski, *Bestand und Wandel*, Oldenbourg, München, 1950.

8a. S. Lie, *Zur Theorie des Integrabilitetsfaktors*, Christiania Forh. 1874, 242–254; reprint, *Gessammelte Abhandlungen*, Vol. III, no. XIII, pp. 176–187.

8b. ______, *Verallgemeinerung und neue Verwertung der Jacobischen Multiplikatortheorie*, Christiania Forh. 1874, 255–274; reprint, *Gesammelte Abhandlungen*, Vol. III, no. XIV, pp. 188–206.

8c. ______, *Theorie der Transformationsgruppen*. II, Arch. für Math. I, 152–193, Christiania 1876; reprint, *Gesammelte Abhandlungen*, Vol. V., no. III, pp. 42–75.

8d. ______, *Theorie der Transformationsgruppen*. I, Math. Ann. **16** (1880), 441–528; reprint, *Gesammelte Abhandlungen*, Vol. V, no. I.

8e. ______, *Untersuchungen über Transformationsgruppen*. I, *Gesammelte Abhandlungen*, Vol. 5, No. XIX, pp. 453–498.

8f. ______, *Gesammelte Abhandlungen*, 7 vols., Teubner, Leipzig, 1922–1960.

9. S. Lie and F. Engel, *Theorie der Transformationsgruppen*, 3 vols., Teubner, Leipzig, 1888–1893.

10. G. D. Mostow, *The extensibility of local Lie groups of transformations and groups on surfaces*, Ann. of Math. (2) **52** (1950), 606–636. MR **14**, 18.

11. J. M. Page, *On the primitive groups of transformations in space of four dimensions*, Amer. J. Math. **10** (1888), 293–346.

12. E. Picard, *Sur les groupes de transformation des équations différentiels linéaires*, C. R. Acad. Sci. Paris **96** (1883), 1131–1134.

13. B. L. van der Waerden, *Die Klassifizierung der einfachen Lie'schen gruppen*, Math. Z. **37** (1933), 446–462.

14. E. Vessiot, *Sur les équations différentielles linéaires*, C. R. Acad. Sci. Paris **112** (1891), 778–780.

15. E. Witt, *Spiegelungsgruppen und Aufzählung halbeinfacher Liescher Ringe*, Abh. Math. Sem. Univ. Hamburg, **14** (1941), 289–322. MR **3**, 100.

2. Invariant differential operators. In our days when Lie group theory has been so highly developed, it is reasonable to reverse the viewpoint in Lie's problem in §1, that is, consider the group as the given object and investigate differential operators invariant under a given group.

Let X be a manifold and D a differential operator on X. Let ϕ be a diffeomorphism of X onto itself. For a smooth function f on X we put $f^{\phi} = f \circ \phi^{-1}$ and define the operator D^{ϕ} by

$$D^{\phi}: f \to \left(Df^{\phi^{-1}}\right)^{\phi};$$

then D^{ϕ} is another differential operator. The operator D is said to be *invariant* under ϕ if

$$D^{\phi} = D. \tag{1}$$

This is a very natural condition on a differential operator and examples abound; the polynomials $P(L)$ in the Laplacian L on $\mathbf{R}^n$ (or on any two-point homogeneous space) are precisely the differential operators invariant under all isometries; similarly the wave operator on $\mathbf{R}^4$ (or on an isotropic Lorentz space) is characterized by its invariance under the Poincaré group (respectively, its isometry group), cf. [**10a**].

We shall now discuss differential operators on a manifold invariant under a transitive Lie group of diffeomorphisms. To be specific, let G be a Lie group, $K \subset G$ a closed subgroup, G/K the manifold of left cosets gK ($g \in G$) and $\mathbf{D}(G/K)$ the algebra of differential operators on G/K invariant under all the transformations $\tau(g)$: $xK \to gxK$ of G/K onto itself. We write $\mathbf{D}(G)$ for $\mathbf{D}(G/e)$, the algebra of left invariant differential operators on G. Let $\mathfrak{g}$ and $\mathfrak{k}$, respectively, denote the Lie algebras of G and K, let $U(\mathfrak{g})$ denote the universal enveloping algebra of $\mathfrak{g}$ and $U(\mathfrak{g})^{\mathfrak{k}}$ the centralizer of $\mathfrak{k}$ in $U(\mathfrak{g})$. As noted by Schwartz and proved in [**8a**, p. 111] we have the canonical isomorphism

$$\mathbf{D}(G) \approx U(\mathfrak{g}). \tag{2}$$

More generally [**10a**, Chapter III], if the coset space G/K is reductive and K connected, we have the isomorphism

$$\mathbf{D}(G/K) \approx U(\mathfrak{g})^{\mathfrak{k}} / \left(U(\mathfrak{g})^{\mathfrak{k}} \cap U(\mathfrak{g})\mathfrak{k}\right) \tag{3}$$

expressing the algebra of invariant differential operators in Lie algebra terms.

Of the many problems one can contemplate for these operators we will discuss the following.

A. ***Solvability***. Given $D \in \mathbf{D}(G/K)$, is the differential equation $Du = f$, for $f \in C^{\infty}(G/K)$ arbitrary, globally solvable (respectively, locally solvable)? If so we say that D is globally solvable (respectively, locally solvable).

For the simplest case $G = \mathbf{R}^n$, $K = (0)$, the operators in $\mathbf{D}(G/K)$ are those of constant coefficients and the global solvability is well known (Ehrenpreis and Malgrange).

B. ***Joint eigenfunctions***. Determine the functions on G/K which are eigenfunctions of each $D \in \mathbf{D}(G/K)$. Similar problems for joint eigendistributions.

C. ***Eigenspace representations***. Let μ: $\mathbf{D}(G/K) \to \mathbf{C}$ be a homomorphism and let E_{μ} denote the corresponding joint eigenspace, i.e.,

$$E_{\mu} = \{f \in C^{\infty}(G/K) \mid Df = \mu(D)f \text{ for all } D \in \mathbf{D}(G/K)\}$$

and let T_μ denote the natural representation of G on this eigenspace, i.e.,

$$(T_\mu(g)f)(xK) = f(g^{-1}xK) \quad \text{for } g, x \in G,$$

$f \in E_\mu$. For which μ is this "eigenspace representation" T_μ irreducible and what representations of G are so obtained?

D. ***Extensions to vector bundles***. In the following sections we survey various results concerning A, B, and C for certain important classes of homogeneous spaces (symmetric spaces and their duals). First we explain how invariance condition (1) can be generalized in a natural way to a smooth vector bundle E over a manifold X (cf. Bott [**1**]). If $p: E \to X$ is the projection map, a smooth section is a C^∞ map $s: X \to E$ such that $p(s(x)) = x$ for all $x \in X$. The smooth sections form a vector space $\Gamma(E)$. Let D be a differential operator on E, that is, a linear operator from $\Gamma(E)$ to $\Gamma(E)$ which via arbitrary local trivializations of E is expressed by means of ordinary partial differential operators with linear transformations of the fibers E_x as coefficients (cf. [**29**, p. 66]). Let $\phi: E \to E$ be a diffeomorphism commuting with p such that for each $x \in X$ the restriction map $E_x \to E_{\phi(x)}$ is a vector space isomorphism. Then ϕ acts on $\Gamma(E)$; if $s \in \Gamma(E)$ the map s^ϕ given by

$$s^\phi(x) = \phi\big(s(\phi^{-1}x)\big) \qquad (x \in X)$$

obviously belongs to $\Gamma(E)$, and we can define the differential operator

$$D^\phi: s \to \big(Ds^{\phi^{-1}}\big)^\phi, \qquad s \in \Gamma(E).$$

Again we call D *invariant under* ϕ if $D^\phi = D$. For the trivial bundle $E = X \times \mathbf{R}$ sections become functions, and this invariance notion reduces to (1).

As an example let $X = \mathbf{R}^4$, and $\tilde{G}$ the universal covering group of the Lorentz group G, and π the natural mapping of $\tilde{G}$ onto G; G acts on $\mathbf{R}^4$ as the Lorentz group and $\tilde{G}$ acts on $\mathbf{R}^4$ via the spinor respresentation. Each $\tilde{g} \in \tilde{G}$ acts on the trivial bundle $E = \mathbf{R}^4 \times \mathbf{R}^4$ by

$$\tilde{g} \cdot (x, y) = (\pi(\tilde{g}) \cdot x, \tilde{g} \cdot y)$$

and the Dirac operator is invariant under this action.

Now suppose $X = G/K$ and δ a representation of K on a finite-dimensional vector space V. Let $E = G \times_K V$ denote the product $G \times V$ modulo the equivalence relation $(gk, v) \sim (g, \delta(k)v)$ for $g \in G$, $k \in K$, $v \in V$. If $[g, v]$ denotes the equivalence class of the element $(g, v) \in G \times V$, then the mapping $p: [g, v] \to gK$ turns E into a vector bundle over $X = G/K$. The group G acts on E via the map $g_0 \cdot [g, v] = [g_0 g, v]$ and this action commutes with p. If for a section $s \in \Gamma(E)$ we put

$$\tilde{s}(g) = g^{-1} \cdot s(gK),$$

then the mapping $s \to \tilde{s}$ is a bijection of $\Gamma(E)$ onto the set of smooth V-valued functions f on G satisfying $f(gk) \equiv \delta(k^{-1})(f(g))$. Let $\mathbf{D}^V(G)$ denote the set of left invariant differential operators on G with coefficients in $\operatorname{Hom}(V, V)$ and let $\mathbf{D}^V(G)^{\mathfrak{k}}$ denote the centralizer of the set $\{T + \delta(T) | T \in \mathfrak{k}\}$ in $\mathbf{D}^V(G)$. Assuming K compact and connected we have, in analogy with (3), that the mapping which sends $D \in D^V(G)^{\mathfrak{k}}$ to the differential operator $\mu(D)$ given by

$$(\mu(D)s)^{\sim} = D\tilde{s}, \qquad s \in \Gamma(E),$$

is a homomorphism of $\mathbf{D}^V(G)^{\mathfrak{k}}$ onto the algebra $\mathbf{D}^\delta(G/K)$ of G-invariant differential operators on E.

Problems A, B, and C are meaningful for the invariant differential operators on E; problem C is particularly interesting for the algebra $\mathbf{D}^\delta(G/K)$ and for subalgebras of it. As will be explained later it seems that the known irreducible representations of semisimple Lie groups and of nilpotent Lie groups can thus be realized as eigenspace representations.

For a manifold X we shall use Schwartz's notation $\mathfrak{D}(X)$ and $\mathfrak{E}(X)$ for the spaces $C_c^\infty(X)$ and $C^\infty(X)$, respectively, with their customary topologies. Their strong duals $\mathfrak{D}'(X)$ and $\mathfrak{E}'(X)$ then consist of the distributions on X and the distributions of compact support, respectively.

3. Solvability. For the solvability question A we consider first one of the best known class of coset spaces, the *symmetric spaces* $X = G/K$ *of the noncompact type* (G a connected semisimple Lie group with finite center, K a maximal compact subgroup). Here we have for each $D \in \mathbf{D}(G/K)$ the global solvability

$$DC^\infty(X) = C^\infty(X) \tag{1}$$

as was proved in [**10e**] using the Fourier transform on X. We recall the original definition of the Fourier transform on X from [**10c**] since we shall state some new results for it below. The Fourier transform of a function $F(y)$ on $\mathbf{R}^n$ can be written

$$\tilde{F}(\eta\omega) = \int_{\mathbf{R}^n} F(y)e^{-i\eta(y,\omega)}\,dy \tag{2}$$

where $\eta \geqslant 0$, $|\omega| = 1$ and (y, ω) is the usual inner product. Geometrically, (y, ω) is the (signed) distance from the origin to the hyperplane through y perpendicular to ω. It turns out to have an analogue for the symmetric space $X = G/K$. To define it let $G = KAN$ be an Iwasawa decomposition of the group G; here A is an abelian subgroup and N a nilpotent subgroup. (In the case $G = \mathbf{SL}(n, \mathbf{R})$, $K = \mathbf{SO}(n)$ the group A is the group of positive diagonal matrices with determinant 1, N the set of supertriangular matrices with diagonal 1; here the decomposition amounts to the usual Gram-Schmidt orthonormalization process.) Let M denote the centralizer of A in K and B the coset space K/M.

A *horocycle* in X is an orbit in X of a group of the form gNg^{-1}. The group G permutes the horocycles transitively. More precisely let $o = \{K\}$ (the origin in X) let ξ_0 denote the horocycle $N \cdot o$. Then each horocycle ξ can be written $\xi = ka \cdot \xi_o$, where kM is unique in B (called the *normal* to the horocycle), and $a \in A$ is unique (called the *complex distance* from o to ξ). This representation of ξ is obvious in the case of the Poincaré disk where the horocycles are the circles tangential to the unit circle. Given $x \in X$, $b \in B$ there is a unique horocycle through x with normal b; let $a(x, b)$ denote its complex distance from o and let $A(x, b)$ be the element in the Lie algebra $\mathfrak{a}$ of A satisfying $\exp A(x, b) = a(x, b)$. This vector-valued function $A(x, b)$ on $X \times B$ is the desired analog of the scalar product (y, ω) for $\mathbf{R}^n \times \mathbf{S}^{n-1}$.

The space X has a certain canonical Riemannian metric and we let dx denote the volume element on X with respect to this metric. Let $\mathfrak{a}^*$ denote the dual of the vector space $\mathfrak{a}$.

Given a function $f(x)$ on X we define its *Fourier transform* in analogy with (2) by

$$\tilde{f}(\lambda, b) = \int_X f(x) e^{(-i\lambda+\rho)(A(x,b))}\, dx \qquad (\lambda \in \mathfrak{a}^*, b \in B). \tag{3}$$

Here ρ is a certain fixed element of $\mathfrak{a}^*$: $2\rho(H)$ is the Jacobian of the automorphism $n \to \exp(H)n\exp(-H)$ of N. It turns out that the transform (3) is one-to-one on $\mathfrak{D}(X)$. There is an inversion formula of the same type as well as a Plancherel formula relating the L^2 norms of f and $\tilde{f}$. But for the proof of (1) the important result to have is a theorem of Paley-Wiener type, that is, an intrinsic characterization of the space $\mathfrak{D}(X)^\sim$. To describe it let W denote the Weyl group of $\mathfrak{a}$, that is, the (finite) group of linear transformations of $\mathfrak{a}$ induced by those elements of K which normalize $\mathfrak{a}$. Then we have the following result [**10e**].

THEOREM 3.1. *The space $\mathfrak{D}(X)^\sim$ of Fourier transforms consists of the smooth functions $\phi(\lambda, b)$ on $\mathfrak{a}^* \times B$ satisfying*:

(i) $\lambda \to \phi(\lambda, b)$ *extends to an entire function of exponential type on the complexification $\mathfrak{a}_c^*$, the exponential type being uniform in $b \in B$.*

(ii) *For each $s \in W$ and each $\lambda \in \mathfrak{a}^*$,*

$$\int_B e^{(is\lambda+\rho)(A(x,b))}\phi(s\lambda, b)\, db = \int_B e^{(i\lambda+\rho)(A(x,b))}\phi(\lambda, b)\, db,$$

where db is a K-invariant measure on B.

The application of the Fourier transform (3) to prove (1) is based on the fact that the kernel in (3) is an eigenfunction of D, i.e.,

$$D_x\left(e^{(i\lambda+\rho)(A(x,b))}\right) = P_D(\lambda) e^{(i\lambda+\rho)(A(x,b))}, \tag{4}$$

where the eigenvalue is a polynomial $P_D(\lambda)$, independent of b. After proving first local solvability of D and using functional analysis tools familiar from the constant coefficient theory, (1) is reduced to proving the implication,

$$f \in \mathfrak{D}(X), \qquad \operatorname{supp}(Df) \subset V \Rightarrow \operatorname{supp}(f) \subset V \tag{5}$$

for any closed ball V in X, supp denoting support. But using Theorem 3.1, statement (5) is translated into the following known property for holomorphic functions. If F is an entire function of exponential type, $P \neq 0$ a polynomial and PF of exponential type $\leqslant R$, then F is of exponential type $\leqslant R$.

In order to have an analog of (1) for distributions it would be sufficient to have a topological version of Theorem 3.1. We will now describe a special result of this type. For this let δ be an irreducible unitary representation of K on a vector space V_δ of dimension $d(\delta)$, and let δ' denote the contragredient representation. Let V_δ^M denote the set of vectors $v \in V_\delta$ fixed under $\delta(M)$ and $l(\delta) = \dim V_\delta^M$. We assume $l(\delta) > 0$. Let $\mathfrak{p}$ be the orthogonal complement of $\mathfrak{k}$ in $\mathfrak{g}$ with respect to the Killing form and $\mathfrak{p}_c$, $\mathfrak{k}_c$ and $\mathfrak{g}_c$ their respective complexifications. Let H be the set of K-harmonic polynomials in the symmetric algebra $S(\mathfrak{p}_c)$ and put $H^* = \lambda(H)$, where λ is the canonical symmetrization map of the symmetric algebra over $\mathfrak{g}_c$ onto the complex universal enveloping algebra $U(\mathfrak{g})$. Let (v_i) be a basis of V_δ^M and (ε_j) a basis of the space of linear maps of V_δ into H^* commuting with the action of K. Since $U(\mathfrak{g}) = (U(\mathfrak{g})\mathfrak{k} + \mathfrak{n}U(\mathfrak{g})) \oplus U(\mathfrak{a})$ (where $\mathfrak{n}$ is the Lie algebra of N) we have a mapping $u \to q^u$ of $U(\mathfrak{g})$ into $U(\mathfrak{a})$ given by $u - q^u \in U(\mathfrak{g})\mathfrak{k} + \mathfrak{n}U(\mathfrak{g})$. As in Kostant [**22a**] we consider for $\lambda \in \mathfrak{a}_c^*$ the complex $l(\delta) \times l(\delta)$ matrix

$$Q^\delta(\lambda) = \left[q^{\varepsilon_j(v_i)}(\rho - i\lambda)\right]_{1 \leq i,j \leq l(\delta)}.$$

The space $\mathcal{H}(\mathfrak{a}^*)$ of entire functions of exponential type on $\mathfrak{a}_c^*$ has a natural topology with respect to which the Euclidean Fourier transform

$$F(a) \to F^*(\lambda) = \int_A F(a) e^{-i\lambda(\log a)}\, da$$

is a homeomorphism of $\mathfrak{D}(A)$ onto $\mathcal{H}(\mathfrak{a}^*)$ [**6b**]. The same holds for the space $\mathcal{H}(\mathfrak{a}^*, V)$ of holomorphic functions of exponential type with values in a finite-dimensional vector space V. Consider now the subspace

$$(6) \qquad \mathcal{H}^\delta(\mathfrak{a}^*) = \left\{F \in \mathcal{H}(\mathfrak{a}^*, \operatorname{Hom}(V_\delta, V_\delta)) | (Q^{\delta'})^{-1} F \ W\text{-invariant}\right\}$$

with the relative topology, and the closed subspace $\mathfrak{D}_{\delta'}(X) \subset \mathfrak{D}(X)$ consisting of those functions in $\mathfrak{D}(X)$ which are K-finite of type δ'. Finally let $\Phi_{\lambda,\delta}$ denote the *generalized spherical function*

$$\Phi_{\lambda,\delta}(x) = \int_K e^{(i\lambda+\rho)(A(x,kM))}\delta(k)\, dk.$$

Then we have the following refinement of Theorem 3.1 (cf. [**10g**]).

THEOREM 3.2. *The mapping $f \to \tilde{f}$ where*

$$\tilde{f}(\lambda) = d(\delta)\int_X f(x)\Phi_{\bar{\lambda},\delta}(x)^*\, dx \qquad (* = \textit{adjoint})$$

is a homeomorphism of $\mathfrak{D}_{\delta'}(X)$ onto $\mathcal{H}^\delta(\mathfrak{a}^*)$.

In addition to Theorem 3.1, the proof uses the relationship of $Q^\delta(\lambda)$ to the intertwining operators for the principal series found by Johnson and Wallach [**17**]. From Theorem 3.2 we have immediately the following consequence.

COROLLARY 3.3. *Let $D \neq 0$ in $\mathbf{D}(G/K)$ and let $\mathfrak{D}_0'(X)$ denote the space of K-finite distributions on X. Then*

$$(7) \qquad D\mathfrak{D}_0'(X) = \mathfrak{D}_0'(X).$$

For the space $\mathcal{E}'(X)$ the analog to (7), of course, fails; as a consequence of Theorem 3.1 the range $D\mathcal{E}'$ consists of those distributions $T \in \mathcal{E}'(X)$ whose Fourier transform $\tilde{T}(\lambda, b)$ is divisible by the polynomial $P_D(\lambda)$ in (4) (cf. [**10e**, §8]).

Next let us consider the solvability question for a Lie group L viewed as a homogeneous space under left translations. The operator $\partial/\partial x + i\, \partial/\partial y +$

$ix\,\partial/\partial z$ on $\mathbf{R}^3$ is a left invariant differential operator on the Heisenberg group

$$\begin{bmatrix} 1 & x & z \\ 0 & 1 & y \\ 0 & 0 & 1 \end{bmatrix}, \qquad x, y, z \in \mathbf{R},$$

yet it is essentially the same as H. Levy's operator which he proved was not locally solvable (cf. [**3**]). Thus the operators in $\mathbf{D}(L)$ are not, in general, locally solvable.

The contrast between this negative result for L and the positive results for the symmetric space G/K disappears when we view L as a symmetric space. A coset space B/C (B a Lie group, C a closed subgroup) is called a *symmetric coset space* if there exists an involutive automorphism σ of B with fixed point set C. The spaces G/K considered above have this property. Now we consider L as a homogeneous space under left and right translations simultaneously, i.e., we let the product group $L \times L$ act on L by

$$(g_1, g_2)\colon g \to g_1 g g_2^{-1}, \qquad g \in L.$$

The subgroup leaving e fixed is the diagonal $L^* \cong L$ so we have the coset space representation

$$L = (L \times L)/L^*. \tag{8}$$

Then the algebra $\mathbf{D}(L \times L/L^*)$ is canonically isomorphic to the algebra $\mathbf{Z}(L)$ of bi-invariant differential operators on L, and the natural problem becomes: Given $D \in \mathbf{Z}(L)$, is it globally (respectively, locally) solvable? In [**5**], Duflo and Raïs proved

THEOREM 3.4. *Let L be a solvable Lie group. Then each bi-invariant differential operator $D \neq 0$ is locally solvable.*

This had been proved in Raïs [**31**] for the special case of a nilpotent Lie group. Their proof is based on a detailed description of the operators $D \in \mathbf{Z}(L)$ by means of harmonic analysis on L, whereby a local fundamental solution can be constructed, giving local solvability. An entirely different proof was given recently by F. Rouvière [**33**]. He proves that given $D \in \mathbf{Z}(L)$ there exists a bi-invariant differential operator $E \neq 0$ on the derived group L' such that for L^2 norms on L,

$$\|Eu\| \leqslant \|Du\| \tag{9}$$

for $u \in C^\infty(L)$ of sufficiently small support. On the other hand Hörmander showed [**15**, p. 157] that local solvability of E on L' implies certain sup norm inequalities which, using the invariance of E, Rouvière converts into norm inequalities,

$$\left|\int_{L'} u(x)v(x)\,dx\right| \leqslant C\|u\|_k \|{}^tEv\|_l. \tag{10}$$

Here dx is a right invariant Haar measure, C a constant, ${}^tE$ the transpose of E, $\|\ \|_k$ a Sobolev-type norm (involving L^2 norm of derivatives up to the kth order) and u and v have sufficiently small support. On the other hand it is well known (cf. [**15**, p. 178] or [**38**, p. 142]) that (10) implies local solvability of E on L'. Now (9) and (10) imply by elementary estimates that (10) holds for the operator D on L. Thus local solvability of E implies that of D so the theorem follows by induction.[5]

THEOREM 3.5. *Let L be a semisimple Lie group. Then each bi-invariant differential operator $D \neq 0$ is locally solvable.*

This is proved in [**10e**] using Lemma 24 in [**8a**] which relates explicitly the action of D on invariant distributions with the action of a constant coefficient operator on the Lie algebra, together with a result of [**31**] that all invariance properties of a constant coefficient operator are preserved in a suitable fundamental solution.

Global solvability does not, in general, hold in Theorem 3.5, as shown by the following interesting example due to Cerèzo and Rouvière [**3**]. Let G and K be as above and assume that $\mathfrak{g}$ has a complex structure J; then $\mathfrak{g} = \mathfrak{k} + J\mathfrak{k}$. Let (T_i) be a basis of $\mathfrak{k}$, orthonormal with respect to the negative of the Killing form on $\mathfrak{g}$. Then the operator $\omega = \sum_i (JT_i)T_i$ lies in $\mathbf{Z}(G)$ but annihilates all functions on G which are right invariant under K, hence is not globally solvable.

Nevertheless we have the global solvability

$$\Omega C^\infty(G) = C^\infty(G) \tag{11}$$

for the Casimir operator Ω on any connected noncompact semisimple Lie group G. This was proved in [**3**] for complex G using harmonic analysis on G. For general semisimple G, (11) was proved by Rauch and Wigner [**32**]. The main ingredient in their proof is the verification that no null bicharacteristics of Ω lie over a compact subset of G. They also prove that Ω is injective on $\mathcal{D}(G)$ and has a property similar to (4). Some general results of that nature have been proved by K. Johnson [**16**].

Since a general Lie group L decomposes into a semisimple Lie group and a solvable Lie group (strict semidirect product if the big group is simply connected) one could hope for a joint generalization of Theorems 3.4 and 3.5 to all Lie groups.[6] But taking (1) into account as well as the representation (8) of L as a symmetric coset space, we are led to an even more general question:

Let B/C be a symmetric coset space and D a B-invariant differential operator on it. Is D necessarily locally solvable?

4. Joint eigenfunctions. We shall now give a survey of relatively recent work towards determining the joint eigenfunctions of the operators in $\mathbf{D}(G/K)$. First we consider the case when G/K is a symmetric space X of the noncompact type. The joint eigenfunctions can be characterized (cf. [**10b**, p. 439]) by the functional equation

$$\phi(o)\int_K \phi(gk\cdot x)\,dk = \phi(g\cdot o)\int_K \phi(k\cdot x)\,dk, \tag{1}$$

[5] After this was written I received a preprint [**4**] from M. Duflo establishing this local solvability in general. His proof combines new algebraic results about $\mathbf{Z}(L)$ with the method of Theorem 3.4.

[6] ADDED IN PROOF. In a recent preprint [**42**], Wigner proves global solvability for a simply connected nilpotent Lie group.

$g \in G$, $x \in G/K$. By (4) in §3 the functions $x \to e^{\nu(A(x,b))}$ ($\nu \in \mathfrak{a}_c^*$) are joint eigenfunctions and the inversion formula for the Fourier transform (3) suggests them as building blocks for all joint eigenfunctions. Harish-Chandra's integral formula

$$\varphi_\nu(x) = \int_B e^{\nu(A(x,b))}\, db \qquad (\nu \in \mathfrak{a}_c^*) \tag{2}$$

for the K-invariant eigenfunctions on X (*the spherical functions*) is the prototype of the integral representations one might expect for the solutions of (1). In Furstenberg [**7b**] and Karpelevič [**18**] the following result is proved.

THEOREM 4.1. *The functions*

$$x \to \int_B e^{\nu(A(x,b))}\, d\mu(b), \qquad x \in X, \tag{3}$$

where $\nu \in \mathfrak{a}^*$ and μ a positive measure on B constitute all the positive joint eigenfunctions of $\mathbf{D}(G/K)$.

Furstenberg's proof is purely measure-theoretic. It consists of proving, using methods of Choquet and Deny, that the convex cone of positive solutions to (1) is generated by the extremal rays, which then are related to the integrand in (3). Karpelevič's method goes further and he gives an integral representation of the *positive* eigenfunctions of the Laplace-Beltrami operator L on G/K. A fundamental solution for the operator L-const is constructed rather explicitly via the heat equation on G/K and the minimal solutions are constructed from the fundamental solution by means of the method of Martin [**26**]. Since any positive eigenfunction is a measure-theoretic superposition of minimal eigenfunctions, the integral representation for positive eigengenfunctions of L follows; formula (3) is a simple corollary.

The problem of determining *all* joint eigenfunctions was raised in [**10d**, p. 139] and it was proved for the hyperbolic disk that *analytic functionals* (alias *hyperfunctions*) on the boundary are what is needed to give (by superposition) all the eigenfunctions of L.

THEOREM 4.2. *The eigenfunctions of the Laplace-Beltrami operator on the hyperbolic space X are precisely the functions*

$$x \to \int_B e^{\nu(A(x,b))}\, dT(b) \tag{4}$$

where $\nu \in \mathfrak{a}_c^*$ and T an analytic functional on B.

The method amounts to first determining the K-finite eigenfunctions of L in terms of hypergeometric functions, and then showing, using asymptotic properties of the hypergeometric functions, that the convergence of the expansion of an arbitrary eigenfunction on X into K-finite ones implies convergence of the boundary values in the sense of analytic functionals. This method gives the following partial solution for X of rank one.

THEOREM 4.3. *Let X be symmetric of rank one. The functions* (4) *for ν real (i.e., $\nu \in \mathfrak{a}^*$) constitute all the eigenfunctions of L with eigenvalue $\geqslant -\langle \rho, \rho \rangle$.*

For the proof see [**10f**]. Intermediary results extending Theorem 4.2, using a similar method, were given by Hashizume, Minemura and Okamoto (cf. [**9b**], [**27a, b, c**]).

The K-finite joint eigenfunctions of $\mathbf{D}(G/K)$ are determined in [**10d, g**] as follows, generalizing (2).

THEOREM 4.4. *Let G/K be arbitrary. The K-finite joint eigenfunctions of $\mathbf{D}(G/K)$ are precisely the functions*

$$\Phi_\nu(x) = \int_B e^{\nu(A(x,b))} F(b)\, db \tag{5}$$

where $\nu \in \mathfrak{a}_c^*$ and F is a K-finite function on B. Moreover, there exists a right invariant differential operator D on G such that

$$\Phi_\nu(gK) = D_g(\psi_\nu(gK)), \qquad g \in G.$$

The proof is primarily based on the Paley-Wiener theorem for K-finite functions (Theorem 3.2). As pointed out in [**10d**, p. 138] for X of rank one, Theorem 4.4 implies that each joint eigenfunction of $\mathbf{D}(G/K)$ has the form

$$f(x) = \int_B e^{\nu(A(x,b))}\, dT(b),$$

where T is a "functional" on B (possibly depending on ν and f). The conjecture, suggested by Theorems 4.2–4.3, that these functionals are just the analytic functionals (hyperfunctions) on B was taken up in a remarkable cooperation of Kashiwara, Kowata, Minemura, Okamoto, Oshima, and Tanaka (cf. [**19**]), and they proved the following general result.

THEOREM 4.5. *Assume $\lambda \in \mathfrak{a}_c^*$ satisfies*[7]

$$2\langle i\lambda, \alpha\rangle/\langle \alpha, \alpha\rangle \notin \mathbf{Z} \tag{6}$$

for all restricted roots α of $\mathfrak{g}$ with respect to $\mathfrak{a}$ and let μ: $\mathbf{D}(G/K) \to \mathbf{C}$ be the homomorphism given by $\mu(D) = P_D(\lambda)$ in the notation of (4), §3. *Then the functions*

$$f(x) = \int_B e^{(i\lambda+\rho)(A(x,b))}\, dT(b), \tag{7}$$

where T is a hyperfunction on B constitute all the elements in the joint eigenspace E_μ.

The proof uses many results from the extensive theory of hyperfunctions while Theorem 4.3 only requires a characterization of their spherical harmonics expansion.

It is of interest to observe that if X has rank 1 and if condition (6) fails, the condition of Theorem 4.3 is satisfied, so we have

THEOREM 4.6. *If X has rank one the functions* (7) *constitute all the eigenfunctions of the Laplacian on X.*

In his paper [**24**], J. Lewis investigates the functions (7) as T runs through the space $\mathcal{D}'(B)$ of *distributions* on B, and proves the following result.

---

[7] ADDED IN PROOF. In a recent communication Minemura and Oshima avoid this restriction.

THEOREM 4.7. *If $T \in \mathfrak{D}'(B)$ then the function $f(x)$ in (7) grows no faster than $e^{Cd(0,x)}$ (C = constant, d distance). The converse holds for X of rank one provided (6) is satisfied.*

In the case when G/K is a Hermitian symmetric space there are integral representations of the type (3) (with B replaced by the Bergman-Silov boundary) for bounded holomorphic functions **[13]**, **[25]**, **[7a]**, **[28]**, and for solutions of a certain overdetermined system **[21]**.

We can also consider the case when $G/K = \mathbf{R}^n$ ($n > 1$) and G the group of all isometries. Here $\mathbf{D}(G/K)$ is generated by the Laplacian L; from general results of Ehrenpreis **[6a]** we have that the solutions to $Lu = -\lambda^2 u$ ($\lambda \in \mathbf{C}$) are given by

$$u(x) = \int e^{i(x,\zeta)}\, d\mu(\zeta)$$

where μ is a measure with support on $\{\zeta \in \mathbf{C}^n \mid \zeta_1^2 + \cdots + \zeta_n^2 = \lambda^2\}$ satisfying

$$\int e^{-(x,\operatorname{Im}\zeta)}(1 + |\zeta|)^N\, d\mu(\zeta) < \infty$$

for all N and all $x \in \mathbf{R}^n$; here, however, μ is not unique. One can also prove (for $\lambda \neq 0$) an integral formula

$$u(x) = \int_{\mathbf{S}^{n-1}} e^{i\lambda(x,\omega)}\, dT(\omega)$$

for the solutions to $Lu = -\lambda^2 u$ in analogy with (4) (cf. **[9a]**, **[10f]**), but then T is more general than an analytic functional: for $n = 2$ the T are the continuous linear functionals on the space of analytic functions on $\mathbf{S}^1$ which extend to holomorphic functions f on $\mathbf{C} - (0)$ such that $f(z)$ and $f(z^{-1})$ have exponential type at $z = \infty$ (cf. **[10f]**).

Next we consider the case of a simply connected symmetric space U/K of the compact type. We assume U/K dual to the noncompact space G/K in Theorem 4.4, that is, G is a subgroup of the simply connected complexification G_c, whose Lie algebra $\mathfrak{g}_c$ is the complexification of $\mathfrak{g}$ as well as of the Lie algebra $\mathfrak{u}$ of U. We also take U to be the analytic subgroup of G_c with Lie algebra $\mathfrak{u}$; then U and U/K are both simply connected.

Let $A(g) = A(gK, eM)$ ($g \in G$) so $g = n \exp A(g)k$ with $n \in N$, $k \in K$. The mapping $A: G \to \mathfrak{a}$ can be extended to a holomorphic mapping A: $G_c^0 \to \mathfrak{a}_c$ of a neighborhood G_c^0 of the identity in G_c into the complexification $\mathfrak{a}_c$ of $\mathfrak{a}$ as follows (Stanton **[36]**, Sherman **[35b]**). For the complexified Iwasawa decomposition $\mathfrak{g}_c = \mathfrak{n}_c + \mathfrak{a}_c + \mathfrak{k}_c$ the mapping

$$(X, H, T) \to \exp X \exp H \exp T$$

is a holomorphic diffeomorphism of a neighborhood of (0, 0, 0) in $\mathfrak{n}_c \times \mathfrak{a}_c \times \mathfrak{k}_c$ onto a neighborhood G_c^0 of e in G_c. The mapping

$$\exp X \exp H \exp T \to H$$

is then the desired holomorphic map A: $G_c^0 \to \mathfrak{a}_c$. Taking $\mathfrak{g}_c$ with the customary Hilbert space inner product $(X, Y) \to -B(X, \tau Y)$, where τ is the conjugation of $\mathfrak{g}_c$ with respect to $\mathfrak{u}$, we may take G_c^0 as the diffeomorphic image (under exp) of an open ball $B_0 \subset \mathfrak{g}_c$ with center 0. Let

$$U_0 = \exp(B_0 \cap \mathfrak{u}).$$

It is clear that for each $\nu \in \mathfrak{a}_c^*$ the function $u \to e^{\nu(A(u))}$ on U_0 is an eigenfunction of each $D \in \mathbf{D}(U/K)$; moreover, as noted in Stanton [**36**] and Sherman [**35a, b**], the spherical functions on U/K are (by analytic continuation) given by the analog of Harish-Chandra's formula (2), namely

$$\phi(uK) = \int_K e^{-\mu(A(kuk^{-1}))}\, dk, \qquad u \in U_0,$$

where μ is the highest weight of the spherical representation associated with ϕ. More generally we have (cf. [**10h**])

THEOREM 4.8. *Each joint eigenfunction of* $\mathbf{D}(U/K)$ *has the form*

$$f(uK) = \int_{K/M} e^{-\mu(A(k^{-1}uk))} F(kM)\, dk_M \qquad (u \in U_0)$$

where $dk_M = db$, $F \in C^\infty(K/M)$ *and* $\mu \in \mathfrak{a}^*$ *satisfies* $\langle \mu, \alpha\rangle/\langle \alpha, \alpha\rangle \in \mathbf{Z}^+$ *for all restricted roots* $\alpha > 0$. *Conversely, if* μ *satisfies this condition the function* f *extends uniquely to an analytic function on* U/K *which is a joint eigenfunction of* $\mathbf{D}(U/K)$.

Next we consider for a symmetric space G/K of the noncompact type the space $\Xi = G/MN$ (§3), which is identified with the space of horocycles in G/K, and has many properties analogous to those of G/K. In particular, the G-invariant differential operators on Ξ are all of the form

$$(8) \qquad (D_P\phi)(gMN) = \left\{ P\left(\frac{\partial}{\partial t_1}, \ldots, \frac{\partial}{\partial t_l}\right)\phi(g\exp(t_1H_1 + \cdots + t_lH_l)MN)\right\}_{t=0}$$

where $P \in S(\mathfrak{a}_c)$ and the mapping $P \to D_P$ is an isomorphism of $S(\mathfrak{a}_c)$ onto $\mathbf{D}(G/MN)$ [**10d**]. Thus the solvability questions of §3 reduce to constant coefficient questions. Noting that $G/MN = (K/M) \times A$ we find easily the following answer to Problem B, §2.

PROPOSITION 4.9. *The joint eigendistributions of* $\mathbf{D}(G/MN)$ *are given by*

$$(9) \qquad \Psi(\phi) = \int_{K/M}\int_A \phi(kaMN)e^{(i\lambda+\rho)(\log a)}\, da\, dS(kM) \qquad (\phi \in \mathfrak{D}(\Xi))$$

where $\lambda \in \mathfrak{a}_c^*$ *and* $S \in \mathfrak{D}'(B)$. *The eigenvalues are polynomials in* λ, *independent of* S.

Let $\mathfrak{D}'_\lambda(\Xi)$ denote the space of the distributions Ψ in (9). Of particular interest are the analogs of the spherical functions on G/K, namely the MN-invariant distributions Ψ in (9), the so-called *conical distributions*. In contrast to the situation for spherical functions it was natural to conjecture that the set of conical distributions in $\mathfrak{D}'_\lambda$ would be parametrized by W. Theorems 4.10, 4.12 give a partial confirmation of this. In order to determine these distributions we decompose Ξ according to the Bruhat decomposition of G. For each $s \in W$ fix $m_s \in K$ such that $\mathrm{Ad}(m_s)|\mathfrak{a} = s$. Then if $\xi_s = m_sMN$, $\Xi_s = MNA \cdot \xi_s$, we have $\Xi = \bigcup_{s\in W}\Xi_s$ (disjoint union). If $\xi \in \Xi_s$ we write $a(\xi)$ for the (unique) A factor in $\xi = mna(\xi)\cdot\xi_s$. Let Σ^+ (respectively

Σ^-) denote the set of positive (negative) roots of $\mathfrak{g}$ with respect to $\mathfrak{a}$, Σ_0^+ (Σ_0^-) the corresponding sets of reduced roots. If

$$(10) \qquad \mathrm{Re}\langle i\lambda, \alpha\rangle > 0 \quad \text{for } \alpha \in \Sigma^+ \cap s^{-1}\Sigma^-,$$

then the function $\xi \to e^{(is\lambda + s\rho)(\log a(\xi))}$ on Ξ_s, viewed as a distribution on Ξ, is a conical distribution $\Psi'_{\lambda,s} \in \mathfrak{D}'_\lambda$. Moreover, if Γ is the Gamma function and we put $\alpha_0 = \alpha/\langle\alpha, \alpha\rangle$, the mapping

$$\lambda \to \Psi_{\lambda,s} = \Big(\prod_{\alpha \in \Sigma_0^+ \cap s^{-1}\Sigma^-} \Gamma(\langle i\lambda, \alpha_0\rangle)\Big)^{-1} \Psi'_{\lambda,s}$$

extends to a holomorphic function on $\mathfrak{a}_c^*$ with values in $\mathfrak{D}'(\Xi)$ and each $\Psi_{\lambda,s}$ is a conical distribution in $\mathfrak{D}'_\lambda(\Xi)$ [**10d**, p. 88]. Let

$$\pi(\lambda) = \prod_{\alpha \in \Sigma_0^+} \langle\lambda, \alpha\rangle$$

and let $\mathbf{e}(\lambda)^{-1}$ denote the denominator in Harish-Chandra's **c**-function,

$$(11) \ \mathbf{e}(\lambda)^{-1} = \prod_{\alpha \in \Sigma_0^+} \Gamma\Big(\tfrac{1}{2}\big(\tfrac{1}{2}m_\alpha + 1 + \langle i\lambda, \alpha_0\rangle\big)\Big)\Gamma\Big(\tfrac{1}{2}\big(\tfrac{1}{2}m_\alpha + m_{2\alpha} + \langle i\lambda, \alpha_0\rangle\big)\Big),$$

m_α and $m_{2\alpha}$ denoting the respective multiplicities. Then we have, by [**10d**, p. 96] and [**10g**, Theorem 6.1],

THEOREM 4.10. *Let $\lambda \in \mathfrak{a}_c^*$ and assume*

$$(12) \qquad \pi(\lambda)\mathbf{e}(\lambda) \neq 0.$$

Then the linear combinations $\sum_{s \in W} c_s \Psi_{\lambda,s}$, $c_s \in \mathbf{C}$, constitute all the conical distributions in $\mathfrak{D}'_\lambda(\Xi)$.

Of the orbits Ξ_s above, exactly one, say Ξ_{s^*}, is an open subset of Ξ, the others have lower dimension. For the corresponding distribution Ψ_{λ,s^*} we have

THEOREM 4.11. *The distribution Ψ_{λ,s^*} is given by a locally integrable function on Ξ if and only if*

$$\mathrm{Re}(\langle i\lambda, \alpha\rangle) > 0 \quad \textit{for } \alpha \in \Sigma^+,$$

and by a C^∞ function on Ξ if and only if

$$\langle i\lambda - \rho, \alpha_0\rangle \in \mathbf{Z}^+ \quad \textit{for } \alpha \in \Sigma^+.$$

For X of rank one, Hu [**14a, b**] managed to eliminate the discrete set of λ which violates condition (12). Since all $\Psi_{0,s}$ ($s \in W$) are proportional, an additional conical distribution $\Psi_0 \in \mathfrak{D}'_0$ is needed. It is defined as follows. The function $e^{\rho(\log a(\xi))}$ on the open dense subset Ξ_{s^*} is not locally integrable on Ξ, but the following regularization

$$\Psi_0(\phi) = \int_\Xi (\phi(\xi) - \phi_0(\xi))e^{\rho(\log a(\xi))}\, d\xi,$$

where for $\phi \in \mathfrak{D}(\Xi)$, ϕ_0 is given by $\phi_0(kaMN) = \phi(aMN)$, gives a well-defined conical distribution in $\mathfrak{D}'_0$. Defining conical distributions by means of the full isometry group $\tilde{G}$ of X instead of G (this makes a difference only for $G = \mathbf{SL}(2, \mathbf{R})$), Hu proved the following result.

THEOREM 4.12. *If X has rank one the linear combinations $c\Psi_{\lambda,s^*} + c'\Psi_{\lambda,e}$ ($c\Psi_0 + c'\Psi_{0,e}$ for $\lambda = 0$) constitute all the conical distributions in $\mathfrak{D}'_\lambda(\Xi)$.*

At the poles of Ψ'_{λ,s^*} the residues are conical distributions which are certain transversal derivatives of $\Psi_{\lambda,e}$. For an algebraic approach giving part of Theorem 4.12 see Lepowsky [**23**]. In view of Theorems 4.5–4.6 and the fact that the dual Radon transform maps joint eigenspaces of $\mathbf{D}(G/MN)$ into joint eigenspaces of $\mathbf{D}(G/K)$ [**10d**, p. 93] it would be natural to consider the joint eigenhyperfunctions of $\mathbf{D}(G/MN)$. While the necessary adjustment in Theorem 4.9 is obvious, one might wonder whether a conical hyperfunction is not necessarily a distribution?

5. Eigenspace representations. In this section we state without proof some results on the eigenspace representations for certain classes of homogeneous spaces G/K (Problem C in §2). Again we start with the case when G/K is a symmetric space of the noncompact type. Since each joint eigenspace contains a spherical function ((2), §4) it is clear that each joint eigenspace has the form

$$\mathcal{E}_\lambda(X) = \{f \in C^\infty(X) | Df = P_D(\lambda)f \text{ for } D \in \mathbf{D}(G/K)\}.$$

Here $\lambda \in \mathfrak{a}_c^*$ is arbitrary and P_D is the polynomial in (4), §3. Let T_λ denote the representation of G on $\mathcal{E}_\lambda(X)$ given by $(T_\lambda(g)f)(x) = f(g^{-1}\cdot x)$. Since P_D is W-invariant we have $\mathcal{E}_{s\lambda} = \mathcal{E}_\lambda$, $T_{s\lambda} = T_\lambda$ for all $s \in W$. Using many different tools the following result is proved in [**10g**].

THEOREM 5.1. *The eigenspace representation T_λ is irreducible if and only if*

$$\mathbf{e}(\lambda)\mathbf{e}(-\lambda) \neq 0 \tag{1}$$

in the notation of (11), §4.

For the non-Euclidean disk with the Riemannian metric

$$ds^2 = (1 - x^2 - y^2)^{-2}(dx^2 + dy^2)$$

the criterion (1) gives the following: The eigenspace of the Laplacian L with eigenvalue $4c(c-1)$ is irreducible if and only if c is not an integer.

Next we consider the case when $G/K = \mathbf{R}^n$ ($n > 1$) and G is the group of all isometries. For $\lambda \in \mathbf{C}$ let $\mathcal{E}_\lambda(\mathbf{R}^n)$ denote the eigenspace

$$\mathcal{E}_\lambda(\mathbf{R}^n) = \{f \in C^\infty(\mathbf{R}^n) | Lf = -\lambda^2 f\},$$

L being the Laplacian.

THEOREM 5.2 [**10f**]. *The natural action of G on $\mathcal{E}_\lambda(\mathbf{R}^n)$ is irreducible if and only if $\lambda \neq 0$.*

For $\lambda = 0$ the harmonic polynomials of degree $\leqslant k$ clearly form a closed, proper invariant subspace. But for eigenvalue 0 the eigenspace of L is mapped into itself by a transformation τ of $\mathbf{R}^n$ provided L is just *quasi-invariant* under τ, i.e., satisfies $L^\tau = \phi L$ where ϕ is a function (this amounts to stability in Lie's sense (§1)). For example the conformal map

$$\tau: z \to (az + b)(cz + d)^{-1}$$

of $\mathbf{R}^2 \cup \{\infty\}$ satisfies $L^\tau = |cz - a|^4 L$. More generally, Ørsted **[40]** shows that if X is a conformal vector field on $\mathbf{R}^n$ then the operator

$$\eta(X)f = Xf - \frac{n-2}{2n}(\operatorname{div} X)f, \qquad f \in C^\infty(\mathbf{R}^n), \tag{2}$$

satisfies

$$L\eta(X)f - \eta(X)Lf = -(2/n)(\operatorname{div} X)Lf \tag{3}$$

and $X \to \eta(X)$ is a representation of the Lie algebra τ of conformal vector fields on $\mathbf{R}^n$ on $C^\infty(\mathbf{R}^n)$. By (3) the space $\mathfrak{E}_0(\mathbf{R}^n)$ of harmonic functions is η-invariant. We have then the following complement to Theorem 5.2 (cf. **[10h]**).

THEOREM 5.3. *The representation $X \to \eta(X)$ of the conformal algebra τ on the space of harmonic functions is scalar irreducible* (*that is, the only commuting operators are the scalars*).

For a coset space U/K of compact groups K and U the irreducibility question for the eigenspace representations is quite easy.

PROPOSITION 5.4. *For compact Lie groups $K \subset U$ the eigenspace representations for U/K are all irreducible under the action of U.*

In fact, each joint eigenspace is finite dimensional and can be decomposed into a direct sum of U-invariant subspaces. Each of these contains a spherical function so the irreducibility follows from the fact that a spherical function on U/K is completely determined by the eigenvalues of $\mathbf{D}(U/K)$.

Next we consider the horocycle space G/MN associated with the noncompact symmetric space G/K. The joint eigenspaces (of distributions) are the spaces $\mathfrak{D}'_\lambda(\Xi)$ ($\lambda \in \mathfrak{a}_c^*$) defined in §4. Let τ_λ denote the representation of G on $\mathfrak{D}'_\lambda$. A representation τ is called *conical* if there is a fixed vector under $\tau(MN)$; τ is called *spherical* if there is a fixed vector under $\tau(K)$. The following result describes the finite-dimensional subrepresentations of τ_λ (cf. **[10d**, p. 144]).

THEOREM 5.5. *Let $\mathfrak{D}'_{\lambda,G}$ denote the space of G-finite vectors in $\mathfrak{D}'_\lambda$. Then*:

(i) $\mathfrak{D}'_{\lambda,G} \neq \{0\} \Leftrightarrow \langle i\lambda - \rho, \alpha_0\rangle \in \mathbf{Z}^+$ *for* $\alpha \in \Sigma^+$ $(\alpha_0 = \alpha/\langle\alpha, \alpha\rangle)$.

(ii) *The representation σ_λ of G on $\mathfrak{D}'_{\lambda,G}$ is finite dimensional and irreducible.*

(iii) *The representations σ_λ are precisely the finite-dimensional irreducible conical representations of G.*

(iv) *The representations σ_λ are precisely the finite-dimensional irreducible spherical representations of G.*

(v) *The lowest weight of σ_λ has restriction to $\mathfrak{a}$ given by $-i\lambda + \rho$.*

For the case when G is complex this can be stated in a somewhat sharper form.

THEOREM 5.6. *Let G be complex semisimple and simply connected and for each homomorphism χ: $\mathbf{D}(G/N) \to \mathbf{C}$ let $\mathcal{H}_\chi$ be the space of holomorphic functions f on G/N satisfying*

$$Df = \chi(D)f \quad \text{for } D \in \mathbf{D}(G/N).$$

Let π_χ denote the representation of G on $\mathcal{H}_\chi$. As χ varies, π_χ runs through all the irreducible finite-dimensional holomorphic representations of G.

For the eigenspace representation τ_λ of the real semisimple G on $\mathcal{D}'_\lambda(\Xi)$ the following irreducibility criterion holds [**22a**, p. 631], [**10g**, §12]

THEOREM 5.7. *τ_λ is irreducible if and only if* $\mathbf{e}(\lambda)\mathbf{e}(-\lambda) \neq 0$.

For simply connected nilpotent Lie groups L the eigenspace representations have been investigated by A. Hole [**11**]. For a linear function $\lambda \neq 0$ on the Lie algebra $\mathfrak{l}$ of L, let $\mathfrak{f} \subset \mathfrak{l}$ be a subalgebra of maximal dimension satisfying $\lambda([\mathfrak{f}, \mathfrak{f}]) = 0$ and put $\mathfrak{h} = \mathfrak{f} \cap \text{kernel}(\lambda)$. Let $H \subset L$ be the analytic subgroup corresponding to $\mathfrak{h}$.

THEOREM 5.8 [**11**]. *The eigenspace representations for L/H are irreducible for all nonzero homomorphisms μ: $\mathbf{D}(L/H) \to \mathbf{C}$. The unitary irreducible representations of L (cf. [**20**]) can be realized in this way.*

We finally indicate how the principal series of a semisimple Lie group G arises from eigenspace representations on suitable vector bundles over G/MN and how the discrete series arises from eigenspace representations on suitable vector bundles over G/K.

Let δ be an irreducible unitary representation of M on V_δ and $\lambda \in \mathfrak{a}_c^*$. Let $\Gamma_{\delta,\lambda}$ denote the space of C^∞ functions f: $G \to V_\delta$ satisfying

$$f(gman) \equiv \delta(m)^{-1} e^{(i\lambda - \rho)(\log a)} f(g)$$

and $\pi_{\delta,\lambda}$ the natural representation of G on $\Gamma_{\delta,\lambda}$. These representations form the *principal series* for G.

Extending δ to a representation of MN on V_δ by $\delta(mn) = \delta(m)$, we consider the associated vector bundle E_δ over G/MN. It is clear from the structure of $\mathbf{D}(G/MN)$ ((8), §4), that *the members of the principal series are the eigenspace representations of $\mathbf{D}(G/MN)$ on the section space $\Gamma(E_\delta)$.*

In conclusion we recall that if G has a compact Cartan subgroup it was proved by Takahashi [**37**] for the de Sitter group, by Hotta [**12**] and Wallach [**39**], in general, that the *discrete series of G can be realized on eigenspaces of the Casimir operator on the space of square integrable sections of vector bundles over G/K.* It would be of interest to know the action of the bigger algebra $\mathbf{D}^\delta(G/K)$ on these sections as well as of the algebra $\mathbf{D}^\delta(G/MN)$ on $\Gamma(E_\delta)$ above.

Related realizations of the discrete series on nullspaces of the Dirac operator on certain vector bundles over G/K were given by Parthasarathy [**30**] and Schmid [**34**]; this is extended by Wolf in [**41**] to the representations of G which make up the Plancherel measure of G.

BIBLIOGRAPHY

1. R. Bott, *The index theorems for homogeneous differential operators*, Differential and Combinatorial Topology (S. S. Cairns, Editor), Princeton Univ. Press, Princeton, N. J., 1965, pp. 167–186. MR **31** #6246.

2. É. Cartan, *Sur la détermination d'un système orthogonal complèt dans un espace de Riemann symétrique clos*, Rend. Circ. Mat. Palermo **53** (1929), 217–252.

3. A. Cerèzo and F. Rouvière, *Opérateurs différentiels invariants sur un groupe de Lie*, Séminaire Goulaouic-Schwartz 1972–1973: Équations aux dérivées partielles et analyse fonctionnelle, Exposé No. 10, École Polytechnique, Paris, 1973. MR **52** #8963.

4. M. Duflo, *Opérateurs differentiels bi-invariants sur un groupe de Lie* (preprint).

5. M. Duflo and M. Raïs, *Sur l'analyse harmonique sur les groupes de Lie résolubles*, Ann. Sci. École Norm. Sup. **9** (1976), 107–144.

6a. L. Ehrenpreis, *A fundamental principle for systems of linear differential equations with constant coefficients, and some of its applications*, Proc. Internat. Sympos. Linear Spaces, Jerusalem, 1960, pp. 161–174. MR **24** #A3420.

6b. ______, *Fourier analysis in several complex variables*, Wiley-Interscience, New York, 1970. MR **44** #3066.

7a. H. Furstenberg, *A Poisson formula for semi-simple Lie groups*, Ann. of Math. (2) **77** (1963), 335–386. MR **26** #3820; **28** #1246.

7b. ______, *Translation-invariant cones of functions on semi-simple Lie groups*, Bull. Amer. Math. Soc. **71** (1965), 271–326. MR **31** #1326.

8a. Harish-Chandra, *The characters of semisimple Lie groups*, Trans. Amer. Math. Soc. **83** (1956), 98–163. MR **18**, 318.

8b. ______, *Invariant eigendistributions on a semisimple Lie group*, Trans. Amer. Math. Soc. **119** (1965), 457–508. MR **31** #4862d.

9a. M. Hashizume, A. Kowata, K. Minemura and K. Okamoto, *An integral representation of an eigenfunction of the Laplacian on the Euclidean space*, Hiroshima Math. J. **2** (1972), 535–545.

9b. M. Hashizume, K. Minemura and K. Okamoto, *Harmonic functions on hermitian hyperbolic spaces*, Hiroshima Math. J. **3** (1973), 81–108. MR **50** #14073.

10a. S. Helgason, *Differential operators on homogeneous spaces*, Acta Math. **102** (1959), 239–299. MR **22** #8457.

10b. ______, *Differential geometry and symmetric spaces*, Academic Press, New York, 1962. MR **26** #2986.

10c. ______, *Radon-Fourier transforms on symmetric spaces and related group representations*, Bull. Amer. Math. Soc. **71** (1965), 757–763. MR **31** #3543.

10d. ______, *A duality for symmetric spaces with applications to group representations*, Advances in Math. **5** (1970), 1–154. MR **41** #8587.

10e. ______, *The surjectivity of invariant differential operators on symmetric spaces*. I, Ann. of Math. (2) **98** (1973), 451–479. MR **51** #3804.

10f. ______, *Eigenspaces of the Laplacian; integral representations and irreducibility*, J. Functional Analysis **17** (1974), 328–353. MR **51** #3353.

10g. ______, *A duality for symmetric spaces with applications to group representations*. II, Advances in Math. **22** (1976), 187–219.

10h. ______, *Some results on eigenfunctions on symmetric spaces and eigenspace representations*, Math. Scand. (to appear).

11. A. Hole, *Representations of nilpotent Lie groups on eigenspaces* (preprint).

12. R. Hotta, *On the realization of the discrete series for semisimple Lie groups*, J. Math. Soc. Japan, **23** (1971), 384–407. MR **46** #5531.

13. L.-K. Hua, *Harmonic analysis of functions of several complex variables in classical domains*, Science Press, Peking, 1958; English transl., Transl. Math. Monographs, vol. 6, Amer. Math. Soc., Providence, 1963. MR **23** #A3277; **30** #2162.

14a. M.-C. Hu, *Determination of the conical distribution for rank one symmetric spaces*, Thesis, Mass. Institute of Technology, 1973.

14b. ______, *Conical distributions for rank one symmetric spaces*, Bull. Amer. Math. Soc. **81** (1975), 98–100. MR **51** #6297.

15. L. Hörmander, *On the theory of general partial differential operators*, Acta Math. **94** (1955), 161–248. MR **17**, 853.

16. K. Johnson, *Differential equations and an analog of the Paley-Wiener theorem for linear semisimple Lie groups*, Nagoya Math. J. **64** (1976), 17–29.

17. K. D. Johnson and N. Wallach, *Composition series and intertwining operators for the spherical principal series*. I, Trans. Amer. Math. Soc. (to appear).

18. F. I. Karpelevic, *The geometry of geodesics and the eigenfunctions of the Beltrami-Laplace operator on symmetric spaces*, Trudy Moskov. Mat. Obšč. **14** (1965), 48–185 = Trans. Moscow Math. Soc., Amer. Math. Soc., Providence, R. I., 1967, pp. 51–199. MR **37** #6876.

19. M. Kashiwara, A. Kowata, K. Minemura, K. Okamoto, T. Oshima and M. Tanaka, *Eigenfunctions of invariant differential operators on a symmetric space* (preprint).

20. A. A. Kirillov, *Unitary representations of nilpotent Lie groups*, Uspehi Mat. Nauk **17** (1962), 57–110 = Russian Math. Surveys **17** (1962), 53–104. MR **25** #5396.

21. A. Korányi and P. Malliavin, *Poisson formula and compound diffusion associated to an overdetermined elliptic system on the Siegel half plane of rank two*, Acta Math. **134** (1975), 185–209.

22a. B. Kostant, *On the existence and irreducibility of certain series of representations*, Bull. Amer. Math. Soc. **75** (1969), 627–642; and in Lie Groups and Their Representations (I. M. Gel'fand, Editor), Halsted Press (Wiley), New York, 1975, pp. 231–329. MR **39** #7031; **51** #5845.

22b. ______, *Verma modules and the existence of quasi-invariant differential operators* (Actes Colloq., Marseilles-Luminy, 1974), Lecture Notes in Math., vol. 466, Springer-Verlag, Berlin and New York, 1975, pp. 101–128.

23. J. Lepowsky, *Conical vectors in induced modules*, Trans. Amer. Math. Soc. **208** (1975), 219–272. MR **51** #12961.

24. J. B. Lewis, *Eigenfunctions on symmetric spaces with distribution-valued boundary forms*, J. Functional Analysis (to appear).

25. D. B. Lowdenslager, *Potential theory in bounded symmetric homogeneous complex domains*, Ann. of Math. (2) **67** (1958), 467–484. MR **21** #2836.

26. R. S. Martin, *Minimal positive harmonic functions*, Trans. Amer. Math. Soc. **49** (1941), 137–172. MR **2**, 292.

27a. K. Minemura, *Harmonic functions on real hyperbolic spaces*, Hiroshima Math. J. **3** (1973), 121–151. MR **48** #11928.

27b. ______, *Eigenfunctions of the Laplacian on a hermitian hyperbolic space*, Hiroshima Math. J. **4** (1974), 441–457.

27c. ______, *Eigenfunctions of the Laplacian on a real hyperbolic space*, J. Math. Soc. Japan, **27** (1975), 82–105.

28. C. C. Moore, *Compactifications of symmetric spaces*. II. *The Cartan domains*, Amer. J. Math. **86** (1964), 358–378. MR **28** #5147.

29. R. S. Palais, *Seminar on the Atiyah-Singer index theorem*, Princeton Univ. Press, Princeton, N. J., 1965. MR **33** #6649.

30. R. Parthasarathy, *Dirac operator and the discrete series*, Ann. of Math. (2) **96** (1972), 1–30. MR **47** #6945.

31. M. Raïs, *Solutions élémentaires des opérateurs différentiels bi-invariants sur un groupe de Lie nilpotent*, C. R. Acad. Sci. Paris Sér. **273** (1971), A495–A498. MR **44** #6908.

32. J. Rauch and D. Wigner, *Global solvability of the Casimir operator*, Ann. of Math. (2) **103** (1976), 229–236.

33. F. Rouvière, *Sur la résolubilité locale des opérateurs bi-invariants*, Ann. Scuola Norm. Sup. Pisa **3** (1976).

34. W. Schmid, *Some properties of square-integrable representations of semisimple Lie groups*, Ann. of Math. (2) **102** (1975), 535–564.

35a. T. O. Sherman, *Fourier analysis on the sphere*, Trans. Amer. Math. Soc. **209** (1975), 1–31.

35b. ______, *Fourier analysis on compact symmetric spaces*; *Local theory* (preprint).

36. R. J. Stanton, *On mean convergence of Fourier series on compact Lie groups*, Trans. Amer. Math. Soc. **218** (1976), 61–87.

37. R. Takahashi, *Sur les représentations unitaires des groupes de Lorentz généralisés*, Bull. Soc. Math. France **91** (1963), 289–433. MR **31** #3544.

38. F. Trèves, *Linear partial differential equations with constant coefficients: Existence, approximation and regularity of solutions*, Gordon and Breach, New York, 1966. MR **37** #557.

39. N. Wallach, *On the Enright-Varadarajan modules*, Ann. Sci. École Norm. Sup. **9** (1976), 81–101.

40. B. Ørsted, *Wave equations, particles and chronometric geometry*, Thesis, Mass. Institute of Technology, 1976.

41. J. A. Wolf, *Partially harmonic spinors and representations of reductive Lie groups*, J. Functional Analysis **15** (1974), 117–154.

42. D. Wigner, *Biinvariant operators on nilpotent Lie groups* (preprint)

DEPARTMENT OF MATHEMATICS, INSTITUTE OF TECHNOLOGY, CAMBRIDGE, MASSACHUSETTS 02139

BULLETIN (New Series) OF THE
AMERICAN MATHEMATICAL SOCIETY
Volume 3, Number 2, September 1980

ON THE ROLE OF THE HEISENBERG GROUP IN HARMONIC ANALYSIS

BY ROGER HOWE

In this article I want to popularize the Heisenberg group, which is remarkably little known considering its ubiquity. I use the word ubiquity advisedly. To justify it, let me give a sample of the many apparently diverse topics where the Heisenberg group reveals itself as an important factor.

(1) Representation Theory of Nilpotent Lie Groups
(2) Foundations of Abelian Harmonic Analysis
(3) Moduli of Abelian Varieties
(4) Structure Theory of Finite Groups
(5) Theory of Partial Differential Equations
(6) Quantum Mechanics
(7) Homological Algebra
(8) Ergodic Theory
(9) Representation Theory of Reductive Algebraic groups
(10) Classical Invariant Theory

This list could easily be lengthened both by adding new topics and making these more specific, for sometimes the applications are multiple. In fact, one of the most important areas of application I have not mentioned at all, to avoid name-dropping. Why has an object with such wide application gone relatively unnoticed until recently? One can only speculate. One reason might be that the role of the Heisenberg group in many situations is relatively subtle. An investigator might be able to get what he wanted out of a situation while overlooking the extra structure imposed by the Heisenberg group, structure which might enable him to get much more. Such may have been the case with Hermann Weyl, one of the pioneers in introducing the Heisenberg group into Quantum Mechanics [**Wy1**]. Indeed, many physicists still call the Heisenberg group the "Weyl group". When Weyl wrote his book *The classical groups*, [**Wy2**] he overlooked the natural occurrence of the Heisenberg group, exploitation of which yields results which one feels Weyl would have liked very much. Again, I gather that an appreciation of the role of the Heisenberg group in rigidifying abelian varieties was an important aspect of Mumford's [**Mm**] fundamental contributions to their study. Another obstacle to the appreciation of the common underlying structure may have been the very diversity of the topics I listed above, for detection of its presence in one place need not suggest its presence elsewhere. Indeed, investigators in one field may very well never have been aware that the Heisenberg group had been found in some field not seemingly related to theirs. Another factor certainly contrib-

Received by the editors December 15, 1979.
AMS (MOS) subject classifications (1970). Primary 42-02, 35-02, 22-02.
Key words and phrases. Heisenberg group, Fourier transform, harmonic analysis, symbols.

utory to its relative obscurity is that what I call "the Heisenberg group" is not in fact one object, but a collection of similar objects, rather like a functor, or a scheme in algebraic geometry, or even a combination of several overlapping functors. Thus one has to look with a certain pair of spectacles in order to see the topics on the list as being united via a single common phenomenon. But I think this is possible, and I want to give some examples of this point of view here. For people unacquainted with the Heisenberg group, this can be an introduction. Hopefully those already acquainted with it will also learn something.

These examples will be taken mostly from harmonic analysis. More particularly, except for the last, they are mostly an attempt to bridge the gap between abelian and nonabelian harmonic analysis by showing how various important results in abelian harmonic analysis may be enriched through an interpretation involving the Heisenberg group. Because of this choice of theme, the examples will not reveal the breadth of the range of applications of the Heisenberg group. But they do hint at more varied applications, and since they treat material that should be relatively familiar, they will perhaps be accessible to a reasonably wide audience. The topics are

(1) Plancherel $\Leftrightarrow$ Stone-von Neumann,

(2) distributional viewpoint towards Fourier Inversion and Poisson Summation,

(3) Bochner's Formula, $\Big\}$ $(Sl_2, O_{p,q})$ duality,
(4) Huyghens' Principle

(5) symbols in the theory of pseudo-differential operators.

I take as my text for this discussion the article by E. M. Stein in *Studies in harmonic analysis*, J. Marshall Ash, ed., published by MAA (1976), entitled *Harmonic analysis on* $\mathbf{R}^n$. Before he gets to the more specialized recent developments, Stein lists a number of facts and formulas he considers basic to the subject of harmonic analysis on $\mathbf{R}^n$. These are

(a) the Fourier Inversion Theorem, (1)

(b) the Schwartz Space,

(c) the Plancherel Theorem,

(d) the fact that Fourier transform interchanges d/dx and multiplication by $2\pi ix$, and translation by y and multiplication by $e^{2\pi ix\cdot y}$. (One might just say, that it interchanges convolution and multiplication.)

(e) that Fourier transform commutes with rotations, and more or less normalizes scalar dilations. (More generally, Fourier transform more or less normalizes the transformations coming from GL_n. If they are unitarized, it does do so.)

(f) that $(e^{-\pi x^2})^\wedge = e^{-\pi x^2}$. Also that the Fourier transform on radial functions is given by a Hankel transform:

$$\hat{f}(r) = (2\pi r)^{-(n-2)/2}\int_0^\infty f(s)J_{(n-2)/2}(2\pi rs)s^{n/2}\,ds.$$

Also Hecke's formula: $(P(x)e^{-\pi x\cdot x})^\wedge = i^kP(x)e^{-\pi x\cdot x}$ if P is a spherical harmonic on $\mathbf{R}^n$ of degree k; and the formula of

Bochner, that $(P(x)f(r))^\wedge = P(x)T(f(r))$, where T is the Hankel transform of degree $n/2 + k - 1$.

These facts are presented by Stein, I think it is accurate to say, as interesting and important facts which happen to be true. I want to cast them in the role of clues pointing to a larger truth. I will try to do so without getting too doctrinaire or theological.

The usual point of view is that Fourier transform (herein denoted $^\wedge$) takes (hard to understand) convolution operators to (easy to understand) multiplication operators, and this is certainly a vital function of $^\wedge$. However, both multiplications and translations are special operators, and to get general operators we must combine them. Thus it might be worthwhile to take a more symmetrical point of view. This can be effected in a group theoretical context as follows. For $x, y \in \mathbf{R}^n$, define

$$T_x f(x') = f(x' - x), \qquad M_y f(x') = e^{2\pi i y \cdot x'} f(x') \tag{2}$$

where $y \cdot x'$ is the usual inner product. Also for $x \in \mathbf{T}$, consider the scalar operators, $S_z f(x') = z f(x')$. Then the set $\{M_y T_x S_z : y \in \mathbf{R}^n, x \in \mathbf{R}^n, z \in \mathbf{T}\}$ is in fact a group of unitary operators on $L^2(\mathbf{R})$. Somewhat more abstractly, if we consider the set $H = \mathbf{R}^n \times \mathbf{R}^n \times \mathbf{T}$, and define on it a law of combination

$$(y_1, x_1, z_1)(y_2, x_2, z_2) = \left(y_1 + y_2, x_1 + x_2, e^{-2\pi i y_2 \cdot x_1} z_1 z_2\right) \tag{3}$$

then H becomes a two step nilpotent group we will call the (reduced) Heisenberg group of order n. Then the map

$$\rho\colon (y, x, z) \to M_y T_x S_z \tag{4}$$

is a faithful unitary representation of H. It is very easy to see that ρ is irreducible. Since the M_y's separate points, any operator commuting with them is a multiplication [L1]. Since the T_x's act transitively, this multiplication operator must be constant, i.e., scalar.

The basic formulas for Fourier transform say that $^\wedge$ normalizes $\rho(H)$. More precisely, let us define an automorphism r of H by the rule

$$r(y, x, z) = (-x, y, e^{2\pi i x \cdot y} z). \tag{5}$$

Then it is immediate to verify that

$$^\wedge\rho(h) = \rho(r(h))^\wedge. \tag{6}$$

That is to say, $^\wedge$ intertwines the two representations ρ and $\rho \circ r$ of H with each other. When you have an intertwining operator between irreducible representations, it has to be terrible not to be unitary. In particular, if densely defined and closable, so if its adjoint is densely defined, it will be unitary (up to multiples). This is the general Schur's lemma [L2, p. 180]. It is relatively obvious that $^\wedge$ has these weak properties, so it is unitary up to multiples. Thus computing one Fourier transform, any one, serves to show that in its standard form,

$$\hat{f}(y) = \int_{\mathbf{R}^n} f(x) e^{-2\pi i x \cdot y}\, dx, \tag{7}$$

$^\wedge$ is unitary.

Fourier inversion can be dealt with similarly. If we repeat $\wedge$ twice we know $\wedge\wedge\rho(h) = \rho(r^2(h))\wedge\wedge$, where $r^2(y, x, z) = (-y, -x, z)$. It is trivial to check that $\omega(-1)$, defined by $\omega(-1)f(x) = f(-x)$ accomplishes the same intertwining. Thus by irreducibility of ρ, $\wedge^2 = \omega(-1)$ up to multiples, and again a single computation suffices to establish that the normalization is correct. With regard to these computations, let us note they are both accomplished by the single formula, listed by Stein (see (1)(f)).

$$(e^{-\pi x\cdot x})^\wedge = e^{-\pi x\cdot x}. \tag{8}$$

For the proof of this, Stein says you make an obvious reduction to the one dimensional case, where it follows by a contour integral argument. A less sophisticated procedure in the one dimensional case is to observe that $e^{-\pi x^2}$ solves the differential equation

$$(d/dx + 2\pi x)f = 0. \tag{9}$$

Since this equation is invariant under Fourier transform, we see $e^{-\pi x^2}$ must be its own Fourier transform up to multiples. Thus the whole business is reduced to the famous formula

$$\int_{-\infty}^{\infty} e^{-\pi x^2}\, dx = 1. \tag{10}$$

The significance of this proof of (8) will be discussed later.

Thus simply the observation that the reduced Heisenberg group exists virtually suffices to establish the basic Plancherel and Fourier Inversion formulas. Reciprocally, the Plancherel Theorem serves to prove and strengthen a celebrated fact about the Heisenberg group, the Stone-von Neumann Theorem, which lies at the foundation of quantum mechanics. Thus the Plancherel and Stone-von Neumann Theorems are revealed as essentially equivalent. We proceed to detail this, bringing in the Schwartz space along the way.

When you have a unitary representation σ of a Lie group G, it is natural to differentiate it and integrate it, and one has a notion of smooth vectors. Let $\mathfrak{H}$ denote the Lie algebra of H. Let $e\colon \mathfrak{H} \to H$ be the exponential map. We can choose a basis X_j, Y_j, T for $\mathfrak{H}$ such that

$$e(\Sigma y_j Y_j + \Sigma x_i X_i + tT) = (y, x, e^{2\pi i(t - y\cdot x/2)}) \tag{11}$$

where $y = (y_1, \ldots, y_n)$, $x = (x_1, \ldots, x_n)$. Then we find easily that

$$\rho(X_j) = -\partial/\partial x_j, \qquad \rho(Y_j) = 2\pi i x_j, \qquad \rho(T) = 2\pi i. \tag{12}$$

The equations (12) are significant in several regards. For one thing, we can compute that

$$[Y_j, X_i] = Y_j X_i - X_i Y_j = 2\pi i \delta_{ij}. \tag{13}$$

The equations (13) are the celebrated canonical commutation relations of Heisenberg. The also famous Stone-von Neumann Theorem asserts that under mild assumptions the operators of (12) are, up to unitary equivalence, the only realization of the canonical commutation relations. We will formulate the group theoretical version of this result. This is essentially the formulation of von Neumann **[vN]**. It is an interesting historical note that Hermann Weyl realized very early **[Wy1]** that the canonical commutation relations were the structural equations for a Lie algebra, and that there was an essentially equivalent formulation of them in terms of the corresponding Lie group. For this reason, among physicists, the group H is often called the Weyl group. This terminology would obviously cause too much confusion among mathematicians, and also probably exemplifies some of the differences in psychology between mathematicians and physicists.

In any case, we have the following result.

THEOREM (STONE-VON NEUMANN). *The representation ρ of H is, up to unitary equivalence, the only irreducible unitary representation of H such that*

$$\rho((0, 0, z)) = zI. \tag{14}$$

(To express (14), we say the central character of ρ is the identity.)

Just as introducing the Heisenberg group facilitated proof of the Plancherel Theorem, the flow can be reversed, and the Plancherel Theorem used to prove the Stone-von Neumann Theorem. This comes about simply by integrating ρ in a reasonably explicit way. The argument comes from I. Segal **[Sg]**, and is almost to be found in von Neumann's original (1931) paper **[vN]**. A great virtue of it is that it proves a meaningfully stronger result than the one we have just stated.

Consider $f \in L^1(G)$. One defines $\rho(f)$ in the usual way:

$$\rho(f) = \int_H f(g)\rho(g)\, dg. \tag{15}$$

Defining left translation in the usual way,

$$L_h(f)(h') = f(h^{-1}h') \tag{16}$$

we find from (15) the relation

$$\rho(L_z f) = z\rho(f) \quad \text{for } z \in \mathbf{T} \subseteq H. \tag{17}$$

Thus, let us define

$$°f = \int_{\mathbf{T}} z^{-1} L_z f\, dz. \tag{18}$$

Then it is easy to see that

$$\text{(a)} \quad \rho(°f) = \rho(f), \qquad \text{(b)} \quad L_z(°f) = z(°f). \tag{19}$$

The subspace $L^1(H, e)$ of L^1 consisting of f satisfying (19)(b) is a two-sided convolution ideal in L^1 and $f \to °f$ is a projection of $L^1(H)$ onto $L^1(H, e)$. Equation (19)(a) says ρ factors through this projection. Hence in computing (15) it will suffice to consider only $f \in L^1(H, e)$. Such f are determined by their restrictions $r(f)$ to the set $(y, x, 1) \subseteq H$. Let $\mathcal{S}(\mathbf{R}^n)$ be the Schwartz space of rapidly decreasing smooth functions on $\mathbf{R}^n$ **[L2]**, **[T]**. Take $\phi \in \mathcal{S}(\mathbf{R}^n)$

and compute

$$\rho(f)(\phi)(x') = \int_{\mathbf{R}^n \oplus \mathbf{R}^n} r(f)(y, x) M_y(T_x \phi)(x')\, dy\, dx$$

$$= \int_{\mathbf{R}^n \oplus \mathbf{R}^n} r(f)(y, x) e^{2\pi i y \cdot x'} \phi(x' - x)\, dy\, dx$$

$$= \int_{\mathbf{R}^n} r(f)^{\wedge_1^{-1}}(x', x) \phi(x' - x)\, dx \tag{20}$$

(where $\wedge_1^{-1}$ indicates inverse Fourier transform in the first variable)

$$= \int_{\mathbf{R}^n} r(f)^{\wedge_1^{-1}}(x', x' - x) \phi(x)\, dx$$

$$= \int_{\mathbf{R}^n} \kappa_f(x', x) \phi(x)\, dx$$

where

$$\kappa_f(x', x) = r(f)^{\wedge_1^{-1}}(x', x' - x). \tag{21}$$

Segal calls formula (21) the Weyl transform. Thus $\rho(f)$ is an integral kernel operator. We recall that the integral operators with square integrable kernel form the algebra of Hilbert-Schmidt operators $\mathcal{H}.\mathcal{S}$. Also, we should note that there is a notion of smooth operator just as well as smooth vector. A smooth operator T is one such that $\rho(u_1) T \rho(u_2)$ is bounded for all u_1, u_2 in the universal enveloping algebra of H. It is easy to see from formulas (12) that the smooth operators for ρ are just the integral kernel operators with Schwartz functions as kernels. These clearly arise as Weyl transforms from $\mathcal{S}(H, e)$, the functions f in $L^1(H, e)$ with $r(f)$ in $\mathcal{S}(\mathbf{R}^{2n})$. From our acquaintance with the Fourier transform, we may use formula (21) to conclude:

THEOREM (SEGAL). *The representation map ρ from $L^1(H, e)$ to bounded operators on $L^2(\mathbf{R}^n)$ extends to an isometric isomorphism from $L^2(H, e)$ to $\mathcal{H}.\mathcal{S}.$, and restricts to a topological isomorphism from $\mathcal{S}(H, e)$ to the smooth operators for ρ.*

This result makes it plausible that ρ is unique, for there is in some sense nothing left from $L^1(H, e)$ out of which to make another representation. Indeed a few exercises in the yoga of matrix coefficients suffice to derive the Stone-von Neumann Theorem from this one.

The fact that ρ extends from L^1 to L^2 means that ρ is what is called a square-integrable representation. It is increasingly appreciated these days that square-integrability is an important property of representations.

We may combine the above results to obtain

THEOREM (STONE-VON NEUMANN-SEGAL). *Up to unitary equivalence there is a unique representation ρ of H with the identity on $\mathbf{T}$ as central character, and this representation is square integrable.*

In the light of recent results in representation theory, in particular those of Moore and Wolf [MW] on square-integrable representations, we can say that square-integrability is a natural thing to include in the description of ρ because

THEOREM (MOORE-WOLF). *An irreducible representation σ of a nilpotent Lie group N is square-integrable (modulo the center of N) if and only if it is the unique irreducible representation of N with its central character.*

From the uniqueness and square-integrability we can conclude immediately that ρ is isometric from $L^2(H, e)$ to $\mathcal{H}.\mathcal{S}$. Reading formula (21) backwards, we conclude that Fourier transform is isometric. Thus we conclude:

The Stone-von Neumann-Segal and the Plancherel Theorems are equivalent theorems.

Another aspect of the formulas (12) is that they imply immediately that the space of smooth vectors of ρ is the Schwartz space $\mathcal{S}(\mathbf{R}^n)$. If we combine this with the Stone-von Neumann Theorem, we see that the Schwartz space is characterized abstractly as the smooth subspace ρ^∞ of a certain irreducible representation ρ of H. Thus all questions of smoothness, growth, etc., of tempered distributions on $\mathbf{R}^n$ could in principle be formulated purely in terms of the structure of ρ. Although this procedure would often be counterproductive, it can be occasionally useful and is something to keep in mind. We will give an example as our last topic below.

Although H has only one representation with central character the identity up to unitary equivalence, there are many rather different looking ways of realizing this representation. This rather Hinduish multiplicity-in-one is characteristic of the theory of H and adds greatly to its richness. We have already seen that the theory of the Fourier transform can be viewed as studying the intertwining between two realizations of ρ. The study of the intertwining between all realizations of ρ leads in many directions, of which we can mention only a few.

Many realizations of ρ are obtained by inducing. Let $A \subseteq H$ be any maximal abelian subgroup, and ψ any unitary character of A restricting to the identity on $\mathbf{T}$. We may form the induced representation

$$\operatorname{ind}_A^H \psi = \rho_{A,\psi}$$

defined by left translation

$$\rho_{A,\psi}(h)f(h') = f(h^{-1}h') \tag{22}$$

on the space of functions f such that

$$f(h'a) = \psi(a)^{-1}f(h'), \qquad \int_{H/A} |f|^2\, dh < \infty. \tag{23}$$

It is clear that the central character of $\rho_{A,\psi}$ is the identity, and since A is specified as maximal abelian, it is not too hard to show, for example via the general results of Mackey that

THEOREM. *$\rho_{A,\psi}$ is irreducible, hence a realization of the unique irreducible representation ρ of H with the identity central character.*

For example, if $A = \{(y, 0, z)\}$ and $\psi(y, 0, z) = z$, then we just get the original action of H defined by formulas (2). In this case, since if $X =$

$\{(0, x, 1)\} \subseteq H$, we have $H = A \cdot X$, we can identify a function in the space of $\rho_{A,\psi}$ with a function on $X \simeq \mathbf{R}^n$, and then the action is just that given in (2). This is called the Schrödinger model of ρ. Since H has a lot of symmetry, this gives a lot of realizations of ρ. We will see further examples below.

Corresponding to the realizations just described there is a distributional version of Frobenius reciprocity. Let $\rho^{\infty *}$ be the dual space of the smooth vectors ρ^{∞} of ρ. In the Schrödinger model, $\rho^{\infty *}$ is of course just the space of tempered distributions. On $\rho^{\infty *}$ one has the representation contragredient to the restriction of ρ to ρ^{∞}.

THEOREM. *For maximal abelian $A \subseteq H$ and unitary character ψ of A restricting to the identity there is a linear functional $\mathfrak{L}(A, \psi)$, unique up to multiples, such that*

$$\rho^{\infty *}(a)\mathfrak{L}(A, \psi) = \psi(a)^{-1}\mathfrak{L}(A, \psi), \qquad a \in A. \tag{24}$$

That is,

$$\mathfrak{L}(A, \psi)(\rho(a)(v)) = \psi(a)\mathfrak{L}(A, \psi)(v), \qquad a \in A, v \in \rho^{\infty}. \tag{25}$$

In the realization $\rho_{A,\psi}$ of ρ, the functional $\mathfrak{L}(A, \psi)$ is just evaluation at the identity in H.

This theorem is probably due more to Cartier [**Cr**] than anyone else.

Let us consider some concrete instances of this theorem. Let $A = \{(y, 0, z)\}$ as before and $B = \{(0, x, z)\}$. Then if $\psi_y(y, 0, z) = z$ and $\psi_x(0, x, z) = z$, we can see in the Schrödinger model that $\mathfrak{L}(A, \psi_y) = \delta_0$ is the Dirac delta at the origin in $\mathbf{R}^n$, and $\mathfrak{L}(B, \psi_x) = dx$ is Lebesgue measure on $\mathbf{R}^n$. These particular results may be proved in a down-to-earth fashion by the following advanced calculus lemmas.

LEMMA. *If $f \in \mathcal{S}(\mathbf{R}^n)$ and $f(0) = 0$, then we may write*

$$f = \sum_{i=1}^{n} x_i f_i \tag{26}$$

with $f_i \in \mathcal{S}(\mathbf{R}^n)$.

LEMMA. *If $f \in \mathcal{S}(\mathbf{R}^n)$ and $\int_{\mathbf{R}^n} f\, dx = 0$, then we may write*

$$f = \sum_{i=1}^{n} \frac{\partial f}{\partial x_i} \tag{27}$$

with $f_i \in \mathcal{S}(\mathbf{R}^n)$.

Of course these lemmas are just Fourier transforms of one another, but we want to take the opposite point of view. Suppose we have the general result on $\mathfrak{L}(A, \psi)$, and let α be an automorphism of H which acts trivially on $\mathbf{T}$. By the uniqueness of ρ we know from general considerations that there is a unitary operator $\omega(\alpha)$, defined up to scalar multiples, such that

$$\omega(\alpha)\rho(h)\omega(\alpha)^{-1} = \rho(\alpha(h)). \tag{28}$$

If A is a maximal abelian subgroup of H, then so is $\alpha(A)$. Let $\alpha^*(\psi)$ be the character on $\alpha(A)$ defined by

$$\alpha^*(\psi)(\alpha(a)) = \psi(a). \tag{29}$$

It is clear that $\omega(\alpha)$ preserves ρ^∞ since this is defined in terms of the action of H. Therefore we can define $\omega^{\infty *}(\alpha)$ acting on $\rho^{\infty *}$ in the usual manner:

$$\omega^{*\infty}(\alpha)(\lambda)(v) = \lambda(\omega(\alpha^{-1})v). \tag{30}$$

Putting these formulas together, we find that

$$\omega^{*\infty}(\alpha)\mathcal{Q}(A, \psi) = \mathcal{Q}(\alpha(A), \alpha^*(\psi)). \tag{31}$$

It is to be understood that (31) is an equation in projective space. If we apply equation (31) to $A = \{(y, 0, z)\}$ and $\alpha = r$ as in formula (5), then we find by formula (6) and our lemmas that Lebesgue measure is the Fourier transform of the Dirac delta, which is simply the Plancherel Theorem from the distributional point of view.

A nice feature of this approach to Fourier transform is that it puts it on the same footing as the Poisson Summation Formula. Thus let $\Gamma = \{(a, b, z)\}$ where $a, b \in \mathbf{Z}^n$. It is easily checked that Γ is a maximal abelian subgroup of H. Let ψ be the character of Γ such that $\psi(a, 0, 1) = 1 = \psi(0, b, 1)$. It is easy to see that the automorphism r of formula (5) fixes Γ and ψ. In the Schrödinger model we conclude that $\mathcal{Q}(\Gamma, \psi)$ is an eigendistribution for Fourier transform. It is easy to check that Haar (counting) measure on $\mathbf{Z}^n$ is invariant under $\rho(a, 0, 1)$ and $\rho(0, b, 1)$, hence must be $\mathcal{Q}(\Gamma, \psi)$. (A direct proof that it is the unique distribution with these invariance properties is another advanced calculus exercise.) Thus counting measure on $\mathbf{Z}^n$ is, up to multiples, invariant under Fourier transform. Again the formula (8) allows us to fix the multiple as 1. Thus we have the Poisson Summation Formula.

The gentle reader will hopefully also have been reminded by the above discussion of our argument for formula (8). Indeed, formula (8) is related to another class of realizations of ρ, not induced in the usual sense, but sometimes called holomorphically induced, and also called Fock models or complex wave models. Study of the Fock model leads into the study of θ-functions, Toeplitz operators and other topics. Indeed, study of the $\mathcal{Q}(A, \psi)$ leads in many directions. They are a central concern in the theory of the Heisenberg group.

We have now discussed most of the topics on Stein's list. There remains the relation of Fourier transform to GL_n and Bochner's and Hecke's formulas. These topics lead deeper into the theory, to the construction of a certain unitary representation with many names. I shall call it the oscillator representation because of its connection with the quantum harmonic oscillator and its analogy with the spin representation. Once having gone so far we will find other topics are also approachable. We will discuss a very classical one–Huyghens' Principle for the wave equation.

Of course GL_n is the group of automorphisms $\mathbf{R}^n$. We see further that if $g \in GL_n$, then

$$\alpha(g)\colon (y, x, z) \to (g^{t^{-1}}(y), g(x), z), \tag{32}$$

where $g^{t^{-1}}$ is the inverse transpose of g, defines an automorphism of H. We more or less began this discussion by observing that Fourier transform also defined an automorphism r of H. It can be observed from (5) and (32) that r

normalizes $\alpha(GL_n)$. Precisely

$$r\alpha(g)r^{-1} = \alpha(g^{t^{-1}}). \tag{33}$$

This has the following operator theoretic consequence.

For $g \in GL_n$, we define $\omega(g)$ on $L^2(\mathbf{R}^n)$ by

$$\omega(g)f(x) = |\det g|^{-1/2} f(g^{-1}(x)). \tag{34}$$

Then $\omega(g)$ is unitary and satisfies (28). We conclude from (33) and (6) that, up to a multiple which one again checks to be 1,

$$^{\wedge}\omega(g) = \omega(g^{t^{-1}})^{\wedge}. \tag{33'}$$

Formula (33′) is of course very easy to verify directly.

Thus (33′) arises because $\wedge$ normalizes $\alpha(GL_n)$ in the automorphisms of the Heisenberg group. At this point it is natural to ask for all the automorphisms of H acting trivially on $\mathbf{T}$. This is most easily attacked by transferring the problem to the Lie algebra $\mathfrak{H}$ of H, as described in (11) and (12). Put $W = \operatorname{span}\{X_j, Y_j\}$, so that

$$\mathfrak{H} = W \oplus \mathbf{R}. \tag{35}$$

First there are inner automorphisms. These are easily checked to act simply transitively on the linear complements to $\mathbf{R}$ in $\mathfrak{H}$. Therefore we may identify the outer automorphism group of $\mathfrak{H}$ with the group of automorphisms fixing W. From formulas (12) it is evident that Lie bracket on $\mathfrak{H}$ induces a symplectic form $\langle\ ,\ \rangle$ on W. Evidently any automorphism of $\mathfrak{H}$ fixing $\mathbf{R}$ pointwise and leaving W invariant must preserve $\langle\ ,\ \rangle$. And from (12) one can also see that conversely any isometry of $\langle\ ,\ \rangle$ on W extends in obvious fashion to an automorphism of $\mathfrak{H}$. We will denote the isometry group of $\langle\ ,\ \rangle$ by $\operatorname{Sp}(W, \langle\ ,\ \rangle) = \operatorname{Sp}$ and will consider it as a group of automorphisms of H. If we note that $W = \mathbf{R}^n \oplus \mathbf{R}^n$, where the first factor is the span of the Y's and the second the span of the X's, then $\alpha(GL_n)$ is the subgroup of Sp preserving this decomposition, and the transformation r of (5) is precisely the nontrivial element of the normalizer of $\alpha(GL_n)$ in Sp.

As we noted above we get for each $g \in \operatorname{Sp}$ a unitary operator $\omega(g)$ defined up to multiples and satisfying (28). We thus get a well-defined action ω of Sp on the projective space $\mathbf{P}L^2(\mathbf{R}^n)$. Such an action is called a projective unitary representation. An ordinary, linear, unitary representation gives rise to a projective representation by just looking at the action on lines. A natural question is the extent to which the reverse is true, and this has been considered in a general context by Mackey **[Ma]**. In the present situation one knows **[Sh]**, **[Wi]**.

THEOREM (SHALE-WEIL). *Let* $\widetilde{\operatorname{Sp}}$ *be the 2-fold cover of* Sp. *Let* $\tilde{g} \to g$ *be the projection map. Then there is a unitary representation* ω *of* $\widetilde{\operatorname{Sp}}$ *on* L^2 *such that*

$$\omega(\tilde{g})\rho(h)\omega(\tilde{g})^{-1} = \rho(g(h)).$$

We will call ω *the oscillator representation.*

We will use this theorem to abuse notation and not distinguish between ω as a projective representation of Sp and as a linear representation of $\widetilde{\operatorname{Sp}}$.

We should describe ω a little bit. We have already described $\omega(\alpha(GL_n))$ and $\omega(r)$. Let N be the subgroup of Sp which leaves the Y_j's fixed. Then N is isomorphic to the space of symmetric $n \times n$ matrices in such a way that $\omega(N)$ consists of multiplications by functions

$$e^{\pi i Q(x, x)}, \qquad Q \text{ symmetric } n \times n. \tag{36}$$

Since $\alpha(Gl_n)$, N and r generate Sp, this in principle describes all $\omega(g)$, and we will leave the matter there. It is a property of ω not to be transparently describable in any realization. However, letting $\mathfrak{sp}$ denote the Lie algebra of Sp, we can easily describe $\omega(\mathfrak{sp})$. It consists of the differential operators of "total order" 2, namely

$$\pi i x_i x_j, \quad \frac{i}{4\pi}\frac{\partial^2}{\partial x_i \partial x_j} \quad \text{and} \quad \frac{1}{2}\left(x_i \frac{\partial}{\partial x_j} + \frac{\partial}{\partial x_j} x_i\right) = x_i \frac{\partial}{\partial x_j} + \frac{\delta_{ij}}{2}. \tag{37}$$

It is the necessity of adding $\frac{1}{2}$ trace to the linear vector fields in order to unitarize the action of GL_n that leads to the order 2 obstruction to linearizing ω as a representation of Sp (as opposed to $\widetilde{\mathrm{Sp}}$).

The oscillator representation has a number of remarkable properties. We are going to give an example of one of the most remarkable, then show how it relates to the topics mentioned above. Fix an inner product (,) on $\mathbf{R}^n$. For convenience say $n = p + q$ and

$$(x, y) = x_1 y_1 + \cdots + x_p y_p - x_{p+1} y_{p+1} \cdots - x_n y_n. \tag{38}$$

Let $O_{p,q} = O$ be the isometry group of (,). Put

$$\rho_{p,q}^2 = \rho^2(x, x). \tag{39}$$

(We will no longer be referring to the representation ρ of H, though, of course, it is there.) Also put

$$\Delta_{p,q} = \Delta = \sum_{j=1}^{p} \frac{\partial^2}{\partial x_j^2} - \sum_{j=p+1}^{p+q} \frac{\partial^2}{\partial x_j^2}. \tag{40}$$

Then ρ^2 and Δ, as differential operators, obviously commute with $\omega(O)$. Also, ρ^2 and Δ belong to the Lie algebra $\omega(\mathfrak{sp})$. We compute

$$[\Delta, \rho^2] = 4\tilde{E} = 4\left(\sum_{j=1}^{p+q} x_j \frac{\partial}{\partial x_j} + \frac{p+q}{2}\right). \tag{41}$$

It may easily be verified that Δ, ρ^2 and $\tilde{E}$ span a Lie algebra $\mathfrak{g}$, with commutation relations

$$\begin{aligned} [\tilde{E}, i\pi\rho^2] &= 2\pi i \rho^2 = 2(i\pi\rho^2), \\ \left[\tilde{E}, \frac{i\Delta}{4\pi}\right] &= -\frac{i\Delta}{2\pi} = -2\left(\frac{i\Delta}{4\pi}\right), \\ \left[i\pi\rho^2, \frac{i\Delta}{4\pi}\right] &= \tilde{E}. \end{aligned} \tag{42}$$

These will be recognized as the commutation relations of a standard basis for $\mathfrak{sl}_2$. It is easy to check further that $\mathfrak{g}'$ is the full Lie subalgebra of $\mathfrak{sp}$ commuting with $\omega(O)$. Let G' be the subgroup of Sp generated by $\mathfrak{g}'$. Then

G' is the centralizer of O in Sp, and in fact, the relation is mutual: (O, G') is a pair of groups in Sp, each of which is the centralizer of the other. We call such a pair a *dual pair*. Let $\tilde{O}$, $\tilde{G}'$ be the inverse images of O and G' in $\tilde{\mathrm{S}}\mathrm{p}$. Then $(\tilde{O}, \tilde{G}')$ are a dual pair in $\tilde{\mathrm{S}}\mathrm{p}$. Hence if we restrict ω to $\tilde{O}$ or $\tilde{G}'$, each group gives us operators commuting with the other, and thus helps us to decompose the restriction of ω to the other group. It turns out that the situation in this regard is as nice as it could be **[Hw1]**.

THEOREM D. *The groups $\omega(\tilde{O})$ and $\omega(\tilde{G}')$ of operators generate each other's commutants, in the sense of von Neumann algebras. Thus*

$$\omega|\tilde{O} \times \tilde{G}' \simeq \int_{\hat{\tilde{G}}'} \sigma \otimes \sigma' \, d\omega(\sigma') \tag{43}$$

where $\hat{\tilde{G}}'$ is the unitary dual of $\tilde{G}'$, and $\sigma \in \hat{\tilde{O}}$, and the map $\sigma' \to \sigma$ is an injection almost everywhere with respect to the spectral measure $d\omega(\sigma')$.

We will consider this result in the cases $q = 0$ and $q = 1$. When $q = 0$ we discover Bochner's and Hecke's formulas. When $q = 1$ we discover Huyghens' principle. Before beginning the details we note two general properties of $\omega|\tilde{G}'$.

(1) From the general considerations about ω it is easy to see that if $p + q$ is even, then $\tilde{G}'$ is just Sl_2, but if $p + q$ is odd, then $\tilde{G}'$ is the nontrivial 2-fold cover of Sl_2, with the nontrivial element of the kernel of the covering map acting by minus the identity on $L^2(X)$.

(2) The generator of SO_2 (or its cover) in $\tilde{G}'$ is $(i/2)(2\pi\rho^2 - \Delta^2/2\pi)$ which is just an indefinite Hermite operator. Hence $e^{(\pi i/2)(2\pi\rho^2 - \Delta^2/2\pi)}$ is essentially just the Fourier transform "adapted to (,)" that is,

$$\wedge^{p,q}(f)(y) = \int_{\mathbf{R}^{p+q}} f(x) e^{-2\pi i (x,y)} \, dx. \tag{44}$$

In other words

$$\exp\left(\frac{\pi}{2} i \left(\frac{\pi\rho^2}{2} - \frac{\Delta^2}{4\pi}\right)\right) = i^{(p-q)/2} \wedge^{p,q} = \pm \omega \widetilde{\begin{pmatrix} 0 & 1 \\ -1 & 0 \end{pmatrix}}$$

$$\text{where} \begin{pmatrix} 0 & 1 \\ -1 & 0 \end{pmatrix} \in Sl_2 \simeq G'. \tag{45}$$

If $q = 0$, then $O_{p,0} = O_p$ is compact, and by general considerations **[L1]** we know we can write $L^2(\mathbf{R}^n) = \bigoplus_{j=0}^{\infty} H_j$ where the H_j are isotypic spaces for O_p. The duality theorem above then says that each H_j is irreducible as an $O_p \times \tilde{G}'$ module. Thus

$$H_j \simeq \sigma_j \otimes \sigma_j' \tag{46}$$

where σ_j is a finite-dimensional representation of O_p, and σ_j' is the corresponding representation of $\tilde{G}'$. As we have mentioned above, the generator of a maximal compact subgroup K of $\tilde{G}'$ is $(i/2)(2\pi\rho^2 - \Delta/2\pi)$. The Hermite operator $\rho^2 - \Delta/4\pi^2$ is obviously nonnegative. Thus $\omega|K$ will have only one-sided spectrum. The representation of $\tilde{G}'$ in which K has one-sided spectrum are called holomorphic representations and are very well understood **[L1]**. This is what they look like. Put

$$\mathfrak{k} = i(\pi\rho^2 - \Delta/4\pi)$$
$$n^+ = \frac{\tilde{E} - \pi\rho^2 - (\Delta/4\pi)}{2} = -\frac{1}{2}e^{\pi\rho^2}(\Delta/4\pi)e^{-\pi\rho^2}$$
$$n^- = \frac{\tilde{E} + \pi\rho^2 + (\Delta/4\pi)}{2} = \frac{1}{2}e^{-\pi\rho^2}(\Delta/4\pi)e^{\pi\rho^2}. \tag{47}$$

Then from (42) we find

$$[\mathfrak{k}, n^+] = 2in^+, \qquad [\mathfrak{k}, n^-] = -2in^-, \qquad [n^+, n^-] = -i\mathfrak{k}. \tag{48}$$

Any unitary representation of $\tilde{G}'$ in which $-i\mathfrak{k}$ has positive spectrum contains a unique vector v_0 such that $n^-v_0 = 0$ and $\mathfrak{k}v_0 = ilv_0/2$ for some positive integer l. The vector v_0 is called the lowest weight vector for the representation, and $l/2$ is the lowest weight. The space of the representation then has an orthogonal basis consisting of the vectors $v_m = n^{+^m}v_0$. These are eigenvectors for $\mathfrak{k}$ with eigenvalues $i(l/2 + 2m)$. Denote by $D_{l/2}$ the representation with lowest weight $l/2$.

From this description and (46) we see that the kernel of n^- will contain exactly one copy of each representation of O_n that occurs in $L^2(\mathbf{R}^n)$, and the irreducible O_n modules in ker n^- will just be the eigenspaces of $\mathfrak{k}$. From the second expression for n^- in (47) we see that ker n^- consists of functions of the form $\varphi e^{-\pi\rho^2}$ where $\Delta\varphi = 0$, i.e., φ is harmonic. Since φ is harmonic it is analytic and we may expand it in a power series

$$\varphi = \sum \varphi_j \tag{49}$$

where φ_j is a polynomial of degree j. Each φ_j will of course also be harmonic. From the formulas (47) we can easily compute that

$$\mathfrak{k}(\varphi e^{-\pi\rho^2}) = i(\tilde{E}\varphi)e^{-\pi\rho^2}. \tag{50}$$

Hence $\varphi e^{-\pi\rho^2}$ is an eigenvector for $\mathfrak{k}$ if and only if φ is a homogeneous polynomial. If φ has degree m, then

$$\mathfrak{k}(\varphi e^{-\pi\rho^2}) = i(m + p/2)\varphi e^{-\pi\rho^2}. \tag{51}$$

Let $\mathcal{H}_m$ be the harmonic polynomials of degree m. Since $(x_1 + ix_2)^m$ is harmonic, we see $\mathcal{H}_m \neq \{0\}$ if $p \geqslant 2$. If $p = 1$, then $\mathcal{H}_0 = \mathbf{C}$, $\mathcal{H}_1 = \mathbf{C}x$, $\mathcal{H}_i = 0$ for $i \geqslant 2$.

A computation shows that the space of functions $\varphi f(r)$ is invariant by n^+ if φ is homogeneous and harmonic. We summarize the consequences of the above computations in the light of Theorem D. We restrict to $p \geqslant 2$ for simplicity.

THEOREM. *For $p \geqslant 2$, we have $\mathcal{S}(\mathbf{R}^n) = \sum_{m=0}^{\infty} \mathcal{H}_m \otimes \mathcal{I}$ where $\mathcal{I} \subseteq \mathcal{S}$ are rotationally invariant functions and $\mathcal{H}_m$ is the space of harmonic polynomials of degree m. Each $\mathcal{H}_m$ is irreducible as O_p module. Each space $\mathcal{H}_m \otimes \mathcal{I}$ is* *irreducible under $\omega|O_p \times \tilde{G}'$, and as $O_p \times \tilde{G}'$ module has the form*

$$\mathcal{H}_m \otimes D_{m+p/2}. \tag{52}$$

Thus we recover the theory of spherical harmonics with extra structure. Furthermore, the formulas of Hecke and Bochner are now clear. Indeed Hecke's formula is immediate since if $\varphi \in \mathcal{H}_m$, then $\varphi e^{-\pi\rho^2}$ is an eigenvector for $\mathfrak{k}$ of eigenvalue $m + p/2$ and $e^{\pi i \mathfrak{k}/2} = i^p \wedge$, as we remarked at the outset. As for Bochner's formula, the curious thing about it is not its specific form–that is what it is–but that restriction of $\wedge$ to $\mathcal{H}_m \otimes \mathcal{G}$ depends only on $m + p/2$. From the viewpoint elaborated above, we can say that all the ingredients in Bochner's formula are defined in terms of $\tilde{G}'$ alone, and therefore the form of the operator will depend only on which $\tilde{G}'$ module is involved–and that, from (51) and (52) depends only on $m + p/2$. This fact in itself can be predicted from the internal structure of the representation theory of $\tilde{G}'$, and general considerations about ω, but we won't go into that.

We now turn to the Lorentz case $q = 1$. One can decompose $L^2(\mathbf{R}^{p+1})$ under $\omega|O_{p,1} \times \tilde{G}'$ explicitly as we did above when $q = 0$, and this has been done by Strichartz **[St]** and by Rallis and Schiffman **[RS]**. Here, however, we will discuss a more special topic, Huyghens' Principle **[Hw2]**.

The situation is this. One considers the wave equation

$$\sum_{j=1}^{p} \frac{\partial^2\varphi}{\partial x_j^2} = \frac{\partial^2\varphi}{\partial t^2}. \tag{53}$$

One wants to solve the initial value problem: given $\varphi(x, 0) = \varphi_0(x)$ and $\partial\varphi(x, 0)/\partial t = \varphi_0'(x)$, find φ solving (53). Standard techniques involving the Fourier transform show there exist distributions P_1 and P_2 on $\mathbf{R}^{p+1}$ such that

$$\varphi(x, t) = P_1 * \varphi_0 + P_2 * \varphi_0'$$

where the convolution is in the x-variables only. The distributions P_i are called propagators. They have been computed classically and the following results are known.

(a) $\varphi(x_1, t)$ depends only on the values of $\varphi_0(x)$ and $\varphi_0'(x)$ for $\|x - x_1\| \leq |t|$. In other words $P_j(\ , t)$ is supported in $\{x: \|x\| \leq |t|\}$. In other words P_j is supported inside the light cone, i.e., in the set $C^- = \{(x, t), \Sigma x_j^2 - t^2 \leq 0\}$.

(b) If p is odd and > 1, then $\varphi(x_1, t)$ depends only on the values of $\varphi_0(x)$ and $\varphi_0'(x)$ (and their derivatives) for $\|x - x_1\| = |t|$. In other words, $P_j(\ , t)$ is supported on $\{x: \|x\| = |t|\}$. In other words, P_j is supported on the light cone $C_0 = \{(x, t): \Sigma x_j^2 - t^2 = 0\}$.

The circumstances of (b), that signals are extinguished suddenly in odd (e.g., (3)) space dimensions is what is known as Huyghens' Principle today. We want to see how it comes out of the group theory.

We begin by observing that the wave equation (53) or more to the point, the initial value problem for the wave equation, has certain symmetries, and these symmetries are reflected in symmetries of P_1 and P_2. (Although the wave equation itself is invariant under the full Lorentz group $O_{n,1}$ as well as dilations of space-time, the initial value problem has somewhat less symmetry since one must also worry about preserving the initial conditions.) Investigating these symmetries, we find

(i) $\omega(R)P_j = P_j$ where $R \in O_n$ is a rotation in the x's.

(ii) $\omega(sI)P_1 = |s|^{(p+1)/2}P_1$, $\omega(sI)P_2 = |s|^{(p-1)/2}\operatorname{sgn}(s)\, P_2$

where sI is a dilation by s on all $\mathbf{R}^{p+1}$.

(iii) P_j satisfies the wave equation. (54)

Although (iii) might not usually be considered a symmetry of P_j, it is appropriate to so call it in our present context since the wave operator is just one of the infinitesimal generators of the $\tilde{G}'$ commuting with $O_{n,1}$.

The conditions (54) can be shown to characterize the P_j up to multiples. Indeed, let $\wedge_{p,1} = \wedge$ be the Fourier transform adapted to $\Sigma x_j^2 - t^2$ on $\mathbf{R}^{p+1}$. Then we see from (54) that

(i) $\omega(R)\hat{P}_j = \hat{P}_j$.

(ii) $\omega(sI)\hat{P}_1 = |s|^{-(p+1)/2}\hat{P}_1$, $\omega(sI)\hat{P}_2 = |s|^{-(p-1)/2}\operatorname{sgn}(s)P_2$.

(iii) $(r^2 - t^2)\hat{P}_j = 0$. (55)

Equation (55)(iii) says $\hat{P}_j$ is supported on the light cone C^0. (Indeed, away from zero, it must factor through the restriction map to the light cone.) Then since O_p and dilations together act transitively on C^0, we find, taking degrees of homogeneity into account, that

$$\hat{P}_1 = b_1 t^2 d\mu, \qquad \hat{P}_2 = b_2 t d\mu \tag{56}$$

where $d\mu$ is the unique $O_{p,1}$ invariant measure on $C^0 - \{0\}$.

Consider now the matter of Huyghens' Principle. Equations (54) say that P_j, aside from being O_n-invariant, is killed by the wave operator and is an eigenvector for the Euler operator E. Recalling formulas (42), we see this says P_j is a highest weight vector for the $\tilde{G}'$ module it generates. We have seen Huyghens' Principle holds exactly when P_j is supported on C^0. Since C^0 is just the locus of zeroes of $\Sigma x_j^2 - t^2$, we know P_j is supported on C^0 if and only if

$$\left(\Sigma x_k^2 - t^2\right)^l P_j = 0 \tag{57}$$

for some l. But in the light of (54)(ii) and (iii) this is equivalent to saying P_j generates a finite-dimensional module for $\tilde{G}'$. That is, we have the

CRITERION. *Huyghens' Principle holds if and only if P_j generates a finite-dimensional module for $\tilde{G}'$.*

As an immediate corollary we see that Huyghens' Principle must fail in odd space-time (even space) dimensions, because then, as we remarked above, $\tilde{G}'$ is not Sl_2 but the two-fold cover, which has no (faithful) finite-dimensional representations.

It is possible, however, for Huyghens' Principle to hold in even space-time dimensions. Taking Fourier transform, we see Huyghens' Principle will hold if and only if $\hat{P}_j$ generates a finite-dimensional $\tilde{G}'$ representation of which it will then be the lowest weight vector. This is guaranteed by the following result, which is a distributional partial analogue to the L^2-duality theorem.

THEOREM D*. *If* $p+1$ *is even, the distribution* $Q(x, t)d\mu$, *with* $d\mu$ *as in* (56) *and* Q *a harmonic polynomial of degree* m, *generates the* $\tilde{G}'$ *module* $V_{m+(p-1)/2}$ *of dimension* $m+(p-1)/2$. *The subspace of* $\mathcal{S}^*(\mathbf{R}^{p+1})$ *transforming according to* $V_{m+(p-1)/2}$ *under* $\tilde{G}'$ *is finite dimensional and irreducible for the joint action of* $O_{p,1}\times\tilde{G}'$ *and isomorphic to the tensor product*

$$\mathcal{H}_m \otimes V_{m+(p-1)/2}. \tag{58}$$

For $m = 1, 2$, we find the $\hat{P}_j$ generates a finite-dimensional $\tilde{G}'$ module, since t is harmonic, and

$$t^2 = \frac{p}{p-1}\left(t^2 - \frac{1}{p}\sum_{j=1}^{p} x_i^2\right) - \frac{1}{p-1}\left(t^2 - \sum_{j=1}^{p} x_i^2\right) \tag{59}$$

where the first term is harmonic, and the product of the second term with $d\mu$ is zero. Thus Huyghens' Principle is verified in this case. In fact, from formula (45) and the well-known structure **[L1]** of finite-dimensional Sl_2 modules, we can write

$$\begin{aligned} P_1 &= \beta_1\left(\sum \frac{\partial^2}{\partial x^2} - \frac{\partial^2}{\partial t^2}\right)^{(p+1)/2}(t^2\,d\mu) \\ P_2 &= \beta_2\left(\sum \frac{\partial^2}{\partial x^2} - \frac{\partial^2}{\partial t^2}\right)^{(p-1)/2}(t\,d\mu) \end{aligned} \tag{60}$$

where the β_i are constants. Thus the propagator is computed explicitly and is by inspection supported on C^0.

The wave equation is of course one very special partial differential equation. Over the past 20 or so years, a major theme in the theory of partial differential equations has been the development of more or less explicit operational calculi to treat various general classes of equations. These calculi are of two main types, generally referred to as classes of pseudo-differential operators, and of Fourier integral operators. Since these theories grew in part out of abelian Fourier analysis and rely heavily on the Fourier transform, with hindsight we may not find it surprising that they may also be developed using the Heisenberg group. In fact some aspects of pseudo-differential operators, especially symbols, fit very naturally into the group-theoretic framework. For a detailed development of the discussion below, see **[Hw 3]**; the ideas were also partially anticipated in **[GLS]**, and indeed some go back to Weyl **[Wy 1]** and his original construction of the Heisenberg group.

The formula (21) tells us, as already noted, that $\mathcal{S}(H, e)$ is taken by the representation ρ, in the Schrödinger model, to the algebra of all integral operators on $\mathcal{S}(\mathbf{R}^n)$ with kernels in $\mathcal{S}(\mathbf{R}^{2n})$. Since ρ is also unitary, we can extend it to a linear isomorphism, still denoted ρ, from $\mathcal{S}^*(H, e)$ to $\mathcal{S}^*(\mathbf{R}^{2n})$. The extended ρ will still be an algebra isomorphism to the extent this makes sense. The Schwartz Kernel Theorem [T] identifies the space $\mathcal{S}^*(\mathbf{R}^{2n})$ to the space of all continuous maps from $\mathcal{S}(\mathbf{R}^n)$ to $\mathcal{S}^*(\mathbf{R}^n)$. Hence *any* operator on $\mathcal{S}(\mathbf{R}^n)$, or even from $\mathcal{S}(\mathbf{R}^n)$ to $\mathcal{S}^*(\mathbf{R}^n)$ may be viewed as coming, via ρ, from a distribution on H. In particular, pseudo-differential operators on $\mathcal{S}(\mathbf{R}^n)$ will arise in this way. Let us consider how the symbol of such a pseudo-differential operator looks from this viewpoint.

Actually we will not discuss the most common symbol, that of Kohn-Nirenberg [**KN**], but a slightly different one, which happens to be the one written down by Weyl [**Wy 1**]. This symbol was also considered in [**GLS**], and used in the very general pseudo-differential calculus of [**Hr**]. We will call it the *isotropic symbol*. It differs in relatively minor ways from the Kohn-Nirenberg symbol, and explicit formulas relating the two can be found in [**Hw 3**] and [**GLS**].

The differences between the two symbols can be understood as follows. Functions in $\mathbb{S}(H, e)$ behave in a known way in the direction of the center **T** of H, so that to know f in $\mathbb{S}(H, e)$ one needs only know it on a cross-section C of the **T**-cosets in H. Thus by restriction one obtains an isomorphism from $\mathbb{S}(H, e)$ to some space of functions on C. By transport of structure one can define a convolution product on these functions, and one can map them to operators via ρ. If one coordinatizes C by means of the projection of H to $H/T \simeq \mathbf{R}^{2n}$, then the convolution product on $\mathbb{S}(H, e)$ transferred to C looks like convolution on $\mathbf{R}^{2n}$ "twisted" by the 2-cocycle defined by the cross-section C. For example, if one uses the cross-section $C_1 = \{(y, x, 1)\}$ as in formula (20), then the convolution (denoted by #) can be written

$$f_1 \# f_2(y', x') = \int_{\mathbf{R}^{2n}} f_1(x, y) f_2(x' - x, y' - y) e^{2\pi i x \cdot (y' - y)}\, dx\, dy. \tag{61}$$

On the other hand, if one uses the cross-section $C_0 = e(W)$, where $e\colon \mathfrak{H} \to H$ is the exponential map and W (and $\mathfrak{H}$) is as in (35), then one gets a slightly different convolution (denoted $\natural$) and writable as

$$f_1 \natural f_2(w') = \int_W f_1(w) f_2(w' - w) e^{-\pi i \langle w, w' \rangle}\, dw. \tag{62}$$

Here $\langle\ ,\ \rangle$ is the symplectic form on W (cf. formula (35) et infra). Other cross-sections give still different convolution product formulas.

The cross-sections C_1 and C_0 have some particular virtues which make them more useful than other cross-sections, and for most purposes they are the only cross-sections one need consider. (In the theory of θ-functions, another cross-section, not exactly of H, but of H extended by the real scalar operators, plays a role.) The cross-section C_1 is most closely related to the conventional manner of writing differential operators, with the differentiations coming first, followed by the multiplications. The cross-section C_0, which we call the *isotropic cross-section* treats differentiations and multiplications in symmetrical fashion. Its virtue is that it is invariant under the full symplectic group $\mathrm{Sp}(W, \langle\ ,\ \rangle) = \mathrm{Sp}$, and it is the only such cross-section. The cross-section C_1 is invariant only under the subgroup $\alpha(GL_n)$ of Sp (as defined in (32)).

With each cross-section we may associate a symbol map. The symbol associated to C_1 is the Kohn-Nirenberg symbol, which was built to conform with the traditional "derivatives first" manner of writing differential operators. The isotropic symbol which we will discuss below is the symbol associated to the isotropic section C_0, and so is more natural in the group-theoretical framework.

The isotropic symbol is defined via the symplectic Fourier transform on W. For $f \in \mathcal{S}(W)$ write

$$\widehat{f}(w') = 2^{-n} \int_W f(w) e^{-\pi i \langle w, w' \rangle} \, dw. \tag{63}$$

Our normalization of dw makes $\frown$ unitary. Since it is defined via a symplectic rather than, as is usual, a symmetric form, the map $\frown$ has order 2 rather than order 4; that is,

$$\widehat{\widehat{f}} = f. \tag{64}$$

Comparing formulas (62) and (63) we note the intriguing fact that $\frown$ amounts to convolution on the right by the constant function 2^{-n}. Furthermore, the function 2^{-n} on W turns out to correspond via ρ to the operator $\omega(-1)$, the image under the oscillator representation of minus the identity in Sp. Finally, we note that of course $\frown$ can be extended to $\mathcal{S}^*(W)$ by continuity. The extended $\frown$ is then the isotropic symbol, and will be denoted also by σ. Thus we have

$$\sigma(D) = \widehat{D} = D \natural 2^{-n}, \qquad \rho(\sigma(D)) = \rho(D)\omega(-1). \tag{65}$$

This symbol has, as it should, a symbolic calculus. To write it we require the usual notation. Set up coordinates in W via the basis of formulas (11) and (12). Let $\alpha = (\alpha_1, \ldots, \alpha_n)$ denote a multi-index of nonnegative integers. We associate various quantities with α:

$$|\alpha| = \sum_{i=1}^{n} |\alpha_i|, \qquad \alpha! = \prod_{i=1}^{n} \alpha_i!, \qquad \partial_x^\alpha = \prod_{i=1}^{n} \frac{\partial^{\alpha_i}}{\partial x_i^{\alpha_i}}. \tag{66}$$

From (65) we see that, given say two distributions D and E,

$$\sigma(D \natural E) = D \natural \sigma(E). \tag{67}$$

From (67) it is straightforward to compute that

$$\begin{aligned} \sigma(D \natural E) &= 2^n \sum_{\alpha,\beta} \left(\frac{i^{|\alpha|-|\beta|}}{\pi^{|\alpha|+|\beta|}} \alpha! \beta! \right) \left(\partial_x^\alpha \partial_y^\beta \sigma(D) \right) \left(\partial_x^\alpha \partial_y^\beta \sigma(E) \right) \\ &= 2^n \left(\sigma(D)\sigma(E) + \frac{i}{\pi} \{ \sigma(D), \sigma(E) \} + \cdots \right). \end{aligned} \tag{68}$$

Ignoring factors of 2^n, π, etc., which are essentially artifacts of our choices of normalization, we see that the leading term of $\sigma(D \natural E)$ is just the pointwise product of $\sigma(D)$ and $\sigma(E)$, as it should be. The second term is the Poisson bracket of $\sigma(D)$ and $\sigma(E)$, and the higher terms are transvectants, certain classical differential invariants of the symplectic group.

An important use of the symbol is to estimate the size of the corresponding operator. A work horse estimate **[Hr]**, **[B]** which can be proved very naturally in this group-theoretical context is the Calderón-Vaillancourt (0, 0) estimate **[CV]**.

THEOREM (CALDERÓN-VAILLANCOURT). *If $\sigma(D)$ and all its derivatives are bounded, then $\rho(D)$ is a bounded operator, with operator norm estimable in terms of the supremum norms on the derivatives.*

This is actually a consequence of a general estimate for representations of locally compact groups, an estimate simple enough to derive. Let G be a locally compact group, and σ a representation of G on a Hilbert space $\mathcal{H}$. Let $\mathcal{H}.\mathcal{S}.$ be the Hilbert-Schmidt operators on $\mathcal{H}$. On $\mathcal{H}.\mathcal{S}.$ is defined the unitary representation $\sigma \otimes \sigma^*$ of G by the rule

$$(\sigma \otimes \sigma^*)(g)(T) = \sigma(g)T\sigma(g)^{-1}, \qquad T \in \mathcal{H}.\mathcal{S}., g \in G. \tag{69}$$

Let u, v be vectors in $\mathcal{H}$, and define the dyad $E_{u,v}$ in $\mathcal{H}.\mathcal{S}.$ by

$$E_{u,v}(w) = (w, v)u, \qquad w \in \mathcal{H} \tag{70}$$

where $(\ ,\ )$ is the inner product in $\mathcal{H}$.

One computes easily that

$$E_{u,v}E_{u',v'} = (u', v)E_{u,v'}$$
$$\sigma(g)E_{u,v} = E_{\sigma(g)u,v}, \qquad E_{u,v}\sigma(g)^{-1} = E_{u,\sigma(g)v}. \tag{71}$$

If v is a unit vector, then the map $\alpha\colon \mathcal{H} \to \mathcal{H}.\mathcal{S}.$ defined by

$$\alpha(u) = E_{u,v} \tag{72}$$

is an isometric embedding, and from (71) we see that

$$\alpha(\sigma(g)u) = \sigma(g)\alpha(u) \qquad \alpha(u)\sigma(g)^{-1}E_{v,v} = \varphi(g)\alpha(u) \tag{73}$$

where

$$\varphi(g) = (v, \sigma(g)v) \tag{74}$$

is the matrix coefficient of σ defined by v. Combining the parts of (73) gives, assuming $\varphi(g) \neq 0$,

$$\alpha(\sigma(g)(u)) = \varphi(g)^{-1}(\sigma \otimes \sigma^*)(\alpha(u))E_{v,v}. \tag{75}$$

Thus if $f \in L^1(G)$ is supported away from the zero set of φ, we may integrate both sides of (75) to obtain

$$\alpha\left(\left(\int_G f(g)\sigma(g)\,dg\right)(u)\right) = \left(\left(\int_G f(g)\varphi(g)^{-1}(\sigma \otimes \sigma^*)(g)\,dg\right)(\alpha(u))\right)E_{v,v} \tag{76}$$

or more briefly

$$\alpha(\sigma(f)(u)) = (\sigma \otimes \sigma^*)(f\varphi^{-1})(\alpha(u))E_{v,v}. \tag{77}$$

Since α is an isometry and right multiplication by $E_{v,v}$ on $\mathcal{H}.\mathcal{S}.$ has norm one, from (77) we get the estimate

$$\|\sigma(f)\| < \|\sigma \otimes \sigma^*(f\varphi^{-1})\| \tag{78}$$

where $\|A\|$ denotes the operator norm of A.

REMARK. Formula (77) reminds one of the theory of Toeplitz operators **[D]**, **[V]**. One might say it exhibits $\sigma(f)$ as a Toeplitz operator constructed from $\sigma \otimes \sigma^*(f\varphi^{-1})$.

Consider now the case when $G = H$, the Heisenberg group, and $\sigma = \rho$, the standard irreducible representation of H. Since the center $\mathbf{T}$ of H is sent by ρ to scalar operators, the representation $\rho \otimes \rho^*$ of H will in fact factor to the abelian group $H/\mathbf{T}$. Since representations of abelian groups are completely analyzable by means of the Fourier transform, and since the isotropic symbol amounts essentially to Fourier transform, it is perhaps plausible that (78) for H and ρ would imply a result like the (0, 0) estimate. At any rate, it does.

Although the isotropic symbol does what it should and its definition (65) is clean, it may still seem less than completely natural. As a final point, we will attempt to further motivate (65) by showing it is consistent with the Kirillov orbit theory for nilpotent groups, as specialized to H. Such a motivation is entirely unhistorical, but it may suggest possibilities for future research.

Let $\tilde{H}$ be the universal covering group of H. The group $\tilde{H}$ has the same Lie algebra $\mathfrak{H}$ as does H, and the exponential map

$$\tilde{e}: \mathfrak{H} \to \tilde{H} \tag{79}$$

is now a diffeomorphism. Let $\mathfrak{Z}$ denote the center of $\mathfrak{H}$ and $Z = \exp \mathfrak{Z}$ the center of $\tilde{H}$. Above in equation (35) we have identified $\mathfrak{Z}$ with $\mathbf{R}$. We will identify Z to $\mathbf{R}$ also in such a way that exp: $\mathfrak{Z} \to Z$ is the identity map on $\mathbf{R}$. Then we may write

$$\tilde{H} = \{(\exp w, t): w \in W, t \in \mathbf{R}\} \tag{80}$$

with W also as in (35). The group law in $\tilde{H}$ then reads

$$(\exp w_1, t_1)(\exp w_2, t_2) = \left(\exp(w_1 + w_2), t_1 + t_2 + \tfrac{1}{2}\langle w_1, w_2\rangle\right). \tag{81}$$

According to Kirillov [**Kr**], the irreducible unitary representations of $\tilde{H}$ are in natural bijection with the orbits of $\tilde{H}$ (or H, since the center acts trivially) in $\mathfrak{H}^*$, the dual of $\mathfrak{H}$, under the co-adjoint action. In accordance with the Stone-von Neumann Theorem, the orbits which are not single points are hyperplanes of the form $\lambda + \mathfrak{Z}^\perp$, where $\mathfrak{Z}^\perp$ is the annihilator in $\mathfrak{H}^*$ of $\mathfrak{Z}$ and λ is some element in the orbit. An irreducible representation τ of $\tilde{H}$ determines a central character χ_τ on Z, and is determined by χ_τ if this is nontrivial. If τ corresponds to the orbit $\lambda + \mathfrak{Z}^\perp$, then

$$e^{2\pi i\lambda(z)} = \chi_\tau(\exp z), \qquad z \in \mathfrak{Z}. \tag{82}$$

There will be a unique point λ_0 in $(\lambda + \mathfrak{Z}^\perp) \cap W^\perp$. Thus having a cross-section W to $\mathfrak{Z}$ is equivalent to having a base point λ_0 in $\lambda + \mathfrak{Z}^\perp = \lambda_0 + \mathfrak{Z}^\perp$. From now on we will take $\lambda = \lambda_0$. Note that $\mathfrak{Z}^\perp$ is naturally dual to W. On the other hand W is self-dual by virtue of the symplectic form $\lambda([\ ,\ ])$ defined on it. Therefore we have an isomorphism

$$\beta: W \to \mathfrak{Z}^\perp$$

defined by

$$\beta(w')(w) = \lambda([w', w]). \tag{83}$$

We can define in a natural manner a Fourier transform ^ from $\mathcal{S}(\mathfrak{H})$ to $\mathcal{S}(\mathfrak{H}^*)$ by

$$\hat{f}(\mu) = \int_{\mathfrak{H}} f(x) e^{2\pi i\mu(x)}\, dx, \qquad f \in \mathcal{S}(\mathfrak{H}), \mu \in \mathfrak{H}^*. \tag{84}$$

Here $dx = dwdz$ where dw is self-dual Haar measure on W and dz is Lebesgue measure on $\mathfrak{Z} \simeq \mathbf{R}$. We may extend (84) to tempered distributions by continuity.

If the representation τ corresponds to the plane $\lambda_0 + \mathfrak{Z}^{\perp}$ in $\mathfrak{H}^*$, then the analogue for $\tilde{H}$, with its noncompact center, of the space $\mathbb{S}(H, e)$ is the space $\mathbb{S}(\tilde{H}, \tau)$ of functions f on $\tilde{H}$ such that

$$L_{\exp z}(f) = e^{2\pi i \lambda_0(z)} f, \quad z \in \mathfrak{Z}, (f \circ \exp)|W \in \mathbb{S}(W). \tag{85}$$

If, under the identification of (35) of $\mathfrak{Z}$ with $\mathbf{R}$, we have $\lambda_0(z) = z$, then τ is simply the lift to $\tilde{H}$ of the representation ρ of H, and the space $\mathbb{S}(\tilde{H}, \tau)$ is simply the pullback to $\tilde{H}$ of $\mathbb{S}(H, e)$. Fix this τ. Then for f in $\mathbb{S}(\tilde{H}, \tau)$ we may speak of the isotropic symbol $\sigma(f)$ as a function on W. On the other hand f may be regarded as a distribution on H, and its pullback to $\mathfrak{H}$ via exp is a distribution on $\mathfrak{H}$. Thus we may consider

$$(f \circ \exp)^{\wedge}$$

which will be a distribution on $\mathfrak{H}^*$, and will have the form

$$(f \circ \exp)^{\wedge} = \tilde{f}\, d\nu \tag{86}$$

where $\tilde{f}$ is a smooth function on $\lambda_0 + \mathfrak{Z}^{\perp}$ and $d\nu$ is an appropriately normalized Lebesgue measure on $\lambda_0 + \mathfrak{Z}^{\perp}$. Comparing (84) and (86) with the definition (63) of the isotropic symbol, and using (83) we see that

$$\sigma(f)(w) = 2^{-n}\tilde{f}(\lambda_0 + \beta(w/2)). \tag{87}$$

Again, the factors of 2 here are essentially irrelevant.

Formula (87) tells us that the symbol as it is understood in P.D.E. is consistent with the map

$$f \to (f \circ \exp)^{\wedge} \tag{88}$$

from $\mathbb{S}(H)$ to $\mathbb{S}(\mathfrak{H}^*)$. But the map (88) has been a basic part of harmonic analysis on nilpotent groups since Kirillov's original paper **[Kr]** where it was used to construct the characters of irreducible representations. This suggests that the map (88) might be usefully construed as a symbol map for a general nilpotent group. It might be used, for example, in deriving norm estimates for the images of functions under representations of the group, including the regular representation. There has developed over the past few years a theory of "singular integral operators" on nilpotent groups with dilations, but this theory has used nilpotent harmonic analysis in a minimal way. The motivating example for this theory was again provided by the Heisenberg group, and appeared in a paper of Knapp and Stein **[KS]**. Because of the dilation-invariance of the Knapp-Stein operators, formula (87) implies that the Knapp-Stein result follows from the Calderón-Vaillancourt (0, 0) estimate. On the other hand, it is a simple matter to dilate the Calderón-Vaillancourt (C-V) result. We will do this explicitly because it further illuminates the group-theoretic basis of this estimate and because it suggests generalization.

Since H acts on $\mathfrak{H}^*$ via the co-adjoint action Ad*, by applying Ad* to $\mathfrak{H}$ we have a Lie algebra of vector fields on $\mathfrak{H}^*$. Let $\{X_j, Y_j, T\}$ be the basis for $\mathfrak{H}$ defined in (11), and let $\{X_j^*, Y_j^*, T^*\}$ be the dual basis for $\mathfrak{H}^*$. Let x_j^*, y_j^*, z^* be the coordinates on $\mathfrak{H}^*$ with respect to this dual basis. Then

$$\mathrm{Ad}^*(X_j) = z^*\frac{\partial}{\partial y_j^*}, \qquad \mathrm{Ad}^*(Y_j) = -z^*\frac{\partial}{\partial x_j^*}, \qquad \mathrm{Ad}^*(T) = 0. \quad (89)$$

Thus, restricted to any hyperplane $\lambda_0 + \mathfrak{Z}^\perp$, the vector fields $\mathrm{Ad}^*(W)$ are just the constant coefficient vector fields, i.e., just give first order partial derivatives, and it is in terms of the symbol and its derivatives that the C-V estimates is couched. Thus the formula (89) together with an inspection of how the constants in C-V depend on Planck's constant allow one to derive results like this.

THEOREM. *Suppose $D \in \mathcal{S}^*(\tilde{H})$ is such that $(D \circ \exp)\hat{}$ is a smooth function and the functions obtained by applying any product of the operators $z^{*-1/2}\mathrm{Ad}^*(w)$, $w \in W$, are bounded. Then for any unitary representation τ of $\tilde{H}$, the operator $\tau(D)$ is bounded. The bound is independent of τ and depends only on the ∞-norms of the functions described above. In particular, left convolution by τ is bounded.*

It is not difficult to extend a result like this to a general 2-step nilpotent group. Then formula (78) would permit use of the 2-step result to get estimates for *irreducible* representations of 3-step nilpotent groups. Thus one might envisage a bootstrap argument yielding analogues of C-V for general nilpotent groups.

To conclude, this paper has tried to illustrate how it can be helpful and pleasing to consider various basic or classical topics in harmonic analysis in terms of the Heisenberg group. It does not begin to exhaust the list of possible topics, even in this relatively restricted domain of discourse. The topics chosen reflect the author's predilections and limitations more than the bounds of known applicability of the Heisenberg group to analysis. In fact, the article of Stein which framed the first part of this article ends with a discussion of some applications of the Heisenberg group, and Stein's work has often involved H. And it should also be clear that the perspective propagandized here has many possibilities for future development.

I would like to thank the University of Maryland and Ron Lipsman, whose invitation to deliver lectures for their Special Year in Harmonic Analysis provided the occasion for writing a preliminary version of this article.

REFERENCES

[B] R. Beals, *A general calculus of pseudodifferential operators*, Duke Math. J. **42** (1975), 1–42.

[CV] A. Calderón and R. Vaillancourt, *On the boundedness of pseudo-differential operators*, J. Math. Soc. Japan **23** (1971), 374–378.

[Cr] P. Cartier, *Quantum mechanical commutation relations and theta functions*, Proc. Sympos. Pure Math., vol. 9, Amer. Math., Soc., Providence, R. I., 1966, pp. 361–383.

[D] R. G. Douglas, *Banach algebra techniques in operator theory*, Academic Press, New York, 1972.

[GLS] A. Grossman, G. Loupias and E. Stein, *An algebra of pseudo-differential operators and quantum mechanics in phase space*, Ann. Inst. Fourier (Grenoble) **18** (1968), 343–368.

[Hr] L. Hörmander, *The Weyl calculus of pseudodifferential operators*, Comm. Pure Appl. Math. **32** (1979), 359–443.

[Hw1] R. Howe, *On some results of Strichartz and of Rallis and Schiffmann*, J. Functional Anal. **32** (1979), 297–303.

[Hw2] ______, *Remarks on Huyghens' principle* (preprint).

[Hw3] ______, *Quantum mechanics and partial differential equations*, J. Functional Anal. (to appear).

[Kr] A. A. Kirillov, *Unitary representations of nilpotent Lie groups*, Uspehi Mat. Nauk. **17** (1962), 57–110; Russian Math. Surveys **17** (1962), 53–104.

[KS] A. Knapp and E. Stein, *Intertwining operators for semisimple groups*, Ann. of Math. (2) **93** (1971), 489–578.

[KN] J. Kohn and L. Nirenberg, *An algebra of pseudodifferential operators*, Comm. Pure Appl. Math. **18** (1965), 269–305.

[L1] S. Lang, $SL_2(R)$, Addison-Wesley, Reading, Mass., 1975.

[L2] ______, *Real analysis*, Addison-Wesley, Reading, Mass., 1969.

[Ma] G. Mackey, *Unitary representations of group extensions*. I, Acta Math. **99** (1958), 265–311.

[MW] C. Moore and J. Wolf, *Square integrable representations of nilpotent groups*, Trans. Amer. Math. Soc. **185** (1973), 445–462.

[Mm] D. Mumford, *On the equations defining abelian varieties*. I, II, III, Invent. Math. **2** and **3**, (1966) and (1967).

[RS] S. Rallis and G. Schiffmann, *Weil representation*. I. *Intertwining distributions and discrete spectrum*, Mem. Amer. Math. Soc. No. 231 (1980).

[Sg] I. Segal, *Transforms for operators and symplectic automorphisms over a locally compact abelian group*, Math. Scand. **13** (1963), 31–43.

[Sh] D. Shale, *Linear symmetries of free boson fields*, Trans. Amer. Math. Soc. **103** (1962), 149–167.

[St] R. Strichartz, *Fourier transforms and non-compact rotation groups*, Indiana Univ. Math. J. **24** (1974), 499–526.

[T] F. Treves, *Topological vector spaces, distributions and kernels*, Academic Press, New York, 1967.

[V] U. Venugopalkrishna, *Fredholm operators associated with strongly pseudoconvex domains in* $\mathbf{C}^n$, J. Functional Anal. **9** (1972), 349–372.

[vN]. J. von Neumann, *Die eindeutigkeit der Schröderschen Operatoren*, Math. Ann. **104** (1931), 570–578.

[Wi] A. Weil, *Sur certains groupes d'operateurs unitaires*, Acta. Math. **111** (1964), 145–211.

[Wy1] H. Weyl, *The theory of groups and quantum mechanics*, Methuen, London, 1931.

[Wy2] ______, *The classical groups*, Princeton Univ. Press, Princeton, N. J., 1939.

DEPARTMENT OF MATHEMATICS, YALE UNIVERSITY, NEW HAVEN, CONNECTICUT 06520

BULLETIN (New Series) OF THE
AMERICAN MATHEMATICAL SOCIETY
Volume 6, Number 3, May 1982

ERGODIC THEORY, GROUP REPRESENTATIONS, AND RIGIDITY

BY ROBERT J. ZIMMER[1]

CONTENTS

These notes represent a mildly expanded version of lectures delivered at the C.I.M.E. summer session on harmonic analysis and group representations in Cortona, Italy, June—July 1980. The author would like to express his thanks and appreciation to the organizers of the conference, Michael Cowling, Sandro Figà-Talamanca, and Massimo Picardello, for inviting him to deliver these lectures and for their most warm and generous hospitality during his stay in Italy. We would also like to thank the other participants of the conference for their interest in these lectures. Finally, we would like to thank Terese S. Zimmer for (among innumerable other things that we need not go into here) helping with the translation of [**29**].

ADDED IN PROOF. Complete and detailed proofs of all the results discussed in these lectures (except those in section 4), as well as further developments, will appear in a forthcoming monograph of the author.

1. Basic notions. In these lectures we discuss some topics concerning the relationship of ergodic theory, representation theory, and the structure of Lie groups and their discrete subgroups.

In studying the representation theory of groups, the assumption of compactness on the group essentially allows one to reduce to a finite dimensional situation, in which case one often can obtain complete information. For noncompact groups, of course, no such reduction is possible and the situation is much more complex. When studying general actions of groups, a somewhat

Received by the editors November 15, 1981.

1980 *Mathematics Subject Classification*. Primary 22D10, 22D40, 22E40, 28D15, 53C35, 57R30, 57S20.

[1]Partially supported by a Sloan Foundation Fellowship and NSF Grant MCS 79-05036.

0273-0979/81/0000-0334/$09.50

similar situation arises. In the compact case every orbit will be closed, the space of orbits will have a reasonable structure, and one can often find nice (with respect to the action) neighborhoods of orbits. A large amount of information about actions of finite and compact groups has been obtained by topological methods. However, once again, if the compactness assumption on the group is dropped, one faces many additional problems. In particular, one can have orbits which are dense (for example, the irrational flow on the torus) and the orbit space may be so badly behaved as to have no continuous functions but constants. Furthermore, moving from a point to a nearby point may produce an orbit which does not follow closely to the original orbit. If one wishes to deal with actions in the noncompact case, this phenomenon of complicated orbit structure must be faced. For many actions, e.g., differentiable actions on manifolds, there are natural measures that behave well with respect to the action. A significant part of ergodic theory is the study of group actions on measure spaces. In particular, ergodic theory aims to understand the phenomenon of bad orbit structure in the presence of a measure.

Throughout these lectures, G will be a locally compact, second countable group. Let (S, μ) be a standard measure space, and assume we have an action $S \times G \to S$ which is a Borel function. Then μ (which is always assumed to be σ-finite) is invariant if $\mu(Ag) = \mu(A)$ for all $A \subset S$ and $g \in G$, and quasi-invariant if $\mu(Ag) = 0$ if and only if $\mu(A) = 0$.

DEFINITION 1.1. The action is called ergodic if $A \subset S$ is G-invariant implies $\mu(A) = 0$ or $\mu(S - A) = 0$.

Clearly any transitive action is ergodic, or, more generally, any essentially transitive action (i.e., transitive on the complement of a null set). We can then write $S = G/G_0$ where $G_0 \subset G$ is a closed subgroup. An ergodic action that is not essentially transitive will be called properly ergodic.

EXAMPLE 1.2. Let $S = \{z \in \mathbf{C} \mid |z| = 1\}$ and $T: S \to S$ be $T(z) = e^{i\alpha}z$ where $\alpha/2\pi$ is irrational. Then T generates a Z-action. If $A \subset S$ is invariant, let $\chi_A(z) = \Sigma a_n z^n$ be the L^2-Fourier expansion of its characteristic function. Then by invariance $\chi_A(z) = \chi_A(e^{i\alpha}z) = \Sigma a_n e^{in\alpha} z^n$. Thus $a_n e^{in\alpha} = a_n$ and so $a_n = 0$ for $n \neq 0$. This implies χ_A is constant, so the action is properly ergodic.

REMARK. If S is a (second countable) topological space and μ is positive on open sets, then proper ergodicity implies almost every orbit is a dense null set. This is one sense in which proper ergodicity is a reflection of complicated orbits. Another is the following.

PROPOSITION 1.3 [**12**]. *Let G act continuously on S where S is metrizable by a complete separable metric. Then the following are equivalent. (We say the action is "smooth" if they hold.)*

(i) *Every G-orbit is locally closed.*

(ii) *S/G is T_0 in the quotient topology.*

(iii) *The quotient Borel structure on S/G is countably separated and generated (i.e., there is a countable family $\{A_i\}$ separating points and generating the Borel structure).*

(iv) *Every quasi-invariant ergodic measure is supported on an orbit.*

PROOF. (i) $\Rightarrow$ (ii) $\Rightarrow$ (iii) are elementary. To see (iii) $\Rightarrow$ (iv), let $p: S \to S/G$ be the projection, and μ an ergodic probability measure on S. Then $\nu = p_*(\mu)$ is a measure on S/G with the property that for any Borel set $B \subset S/G$, $\nu(B) = 0$ or 1. Since S/G is countably separated and generated, ν is supported on a point, so μ is supported on an orbit. The implication (iv) $\Rightarrow$ (i) is difficult (and we will not be using it).

We will be making constant use of the implication (i) $\Rightarrow$ (iv). For example,

COROLLARY 1.4. *Every ergodic action of a compact group is essentially transitive.*

If the action is on a metric space, this follows immediately. However, a theorem of Varadarajan [**45**] implies that any action can be so realized.

COROLLARY 1.5. *Every ergodic algebraic action of a real (or p-adic) algebraic group (more precisely, the real or p-adic points) on an algebraic variety is essentially transitive.*

This follows from the theorem of Borel and Borel-Serre that orbits are locally closed [**3**, **6**].

While the decomposition of a general action into orbits may not be satisfactory there is always a good decomposition into ergodic components.

PROPOSITION 1.6. *Let (S, μ) be a G-space. Then there is a standard measure space (E, ν), a conull G-invariant set $S_0 \subset S$, and a G-invariant Borel map φ: $S \to E$ with $\varphi_*(\mu) = \nu$ such that, writing $\mu = \int^{\oplus} \mu_y \, d\nu(y)$ where μ_y is supported on $\varphi^{-1}(y)$, we have μ_y is quasi-invariant and ergodic under G for almost all y.*

(E, ν) is called the space of ergodic components of the action (and is essentially uniquely determined by the above conditions).

We now discuss some notions of "isomorphism".

DEFINITION 1.7. Let (S, μ), (S', μ') be ergodic G-spaces. Call them conjugate if modulo null sets there is φ: $S \to S'$ with

(i) φ a bijective Borel isomorphism.

(ii) $\varphi_*(\mu) \sim \mu'$ (i.e., same null sets).

(iii) $\varphi(sg) = \varphi(s)g$.

If $A \in \mathrm{Aut}(G)$ and S is a G-space, we have a new G-action on S by defining $s \circ g = s \cdot A(g)$.

DEFINITION 1.8. Call two actions automorphically conjugate if they become conjugate when modified by some automorphism.

An a priori much weaker notion is simply to ask for the orbit pictures to be the same. Here, we can compare actions of different groups.

DEFINITION 1.9. Suppose (S, μ) is a G-space, (S', μ') a G'-space. Call the actions orbit equivalent if (modulo null sets) there exists φ: $S \to S'$ with

(i) φ a bijective Borel isomorphism.

(ii) $\varphi_*(\mu) \sim \mu'$.

(iii) $\varphi(G\text{-orbit}) = G'\text{-orbit}$.

If φ: $(X, \mu) \to (Y, \nu)$ is a measure class preserving G-map of G-spaces we call X an extension of Y or Y a factor of X. Observe that we automatically have $\nu(Y - \varphi(X)) = 0$. If $H \subset G$ is a subgroup, and X is an ergodic G-space, we can restrict to obtain an action of H, which of course no longer need be ergodic. In the other direction, we can induce. Namely, suppose S is an ergodic H-space and $H \subset G$ is a closed subgroup. Then we obtain a naturally associated G-space as follows. Let H act on $S \times G$ by $(s, g)h = (sh, gh)$ and let $X = (S \times G)/H$. Then G acts on $S \times G$ by $(s, g)\tilde{g} = (s, \tilde{g}^{-1}g)$, and this action commutes with the H-action. Hence there is an induced action of G on X which will be ergodic with its natural measure class.

DEFINITION 1.10. X is called the ergodic G-space induced from the G-action, and we denote it by $\operatorname{ind}_H^G(S)$.

For example, if $H = Z$, $G = \mathbf{R}$, then X can be identified with $(S \times [0, 1])/\sim$ where $\sim$ identifies $(s, 1)$ with $(Ts, 0)$. Under the induced $\mathbf{R}$-action a point simply flows up along the line it is in with unit speed.

Given an ergodic G-space X, it is useful to know when it is induced from an action of a subgroup. The following is helpful in this regard.

PROPOSITION 1.11 [**52**]. *If X is an ergodic G-space and $H \subset G$ is a closed subgroup, then $X = \operatorname{ind}_H^G(S)$ for some H-space S if and only if G/H is a factor of X, i.e., there is a measure class preserving G-map $X \to G/H$.*

If X is a G-space, is there a unique (up to conjugacy) smallest closed subgroup from which it is induced? The answer in general is no, but we have the following

PROPOSITION 1.12 [**54**]. *Suppose G is (the real points, or k-points, k a p-adic field) of an algebraic group and X an ergodic G-space. Then there is a unique conjugacy class of algebraic subgroups such that $X = \operatorname{ind}_H^G(S)$ for H algebraic (and some S), if and only if H contains a member of this conjugacy class.*

DEFINITION 1.13 [**54**]. If H is in this class, call H the algebraic hull of the action. If this is all of G, call the action Zariski dense.

If $X = G/G_0$, then the algebraic hull is just usual algebraic hull of the group G_0.

PROOF OF 1.12. There exist minimal such groups from the descending chain condition on algebraic subgroups. Suppose $H_1, H_2 \subset G$ are two such minimal algebraic groups. We have $\varphi_i\colon X \to G/H_i$. Let $\varphi = (\varphi_1, \varphi_2)\colon X \to G/H_1 \times G/H_2$. Then $\varphi_*(\mu)$ is an ergodic quasi-invariant measure on $G/H_1 \times G/H_2$. But the G-action on this product is algebraic, so $\varphi_*(\mu)$ is supported on an orbit. But as a G-space, an orbit is $G/(g_1H_1g_1^{-1} \cap g_2H_2g_2^{-1})$. By minimality assumptions, H_1 and H_2 are conjugate.

THEOREM 1.14 (BOREL DENSITY THEOREM [**4**]). *If G is a connected semisimple real algebraic group with no compact factors, and S an ergodic G-space with finite invariant measure, then S is Zariski dense in G.*

As an example of an ergodic action of such a group, we point out the following example. (One can show there are uncountably many inequivalent actions of such groups [with finite invariant measure].)

EXAMPLE 1.15. Let $SL(n, \mathbf{Z})$ act on $\mathbf{R}^n/\mathbf{Z}^n$ by automorphisms. This is ergodic. The induced $SL(n, \mathbf{R})$ action will be properly ergodic, essentially free (i.e. almost all stabilizers trivial), and have finite invariant measure.

2. Ergodicity theorems. A natural class of actions that arises in a variety of situations are actions on homogeneous spaces. Thus, if $H_1, H_2 \subset G$ are subgroups with H_2 closed, H_1 acts on G/H_2 and the question arises as to when this is ergodic. This is a special case of the following question. Suppose S is an ergodic G-space and $H \subset G$ is a subgroup. When will the restriction to H still be ergodic? In the special case in which S has a finite invariant measure, results about unitary representations can be directly applied. Namely, let $(U_g f)(s) = f(sg)$, where $f \in L^2(S)$. This defines a unitary representation of G on $L^2(S)$ and G is ergodic on S (assuming finite invariant measure) if and only if there are no nonzero invariant vectors in $L^2(S) \ominus C$. Thus, to settle the question about ergodicity of restrictions in this case, we have a representation U of G with no invariant vectors and we ask whether or not $U|H$ has invariant vectors. Let us consider some of the classical examples, when G is transitive on S.

EXAMPLE 2.1. Suppose G is compact, $S = G$. Then $H \subset G$ is ergodic on S if and only if H is dense. This includes Example 1.2.

Now let N be a simply connected nilpotent Lie group, $\Gamma \subset N$ a lattice (i.e., Γ is discrete and N/Γ has finite invariant measure). For example,

$$N = \left\{ \begin{pmatrix} 1 & x & z \\ 0 & 1 & y \\ 0 & 0 & 1 \end{pmatrix} \,\middle|\, x, y, z \in \mathbf{R} \right\}$$

and $\Gamma = N_Z$, the subgroup with $x, y, z \in Z$. Then $[N, N] = \{A \in N \mid x = y = 0\}$, and $N/[N, N]\Gamma$ is a torus. The map $N/\Gamma \to N/[N, N]\Gamma$ exhibits the 3-manifold N/Γ as a circle bundle over the torus. In general $N/\Gamma \to N/[N, N]\Gamma$ will be a bundle over the torus with fiber $[N, N]/[N, N] \cap \Gamma$.

THEOREM 2.2 (L. GREEN [**1**]). *$H \subset N$ is ergodic on N/Γ if and only if it is ergodic on $N/[N, N]\Gamma$.*

As the latter is a torus, ergodicity can be determined as in Example 2.1. The proof of this depends on writing down the representation of N which appear in $L^2(N/\Gamma)$ and examining them with respect to restriction to subgroups. See [**1**] for details.

Results for 1-parameter subgroups acting on compact homogeneous spaces of solvable Lie groups have been obtained by Auslander [**2**] and Brezin and Moore [**7**].

If $G = SL(2, \mathbf{R})$, $\Gamma \subset G$ is a lattice in G and $H \subset G$ is the group of positive diagonal matrices, then G/Γ is in a natural way the unit tangent bundle of the finite volume negatively curved manifold D/Γ where $D = SO(2, \mathbf{R}) \backslash G$ is the Poincaré disk, and H is the geodesic flow. Thus, a classical result of Hedlund [**20**] and Hopf [**21**] says that H is ergodic on G/Γ. C. C. Moore generalized this to allow G to be a very general semisimple Lie group, and H to be an aribtrary subgroup.

THEOREM 2.3 (C. C. MOORE [**32**]). *Let $G = \prod G_i$ where G_i is a noncompact connected simple Lie group with finite center and let $\Gamma \subset G$ be an irreducible latice. Then $H \subset G$ is ergodic on G/Γ if and only if $\overline{H}$ is not compact.*

This theorem was proved by showing the following general result about arbitrary representations (not necessarily one appearing in $L^2(G/\Gamma)$). Let G be a noncompact connected simple Lie group with finite center and π a unitary representation of G with no nonzero invariant vectors. Then for any vector $x \neq 0$, $\{g \in G \mid \pi(g)x = x\}$ is compact. This result easily implies the theorem. A stronger result about such representations that we will need has subsequently come to light.

THEOREM 2.4. *If π is any unitary representation of a connected noncompact simple Lie group with finite center, then the matrix coefficients $f(g) = \langle \pi(g)x \mid y \rangle \to 0$ as $g \to \infty$, assuming there are no $\pi(G)$-invariant vectors.*

A nice proof of this appears in a paper of Howe and Moore [**22**] although the basic idea is present in the work of Sherman [**43**] (see also [**49**]). The idea of the proof is to let $G = KAK$ be a Cartan decomposition. Since K is compact, it suffices to see $f(a) \to 0$ as $a \to \infty$. Consider the example $G = SL(2, \mathbf{R})$, so that A is the positive diagonals. Let P be the upper triangular 2×2 matrices in G with positive diagonal entries. The representation theory of P is well known. There are 1-dimensional representations which factor through $[P, P]$ and 2 infinite dimensional representations induced from $[P, P]$. For the latter, it is clear that the restriction of a representation to A is just the regular representation of A for which it is clear that matrix coefficients vanish at ∞. Thus it suffices to see that $\pi \mid P$ has a spectral decomposition which assigns measure 0 to the 1-dimensional representations. But if it assigned positive measure, $[P, P]$ would have to leave a vector fixed, say v. Then $\varphi(g) = \langle \pi(g)v \mid v \rangle$ would be bi-invariant under $[P, P] = N$. G/N can be identified with $\mathbf{R}^2 - \{0\}$, and the N orbits on G/N are the horizontal lines except for the x-axis, and single points on the x-axis. A continuous function on G/N constant on the orbits must clearly be constant on the x-axis as well. This translates into $\varphi(g) = 1$ for all $g \in P$, and since π is unitary, v is P-invariant. Thus φ is bi-invariant under P, and since P has a dense orbit on G/P, $\varphi(g) = 1$ for all $g \in G$, showing that v is G-invariant.

We thus have good information about some basic examples for the question of ergodicity of actions on homogeneous spaces of finite invariant measure. For the general homogeneous space we make use of the following observation.

PROPOSITION 2.5 [**49**]. *If S is an ergodic G-space (general quasi-invariant measure) and $H \subset G$ is a closed subgroup, then H is ergodic on S if and only if G acts ergodically on the product $G/H \times S$.*

To see this, suppose $A \subset G/H \times S$ is G-invariant. For each $x \in G/H$, let $A_x = \{s \in S \mid (x, s) \in A\}$. By quasi-invariance one easily sees that A and all A_x are simultaneously either null, of null complement, or neither. $A_{[e]}$ is an H-invariant set, and clearly any H-invariant set $B \subset S$ is of the form $B = A_{[e]}$ for some G-invariant A.

COROLLARY 2.6 [**32**]. *If* Γ, $H \subset G$ *are closed subgroups, then* H *is ergodic on* G/Γ *if and only if* Γ *is ergodic on* G/H.

This enables us to use information about ergodicity of restrictions on spaces for which there is a finite invariant measure to obtain results in the case no such measure exists.

COROLLARY 2.7 (MOORE). $G = \prod G_i$, Γ *as in Theorem* 2.3. *If* S *is a transitive* G-*space, then* Γ *is ergodic on* S *if and only if the stabilizers in* G *of points in* S *are not compact.*

EXAMPLE 2.8 (MOORE). $\mathrm{SL}(n, Z)$ is ergodic on $\mathbf{R}^n$, $n \geq 2$. This follows since $\mathrm{SL}(n, \mathbf{R})$ is essentially transitive on $\mathbf{R}^n$ and the stabilizers in the orbit of full measure are not compact.

EXAMPLE 2.9. Consider the action of $\mathrm{SL}(2, \mathbf{R})$ on the Poincaré disk $\mathrm{SL}(2, \mathbf{R})/\mathrm{SO}(2, \mathbf{R})$. This action extends to the boundary circle, and the boundary can be identified with $\mathrm{SL}(2, \mathbf{R})/P$, where P is the upper triangular matrices in G. If $\Gamma \subset \mathrm{SL}(2, \mathbf{R})$ is a torsion free lattice, then Γ acts in a properly discontinuous fashion on the disk, and the quotient space D/Γ is a Riemann surface of finite volume. On the other hand, since P is not compact, the action of Γ on the boundary will be properly ergodic. More generally, if G is any semisimple Lie group and $P \subset G$ is a minimal parabolic subgroup, then G/P is the unique compact G-orbit in the boundary of a natural compactifiation of the symmetric space $X = G/K$, $K \subset G$ maximal compact. Here again, Γ is ergodic on G/P. Thus these ergodic actions of Γ on homogeneous spaces of G arise very naturally in a geometric setting, and the study of these ergodic actions is extremely useful in understanding Γ.

Since this is such an important example, let us point out that for G/P compact (e.g. P a parabolic) that ergodicity of Γ on G/P can be demonstrated in a much less sophisticated fashion. Namely if there is a P-invariant vector in $L^2(G/\Gamma) \ominus \mathbf{C}$ then there is a compact G-orbit in the Hilbert space. As is well known, this implies that there exist finite dimensional subrepresentations, which for G, it is also well known, must be the identity. This is impossible.

Corollary 2.7 deals with the restriction of transitive G-actions to Γ. We now deal with the properly ergodic case.

THEOREM 2.10 [**49**]. *If* $G = \prod G_i$, G_i *connected noncompact simple Lie groups with finite center,* $\Gamma \subset G$ *an irreducible lattice and* S *is a properly ergodic* G-*space, then* Γ *is ergodic on* S.

PROOF. Suppose not. Let $A \subset S \times G/\Gamma$ be invariant. For each s, let $f_s \in L^2(G/\Gamma) \ominus \mathbf{C}$ be the image under orthogonal projection of the characteristic function of $A_s = \{x \in G/\Gamma \mid (s, x) \in A\}$. We can suppose $f_s \neq 0$ on a set of positive measure. Invariance of A is easily seen to imply that if we let $B \subset L^2(G/\Gamma) \ominus \mathbf{C}$ be the unit ball and let G act on the right in G via the unitary representation of G on $L^2(G/\Gamma)$, then $\Phi\colon S \to B$, $\Phi(s) = f_s$ is a G-map. Then $\Phi_*(\mu)$ is a quasi-invariant ergodic measure on B.

But by vanishing of the matrix coefficients (Theorem 2.4), for $w \in B$, $w \cdot g \to 0$ weakly as $g \to \infty$. This implies G-orbits in B are locally closed, i.e.

the action is smooth. It follows that $\Phi_*(\mu)$ is supported on an orbit, so we can suppose $\Phi: S \to G/G_0$ where G_0 is the stabilizer of a point in this orbit. This implies $S = \operatorname{ind}_{G_0}^G(S_0)$ where S_0 is an ergodic G_0 space. But G_0 is compact, so G_0 is transitive on S_0. This implies G is transitive on S, which contradicts our hypotheses.

Similar results can be proven for other groups for which there is a vanishing theorem for matrix coefficients.

THEOREM 2.11 [**50**]. *Let G be an exponential solvable Lie group and S an ergodic G-space. Suppose $[G, G]$ is ergodic on S. Then Γ is also ergodic on S for every cocompact $\Gamma \subset G$.*

The proof uses the result of Howe and Moore [**22**] that for such a group, the matrix coefficients $\langle \pi(g)v \mid w\rangle \to 0$ as $g \to \infty$ in G/P_π where $P_\pi = \{g \mid \pi(g)$ is scalar$\}$. Here π is assumed irreducible.

3. Cocycles. If X is a G-space and Y a Borel space, let $F(X, Y)$ be the space of measurable functions $X \to Y$, two functions being identified if they agree off a null set. G acts on $X \times Y$ by $(x, y) \cdot g = (xg, y)$ and on $F(X, Y)$ by $(g \cdot f)(x) = f(xg)$. If Y is also an H-space for some group H, we can define "twisted" actions. Namely, $(x, y) \cdot g = (xg, y \cdot \alpha(x, g))$ where $\alpha(x, g) \in H$, and for $f \in F(X, Y)$ (where for convenience we usually take H to be acting on the left), $(g \cdot f)(x) = \alpha(x, g)f(xg)$. For these to define actions, we need the following compatibility condition: $\alpha(x, gh) = \alpha(x, g)\alpha(xg, h)$. Such a Borel function $\alpha: X \times G \to H$ will be called a cocycle. (The question as to whether this holds everywhere or almost everywhere is an important technical point which we will not discuss. See [**41**].) When endowed with this action we shall denote $X \times Y$ by $X \times_\alpha Y$. If $Y = \mathcal{H}$ a Hilbert space, $H = U(\mathcal{H})$, the unitary group of the Hilbert space, and the measure on X is invariant, then the α-twisted action on $F(X, Y)$ restricts to $L^2(X; \mathcal{H})$ to yield a unitary representation U^α. If $\alpha, \beta: X \times G \to H$ are cocycles there is a certain relation which immediately implies equivalence of the actions or representations. Namely if we have a Borel map $\varphi: X \to H$ such that $\alpha(x, g) = \varphi(x)\beta(x, g)\varphi(xg)^{-1}$, this will be the case. We then call α and β equivalent, or cohomologous, and write $\alpha \sim \beta$.

To get some further feeling for this notion, consider the case $X = G/G_0$. If $\alpha: G/G_0 \times G \to H$ is a cocycle, then $\alpha \mid [e] \times G_0$ defines a homomorphism $G_0 \to H$. Equivalent cocycles yield conjugate homomorphisms. Furthermore, every homomorphism $G_0 \to H$ arises from a cocycle α in this way. Namely, let $\gamma: G/G_0 \to G$ be a Borel section. Then for $(x, g) \in G/G_0 \times G$, $\gamma(xg)$ and $\gamma(x) \cdot g$ are equal when projected to G/G_0. Thus $\gamma(x)g\gamma(xg)^{-1} \in G_0$. We can suppose $\gamma([e]) = e$, and then $(x, g) \to \gamma(x)g\gamma(xg)^{-1}$ is a cocycle $G/G_0 \times G \to G_0$ which when restricted to $[e] \times G_0$ yields the identity $G_0 \to G_0$. Thus if $\pi: G_0 \to H$ is a homomorphism, $\alpha(x, g) = \pi(\gamma(x)g\gamma(xg)^{-1})$ is the required cocycle. Thus we have

THEOREM 3.1. *$\alpha \to \alpha \mid [e] \times G_0$ defines a bijection between equivalence classes of cocycles $G/G_0 \times G \to H$ and conjugacy classes of homomorphisms $G_0 \to H$.*

We remark that if $H = U(\mathcal{H})$, and $\pi: G_0 \to H$ is a unitary representation, we have an associated cocycle $\alpha: G/G_0 \times G \to H$, and then an associated representation U^α of G on $L^2(G/G_0; \mathcal{H})$. Of course $U^\alpha = \operatorname{ind}_{G_0}^G(\pi)$. See **[45]** for this approach to induced representations.

We now consider some other examples.

EXAMPLE 3.2. (a) If $h: G \to H$ is a homomorphism, X a G-space, then $\alpha(x, g) = h(g)$ is a cocycle. If $X = G/G_0$, this corresponds to a homomorphism $G_0 \to H$, which is simply $h \mid G_0$. Thus in general we shall sometimes call α the restriction of h to $X \times G$ and write $\alpha = h \mid X \times G$.

(b) Suppose X is an ergodic G-space with quasi-invariant measure μ. Let $r_\mu(x, g) = d\mu(xg)/d\mu(x)$, the Radon-Nikodym derivative. The chain rule implies $r_\mu: X \times G \to R^+$ is a cocycle, called the Radon-Nikodym cocycle. If $\mu \sim \nu$, so $d\mu = f\,d\nu, f > 0$, then

$$d\mu(xg)/d\mu(x) = f(x)^{-1}(d\nu(xg)/d\nu(x))f(xg),$$

i.e., $r_\mu \sim r_\nu$. Therefore the cohomology class we obtain does not depend upon the measure, only the measure class. In particular, there is an equivalent σ-finite invariant measure if and only if the cocycle is trivial (i.e. equivalent to the identity $\alpha(x, g) = 1$).

(c) Suppose X is a G-space, X' a free G'-space, and that the actions are orbit equivalent, with $\theta: X \to X'$ the orbit equivalence. Then for $(x, g) \in X \times G$, $\theta(x)$ and $\theta(xg)$ are in the same G'-orbit, say $\theta(x)\alpha(x, g) = \theta(xg)$ for $\alpha(x, g) \in G'$. Then $\alpha: X \times G \to G'$ is a cocycle. If $G = G'$, we have the following.

PROPOSITION 3.3 **[55]**. *If α is equivalent to the restriction of an automorphism $A \in \operatorname{Aut}(G)$ to $X \times G$, then X and X' are automorphically conjugate. If this automorphism is inner, then X and X' are conjugate.*

PROOF. If $\alpha(x, g) = \lambda(x)A(g)\lambda(xg)^{-1}$, then $\theta_1(x) = \theta(x)\lambda(x)$ satisfies $\theta_1(xg) = \theta_1(x)A(g)$, so we have automorphic conjugacy. If $A(g) = hgh^{-1}$, let $\theta_2(x) = \theta_1(x)h$. This is then a G-map.

There are many other naturally arising situations in which cocycles appear, but we shall not have time to discuss them here. Instead, we turn to an important invariant attached to a cocycle, namely the Mackey range. Let α: $S \times G \to H$ where H is also locally compact. Form the twisted G-action $S \times_\alpha H$ where we view H as acting on itself by right translations. H also acts on $S \times_\alpha H$ by $(s, h) \cdot h_0 = (s, h_0^{-1}h)$, and this H action commutes with the G-action. Note that if $i: G \to H$ is an embedding of G into a larger group and $\alpha = i \mid S \times G$, this is exactly the situation in the inducing procedure. As in the latter, we obtain an action of H on the space of G-orbits. But this space may not be a decent measure space, so instead, we let X be the space of G-ergodic components of the action of G of $S \times_\alpha H$. Then H will act on X as well, and this will be an ergodic H-action.

DEFINITION 3.4. If $\alpha: S \times G \to H$ is a cocycle, the associated H-space X will be called the Mackey range of α. This is a cohomology invariant of α.

EXAMPLE 3.5. (a) If $i: G \to H$ is an embedding of G as a closed subgroup of H, and $\alpha(s, g) = i(g)$, i.e. $\alpha = i \mid S \times G$, then the Mackey range of α is $\operatorname{ind}_G^H(S)$.

(b) If θ: $X \to X'$ is an orbit equivalence, α: $X \times G \to G'$ the associated cocycle, then the Mackey range is the G'-space X'.

(c) If $S = G/G_0$ and α corresponds to a homomorphism π: $G_0 \to H$, then the Mackey range of α: $G/G_0 \times G \to H$ is the H-space $H/\overline{\pi(G_0)}$.

Finally, the following relates the Mackey range to the cohomology class of α.

PROPOSITION 3.6. *If* α: $S \times G \to H$, *the following are equivalent.*

(i) $\alpha \sim \beta$ *where* $\beta(S \times G) \subset H_0$, $H_0 \subset H$ *a closed subgroup.*

(ii) H/H_0 *is a factor of the Mackey range.*

(iii) $X = \operatorname{ind}_{H_0}^{H}(S_0)$ *for some* S_0, *where* X *is the Mackey range.*

For a proof, see [**47, 52**].

4. Generalized discrete spectrum. Suppose (S, μ) is an ergodic space with μ finite and invariant. In this lecture we try to see what the algebraic structure of the representation π of G on $L^2(S)$ says about the geometric structure of the action.

DEFINITION 4.1. We say that the action has discrete spectrum if π is the direct sum of finite dimensional irreducible subrepresentations.

EXAMPLE 4.2. Let K be a compact group, H a closed subgroup, and φ: $G \to K$ a homomorphism with $\varphi(G)$ dense in K. Let G act on K/H by $[k] \cdot g = [k\varphi(g)]$. Then this action has discrete spectrum.

THEOREM 4.3 (VON NEUMANN-HALMOS-MACKEY). *These are all the examples. That is, if* S *is a* G*-space with discrete spectrum, then there exists a compact group* K, *a closed subgroup* H, *and a homomorphism* φ: $G \to K$ *with dense range such that* S *and* K/H *are conjugate* G*-spaces.*

This was originally proved by von Neumann and Halmos for $G = Z$ or $\mathbf{R}$, and by Mackey [**26**] for general G. We sketch Mackey's proof.

Let $L^2(S) = \sum^{\oplus} W_i$ where W_i are $\pi(G)$-invariant and finite dimensional. Let $B = \prod U(W_i)$, the product of the associated unitary groups, which is a compact subgroup of $U(L^2(S))$. Further, π: $G \to B$. Let $K = \overline{\pi(G)}$, so that K is also compact. Let M be the abelian von Neumann algebra on $L^2(S)$ consisting of multiplication by elements of $L^\infty(S)$. Then clearly $\pi(g)M\pi(g)^{-1} = M$, and by passing to the strong limit, we obtain $TMT^{-1} = M$ for all $T \in K$. From this one can deduce that each operator T in K is induced by a point transformation of S, and thus the G-action on S extends to an action of K. (There is some delicate measure theory we are ignoring here.) Since the G-action is already ergodic, so is the K-action. Since K is compact, K must act transitively, so we can identify $S \cong K/H$.

Theorem 4.3 can be generalized to extensions. Namely, suppose $X \to Y$ is an extension of ergodic G-spaces with finite invariant measure. The Hilbert space $L^2(X)$ not only has a natural representation of G on it, but $L^2(X)$ is also an $L^\infty(Y)$-module in a natural way. (Namely, lift a function on Y to a function on X and multiply.) Alternatively, we can express this by saying that there is a natural system of imprimitivity for π on $L^2(X)$ based on Y.

DEFINITION 4.4 [**47**]. We say that X has relatively discrete spectrum over Y if $L^2(X) = \overline{\Sigma^{\oplus} W_i}$ where W_i are G-invariant subspaces that are finitely generated as $L^\infty(Y)$-modules.

EXAMPLE 4.5. Suppose Y is an ergodic G-space with finite invariant measure, $\alpha\colon Y \times G \to K$ is a cocycle where K is compact, and $H \subset K$ is a closed subgroup. Then $X = Y \times_\alpha K/H$ is an extension of Y with relatively discrete spectrum. To see this, observe that $L^2(X) = L^2((Y); L^2(K/H))$. Write $L^2(K/H) = \Sigma^\oplus Z_i$ where Z_i are finite dimensional and K-invariant. We then have $L^2(X) = \Sigma^\oplus L^2(Y; Z_i)$ and $L^2(Y; Z_i)$ will be G-invariant since G acts from fiber to fiber in X by an element of K, and Z_i is K-invariant. Clearly $L^2(Y; Z_i) = \overline{L^\infty(Y, Z_i)}$ and the latter is finitely generated over $L^\infty(Y)$.

THEOREM 4.6 [**47**]. *These are all the examples. That is, if $X \to Y$ is an ergodic extension with relatively discrete spectrum, then there exists a compact group K, a closed subgroup $H \subset K$, and a cocycle $\alpha\colon Y \times G \to K$, such that as extensions of Y, $X \cong Y \times_\alpha K/H$.*

Thus Theorem 4.6 tells us how to recognize extensions of the form $Y \times_\alpha K/H$ from information about the unitary representation of the extension. There is now a larger class of actions whose "structure" we know.

DEFINITION 4.7 [**48**]. We say that X has generalized discrete spectrum if X can be built from a point via the operations of taking extensions with relatively discrete spectrum and inverse limits. More precisely, there is a countable ordinal α and for each $\sigma \leqslant \alpha$ a factor X_σ of X such that

(i) $X_0 =$ point,

(ii) $X_{\sigma+1} \to X_\sigma$ is an extension with relatively discrete spectrum, for $\sigma < \alpha$,

(iii) if σ is a limit ordinal, $X_\sigma = \varprojlim \{X_\beta, \beta < \sigma\}$,

(iv) $X_\alpha = X$.

In light of Theorem 4.6, we have an exact picture of the structure of such actions. We would now like to see which actions arise in this fashion.

DEFINITION 4.8. If G acts continuously on a compact metric space X, G is called distal on X if $x, y \in X$, $x \neq y$, implies $\inf_{g \in G} d(xg, yg) > 0$.

Clearly any isometric action is distal. However, not every distal action admits an invariant metric. For example, if N is a nilpotent Lie group and $\Gamma \subset N$ is a lattice, then the action of N on N/Γ is distal. This was first shown in [**1**].

DEFINITION 4.9 (PARRY [**38**]). If (S, μ) is an ergodic G-space, call the action measure distal if there is a decreasing sequence of sets of positive measure $\{A_i\}$ with $\mu(A_i) \to 0$, such that if $x, y \in S$, $xg_i, yg_i \in A_i$ for some sequence $g_i \in G$, then $x = y$. (We have ignored some measure theoretic issues in this definition, which arise if G is not discrete. See [**48**] for a more careful formulation.)

Any distal action with an invariant measure that is positive on open sets is clearly measure distal.

THEOREM 4.10 [**48**]. *A finite measure preserving ergodic action (on a non-atomic measure space) is measure distal if and only if it has generalized discrete spectrum.*

This is an analogue for measure theoretic actions of the Furstenberg structure theorem for minimal (i.e. every orbit dense) distal actions on compact metric spaces [**16**].

Another situation in which actions with generalized discrete spectrum arise is the following.

THEOREM 4.11. *Let N be a nilpotent group. Suppose S is an ergodic N-space for which $L^2(S)$ is a direct sum of irreducible representations (not necessarily finite dimensional). Then S has generalized discrete spectrum (and the ordinal in Definition* 4.7 *can be taken to be finite).*

This theorem is false for solvable groups. Let us give an example of such a properly ergodic N-space. Let N be the Heisenberg group, $N_Z = \Gamma$ the integer points, so that Γ is a lattice. There is an injective homomorphism $\Gamma \to K$ where K is compact. For example, let $K = \prod N_{Z/pZ}$, where the product is taken over all primes. Then K is a Γ-space with discrete spectrum. Let $X = \mathrm{ind}_\Gamma^N(K)$. Since a finite dimensional representation of Γ induced to N decomposes into a direct sum of irreducibles, and π on $L^2(X)$ can be expressed as $\pi = \mathrm{ind}_\Gamma^N(\sigma)$ where σ is the representation of Γ on $L^2(K)$, it follows that $L^2(X)$ is a direct sum of irreducibles.

It is natural to ask which groups have actions of the sort we have been discussing in a nontrivial way, say an effective or essentially free action. A group will have an effective or free action with discrete spectrum if and only if there are enough finite dimensional unitary representations to separate points. In the connected case, such groups are identified by a classical theorem of Freudenthal.

THEOREM 4.12 (FREDUENTHAL [**14**]). *A connected group has a free (or effective) action with discrete spectrum if and only if it is isomorphic to $\mathbf{R}^n \times K$ where K is compact.*

To describe the analogous result for generalized discrete spectrum, we recall that a connected group is said to be of polynomial growth if for any compact neighborhood of the identity, W, the Haar measure $m(W^n)$ grows no faster than a polynomial in n. (If this is true for one compact neighborhood, it is true for all such neighborhoods.) For Lie groups, this condition is equivalent to the group being of type (R) [**19**, **23**]. We recall that this means that every eigenvalue of $\mathrm{Ad}(g)$ lies on the unit circle for all $g \in G$. For example, nilpotent groups and Euclidean motion groups are type (R), while semisimple groups and the $ax + b$ group are not. The following is joint work with C. C. Moore.

THEOREM 4.13 [**34**]. *A connected group has a free (or effective) ergodic action with generalized discrete spectrum if and only if it is of polynomial growth.*

PROOF. We indicate the proof for the $ax + b$ group. The general proof is based on this argument and some structure theory for Lie groups, particularly that of solvable Lie groups.

Let $G = AB$ be a semidirect product where $B = \mathbf{R}$ is normal and $A = \mathbf{R}^+$ acts on B by multiplication. If X is a G-space with generalized discrete spectrum, and X_1 is the factor of X with discrete spectrum, then B must act trivially on X_1 since all finite dimensional unitary representations of G are one dimensional and thus factor through $B = [G, G]$. It therefore suffices to show the following: suppose $\varphi: X \to Y$ is an extension of G-spaces with relatively discrete spectrum and suppose B acts trivially on Y; then B acts trivially on X. To prove this assertion, let $L^2(X) = \Sigma^{\oplus} \overline{W_i}$, W_i a finitely generated $L^\infty(Y)$-module which is G-invariant. Let μ, ν be the given measures on X and Y respectively, and decompose μ with respect to ν over the fibers of φ. Thus, we write $\mu = \int \mu_y \, d\nu(y)$ where μ_y is supported on $\varphi^{-1}(y)$. This gives us a direct integral decomposition $L^2(X) = \int_Y^{\oplus} L^2(\varphi^{-1}(y), \mu_y)\, d\nu$. For each y, g we have $\alpha(y, g)$: $L^2(\varphi^{-1}(yg), \mu_{yg}) \to L^2(\varphi^{-1}(y), \mu_y)$ given by $[\alpha(y, g)f](z) = f(zg)$ for $z \in \varphi^{-1}(y)$. Fix i. Saying that W_i is finitely generated over $L^\infty(Y)$ and G-invariant means that there is $V_y \subset L^2(\varphi^{-1}(y))$, a finite dimensional subspace, such that $W_i = \int^{\oplus} V_y \, d\nu(y)$ and $\alpha(y, g)V_{yg} = V_y$. For $g \in B$, $yg = y$, so $\alpha \mid \{y\} \times B$ is a unitary representation of B on V_y. Say $\dim V_y = n$. Then for each y, we have n elements in $\hat{B}$ = character group of B. Furthermore, G acts on $\hat{B}$ and one can check that the cocycle identity for α implies

(*)

$$(\alpha \mid \{y\} \times B) \cdot g \cong \alpha \mid \{yg\} \times B, \quad \text{where } \cong \text{ means unitary equivalence.}$$

Let $\hat{B}^n/S_n$ (S_n is the symmetric group on n letters) be the set of unordered n-tuples of elements of $\hat{B}$. We have a map Φ: $Y \to \hat{B}^n/S_n$, and (*) implies that Φ is a G-map. The action of G on $\hat{B} \cong \mathbf{R}$ has three orbits, namely the origin and the 2 half lines. From this it is easy to see that every G-orbit in $\hat{B}^n$ (and hence in $\hat{B}^n/S_n$) is locally closed, and that the only compact G-orbit is the identity (i.e. the origin). But $\Phi_*(\mu)$ is a finite invariant ergodic measure on $\hat{B}^n/S_n$. By smoothness, this must be supported on an orbit and by finiteness and invariance, this must clearly be the zero orbit. Thus, $\alpha \mid \{y\} \times B$ is the identity for all y, so B acts trivially on each W_i. Therefore, B is trivial on $L^2(X)$ and hence on X as well.

Finally, we remark that the notion of generalized discrete spectrum yields a type of structure theorem for general actions with finite invariant measure that is sometimes useful. If S is an ergodic G-space, it is not always true that $S \times S$ is also ergodic, where G acts by $(s, t)g = (sg, tg)$. If this additional ergodicity property holds, the action is called weakly mixing. More generally, if $X \to Y$ is an ergodic extension of Y, the fibered product $X \times_Y X$ has a natural G-invariant measure on it **[47]**, but this action no longer need be ergodic. Once again, if this extra ergodicity holds, the extension X is called relatively weak mixing over Y. Given any ergodic G-space X, there is a unique maximal factor Z of X such that Z has generalized discrete spectrum and X is relatively weakly mixing over Z. Thus we break X up into a factor whose structure we know explicitly, and an extension with extra ergodicity properties. Of course, simply by knowing that an action or extension is weak mixing does not say very much about its detailed structure, so for most questions, this is not a satisfactory structure theorem aside from the factor Z. Nevertheless, weak mixing does clearly have some information, and thus one can hope to find this decomposition useful in some circumstances. An example of this appears in recent work of Furstenberg. Szemeredi recently succeeded in proving a conjecture of Erdös which asserts that every set of positive integers of positive upper density contains arithmetic progressions of arbitrary (finite) length. In **[17]**, Furstenberg gave another proof of Szemeredi's theorem, first by converting this to a statement about measure preserving integer actions, and then proving the latter by proving it first for actions with generalized discrete spectrum, and then showing the property is preserved by passing to relatively weakly mixing extensions.

5. Amenability. The notion of an amenable group can be described in a variety of ways. Here, we shall focus on the fixed point property.

Let E be a separable Banach space, E^* the dual, E_1^* the unit ball in E^*, and $\mathrm{Iso}(E)$ the group of isometric isomorphisms of E. Suppose $\pi: G \to \mathrm{Iso}(E)$ is a representation of G on E, and that $A \subset E_1^*$ is a compact convex G-invariant set. (Here G acts on E^* via the adjoint representation, $\pi^*(g) = (\pi(g^{-1}))^*$.)

DEFINITION 5.1. G is amenable if for all π and A as above, there is a fixed point for G in A.

For example, if G is amenable and G acts continuously on a compact metric space X, then there is a G-invariant probability measure on X. We simply apply the definition to $E = C(X)$ where $A \subset C(X)^*$ is the set of probability measures. In fact a standard convexity argument shows that G is amenable if and only if there is a G-invariant measure on every compact metric G-space.

Abelian groups are amenable by the Markov-Kakutani fixed point theorem, and compact groups are easily seen to be amenable. If $0 \to A \to B \to C \to 0$ is an exact sequence, then B is amenable if and only if A and C are amenable. Thus, groups with a cocompact solvable normal subgroup are amenable. Every connected amenable group is of this form, but this is no longer true among all discrete groups **[18]**.

We now wish to define the notion of an amenable ergodic action of a group, originally introduced in **[51]**. This will include all actions of amenable groups, as well as some actions of nonamenable groups. We begin by describing certain classes of G-invariant compact convex sets that arise from an ergodic G-space S.

So suppose S is an ergodic G-space and $\alpha: S \times G \to \mathrm{Iso}(E)$ is a cocycle. We then have the adjoint cocycle $\alpha^*(s, g) = (\alpha(s, g)^{-1})^*$, and the α^*-twisted action on $L^\infty(S, E^*)$, given by $(g \cdot f)(s) = \alpha^*(s, g)f(sg)$, for $f \in L^\infty(S, E^*)$. We observe that $L^\infty(S, E^*) = (L^1(S, E))^*$, so that $L^\infty(S, E^*)$ is a dual space. We want to describe certain G-invariant compact convex sets in the unit ball of this dual space. One natural possibility is to take $A \subset E_1^*$ compact, convex, and satisfying the condition $\alpha^*(s, g)A = A$. Then $F(S, A)$ (= measurable functions $S \to A$) will be a compact convex G-invariant set in $L^\infty(S, E^*)$. However, it is also possible to vary the set A as we move from point to point in S. Thus, suppose $\{A_s\}$ is a collection of compact convex subsets $A_s \subset E_1^*$, which vary measurably in s, and satisfying the condition $\alpha^*(s, g)A_{sg} = A_s$. Then $F(S, \{A_s\}) = \{f \in L^\infty(S, E^*) \mid f(s) \in A_s\}$ is a compact convex G-invariant set.

DEFINITION 5.2 [**51**]. Call a set of the form $F(S, \{A_s\})$ a compact convex set over S. An ergodic action of G on S is called amenable if every compact convex G-invariant set over S has a fixed point.

Thus while amenability of G demands a fixed point in every compact convex set, amenability of the action demands a fixed point only in compact convex sets over the action. We also remark that the condition that one has a fixed point simply means $\alpha^*(s, g)f(sg) = f(s)$ for $f\colon S \to E^*$, $f(s) \in A_s$. As an example of how one can use this condition, suppose S is an amenable G-space and that X is a compact metric G-space. Let $M(X)$ be the space of probability measures on X. We have a representation $\pi\colon G \to \mathrm{Iso}(C(X))$ and hence a cocycle $\alpha\colon S \times G \to \mathrm{Iso}(C(X))$ by restriction, i.e. $\alpha(s, g) = \pi(g)$. $M(X)$ will be a G-invariant compact convex set, and thus we can take $A_s = M(X)$ for all s. (So for this example, we did not have to vary the compact convex set in going from point to point.) Amenability of the action then implies that there is a function $f\colon S \to M(X)$ such that $\alpha^*(s, g)f(sg) = f(s)$, i.e., $\pi^*(g)f(sg) = f(s)$. Switching to a right action on $M(X)$, we obtain that $f(sg) = f(s) \cdot g$. Thus, we conclude that if S is an amenable ergodic G-space, X a compact metric G-space, then there is a measurable G-map $f\colon S \to M(X)$.

We now list some basic properties. Proofs can be found in [**51**, **52**].

PROPOSITION 5.3. (a) *If G is amenable, every ergodic G-space is amenable.*

(b) *If S is an amenable ergodic G-space with finite invariant measure, then G is amenable.*

(c) *If $S = G/H$, then G/H is an amenable G-space if and only if H is amenable.*

(d) *If S is an amenable ergodic G-space, and $\Gamma \subset G$ is a closed subgroup, then the restriction of the action on S to Γ is amenable.*

EXAMPLE 5.4. Let $\Gamma \subset \mathrm{SL}(2, \mathbf{R})$ be a lattice and consider the ergodic action of Γ on the boundary circle of the Poincaré disk. This is just the action of Γ on $\mathrm{SL}(2, \mathbf{R})/P$ where P is the upper triangular subgroup. Since P is amenable, the Γ-action on the boundary circle is amenable by (c) and (d) of the above proposition. More generally, let G be a semisimple Lie group, $\Gamma \subset G$ a lattice, and $P \subset G$ a minimal parabolic subgroup. Then P is amenable and so Γ acting on G/P is amenable.

This example indicates how natural and important examples of actions of nonamenable groups are amenable. Assertion (c) of the above proposition shows that any group has amenable transitive actions. More generally, we have the following.

PROPOSITION 5.5 [**52**]. *If $H \subset G$ is a closed subgroup and S is an ergodic H-space, then S is an amenable H-space if and only if $\mathrm{ind}_H^G(S)$ is an amenable G-space.*

This proposition raises the following question. Although one can have amenable actions of nonamenable groups, does every such action come in a simple way from an action of an amenable subgroup, namely just by inducing? In fact, the answer in general is no. One can show that if $\Gamma \subset \mathrm{SL}(2, \mathbf{C})$ is a lattice, then Γ is amenable on $\mathbf{CP}^1$ (this is just Example 5.4) but that this action is not induced from an action of an amenable subgroup. This latter fact is not trivial, and a proof is given in [**52**]. On the other hand, we do have the following.

THEOREM 5.6 [**52**, **55**]. *Let G be a connected group. Then every amenable ergodic action of G is induced from an action of an amenable subgroup.*

PROOF. We give the proof for G semisimple. Let S be an amenable G-space and $P \subset G$ be a minimal parabolic subgroup. Since G/P is compact, by the remarks following Definition 5.2, there is a measurable G-map $f: S \to M(G/P)$, where the latter is the space of probability measures on G/P. The proof will now follow from two basic results about the action of G on $M(G/P)$. The first is due to the author, the second to C. C. Moore [**33**].

THEOREM 5.7 [**52**, **55**]. *Let G be a connected semisimple Lie group and $P \subset G$ any parabolic subgroup. Then every element in $M(G/P)$ has a locally closed orbit under the G-action.*

THEOREM 5.8 (C. C. MOORE [**33**]). *Let G be a connected semisimple Lie group with trivial center and $P \subset G$ a minimal parabolic subgroup. Let $\mu \in M(G/P)$ and G_μ the stabilizer of μ in G. Then G_μ is an amenable algebraic group. (By algebraic, we mean the intersection of G with an algebraic subgroup of $\tilde{G}$, where the latter is an algebraic subgroup containing G as a subgroup of finite index.)*

To conclude the proof of Theorem 5.6 given these results, we simply observe that Theorem 5.7 asserts the smoothness of the G-action on $M(G/P)$, so if m denotes the measure on S, then $f_*(m)$ is quasi-invariant and ergodic on $M(G/P)$ and hence supported on an orbit. Thus we can consider f as $f: S \to G/H$, where H is the stabilizer of an orbit. But by Theorem 5.8, H is amenable so the result follows from Proposition 1.11.

Without giving a proof of Theorems 5.7 and 5.8, let us at least try to give some indication of why they are true. To this end, we state the following lemma of Furstenberg [**15**].

LEMMA 5.9 (FURSTENBERG). *Suppose $g_n \in \mathrm{SL}(n, \mathbf{R})$, $g_n \to \infty$. Let $\mu \in M(\mathbf{P}^{n-1})$ and suppose $\mu \cdot g_n \to \nu$ for some $\nu \in M(\mathbf{P}^{n-1})$. Then ν is supported on a union of two proper projective subspaces.*

REMARK. (i) Since the set of measures supported on a union of two proper projective subspaces is closed, this shows that the orbit of any μ not so supported is locally closed. This is the beginning of an inductive procedure for proving Theorem 5.7.

(ii) If $H \subset \mathrm{SL}(n, \mathbf{R})$ leaves a measure fixed, then H is either compact or leaves the union of finitely many proper subspaces invariant. Both of these conditions are "algebraic". Proceeding inductively, one should obtain either compactness or further splitting. This should result in an algebraic object which is a compact extension of a solvable group. In this way, Lemma 5.9 suggests Theorem 5.8 as well.

PROOF OF LEMMA 5.9. Let $h_n = g_n/\|g_n\|$. Then $\|h_n\| = 1$, $\det(h_n) \to 0$, so we can assume $h_n \to h$, $h \neq 0$, $\det(h) = 0$. Let $V = \text{range}(h)$, $N = \ker(h)$, $[V]$, $[N]$ the corresponding projective subspaces in $\mathbf{P}^{n-1}$. Write $\mu = \mu_1 + \mu_2$ where $\text{support}(\mu_1) \subset [N]$, $\text{support}(\mu_2) \subset \mathbf{P}^{n-1} - [N]$. If $x \in \mathbf{P}_{n-1} - [N]$, then as $n \to \infty$, $h_n(x) \to h(x) \in [V]$. Passing to a subsequence, we can write

$$\nu = \lim \mu \cdot g_n = \lim \mu_1 \cdot g_n + \lim \mu_2 \cdot g_n,$$

and the previous sentence implies $\lim \mu_2 \cdot g_n$ is supported on $[V]$. We also have $\lim \mu_2 \cdot g_n$ supported on $[W]$ where $\lim[N] \cdot g_n = [W]$.

We now turn to the theory of orbit equivalence for amenable actions. We begin with an observation.

PROPOSITION 5.10. *For free ergodic actions amenability is an invariant of orbit equivalence.*

The circle of ideas concerning orbit equivalence began in the late 1950s with the work of H. Dye [**10**, **11**] and for amenable actions has been brought to complete form very recently. We begin with the fundamental theorem of Dye.

THEOREM 5.11 (DYE [**10**]). *All finite measure preserving (properly) ergodic* **Z***-actions are orbit equivalent.*

This was later extended to the σ-finite case by Krieger [**25**].

THEOREM 5.12 (KRIEGER). *All σ-finite (but not finite) measure preserving* **Z***-actions are orbit equivalent.*

Krieger also extended the theorem to the case of quasi-invariant measure without equivalent invariant measure. He showed that a measurement of the extent to which the action fails to be measure preserving is a complete invariant of orbit equivalence. Namely, let X be an ergodic G-space, and r: $X \times G \to \mathbf{R}^+$ be the Radon-Nikodym cocycle. Let Δ: $G \to \mathbf{R}^+$ be the modular function of G, and let m: $X \times G \to \mathbf{R}^+$ be $m(x, g) = r(x, g)\Delta(g)^{-1}$. We call m the modular cocycle. The Mackey range of this cocycle will be an ergodic $\mathbf{R}^+$-action, which we call the modular flow or the modular range. For unimodular groups, the modular flow will be translation of $\mathbf{R}^+$ on $\mathbf{R}^+$ itself if and only if there is an invariant measure for the action (for then the Radon-Nikodym cocycle is trivial (Proposition 3.6)).

THEOREM 5.13 (KRIEGER). *For Z-actions with quasi-invariant measure, and not possessing (an equivalent) finite invariant measure, the modular flow is a complete invariant of orbit equivalence.*

A good account of Krieger's work is [**44**].

EXAMPLE 5.14. Let us see how to compute the modular flow in some examples. Let $\Gamma \subset SL(2, \mathbf{R})$ be a lattice, P the upper triangular subgroup. The Radon-Nikodym cocycle for the action of $G = SL(2, \mathbf{R})$ on G/P is the cocycle α: $G/P \times G \to \mathbf{R}^+$ corresponding to the homomorphism $P \to \mathbf{R}^+$ given by Δ_P, the modular function of P. Clearly r: $G/P \times \Gamma \to \mathbf{R}^+$ is just $\alpha \mid G/P \times \Gamma$. Now the Γ action on $G/P \times \Gamma \to \mathbf{R}^+$ that appears in the construction of the

Mackey range is just the restriction to Γ of the G-action $G/P \times_\alpha \mathbf{R}^+$. Since Δ: $P \to \mathbf{R}^+$ is surjective, as is well known, this is the G-action $G/P \times_\alpha P/\ker \Delta_P$ on which G acts transitively with stabilizer $\ker \Delta_P$. Thus as a Γ-space, $G/P \times_r \mathbf{R}^+$ is just the action of Γ on $G/\ker \Delta_P$. Since $\ker \Delta_P$ is not compact, by Moore's ergodicity theorem (§2) Γ is ergodic on this space. Thus there is only one ergodic component, and so the modular flow is the action of $\mathbf{R}^+$ on a point. This computation can clearly be carried out on any semisimple noncompact Lie group. If Γ is a lattice in such a group G, and $P \subset G$ is a minimal parabolic, then the modular flow of the action of Γ on G/P will be the action of $\mathbf{R}^+$ on a point.

The Dye-Krieger theorems were extended over the years by a number of persons to include within its framework actions of larger classes of groups. (In fact Dye did not restrict himself to the integers.) This work has recently culminated with the following theorems.

THEOREM 5.15 (CONNES-FELDMAN-ORNSTEIN-WEISS [**8**, **37**]). (i) *A free properly ergodic action of a discrete group is amenable if and only if it is orbit equivalent to a* $\mathbf{Z}$-*action.*

(ii) *The Dye-Krieger Theorems* 5.11–5.13 *hold for the class of amenable properly ergodic actions of discrete groups.*

THEOREM 5.16 (CONNES-FELDMAN-ORNSTEIN-WEISS [**8**, **37**]). (i) *A free properly ergodic action of a continuous group is amenable if and only if it is orbit equivalent to an* $\mathbf{R}$-*action.*

(ii) *For such actions, the modular flow is a complete invariant of orbit equivalence. In particular, any two free properly ergodic actions of continuous amenable unimodular groups with invariant measure are orbit equivalent.*

EXAMPLE 5.17. If G is a semisimple noncompact Lie group, $\Gamma \subset G$ a lattice, $P \subset G$ a minimal parabolic, we saw in Example 5.14 that the modular flow of Γ on G/P is independent of G and Γ. Since these actions are amenable, Theorem 5.15 says that they are all orbit equivalent.

6. Rigidity: The Mostow-Margulis theorem and a generalization to ergodic actions. In this lecture we shall describe the proof of the Mostow-Margulis rigidity theorem for lattices in semisimple Lie groups and indicate how this result can be extended to yield results about orbit equivalence for ergodic actions of semisimple groups and their lattices.

THEOREM 6.1 (MOSTOW-MARGULIS RIGIDITY). *Let G, G' be connected semisimple Lie groups with finite center, no compact factors, and Γ, Γ' irreducible lattices in G, G' respectively. Suppose* $\mathbf{R}$-*rank*$(G) \geqslant 2$. *Then*

(i) *if Γ and Γ' are isomorphic, then G and G' are locally isomorphic.*

(ii) *In the center free case, any isomorphism $\Gamma \to \Gamma'$ extends to a rational isomorphism $G \to G'$.*

This was first proved for cocompact lattices by Mostow [**36**] and for noncocompact lattices by Margulis [**27**]. In an extraordinary and highly original and innovative paper, Margulis then gave an alternate proof in [**28**] which subsumed both cases, gave stronger results on the extension of homomorphisms from Γ to G', and which was powerful enough to prove the arithmeticity of lattices. In **[55]** we showed how Margulis' techniques could be incorporated into a proof of rigidity for ergodic actions. Theorem 6.1 is also true without the **R**-rank assumption as long as $G \neq \mathrm{PSL}(2, \mathbf{R})$. This is due to Mostow **[36]** and Prasad **[39]**. We now describe rigidity for ergodic actions.

DEFINITION 6.2 **[55]**. Suppose G is a semisimple connected Lie group with finite center and no compact factors. An ergodic G-space S is called irreducible if every noncentral normal subgroup of G is also ergodic on S. (Thus if Γ is a lattice, G/Γ is irreducible if and only if Γ is irreducible.)

THEOREM 6.3 **[55]** (RIGIDITY FOR ERGODIC ACTIONS). *Let G, G' be connected semisimple Lie groups with finite center and no compact factors, S, S' free irreducible ergodic G-, G'-spaces, respectively, with finite invariant measure. Let* **R**-*rank*$(G) \geq 2$. *Suppose the actions are orbit equivalent. Then*

(i) *G and G' are locally isomorphic,*

(ii) *in the centerfree case, we can take $G = G'$, and then the actions on S and S' are automorphically conjugate.*

Thus, this theorem asserts that one has behavior that is diametrically opposed to the behavior of actions of amenable groups. Although Theorems 6.1 and 6.3 look rather different, let us show that they are both direct consequences of the following theorem.

THEOREM 6.4 **[55]**. *Let G, G' be connected semisimple Lie groups with trivial center and no compact factors, and let S be an irreducible ergodic G-space with finite invariant measure. Assume* **R**-*rank*$(G) \geq 2$. *Let α: $S \times G \to G'$ be a cocycle whose Mackey range is Zariski dense in G'. Then α is equivalent to a cocycle β that is the restriction of a rational epimorphism π: $G \to G'$.*

To deduce Theorem 6.1 from Theorem 6.4, one observes that when applied to the case $S = G/\Gamma$ and α: $G/\Gamma \times G \to G'$ a cocycle corresponding to a homomorphism $\Gamma \to G'$, Theorem 6.4 yields the following theorem of Margulis.

THEOREM 6.5 (MARGULIS). *G, G' as in Theorem 6.4,* (**R**-*rank*$(G) \geq 2$), *$\Gamma \subset G$ an irreducible lattice. Suppose π $\Gamma \to G'$ is a homomorphism with $\pi(\Gamma)$ Zariski dense in G'. Then π extends to a rational epimorphism $G \to G'$.*

Theorem 6.1 then follows. To deduce Theorem 6.3 from Theorem 6.4, one simply applies Theorem 6.4 to the cocycle α: $S \times G \to G'$ coming from an orbit equivalence. The hypotheses of Theorem 6.4 are satisfied since the Mackey range of α is the G'-space S' and the Borel density theorem implies Zariski density. The conclusion of 6.4 implies that of 6.3 by Proposition 3.3.

Let us give an example of how to apply Theorem 6.3 to some natural examples.

COROLLARY 6.6 **[55]**. *Let G, G' be connected simple noncompact Lie groups with finite center, Γ, $\Gamma' \subset G$, G' lattices and suppose S, S' are free ergodic Γ-,*

Γ'-spaces with finite invariant measure. Suppose $\mathbf{R}$-*rank*$(G) \geqslant 2$, *and that the Γ-action on S and Γ'-action on S' are orbit equivalent. Then G and G' are locally isomorphic.*

PROOF. Let $X = \mathrm{ind}_\Gamma^G(S)$, $X' = \mathrm{ind}_{\Gamma'}^{G'}(S')$. Then one easily checks that the hypothesis of Theorem 6.3 is satisfied.

EXAMPLE 6.7 [**55**]. As we vary n, $n \geqslant 2$, the natural actions of $\mathrm{SL}(n, Z)$ on $\mathbf{R}^n/Z^n$ by automorphisms are mutually nonorbit equivalent.

We now turn to some proofs. We will not prove Theorem 6.4 here, but rather only Theorem 6.5 (which therefore gives us a proof of Theorem 6.1). The first part of the proof we present is different from Margulis' original argument. Instead, we present an argument which generalizes nicely when one attempts to prove Theorem 6.4. It is perhaps also more transparent than the original argument. The second part of the proof will be that of Margulis, although we shall try to give some motivation. For the proof of Theorem 6.4, see [**55**].

PROOF OF THEOREM 6.5. Let $P \subset G$, $P' \subset G'$ be minimal parabolic subgroups. We have a homomorphism $\pi\colon \Gamma \to G'$ and hence G'/P' becomes a compact metrizable Γ-space. On the other hand, as we observed in Example 5.4, the action of Γ on G/P is ergodic and amenable. By the remarks following Definition 5.2, there is a measurable Γ-map $\varphi\colon G/P \to M(G'/P')$, the latter space being the space of probability measures on G'/P'. By Theorem 5.7, the action of G' on $M(G'/P')$ is smooth, so $\hat{M}(G'/P') = [M(G'/P')]/G'$ is countably separated and generated. Since φ is a Γ-map, $\varphi(x\gamma) = \varphi(x)\pi(\gamma)$, so $\varphi(x\gamma) \equiv \varphi(x)$ in $\hat{M}(G'/P')$. By ergodicity of Γ on G/P, the projection of φ into $\hat{M}(G'/P')$ is essentially constant, i.e. $\varphi(G/P)$ can be assumed to lie in one G'-orbit in $M(G'/P')$. Thus, we can view φ as a Γ-map $\varphi\colon G/P \to G'/H'$, and by Theorem 5.8, H' is an amenable algebraic subgroup. What we have done is to obtain a Γ-map φ where the image is no longer an infinite dimensional space $M(G'/P')$ but an algebraic variety G'/H'. The existence of such a measurable map φ is the first main step in the proof. The second step is the following fundamental lemma of Margulis.

LEMMA A. *$\varphi\colon G/P \to G'/H'$ is a rational mapping of algebraic varieties.*

Let us show why this lemma suffices to prove the theorem. Suppose $\varphi\colon G/P \to G'/H'$ is a rational mapping such that $\varphi(x\gamma) = \varphi(x)\pi(\gamma)$. Let $R(G/P, G'/H')$ be the space of rational mappings. Then $G \times G'$ acts on $R(G/P, G'/H')$ by

$$[(g, g') \cdot f](x) = f(xg) \cdot (g')^{-1}.$$

The fact that φ is a Γ-map means φ is fixed under Γ, where Γ is identified with the subgroup of $G \times G'$ given by $\{(\gamma, \pi(\gamma))\}$. Let $\bar{\Gamma}$ be the algebraic hull of Γ in $G \times G'$. Then φ is also fixed under $\bar{\Gamma}$. We claim that $\bar{\Gamma}$ is the graph of homomorphism $G \to G'$. Since Γ is Zariski dense in G, $\bar{\Gamma}$ must project onto all of G. So suppose $(g, h_1), (g, h_2) \in \bar{\Gamma}$. Then $\varphi(xg) = \varphi(x)h_1$ and $\varphi(xg) = \varphi(x)h_2$. Therefore $h_1h_2^{-1}$ leaves $\varphi(G/P)$ pointwise fixed. But $\pi(\Gamma)$ leaves $\varphi(G/P)$ invariant, and since $\pi(\Gamma)$ is Zariski dense in G', $\varphi(G/P)$ must be Zariski dense in G'/H'. Therefore $h_1h_2^{-1}$ leaves all G'/H' pointwise fixed, and since $\bigcap_{g \in G'} gH'g^{-1} = \{e\}$ (since it is an amenable normal subgroup), $h_1 h_2^{-1} = e$. Therefore $\overline{\Gamma}$ is the graph of a function $G \to G'$, which is a homomorphism since $\Gamma \subset G$ is Zariski dense and the map is a homomorphism on Γ.

We now return to the proof of the lemma. We must show that a certain measurable mapping between varieties is actually rational. There is one well known situation in which a measurable map is known to have much stronger properties, namely if the map is a homomorphism. For example, any measurable homomorphism between Lie groups is C^∞, and similarly, any measurable homomorphism between real algebraic groups $R \to R'$, with R reductive, will be rational on all unipotent subgroups of R. Of course our map φ is defined on G/P which is not a group. However, up to a set of measure 0, it is a group. For example, consider $G = \mathrm{SL}(n, \mathbf{R})$, $P =$ upper triangular subgroup. Let U be the lower triangular unipotent matrices. Then the natural map $G \to G/P$ carries U onto an open subset of measure 1. Furthermore, this establishes an isomorphism of U with its image as algebraic varieties. Thus, we can view φ as a map $U \to G'/H'$. Now, although we have φ defined on a group, it is not a homomorphism (as the image is not even a group). We do however, have some sort of algebraic relation, namely the fact that φ is a Γ-map. Thus we might hope to be able to force this algebraic relation to show that φ only depends upon a homomorphism of U. This, however, is not possible. As we shall see, when we try to force the algebra, we shall need some commutativity of U with elements of A, where A is the positive diagonals. But the centralizer of U in A is trivial. In the $\mathbf{R}$-rank 1 case, we can proceed no further, but in higher rank all is not lost. Let us fix an element $t \in A$, and consider the centralizer C_t. For example, in $\mathrm{SL}(3, \mathbf{R})$, let

$$t = \begin{pmatrix} \lambda & 0 & 0 \\ 0 & \lambda & 0 \\ 0 & 0 & \lambda^{-2} \end{pmatrix}, \quad C_t = \left\{ \begin{pmatrix} M & 0 \\ \hline 0 & \alpha \end{pmatrix} \text{ where } \alpha = (\det M)^{-1} \right\}.$$

Let

$$C_t^u = C_t \cap U = \left\{ \begin{pmatrix} 1 & 0 & 0 \\ a & 1 & 0 \\ 0 & 0 & 1 \end{pmatrix} \right\}.$$

Now $U \cong \mathbf{R}^3$, and $C_t^u \cong \mathbf{R}$. Thus C_t^u will give us one direction in U, and C_t is a reductive group that has a centralizer in A. As we shall see, this will be enough to show that $\varphi\colon U \to G'/H'$ depends rationally on C_t^u. But now if we vary $t \in A$, we can pick up the other directions in the same way. The following lemma of Margulis is now clearly relevant.

LEMMA B. *Let φ be a measurable function defined on $\mathbf{R}^n \times \mathbf{R}^k$. If φ is rational in x for almost all $y \in \mathbf{R}^k$ and rational in y for almost all $x \in \mathbf{R}^n$, then φ is rational.*

The above remarks about $\mathrm{SL}(n, \mathbf{R})$ extend to general G. Thus, if we let U be the unipotent radical of the parabolic opposite to P, then $U \to G/P$ is an isomorphism of algebraic varieties with its image, the latter being open and of full measure. Let $A \subset P$ be a maximal abelian $\mathbf{R}$-diagonalizable subgroup and

$t \in A$, $t \neq 0$. Let C_t be the centralizer of t in G. Then C_t is reductive. Letting $C_t^u = U \cap C_t$, U can be built from the various C_t^u by varying t. Thus, using Lemma B, it suffices now to prove the following. View $\varphi\colon G/P \to G'/H'$ as a map $\varphi\colon G \to G'/H'$, with $\varphi(pg) = \varphi(g)$, for $p \in P$.

LEMMA C. *For almost all $g \in G$, $\varphi(cg)$ depends rationally on c for $c \in C_t^u$ (for any $t \in A$).*

PROOF. Let $C_t = C$. We want to study dependence of φ on C, so for each $g \in G$, define $\omega_g\colon C \to G'/H'$ by $\omega_g(c) = \varphi(cg)$.

Thus we have a map $\omega\colon G \to F(C, G'/H')$, the latter being the space of measurable maps $C \to G'/H'$. Let $T = \{t^n\}$. Then $\omega_{tg}(c) = \omega(ctg) = \omega(tcg) = \omega(cg)$ (since $t \in P$), and so we have $\omega_{tg}(c) = \omega_g(c)$. Thus we can view ω as a map $\omega\colon G/T \to F(C, G'/H')$. We now use Γ-invariance of φ:

$$\omega_{g\gamma}(c) = \varphi(cg\gamma) = \varphi(cg)\pi(\gamma) = \omega_g(c) \cdot \pi(\gamma).$$

Thus, $\omega_{g\gamma}$ and ω_g are in the same G'-orbit in $F(C, G'/H')$, where G' acts on the latter pointwise. We now need another smoothness result.

LEMMA D. *Every G'-orbit in $F(X, G'/H')$ is locally closed, where X is a measure space.*

We observe that if X is finite, this is immediate from the fact that $F(X, G'/H')$ would then be a variety. Margulis observed that the lemma is true for any measure space X. For a simple proof, see the appendix of **[55]**.

Returning now to the proof of Lemma C, we have that $\omega_{g\gamma} \equiv \omega_g$ when projected to $[F(C, G'/H')]/G'$. By Lemma D, this latter space is countably generated and separated, and by Moore's theorem Γ is ergodic on G/T. Therefore, all ω_g are equal when projected to $[F(C, G'/H')]/G'$, or equivalently, all ω_g lie in the same G'-orbit. So for $a, g \in G$, we have

$$\omega_{ag} = \omega_g \cdot h(a, g)$$

where $h(a, g) \in G'$ and h is measurable. For any $f \in F(C, G'/H')$, let G'_f be the stabilizer, and N_f the normalizer of G'_f in G'. Clearly for $a \in C$, $G'_{\omega_{ag}} = G'_{\omega_g}$, so for $a \in C$, $h(a, g) \in N_{\omega_g}$. Suppose now that $a_1, a_2 \in C$. Then

$$\begin{aligned}
\omega_g(c) \cdot h(a_1 a_2, g) &= \omega_{a_1 a_2 g}(c) = \varphi(ca_1 a_2 g) \\
&= \omega_{a_2 g}(ca_1) = \omega_g(ca_1) \cdot h(a_2, g) \\
&= \varphi(ca_1 g) \cdot h(a_2, g) = \omega_{a_1 g}(c) \cdot h(a_2, g) \\
&= \omega_g(c) \cdot h(a_1, g) h(a_2, g).
\end{aligned}$$

Thus, for almost all g, $a \to h(a, g)$ is a measurable homomorphism

$$C \to N_{\omega_g}/G'_{\omega_g}.$$

We have $\varphi(cag) = \varphi(cg) \cdot h(a, g)$. Thus for a in any unipotent subgroup of C, $a \to \varphi(cag)$ depends rationally on a. Choosing this subgroup to be $c^{-1}C_t^u c$, we obtain $\varphi(bg)$ depends rationally on $b \in C_t^u$. This completes the proof.

7. Complements to the rigidity theorem for ergodic actions: Foliations by symmetric spaces and Kazhdan's property (T). The rigidity theorem for ergodic actions stated in §6 allowed us to distinguish ergodic actions of lattices on the basis of orbit equivalence if the actions had finite invariant measure (e.g. Corollary 6.6 and Example 6.7). However, some of the most interesting actions of lattices, e.g., the action of $SL(n, Z)$ on $\mathbf{P}^{n-1}$ or other flag and Grassman varieties, do not have finite invariant measure. We now indicate how to extend the rigidity theorem to enable us to deal with this situation. The main step is to first extend the rigidity theorem to actions of general connected groups.

Let H be a connected group. Every locally compact group has a unique maximal normal amenable subgroup N. If H is connected H/N will be a product of noncompact connected simple Lie groups with trivial center. We shall say that an ergodic action of H is irreducible if the inverse image in H of each of these simple factors of H/N is still ergodic.

THEOREM 7.1 [**56**]. *Let H, H' be connected locally compact groups, N, N' the maximal normal amenable subgroups. Suppose $\mathbf{R}$-rank$(H/N) \geq 2$. Let S, S' be free ergodic irreducible H-, H'-spaces with finite invariant measure, and suppose the actions are orbit equivalent. Then H/N and H'/N' are isomorphic, and N is compact if and only if N' is also compact.*

Thus, for connected groups, orbit equivalence implies isomorphism of the semisimple parts of the groups. The proof of this result is an extension of the proof of the rigidity theorem for ergodic actions of semisimple groups. To see how to apply this to obtain results about actions of lattices without invariant measure, observe that the orbit space of Γ acting on G/H can be identified with the orbit space of H acting on G/Γ. Theorem 7.1 deals with the latter situation since now G/Γ has a finite H-invariant measure. so we can try to apply this to the action of Γ on G/H. One can then prove the following precise result.

THEOREM 7.2 [**56**]. *Let G, G' be connected semisimple Lie groups with finite center, Γ, Γ' irreducible lattices. Let $H \subset G$, $H' \subset G'$ be almost connected noncompact subgroups. Assume the actions of Γ on G/H and Γ' on G'/H' are essentially free and orbit equivalent. Let N, $N' \subset H$, H' be the maximal normal amenable subgroups, and suppose $\mathbf{R}$-rank$(H/N) \geq 2$. Then H/N and H'/N' are locally isomorphic.*

EXAMPLE 7.3 [**56**]. As we vary n, $n \geq 2$, the actions of $SL(n, Z)$ on $\mathbf{P}^{n-1}$ are mutually nonorbit equivalent. This follows by simply observing that the semisimple parts of the corresponding maximal parabolics in $SL(n, \mathbf{R})$ are not isomorphic. (Actually, Theorem 7.2 will not apply to compare the cases $n = 2$ and $n = 3$. However, the action of $SL(2, Z)$ on $\mathbf{P}^1$ is amenable, while the action of $SL(3, Z)$ on $\mathbf{P}^2$ is not.) In a similar fashion, one can read off a large number of results about actions of lattices on the flag and Grassman varieties.

A natural question that arises in light of Theorem 7.1 is how sensitive orbit equivalence is to the way in which H is built from N and H/N. For example, what is the relation of actions of $SL(n, \mathbf{R}) \times \mathbf{R}^n$ to that of $SL(n, \mathbf{R}) \circledS \mathbf{R}^n$, where the latter semidirect product just results from the natural action of $SL(n, \mathbf{R})$ on $\mathbf{R}^n$? To answer this question, we recall Kazhdan's notion of property (T) for groups, and then indicate how to define this for actions.

Let G be a locally compact group, and I the one dimensional trivial representation. If π_n, π are unitary representations of G, then recall $\pi_n \to \pi$ means that for any unit vectors $v_1, \ldots, v_k \in H_\pi$ there exist unit vectors $v_1^n, \ldots, v_k^n \in H_{\pi_n}$ such that $\langle \pi_n(g)v_i^n \mid v_j^n \rangle \to \langle \pi(g)v_i \mid v_j \rangle$ uniformly on compact sets in G for each i, j. Kazhdan **[24]** defined a group to have property (T) if $\pi_n \to I$ implies $I \leqslant \pi_n$ for n sufficiently large.

THEOREM 7.4 (KAZHDAN **[24, 9]**). (i) *Semisimple Lie groups with all simple factors having* $\mathbf{R}$*-rank at least* 2 *have property* (T). (*Actually, Kazhdan proved this assuming* $\mathbf{R}$*-rank* $\geqslant 3$. *That one only need assume* $\mathbf{R}$*-rank* $\geqslant 2$ *was observed by a number of persons, e.g.* **[9]**.)

(ii) *Any lattice subgroup of a group with property* (T) *also has property* (T).

We shall also need the following result of Wang.

THEOREM 7.5 (WANG **[46]**). $SL(n, \mathbf{R}) \circledS \mathbf{R}^n$ *has property* (T), *and hence so does* $SL(n, Z) \circledS Z^n$ $(n \geqslant 3)$.

We now define property (T) for ergodic actions. For simplicity, we restrict attention to actions of discrete groups. This notion for actions originally appeared in **[57]**.

Let G be a discrete group, S an ergodic G-space. Let $\alpha: S \times G \to U(H)$ be a unitary group valued cocycle. Let $v, w: S \to H$ be Borel functions with $\|v\|_\infty = \|w\|_\infty = 1$. Let $f_{\alpha,v,w}: S \times G \to \mathbf{C}$ be given by

$$f_{\alpha,v,w}(s, g) = \langle \alpha(s, g)v(sg) \mid w(s) \rangle.$$

We consider $f_{\alpha,v,w}$ as a function $G \to F(S, \mathbf{C})$, and we endow $F(S, \mathbf{C})$ with the topology of convergence in measure. For cocycles α_n, α, we say $\alpha_n \to \alpha$ if given $v_1, \ldots, v_k: S \to H_\alpha$, $\|v_i\|_\infty = 1$, there exist $v_1^n, \ldots, v_k^n: S \to H_{\alpha_n}$ such that $f_{\alpha_n, v_i^n, v_j^n} \to f_{\alpha, v_i, v_j}$ pointwise on G (i.e. in measure on S for each $g \in G$) for all i, j.

DEFINITION 7.6 **[57]**. The action of G on S has property (T) if $\alpha_n \to I$ implies $\alpha_n \geqslant I$ for n sufficiently large. Here I is the one dimensional trivial cocycle and $\alpha \geqslant I$ means $\alpha \sim \beta$ where $\beta(s, g)v = v$ for some nonzero vector v.

We then have the following results.

THEOREM 7.7 **[57]**. (a) *If G has property* (T), *and S has a finite invariant measure then S has property* (T).

(b) *If S has property* (T), *finite invariant measure and is weak mixing* (*i.e. there are no finite dimensional invariant subspaces in* $L^2(S) \ominus \mathbf{C}$), *then G has property* (T).

(c) *For free actions of discrete groups, property* (T) *is an invariant of orbit equivalence.*

Combining 7.5 and 7.7, we have the following, showing that, in fact, orbit equivalence is quite sensitive to the way H is constructed from N and H/N.

COROLLARY 7.8 [**57**]. *Let* $n \geqslant 3$, *and* $\Gamma_1 = \mathrm{SL}(n, Z) \times Z^n$, $\Gamma_2 = \mathrm{SL}(n, Z) \circledS Z^n$. *Then* Γ_1 *and* Γ_2 *do not have free orbit equivalent weakly mixing actions with finite invariant measure.*

We shall now describe a geometric interpretation of the rigidity theorem for ergodic actions. We begin by recalling the geometric formulation of the Mostow-Margulis theorem. Let G be a connected semisimple Lie group with finite center and no compact factors, $K \subset G$ a maximal compact subgroup, and $\Gamma \subset G$ a torsion free lattice. Then G/K is a Riemannian symmetric space (diffeomorphic to Euclidean space), and Γ operates properly discontinuously on $X = G/K$. Thus $\Gamma \backslash X$ is a locally symmetric space of finite volume, and $\pi_1(\Gamma \backslash X) \cong \Gamma$.

THEOREM 7.9 (MOSTOW-MARGULIS RIGIDITY, GEOMETRIC FORM). *Let* M_1, M_2 *be locally symmetric Riemannian manifolds of finite volume whose universal covers,* X_1, X_2 *are symmetric spaces of purely noncompact type, and whose fundamental groups* $\pi_1(M_i)$ *act as irreducible groups of isometries of* X_i. *Suppose further that the rank of* M_1 *is at least* 2. *Then any isomorphism* $\pi_1(M_1) \to \pi_1(M_2)$ *is induced by a diffeomorphism* $M_1 \to M_2$ *that is an isometry modulo normalizing scalar multiples.*

Rougly speaking, this asserts that for a particular class of Riemannian manifolds, i.e. suitable locally symmetric spaces, that a purely topological invariant, namely the fundamental group, determines the Riemannian structure. We now describe an analogous geometric interpretation of the rigidity theorem for actions which will make an assertion about foliations by symmetric spaces.

Let G be a connected semisimple noncompact Lie group with finite center and no compact factors, $K \subset G$ a maximal compact subgroup, and (S, μ) a free ergodic G-space with finite invariant measure. Let $Y = S/K$. Then because K is compact, Y is a standard Borel space, and the orbits in S yield an equivalence relation $\mathfrak{F}$ on Y in which each equivalence class can be identified with G/K. Thus, $(Y, \mathfrak{F})$ is a "Riemannian measurable foliation", i.e., a measure space with an equivalence relation $\mathfrak{F}$ in which each equivalence class (or "leaf") has the structure of a C^∞-Riemannian manifold, so that these structures vary measurably in a suitable sense [**53**] over Y. Given Y, Y', two spaces supporting Riemannian measurable foliations, we call them isometric if there is a measure space isomorphism between Y and Y' that carries leaves onto leaves isometrically (possibly after discarding null sets of leaves). By a transversal for $(Y, \mathfrak{F})$ we mean a Borel set intersecting almost every leaf in a countable set. Such a set T will have a natural equivalence relation on it with countable equivalence classes (namely $\mathfrak{F} | T$), and a natural measure class ν satisfying the condition that for $B \subset T$, $\nu(B) = 0$ if and only if the union of the leaves intersecting B has μ-measure 0 [**13**, **41**]. We call two Riemannian measurable foliations transversally equivalent if they have isomorphic transversals. By an isomorphism of transversals, we mean isomorphism as measure spaces with equivalence relations, i.e., a measure space isomorphism carrying one equivalence relation onto the other. This is a purely measure-theoretic invariant of the foliation. The following is the geometric version of Theorem 6.3.

THEOREM 7.10 [**55**] (RIGIDITY FOR FOLIATIONS BY SYMMETRIC SPACES). *Let G, G', S, S' be as in Theorem* 6.3. *Let $Y = S/K$, $Y' = S'/K'$ where $K, K' \subset G, G'$ are the maximal compact subgroups. Let $(Y, \mathfrak{T})$, $(Y', \mathfrak{T}')$ be the associated Riemannian measurable foliations by symmetric spaces. If the foliations are transversally equivalent, then they are isometric, modulo normalizing scalar multiples (independent of the leaves).*

Thus, roughly speaking, for suitable foliations in which the leaves are symmetric spaces, a purely measure-theoretic invariant, namely the measure theory of the transversal, determines the Riemannian structure on almost every leaf.

As we have already remarked, the rigidity theorem for lattices holds in the $\mathbf{R}$-rank 1 case as well as long as $G \neq \mathrm{PSL}(2, \mathbf{R})$, although the proof we have given in §6 does not apply, and one must use other techniques, for example those of Mostow [**36**] and Prasad [**39**]. It is natural to inquire as to what extent the rigidity theorem for ergodic actions holds in the $\mathbf{R}$-rank 1 case as well. In [**58**] we proved the following result in this direction, applying basic results of Mostow [**35**] on quasi-conformal mappings.

THEOREM 7.11 [**58**]. *Let S, S' be free ergodic $\mathrm{SO}(1, n)/\{\pm, I\}$-spaces with finite invariant measure, and assume $n \geqslant 3$. Let $(Y, \mathfrak{T})$, $(Y', \mathfrak{T}')$ be the associated measurable foliations by hyperbolic space (as in the discussion preceding Theorem* 7.10). *If $(Y, \mathfrak{T})$ and $(Y', \mathfrak{T}')$ are quasi-conformally equivalent, then they are isometric (modulo a normalizing scalar independent of the leaf), and the actions of $\mathrm{SO}(1, n)/\{\pm I\}$ on S and S' are automorphically conjugate.*

Here, of course, quasi-conformal equivalence asserts the existence of a measure space isomorphism taking (almost all) leaves to leaves quasi-conformally. We remark that the analogous statement for $\mathbf{R}^n$ actions can be shown to be false by many counterexamples.

8. Margulis' finiteness theorem. In §6, we saw how the analysis of the ergodic action of Γ on G/P led to Margulis' proof of the rigidity theorem. Margulis has also demonstrated some other deep properties of this ergodic action and used this to obtain very strong results about the structure of Γ. More precisely, he has shown the following.

THEOREM 8.1 (MARGULIS [**30, 31**]). *Let G be a connected semisimple Lie group with finite center and no compact factors, and assume $\mathbf{R}$-rank$(G) \geqslant 2$. Let $\Gamma \subset G$ be an irreducible lattice, and $H = \Gamma/N$ a nonamenable quotient group. Then $N \subset Z/Z(G)$, the center of G, and in particular, is finite.*

If we further assume that the $\mathbf{R}$-rank of every simple factor of G is at least 2, then Γ has property (T) of Kazhdan [**24**], and hence if $H = \Gamma/N$ is an amenable quotient, H must also have property (T) and hence is finite. Thus, we conclude the following.

COROLLARY 8.2. *Let G be a connected semisimple Lie group with finite center and assume* **R**-*rank of each simple factor of G is at least 2. Let $\Gamma \subset G$ be an irreducible lattice. Then every normal subgroup of G is either finite or of finite index.*

Margulis' results are in fact significantly more general, both in terms of taking lattices in products of algebraic groups defined over various local fields and in terms of rank restrictions. The basic difficult step in the proof of Theorem 8.1 is the following result concerning the action of Γ on G/P. Let P' be another parabolic subgroup containing P. Then there is a Γ-map $G/P \to G/P'$, i.e. G/P' is a Γ-space factor of G/P.

THEOREM 8.3 (MARGULIS [**30**]). *Let G, Γ be as in Theorem 8.1, $P \subset G$ a minimal parabolic. Then any measurable factor of the Γ-space G/P is of the form $G/P \to G/P'$ for some parabolic $P' \supset P$.*

In other words, every measurable Γ-factor of G/P is actually also a G-factor. This theorem is difficult and we will not prove it here. Instead, we show how to deduce Theorem 8.1 from it.

Let $H = \Gamma/N$ be a nonamenable quotient. Then there is a compact metric H-space X so that there is no H-invariant measure on X. We can also view X as a compact metric Γ-space. Since the action of Γ on G/P is amenable, by the discussion following Definition 5.2, there is a measurable Γ-map φ: $G/P \to M(X)$, where the latter is the space of probability measures on X. If we let μ be a measure on G/P in the natural measure class, then $(M(X), \varphi_*(\mu))$ is a Γ-space factor of G/P. Thus, there is some parabolic P' so that as Γ-spaces, $(M(X), \varphi_*(\mu))$ is conjugate to G/P'. Since there are no fixed points in $M(X)$ under Γ, $P' \neq G$. But N acts trivially on $M(X)$ by definition, so N is trivial on G/P' which implies $N \subset \bigcap_{g \in G} gP'g^{-1}$, a proper normal subgroup of G. Dividing G by its center, it clearly suffices to observe that if $\Gamma \subset \prod_{i \in I} G_i$ is an irreducible lattice in a product of simple Lie groups with trivial center, that $N = \Gamma \cap \prod_{i \in J} G_i$ is trivial for $J \subset I$ a proper subset. But since N is normalized by Γ and $\prod_{I-J} G_i$, it is normalized by the product of these groups which is dense in G by irreducibility. The result follows.

9. Margulis' arithmeticity theorem. (This section will require a bit more knowledge about algebraic groups than previous sections. We also caution the reader that in this section, by algebraic group, Zariski closure, etc., we shall mean with respect to the algebraically closed field, unless we explicitly declare otherwise in a given instance.)

In this section we describe the proof of Margulis' arithmeticity theorem for lattices in semisimple Lie groups. The proof of the rigidity theorem in §6 was based on a result asserting that under suitable hypotheses, a homomorphism of Γ into a real algebraic group extended to a homomorphism of G. This result is also basic to the proof of the arithmeticity theorem. However, we shall also need results concerning homomorphisms of Γ into complex groups and algebraic groups over local fields. With some additional comments, the proof of Theorem 6.5 can be applied to give us these needed results, so that the bulk of the work of the proof of arithmeticity has in fact already been done. But before passing to these arguments, let us recall the statement of the problem.

The first example of a lattice in a Lie group is the integer lattice $Z^n \subset \mathbf{R}^n$. This is of course not the only lattice in $\mathbf{R}^n$. However if L is any lattice there is an automorphism $A\colon \mathbf{R}^n \to \mathbf{R}^n$ such that $A(Z^n) = L$. Thus, L is "arithmetically" defined.

To get other examples of lattices, suppose $G \subset \mathrm{GL}(n, \mathbf{C})$ is an algebraic group defined over $\mathbf{Q}$, i.e. there is an ideal $I \subset \mathbf{Q}[a_{ij}, \det(a_{ij})^{-1}]$ such that $G = \{a \in \mathrm{GL}(n, \mathbf{C}) \mid p(a) = 0 \text{ for all } p \in I\}$. As usual, if $B \subset \mathbf{C}$ is any subring, we let

$$G_B = \left\{a \in G \mid a_{ij} \in B, \text{ for all } i, j \text{ and } \det(a_{ij})^{-1} \in B\right\}.$$

THEOREM 9.1 (BOREL-HARISH-CHANDRA [**5**]). *If G is semisimple, then G_Z is a lattice in $G_{\mathbf{R}}$.*

For example, for $G = \mathrm{SL}(n, \mathbf{C})$, we have $\mathrm{SL}(n, Z)$ is a lattice in $\mathrm{SL}(n, \mathbf{R})$. The question the arithmeticity theorem answers is to what extent this is a general construction, i.e. to what extent are lattices arithmetically defined? We now exhibit two ways of modifying a given lattice to obtain a new lattice.

DEFINITION 9.2. If Γ, Γ' are discrete groups, then Γ and Γ' are called commensurable if $[\Gamma\colon \Gamma \cap \Gamma'] < \infty$ and $[\Gamma'\colon \Gamma \cap \Gamma'] < \infty$.

PROPOSITION 9.3. *If $\Gamma, \Gamma' \subset G$, Γ is a lattice and Γ, Γ' are commensurable, then Γ' is a lattice.*

For example, given that $\mathrm{SL}(n, Z)$ is a lattice, $\{a \in \mathrm{SL}(n, Z) \mid a = I \bmod p$ for a given prime $p\}$ is a commensurable lattice.

Here is another way to get new lattices.

PROPOSITION 9.4. *If $\Gamma \subset H$ is a lattice, $\varphi\colon H \to G$ a surjective homomorphism with compact kernel then $\varphi(\Gamma)$ is a lattice in G.*

Margulis' theorem says that aside from these two types of rather trivial modifications, every lattice in a semisimple Lie group of higher $\mathbf{R}$-rank arises as in Theorem 9.1. More precisely, let us make the following definition. (If H is a group, H^0 denotes the topologically connected component of the identity.)

DEFINITION 9.5. Let G be a connected semisimple Lie group with trivial center and no compact factors. Let $\Gamma \subset G$ be a lattice. Then Γ is called arithmetic if there exists an algebraic group H defined over $\mathbf{Q}$, and a surjective homomorphism $\varphi\colon H^0_{\mathbf{R}} \to G$ such that

(i) kernel(φ) is compact;

(ii) $\varphi(H_Z \cap H^0_{\mathbf{R}})$ is a lattice in G commensurable with Γ.

THEOREM 9.6 (MARGULIS [**28**]). *Let G be as in Definition 9.5, and assume $\mathbf{R}$-rank$(G) \geq 2$. Then any irreducible lattice in G is arithmetic.*

As we indicated above, the proof is based on two further results about homomorphisms of Γ.

THEOREM 9.7 (MARGULIS [28]). *Let $\Gamma \subset G$ be an irreducible lattice, G as above, $\mathbf{R}$-rank$(G) \geq 2$.*

(i) *If H is a (complex) simple algebraic group, connected and with trivial center, then any homomorphism $\pi: \Gamma \to H$ with $\pi(\Gamma)$ Zariski dense in H either satisfies $\overline{\pi(\Gamma)}$ compact or extends to a rational endomorphism $\tilde{G} \to H$, where $\tilde{G}$ is the Zariski closure of G (embedding G in the linear transformations in the complexified Lie algebra for example).*

(ii) *Any homomorphism $\pi: \Gamma \to H_K$ where H is a semisimple algebraic group over K, and K is a local totally disconnected field of characteristic 0, with $\pi(\Gamma)$ Zariski dense, satisfies $\overline{\pi(\Gamma)}$ is compact.*

The proof we present is in the spirit of the proof we gave of Theorem 6.5 so as to be generalizable to cocycles defined on general ergodic G-spaces. We expect these generalized results to be of use in describing "arithmetic" features of an ergodic action, but we do not discuss this here.

PROOF. (i) The proof we gave of Theorem 6.5 can be applied if we can find a measurable Γ-map $\varphi: G/P \to H/H_0$ where H_0 is an algebraic subgroup of H such that $\bigcap_{h \in H} hH_0h^{-1} = \{e\}$. As in Theorem 6.5, we can let $P' \subset H$ be a minimal parabolic subgroup, use amenability to find a Γ-map $\varphi: G/P \to M(H/P')$ and prove that each orbit in $M(H/P')$ under H is locally closed. Again, as in 6.5, we can then assume $\varphi: G/P \to H/H_1$ where H_1 is the stabilizer of a measure in $M(H/P')$. Unlike the real case however, this stabilizer need not be algebraic. For example, the group may be compact which in the real case implies that it is the real points of an algebraic group, while in the complex case, of course, a compact group will not be algebraic. However, we can suppose H is rationally represented on a finite dimensional complex space in such a way that P' is the stabilizer of a point in projective space. Let μ be the measure on H/P' stabilized by H_1. If H_1 is not compact, then using an argument as in Furstenberg's lemma (Lemma 5.9) we see that μ must be supported on the intersection of H/P' with the union of two proper projective subspaces. Choose a proper subspace V so that $\mu(H/P' \cap [V]) > 0$, and V has minimal dimension among all subspaces with this property. By the minimality property of $[V]$ and H_1-invariance of μ, the H_1-orbit of $[V]$ must clearly be a finite union of projective subspaces. Hence, if we let H_0 be the Zariski closure of H_1, then $H_0 \subset H$ is a proper algebraic subgroup. Since H is simple and with trivial center $\bigcap hH_0h^{-1} = \{e\}$, and as we remarked at the beginning of the proof, this suffices.

We must now consider the case in which H_1 is compact. We then have a Γ-map $\varphi: G/P \to H/H_1$, so that if we let $\nu = \varphi_*(\mu)$, ν is a quasi-invariant ergodic measure for the action of Γ on H/H_1. (Unlike the previous paragraph, μ is now the natural measure class on G/P.) Consider the Γ-map $\varphi \times \varphi$: $G/P \times G/P \to H/H_1 \times H/H_1$. It is well known that on G/P, the P-action is essentially transitive, the conull orbit having $P \cap \overline{P}$ as stabilizer, where $\overline{P}$ is the opposite parabolic to P. Thus as a G-space, $G/P \times G/P$ will be essentially transitive with stabilizer $P \cap \overline{P}$, which is noncompact. By Moore's ergodicity theorem (§2), Γ is therefore ergodic on $G/P \times G/P$. It follows that Γ must also be ergodic on $(H/H_1, \nu) \times (H/H_1, \nu)$. Since H_1 is compact, the H-orbits on $H/H_1 \times H/H_1$ are closed. Since $(\nu \times \nu)$ is ergodic under Γ, and Γ-orbits are of course contained in H-orbits, smoothness of the H-action on $H/H_1 \times H/H_1$ implies that $\nu \times \nu$ must be supported on an H-orbit. From Fubini's theorem, one easily deduces that ν must be supported on an H_2 orbit in H/H_1 where H_2 is a conjugate of H_1, and in particular is compact. Thus, support(ν) is compact. Since ν is quasi-invariant under $\pi(\Gamma)$, support(ν) is $\pi(\Gamma)$-invariant, and since H_1 is also compact, it follows that $\pi(\Gamma)$ is contained in a compact set. This completes the proof of (i).

(ii) Let $P' \subset H$ be a minimal parabolic K-subgroup, so that H_K/P'_K is compact, and P'_K contains no normal algebraic subgroup. We again wish to apply the same type of argument as in the proof of Theorem 6.5. The first step is to prove the analogue of Theorem 5.7 over K. In fact the proof in **[52]** shows that $\mathrm{GL}(n, K)$ acts smoothly on $M(\mathbf{P}^{n-1}(K))$. We can assume that we have a faithful rational representation of H_K on K^n so that H_K/P'_K is an orbit in $\mathbf{P}^{n-1}(K)$. By amenability of the Γ-action on G/P, there is a Γ-map φ: $G/P \to M(H_K/P'_K) \subset M(\mathbf{P}^{n-1}(K))$. By smoothness of the $\mathrm{GL}(n, K)$-action on $M(\mathbf{P}^{n-1}(K))$, we can view φ as a map φ: $G/P \to [\mu \cdot \mathrm{GL}(n, K)] \cap M(H_K/P'_K)$ where $\mu \in M(H_K/P'_K)$. Let S be the stabilizer of μ in $\mathrm{GL}(n, K)$. If S is compact in $\mathrm{PGL}(n, K)$, then by the argument in part (i), $\overline{\pi(\Gamma)}$ will be compact in $\mathrm{PGL}(n, K)$, and so $\overline{\pi(\Gamma)}$ will also be compact in H. If not, then using an argument as in Furstenberg's lemma (Lemma 5.9), we can, as in part (i), assume the Zariski closure L of S is a proper algebraic subgroup. Furthermore, we can clearly assume from the construction of L as in part (i), that for any $g \in \mathrm{GL}(n, K)$, $\dim(H \cap gLg^{-1}) < \dim H$. By the condition of Zariski density, it therefore suffices to see that $\pi(\Gamma) \subset gL_K g^{-1}$ for some $g \in \mathrm{GL}(n, K)$. We have φ: $G/P \to \mathrm{GL}(n, K)/L_K$ a measurable Γ-map. In the real case we showed φ was rational by showing it could be built from homomorphisms of unipotent subgroups of G which had to be rational. In the present situation, we can construct the same type of homomorphism using the argument of Theorem 6.5, but now, since the image group is totally disconnected, these maps must be constant. We thus conclude that φ: $G/P \to \mathrm{GL}(n, K)/L_K$ is essentially constant. Since $\pi(\Gamma)$ leaves $\varphi(G/P)$ fixed, this implies $\pi(\Gamma)$ is contained in a $\mathrm{GL}(n, K)$-conjugate of L_K, and this completes the proof.

We now turn to the proof of Theorem 9.6 itself. We may take the semisimple Lie group to be $G^0_{\mathbf{R}}$, where now $G \subset \mathrm{GL}(n, \mathbf{C})$ represents an algebraic group defined over $\mathbf{Q}$. Thus $\Gamma \subset G^0_{\mathbf{R}}$ is an irreducible lattice. The following lemma is classical, and follows for example from an argument of Selberg **[42]** (see also **[40**, Proposition 6.6] for the same argument). This argument is based on expressing the embeddings of Γ into G as an algebraic variety, and then choosing a real algebraic point of this variety. However, with Theorem 9.7 at hand, we present an alternative argument due to Margulis **[29]**.

LEMMA 9.8. *There is a real algebraic number field k and a rational faithful representation of G such that, identifying G with its image under this representation, $\Gamma \subset G_k$.*

PROOF. The first step is to show that for K the field of real algebraic numbers we have $\mathrm{Tr}(\mathrm{Ad}(\gamma)) \in K$ for all γ. Following Margulis, we let σ be an automorphism of $\mathbf{C}$. Then σ acts on matrices with entries in $\mathbf{C}$ by taking $(z_{ij}) \to (\sigma(z_{ij}))$, and since G is defined over $\mathbf{Q}$, σ induces an automorphism of G. (Of course this is an automorphism of G only as an abstract group, and will in general not be measurable.) Let us assume for the moment that G is simple. Then $\sigma \mid \Gamma$ satisfies either (i) $\overline{\sigma(\Gamma)}$ is compact; or (ii) $\sigma \mid \Gamma$ extends to a rational automorphism of G. This follows from Theorem 9.7. In the first case, all eigenvalues of $\mathrm{Ad}(\sigma(\gamma))$ will have absolute value one, and in the second case, these eigenvalues coincide with those of $\mathrm{Ad}(\gamma)$. (We remark that if $A: G \to G$ is an automorphism we have $dA \circ \mathrm{Ad}(A(g)) \circ (dA)^{-1} = \mathrm{Ad}(g)$, so $\mathrm{Tr}(\mathrm{Ad}\, A(g)) = \mathrm{Tr}(\mathrm{Ad}\, g)$.) Hence for each $\gamma \in \Gamma$, $\{\sigma(\mathrm{Tr}(\mathrm{Ad}(\gamma))) \mid \sigma \in \mathrm{Aut}(\mathbf{C})\}$ is bounded. The same can easily be seen if G is semisimple by examining the composition of $\sigma \mid \Gamma$ with projection on the simple factors. However, since $\mathrm{Aut}(\mathbf{C})$ is transitive on the transcendental numbers, it follows that $\mathrm{Tr}(\mathrm{Ad}(\gamma))$ is algebraic for all $\gamma \in \Gamma$.

Thus, identifying G with $\mathrm{Ad}(G)$, we have $\Gamma \subset G$ with $\mathrm{Tr}(\gamma) \in K$ for all $\gamma \in \Gamma$. The next step, which is classical, is to observe that this implies that there is a faithful rational representation of G, defined over K, such that $\Gamma \subset G_K$, once again, identifying G with its image under this representation. We recall the construction. Consider $\mathrm{Tr}: G \to \mathbf{C}$, and let $V = \mathbf{C}$-linear span of Γ-translates of Tr. By the Borel density theorem V will also be G-invariant. Choose a basis of V of the form $\gamma_i \cdot \mathrm{Tr}$. Then one can verify in a straightforward manner that with respect to this basis, the matrix elements of $\gamma \in \Gamma$ acting on this space are all in K. Since Γ is finitely generated (in the property (T) case this follows easily [**9**]) we can find an algebraic number field k with $\Gamma \subset G_k$.

We now recall the basic operation of restriction of scalars. Suppose G is an algebraic group defined over an algebraic number field k. Then there exists an algebraic group $\tilde{G}$ defined over $\mathbf{Q}$ such that

(i) there is an injective map $\alpha: G_k \to \tilde{G}_{\mathbf{Q}}$; and

(ii) there is a surjective rational homomorphism $p: \tilde{G} \to G$ defined over k such that $p(\tilde{G}_{\mathbf{Q}}) = G_k$ and $p \circ \alpha: G_k \to G_k$ is the identity.

We recall here two ways of describing this construction. We can take $\tilde{G} = \prod_{\sigma} \sigma(G)$ where σ runs through the distinct embeddings of k in $\mathbf{C}$. Then $\alpha: G_k \to \tilde{G}$ is the map $\alpha(g) = (\sigma_1(g), \ldots, \sigma_r(g))$, and $p: \tilde{G} \to G$ is projection onto the factor corresponding to $\sigma = \mathrm{id}$. Alternatively, let $[k: Q] = r$, and choose an identification $k^n \leftrightarrow \mathbf{Q}^{nr}$. Then we have a map $\alpha: G_k \to \mathrm{GL}(nr, \mathbf{Q})$, and we let $\tilde{G}$ be the Zariski closure of $\alpha(G_k)$ in $\mathrm{GL}(nr, \mathbf{C})$. In this formulation, p arises from the fact that the entries of $g \in G_k$ are described by k-linear combinations of the entries of $\alpha(g)$. These linear expressions allow us to define a linear map from $nr \times nr$ $\mathbf{Q}$-matrices to $n \times n$ k-matrices, and thus a map $\tilde{G} \to G$. (Recall that G is the Zariski closure of G_k.) Since this map is clearly a homomorphism on $\alpha(G_k)$ (being the inverse of α), it is also a homomorphism on its Zariski-closure, $\tilde{G}$.

COMPLETION OF PROOF OF THEOREM 9.6. We let k be as in Lemma 9.8, and $\tilde{G}, p, \alpha$ as above. We have $\Gamma \subset G_k$, and we let $H \subset \tilde{G}$ be the Zariski closure of

Γ. We still have $p(H) = G$ by Zariski density of Γ in G. Since G is semisimple, p trivial on the radical of H, and, replacing H by the quotient of H by its radical, we can assume H is a semisimple group defined over $\mathbf{Q}$, and with trivial center.

We now claim that $(\ker p)_{\mathbf{R}}$ is compact. Let F be a simple factor of $\ker p$. Then as algebraic groups defined over $\mathbf{R}$, we can write $H \cong G \times F \times F'$ where F' is the product of the remaining simple factors. Since $\alpha(\Gamma)$ is Zariski dense in H, $(q \circ \alpha)(\Gamma)$ is Zariski dense in F where q is projection of H onto F. We claim $F_{\mathbf{R}}$ must be compact. If not, then $(q \circ \alpha)(\Gamma)$ cannot have compact (topological) closure since compact real matrix groups are real points of algebraic groups. This would imply by Theorem 9 7 that $q \circ \alpha$ extended to a rational homomorphism $h\colon G \to F$. But then $\{(g, h(g), f') \mid g \in G, f' \in F'\}$ would be a proper algebraic subgroup containing $\alpha(\Gamma)$, contradicting Zariski density of $\alpha(\Gamma)$ in H. This verifies compactness of $F_{\mathbf{R}}$, and doing the same for each factor, compactness of $(\ker p)_{\mathbf{R}}$.

Now consider $\alpha\colon \Gamma \to H_{\mathbf{Q}}$. For each prime a, the image of $\alpha(\Gamma)$ in $H_{\mathbf{Q}_a}$ must be bounded by Theorem 9.7(ii). This means that the powers of each prime appearing in the denominators of matrix entries of $\alpha(\gamma) \in H_{\mathbf{Q}}$ are bounded uniformly over $\gamma \in \Gamma$. But Γ is finitely generated, and hence only finitely many primes will appear at all. This is readily seen to imply that $\alpha(\Gamma) \cap H_Z$ is of finite index in $\alpha(\Gamma)$, and hence, applying p, that $\Gamma \cap p(H_Z)$ is of finite index in Γ. This in turn implies that $\Gamma \cap p(H_Z)$ is a lattice in $G_{\mathbf{R}}$. By Theorem 9.1 and Proposition 9.4, $p(H_Z)$ is a lattice in $G_{\mathbf{R}}$, and since $(\Gamma \cap p(H_Z)) \subset p(H_Z)$ is an inclusion of lattices, we also have $[p(H_Z)\colon \Gamma \cap p(H_Z)] < \infty$. This shows commensurability of Γ and $p(H_Z)$, completing the proof.

REFERENCES

1. L. Auslander, L. Green and F. Hahn, *Flows on homogeneous spaces*, Ann. of Math. Studies, no. 53, Princeton Univ. Press, Princeton, N.J., 1963.

2. L. Auslander, *An exposition of the structure of solvmanifolds*, Bull. Amer. Math. Soc. **79** (1973), 227–285.

3. A. Borel, *Linear algebraic groups*, Benjamin, New York, 1969.

4. ______, *Density properties for certain subgroups of semisimple Lie groups without compact factors*, Ann. of Math. **72** (1960), 179–188.

5. A. Borel and Harish-Chandra, *Arithmetic subgroups of algebraic groups*, Ann. of Math. **75** (1962), 485–535.

6. A. Borel and J. P. Serre, *Théorèmes de finitude en cohomologie Galoisienne*, Comment. Math. Helv. **39** (1964), 111–164.

7. J. Brezin and C. C. Moore, *Flows on homogeneous spaces: A new look* (preprint).

8. A. Connes, J. Feldman and B. Weiss, *Amenable equivalence relations are generated by a single transformation* (preprint).

9. C. Delaroche and A. Kirillov, *Sur les relations entre l'espace dual d'un groupe et la structure de ses sous-groupes fermés*, Seminarie Bourbaki, no. 343, 1967/68.

10. H. A. Dye, *On groups of measure preserving transformations*. I, Amer. J. Math. **81** (1959), 119–159.

11. ______, *On groups of measure preserving transformations*. II, Amer. J. Math. **85** (1963), 551–576.

12. E. G. Effros, *Transformation groups and C*-algebras*, Ann. of Math. **81** (1965), 38–55.

13. J. Feldman, P. Hahn and C. C. Moore, *Orbit structure and countable sections for actions of continuous groups*, Adv. in Math. **28** (1978), 186–230.

14. H. Freudenthal, *Topologische Gruppen mit Genugend Vielen Fastperiodishen Funktionen*, Ann. of Math. **37** (1936), 57–77.

15. H. Furstenberg, *A Poisson formula for semisimple Lie groups*, Ann. of Math. **77** (1963), 335–383.

16. ______, *The structure of distal flows*, Amer. J. Math. **85** (1963), 477–515.

17. ______, *Ergodic behavior of diagonal measures and a theorem of Szemeredi on arithmetic progressions*, J. Analyse Math. **31** (1977), 204–256.

18. F. P. Greenleaf, *Invariant means on topological groups*, Van Nostrand, New York, 1969.

19. Y. Guivarc'h, *Croissance polynomiale et periodes des fonctions harmonique*, Bull. Soc. Math. France **101** (1973), 333–379.

20. G. Hedlund, *The dynamics of geodesic flows*, Bull. Amer. Math. Soc. **45** (1939), 241–260.

21. E. Hopf, *Statistik der Lösungen geodätischer probleme vom unstabilen typus*, Math. Ann. **117** (1940), 590–608.

22. R. Howe and C. C. Moore, *Asymptotic properties of unitary representations*, J. Funct. Anal. **32** (1979), 72–96.

23. J. W. Jenkins, *Growth of connected locally compact groups*, J. Funct. Anal. **12** (1973), 113–127.

24. D. Kazhdan, *Connection of the dual space of a group with the structure of its closed subgroups*, Functional Anal. Appl. **1** (1967), 63–65.

25. W. Krieger, *On ergodic flows and the isomorphism of factors*, Math. Ann. **223** (1976), 19–70.

26. G. W. Mackey, *Ergodic transformation groups with a pure point spectrum*, Illinois J. Math. **8** (1964), 593–600.

27. G. A. Margulis, *Non-uniform lattices in semisimple algebraic groups*, *in Lie groups and their representations* (I. M. Gelfand, ed.), Wiley, New York.

28. ______, *Discrete groups of motions of manifolds of nonpositive curvature*, Amer. Math. Soc. Transl., vol. 109 (1977), 33–45.

29. ______, *Arithmeticity of irreducible lattices in semisimple groups of rank greater than* 1, Appendix to Russian translation of M. Ragunathan, *Discrete subgroups of Lie groups*, Mir, Moscow, 1977. (Russian)

30. ______, *Factor groups of discrete subgroups*, Soviet Math. Dokl. **19** (1978), 1145–1149.

31. ______, *Quotient groups of discrete subgroups and measure theory*, Functional Anal. Appl. **12** (1978), 295–305.

32. C. C. Moore, *Ergodicity of flows on homogeneous spaces*, Amer. J. Math. **88** (1966), 154–178.

33. ______, *Amenable subgroups of semisimple groups and proximal flows*, Israel J. Math. **34** (1979), 121–138.

34. C. C. Moore and R. J. Zimmer, *Groups admitting ergodic actions with generalized discrete spectrum*, Invent. Math. **51** (1979), 171–188.

35. G. D. Mostow, *Quasi-conformal mappings in n-space and the rigidity of hyperbolic space forms*, Publ. Math. I.H.E.S. (1967), 53–104.

36. ______, *Strong rigidity of locally symmetric spaces*, Ann. of Math. Studies, no. 78, Princeton Univ. Press, Princeton, N.J., 1973.

37. D. Ornstein and B. Weiss (to appear).

38. W. Parry, *Zero entropy of distal and related transformations*, Topological Dynamics (J. Auslander and W. Gottschalk, eds.), Benjamin, New Yor, 1968.

39. G. Prasad, *Strong rigidity of Q-rank* 1 *lattices*, Invent. Math. **21** (1973), 255–286.

40. M. Ragunathan, *Discrete subgroups of Lie groups*, Springer-Verlag, New York, 1972.

41. A. Ramsey, *Virtual groups and group actions*, Adv. in Math. **6** (1971), 253–322.

42. A. Selberg, *On discontinuous groups in higher dimensional symmetric spaces*, Int. Colloquium on Function Theory, Tata Institute, Bombay, 1960.

43. T. Sherman, *A weight theory for unitary representations*, Canad. J. Math. **18** (1966), 159–168.

44. C. Sutherland, *Orbit equivalence: Lectures on Krieger's theorem*, Univ. of Oslo Lecture Notes.

45. V. S. Varadarajan, *Geometry of quantum theory*. II, Van Nostrand, Princeton, N.J., 1970.

46. S. P. Wang, *On isolated points in the dual spaces of locally compact groups*, Math. Ann. **218** (1975), 19–34.

47. R. J. Zimmer, *Extensions of ergodic group actions*, Illinois J. Math. **20** (1976), 373–409.

48. ______, *Ergodic actions with generalized discrete spectrum*, Illinois J. Math. **20** (1976), 555–588.

49. ______, *Orbit spaces of unitary representations, ergodic theory, and simple Lie groups*, Ann. of Math. **106** (1977), 573–588.

50. ______, *Uniform subgroups and ergodic actions of exponential Lie groups*, Pacific J. Math. **78** (1978), 267–272.

51. ______, *Amenable ergodic group actions and an application to Poisson boundaries of random walks*, J. Funct. Anal. **27** (1978), 350–372.

52. ______, *Induced and amenable ergodic actions of Lie groups*, Ann. Sci. École Norm. Sup. (4) **11** (1978), 407–428.

53. ______, *Algebraic topology of ergodic Lie group action and measurable foliations* (preprint).

54. ______, *An algebraic group associated to an ergodic diffeomorphism*, Compositio Math. **43** (1981), 59–69.

55. ______, *Strong rigidity for ergodic actions of semisimple Lie groups*, Ann. of Math. **112** (1980), 511–529.

56. ______, *Orbit equivalence and rigidity of ergodic actions of Lie groups*, Ergod. Th. and Dynam. Sys. **1** (1981), 237–253.

57. ______, *On the cohomology of ergodic actions of semisimple Lie groups and discrete subgroups*, Amer. J. Math. **103** (1981), 937–950.

58. ______, *On the Mostow rigidity theorem and measurable foliations by hyperbolic space* (preprint).

DEPARTMENT OF MATHEMATICS, UNIVERSITY OF CHICAGO, CHICAGO, ILLINOIS 60637

BULLETIN (New Series) OF THE
AMERICAN MATHEMATICAL SOCIETY
Volume 10, Number 2, April 1984

AN ELEMENTARY INTRODUCTION TO THE LANGLANDS PROGRAM

BY STEPHEN GELBART[1]

TABLE OF CONTENTS

PREFACE

In a recent issue of the Notices of the American Mathematical Society (April 1983, p. 273), as part of a very brief summary of *Progress in Theoretical Mathematics* presented to the Office of Science and Technology of the President of the United States by a briefing panel from the National Academy of Sciences chaired by William Browder, the general mathematical reader will find the following paragraphs:

Based on a lecture delivered at the Conference dedicating the Professor Abe Gelbart Chair in Mathematics at Bar Ilan University, Ramat Gan, Israel, January 1983; received by the editors July 12, 1983.

1980 *Mathematics Subject Classification*. Primary 10D40, 12A67; Secondary 22E55.

[1]Supported in part by a grant from the National Science Foundation.

"The unifying role of group symmetry in geometry, so penetratingly expounded by Felix Klein in his 1872 Erlanger Program, has led to a century of progress. A worthy successor to the Erlanger Program seems to be Langlands' program to use infinite dimensional representations of Lie groups to illuminate number theory.

That the possible number fields of degree n are restricted in nature by the irreducible infinite dimensional representations of $\mathrm{GL}(n)$ was the visionary conjecture of R. P. Langlands. His far-reaching conjectures present tantalizing problems whose solution will lead us to a better understanding of representation theory, number theory and algebraic geometry. Impressive progress has already been made, but very much more lies ahead."

The purpose of this paper is to explain what the Langlands program is about —what new perspectives on number theory it affords, and what kinds of results it can be expected to prove.

To begin with, Langlands' program is a synthesis of several important themes in classical number theory. It is also—and more significantly—a program for future research. This program emerged around 1967 in the form of a series of conjectures, and it has subsequently influenced recent research in number theory in much the same way the conjectures of A. Weil shaped the course of algebraic geometry since 1948.

At the heart of Langlands' program is the general notion of an "automorphic representation" π and its L-function $L(s, \pi)$. These notions, both defined via group theory and the theory of harmonic analysis on so-called adele groups, will of course be explained in this paper. The conjectures of Langlands just alluded to amount (roughly) to the assertion that the other zeta-functions arising in number theory are but special realizations of these $L(s, \pi)$.

Herein lies the agony as well as the ecstasy of Langlands' program. To merely state the conjectures correctly requires much of the machinery of class field theory, the structure theory of algebraic groups, the representation theory of real and p-adic groups, and (at least) the language of algebraic geometry. In other words, though the promised rewards are great, the initiation process is forbidding.

Two excellent recent introductions to Langlands' theory are [**Bo** and **Art**]. However, the first essentially assumes all the prerequisites just mentioned, while the second concentrates on links with Langlands' earlier theory of Eisenstein series.

The idea of writing the present survey came to me from Professor Paul Halmos, and I am grateful to him for his encouragement. Although the finished product is not what he had in mind, my hope is that it will still make accessible to a wider audience the beauty and appeal of this subject; in particular, I shall be pleased if this paper serves as a suitable introduction to the surveys of Borel and Arthur.

One final remark: This paper is *not* addressed to the experts. Readers who wish to find additional information on such topics as the trace formula, θ-series, L-indistinguishability, zeta-functions of varieties, etc., are referred to the (annotated) bibliography appearing after Part IV. I am indebted to Martin Karel and Paul Sally for their help in seeing this paper through to its publication.

I. INTRODUCTION

In this article I shall describe Langlands' theory in terms of the classical works which anticipated, as well as motivated, it. Examples are the local-global methods used in solving polynomial equations in integers, especially "Hasse's principle" for quadratic forms; the use of classical automorphic forms and zeta-functions to study integers in algebraic number fields; and the use of groups and their representations to bridge the gap between analytic and algebraic problems. Thus, more than one half of this survey will be devoted to material which is quite well known, though perhaps never before presented purely as a vehicle for introducing Langlands' program.

To give some idea of the depth and breadth of Langlands' program, let me leisurely describe one particular conjecture of Langlands; the rest of this paper will be devoted to adding flesh (and pretty clothes) to this skeletal sketch (as well as defining all the terms alluded to in this Introduction!).

In algebraic number theory, a fundamental problem is to describe how an ordinary prime p factors into "primes" in the ring of "integers" of an arbitrary finite extension E of $\mathbf{Q}$. Recall that the ring of integers O_E consists of those x in E which satisfy a monic polynomial with coefficients in $\mathbf{Z}$. Though O_E need not have unique factorization in the classical sense, every *ideal* of O_E must factor uniquely into prime *ideals* (the "primes" of O_E). Thus, in particular,

$$(*) \qquad pO_E = \prod \mathfrak{P}_i,$$

with each $\mathfrak{P}_i$ a prime ideal of O_E, and the collection $\{\mathfrak{P}_i\}$ completely determined by p.

Now suppose, in addition, that E is *Galois* over $\mathbf{Q}$, with Galois group $G = \mathrm{Gal}(E/\mathbf{Q})$. This means that E is the splitting field of some monic polynomial in $\mathbf{Q}[x]$, and G is the group of field automorphisms of E fixing $\mathbf{Q}$ pointwise. According to a well-known theorem, each element of G moves around the primes $\mathfrak{P}_i$ "dividing" p, and G acts transitively on this set. Thus the "splitting type" of p in O_E is completely determined by the size of the subgroup of G which fixes any $\mathfrak{P}_i$, i.e., by the size of the "isotropy groups" G_i (which are conjugate in G).

For simplicity, we shall now assume that the primes $\mathfrak{P}_i$ in $(*)$ are *distinct*, i.e., the prime p is *unramified* in E. In this case, the afore-mentioned isotropy groups are cyclic. To obtain information about the factorization of such p, attention is focused on the so-called *Frobenius element* $\mathrm{Fr}_{\mathfrak{P}_i}$ of G, the canonical generator of the subgroup of G which maps any $\mathfrak{P}_i$ into itself. (We shall discuss all these matters in more detail in II.C.2.) To be sure, $\mathrm{Fr}_{\mathfrak{P}_i}$ is an automorphism of E over $\mathbf{Q}$ determined *only up to conjugacy* in G. Nevertheless, the resulting *conjugacy class* $\{\mathrm{Fr}_p\}$ completely determines the factorization type of $(*)$. For example, when $\{\mathrm{Fr}_p\}$ is the class of the identity alone, then (and only then) p *splits completely* in E, i.e., p factors into the maximum number of primes in O_E (namely $r = [E:\mathbf{Q}] = \#G$).

In general, one seeks to describe $\{\mathrm{Fr}_p\}$ (and hence the factorization of p in E) intrinsically in terms of p and the arithmetic of $\mathbf{Q}$. To see what this means, consider the example

$$E = \mathbf{Q}(i) = \{\alpha + \beta i : \alpha, \beta \in \mathbf{Q}\},$$

with $O_E = \mathbf{Z}(i) = \{n + mi\colon\ n, m \in \mathbf{Z}\}$. In this case, $G = \mathrm{Gal}(E/\mathbf{Q}) = \{I, \text{complex conjugation}\}$, and some elementary algebra shows that

$$\mathrm{Fr}_p = \begin{cases} I & \text{if } -1 \text{ is a quadratic residue mod } p, \\ \text{conjugation} & \text{otherwise.} \end{cases}$$

For convenience, let us identify $\mathrm{Gal}(E/\mathbf{Q})$ with the subgroup $\{\pm 1\}$ of $\mathbf{C}^{\times}$ via the obvious isomorphism $\sigma\colon G \to \{\pm 1\}$. Then we have

$$\sigma(\mathrm{Fr}_p) = (-1/p),$$

with $(-1/p)$ the Legendre symbol (equal to 1 or -1 according to whether -1 is, or is not, a quadratic residue mod p). To express this condition in terms of a congruence condition *on p* instead of on -1, we appeal to a part of the quadratic reciprocity law for $\mathbf{Q}$ which states that (for odd p, precisely those p unramified in $\mathbf{Q}(i)$)

$$(-1/p) = (-1)^{(p-1)/2}, \quad \text{i.e.,} \quad \sigma(\mathrm{Fr}_p) = (-1)^{(p-1)/2}.$$

This is the type of intrinsic description of Fr_p we sought; from it, and the fact that

$$(-1)^{(p-1)/2} = 1 \Leftrightarrow p \equiv 1 \ (4),$$

we conclude that the factorization of p in $\mathbf{Z}(i)$ depends only on its residue modulo 4. In particular, all primes in a given arithmetic progression mod 4 have the same factorization type in $\mathbf{Z}(i)$. Moreover, since all the prime ideals of $\mathbf{Z}(i)$ are principal, and of the form (n) or $(n + im)$, we obtain the following:

THEOREM (FERMAT 1640, EULER 1754). *Suppose p is an odd prime. Then p can be written as the sum of two squares $n^2 + m^2$ if and only if $p \equiv 1$ (4).*

PROOF. $p = n^2 + m^2 = (n + im)(n - im)$ if and only if p splits completely in $\mathbf{Z}(i)$.

A major goal of class field theory is to give a similar description of $\{\mathrm{Fr}_p\}$ for arbitrary Galois extensions E. However, this goal is far from achieved and, in general, is probably impossible.

In general, we cannot expect there to be a modulus N such that $\{\mathrm{Fr}_p\} = \{I\}$ if and only if p lies in some arithmetic progression mod N. However, if E is *abelian*, i.e., $G = \mathrm{Gal}(E/\mathbf{Q})$ is abelian, then a great deal can be said. Indeed, suppose E is such an extension, and $\sigma\colon G \to \mathbf{C}^{\times}$ is a homomorphism. Then it is known that there exists an integer $N_\sigma > 0$ and a Dirichlet character

$$\chi_\sigma\colon (\mathbf{Z}/N\mathbf{Z})^{\times} \to \mathbf{C}^{\times} \quad \text{such that} \quad \sigma(\mathrm{Fr}_p) = \chi_\sigma(p)$$

for all primes p (unramified in E). This is E. Artin's famous and fundamental *reciprocity law of abelian class field theory.*[2] It implies—just as in the special case $E = \mathbf{Q}(i)$—that the splitting properties of p in E depend only on its

[2]The more familiar form of this law directly identifies $\mathrm{Gal}(E/\mathbf{Q})$ with the idele class group of $\mathbf{Q}$ modulo the "norms from E"; we stress the "dual form" of this assertion only because its formulation seems more amenable to generalization (i.e., nonabelian E).

residue modulo some *fixed modulus* N (depending on E). To see how this result directly generalizes the classical result of Fermat and Euler, we note that when $E = \mathbf{Q}(i)$ and $\sigma: G \to \{\pm 1\}$ is as before,

$$\sigma(\mathrm{Fr}_p) = \chi_\sigma(p),$$

with χ: $(\mathbf{Z}/4\mathbf{Z})^\times \to \mathbf{C}^\times$ defined as follows:

$$\chi_\sigma(n) = (-1)^{(n-1)/2}.$$

For more general abelian extensions, Artin's theorem not only implies the general quadratic reciprocity law (in place of the supplementary rule $(-1/p) = (-1)^{(p-1)/2}$) but also the so-called higher reciprocity laws of abelian class field theory. For a discussion of such matters, see, for example, [**Goldstein, Tate**, or **Mazur**].

The question remains: for *non*abelian Galois extensions, how can the family $\{\mathrm{Fr}_p\}$ be described in terms of the ground field $\mathbf{Q}$?

Recognizing the utility of studying groups in terms of their matrix representations, Artin focused attention on homomorphisms of the form σ: Gal$(E/\mathbf{Q}) \to \mathrm{GL}_n(\mathbf{C})$, i.e., on n-dimensional *representations* of the Galois group G. In this way he was able to transfer the problem of analyzing certain conjugacy classes in G to an analogous problem inside $\mathrm{GL}_n(\mathbf{C})$ (where such classes as $\{\sigma(\mathrm{Fr}_p)\}$ are completely determined by their characteristic polynomials $\det[I_n - \sigma(\mathrm{Fr}_p)p^{-s}]$). By also introducing the (Artin) L-functions

$$L(s, \sigma) = \prod_p \left(\det\left[I_n - \sigma(\mathrm{Fr}_p)p^{-s}\right]\right)^{-1}$$

(whose exact definition will be given in II.C.2), Artin was further able to reduce this problem to one involving the *analytic* objects $L(s, \sigma)$.

Problem. Can the L-functions $L(s, \sigma)$ be defined in terms of the arithmetic of $\mathbf{Q}$ alone?

It was in the context of this problem that Artin proved his fundamental reciprocity law. Indeed, for abelian E over $\mathbf{Q}$, and one-dimensional σ, Artin proved that his $L(s, \sigma)$ is identical to a Dirichlet L-series

$$L(s, \chi) = \prod (1 - \chi(p)p^{-s})^{-1}$$

for an appropriate choice of character χ: $(\mathbf{Z}/N\mathbf{Z})^\times \to \mathbf{C}^\times$.

For arbitrary E and σ, Artin was able to derive important analytic properties of $L(s, \sigma)$. However, what he was unable to do was discover the appropriate "n-dimensional" analogues of Dirichlet's characters and L-functions. Although some such 2-dimensional "automorphic" L-functions were being studied nearby (and concurrently) by Hecke, it remained for Langlands (40 years later) to see the connection and map out some general conjectures.

Roughly speaking, here is what Langlands did. He isolated the notion of an "automorphic representation of the group GL_n over the adeles of $\mathbf{Q}$" as the appropriate generalization of a Dirichlet character. Furthermore, he associated L-functions with these automorphic representations, generalizing Dirichlet's

L-functions in the case $n = 1$. Finally, he *conjectured* that each n-dimensional Artin L-function $L(s, \sigma)$ is exactly the L-function $L(s, \pi_\sigma)$ for an appropriate automorphic representation π_σ of GL_n. This is discussed—with an arbitrary number field F in place of $\mathbf{Q}$—in Part IV of the present paper; cf. Conjecture 1 in IV.A.

The (conjectured) correspondence $\sigma \to \pi_\sigma$ is to be regarded as a far reaching generalization of Artin's reciprocity map $\sigma \to \chi_\sigma$. In case $n = 2$, when π_σ corresponds to a classical automorphic form $f(z)$ in the sense of Hecke (see I.B), the map $\sigma \to \pi_\sigma$ affords an interpretation of the classes $\{\mathrm{Fr}_p\}$ in terms of certain conjugacy classes in $GL_2(\mathbf{C})$ determined by the Fourier coefficients of the form $f(z)$. In general, the proper formulation of this conjecture (and other conjectures of Langlands) requires a synthesis and further development of all the themes alluded to heretofore: local-global principles, automorphic forms, group representations, etc.

In Part II of this paper, I motivate the use of p-adic numbers and adeles and survey Hecke's theory of automorphic forms, the L-functions of Artin and Hecke, and the use of group representations in number theory. Perforce, this brings us to the theory of *infinite-dimensional* representations of real and p-adic groups.

In Part III these "classical" themes and ingredients are mixed together to produce the all-important notion of an "automorphic representation of GL_n over $\mathbf{Q}$". Finally, in Part IV, I survey the high points of Langlands' general program, with an emphasis on its historical perspective, and a brief description of techniques and known results.

II. CLASSICAL THEMES

A. The local-global principle. One of the major preoccupations of number theory in general has been finding integer solutions of polynomial equations of the form

$$P(x_1, x_2, \ldots, x_n) = 0. \tag{1}$$

For convenience, let us assume that P is actually a homogeneous polynomial, and let us agree that only *nonzero* solutions are of interest. The difficulty in solving (1) is illustrated by Fermat's famous unproved assertion that the particular equation

$$X^n + Y^n - Z^n = 0$$

has no nontrivial solutions in integers for $n > 2$. Indeed, much of the development of the theory of algebraic numbers is linked to attempts by people contemporary with Kummer to solve this problem.

On the other hand, a question which is more easily decided is the existence of integral solutions "modulo m". Clearly a *necessary* condition that integer solutions of (1) exist is that the congruence

$$P(x_1, \ldots, x_n) \equiv 0 \pmod m \tag{2}$$

be solvable for *every* value of the modulus m. This observation leads naturally to the "local methods" we shall now explain.

Suppose $m = NM$ with N and M relatively prime. By the Chinese Remainder Theorem, (2) has a solution if and only if the similar congruences for N and M do. In other words, to solve (2) it is sufficient to solve congruences modulo p^k for any prime p and all positive integers k.

Whenever we focus on a fixed prime p, we say we are working "locally". So suppose we fix a prime p and ask whether the congruence

$$P(x_1,\dots,x_n) \equiv 0 \pmod{p^k} \tag{3}$$

has a solution for all natural numbers k. It was Hensel who reformulated this question in a formal, yet significant, way in 1897. For each prime p he introduced a new field of numbers—the "p-adic numbers"—and he showed that the solvability of (3) *for all* k is equivalent to the solvability of (1) in the p-adic numbers. Thus the solvability of the congruence (2) for all n is equivalent to the solvability of (1) in the p-adic numbers for all p.

Let us return now to the original problem of solving (1) in ordinary integers. In addition to being able to solve (2) modulo all integers m, it is also clearly necessary to be able to find real solutions for (1). The question of when these obviously necessary conditions are also sufficient is much more difficult, since the assertion that "an equation is solvable if and only if it is solvable modulo any integer and has real solutions" is in general false, or at least not known. For example, the Fermat equation has been known to be solvable p-adically for all p since around 1909.

On the other hand, there are important instances where this "local-global principle" is known to work.

THEOREM (HASSE-MINKOWSKI). *Suppose*

$$Q(x_1,\dots,x_n) = \sum_{i,j=1}^{n} a_{ij}X_iX_j$$

is a quadratic form with a_{ij} in $\mathbf{Z}$ and $\det(a_{ij}) \neq 0$. *Then $Q(x_1,\dots,x_n) = 0$ has a nontrivial integer solution if and only if it has a real solution and a p-adic solution for each p.*

In order to give a more symmetric form to this example of the local-global principle, let me recall how the p-adic numbers can be constructed analogously to the real numbers. Fixing a prime p, we can express any fraction x in the form $p^\alpha n/m$, with n and m relatively prime to each other and to p. Then an absolute value is defined on $\mathbf{Q}$ by

$$|x|_p = p^{-\alpha},$$

and the field of p-adic numbers is just the completion of $\mathbf{Q}$ with respect to this metric $|\ |_p$. Note that the integer α (called the p-adic order of x) can be negative, and the integers that are close to zero "p-adically" are precisely the ones that are *highly* divisible by p. Though perhaps jarring at first, this p-adic notion of size is entirely natural given our earlier motivations: the congruence $n \equiv 0\ (p^k)$, with k large, translates into the statement that n is close to zero (p-adically).

Because $\mathbf{R}$ is the completion of $\mathbf{Q}$ with respect to the usual absolute value $|\ |$, it is customary to write $|x|_\infty$ for $|x|$, $\mathbf{Q}_\infty$ for $\mathbf{R}$, and then call $\mathbf{R}$ the completion of $\mathbf{Q}$ at "the infinite prime" ∞. The result is a family of locally compact complete topological fields $\mathbf{Q}_p$ which contain $\mathbf{Q}$, one for each $p \leqslant \infty$. Each $\mathbf{Q}_p$ is called a "local field", and $\mathbf{Q}$ itself is called a "global field". With this terminology the Hasse-Minkowski theorem takes the following symmetric form: a quadratic form over $\mathbf{Q}$ has a global solution if and only if it has a local solution for each prime p.

For the purposes of this article, the significance of the local-global principle is this: global problems should be analyzed purely locally, and with equal attention paid to each of the local "places" $\mathbf{Q}_p$.

Note. For a leisurely discussion of p-adic numbers, and instances of the local-global principle, the reader is urged to browse through the Introduction to [**Bo Shaf** and **Cassels**]. Also highly recommended is the expository article [**Rob 2**].

B. Hecke theory and the centrality of automorphic forms. In the 19th century the arithmetic significance of automorphic forms was clearly recognized, and examples of such forms were used to great effect in number theory.

Around 1830, Jacobi worked with the classical theta-function $\theta(z)$ in order to obtain exact formulas for the representation numbers of n as a sum of r squares. Then 30 years later, Riemann exploited this same function in order to derive the analytic continuation and functional equation of his famous zeta-function $\zeta(s)$.

Before explaining these matters in more detail, let us briefly recall the classical notion of an automorphic form.

1. *Basic notions.* Let H denote the upper half-plane in $\mathbf{C}$, and regard the group

$$\mathrm{SL}_2(\mathbf{R}) = \left\{ \begin{bmatrix} a & b \\ c & d \end{bmatrix} : a, b, c, d \text{ real}, ad - bc = 1 \right\}$$

as the group of fractional linear transformations of H. An automorphic form of weight k is a function $f(z)$ which is holomorphic in H and "almost" invariant for the transformations $\gamma = \left[\begin{smallmatrix} a & b \\ c & d \end{smallmatrix}\right]$ in some *discrete* subgroup Γ of $\mathrm{SL}_2(\mathbf{R})$, i.e.,

$$f\left(\frac{az+b}{cz+d}\right) = (cz+d)^k f(z) \tag{1}$$

for all $\gamma = \left[\begin{smallmatrix} a & b \\ c & d \end{smallmatrix}\right]$ in Γ.

The most famous example of an automorphic form is the classical theta-function

$$\theta(z) = \sum_{n=-\infty}^{\infty} e^{\pi i n^2 z} = 1 + \sum_{n=1}^{\infty} 2e^{\pi i n^2 z}.$$

This is an automorphic form of weight $\frac{1}{2}$ for the group

$$\Gamma(2) = \left\{ \begin{bmatrix} a & b \\ c & d \end{bmatrix} \equiv \mathrm{SL}_2(\mathbf{Z}) \colon b, c \equiv 0\ (2), \quad a, d \equiv 1\ (2) \right\};$$

moreover,

$$\theta(-1/z) = (-iz)^{1/2}\theta(z).$$

More generally, let $Q_r(x_1, \ldots, x_r)$ denote the quadratic form $\Sigma_{i=1}^r x_i^2$, and set

$$\theta_r(z) = \sum_{(n_1, \ldots, n_r)} e^{\pi i Q_r(n_1, \ldots, n_r) z},$$

the sum extending over all "integral" vectors $(n_1, \ldots, n_r)$. Then $\theta_r(z)$ is again an automorphic form, this time of weight $r/2$. This example has special number theoretic significance because the coefficients in the Fourier expansion of this periodic function are the representation numbers of the quadratic form Q_r. Indeed, if $r(n, Q_r)$ denotes the number of distinct ways of expressing n as the sum of r squares, then

$$\theta_r(z) = \theta(z)^r = \sum_{n=0}^{\infty} r(n, Q) e^{\pi i n z}.$$

Here are some more examples of automorphic forms:

(i) Let $\Delta(z)$ denote the function defined in H by

$$\Delta(z) = e^{2\pi i z} \prod_{n=1}^{\infty} (1 - e^{2\pi i n z})^{24} = \sum_{n=1}^{\infty} \tau(n) e^{2\pi i n z}.$$

It is an automorphic form of weight 12 for the full modular group $\Gamma = \mathrm{SL}_2(\mathbf{Z})$, and its Fourier coefficients $\tau(n)$—carefully investigated by Ramanujan in 1916—are closely related to the classical partition function $p(n)$.

(ii) For $k > 1$ the function

$$E_{2k}(z) = \frac{1}{2\zeta(2k)} \sum_{\substack{(c, d) \neq (0,0) \\ \text{in } \mathbf{Z}^2}} \frac{1}{(cz + d)^{2k}}$$

is called the (normalized) *Eisenstein series* of weight $2k$. It is again an automorphic form with respect to the full modular group $\mathrm{SL}_2(\mathbf{Z})$, this time with Fourier expansion

$$E_{2k}(z) = 1 + \frac{(-1)^k 4k}{B_k} \sum_{n=1}^{\infty} \sigma_{2k-1}(n) e^{2\pi i n z},$$

with B_k the so-called nth Bernoulli number, and $\sigma_r(n) = \Sigma_{d|n} d^r$.

From these few examples, it is already clearly indicated that automorphic forms comprise an integral part of number theory. Indeed, invariance of the form with respect to translations of the type $z \to z + h$ implies the existence of a Fourier expansion $\Sigma a_n e^{2\pi i k z/h}$, with the a_n of number-theoretic significance.

In general, the automorphy property (1) implies $f(z)$ is determined by its values on a "fundamental domain" D for the action of Γ in H. More precisely,

D is a subset of H such that every orbit of Γ (with respect to the action $z \to (az + b)/(cz + d)$) has exactly one representative in D. For example, for $\Gamma = SL_2(\mathbf{Z})$, the fundamental domain D looks like this:

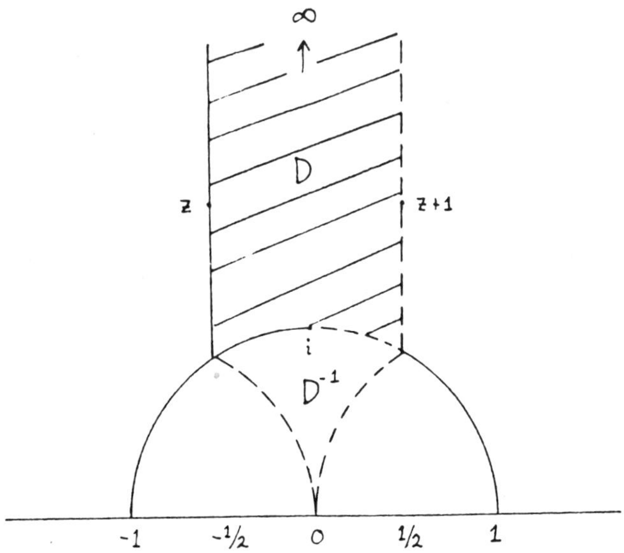

Note that any other fundamental domain must be obtained by applying to this D some $[\begin{smallmatrix} a & b \\ c & d \end{smallmatrix}]$ in Γ. In particular, the domain D^{-1} pictured above is precisely the image of D by the "inversion" element $[\begin{smallmatrix} 0 & 1 \\ -1 & 0 \end{smallmatrix}]$, the point "at infinity" for D being mapped to the "cusp" at 0 in (the boundary of) the fundamental domain D^{-1}.

To be able to apply convenient methods of analysis to the study of automorphic forms, it is customary to impose additional technical restrictions on the regularity of f at "cusps" along the boundary of a fundamental domain, especially "at infinity". This implies in particular that $f(z)$ always has a Fourier expansion of the form

$$f(z) = \sum_{n=0}^{\infty} a_n e^{2\pi i n z/h}. \tag{2}$$

For example, for $\Delta(z)$ or $E_{2k}(z)$ we can take $h = 1$, but for $\theta(z)$, which is an automorphic form only on $\Gamma(2)$ (which does not contain the translation $z \to z + 1$), the period is no longer 1, and we must take $h = 2$.

Let us denote by $M_k(\Gamma)$ the vector space of automorphic forms of weight k for Γ which are "regular at the cusps" of Γ, and by $S_k(\Gamma)$ the subspace of $f(z)$ in $M_k(\Gamma)$ which actually vanish at the cusps. Functions in this latter space are called *cusp forms*; for such functions (like the "modular discriminant" $\Delta(z)$), the constant term a_0 in the expansion (2) is zero.

We have already remarked that automorphic forms in general have number-theoretic interest because their Fourier coefficients involve solution numbers of number-theoretic problems. For example, by relating $\theta_4(z)$ to certain Eisenstein series *on* $\Gamma(2)$, we obtain Jacobi's remarkable formula

$$r(n, Q_4) = 8 \sum_{\substack{d|n \\ 4\nmid d}} d.$$

Thus the need for analyzing this space $M_k(\Gamma)$ is clearly indicated.

As we shall soon see, the subsequent theory developed by Hecke was so successful that it suggested new ways to look at automorphic forms in number theory as well as immediately providing the tools to solve existing classical problems.

2. *Hecke's theory*. Hecke's key idea was to characterize the properties of an automorphic form in terms of a corresponding Dirichlet series. The most famous Dirichlet series around is, of course, Riemann's zeta-function

$$\zeta(s) = \sum_{n=1}^{\infty} \frac{1}{n^s} = \prod_{p<\infty} (1 - p^{-s})^{-1}.$$

So let us first sketch Riemann's original analysis of $\zeta(s)$ which Hecke so brilliantly generalized.

Recall the gamma function identity,

$$\Gamma(s) = \int_0^{\infty} t^s e^{-t} \frac{dt}{t},$$

valid for $\mathrm{Re}(s) > 0$. (In modern parlance, we say that $\Gamma(s)$ is the Mellin transform of e^{-t} at s.) With this identity, we derive the relation

$$\pi^{-s}\Gamma(s)\zeta(2s) = \int_0^{\infty} \left[\frac{\theta(it) - 1}{2}\right] t^s \frac{dt}{t},$$

with θ the classical theta-function already encountered. In other words, $\zeta(2s)$ is essentially the Mellin transform of $\theta(it)$. From this fact, it is a simple matter to derive the desired analytic properties of $\zeta(s)$ in terms of the automorphic properties of $\theta(z)$, and conversely! Here are the key steps:

$$\begin{aligned}
\pi^{-s}\Gamma(s)\zeta(2s) &= \int_1^{\infty} t^{s-1}\left(\frac{\theta(it) - 1}{2}\right) dt \\
&\quad - \frac{1}{2}\frac{t^s}{s}\Big|_0^1 + \frac{1}{2}\int_0^1 t^{s-1}\theta(it)\, dt \\
&= \int_1^{\infty} t^{s-1}\frac{(\theta(it) - 1)}{2} dt - \frac{1}{2s} + \frac{1}{2}\int_1^{\infty} t^{-s-1}\theta\left(\frac{i}{t}\right) dt
\end{aligned}$$

(using the change of variable $t \to 1/t$)

$$= \int_1^{\infty} (t^{s-1} + t^{1/2-(s+1)})\frac{(\theta(it) - 1)}{2} dt - \frac{1}{2s} - \frac{1}{1 - 2s}$$

$\left(\text{using the automorphy property } \theta\left(\frac{i}{t}\right) = t^{1/2}\theta(it)\right).$

Note that invariance with respect to the substitution $s \to \frac{1}{2} - s$ is already obvious. To reverse the process and derive the "functional equation", i.e., automorphy condition of $\theta(z)$ from that of $\zeta(s)$, we require "Mellin inversion":

$$\frac{\theta(it) - 1}{2} = \frac{1}{2\pi i} \int_{\mathrm{Re}(s)=\sigma>0} t^{-s} \pi^{-s} \Gamma(s) \zeta(2s)\, ds.$$

By generalizing this proof, Hecke was able to "explain" the symmetry of a large number of Dirichlet series and also pave the way towards finding automorphic forms seemingly everywhere in number theory.

Given a sequence of complex numbers $a_0, a_1, \ldots, a_n, \ldots$ with $a_n = O(n^c)$ for some $c > 0$, and given $h > 0$, $k > 0$, $C = \pm 1$, consider the series

$$\phi(s) = \sum_{n=1}^{\infty} \frac{a_n}{n^s}$$

and

$$\Phi(s) = (2\pi/\lambda)^{-s} \Gamma(s) \phi(s),$$

and the function defined in H by

$$f(z) = \sum_{n=0}^{\infty} a_n e^{2\pi i n z/h}.$$

THEOREM 1 (HECKE). *The following two conditions are equivalent*:

(A) $\Phi(s) + a_0/s + C/(k - s)$ *is entire, bounded in every vertical strip, and satisfies the functional equation* $\Phi(k - s) = C\Phi(s)$;

(B) $f(-1/z) = C(z/i)^k f(z)$.

In other words, the holomorphic function $f(z)$ is automorphic of weight k (for the group of transformations generated by $z \to z + h$ and $z \to -1/z$) if and only if its associated Dirichlet series $\Sigma a_n/n^s$ is "nice". (We shall often use the term "nice" to describe a Dirichlet series satisfying certain analytic properties similar to $\zeta(s)$.)

The second part of Hecke's theory answers the question: when does $\phi(s) = \Sigma a_n/n^s$ have an Euler product expansion of the form $\phi(s) = \prod_{p<\infty} L_p(s)$, with $L_p(s)$ a power series in p^{-s}? A formal computation shows that $\phi(s)$ factors as

$$\prod_{p \text{ prime}} \sum_{m \geq 0} \frac{a_{p^m}}{p^{ms}}$$

whenever the coefficients a_n are *multiplicative*, i.e., $a_{nm} = a_n a_m$ if n and m are relatively prime.

Characterizing such multiplicativity is crucial. Indeed, since the coefficients a_n always have number-theoretic significance, it is of great interest to know when knowledge of these a_n's can be reduced to knowing a_p for p prime.

Note that when the a_n's are *completely* multiplicative, i.e., $a_{nm} = a_n a_m$ for *all* n and m, the Euler product expansion above reduces to the familiar expression

$$\sum \frac{a_n}{n^s} = \prod_p (1 - a_p p^{-s})^{-1}. \tag{3}$$

Such is, of course, the case for the Riemann zeta-function

$$\zeta(s) = \sum \frac{1}{n^s} = \prod (1 - p^{-s})^{-1}$$

where the Euler product expansion (discovered appropriately by Euler) is tantamount to the fundamental theorem of arithmetic. Another example is provided by the coefficients $a_n = \chi(n)$ with χ a "character of the integers modulo some N", i.e., χ is completely multiplicative, of period N, and $\chi(0) = 0$; such a character is called a "Dirichlet character" mod N, and the corresponding series

$$\sum \frac{a_n}{n^s} = \sum \frac{\chi(n)}{n^s} = \prod (1 - \chi(p)p^{-s})^{-1}$$

is a "Dirichlet L-series".

In both these examples, the Euler factors $L_p(s)$ are of degree 1 in p^{-s}. In general, such an expansion as (2) is too much to ask for; usually we ask for only ordinary multiplicativity, and then the factors $L_p(s)$ turn out to be of the second (not first) degree in p^{-s}.

Hecke's contribution was to characterize the multiplicativity of the a_n (or $\phi(s)$) intrinsically (and even "locally") in terms of $f(z)$ by introducing a certain ring of "Hecke operators" $T(p)$ defined in a space of automorphic forms of fixed weight.

THEOREM 2 (HECKE). *Assume, for convenience, that*

$$f(z) = \sum_{n=1}^{\infty} a_n e^{2\pi i n z}$$

belongs to $S_k(\mathrm{SL}_2(\mathbf{Z}))$ *and* $a_1 = 1$. *Then the* a_n*'s are multiplicative (and* $\phi(s)$ *has an Euler product expansion) if and only if* f *is an eigenfunction for all the Hecke operators* $T(p)$, *with* $T(p)f = a_p f$. *In this case,*

$$\phi(s) = \prod L_p(s)$$

with

$$L_p(s) = \left(1 - a_p p^{-s} + p^{k-1-2s}\right)^{-1}.$$

EXAMPLE. Since $S_{12}(\mathrm{SL}_2(\mathbf{Z}))$ is one dimensional, and $T(p)$ preserves this space, the condition $T(p)\Delta = \lambda_p\Delta$ is automatic. Thus one obtains the multiplicativity of the coefficients $\tau(n)$ (conjectured by Ramanujan and first proved by L. J. Mordell).

REMARKS. (1) Hecke's Theorem 1 really says that an automorphic eigenform of weight k on $\mathrm{SL}_2(\mathbf{Z})$ is indistinguishable from an Euler product of degree 2 with prescribed analytic behavior (and functional equation involving the substitution $s \to k - s$). This observation sheds a new light on the theory of automorphic forms, since there are many number-theoretical situations where data $\{a_n\}$ leads to a nice L-function, hence by Hecke to an automorphic form. Following up on this idea, A. Weil in 1967 completed Hecke's theory by similarly characterizing automorphic forms not just on $SL_2(\mathbf{Z})$, but also on the so-called congruence subgroups such as

$$\Gamma_0(N) = \left\{ \begin{bmatrix} a & b \\ c & d \end{bmatrix} \in SL_2(\mathbf{Z}) \colon c \equiv 0 \right\}.$$

(These subgroups have in general many generators, whereas Hecke's theorem deals with automorphic forms only for the groups generated by $z \to z + \lambda$ and $z \to -1/z$.)

In this way, Weil was led to an extremely interesting conjecture. By carefully analyzing the zeta-function attached to an elliptic curve E over Q (with $L_p(s) = (1 - a_p p^{-s} + p^{1-2s})^{-1}$, and $1 + p - a_p$ the number of points on the "reduced curve modulo p"), Weil was able to conjecture that such a zeta-function is the Dirichlet series attached to an automorphic form in some $S_2(\Gamma_0(N))$; cf. [**We 1**]. In other words, the study of these curves might (perversely) be regarded as a special chapter in the theory of automorphic forms!

(2) Perhaps it is now clear to the reader that an automorphic form $f(z)$ (like an elliptic curve or a quadratic form) should be regarded as a "global object" over $\mathbf{Q}$, and that the a_p (or the Euler factors $L_p(s)$) comprise local data for f in much the same way that p-adic solutions comprise local data for rational (i.e. "global") solutions of Diophantine equations. This turns out to be the case, but must remain a fuzzy notion until the language of automorphic representations is introduced in Part III.

Note. Two excellent sources on Hecke theory, which we have followed closely, are [**Ogg** and **Rob 3**].

C. Artin (and other) L-functions. Around 1840, Dirichlet succeeded in proving the existence of infinitely many primes in an arithmetic progression by replacing (Euler's) analysis of the series $\Sigma 1/n^s$ by his own analysis of the "Dirichlet L-functions" $L(s, \chi) = \Sigma \chi(n)/n^s$. Soon afterwards, Riemann focused on such Dirichlet series as functions of a *complex* variable, thereby inspiring a spate of applications of Dirichlet series to number theory in general, and the theory of prime numbers in particular. Finally, in 1870, Dedekind introduced a new kind of zeta-function to study the integers in an arbitrary number field E, i.e., any finite extension of $\mathbf{Q}$. This kind of zeta-function, now called a Dedekind zeta-function and denoted $\zeta_E(s)$, made it possible to relate the primes of $\mathbf{Q}$ to those of E and to analyze the distribution of primes within E alone.

Despite this widespread use of "L-series" in the nineteenth century, and the concomitant need to generalize these functions further, a full understanding of the arithmetic significance of L-functions awaited twentieth century developments.

1. *Abelian L-functions*. In 1916, Hecke was able to establish the analytic continuation and functional equation of Dirichlet's L-functions and to generalize them to the setting of an arbitrary number field. To describe Hecke's achievement properly, we must first recall how to generalize the family of p-adic fields $\mathbf{Q}_p$ considered in II.A.

Fix a finite extension E of $\mathbf{Q}$ and let O_E denote the ring of integers of E. By a finite "place" or prime v of E we understand a prime ideal $\mathcal{P}$ in O_E (and we often confuse the notations v and $\mathcal{P}$); by a "fractional ideal" of O_E we understand an O_E-submodule of E with the property that $x\mathfrak{A} \subset O_E$ for some $x \in E^\times$. It is a basic fact that the prime ideals are invertible (in the sense that $\mathcal{P} \cdot \mathcal{P}^{-1} = O_E$ for some fractional ideal $\mathcal{P}^{-1}$) and that every fractional ideal factors (uniquely) into powers of prime ideals.

Now if x is in $E^\times$, we define $\text{ord}_{\mathcal{P}}(x)$ to be the (positive or negative) power of $\mathcal{P}$ appearing in the factorization of the principal ideal (x), and we set

$$|x|_v = |x|_{\mathcal{P}} = N\mathcal{P}^{-\text{ord}_{\mathcal{P}}(x)},$$

with $N\mathcal{P}$ the cardinality of the field $O_E/\mathcal{P}$. By analogy with the case of $\mathbf{Q}$, we also define a "real" place v of E to be a norm $|x|_v = |\sigma(x)|$, with $\sigma\colon E \to \mathbf{R}$ a real embedding. ("Complex" infinite places are defined analogously.) The result is a family of completions E_v of E, one for each prime (or place) v of E. Following the lead of the local-global principle for $\mathbf{Q}$, we treat all these "finite" or "infinite" places equally.

Recall that a classical Dirichlet character is just a homomorphism of $(\mathbf{Z}/N\mathbf{Z})^\times$ into $\mathbf{C}^\times$ "extended" to $\mathbf{Z}$ by composition with the natural homomorphism $\mathbf{Z} \to \mathbf{Z}/N\mathbf{Z}$ (and with the convention that $\chi(n) = 0$ if $(n, N) > 1$). The appropriate generalization of such a character to the number field E is called a *Hecke character* (or grossencharacter) χ. This is a family of homomorphisms $\chi_v\colon E_v^\times \to \mathbf{C}^\times$, one for every place v of E, such that for any x in $E^\times$ (regarded as embedded in each $E_v^\times$), $\prod_{\text{all } v} \chi_v(x) = 1$. I.e., Hecke characters are "trivial" on $E^\times$. Implicit here is the fact that all but finitely many of the χ_v are *unramified*, i.e., $\chi_v(x_v) = 1$ for all x_v in $E_v^\times$ such that $|x|_v = 1$. (The fact that Dirichlet characters give rise to such Hecke characters is spelled out in [**We3**, p. 313]; we shall return to these matters in II.D.1.)

Now we can define Hecke's abelian L-series attached to $\chi = (\chi_v)$ by

$$L_E(\chi, s) = \sum \frac{\chi(\mathfrak{A})}{N(\mathfrak{A})^s} = \prod_{\mathcal{P}} (1 - \chi(\mathcal{P})N\mathcal{P}^{-s})^{-1}.$$

Here $\mathfrak{A}$ is an (ordinary) ideal of O_E, and $\chi(\mathfrak{A})$ is defined multiplicatively on the ideals relatively prime to those such that χ_v is ramified (the "conductor" of χ). In particular, $\chi(\mathcal{P}) = 0$ whenever χ_v is ramified; otherwise, $\chi(\mathcal{P}) = \chi_v(\tilde{\omega}_v)$ with $\tilde{\omega}_v$ in E_v such that $|\tilde{\omega}_v|_v = N\mathcal{P}^{-1}$ (a "uniformizing" variable for E_v).

If χ is the unit character, i.e., $\chi_v \equiv 1$ for all v, then $L_E(\chi, s)$ reduces to Dedekind's zeta-function $\zeta_E(s)$. On the other hand, when $E = \mathbf{Q}$ and χ is of finite order, $L_E(\chi, s)$ reduces to a familiar Dirichlet L-series $L(s, \chi)$. In general, using ingenious and complicated arguments, Hecke was able to derive the analytic continuation and functional equation for all these L-series in 1917.

This settled, a natural arithmetic question was: how does a series like $\zeta_E(s)$ factor into L-series involving *only* the field $\mathbf{Q}$? Partly in an attempt to solve this problem, E. Artin was led (around 1925) to define yet another new L-series.

2. *Nonabelian L-functions*. Suppose K is a number field, and E is a finite Galois extension of K with Galois group $G = \text{Gal}(E/K)$. By a *representation* of G we understand a homomorphism σ of G into $\text{GL}(V)$, the group of invertible linear transformations of a complex vector space of dimension n.

Given a finite place $\mathfrak{P}$ of K, we say a prime $\mathfrak{B}$ of O_E *lies over* $\mathfrak{P}$ (or *divides* $\mathfrak{P}$) if $\mathfrak{B}$ appears in the factorization of $\mathfrak{P}O_E$ into prime ideals of O_E. Given such a pair $\mathfrak{B}/\mathfrak{P}$, the following "local" objects are defined, each depending on $\mathfrak{B}$ only up to conjugation in G:

(1) The *decomposition group* $D_{\mathfrak{B}} = \{g \in G: g(\mathfrak{B}) = \mathfrak{B}\}$;

(2) The *inertia* subgroup

$$I_{\mathfrak{B}} = \{g \in D_{\mathfrak{B}}: g(x) = x \ (\mathrm{mod}\ \mathfrak{B}) \text{ for all } x \in O_E\};$$

(3) The *Frobenius automorphism* $\mathrm{Fr}_{\mathfrak{B}}$ generating the cyclic group $D_{\mathfrak{B}}/I_{\mathfrak{B}} \approx \mathrm{Gal}(O_E/\mathfrak{B}\colon O_K/\mathfrak{P})$. When $I_{\mathfrak{B}} = \{1\}$, we call $\mathfrak{P}$ *unramified* in E. In this case, $\mathrm{Fr}_{\mathfrak{B}}$ represents a conjugacy class in G. Because almost all $\mathfrak{P}$ are unramified in this sense, and because knowledge of $\mathrm{Fr}_{\mathfrak{B}}$ completely determines the factorization type of such $\mathfrak{P}$ (cf. our Introduction), it is natural to focus attention on these $\mathrm{Fr}_{\mathfrak{B}}$. In general, if E is *abelian* over K, i.e., $G = \mathrm{Gal}(E/K)$ is abelian, than $D_{\mathfrak{B}}$, $I_{\mathfrak{B}}$ and $\mathrm{Fr}_{\mathfrak{B}}$ depend only on $\mathfrak{P}$ (and are denoted $D_{\mathfrak{P}}$, $I_{\mathfrak{P}}$ and $\mathrm{Fr}_{\mathfrak{P}}$ accordingly). In particular, $\mathrm{Fr}_{\mathfrak{P}}$ reduces to a distinguished element (as opposed to a conjugacy class) of G.

Returning to our representation $\sigma\colon G \to \mathrm{GL}(V)$, let $V_{\mathfrak{B}}$ denote the subspace of V formed by vectors invariant by $\sigma(I_{\mathfrak{B}})$. Then $\sigma(\mathrm{Fr}_{\mathfrak{B}})$ is defined unambiguously on $V_{\mathfrak{B}}$, and the "Euler factor"

$$L_{\mathfrak{P}}(\sigma, s) = \left[\det\left(I - \sigma(\mathrm{Fr}_{\mathfrak{B}})N\mathfrak{P}^{-1}|_{V_{\mathfrak{B}}}\right)\right]^{-1}$$

depends only on $\mathfrak{P}$, not $\mathfrak{B}$. Note that for almost all $\mathfrak{P}$, $I_{\mathfrak{B}} = \{1\}$, $V_{\mathfrak{B}} = V$, and hence $L_{\mathfrak{P}}(\sigma, s)$ is of degree n in $(N\mathfrak{P})^{-s}$.

Now we can finally define *Artin's L*-function attached to σ by the product

$$L(\sigma, s) = L_{E/K}(\sigma, s) = \prod_{\mathfrak{P}} L_{\mathfrak{P}}(\sigma, s),$$

convergent for $\mathrm{Re}(s) > 1$.

THEOREM 1 (ARTIN). *Suppose E is abelian over K, and $\sigma\colon G \to \mathbf{C}^x$ is a character. Then there exists a Hecke character $\chi = \{\chi_v\}_v$ of K (not E) such that $L_{E/K}(\sigma, s) = L(\chi, s)$ (Hecke's abelian L-series for K).*

In order to prove this theorem, Artin formulated and proved his celebrated "reciprocity law" for abelian extensions of number fields. For a leisurely account of these matters, see [**Tate**]. Because Artin and Hecke L-series coincide in the case of abelian extensions, the terminology "abelian L-functions" for Hecke's $L(s, \chi)$ seems apt.

THEOREM 2 (ARTIN-BRAUER). *In general, $L(\sigma, s)$ extends to a meromorphic function of $\mathbf{C}$ with functional equation*

$$L(\sigma, s) = \varepsilon(\sigma, s)L(\sigma^v, 1 - s),$$

σ^v the contragredient *representation* $\sigma(g^{-1})^t$.

The proof of this theorem relies on Theorem 1 together with R. Brauer's (1947) factorization

$$L_{E/K}(\sigma, s) = \prod L_{E/K_i}(\sigma_i, s)^{n_i};$$

here each K_i is intermediate between K and E, σ_i is a representation of $\mathrm{Gal}(E/K_i)$ of dimension 1, and n_i is an integer.

THEOREM 3 (ARTIN). *Let H denote the (finite) set of all irreducible representations of G (listed up to isomorphism). Then*

$$\zeta_E(s) = \prod_{\sigma\in H} L_{E/K}(\sigma, s)^{\dim(\sigma)}.$$

This is the sought-after factorization of $\zeta_E(s)$ alluded to at the end of paragraph II.C.1. Note that when E is *abelian* over K, and $K = \mathbf{Q}$, Theorems 3 and 1 together imply that $\zeta_E(s)$ factors as the product of Hecke L-series, a nontrivial assertion involving the theory of cyclotomic fields (in particular, the Kronecker-Weber theorem asserting that every abelian extension of $\mathbf{Q}$ embeds in some cyclotomic field).

Question. When $\dim(\sigma) > 1$, what is the nature of these *nonabelian* L-series $L(\sigma, s)$? In particular, if $\dim(\sigma) = 2$, does $L(\sigma, s)$ have any relation to Hecke's "automorphic" Dirichlet series $\phi_f(s) = \Sigma a_n/n^s$ (if not with the abelian L-series $L(\chi, s)$)?

CONJECTURE (ARTIN). Suppose $\dim(\sigma) > 1$, and σ is irreducible, i.e., V has no proper invariant subspaces under $\sigma(G)$. Then $L_{E/K}(\sigma, s)$ is actually entire.

Note that the truth of Artin's Conjecture is at least consistent with the known analytic behavior of the automorphic L-functions $L(s, f)$ (cf. Hecke's Theorem 2 in II.B.2). We shall return to these matters in earnest in Parts III and IV.

CONCLUDING REMARK. The "constant" $\varepsilon(\sigma, s)$ appearing in the functional equation for $L(\sigma, s)$ is defined by piecing together some global notions (the Artin "conductor of σ") with a product of gamma functions. But following the lead of our local-global principle, one should attempt to define $\varepsilon(\sigma, s)$ instead as the product of purely local factors $\varepsilon(\sigma_v, s)$. Eventually this was accomplished by Langlands (with some finishing touches by Deligne). As we shall see, this accomplishment made possible a serious attack on the above questions (at least for $\dim(\sigma) = 2$) and helped to develop the program this article eventually describes.

Note. Our exposition here closely follows the first few sections of Cartier's Bourbaki talk [**Cart 1**].

D. Group representations in number theory. We come at last to the fourth and final ingredient for the soup we shall mix up in Part III.

If G is a group, a (unitary) representation of G is a homomorphism π from G to the group of invertible (unitary) operators on some (Hilbert) space V, not necessarily finite dimensional. If G is a topological group, continuity assumptions are also imposed on π, but we shall ignore them here. Examples of representations encountered thus far in this survey include:

(1) The case when V is the one-dimensional space $\mathbf{C}$; in this case, a representation on V is simply a character—a homomorphism from G to (the torus in) $\mathbf{C}^\times$. A Hecke character, for example, is a collection (χ_v) of 1-dimensional representations of $E_v^\times$.

(2) The case when V is finite dimensional and G is a finite group; the Artin L-functions $L(\sigma, s)$, for example, are attached to n-dimensional representations σ of $G = \mathrm{Gal}(E/K)$.

The use of group representations in systematizing and resolving diverse mathematical problems is certainly not new, and the subject has been ably surveyed in several recent articles, notably [**Gross** and **Mackey**]. The reader is strongly urged to consult these articles, especially for their reformulation of harmonic analysis as a chapter in the theory of group representations.

In harmonic analysis, as well as in the theory of automorphic forms, the fundamental example of a (unitary) representation is the so-called "right regular" representation of G, defined in a space of functions $\{\phi\}$ on G by the formula $(R(g)\phi)(x) = \phi(xg)$.

Our interest here is in the role representation theory has played in the theory of automorphic forms. We focus on two separate developments, both of which are eventually synthesized in the Langlands program, and both of which derive from the original contributions of Hecke already described.

1. *Tate's adelic treatment of Hecke's L-series.* In his 1950 Princeton thesis, J. Tate showed how to use representations (or rather characters) of adele groups to reformulate and reprove Hecke's (complicated) results on the functional equation of his "abelian" L-series.

What is an adele group? It is really just a device for simultaneously considering all the completions of an object such as the rational field $\mathbf{Q}$. In this context, the adele ring $\mathbf{A}$ is the subgroup of the infinite product $\prod_{p \leq \infty} \mathbf{Q}_p$ consisting of those sequences $\{x_p\}$ with $|x_p|_p \leq 1$ for all but a finite number of primes p. The *idele* group $\mathbf{A}^x$ is defined similarly as the subgroup of the (multiplicative) group $\prod_{p<\infty} \mathbf{Q}_p^x$ consisting of sequences $\{y_p\}$ with $|y_p|_p = 1$ for all but finitely many places p. In other words, the adeles and ideles are *restricted* direct products of p-adic objects. For a general number field in place of $\mathbf{Q}$, similar definitions apply with v in place of p.

Note that the assignment of an idele or adele group to a number field F is entirely consistent with the local-global principle of II.A in that all local objects are introduced on an equal footing and the goal is to analyze the global object F. Pushing this principle to the extreme, Tate used this notion to define and analyze Hecke's "abelian L-functions" purely in local terms, as we shall now explain.

Let $\mathbf{A}_F$ (resp. $\mathbf{A}_F^x$) denote the adeles (resp. ideles) of F. The topology on $\mathbf{A}$ (resp. $\mathbf{A}^x$) is defined in such a way that every continuous character ψ of $\mathbf{A}$ is of the form $\psi(x) = \prod_v \psi_v(x_v)$ with ψ_v a continuous character of F_v for each place v, and for all but finitely many v, $\psi_v(x_v) = 1$ whenever $|x_v|_v \leq 1$. Similarly, any character χ of $\mathbf{A}^x$ is of the form $\prod_v \chi_v(x_v)$, with χ_v *unramified* for almost all v, i.e., $\chi_v(x_v) = 1$ whenever $|x_v|_v = 1$. Thus a character χ which is trivial on the so-called principle ideles F^x (embedded diagonally in $\mathbf{A}^x$) is the same thing as a Hecke character (χ_v) of E. In this case, we also call χ an *idele class* character, since it descends to a character of the so-called idele class group $\mathbf{A}^x/F^x$.

Now fix a character $\psi = \prod \psi_v$ of $\mathbf{A}$ which is trivial on F (again embedded diagonally in $\mathbf{A}$). Given a Hecke character $\chi = \prod \chi_v$, Tate defined, for each nice "Schwartz function" f_v on the field F_v, a "local zeta-function"

$$\zeta(f_v, \chi_v, s) = \int_{F_v^*} f_v(x)\chi_v(x)|x|^{s-1}\,dx.$$

Using a local Fourier transform $\hat{f}_v$ (defined in terms of the local character ψ_v), Tate then proved that each $\zeta(f_v, \chi_v, s)$ has an analytic continuation and functional equation of the form

$$\zeta(f_v, \chi_v, s) = \gamma(\chi_v, \psi_v, s)\zeta(\hat{f}_v, \chi_v^{-1}, 1-s),$$

with $\gamma(\chi_v, \psi_v, s)$ a meromorphic function of s which is independent of f_v. In fact, let

$$L(\chi_v, s) = (1 - \chi_v(\tilde{\omega}_v)N\mathcal{P}^{-s})^{-1}$$

if χ_v is unramified, and 1 otherwise. Then $\zeta(f_v, \chi_v, s)/L(\chi_v, s)$ is entire for all f_v, and equals 1 for appropriately chosen f_v; moreover,

$$\gamma(\psi_v, s) = \frac{\varepsilon(\chi, \psi_v, s)^{-1}L(\chi_v^{-1}, 1-s)}{L(\chi_v, s)}$$

with $\varepsilon(\chi_v, \psi_v, s) = 1$ whenever χ_v (also ψ_v) is unramified.

Now return to the global setting and consider the (global) Hecke L-series

$$L(\chi, s) = \prod_v L(\chi_v, s).$$

By considering global zeta-functions of the form

$$\zeta(f, \chi, s) = \int_{\mathbf{A}^\times} f(x)\chi(x)(x)^{s-1}\,dx = \prod_v \zeta(f_v, \chi_v, s),$$

Tate was *easily* able to prove that $L(\chi, s)$ has an analytic continuation and functional equation of the form

$$L(s, \chi) = \varepsilon(s, \chi)L(1-s, \chi^{-1}) \quad \text{with } \varepsilon(s, \chi) = \prod_v \varepsilon(\chi_v, \psi_v, s).$$

In addition to clarifying and simplifying Hecke's work, Tate was thus also able to give a purely local interpretation to the "constant" ε appearing in Hecke's functional equation.

These ideas and others are vastly generalized by Langlands in the "Langlands program". The immediate impetus and inspiration for this program actually came from Langlands' general theory of Eisenstein series (cf. **[Art]**). The sequence of events seems to have unfolded like this: a careful analysis of the "constant term" of these general Eisenstein series first suggested the definition of the general automorphic L-functions $L(s, \pi)$ (discussed in Part III), then the general conjectures (discussed in Part IV), and finally the purely local definition of the constants in the functional equations of Artin's *non-abelian* L-series (analogous to Tate's treatment of the abelian case). In fact, the existence and properties of the local ε-factors "on the Galois side" were predicted by the newly found properties of the ε-factors "on the representation theory side"; cf. Part III. It is interesting to note that here, as in Tate's work and its generalizations, most of the work in proving the global results goes into the local theory; for earlier important work on the "Galois" ε-factors, see **[Dwork]**.

Finally, as we shall see in Part III, Hecke characters are themselves examples of automorphic representations. Thus most of these results of Tate, Hecke and Artin are subsumed simultaneously in Langlands' theory.

Notes. There are now several good expositions of Tate's original work, **[Rob 2]** being one of them. Tate's thesis itself makes for surprisingly pleasant reading; cf. **[Cas Fro]**.

2. *Automorphic forms as group representations.* How do classical automorphic forms amount to special examples of infinite-dimensional representations—first of the group $\mathrm{SL}_2(\mathbf{R})$, then of the so-called adele group of GL_2?

For convenience, we shall consider only automorphic forms of even integral weight for the full modular group $\Gamma = \mathrm{SL}_2(\mathbf{Z})$; for details and more examples, the reader is referred to **[Ge 1]**.

The connection between automorphic forms and infinite-dimensional representations seems first to have been made explicit in **[Ge Fo]**. It starts with the observation that the stabilizer in $\mathrm{SL}_2(\mathbf{R})$ of the point i in the upper halfplane H is the rotation group $\mathrm{SO}_2(\mathbf{R})$, whence the identification

$$H \approx \mathrm{SL}_2(\mathbf{R})/\mathrm{SO}_2(\mathbf{R}).$$

If $f(z)$ is an automorphic form of weight k for $\mathrm{SL}_2(\mathbf{Z})$ it defines a function ϕ_f on $\mathrm{SL}_2(\mathbf{R})$ through the formula

$$(1) \qquad \phi_f(g) = (ci+d)^{-k} f\left(\frac{ai+b}{ci+d}\right) \quad \text{if } g = \begin{bmatrix} a & b \\ c & d \end{bmatrix}.$$

This function can be shown to satisfy the following properties:

$$(0) \qquad \phi(g\gamma(\theta)) = e^{-ik\theta}\phi(g) \quad \text{for } r(\theta) = \begin{bmatrix} \cos\theta & \sin\theta \\ -\sin\theta & \cos\theta \end{bmatrix};$$

(i) $\phi(\gamma g) = \phi(g)$ for all ϕ in $\mathrm{SL}_2(\mathbf{Z})$;

(ii) there is a second-order differential operator Δ—called the Laplace or Casimir operator on $\mathrm{SL}_2(\mathbf{R})$—such that

$$\Delta\phi = -\tfrac{1}{4}k(k-2)\phi.$$

Indeed the automorphy property of $f(z)$ translates into condition (i), and the holomorphy of f translates into the eigenfunction condition (ii). The technical regularity condition on f "at the cusps" corresponds to a boundedness condition on ϕ, and if f actually "vanishes" at the cusps, then ϕ satisfies a "cuspidal" condition which puts it in a distinguished subspace L_0^2 of $L^2(\mathrm{SL}_2(\mathbf{Z})\backslash\mathrm{SL}_2(\mathbf{R}))$. Here L^2 denotes square-integrable functions with respect to the natural Haar measure on $\mathrm{SL}_2(\mathbf{R})$ pushed down to the quotient space $\mathrm{SL}_2(\mathbf{Z})\backslash\mathrm{SL}_2(\mathbf{R})$; the "cuspidal" condition is included in the following hypothesis:

(iii) $\phi \in L^2$ and the "zeroth Fourier coefficient"

$$\int_{\mathbf{Z}\backslash\mathbf{R}} \phi\left(\begin{bmatrix} 1 & x \\ 0 & 1 \end{bmatrix} g\right) dx$$

vanishes.

Conversely, if $\phi(g)$ is any function on $SL_2(\mathbf{R})$ satisfying these four conditions, then ϕ is of the form ϕ_f for some f in $S_k(\Gamma)$. In other words, $S_k(\Gamma)$ is isomorphic to a certain subspace $A_k(\Gamma)$ of the space of "cusp" forms $L_0^2(SL_2(\mathbf{Z})\backslash SL_2(\mathbf{R}))$.

To obtain a more general notion of "automorphic cusp form on $SL_2(\mathbf{R})$" it is natural to weaken (ii) to the condition that ϕ be *any* eigenform of Δ. In this way we recover, for example, the "new" real-analytic "wave forms" which H. Maass discovered in 1949 in connection with his analysis of the functional equation of the Dedekind zeta-function of real (as opposed to imaginary) quadratic extensions of $\mathbf{Q}$; cf. §2 of **[Ge 1]** for more details.

Having brought into play the group $SL_2(\mathbf{R})$, the next step is to bring into play its representation. (It is also natural to weaken condition (0) and simply require $\varnothing$ to be "right SO(2)-finite".)

The *right regular representation* R of $SL_2(\mathbf{R})$ on the space L_0^2 is defined by

$$R(g)\phi(h) = \phi(hg), \qquad g, h \text{ in } SL_2(\mathbf{R}).$$

We recall this is a unitary representation of G because the mapping $g \mapsto R(g)$ is a homomorphism from G to the group of unitary operators on L_0^2. Moreover, this representation is related to the differential operator Δ since it commutes with Δ, i.e.,

$$\Delta R\phi = R\Delta\phi, \quad \text{for all nice } \phi.$$

By a general principle of representation theory, such *invariant* operators Δ may be used to decompose the representation R into invariant subspaces. More precisely, the closure of each eigenspace $\mathfrak{L}$ of Δ is then *invariant* for the action of R, i.e., $R(g)(\mathfrak{L}) \subset \mathfrak{L}$ for all g in $SL_2(\mathbf{R})$. In particular, R is the direct sum of representations of $SL_2(\mathbf{R})$ in the various eigenspaces $\mathfrak{L}$.

Since all the irreducible unitary representations of $SL_2(\mathbf{R})$ are known, it is natural to ask which of them occurs in these eigenspaces, and with what multiplicity.

EXAMPLE. Fix f in $S_k(SL_2(\mathbf{Z}))$, and let $\mathfrak{L}_f$ denote the subspace of $L_0^2(SL_2(\mathbf{Z})\backslash SL_2(\mathbf{R}))$ generated by ϕ_f and its right translates by elements of $SL_2(\mathbf{R})$. Suppose, in addition, that f is an eigenfunction for all the Hecke operators $T(p)$. Then $\mathfrak{L}_f$ is an irreducible subspace of L_0^2 realizing the so-called discrete series representation of $SL_2(\mathbf{R})$ of weight k.

This example shows that automorphic cusp forms are one and the same thing as irreducible unitary representations of $SL_2(\mathbf{R})$ which are "automorphic cuspidal", i.e., equivalent to some subrepresentation of R in L_0^2. In other words, the construction of automorphic forms amounts to knowing how the right regular representation R decomposes. Thus it is natural to generalize this notion further to an arbitrary Lie group G and discrete subgroups Γ such that $\Gamma\backslash G$ has finite volume. In practice, one deals primarily with "semisimple" groups G and "arithmetic" subgroups Γ. For example, if G is the real symplectic group $Sp_n(\mathbf{R})$, and $\Gamma = Sp_n(\mathbf{Z})$, a generalization of the so-called "Siegel modular forms" (studied by C. L. Siegel) is obtained; cf. **[Langlands 1]**. In general, much progress has been made in analyzing qualitative features of the decomposition of the regular representation R of G in the space L_0^2. In particular, one knows there are only finitely many invariant subspaces of L_0^2

which are equivalent (as G-modules) to a given "automorphic cuspidal representation" π (similarly any space of generalized cusp forms, of "fixed weight" for Γ, is finite dimensional).

Despite the appeal of the program just described, recalling the local-global principle should give us pause for thought. By working with a group like $\mathrm{SL}_2(\mathbf{R})$, we have put special emphasis on $\mathbf{R}$ as a completion of $\mathbf{Q}$, whereas we really should be paying equal attention to all the p-adic fields $\mathbf{Q}_p$. Indeed by dealing only with $\mathrm{SL}_2(\mathbf{R})$, we cannot hope to naturally view $f(z)$ as a "global object", or to throw light on the Euler product expansion which can exist for the Dirichlet series of $f(z)$.

What is clearly indicated now is that we view groups like $\mathrm{SL}_2(\mathbf{R})$ as the infinite component of some *global* object. Following the lead of imbedding $\mathbf{R}$ as the infinite component of $\mathbf{A}_{\mathbf{Q}}$, we set, for each prime p,

$$G_p = \mathrm{GL}_2(\mathbf{Q}_p), \qquad K_p = \mathrm{GL}_2(O_p),$$

where $O_p = \{x \in \mathbf{Q}_p\colon |x|_p \leq 1\}$, and we write G_∞ for $\mathrm{GL}_2(\mathbf{R})$. Then the corresponding adelic group $G_{\mathbf{A}}$ is defined to be the restricted direct product

$$G_{\mathbf{A}} = \mathrm{GL}_2(\mathbf{A}) = \prod_{p \leq \infty}' G_p,$$

where "restricted" here means we are including only the sequences (g_p) for which g_p belongs to K_p for all but finitely many p. The group $G_{\mathbf{Q}} = \mathrm{GL}_2(\mathbf{Q})$ is embedded as a discrete subgroup of $G_{\mathbf{A}}$ via the diagonal embedding.

Now suppose, as before, that $f(z)$ belongs to $S_k(\mathrm{SL}_2(\mathbf{Z}))$. By modifying formula (1), we can "lift" $f(z)$ all the way up to a function $\phi_f(g)$ on $G_{\mathbf{A}}$ such that

$$\phi(\gamma g) = \phi(g), \quad \text{for all } \gamma \text{ in } G_{\mathbf{Q}},$$

and $\phi(zg) = \phi(g)$, for all z in the center $Z_{\mathbf{A}} = \{[\begin{smallmatrix} a & 0 \\ 0 & a \end{smallmatrix}]\colon a \in \mathbf{A}^x\}$ of $G_{\mathbf{A}}$. Moreover, $\phi_f(g)$ satisfies a cuspidal condition like (iii); we refer the reader to **[Ge 1]** for a careful discussion of the approximation theorem which makes this lifting possible and well defined. The switch from SL_2 to GL_2 was made for reasons of convenience which need not concern the nonexpert.

Where do we stand now? We again have a right regular representation R of $\mathrm{GL}_2(\mathbf{A})$, this time on $L_0^2(Z_{\mathbf{A}}G_{\mathbf{Q}}\backslash G_{\mathbf{A}})$, and the functions ϕ_f just described lie in this space. So we may again ask, for a given f, which irreducible unitary representations of $G_{\mathbf{A}}$ occur in the representation R restricted to the subspace generated by each ϕ and its right translates. It is known that:

(i) Every irreducible unitary representation π of $G_{\mathbf{A}}$ is of the form $\otimes_p \pi_p$, with π_p an irreducible unitary representation of G_p for each p, and for almost all p, π_p has a vector fixed for the action of $\pi_p(K_p)$, i.e., π_p is *unramified*. (This is in complete analogy to the description of an idele class character χ as a Hecke character $\prod \chi_p$.)

(ii) The classical Hecke operators $T(p)$ can be described purely *locally* as convolution operators on the G_p-component of the function ϕ_f on $G_{\mathbf{A}}$.

(iii) If $T(p)f = a_p f$ for all p, then ϕ_f generates an *irreducible* subspace $\mathcal{L}_f$ of $L_0^2(Z_{\mathbf{A}}G_{\mathbf{Q}}\backslash G_{\mathbf{A}})$, and the representation $\pi = \otimes \pi_p$ realized in this subspace can be described purely locally in terms of the eigenvalues a_p (and the weight k).

The point is that we have finally succeeded in viewing automorphic forms as global objects which can be analyzed by "local objects" (Hecke operators, for example). Moreover, we have indicated that these notions might make sense for groups more general than GL_2, and that all "places" should be treated on an equal footing. Thus we are ready—at last—to introduce the general notion of an "automorphic representation" and to describe Langlands' theory.

We note that classical automorphic forms on all the so-called congruence subgroups of $SL_2(\mathbf{Z})$ can also be lifted to functions on $G_{\mathbf{Q}}\backslash G_{\mathbf{A}}$. So by working "adelically" with GL_2, we are actually simultaneously treating the theory of automorphic forms for *all* these discrete subgroups. For a thorough discussion of these and related matters, see [**Ge 1, Rob 1** and **G G PS**].

III. AUTOMORPHIC REPRESENTATIONS

In this part, G will denote the group GL_n regarded as an "algebraic group over the number field F". For all practical purposes, this simply means we can "complete" G at any place v of F to form the groups $G_v = GL_n(F_v)$ and then "adelize" G by forming the restricted direct product $G_{\mathbf{A}} = \prod'_v G_v$, just as we did for GL_2 and GL_1 $(= F^x)$. In the case of GL_n, the product $\prod_v G_v$ is restricted with respect to the subgroups $K_v = GL_n(O_v)$, with $O_v = \{x \in F_v\colon |x|_v \leq 1\}$. By G_F we denote the group $G_F = GL_n(F)$ regarded as a discrete subgroup of $G_{\mathbf{A}}$ via the diagonal embedding $\gamma \to (\gamma, \ldots, \gamma, \ldots)$.

Although GL_n is almost sufficient for the purposes of this expository article, it should be stressed that we could just as well be dealing with an arbitrary "connected reductive algebraic group G defined over F" (as we shall in fact do at the end of Part IV).

A. Some definitions. Let L_0^2 denote the Hilbert space of square-integrable functions ϕ on $G_F\backslash G_{\mathbf{A}}$ satisfying a certain cuspidal condition. When $n = 1$, this condition is vacuous; when $n = 2$, it generalizes the condition $a_0 = 0$ for functions ϕ corresponding to classical automorphic forms $f(z) = \sum_{n=0}^{\infty} a_n e^{2\pi i n z}$ (cf. condition (iii) in II.D.2). By R_0 we denote the right regular representation of $G_{\mathbf{A}}$ in L_0^2 given by $R_0(g)\phi(h) = \phi(hg)$.

Suppose π is an irreducible unitary representation of $G_{\mathbf{A}}$ in some space H_π. Then π can be factored as a restricted direct product $\pi = \otimes \pi_v$, with each π_v an irreducible unitary representation of $G_v = GL_n(F_v)$. This is analogous to the factorization $\prod X_v$ encountered for Hecke characters, and to the factorization $\pi_f = \otimes \pi_p$ discussed in Part II.D.2. In particular, for almost all v, π_v is *unramified*; this means that its space H_v contains (a one-dimensional space of) vectors which are left fixed by the group of operators $\pi(k_v)$, k_v in K_v. The product $\otimes_v \pi_v$ is restricted in the sense that its space consists of vectors $w = \otimes w_v$ with the property that w_v is K_v-fixed for almost all v. Thus the representation $\pi(g)$ is well defined by the formula

$$\pi(g)w = \prod_v \pi_v(g_v)w_v$$

since $\pi_v(g_v)w_v$ belongs to K_v for almost every v.

If there exists an isomorphism A between H_π (the space of π) and some subspace V_π of L_0^2 such that $A\pi(g) = R(g)A$ for all g in $G_{\mathbf{A}}$, then we say π is equivalent to some subrepresentation of R_0, or more simply, that π "appears" in R_0.

DEFINITION. An irreducible unitary representation of $G_{\mathbf{A}}$ is called a *cuspidal automorphic representation of* G if it appears in the right regular representation R_0 of G.

REMARKS. (i) In order to avoid further technical discussion, we have overlooked the fact that $G_F \backslash G_{\mathbf{A}}$ does not quite have finite volume, and thus that L_0^2 does not quite possess minimally invariant subspaces V_π. These technical problems are overcome by considering invariance with respect to the central subgroup

$$Z = \left\{ \begin{bmatrix} z & & 0 \\ & \ddots & \\ 0 & & z \end{bmatrix} : z \in \mathbf{A} \right\};$$

the reader can consult [**Ge 1** or **Art**] for details.

(ii) For similar reasons we have made our notion of automorphic representation more restrictive than it should be. Indeed, we should not be restricting ourselves to cuspidal functions, and we also should not be insisting that π necessarily occur "discretely" in any right regular representation. The correct notion of automorphic representation, together with a discussion of the technical problems which ensue, is discussed in [**Langlands 5** and **Bo Ja**].

EXAMPLES OF AUTOMORPHIC REPRESENTATIONS. (1) An automorphic representation of GL_1 is just a character $\chi = \prod \chi_v$ on the idele class group $F \backslash \mathbf{A}$, and $L(s, \chi) = \prod (1 - \chi_v(\tilde{\omega}_v) N v^{-s})^{-1}$ is the L-function attached to this automorphic representation. Here, and henceforth, Nv denotes the cardinality of the finite field O_v / P_v, the ring of integers in F_v modulo the prime ideal $P_v = \{x \in O_v : |x| \leq 1\}$.

(2) In II.D.2, we saw that a classical automorphic form in $S_k(\mathrm{SL}_2(\mathbf{Z}))$, with

$$T(p) f = a_p f \quad \text{for all } p,$$

uniquely determines an automorphic cuspidal representation $\pi_f = \otimes \pi_p$ of GL_2 *over* $\mathbf{Q}$. Conversely, every automorphic cuspidal representation $\pi = \otimes \pi_p$ determines some classical automorphic form $f(z)$, though not necessarily for the full modular group $\mathrm{SL}_2(\mathbf{Z})$, and not necessarily holomorphic (as opposed to real analytic). Moreover, for f as above, the pth local factor in the Euler product expansion of $\phi_f(s) = \sum a_n n^{-s}$ is

$$L_p(s) = \left(1 - a_p p^{-s} + p^{k-1-2s}\right)^{-1} = \left[\left(1 - \alpha_p p^{-s}\right)\left(1 - \beta_p p^{-s}\right)\right]^{-1}.$$

In both these examples, important information about global objects (χ or $f(z)$) is stored by local objects (χ_v or a_p). Indeed, it is precisely this fact which makes it possible to introduce an L-function in local terms and in a uniform manner independent of the underlying group.

Let us now explain how this works in general.

B. Local invariants. Given an arbitrary irreducible unitary representation $\pi = \otimes \pi_v$, how can we attach to π_v local data which will simultaneously characterize π_v and feed neatly into a local Euler factor $L(s, \pi_v)$—even when π is not necessarily automorphic? The answer to this question involves the so-called theory of spherical functions and unramified representations of G_v.

Fix a finite place v of F, and suppose π_v is an unramified representation of G_v. Then the theory of spherical functions assigns to π_v an element A_v of the complex torus T^n which is unique up to the natural action of the permutation group S_n on T^n.

Roughly speaking, the procedure is this: let H_v^0 denote the space of $\mathrm{GL}_n(O_v)$-fixed vectors w_v in the space of π_v, and let $\mathcal{H}_v$ denote the convolution algebra of locally constant, $\mathrm{GL}_n(O_v)$-bi-invariant, and compactly supported complex-valued functions on G_v (called the Hecke algebra); then the formula

$$\pi_v(f)w_v = \int_{G_v} f(g)\pi(g)w_v\,dg = \chi_{\pi_v}(f)w_v$$

defines a representation of $\mathcal{H}_v$ on the one-dimensional space H_v^0, and the resulting character $f \to \chi_{\pi_v}(f)$ defines (via the so-called Satake isomorphism) an element A_v of T^n (unique up to permutation); details are discussed in §IV of **[Cart 2]** and §§6, 7 of **[Bo]**. In fact, π_v is itself determined by (the class of) A_v. Thus there results an injection $\pi_v \mapsto A_v$ taking unramified representations of G_v to "semisimple" (i.e., diagonalizable) conjugacy classes in the group $\mathrm{GL}_n(\mathbf{C})$.

Let us return now to the global situation. Given $\pi = \otimes\pi_v$, we let S_π denote the finite set of places v outside of which π_v is unramified, and we consider the family of conjugacy classes $\{A_v\}$, $v \notin S_\pi$. The point is that this construction generalizes the assignment of conjugacy classes which is implicit in the classical theory of Hecke, and yet it makes sense for all (not necessarily automorphic) representations π of $\mathrm{GL}_n(\mathbf{A})$.

For example, if $n = 1$ and π_v is an unramified character χ_v of F_v^x, then A_v is simply $\chi_v(\tilde{\omega}_v)$, the value of χ_v at any local uniformizing variable. Also, if $\chi = \prod\chi_v$ is actually automorphic, i.e., trivial on F^x, then the family $\{A_v\}$, $v \notin S_\pi$, actually determines χ uniquely.

Now suppose $n = 2$, and $\pi_f = \otimes\pi_p$ is an automorphic cuspidal representation of $\mathrm{GL}_2(\mathbf{A})$ corresponding to the classical cusp form $f(z) = \sum a_n e^{2\pi i n z}$ in $S_k(\mathrm{SL}_2(\mathbf{Z}))$. If $T(p)f = a_p f$ for all p, then

$$A_p = \begin{bmatrix} \alpha_p & 0 \\ 0 & \beta_p \end{bmatrix} \quad \text{in } \mathrm{GL}_2(\mathbf{C}),$$

with $\alpha_p\beta_p = p^{k-1}$ and $\alpha_p + \beta_p = a_p$. In particular, the eigenvalues a_p completely determine the local representations π_p, and it turns out that the *family* of classes $\{A_p\}$ completely determines f (or π).

In general, for GL_n, it still turns out that the local data $\{A_v\}$ determine the global representation π uniquely *provided* π is automorphic. (This is a nontrivial recent result of Jacquet, Piatetski-Shapiro and Shalika **[J PS S 2]**.) Thus we are witnessing a powerful variation of the local-global principle at work in the context of general automorphic representations. Simultaneously, we are obtaining a handle on introducing general L-functions *and* comparing automorphic representations of different groups, both cornerstones of the Langlands program we shall now finally describe.

IV. THE LANGLANDS PROGRAM

How does Langlands' theory not only tie together the strands woven in Part II but also develop the general notions of Part III (automorphic representations and L-functions) beyond their classical origins?

A. Preliminary L-functions. Langlands' theory begins by attaching an L-function to an arbitrary irreducible unitary representation $\pi = \otimes \pi_v$ of $G_{\mathbf{A}}$. (Here, as in Part III, G is the group GL_n, but the more adventurous reader can imagine it to be an arbitrary reductive algebraic group.)

Given π, let S_π again denote the finite set of places v of F outside of which π_v is unramified, and for each $v \notin S_\pi$ let A_v denote the semisimple conjugacy class

$$A_v = \begin{bmatrix} \alpha_1 & & 0 \\ & \ddots & \\ 0 & & \alpha_n \end{bmatrix}$$

in $\mathrm{GL}(n, \mathbf{C})$ corresponding to π_v as in III.B. Then consider the Euler product

$$L_{S_\pi}(s, \pi) = \prod_{v \notin S_\pi} L(s, \pi_v)$$

with

$$L(s, \pi_v) = \left[\det(I - [A_v] Nv^{-s})\right]^{-1}.$$

This is an Euler product *of degree n* in the sense that each Euler factor $L(s, \pi_v)$ is of the form $P^{-1}(Nv^{-s})$ with P a polynomial of degree n and $P(0) = 1$. The infinite product can be shown to converge for $\mathrm{Re}(s)$ sufficiently large.

THEOREM. *Suppose $\pi = \otimes \pi_v$ is an arbitrary irreducible representation of $G_{\mathbf{A}}$. Then for all v one can define Euler factors $L(s, \pi_v)$ of degree $\leqslant n$, and local factors $\varepsilon(s, \pi_v)$, such that $\varepsilon(s, \pi_v)$ is* 1 *for almost all v, and*

$$L(s, \pi_v) = \det(I - [A_v] Nv^{-s})^{-1}$$

whenever v is unramified. Moreover, if π is actually automorphic cuspidal, then the Euler product

$$L(s, \pi) = \prod_v L(s, \pi_v),$$

initially defined only in some right half-plane, satisfies the following properties:

(i) *it extends to an entire function of* $\mathbf{C}$ (*unless $n = 1$ and π is the trivial character, in which case $L(s, \pi)$ has a pole*);

(ii) *it satisfies the functional equation*

$$L(s, \pi) = \varepsilon(s, \pi) L(1 - s, \tilde{\pi}),$$

with $\tilde{\pi}$ the representation "contragredient" to π, and

$$\varepsilon(s, \pi) = \prod_v \varepsilon(s, \pi_v)$$

REMARKS. (i) This theorem was first proved by Jacquet and Langlands in **[J L]** for the case $n = 2$. The general case appears in **[Go Ja]**. We note that when $n = 1$, this theorem reduces to the work of Tate described in Part II.D.1. In case $n = 2$ it represents a vast reformulation and generalization of Hecke's work; in particular, it finally sheds light on the Euler product expansion of Hecke's Dirichlet series

$$\phi_f(s) = \sum \frac{a_n}{n^s} = (2\pi)^s \Gamma(s)^{-1} L(s', \pi_f)$$

and interprets the constant in the functional equation of $\phi_f(s)$ in terms of the *local* groups G_v. This last point underscores one of the key contributions of the original work of Jacquet and Langlands and will be discussed presently.

(ii) It is natural and important to ask if this theorem has a converse. In other words, suppose that $\{A_v^*\}$, v outside some finite set of places S, is a family of semisimple conjugacy classes in $\mathrm{GL}_n(\mathbf{C})$. Suppose, in addition, that

$$\prod_{v \notin S} \det\left(I - A_v^*(Nv)^{-s}\right)^{-1}$$

converges in some right half-plane to an analytic function which has properties similar to those established in the theorem above. Is there then an automorphic cuspidal representation π of $G_{\mathbf{A}}$ such that $A_v(\pi_v) = A_v^*$ for each v outside S? The answer in general is no. However, an appropriate characterization of "automorphic" families $\{A_v\}$ was given in **[J L]** for the case $n = 2$ by generalizing the classical converse theorems of Hecke and Weil already described; for $n = 3$, a "converse theorem" is given in **[J PS S 1]**.

Let us briefly discuss some corollaries of the Converse Theorem for GL_2. As mentioned earlier, Weil's characterization of the L-functions attached to classical automorphic forms led him to conjecture that the zeta-function

$$\zeta(E, s) = \prod \left\{\det\left[I - A_p^* p^{-s}\right]\right\}^{-1}$$

of an elliptic curve E over $\mathbf{Q}$ is really the L-function of an automorphic cusp form of weight 2. The point is that once the zeta-function of such a curve is conjectured to have nice analytic properties, it will follow that the semisimple conjugacy classes $\{A_p^*\}$ in $\mathrm{GL}_2(\mathbf{C})$—which one obtains by counting points on the "reduced curve mod p"—should comprise an "automorphic" family of conjugacy classes for GL_2.

On the other hand, Langlands' local analysis of the constants in the functional equation of Artin's nonabelian L-functions led Jacquet and Langlands to relate these L-functions as well to "automorphic" L-functions. Indeed, modulo Artin's conjecture on the entirety of his L-functions, the nonabelian L-functions of degree 2 have been shown to satisfy all the hypotheses of the "Converse theorem" for GL_2 (cf. **[Deligne 2]**). For arbitrary n, there is the following remarkable "Reciprocity Conjecture".

CONJECTURE 1 (LANGLANDS). Suppose E is a finite Galois extension of F with Galois group $G = \mathrm{Gal}(E/F)$, and $\sigma\colon G \to \mathrm{GL}_n(C)$ is an irreducible representation of G. Then there exists an automorphic cuspidal representation π_σ on GL_n *over* F such that $L(s, \pi_\sigma) = L(s, \sigma)$.

Note that when $n = 1$ and E over F is abelian, this conjecture reduces to Artin's celebrated "reciprocity law" relating $L(s, \sigma)$ to Hecke's abelian L-series $L(s, \chi)$. Moreover, for arbitrary n, the truth of this conjecture implies the truth of Artin's conjecture on the entirety of his $L(s, \sigma)$, since $L(s, \pi)$ is entire for any (nontrivial) cuspidal representation π of GL_n.

Note also that we have at last succeeded in showing how Langlands' theory (at least conjecturally) subsumes all the classical themes and results discussed in Part II.

B. L-groups and the functoriality of automorphic representations. To proceed deeper into Langlands' program, it is necessary to deal finally with more general groups than GL_n and to introduce the notion of an "L-group". This latter notion is already implicit in the definition of L-functions for GL_n, but needs to be made explicit before these functions can be generalized.

Recall that if $\pi = \otimes \pi_v$ is a representation of $GL_n(\mathbf{A})$, the theory of spherical function assings to each $v \notin S_\pi$ a well-defined semisimple conjugacy class in $GL_n(\mathbf{C})$. Actually, to be more precise, the theory of spherical functions establishes a bijection between unramified representations π_v and orbits of unramified homomorphisms of the local torus

$$T_v = \left\{ \begin{bmatrix} a_1 & & & \bigcirc \\ & a_2 & & \\ & & \ddots & \\ \bigcirc & & & a_n \end{bmatrix} : a_i \text{ in } F_v^x \right\}.$$

Here "unramified" means that the restriction to

$$T_v(O_v) = \left\{ \begin{bmatrix} \alpha_1 & & \bigcirc \\ & \ddots & \\ \bigcirc & & \alpha_n \end{bmatrix} : \alpha_i \text{ in } O_v^x \right\}$$

is trivial, and orbits are understood taken with respect to the action of the so-called Weyl group (which in this case is just S_n).

Now the structure theory of GL_n is such that these sets of orbits of "quasi-characters" are in natural 1-1 correspondence with semisimple conjugacy classes of the complex group $GL_n(\mathbf{C})$. What the notion of an L-group does is systematize this analysis and extend it to more general reductive groups. Since we are ignoring the definition and structure theory of a general reductive group, we refer the reader to [**Cart 2, Springer, Humph** or **Ti**] for details and examples.

Suppose first that G is a "split" such group—an arbitrary connected reductive group with a maximal torus split over F. Guided by the facts just described for GL_n, Langlands (in [**Langlands 2**]) constructed a complex reductive Lie group ${}^LG^0$—the "L-group of G". For $G = \mathrm{GL}_n$, of course ${}^LG^0 = \mathrm{GL}_n(\mathbf{C})$. Here are some other examples:

G	${}^LG^0$
SL_n	$\mathrm{PGL}_n(\mathbf{C})$
PGL_n	$\mathrm{SL}_n(\mathbf{C})$
Sp_{2n}	$\mathrm{SO}_{2n+1}(\mathbf{C})$
SO_{2n+1}	$\mathrm{Sp}_{2n}(\mathbf{C})$

In general, the construction of ${}^LG^0$ uses the notions of maximal tori, root systems, etc., concepts originally introduced to classify the complex Lie algebras and their finite-dimensional representations. The result is a bijection between the unramified representations of G_v and the semisimple conjugacy classes of ${}^LG^0$, i.e., orbits of the maximal torus ${}^LT^0$ with respect to the "Weyl group" of $({}^LG^0, {}^LT^0)$; note that if $G = \mathrm{GL}_n$, then ${}^LT^0 = T^n$, and the Weyl group is just (isomorphic to) S^n. In any case, given an irreducible unitary representation $\pi = \otimes \pi_v$ of $G_{\mathbf{A}}$, with π_v unramified for all v outside S_π, there is defined a collection of conjugacy classes $\{A_v\}$ in ${}^LG^0$.

DEFINITION. If r is any finite-dimensional complex analytic representation of ${}^LG^0$, define

$$L_v(s, \pi, r) = \det[I - r(A_v)Nv^{-s}]^{-1}$$

for $v \notin S_\pi$, and

$$L(s, \pi, r) = \prod_{v \notin S_\pi} L_v(s, \pi, r).$$

Here, as before, Nv denotes the cardinality of the finite field O_v/P_v.

These L-functions generalize those already defined for GL_n in Part III. Indeed, if $G = \mathrm{GL}_n$, and r: $\mathrm{GL}_n(\mathbf{C}) \to \mathrm{GL}_n(\mathbf{C})$ is the obvious "standard" representation, then $L(s, \pi, r) = L(s, \pi)$. On the other hand, fixing $G = \mathrm{GL}_n$, and letting r vary over all possible representations (and dimensions), we obtain a *family* of L-functions for GL_n.

CONJECTURE 2′. Suppose π is actually automorphic. Then $L(s, \pi, r)$, initially defined only in some right half-plane, continues meromorphically to $\mathbf{C}$ with a functional equation relating $L(s, \pi, r)$ to $L(1 - s, \tilde{\pi}, r)$.

Unfortunately, there are very few examples of verifications of Conjecture 2′. One example, however, is particularly closely related to the "principle of functoriality" which we shall soon discuss.

Suppose $G = \mathrm{GL}_2$ (so ${}^LG^0 = \mathrm{GL}_2(\mathbf{C})$), and let r denote the 3-dimensional adjoint representation of $\mathrm{GL}_2(\mathbf{C})$ obtained by composing the natural adjoint action of $\mathrm{PGL}_2(\mathbf{C})$ (on the 3-dimensional Lie algebra of trace zero 2×2

matrices) with the natural projection map $GL_2(\mathbf{C}) \to PGL_2(\mathbf{C})$. Thus we have the diagram

$$\begin{array}{ccc} {}^LG^0 = GL_2(\mathbf{C}) & \xrightarrow{r} & GL_3(\mathbf{C}) \\ & \searrow \quad \nearrow \mathrm{Ad} & \\ & PGL_2(\mathbf{C}) & \end{array}$$

In **[Ge Ja]** it is shown that Conjecture 2′ is true for this G and r. Note, in this case, that

$$L_v(s, \pi, r) = \det[I - r(A_v)N_v^{-s}]^{-1} = \det[I - [A_v']Nv^{-s}]^{-1}$$

is an Euler factor of degree 3. What is shown in **[Ge Ja]**, using the Converse Theorem for GL_3, is that the family of conjugacy classes $\{r(A_v)\} = \{A_v'\}$ in $GL_3(\mathbf{C}) = {}^L(GL_3)^0$ is actually "automorphic", i.e. belongs to an automorphic representation $\Pi = \otimes \Pi_v$ of GL_3. This result is predicted by, and indeed simply a special realization of the principle embodied in Conjecture 3′ below.

Suppose G and G' are split groups, and ρ: ${}^LG^0 \to {}^LG'^0$ is an analytic homomorphism. If r' is any finite-dimensional representation of ${}^LG'^0$, then $r = r' \circ \rho$ is a finite-dimensional analytic representation of ${}^LG^0$. If π is an automorphic cuspidal representation of G, and $v \notin S_\pi$, let A_v' denote the semisimple conjugacy class *in* ${}^LG'^0$ which contains $\rho(A_v)$. Then

$$L(s, \pi, r) = \prod_{v \notin S} \det[I - r'(A_v')Nv^{-s}]^{-1},$$

and the analytic continuation and functional equation of the function *on the right*, i.e., for the family A_v', would follow from Conjecture 2′. Thus we are led to the following "functoriality principle" of Langlands:

CONJECTURE 3′. Given an analytic homomorphism ρ: ${}^LG^0 \to {}^LG^0$, and an automorphic representation $\pi = \otimes \pi_v$ of G, there is an automorphic representation π' of G' such that $S_\pi = S_{\pi'}$, and such that for each $v \notin S_{\pi'}$, A_v' is the conjugacy class in ${}^LG'^0$ which contains $\rho(A_v)$. In particular,

$$L(s, \pi', r') = L(s, \pi, r' \circ \rho)$$

for each finite-dimensional representation r' of ${}^LG'^0$.

In order to minimize the brain strain that results from the unfolding of these conjectures, it is helpful to note that each is actually contained in an appropriately formulated functoriality conjecture. For example, suppose we take $G' = GL_n$ and r' to be the standard representation of $GL_n(\mathbf{C})$. Then $L(s, \pi, r' \circ \rho) = L(s, \pi, r) = L(s, \pi')$, and the functional equation and analytic continuation of $L(s, \pi, r)$ would be assured by Theorem 1 if π' were indeed automorphic. In other words, Conjecture 3′ implies Conjecture 2′.

In order to also incorporate Conjecture 1 (Langlands' reciprocity conjecture) it is necessary to slightly reformulate Conjecture 3′ by taking into account the fact that, unlike GL_n, not all groups are "split" over F. Without going into details (to which we refer the reader to **[Bo]**), we collect some fundamental facts below.

Suppose G is an arbitrary connected reductive group defined over F, and K is a (sufficiently large) Galois extension of F. Then one can define a complex reductive Lie group ${}^LG^0$, together with an action of $\mathrm{Gal}(K/F)$ on ${}^LG^0$, such that the resulting semidirect product, ${}^LG = {}^LG^0 \rtimes \mathrm{Gal}(K/F)$, satisfies the following properties:

(i) In case G is "split" over F (in particular, when $G = \mathrm{GL}_n$), ${}^LG$ reduces to a direct product (the action of $\mathrm{Gal}(K/F)$ on ${}^LG^\circ$ being trivial);

(ii) In general, if v is a prime of F unramified in K, with corresponding Frobenius automorphism Fr_v, there is a 1-1 correspondence between "unramified" representations π_v of $G_v = G(F_v)$ and conjugacy classes $t(\pi_v)$ in ${}^LG$ such that the projection of $t(\pi_v)$ onto $\mathrm{Gal}(K/F)$ is the class of Fr_v; see **[Bo]** for more details.

This group ${}^LG$, which plays the same role for arbitrary G as $\mathrm{GL}_n(\mathbf{C})$ plays for the group G, is called the (Galois form of the) *L-group of G*. By a *representation* of ${}^LG$ we understand a homomorphism r: ${}^LG \to \mathrm{GL}_d(\mathbf{C})$ whose restriction to ${}^LG^0$ is complex analytic. By an *L-homomorphism* of L-groups ${}^LG$ and ${}^LG'$ we understand a continuous homomorphism

$$\begin{array}{ccc} \rho\colon {}^LG & \rightarrow & {}^LG' \\ & \searrow \quad \swarrow & \\ & \mathrm{Gal}(K/F) & \end{array}$$

which is compatible with the natural projections of each group onto $\mathrm{Gal}(K/F)$ (and whose restriction to ${}^LG^0$ is a complex analytic map of ${}^LG^0$ to ${}^LG'^0$).

In terms of these concepts we can finally formulate the ultimate generalizations of Conjectures 2′ and 3′.

CONJECTURE 2. Suppose $\pi = \otimes \pi_v$ is an automorphic representation of G and r is a finite-dimensional representation of ${}^LG$. Then the Euler product

$$L(s, \pi, r) = \prod_{\substack{v \\ \text{unramified}}} \det\left[I - r(t(\pi_v))Nv^{-s}\right]^{-1},$$

initially defined in a right half-plane of s, continues meromorphically to all of C with functional equation relating $L(s, \pi, r)$ to $L(1 - s, \tilde{\pi}, r)$.

CONJECTURE 3. (Functoriality of automorphic forms with respect to the L-group). Suppose G and G' are reductive groups and ρ: ${}^LG \to {}^LG'$ is an L-homomorphism. Then to each automorphic representation $\pi = \otimes \pi_v$ of G there is an automorphic representation $\pi' = \otimes \pi_v'$ of G' such that for all $v \notin S_{\pi'}$ (i.e. unramified v), $t(\pi_v')$ is the conjugacy class in ${}^LG'$ which contains $t(\pi_v)$. Moreover, for any finite-dimensional representation r' of ${}^LG'$,

$$L(s, \pi', r') = L(s, \pi, r' \circ \rho)$$

CONCLUDING REMARKS. This last conjecture of Langlands' really does imply all the preceding conjectures discussed heretofore.

First take $G' = \mathrm{GL}_n$ and suppose r is any n-dimensional representation of ${}^LG$ (G arbitrary but fixed). Let ρ: ${}^LG \to {}^LG'$ be such that the following diagram commutes:

$$\begin{array}{ccc} {}^LG & \xrightarrow{r} & \mathrm{GL}_n(\mathbf{C}) \\ \rho\searrow & & \nearrow \mathrm{St} \\ & {}^LG' = \mathrm{GL}_n(\mathbf{C}) \times \mathrm{Gal}(K/F) & \end{array}$$

(Here St: ${}^LG \to \mathrm{GL}_n(\mathbf{C})$ is the standard representation of ${}^LG$, namely projection onto the component ${}^LG^0 = \mathrm{GL}_n(\mathbf{C})$.) Then *assuming* the truth of Conjecture 3, we have a lift of automorphic forms $\pi \to \pi'$ between G and G' with

$$L(s, \pi, r) = L(s, \pi, \mathrm{St} \circ \rho) = L(s, \pi', \mathrm{St}) = L(s, \pi'),$$

with the last L-function (on GL_n) "nice" by **[Go Ja]**. Thus Conjecture 3 not only implies Conjecture 2, but also reduces the study of generalized L-functions for arbitrary G to the known theory for GL_n!

Now suppose $G = \{e\}$ (the trivial group) and again take $G' = \mathrm{GL}_n$. Then the only possible automorphic representation of G is the trivial one, and an L-homomorphism ρ_σ: ${}^LG \to {}^LG'$ amounts to specifying a representation σ: $\mathrm{Gal}(K/F) \to \mathrm{GL}_n(\mathbf{C})$ such that $\rho_\sigma(1 \times \gamma) = \sigma(\gamma) \times \gamma$ for all $\gamma \in \mathrm{Gal}(K/F)$. Thus Conjecture 3 amounts to the assertion that there is an automorphic representation π_σ of GL_n (associated to the trivial representation of G via $\rho = \rho_\sigma$) such that for all unramified primes v, the projection of $t(\pi_v)$ on $\mathrm{GL}_n(\mathbf{C})$ is just $\sigma(\mathrm{Fr}_v)$. In particular,

$$L(s, \pi_\sigma) = L(s, \sigma),$$

which means Conjecture 3 indeed implies Conjecture 1 (Langlands' Reciprocity Conjecture).

Note. A careful and detailed exposition of Langlands' general program is found in **[Bo]**; our brief sketch of the theory follows **[Art]**.

C. What's known? Though many specific cases of the functoriality conjecture (Conjecture 3) have been verified, it is far from being solved. The most comprehensive survey of known results (and work in progress) is found in **[Bo]**. The reader is also referred to **[Langlands 3]** for a concise, illuminating discussion.

In the paragraphs below, we shall comment briefly only on a small part of the recent work in this area.

1. *Artin's Conjecture and Langlands' Reciprocity Conjecture*. Here we follow **[Art]** quite closely. Suppose K is a Galois extension of F, and $F = \mathbf{Q}$ (for simplicity). We think of K as the splitting field of some monic polynomial $f(x)$ with integer coefficients. For almost all primes p, namely those "unramified" in K, we let Fr_p denote the (conjugacy class of a) Frobenius automorphism in $\mathrm{Gal}(K/\mathbf{Q})$. Recall that the prime p "splits completely" in K (i.e., the ideal it generates in the ring of integers O_K factors into $[K:\mathbf{Q}]$ distinct prime ideals of O_K) if and only if $\mathrm{Fr}_p = \mathrm{Id}$. In terms of the polynomial $f(x)$, this means (in general) that $f(x)$ splits into linear factors "mod p".

Let $S(K)$ denote the set of primes p which split completely in K. For example, if $K = \mathbf{Q}(\sqrt{-1})$, then $S(K) = \{p\colon p \equiv 1 \pmod 4\}$. In general, it is known that the map $K \to S(K)$ is an *injective* order reversing map from finite Galois extensions of $\mathbf{Q}$ into subsets of prime numbers. In other words, the set of splitting primes determines K uniquely. Thus it is natural to pose the following:

Problem. What is the image of this map? I.e., what sets of prime numbers are of the form $S(K)$?

A solution to this problem would constitute some kind of "nonabelian class field theory" since we would be able to parametrize all the finite Galois extensions K of $\mathbf{Q}$ by the collections $S(K)$ (which are intrinsic to $\mathbf{Q}$). In case we restrict attention to *abelian* extensions, a solution is known in terms of congruence conditions like those for the example $K = \mathbf{Q}(\sqrt{-1})$; this is the "abelian class field theory" discussed in Parts I and II. In general, such a neat solution cannot be expected, but any intrinsic characterization of these sets $S(K)$ certainly deserves to be called a reciprocity law.

What light do Langlands' ideas shed on this fundamental problem?

Let $\overline{\mathbf{Q}}$ denote an algebraic closure of $\mathbf{Q}$. Given a Galois extension K of $\mathbf{Q}$ as above, there exists a homomorphism $\sigma\colon \mathrm{Gal}(\overline{\mathbf{Q}}/\mathbf{Q}) \to \mathrm{GL}_n(\mathbf{C})$ with the property that $\mathrm{Gal}(\overline{\mathbf{Q}}/K)$ is the kernel of r. Thus we get an *injective* homomorphism $\sigma\colon \mathrm{Gal}(K/\mathbf{Q}) \to \mathrm{GL}_n(\mathbf{C})$ to which we can attach the Artin L-function $L(s, \sigma)$ discussed in II.C.2. Moreover, the definitions are such that

$$S(K) = \{p\colon \sigma(\mathrm{Fr}_p) = I\}.$$

Now consider again Langlands' Conjecture 1. It asserts that the family $\{\sigma(\mathrm{Fr}_p)\}$ is automorphic, i.e., that there exists an automorphic representation $\pi = \bigotimes \pi_p$ of GL_n such that for all p outside S_π, $A_p = \sigma(\mathrm{Fr}_p)$. In particular,

$$S(K) = \{p\colon A_p = I\}.$$

Thus (the truth of) Conjecture 1 reduces the reciprocity problem above to the study of automorphic representations of $\mathrm{GL}_n(\mathbf{A})$. Moreover, this relation is typical of the perspectives which Langlands' program brings to classical number theory.

The fact that the collections $S(K)$—which classify Galois extensions of $\mathbf{Q}$—might be recovered from data obtained analytically from the decomposition of the right regular representation R into irreducibles is not only abstractly satisfying, it also gives us a handle on solving the original problem. Indeed, when $n = 2$, Langlands has already applied the theory of representations to prove Conjecture 1 for a wide class of irreducible representations σ of $\mathrm{Gal}(K/F)$. For a discussion of these matters, see [**Ge 3, Ger Lab**] and the original sources [**Langlands 3, 4**]; the most recent results are described in [**Tun**].

2. *Other examples.* Below we provide a partial list of recent verifications of the functoriality principle; for more complete discussions and references the reader is again referred to [**Bo**].

(a) *Base-change.* Take E to be a cyclic Galois extension of F of prime degree. Then automorphic representations of GL_2 over F "lift" to automorphic representations of GL_2 *over* E corresponding to the following homomorphism of L-groups. If $G = \mathrm{GL}_2$, then the group $G' = \mathrm{Res}_F^E G$ (obtained "by restriction of scalars") is such that G'_F (the points of G' over F) $= \mathrm{GL}_2(E)$; its L-groups is ${}^LG' = \mathrm{GL}_2(\mathbf{C}) \times \cdots \times \mathrm{GL}_2(\mathbf{C}) \rtimes \mathrm{Gal}(E/F)$, the Galois group acting on ${}^LG'^0 = \mathrm{GL}_2(\mathbf{C}) \times \cdots \times \mathrm{GL}_2(\mathbf{C})$ by permuting coordinates. The L-group homomorphism ρ which gives rise to "base change lifting" is then given by simply imbedding ${}^LG^0$ diagonally in ${}^LG^{0\prime}$. The resulting "lifting theorems" (due to Saito, Shintaini and Langlands) play a fundamental role in Langlands' proof of his Conjecture 1 for (certain) two-dimensional Galois representations σ.

(b) *Zeta-functions of algebraic varieties.* We have already alluded to Weil's conjecture asserting that the zeta-function of an elliptic curve is "automorphic". Similar results have been conjectured more generally for the zeta-functions (counting points mod p) for general algebraic varieties over F. The classical results center around "Eichler-Shimura theory", and a general theory has been developed by Langlands, Deligne, Milne, Shih, Shimura and others. For an elementary introduction, see [**Ge 2**]; for a survey of recent developments, see [**Bo, Cas**], and the references therein.

(c) *Local results.* Although we have not emphasized this fact, nearly all the assertions thus far treated have local counterparts which are part and parcel of the global theory. For example, the local part of Langlands' Reciprocity Conjecture amounts to a (conjectured) parametrization of the irreducible representations of $G_v = \mathrm{GL}_n(F_v)$ by n-dimensional representations of the Galois group of F_v (such that ε-factors are preserved.) For $n = 2$ this is already a highly nontrivial and very interesting assertion, the truth of which has just recently been verified by P. Kutzko; see [**Cart 1**] for a complete exposition of the problem. For GL_n over a p-adic field, especially when $p \nmid n$, see [**Moy** and **Hen**].

3. *Related questions.* There are other important directions in the theory of automorphic forms (cf. [**Mazur Wiles, Gross B** and **Rib**]) which do not as yet fit in neatly with the general Langlands program. Clearly it would be profitable to pursue the connections with these works and other purely diophantine investigations.

D. Methods of proof. Thus far, three general methods have been used to attack automorphic problems such as the functoriality conjecture. We shall merely sketch the barest outline of these methods and some representative examples of their successes. Undoubtedly, totally new methods are called for as well.

(1) *θ-series.* This is perhaps the oldest, most venerable approach, which began with the classical discovery that

$$\theta(z) = \sum_{-\infty}^{\infty} e^{\pi i n^2 z}$$

defines an automorphic form; more generally, as we already remarked in II.B, similar (automorphic) theta-series can be attached to quadratic forms in any number of variables.

The representation-theoretic construction of automorphic forms via theta-series has as its point of departure the fundamental paper [**We 2**] published in 1964. In that paper, Weil found a proper group-theoretic home for general theta-functions $\theta(g)$, namely the symplectic groups $\mathrm{Sp}_{2n}(A)$ and their two-fold covering groups (the so-called metaplectic groups). Thus Weil was able to reinterpret the extensive earlier works of C. L. Siegel on quadratic forms and to open the way for group representation theory to be used in the construction of (automorphic) theta-series.

According to Weil, the proper generalization of the classical theta-series $\theta(z)$ is a certain *automorphic* representation of $G_{\mathbf{A}} = \mathrm{Sp}_{2n}(\mathbf{A})$, called *Weil's representation*. This representation acts by right translation in a space of generalized theta-functions $\theta(g)$; by analyzing (in particular, decomposing) this representation, a great deal of interesting information about theta-series can be obtained. For example, in order to understand Hecke's construction of theta-series attached to grossencharacters of a quadratic extension of $\mathbf{Q}$, one simply decomposes (an appropriate tensor product of) Weil's representation of $\mathrm{SL}_2(\mathbf{A}) = \mathrm{Sp}_2(\mathbf{A})$ into irreducible (automorphic) representations; this is what was carried out in [**Sha Tan**]. For a survey of similar applications of Weil's representation to the construction of automorphic forms, see [**Ge 4**].

Now what is the connection between these constructions and the functoriality conjecture of Langlands? The best way to answer this question is to bring into play R. Howe's theory of "dual reductive pairs", yet another simple principle of great beauty and consequence.

Suppose G and G' are subgroups of Sp_{2n} which are each others' centralizers, i.e., (G, G') comprises a "dual reductive pair" in the sense of [**Ho 1**].

The decomposition of Weil's representation restricted to $G \times G'$ *should* then give a correspondence $\pi \to \pi'$ attached to some L-group homomorphism ${}^LG \to {}^LG'$. Indeed, this restriction should decompose as a sum of representations $\pi \otimes \pi'$ with π' an irreducible representation of G' uniquely determined by π (a representation of G); moreover, π' should be automorphic if and only if π is. For a careful description of this "duality correspondence", see [**Ho 1**]. In practice, one can usually construct the (global) correspondence $\pi \to \pi'$ directly by using a formula like

$$(1) \qquad f(h) \to \phi_f(g) = \int_{G_F \backslash G_{\mathbf{A}}} \theta(g, h) f(h)\, dh;$$

here f is an automorphic form on $G_{\mathbf{A}}$ in the space of π, $\theta(g, h)$ is a theta-function on $\mathrm{Sp}_{2n}(\mathbf{A})$ restricted to $G_{\mathbf{A}} \times G'_{\mathbf{A}}$, and ϕ_f is a function on $G'_{\mathbf{A}}$ which generates the automorphic representation π'. (This operator essentially projects the restriction of Weil's representation onto the isotypic component "belonging to the irreducible representation π".)

EXAMPLES FOR (1). (i) Take $(G, G') \subset \mathrm{Sp}_4$ with G the norm 1 group of a quadratic extension E of $\mathbf{Q}$, and $G' = \mathrm{SL}_2 = \mathrm{Sp}_2$. Then the corresponding lift $\chi \to \pi_\chi$ defined by (1) generalizes the construction of Hecke's just alluded to. In particular, if E over $\mathbf{Q}$ is *real*, one obtains Maass' construction of nonnolomorphic modular forms; cf. [**Ge 1**] for details.

(ii) Let G (resp. G') denote the unitary group of an isotropic Hermitian space in two (resp. three) variables over a quadratic extension E of F. Then (G, G')

naturally embeds as a dual reductive pair in Sp_{12}, and the corresponding lift $\pi \to \pi'$ has been analyzed via (1) in **[Ge PS 1]**.

We note that the *exact* relation between the duality correspondence just sketched and the Langlands lifting predicted by Conjecture 3 is not at all transparent, and in fact is a bit delicate; the interested reader is referred to the Introduction of **[Ra 1]** for more discussion of "Langlands functoriality for the Weil representation".

(2) *L-functions.* This method has already been discussed at some length in this paper. Its success in constructing automorphic representations (and thereby verifying the functoriality conjecture) is based mainly on the "Converse Theorem", which asserts that a representation of $G_{\mathbf{A}}$ is automorphic if and only if enough of its L-functions are "nice". To be sure, since this "Converse Theorem" has been proved in a useful form only for GL_2 and GL_3, the range of applicability of this method is somewhat limited (see, however, the remarks following the examples below).

EXAMPLES FOR (2). (i) Take E to be a cubic (not necessarily Galois) extension of a number field F, $G = E^x$ (actually $\mathrm{Res}_F^E GL_1$), and $G' = GL_3$. Then there is a natural L-homomorphism ρ: ${}^LG \to {}^LG'$ (corresponding roughly to the "toral" embedding of ${}^LG^0 = C^x \times C^x \times C^x$ into $GL_3(\mathbf{C})$), and the corresponding lift $\chi \to \pi_\chi$ between grossencharacters of E^x and automorphic representations of GL_3 is such that $L(s, \pi_\chi) = L(s, \chi)$. The converse theorem works here because $L(s, \chi)$ (and hence $L(s, \pi_\chi)$) is known to be "nice" by Hecke's theory of abelian L-functions (cf. Parts II.C.1 and II.D.1); analogous arguments for quadratic extensions and GL_2 give a different approach to Example (1)(i).

(ii) If $G = GL_2$, $G' = GL_3$, and ρ: $GL_2(\mathbf{C}) \to GL_3(\mathbf{C})$ is the adjoint representation described in Example III.B, then the corresponding lift $\pi \to \Pi$ has already been discussed.

(iii) Take G equal to the unitary group in three variables over E introduced in Example (1)(ii), and $G' = \mathrm{Res}_F^E G$ (so $G'_F = GL_3(E)$). Then one can attach to each automorphic cuspidal representation π of $G_{\mathbf{A}}$ an L-function (of degree 6 over F) which is meromorphic with functional equation and (by the converse theorem for GL_3 *over* E) belongs to an automorphic representation π' of $GL_3(\mathbf{A}_E)$); cf. **[Ge PS 1]**. The resulting correspondence $\pi \to \pi'$ then defines a "base change lift" for U_3.

There are more subtle (and recent) applications of the theory of L-functions, especially to the functoriality principle, which go beyond the confines of the converse theorem. I have in mind mostly the use of L-functions in characterizing the image of the liftings which come from dual reductive pairs, for example the deep work of **[Wald]** characterizing the image of the so-called Shimura correspondence, and the work of **[PS]** on the Saito-Kurokawa lifting between PGL_2 and Sp_4; this work promises to shed light on a wide variety of examples, in particular, the (as yet undeveloped) theories of base change for the metaplectic group and $PGSp_4$.

Equally worthy of mention is the recent work of **[J PS S 3]** on base change for GL_2 to a cubic *nonnormal* extension E of F. As mentioned in IV.C.2(a), "base-change" was fully developed for GL_2 (cf. **[Langlands 4]**) for cyclic Galois extensions of F of prime degree. To prove the theorem for nonnormal E an appeal is made to the theory of L-functions on $\mathrm{GL}_n \times \mathrm{GL}_m$ as developed in [**J PS S 2**]; moveover, to date at least, the trace formula proof used for Galois extensions (see below) has not been made to work for nonnormal E.

(3) *The trace formula.* This method is both the newest and "hottest" approach to studying automorphic representations. Without pretending to go to the heart of the matter we shall give a very rough idea of how this method works. Unfortunately, the subject is in such a state of flux and development that even the experts can get confused and frustrated.

What is "the trace formula"? For a given group G, let us consider our friend the right regular representation R (cf. II.D) of $G_{\mathbf{A}}$ acting in $L_0^2(G_F \backslash G_{\mathbf{A}})$. It decomposes as a sum of irreducible (unitary) automorphic representations π of G, each with finite multiplicity m_π, i.e.,

$$R = \oplus m_\pi \pi. \tag{2}$$

Since one goal of the theory of automorphic forms is to understand which π occur in (2), and since an irreducible representation π is determined by its "character", it is natural to want to compute the character of R. This is what "the trace formula for G" does. Of course we can't just take the trace of a unitary representation (it doesn't exist!), so we have first to integrate the representation against a "nice" compactly supported function f on $G_{\mathbf{A}}$ (getting the operator $R(f) = \int f(g)R(g)\,dg$ which can be shown to be of trace class). On the one hand, we have

$$\operatorname{tr} R(f) = \Sigma m_\pi \chi_\pi(f), \tag{3}$$

where

$$\chi_\pi(f) = \operatorname{trace}\{\int f(g)\pi(g)\,dg\}.$$

On the other hand, using the explicit form of $R(f)$, as an integral operator in $L^2(G_{\mathbf{Q}} \backslash G_{\mathbf{A}})$, we get a second (more complicated) expression for the trace which makes no reference to the decomposition (2); instead it involves expressions intrinsic to the geometry of the group, for example "orbital integrals" of f over "rational" conjugacy classes in $G_{\mathbf{A}}$, etc.

The idea is that by carefully examining this "second form of the trace formula" one should be able to conclude something about the expression (3), i.e., about the automorphic representations π of $G_{\mathbf{A}}$. In practice, however, this turns out to be nearly impossible, i.e., it is difficult to show that a given representation π of $G_{\mathbf{A}}$ occurs in (2) by just analyzing an explicit formula for $\operatorname{tr} R(f)$. What *does* seem to work, however, is to *compare* the (second forms of the) trace formula for two *different* groups G and G' and then conclude that *if* π occurs in $(R, G_{\mathbf{A}})$ then π' *will* occur in $(R', G'_{\mathbf{A}})$.

For example, suppose G is the multiplicative group of a division quaternion algebra and $G' = \mathrm{GL}_2$. In §16 of [**JL**] the trace formulas for these groups are compared, and the conclusion is that there is a correspondence $\pi \to \pi'$ between the (greater than one-dimensional) automorphic representations of G and a certain subset of the automorphic π' on $\mathrm{GL}_2(\mathbf{A})$. In fact, this correspondence is consistent with the *identity* homomorphism between the L-groups of G and G'; in particular, $L(s, \pi) = L(s, \pi')$.

As already suggested, the prototype example of this approach to the functoriality principle is the base-change lifting introduced in Example IV.C.2(a). Other examples include:

(i) A new treatment of the "Adjoint" lifting $GL_2 \to GL_3$ due to Flicker [**Flick 1**]; here, as in all such applications of the trace formula, one can characterize the image of the lifting as well;

(ii) A different proof of base change and lifting for the unitary group U_3; cf. [**Flick 2**].

Warning. Most applications of the trace formula involve pairs of groups G, G' whose conjugacy classes are not so earily compared as suggested in the examples above; this leads to the thorny notions of "instability" for the trace formula, "L-indistinguishability" and the like. Though these notions arose initially as impediments to a direct application of the trace formula, Langlands figured out how to turn them into powerful weapons for *proving* his Functoriality Conjecture. Indeed, these suggestions have already been followed in some of the works just mentioned; for a comprehensive but difficult introduction to this increasingly active research domain, see [**Langlands 6, Shel 5** and **Flick 1, 2**].

CONCLUDING REMARKS. It is a happy circumstance that the three methods of proof just sketched complement each other remarkably well. For example, sometimes a deep theorem can be proved *only* by using a mixture of two or three of these approaches. An example of this is Langlands' Reciprocity Conjecture for "tetrahedral" two-dimensional representations of $\mathrm{Gal}(K/F)$; cf. [**Langlands 3** or **Ge 3**] for a leisurely discussion. On the other hand, sometimes a result can be proved by using *any* one of these approaches, but each method affords its own particular advantages. An example of this is the lifting of automorphic representations from a division quaternion algebra to GL_2; proofs using L-functions or the trace formula are found in [**JL**], and a direct proof using theta-series is the subject matter of [**Shimizu**]. A slightly different kind of example is the correspondence between automorphic representations of the metaplectic group and GL_2 (Shimura's correspondence); cf. [**Shimura, Ge PS 1, Flick 3** and **Wald**]. Thus, as already suggested by our discussion of base change, no one of these methods has a monopoly on proving interesting theorems.

E. A few last words. It should be clear by now that the strength of Langlands' program lies as much (or more) in suggesting new problems as well as in resolving old ones.

Given two groups G and G', and an L-homomorphism between them, what is the relation between the automorphic representations of G and G'?

Given a number-theoretically defined family of conjugacy classes in some complex group like $GL_n(\mathbf{C})$, what is the "automorphic nature" of this collection?

Questions like these will undoubtedly keep mathematicians busy for a long time to come.

BIBLIOGRAPHY

Code for Superscripts

a = General reference
b = Langlands program
c = Representation theory (real or p-adic groups)
d = Automorphic L-functions (local or global theory)
e = θ-series
f = Trace formula
g = Zeta-functions of varieties
h = L-indistinguishability
i = Local global principle
j = Related developments in number theory

[d] **[And]** A. N. Andrianov, *Zeta functions and the Siegel modular forms*, Lie Groups and Their Representations (I. M. Gelfand, ed.), Wiley, New York, 1975, pp. 9–20.

[b] **[Art]** J. Arthur, *Automorphic representations and number theory*, Canad. Math. Soc. Conf. Proc. Vol. 1 1981, pp. 3–54.

[f] **[Art 2]** ______, *Eisenstein series and the trace formula*, Proc. Sympos. Pure Math., Vol. 33, part 1, Amer. Math. Soc. Providence, R. I., 1979, pp. 253–274.

[f] **[Art 3]** ______, *A trace formula for reductive groups*. I, II, Duke Math. J. **45** (1978), 911–952; Compositio Math. **40** (1980), 87–121.

[f] **[Art 4]** ______, *The trace formula in invariant form*, Ann. of Math. (2) **114** (1981), 1–74.

[d] **[Asai]** T. Asai, *On certain Dirichlet series associated with Hilbert modular forms and Rankin's method*, Math. Ann. **226** (1977), 81–94.

[c] **[Be Ze]** I. N. Bernstein and A. Zelevinskii, *Representations of the group* GL(n, F) *when F is a nonarchimedean local field*, Russian Math. Surveys **31** (1976), no. 3, 1–68.

[g] **[Bi SD]** B. J. Birch and H. P. F. Swinnerton-Dyer, *Elliptic curves and modular functions of one variable*, Lecture Notes in Math., Vol. 476, Springer, New York, 1975, pp. 2–32.

[b] **[Bo]** A. Borel, *Automorphic L-functions* Proc. Sympos. Pure Math., Vol. 33, part 2, Amer. Math. Soc., Providence, R. I., 1979, pp. 27–61.

[a,b] **[Bo 2]** ______, *Formes automorphes et series de Dirichlet*, Sém. Bourbaki, No. 466, Lecture Notes in Math., Vol. 514, Springer, New York, pp. 189–222.

[a] **[Bo Ja]** A. Borel and H. Jacquet, *Automorphic forms and automorphic representations*, Proc. Sympos. Pure Math., Vol. 33, part 1, Amer. Math. Soc., Providence, R. I., 1979, pp. 189–202.

[a] **[Bo Shaf]** Z. I. Borevich and I. R. Shafarevich, *Number theory*, Academic Press, 1967.

[a,b,c] **[Cart 1]** P. Cartier, *La conjecture locale de Langlands pour* GL(2) *et la demonstration de Ph. Kutzko*, Sem. Bourbaki, 1979/80, No. 550, Lecture Notes in Math., Vol. 842, Springer, 1981.

[c] **[Cart 2]** ______, *Representations of p-adic groups: A survey*, Proc. Sympos. Pure Math., Vol. 33, part 1, Amer. Math. Soc., Providence, R. I. 1979, pp. 111–156.

[g] **[Cas]** W. Casselman, *The Hasse-Weil ζ-function of some moduli varieties of dimension greater than one*, Proc. Sympos. Pure Math., Vol. 33, part 2, Amer. Math. Soc., Providence, R. I., 1979, pp. 141–164.

[b,c] **[Cas 2]** ______, "GL_n", Proc. Durham Sympos. Algebraic Number Fields (London), Academic Press, New York, 1977.

[a] **[Cassels]** J. W. Cassels, *Rational quadratic forms*, Academic Press, London and New York, 1978.

[a] **[Cas Fro]** J. W. Cassels and A. Frölich (Editors), *Algebraic number theory*, Proc. Instructional Conf. (Brighton), Thompson Book, Washington, D. C., 1967.

[c] **[Cloz]** L. Clozel, *Base change for real groups*, preprints, 1980-1982.

[j] **[Coates]** J. Coates, *The work of Mazur and Wiles on cyclotomic fields*, Sém. Bourbaki 1980/81, Lecture Notes in Math., Vol. 575, Springer-Verlag, 1982, pp. 220–242.

[a] **[Deligne]** P. Deligne, *Formes modulaires et representations de* GL_2, Modular Functions of One Variable, Lecture Notes in Math., Vol. 349, Springer, New York, 1973, pp. 55–106.

[a,b] **[Deligne 2]** ______, *Les constantes des équations fonctionelles des fonctions L*, Modular Functions of One Variable. II, Lecture Notes in Math., Vol. 249, Springer-Verlag, 1973, pp. 501–597.

[a][De Se] P. Deligne and J.-P. Serre, *Formes modulaires de poids* 1, Ann. Sci. École Norm. Sup. (4) **7** (1974), 507–530; also article in Proc. Durham Sympos. Algebraic Number Fields (London), Academic Press, New York, 1977, pp. 193–268.

[b][Drin] V. G. Drinfeld, *Langlands conjecture for* GL(2) *over function fields*, Proc. Internat. Congress. Math., Helsinki, 1978.

[a,j][Dwork] D. Dwork, *On the Artin root number*, Amer. J. Math. **78** (1956), 444–472.

[f][Flath] D. Flath, *A comparison of the automorphic representations of* GL(3) *and its twisted forms*, Pacific J. Math. **97** (1981), 373–402.

[f][Flick 1] Y. Flicker, *The adjoint lifting from* SL(2) *to* PGL(3), I.H.E.S., 1981, preprint.

[f][Flick 2] ______, *L-packets and liftings for* $U(3)$, Princeton, 1982, preprint.

[f][Flick 3] ______, *Automorphic forms on covering groups of* GL_2, Invent. Math. **57** (1980), 119–182.

[f][Flick 4] ______, *The trace formula and base change for* GL(3), Lecture Notes in Math., Vol. 927, Springer, New York, 1982.

[a][Ge 1] S. Gelbart, *Automorphic forms on adele groups*, Ann. of Math. Stud., No. 83, Princeton Univ. Press, Princeton, N. J., 1975.

[g][Ge 2] ______, *Elliptic curves and automorphic representations*, Adv. in Math. **21** (1976), 235–292.

[b,d][Ge 3] ______, *Automorphic forms and Artin's conjecture*, Lecture Notes in Math., Vol. 627, Springer-Verlag, 1977.

[c][Ge 4] ______, *Examples of dual reductive pairs*, Proc. Sympos. Pure Math., Vol. 33, part 1, Amer. Math. Soc., Providence, R. I., 1979, pp. 287–296.

[d][Ge Ja] S. Gelbart and H. Jacquet, *A relation between automorphic representations of* GL(2) *and* GL(3), Ann. Sci. École. Norm. Sup. (4) **11** (1978), 471–552.

[f][Ge Ja 2] ______, *Forms of* GL(2) *from the analytic point of view*, Proc. Sympos. Pure Math., Vol. 33, part 1, Amer. Math. Soc., Providence, R. I., 1979, pp. 213–251.

[c,h][Ge Kn] S. Gelbart and A. Knapp, *L-indistinguishability and R groups for the special linear groups*, Adv. in Math. **43** (1982), 101–121.

[d][Ge PS 1] S. Gelbart and I. Piatetski-Shapiro, *On Shimura's correspondence for modular forms of half-integral weight*, Automorphic Forms, Representation Theory and Arithmetic (Bombay, 1979), Tata Inst. Fund. Res. Stud. Math., No. 10, Bombay, 1981, pp. 1–39.

[d][Ge PS 2] ______, *Automorphic forms and L-functions for the unitary group*, Proc. Maryland Special Year Conf. Lie Groups, 1982-1983, Lecture Notes in Math., Springer-Verlag (to appear).

[a][Ge Fo] I. M. Gelfand and S. Fomin, *Geodesic flows on manifolds of constant negative curvature*, Uspekhi Mat. Nauk 7 (1952), no. 1, 118–137; English transl., Amer. Math. Soc. Transl. (2) **1** (1965), 49–65.

[a][G G PS] I. M. Gelfand, M. Graev and I. Piatetski-Shapiro, *Representation theory and automorphic functions*, Saunders, Philadelphia, 1969.

[c][Ge Ka] I. M. Gelfand and D. Kazhdan, *Representations of* $GL(n, K)$ *where* K *is a local field*, Lie Groups and their Representations, Wiley, 1975, 95–118.

[c,d][Ger] P. Gerardin, *Cuspidal unramified series for central simple algebras over local fields*, Proc. Sympos. Pure Math., Vol. 33, part 1, Amer. Math. Soc., Providence, R. I., 1979, pp. 157–169.

[b,f][Ger Lab] P. Gerardin and J.-P. Labesse, *The solution of a base change problem for* GL(2) (*following Langlands, Saito, Shintani*), Proc. Sympos. Pure Math., Vol. 33, part 2, Amer. Math. Soc., Providence, R. I., 1979, pp. 115–134.

[b][God 1] R. Godement, *Fonctions automorphes et produits euleriens*, Sém. Bourbaki 1968/69, No. 349, Lecture Notes in Math., Vol. 179, Springer-Verlag, New York, 1970, pp. 37–53.

[f][God 2] ______, *Introduction aux travaux de Selberg*, Sem. Bourbaki, 9[e] année: 1956/1957, No. 144.

[d][Go Ja] R. Godement and H. Jacquet, *Zeta-functions of simple algebras*, Lecture Notes in Math., Vol. 260, Springer, New York, 1972.

[a][Goldstein] L. Goldstein, *Analytic number theory*, Prentice-Hall, 1971.

[a][Gross] K. I. Gross, *On the evolution of non-commutative harmonic analysis*, Amer. Math. Monthly **85** (1978), 525–548.

[j][Gross B] B. Gross, *On the conjecture of Birch and Swinnerton-Dyer for elliptic curves with complex multiplication*, Number Theory Related to Fermat's Last Theorem, Progress in Math., Vol. 26, Birkhauser, Boston, 1982.

[a][**Harder**] G. Harder, *Chevalley groups over function fields and automorphic forms*, Ann. of Math. (2) **100** (1974), 249–306.

[a][**HC**] Harish-Chandra, *Automorphic forms on semi-simple Lie groups*, Lecture Notes in Math., Vol. 62, Springer, New York, 1968.

[b,d][**Hen**] G. Henniart, *La conjecture de Langlands locale pour* GL(3), I.H.E.S., 1982, preprint.

[c][**Ho 1**] R. Howe, *θ-series and invariant theory*, Proc. Sympos. Pure Math., Vol. 33, part 1, Amer. Math. Soc., Providence, R. I. 1979, pp. 275–286.

[c][**Ho 2**] ______, *Tamely ramified supercuspidal representations of* $\mathrm{GL}_n(F)$, Pacific J. Math. **73** (1977), 437–460.

[c][**Ho 3**] ______, *Some qualitative results on the representation theory of* GL_n *over a p-adic field*, Pacific J. Math. **73** (1977), 479–538.

[a][**Humph**] J. E. Humphreys, *Linear algebraic groups*, Springer-Verlag, 1975.

[c][**Jacquet 1**] H. Jacquet, *Representations des groupes lineares p-adiques*, Theory of Group Representations and Fourier Analysis, (Centro Internaz. Mat. Estivo (C.I.M.E.), II Ciclo, Montecatini Terme, 1970), Edizioni Cremonese, Rome, 1971, pp. 119–220.

[d][**Jacquet 2**] ______, *Automorphic forms on* GL_2. II, Lecture Notes in Math., Vol. 278, Springer, New York, 1972.

[d][**Jacquet 3**] ______, *Principal L-functions of the linear group*, Proc. Sympos. Pure Math., Vol. 33, part 2, Amer. Math. Soc., Providence, R. I., 1979, pp. 63–86.

[b,c,d][**J L**] H. Jacquet and R. P. Langlands, *Automorphic forms on* GL(2), Lecture Notes in Math., Vol. 114, Springer-Verlag, New York, 1970.

[d][**J PS S1**] H. Jacquet, I. Piatetski-Shapiro and J. Shalika, *Automorphic forms on* GL(3). I, II, Ann. of Math. (2) **109** (1979), 169–258.

[d][**J PS S2**] ______, *On Euler products and the classification of automorphic representations*. I, II, Amer. J. Math. **103** (1981), 499–558, 777–815.

[d][**J PS S3**] ______, *Relèvement cubique non normal*, C. R. Acad. Sci. Paris Sér. A **292** (1981), 567–571.

[c][**Kaz**] D. Kazhdan, *Some applications of the Weil representation*, J. Analyse Math. **32** (1977), 235–248.

[c][**Kn Zn**] A. W. Knapp and G. Zuckerman, *Normalizing factors, tempered representations, and L-groups*, Proc. Sympos. Pure Math., Vol. 33, part 1, Amer. Math. Soc., Providence, R. I., 1979, pp. 93–105.

[c][**Kott**] R. Kottwitz, *Unstable orbital integrals on* SL(3), Duke Math. J. **48** (1981).

[c][**Kudla**] S. Kudla, *Relations between automorphic forms produced by theta-functions*, Modular Functions of One Variable. VI, Lecture Notes in Math., vol. 627, Springer, New York, 1977.

[h][**Lab Lang**] J.-P. Labesse and R. P. Langlands, *L-indistinguishability for* SL(2), Canad. J. Math. **31** (1979), 726–785.

[b][**Langlands 1**] R. P. Langlands, *Euler products*, Yale Univ. Press, New Haven, Conn., 1967.

[b][**Langlands 2**] ______, *Problems in the theory of automorphic forms*, Lecture Notes in Math., Vol. 170, Springer-Verlag, New York, 1970, pp. 18–86.

[b,d][**Langlands 3**] ______, *L-functions and automorphic representations*, Proc. Internat. Congress Math., Helsinki, 1978.

[b,f][**Langlands 4**] ______, *Base change for* GL_2, Ann. of Math. Stud., No. 96, Princeton Univ. Press, Princeton, N. J., 1980.

[a][**Langlands 5**] ______, *On the notion of an automorphic representation*, Proc. Sympos. Pure Math., Vol. 33, part 1, Amer. Math. Soc., Providence, R. I., 1979, pp. 203–208.

[f][**Langlands 6**] ______, *Les débuts d'une formule des traces stables*, Lecture Notes, L'École Normale Supérieure de Jeunes Filles, Paris, 1980.

[f,g][**Langlands 7**] ______, *Shimura varieties and the Selberg trace formula*, Canad. J. Math. **29** (1977), 1292–1299.

[f,g][**Langlands 8**], ______, *Automorphic representations, Shimura varieties and motives*, Proc. Sympos. Pure Math. Vol. 33, part 2, Amer. Math. Soc., Providence, R. I., 1979, pp. 205–246.

[h][**Langlands 9**] ______, *Stable conjugacy: definitions and lemmas*, Canad. J. Math. **31** (1979), 700–725.

[c,f,d][**Li**] W. Li, *On the representations of* GL(2). I, II, J. Reine Angew. Math **313** (1980), 27–42; ibid **314** (1980), 3–20; also *Hecke-Weil-Jacquet-Langlands theorem revisited*, Number Theory (Carbondale, 1979), Lecture Notes in Math., Vol. 751, Springer, Berlin, 1979.

[c][**Lips**] R. Lipsman, *Group representations*, Lecture Notes in Math., No. 388, Springer, New York, 1974.

[c][**Mac**] I. G. Macdonald, *Spherical functions on a group of p-adic type*, Ramanujan Inst. for Advanced Study in Math., Univ. of Madras, Madras, India, 1971.

[a][**Mackey**] G. Mackey, *Harmonic analysis as the exploitation of symmetry—a historical survey*, Bull. Amer. Math. Soc. (N.S.) **3** (1980), 543–698.

[a,j][**Mazur**] B. Mazur, Review of *Ernst Edward Kummer, Collected papers*, Bull. Amer. Math. Soc. **83** (1977), 976–988.

[j][**Mazur Wiles**] B. Mazur and A. Wiles, *Class fields of abelian extensions of* **Q**, Invent. Math. (submitted)

[g][**Milne Shih**] J. Milne and K-y. Shih, *The action of complex conjugation on a Shimura variety*, Ann. of Math. (2) **113** (1981), 569–599.

[b,c][**Moy**] A. Moy, *Local constants and the tame Langlands correspondence*, Thesis, Univ. of Chicago, 1982.

[d][**Novod**] M. Novodvordsky, *Automorphic L-functions for the symplectic group* $GSp(4)$, Proc. Sympos. Pure Math. Vol. 33, part 2, Amer. Math. Soc., Providence, R. I., 1979, pp. 87–95.

[a][**Ogg**] A. Ogg, *Modular forms and Dirichlet series*, Benjamin, New York, 1969.

[f][**Os Wa**] M. S. Osborne and G. Warner, *The Selberg trace formula*, I–V, Univ. of Washington, 1980–1983, preprints.

[d,e][**PS**] I. Piatetski-Shapiro, *On the Saito-Kurokawa lifting*, Invent. Math. **71** (1983), 309–338.

[d][**PS 2**] ______, *L-functions for* GSp_4, preprint, 1982.

[d][**PS 3**] ______, *Tate theory for reductive groups and distinguished representations*, Proc. Internat. Congress Math., Helsinki, 1978.

[**PS S**] I. Piatetski-Shapiro and D. Soudry, *L and* ε *factors for* GSp(4), J. Fac. Sci. Univ. Tokyo Sect. IA Math. **28** (1982), 505–530.

[i,a][**Plat**] V. P. Platonov, *The arithmetic theory of algebraic groups*, Russian Math. Surveys **37** (1982), no. 3, 1–62.

[e][**Ral**] S. Rallis, *Langlands' functorially and the Weil representation*, Amer. J. Math. **104** (1982), 469–515.

[e][**R S**] S. Rallis and G. Schiffmann, *Automorphic forms constructed from the Weil representation: Holomorphic case*, Amer. J. Math. **100** (1978), 1049–1122.

[g][**Rap Z**] M. Rapoport and T. Zink, *Uber die lokale Zeta-funktion von Shimuravarietaten*, Invent. Math. **68** (1982), 21–101.

[c][**Rep**] J. Repka, *Base change lifting and Galois invariance*, Pacific J. Math. **95** (1981), 205–212; ibid. **93** (1981), 193–200.

[j][**Rib**] K. Ribet, *A modular construction of unramified p-extensions of* $\mathbf{Q}(\mu_p)$, Invent. Math. **34** (1976), 151–162.

[b][**Rob 1**] A. Robert, *Formes automorphes sur* GL_2 [*Travaux de H. Jacquet et R. P. Langlands*], Semi. Bourbaki 1971/1972, no. 415, Lecture Notes in Math., Vol. 317, Springer, 1973.

[a][**Rob 2**] ______, *Des adeles: pourquoi?*, L'Enseigenment Math. (2) **20** (1974), 133–146.

[a][**Rob 3**] ______, *Lectures on automorphic forms*, Queens Univ., Ontario, 1976.

[c,f][**Rog**] J. Rogawski, *Representations of* $GL(n)$ *and division algebras over a p-adic field*, Duke Math. J. **50** (1983), 161–196.

[f][**Saito**] H. Saito, *Automorphic forms and algebraic extensions of number fields*, Lectures in Math., Kyoto Univ., 1975.

[d][**Sa Sh**] P. Sally and J. Shalika, *Characters of the discrete series of representations of* SL(2) *over a local field*, Proc. Nat. Acad. Sci. U.S.A. **61** (1968), 1231–1237; also *the Plancherel formula for* SL(2) *over a local field*, Proc. Nat. Acad. Sci. U.S.A. **63** (1969), 661–667.

[f][**Selberg**] A. Selberg, *Harmonic analysis and discontinuous groups* (= *the original "trace formula"*!), J. Indian Math. Soc. **20** (1956), 47–87.

[a,b][**Serre**] J.-P. Serre, *Modular forms of weight one and Galois representations*, Algebraic Number Fields (A. Fröhlich, ed.), Academic Press, 1977.

[d][**Sha**] F. Shahidi, *On certain L-functions*, Amer. J. Math. **103** (1981), 297–356.

[d][**Shalika**] J. Shalika, *Some conjectures in class field theory*, Proc. Sympos. Pure Math., Vol. 20, Amer. Math. Soc., Providence, R. I., 1971, pp. 115–122.

[c][**Sha Tan**] J. Shalika and S. Tanaka, *On an explicit construction of a certain class of automorphic forms*, Amer. J. Math. **91** (1969), 1049–1076.

[h][**Shel 1**] D. Shelstad, *Notes on L-indistinguishability (based on a lecture of R. P. Langlands)*, Proc. Sympos. Pure Math. Vol. 33, part 2, Amer. Math. Soc., Providence, R. I., 1979, pp. 193–203.

[c][**Shel 2**] ______, *Characters and inner forms of a quasi-split group over* **R**, Compositio Math. **39** (1979), 11–45.

[c][**Shel 3**] ______, *Orbital integrals and a family of groups attached to a real reductive group*, Ann. Sci. École Norm. Sup. **12** (1979), 1–31.

[b,h][**Shel 4**] ______, *Embeddings of L-groups*, Canad. J. Math. **33** (1981), 513–558.

[h][**Shel 5**] ______, *L-indistinguishability for real groups*, Math. Ann. **259** (1982), 385–430.

[e][**Shimizu**] H. Shimizu, *Theta-series and automorphic forms on GL_2*, J. Math. Soc. Japan **24** (1972), 638–683.

[a][**Shimura**] G. Shimura, *On modular forms of half-integral weight*, Ann. of Math. (2) **97** (1973), 440–481.

[g][**Shimura 2**] ______, *Correspondence modulaires et les fonctions ζ de courbes algébriques*, J. Math. Soc. Japan **10** (1958), 1–28.

[a][**Shimura 3**] ______, *Introduction to the arithmetic theory of automorphic functions*, Publ. Math. Soc. Japan, No. 11, Princeton Univ. Press, Princeton, N. J., 1971.

[f][**Shintani**] T. Shintani, *On liftings of holomorphic cusp forms*, Proc. Sympos. Pure Math., Vol. 33, part 2, Amer. Math. Soc., Providence, R. I., 1979, pp. 97–110.

[c][**Sil**] A. Silberger, *Discrete series and classification for p-adic groups*. I, Amer. J. Math. **103** (1981), 1241–1321.

[a][**Springer**] T. A. Springer, *Reductive groups* Proc. Sympos. Pure Math., Vol. 33, part 1, Amer. Math. Soc., Providence, R. I., 1979, pp. 3–28.

[a][**Tate**] J. Tate, *Problem* 9: *the general reciprocity law*, Proc. Sympos. Pure Math., Vol. 28, Amer. Math. Soc., Providence, R. I., 1976, pp. 311–322.

[a][**Tate 2**] ______, *Number theoretic background*, Proc. Symos. Pure Math., Vol. 33, part 2, Amer. Math. Soc., Providence, R. I., 1979, pp. 3–26.

[a][**Ti**] J. Tits, *Reductive groups over local fields* Proc. Sympos. Pure Math., Vol. 33, part I, Amer. Math. Soc., Providence, R. I., 1979, pp. 29–70.

[b,d][**Tun**] J. Tunnell, *Artin's conjecture for representations or octahedral type*, Bull. Amer. Math. Soc. (N.S.) **5** (1981), 173–175.

[c][**Vigneras**] M.-F. Vigneras, *Charactérisation des integrales orbitales sur un groupe réductife p-adique*, J. Fac. Sci. Univ. Tokyo Sect. IA Math. **28** (1981), no. 3, 945–961 (1982).

[e][**Wald**] J.-L. Waldspurger, *Correspondance de Shimura*, J. Math. Pures Appl. **59** (1980), 1–133.

[c][**Warner**] G. Warner, *Harmonic analysis on semi-simple Lie groups*. I, II, Springer-Verlag, Berlin, 1972.

[a][**We 1**] A. Weil, *Über die bestimmung Dirichletscher Rieben durch funktionelgleichungen*, Math. Ann. **168** (1967), 149–156; also *Dirichlet series and automorphic forms*, Lecture Notes in Math., Vol. 189, Springer-Verlag, New York, 1971.

[e][**We 2**] ______, *Sur certaines groups d'operateurs unitaires*, Acta Math. **111** (1964), 143–211.

[a][**We 3**] ______, *Basic number theory*, 3rd ed., Springer-Verlag, New York, 1974.

[f][**Zagier**] D. Zagier, *Eisenstein series and the Selberg trace formula*. I, Automorphic Forms, Representation Theory and Arithmetic (Bombay, 1979), Springer-Verlag, Berlin, 1981, pp. 275–301.

Department of Mathematics, Cornell University, Ithaca, New York 14853

Current address: Department of Mathematics, Tel-Aviv University, Ramat-Aviv, Israel